Gene-Edited Crops

CRISPR-based genome editing has revolutionized the precise genetic improvement of agriculturally important crops, offering unprecedented precision and efficiency in crop improvement. This powerful technology holds immense potential to reshape the future of food security, enabling scientists and agriculturalists to develop resilient, high yield crops resistant to diseases, pests, and environmental stresses. *Gene-Edited Crops: The CRISPR Solution for Global Food Security* offers a comprehensive guide to the latest research, advancements, and real-world applications of CRISPR in crop genetic improvement for sustainable agriculture and food security.

Through in-depth chapters, this book provides an accessible yet rigorous exploration of CRISPR technology, CRISPR reagents and their potential delivery methods, recent developments in CRISPR, and its transformative role in agriculture and food security. With a focus on practical implementation and future potential, this book examines the science behind genome editing and the strategic pathways to address key challenges in crop development—from enhancing nutritional profiles to improving environmental adaptability.

Features include:

- **Comprehensive overview**: An introduction to the fundamentals of CRISPR technology and its application in crop science.
- **Computational tools and CRISPR reagents**: Bioinformatic tools for designing precise gRNA, CRISPR reagents and their effective delivery methods.
- **Recent developments in CRISPR**: Emerging developments in CRISPR for precise gene editing in plants such as base editing, prime editing, Fanzor, and multiplex gene editing.
- **Real-world applications**: Application of CRISPR for developing future crops, including projects targeting enhanced disease resistance, drought tolerance, and improved nutritional content.
- **Regulatory landscape**: Insightful discussion on the evolving global regulatory framework governing gene-edited crops, including the transformative role of artificial intelligence in navigating these regulations.
- **Future directions**: Exploration of ongoing advances and the outlook for CRISPR technology in addressing food security.

With detailed examples, data, and resources, this book serves as an essential reference for scientists, agriculturalists, and policymakers invested in the future of food security. It also doubles as an educational resource for advanced courses in genetics, plant sciences, and agricultural biotechnology, making it invaluable to both seasoned researchers and those newly interested in the potential of CRISPR.

Gene-Edited Crops

The CRISPR Solution for Global Food Security

Edited by
Aftab Ahmad, Nayla Munawar, and Baohong Zhang

CRC Press
Taylor & Francis Group
Boca Raton London New York

CRC Press is an imprint of the
Taylor & Francis Group, an **informa** business

Designed cover image: Shutterstock

First edition published 2025
by CRC Press
2385 NW Executive Center Drive, Suite 320, Boca Raton FL 33431

and by CRC Press
4 Park Square, Milton Park, Abingdon, Oxon, OX14 4RN

CRC Press is an imprint of Taylor & Francis Group, LLC

ISBN: 978-1-032-81693-7 (hbk)
ISBN: 978-1-032-81694-4 (pbk)
ISBN: 978-1-003-50093-3 (ebk)

DOI: 10.1201/9781003500933

Typeset in Times
by Newgen Publishing UK

Contents

Preface

The need to secure global food and nutrition for an ever-growing population in the face of climate change has never been more urgent. Agricultural systems today face unprecedented challenges such as limited arable land, water scarcity, climate change, and increased demand to produce more food without further environmental degradation. Conventional tools for plant breeding are still helpful but inadequate in meeting the demands of modern agriculture. In this context, the CRISPR/Cas gene-editing technology provides a versatile toolset to manipulate the genomic landscape of agriculturally important crops, offering new opportunities for enhancing crop yields, improving nutrition, and strengthening resilience against environmental stresses.

With their advanced variants, the evolving CRISPR-Cas systems, such as Cas9, Cas12, and Cas13, provide unparalleled precision and efficiency and represent a paradigm shift in how we can engineer crops. CRISPR applications such as CRISPRi, CRISPRa, base editing, prime editing, and RNA editing provide potent tools for introducing precise modifications in the genome. The use of RNPs as CRISPR reagents provides a powerful tool for developing transgene-free plants without inserting any foreign genes. CRISPR-edited plants, especially with SDN1 and SDN2 modifications, are classified as non-GMOs by different countries as these modifications can be introduced by conventional breeding as well. CRISPR has accelerated breeding, enabling scientists to develop improved crop varieties in a few years rather than decades like traditional breeding. In addition, it also minimizes unintended consequences by precise targeting of the genome that has often been associated with other genetic modification techniques. The potential benefits of CRISPR technology in agriculture include functional genomics, developing pest-resistant crops, and improving plant tolerance to environmental stresses such as drought, heat, heavy metals, and salinity, thus addressing the fundamental challenges posed by climate change.

In this book, we explore the immense potential and practical applications of CRISPR/Cas systems for precise genetic improvement of crop plants to address the global challenges of food security and climate change. We begin with an introduction to the CRISPR-Cas technology, the classification of CRISPR systems, the computational tools for gRNA design, and the CRISPR reagents with their potential delivery methods that have made gene editing easy and accessible. From there, we cover various applications of CRISPR-Cas technology such as CRISPRi, CRISPRa, base editing, prime editing, and RNA editing for genetic improvements of crops. We also highlight emerging technologies such as CRISPRASE, Retrons, OMEGA, and Fanzor that are pushing the boundaries of plant genome editing even further. Finally, we cover applications of CRISPR-Cas systems in plant science, ranging from biotic and abiotic stress resistance to yield and nutritional improvement.

However, like any other technology, the immense power of CRISPR also comes with great responsibility. Although CRISPR crops are touching world markets, the world community is divided over regulatory frameworks of these crops. For example, the USA has been deregulating CRISPR-edited crops, especially SDN1 and SDN2-edited crops, while the EU has been regulating all CRISPR-edited crops as GM. This book does not shy away from addressing the ethical, social, and regulatory concerns surrounding gene-editing technologies in agriculture. We recognize that public perception and regulatory frameworks will play crucial roles in determining the future of CRISPR-edited crops. Therefore, this volume discusses the global regulatory landscape of CRISPR crops and the governance structures necessary for their safe and equitable use.

Our main objective in writing this book is to provide a comprehensive resource for students, researchers, educators, policymakers, and anyone interested in the future of food security through sustainable agricultural practices. Whether you are a biotechnologist eager to apply CRISPR in your research, a policymaker navigating the regulatory landscape, or a student hoping to understand this powerful technology, we hope this book will serve as an invaluable guide.

We live in a time when the convergence of AI, scientific innovation, and agricultural necessity has the potential to reshape the future. We hope the knowledge shared within these pages will help foster further advances in CRISPR technology, promote food security, and contribute to a more sustainable agricultural future for future generations.

Aftab Ahmad
Nayla Munawar
Baohong Zhang

About the Editors

Aftab Ahmad is currently working as an Associate Professor in the Natural and Medical Sciences Research Center (NMSRC), University of Nizwa, Oman. Dr Aftab's research interests include developing gene edited plants especially virus-resistant and salt-tolerant plants using CRISPR/Cas technology. He has published more than 50 research articles in international peer-reviewed journals including *Frontiers in Plant Sciences*, *Journal of Biomedical Sciences*, *Scientific Reports-Nature*, *PLOS One*, *Journal of Translational Medicine*, *Plant Molecular Biology Reporter*, *Molecules*, *Journal of Cereal Science*, and *Genetic Resources and Crop Evolution*. Dr Aftab pursued his PhD and postdocs in Plant Molecular Biology from the University of Shizuoka, Japan, under the Center of Excellence (COE, 21st Century) program. Dr Aftab also served as a visiting scientist in the Department of Plant Sciences, University of California, Davis, USA. Dr Aftab has 10 years of teaching experience in biochemistry and molecular biology at University of Agriculture, Faisalabad (UAF), Pakistan and his research focuses in the area of CRISPR/Cas-based genome editing in plants.

Nayla Munawar is currently working as an Associate Professor of Biochemistry in the Department of Chemistry, College of Science, United Arab Emirates University (UAEU), Al-Ain, UAE. Dr. Munawar's research interests include the synthesis from plants, whole microbial cells, and/or purified enzymes, analysis, and applications of biomaterials. She is interested in different protein and carbohydrate chemistry aspects, especially in the engineering of natural enzymes, molecules interactions, and epigenetic regulatory protein complexes. Her current research motivation is hydrolases, transferases, and dehydrogenase enzymes for the production of novel biomaterials for medical, agriculture, and biotechnological applications. She has published 30 articles in international peer-reviewed journals and 9 book chapters. She is a certified reviewer of MDPI journals and serves as guest editor of *Frontiers in Genome Editing* and *Frontiers in Biotechnology* journals. Dr. Nayla pursued her Ph.D. in Biochemistry from Conway Institute, University College Dublin, Ireland, in 2011 and then been awarded a postdoc position in the same institute in the field of proteomics and epigenetics. She has served as an assistant lecturer at the Dublin Institute of Technology, Ireland, and then joined the Department of Biochemistry at the University of Agriculture, Faisalabad, Pakistan as an assistant professor. She was offered another postdoc position at the Royal College of Surgeons, Ireland, in 2016, and then she joined the Department of Chemistry at the United Arab Emirates University, Al-Ain, UAE in 2019.

Baohong Zhang is a well-known scientist and distinguished Professor in the Department of Biology at East Carolina University, North Carolina, USA. Dr. Zhang received his bachelor's degree in plant genetics and breeding from the China Agricultural University and his Ph.D. in environmental toxicology from Texas Tech University, USA. Dr. Zhang not only provides leadership in research activities but also training the next generation scientists and serving the scientific community. He has been working in the field of genomics and biotechnology, including transgenic and genome editing, for more than 20 years. His work has been supported by federal funding agents (NSF, NIH, DoD and USDA) and industries, such as North Carolina Biotechnology Center and DuPont. Dr. Zhang is frequently invited to international conferences to give plenary or keynote lectures. During his long and distinguished career, he has published 10 scientific books, including *MicroRNAs in Stem* Cell, and *Transgenic Cotton*, published >200 papers at the peer-reviewed international journals. His papers have been cited more than 19,600 times; many of them were listed as highly cited papers by the ISI Web of Sciences. He is frequently invited to serve on grant reviewing panels, including NSF, USDA, DoE, and the European Research Council Executive Agency. Dr. Zhang is also serving as Editor-in-Chief, associate editor or guest editor for 10 international journals, including *Functional & Integrative Genomics*, *Plants*, *Scientific Reports*, *Plant Biotechnology Journal*, *Molecular Biotechnology*, *Plant Molecular Biology*, and *Journal of Cotton Research*. Dr. Zhang frequently reviews manuscripts for more than 100 international journals, including *Nature* and *Nature Biotechnology*. Dr. Zhang has won many prestigious awards, including not only the Cotton Researcher of the year 2018, but also Science/AAAS Excellence in Science (2006), the ECU Five Year Research Achievement Award (2013), and THCAS Distinguished Professorship (2018). He has collaborated with more than 1,000 scientists from >20 countries and won the inaugural ECU Achievement in International Research and Creative Activity Award (2017). In 2018, he was elected to the Fellow of the American Association for the Advancement of Science (AAAS) and won the Lifetime Achievement Research Award from ECU (2019). In 2020, he also won the Cotton Biotechnology Award.

Contributors

Kamel Abd Elsalam
Plant Pathology Research Institute
Agricultural Research Center
Giza, Egypt

Rabail Afzal
Department of Chemistry
United Arab Emirates University
Al-Ain, United Arab Emirates

Aftab Ahmad
Natural and Medical Sciences Research Center (NMSRC)
University of Nizwa
Oman

Muhammad Sarfraz Ahmed
Department of Zoology, Wildlife and Fisheries
University of Agriculture
Faisalabad, Pakistan

Emre Aksoy
Department of Biological Sciences
Faculty of Arts and Science
Middle East Technical University
Ankara, Türkiye

Ahmed Al-Harrasi
Natural & Medical Sciences Research Center (NMSRC)
University of Nizwa, Oman

Florencia Alvarez
Universidad de Buenos Aires
Facultad de Agronomía, Cátedra de Bioquímica
Buenos Aires, Argentina

Burcu Arıkan
Department of Molecular Biology and Genetics
Faculty of Science
Istanbul University
Istanbul, Turkey

Ruben D. Arroyo-Olarte
Unidad de Biomedicina
Facultad de Estudios Superiores Iztacala
Universidad Nacional Autónoma de México
Tlalnepantla, México

Sidra Ashraf
Department of Biochemistry
University of Agriculture
Faisalabad, Pakistan

Noor Asma
Department of Biochemistry
University of Agriculture
Faisalabad, Pakistan

Asma Ayub
Department of Botany
University of Agriculture
Faisalabad, Pakistan

Faheem Shahzad Baloch
Department of Plant Protection
Faculty of Agricultural Sciences and Technologies
Sivas University of Science and Technology
Sivas, Turkey

Özgür Çakır
Department of Molecular Biology and Genetics
Faculty of Science
Istanbul University
Istanbul, Turkey

Selime Çelik
Department of Biological Sciences
Faculty of Arts and Science
Middle East Technical University
Ankara, Türkiye

Ricardo Reyes-Chilpa
Laboratorio de Productos Naturales
Instituto de Química
Universidad Nacional Autónoma de México
México

Medha Dange
Department of Biotechnology
MES Abasaheb Garware College
Pune, Maharashtra, India

Utku Deniz
Department of Biological Sciences
Faculty of Arts and Science
Middle East Technical University
Ankara, Türkiye

Priyanjali Dixit
Department of Biotechnology
MES Abasaheb Garware College
Pune, Maharashtra, India

Selin Doğru
Department of Biotechnology
Faculty of Science
Mersin University
Mersin, Turkey

Alman Fatima
Department of Biochemistry
McGill University
Montreal, Quebec, Canada

Noreen Fatima
Department of Plant Breeding and Genetics
University of Agriculture
Faisalabad, Pakistan

Muhammad Faheem
Department of Biochemistry
University of Agriculture
Faisalabad, Pakistan

Abdul Ghaffar
Department of Biochemistry
Government College University
Faisalabad, Pakistan

Rafaqat Ali Gill
Lushan Botanical Garden
Chinese Academy of Sciences
Jiujiang, China

Rana Ali Haider
Department of Biochemistry
University of Agriculture
Faisalabad, Pakistan

Qiong Hu
Key Laboratory for Biology and Genetic Improvement of Oil Crops
Ministry of Agriculture and Rural Affairs
Oil Crops Research Institute, Chinese Academy of Agricultural Sciences
Wuhan, China

Noor ul Huda
Department of Biochemistry
University of Agriculture
Faisalabad, Pakistan

Muhammad Ibrar
Department of Computer and Mathematical Sciences at New Mexico Highlands University in Las Vegas
New Mexico, USA

Adeen Ijaz
Department of Biotechnology
Forman Christian College
Lahore, Pakistan

Annena Ijaz
Department of Biochemistry
Government College University
Faisalabad, Pakistan

Neslihan Turgut Kara
Department of Molecular Biology and Genetics
Faculty of Science
Istanbul University
Istanbul, Turkey

Ceyhun Kayıhan
Department of Molecular Biology and Genetics
Faculty of Arts and Science
Baskent University
Ankara, Türkiye

Doğa Selin Kayıhan
Department of Pharmacognosy
Faculty of Pharmacy
Baskent University
Ankara, Türkiye

Darya Khan
Hainan Yazhou Bay Seed Laboratory
Sanya Nanfan Research Institute of Hainan University
Sanya, China

Hammad Ahmad Khan
Department of Zoology, Wildlife and Fisheries
University of Agriculture
Faisalabad, Pakistan

Muhammad Naeem Khan
School of Pharmaceutical Science and Technology
Tianjin University
Tianjin, China

Maya Khater
Department of Biotechnology
MES Abasaheb Garware College
Pune, Maharashtra, India

Hira Khursheed
Fuji Foundation Medical College
Islamabad, Pakistan

Nidhi Kumari
College of Forestry
Sam Higginbottom University of Agriculture, Technology and Sciences (SHUATS), Prayagraj, Uttar Pradesh, India

Waqar Majeed
Department of Zoology, Wildlife and Fisheries
University of Agriculture
Faisalabad, Pakistan

Sayanti Mandal
Department of Chemistry & Biochemistry Sharda School of Engineering & Sciences Sharda University
Greater Noida
Uttar Pradesh, India

Ayesha Mobeen
Department of Biochemistry
University of Agriculture
Faisalabad, Pakistan

Nayla Munawar
Department of Chemistry
United Arab Emirates University
Al-Ain, United Arab Emirates

Ahmad Munir
Department of Biochemistry
University of Agriculture
Faisalabad, Pakistan

Muhammad Naeem
College of Life Science
Hebei Normal University
Shijiazhuang, China

Anam Niamat
Department of Biochemistry
University of Agriculture
Faisalabad, Pakistan

Karen Betsabe Miranda Ortega
Laboratorio de Productos Naturales
Instituto de Química
Universidad Nacional Autónoma de México
México

Aman Prakash
Department of Marine Biology
Leon H. Charney School of Marine Sciences
University of Haifa
Haifa, Israel

Elif Pulat
Program of Molecular Biotechnology and Genetics
Institute of Science
Istanbul University
Istanbul, Turkey

Sameer H. Qari
Department of Biology
Al-Jumum University College
Umm Al-Qura University
Makkah, Saudi Arabia

Uzma Ramzan
Institute of Zoology
University of the Punjab
Lahore, Pakistan

Naureen Rana
Department of Zoology, Wildlife and Fisheries
University of Agriculture
Faisalabad, Pakistan

Shashi Ranjan
Amrita Centre for Nanosciences and Molecular Medicine
Amrita Vishwa Vidyapeetham
Kochi, Kerala, India

Muhammad Rasheed
Department of Biochemistry
University of Agriculture
Faisalabad, Pakistan

Muhammad Zawwad Raza
Fuji Foundation Medical College
Islamabad, Pakistan

İlayda Reis
Department of Biotechnology
Faculty of Science
Mersin University
Mersin, Turkey

Aqsa Riaz
Department of Zoology, Wildlife and Fisheries
University of Agriculture
Faisalabad, Pakistan

Muhammad Riaz
Yale University School of Medicine
New Haven, CT, USA

Karla Daniela Rodríguez-Hernández
Laboratorio de Productos Naturales
Instituto de Química
Universidad Nacional Autónoma de México
México

Sehrish Sarwar
Department of Biochemistry
University of Agriculture
Faisalabad, Pakistan

Vijay Sheri
Department of Biology
East Carolina University
Greenville, USA

Raj Bahadur Singh
Department of Botany
Institute of Science
Banaras Hindu University
Lanka Varanasi, India

Sanjana Singh
Department of Botany
Institute of Science
Banaras Hindu University
Lanka Varanasi, India

Krishna Ravi Srinivas
NALSAR University of Law
Hyderabad, India
and
RIS
New Delhi, India

Dilek Tekdal
Department of Biotechnology
Faculty of Science
Mersin University
Mersin, Turkey

Pragalbh Tiwari
College of Forestry
Sam Higginbottom University of Agriculture, Technology and Sciences (SHUATS)
Prayagraj, Uttar Pradesh, India

Hua Feng Wang
Hainan Yazhou Bay Seed Laboratory
Sanya Nanfan Research Institute of Hainan University
Sanya, China

Revati Wanikar
Department of Biotechnology
MES Abasaheb Garware College
Pune, Maharashtra, India

Neslihan Yaşar
Department of Molecular Biology and Genetics
Faculty of Arts and Science
Baskent University
Ankara, Türkiye

Rohini Yevale
Department of Biotechnology
MES Abasaheb Garware College
Pune, Maharashtra, India

Şükran Yıldız
Department of Biotechnology
Faculty of Science
Mersin University
Mersin, Turkey

Hira Yousaf
Department of Biochemistry
University of Agriculture
Faisalabad, Pakistan

Hassan Zafar
Department of Pharmacology
University of Vermont
Vermont, USA

Muhammad Kashif Zahoor
Department of Zoology
Government College University
Faisalabad, Pakistan

Paulo Zaini
Department of Plant Sciences
University of California, Davis
California, USA

Qamar U. Zaman
Hainan Yazhou Bay Seed Laboratory
Sanya Nanfan Research Institute of Hainan University
Sanya, China

Jorge Alberto Zavala
Universidad de Buenos Aires
Facultad de Agronomía, Cátedra de Bioquímica
Buenos Aires, Argentina

Baohong Zhang
Department of Biology
East Carolina University
Greenville, USA

Part I

CRISPR Technology

Introduction, Design, and Delivery

1 Genome Editing

A Renewed Opportunity for Food Security

Nayla Munawar and Aftab Ahmad

1.1 GLOBAL FOOD SECURITY CHALLENGES

Thomas Malthus, an English scholar, warned that "population grows geometrically, whereas food supplies grow arithmetically" in his book, *An Essay on the Principle of Population*, published in 1798. Despite the reasons behind this, such as increased birth rate and advances in disease treatment, the predictions of an increase in the global population to almost 10 billion people by 2050 (FAO, 2009) have highlighted the insecurity of the food supply to feed such large numbers of people using the same area of cultivated land. According to the Food and Agricultural Organization of the United Nations (FAO), the production of food should be increased swiftly to meet the consumption demand (FAO, 2009). Regardless of the debate surrounding the assumptions behind the population increase, the demand for high-quality food continues to rise daily. Thus, food production must be increased efficiently and sustainably by 70% by the year 2050. However, climate change, water scarcity, disease outbreaks, pests, and land degradation are the main hindrances to achieving this goal.

Global environmental change includes all the changes within the Earth's system, from natural processes and human activities, with regard to ecosystems and climate. At the end of this century, the global temperature is most likely to rise up to 2.6°C to 4.8°C (Scafetta, 2024). The significant rise in temperature, along with an increased frequency of droughts, will affect vast areas of agricultural land used for food production. Changes to land use and continued climate change make up some of the key challenges to agriculture, mainly from a quantity and quality standpoint regarding water and land fertility properties. Climate change is more likely to shift the geographic distributions of pests and pathogens, which would result in new plant diseases and pest outbreaks posing significant threats to food security and public health. This will prospectively lead to major economic impacts on agriculture, further intensifying the challenges faced by agricultural systems (Anderson et al., 2004). The global loss of food crops like wheat (21.5%), maize (22.6%), soybean (21.4%), rice (30.3%), and potato (17.2%) is already high due to diseases and crop pests (Savary et al., 2019). Plant diseases are known to be the reason for significant pre-harvest losses in beans, maize, and potatoes, which result in low yield, high mitigation costs, and also affect human health (Delgado et al., 2017). Food security is a complex multidimensional challenge that needs urgent attention in order to integrate intense transdisciplinary research and innovations to enhance crop yield, promote disease resistance, and minimize climate impact.

1.2 BIOTECHNOLOGY IN AGRICULTURE

Plant breeding in agriculture is as old as human civilization. Traditional plant breeding was considered successful on the basis of improved quality and yield of edible parts, pests' resistance, and environmental stress as compared to the parent plants. Before the first plant hybridization by Kölreuter in the 1760s, natural plant breeding was the only technique used to obtain the desired trait in the plants. However, in the nineteenth and twentieth centuries, after the discovery of hereditary laws, the significance of hybridization was recognized, and it is still used in many plant breeding programs (Xu, 2010). Humans have been trying to improve plant traits for hundreds of years, however, the biotechnology revolution in the mid-1970s, with new advances in genetics and the production or alteration of products with the help of living organisms, sped up the process of incorporating new traits from any species enormously. Further, the addition of bioinformatics tools enhanced the understanding of genetics and improved the ability to transfer genes across species, thus making useful modifications possible. However, conventional plant breeding and hybridization require more than a decade to produce the desired traits or new variants and bring them to market. Breeding goals could still be achieved with these techniques but with the sacrifice of precision, genetic variability, and speed. Genetic modification of crops by inserting a gene between species faces huge public skepticism and regulatory hurdles. Genome editing offers an alternative approach for crop improvement with faster time to market.

1.3 GENOME-EDITING TECHNOLOGIES: OVERVIEW AND ADVANTAGES

Generally, genetically modified (GM) refers to the changes that may not occur naturally or through reproduction but necessarily through biotechnological processes (Ahmad et al., 2023).

DOI: 10.1201/9781003500933-2

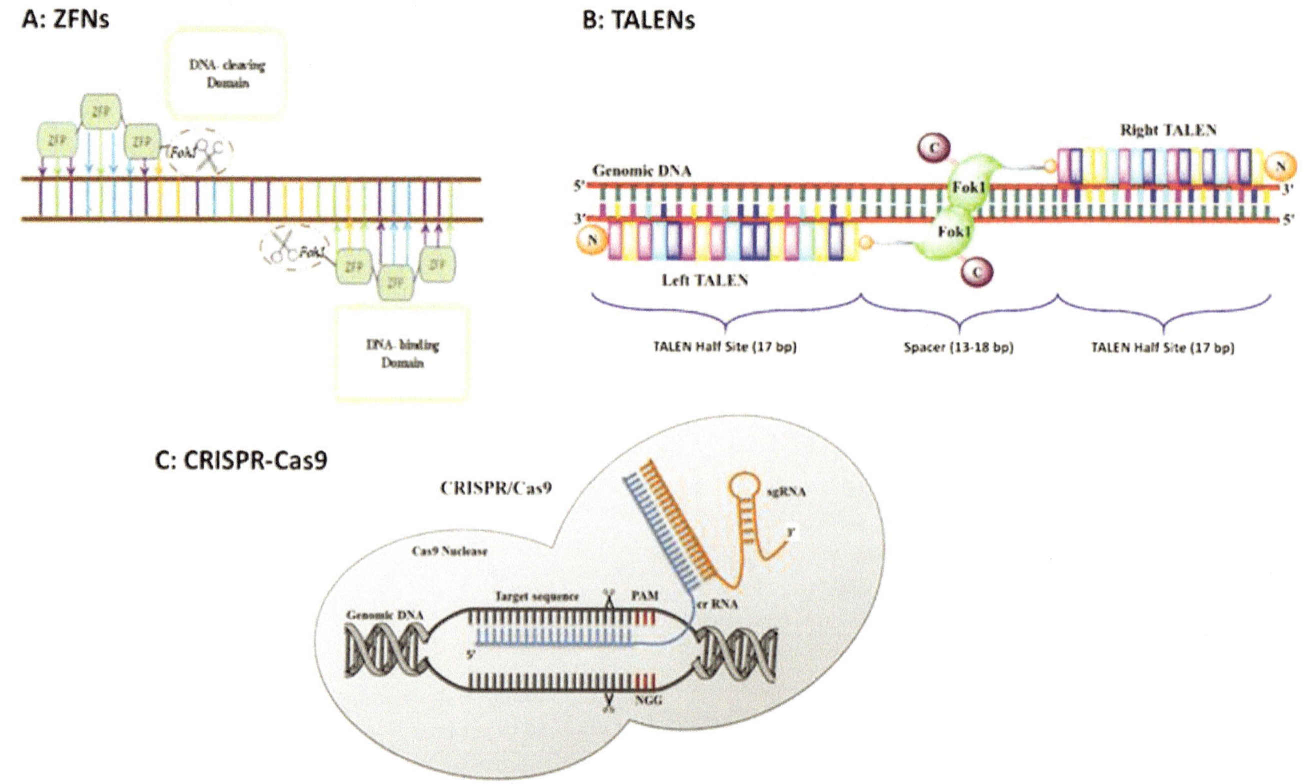

FIGURE 1.1 Endonucleases-based genome editing tools and their components. A. Each zinc finger recognizes three nucleotides in the DNA and is customizable. Zinc figures are fused to FokI nuclease; B. Customizable TALENs for each target are fused with FokI nuclease through a spacer; C. Cas 9 nuclease is guided by a customizable guide RNA (sgRNA) complementary to each target DNA. (Adapted from Munawar and Ahmad, 2022).

"Genome editing" refers to changes in the original DNA without the integration of exogenous gene sequences of interspecies. Advanced genome-editing techniques make use of designer DNA nucleases and the cell's own DNA repair machinery to make precise changes in the genome. The three most powerful genome-editing tools are (1) zinc finger nucleases (ZNFs), (2) transcription activator-like effector nucleases (TALENs), and (3) Clustered Regularly Interspaced Short Palindromic Repeats (CRISPR)-Cas9, as shown in Figure 1.1, and their advantages and disadvantages are briefly discussed below.

1.3.1 ZFNs

The discovery of zinc finger nucleases (ZFNs) in the 1990s represented a major breakthrough in genome editing. ZFNs are engineered DNA restriction enzymes having a zinc finger DNA-binding domain fused with a DNA-cleavage domain through a linker. One zinc figure domain recognizes three nucleotides in the target DNA, therefore, connecting multiple domains facilitates binding up to 9 to 18 base pairs of the target DNA. The linker region connects the zinc finger DNA-binding domains to the DNA cleavage domain, which is derived from the Fok1 restriction enzyme. Since FokI requires dimerization for activity, two FokI domains are fused with zinc finger domains to allow dimerization when two ZFNs bind to the target site on opposite DNA strands. The linker not only serves as a spacer between two domains but also provides a proper orientation to bind to the DNA and facilitates the cutting of DNA by FokI restriction enzymes on both sides to generate double-stranded breaks (DSBs) at the target site. The DSBs in the DNA provoke the natural DNA repair systems, non-homologous end joining (NHEJ) and homology-directed repair (HDR), in the cell. NHEJ is an error-prone process that naturally leads to deletions or insertions at the break site, resulting in gene disruption, whereas HDR allows precise insertion of a homologous DNA template, provided to the cell, to repair the break accurately.

The customized designing of zinc finger DNA binding domains for unique targets makes this tool versatile, having the capability to bind any region of DNA in the genome. However, the complexity of designing for every target, the high cost, the labor-intensive, off-target binding, and the low and difficult delivery in the cell have restricted the widespread use of this tool. Nonetheless, an understanding of the designing and targeted genome-editing procedure through ZFNs paved the way to create new, precise, and efficient tools like TALENs and CRISPR-Cas for DNA editing.

1.3.2 TALENs

In the early 2000s, the next breakthrough in the genome-editing technology came with the introduction of transcription activator-like effector nucleases (TALENs). FokI nuclease fused with a DNA-binding domain derived from transcription activator-like effectors to create an alternative tool called transcription activator-like effector nucleases (TALENs). The DNA-binding domain in TALENs is comprised of repeating motifs that recognize specific nucleotide sequences, offering improved specificity, straightforward customization, and ease of design compared to ZFNs. The binding of a pair of TALEN DNA-binding domains to their respective target site on opposite DNA strands provides an opportunity for dimerization of FokI nuclease to introduce double-stranded breaks in DNA that could be repaired through NHEJ or HDR, offering targeted genome editing. Because of high specificity and flexibility, TALENs have been applied in various fields, including agriculture, to target a wide range of DNA sequences successfully (Jankele and Svoboda, 2014). It was the most popular technique until 2013, followed by ZFNs (Christian et al., 2010). In both techniques, the DNA binding domains are fused to a non-specific FokI nuclease domain, which uses sticky ends of various lengths to cut the DNA, depending on the linker and spacer used, making the construct complex with less target specificity and low cutting efficiency. Moreover, the requirement of new proteins according to the verified target DNA makes construction more labor-intensive and time-consuming, which encouraged researchers to explore alternative tools for precise genome-editing applications.

1.3.3 CRISPR-Cas

The discovery of CRISPR/Cas technology in 2012 (Jinek et al., 2012) brought the era of genome editing to new heights and resulted in the fastest Nobel Prize in history being awarded to Emmanuelle Charpentier and Jennifer Doudna for their discovery in 2020. CRISPR/Cas solved the practical challenges posed by ZFNs and TALENs through its easy design, high target specificity, high transformation efficiency, low cost, and ability to target multiple genes simultaneously. "CRISPR" stands for "Clustered Regularly Interspaced Short Palindromic Repeats," initially observed by Dr. Nakata's group in *Escherichia coli* DNA in 1987 (Ishino et al., 1987) as part of the adaptive immune system in prokaryotes. After understanding the components required for its function and the mechanism of action of this natural system by various research groups over years (Bolotin et al., 2005; Mojica et al., 2005; Pourcel et al., 2005; Barrangou et al., 2007; Brouns et al., 2008; Deveau et al., 2008; Garneau et al., 2010; Deltcheva et al., 2011), scientists modified it and converted it into a powerful reprogrammable gene-editing tool (Jinek et al., 2012). Unlike ZFNs and TALENs, where the DNA binding and DNA cleavage (FokI nuclease) are two different proteins needing to be fused together, CRISPR has Cas9, one multidomain protein having endonuclease (NUC) and DNA binding/recognition (REC) lobes in its structure. The Cas protein can be guided to the target DNA by an RNA sequence called guide RNA (sgRNA), which is complementary to the target site. The NUC lobe of the Cas protein contains HNH and RuvC endonuclease domains along with a protospacer adjacent motif (PAM) interacting domain (Babu et al., 2021). PAM is a very specific 2–8 nucleotides (nt) sequence, which is highly desirable for Cas protein in some CRISPR types to recognize and cut the target DNA. Researchers only need to select a target site with a PAM sequence and construct a complementary RNA sequence to guide Cas 9 nuclease to the target where it introduces double-stranded DNA breaks through its nuclease activity. Specific genetic changes could be introduced by the cell's natural repair mechanisms, NHEJ or HDR (Yuan et al., 2024). Despite the off-target concerns by the scientific community, the simple design and easy programmability of CRISPR have made it the first choice of researchers worldwide for genome-editing applications in all fields, including agriculture and beyond.

Compared to conventional breeding and genetic alteration, genome editing has a number of significant advantages. First, it helps allay public fears about genetically modified organisms by enabling precise gene-level modifications without the introduction of foreign DNA into the genome. Second, compared to traditional breeding, which might take years or even decades to obtain comparable results, genome editing can yield results much more quickly. Third, genome editing makes it possible to change several genes at once, leading to the creation of crops with a variety of advantageous characteristics, including increased nutritional value, disease resistance, and drought tolerance.

1.4 APPLICATIONS OF CRISPR/CAS GENOME EDITING IN AGRICULTURE

CRISPR holds immense potential to transform the genetic makeup of agricultural crops with the possibility of fulfilling the early promise of genetic engineering of producing plants with better abiotic and biotic resistance, higher yield of edible parts with more nutrition (Figure 1.2) (Ahmad et al., 2021; Kumar et al., 2023).

1.5 IMPROVING CROP YIELD

The CRISPR/Cas-based gene editing in plants includes target site selection, sgRNA design and synthesis, selection of transformation vectors or ribonucleoprotein (RNP) in plant cells, and transformation and screening of gene-edited plants. Many genome-editing techniques, such as gene knock-in, knockout, knockdown, and stimulation of gene expression, have been proven by the CRISPR/Cas9 system and its variants in plants to result in improved crop yield. The literature review indicated that the number of publications on crop improvement using CRISPR/Cas has increased enormously in the last five years. The CRISPR/Cas 9-mediated gene editing in 41 food crops (Apple, Banana, Barley, Basil, Blueberry, Cabbage, Carrot, Cassava, Chickpea, Chilli, Citrus, Coconut, Cowpea, Cucumber, Date Palm, Grapefruit, Grapes, Kale, Kiwifruit, Lactuca sativa, Lemon, Lettuce, Lychee, Maize, Melon, Oats, Orange, Papaya, Pear, Pepper, Potato, Pumpkin, Rice, Saffron, Strawberry, Sugar beet, Sweet potato, Tomato, Watermelon, Wheat, Yam), 14 industrial crops

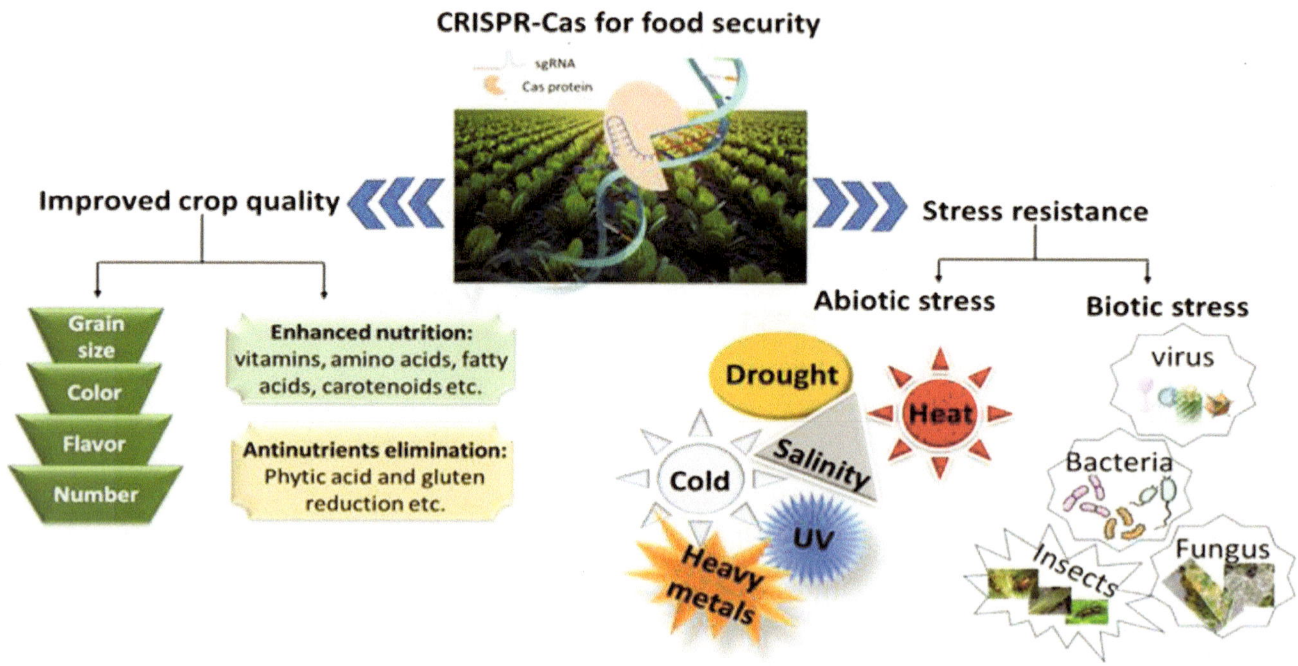

FIGURE 1.2 Potential role of genome editing by CRSIPR/Cas in agriculture for food security.

(Cichorium intybus, Coffee, Dandelion, Hevea brasiliesis, Jatropha curcas, Millet, Papaver, Parasponia, Salvia miltiorrhiza, Sorghum, Sugarcane, Switchgrass, Tragopogon, Tripterygium wilfordii), 6 oil crops (Canola, Flax, Oil palm, Oilseed rape, Soybean, Sunflower), 8 ornamental crops (Lily, Lotus, Petunia, Poplar, Rose, Sedum, Snapdragon, Torenia fournieri), 1 fiber crop (Cotton), and 1 feed crop (Alfalfa) have been reported (Liu et al., 2021). The grain size and shape of food crops affect the quality and weight of the edible part. The size and shape of the fruit have been regulated by gene knockouts in rice and tomato. The knockout of *GRAIN SIZE 3 (GS3)*, a quantitative trait loci (QTLs), in five japonica rice varieties, increased the grain length, while improved grain weight has been observed by targeting multiple, GW2, GW5, and GW6, negative regulators of grain weight (Xu et al., 2016). Knockout of *TaGW7* in wheat resulted in an increased grain width and weight (Wang et al., 2019). Modifications in the expression of *ENO* (Yuste-Lisbona et al., 2020), *OVATE, CLV* (Zsögön et al., 2018), and *fas* and *lc* (Rodríguez-Leal et al., 2017) positively modulated tomato fruit shape and size. RING-type E3 ubiquitin ligase encoded by *GRAIN WIDTH and WEIGHT2* (*GW2*) genes determine grain weight in cereal crops. A GW2 knockout in rice (var. MTU1010) by CRISPR in India showed improved grain filling architecture and nutritional quality, suggesting the possibility of improvement in other cereal crops through GW2 gene knockout (Achary and Reddy, 2021). Mutations by CRISPR/Cas9 in *GS3*, *TGW3*, and *GW8* grain size genes in three-line hybrid rice showed significant effects on the grain size with minimal impact on agronomic traits of the crop (Huang et al., 2024), indicating the effective use of genome editing to obtain high yield and quality for food security.

1.6 ABIOTIC STRESS TOLERANCE

Abiotic stresses such as drought, heat, cold, salt, UV rays, heavy metals, and oxidative stress significantly reduce crop output and agricultural yields. CRISPR/Cas9 tools to target particular genes involved in stress responses provide precise editing that could improve crop grains' tolerance against abiotic stresses and enhance agricultural sustainability.

1.6.1 Drought

A major obstacle to food security is the susceptibility of staple grain crops like wheat, rice, and maize to water constraints. A precise genetic alteration with CRISPR-Cas in ZmHDT103 gene, which is a component of the ABA signaling pathway, was demonstrated to increase maize's resistance to drought by strengthening the plant's ability to withstand water scarcity without impairing growth or yield in typical circumstances (Wang et al., 2024). CRISPR-Cas editing of the Sal1 gene in wheat improved the production of osmoprotectants like proline and enabled the survival of the crop during a dry period, showing the potential of improved drought resistance through CRISPR technology (Mohr et al., 2022). Alteration in the TaRPK1 gene in wheat enhanced the water absorption capacity of the plants to contest drought stress (Abdul Rahim et al., 2024).

1.6.2 Salinity

Soil salinization, due to poor irrigation practices and rising sea levels, is a growing agricultural problem. CRISPR-Cas offers a solution through tailoring the genes involved in ion homeostasis and the osmotic balance in crops. For instance, modification in sodium uptake and compartmentalization transporters can reduce sodium accumulation in the cell and improve salt tolerance, as experienced in soybean and barley. Enhanced salt tolerance has been observed in many crops, including rice and soybeans, by upregulation of the DREB2A transcription factor (Farhat et al., 2019). Moreover, the knockout of WRKYs and AKT1 genes enhanced salt tolerance while maintaining growth and yield in barley and soybean, as reported by Price (2022) and Feng et al. (2023), respectively.

1.6.3 Temperature

Global warming is changing the climate to extreme heat and cold temperatures, producing challenging environments for plants to maintain photosynthetic efficiency, thus impairing plant growth and decreasing yield. By targeting heat shock proteins (HSPs) and factors (HSFs) genes, CRISPR/Cas has been used to develop heat tolerance in plants. For example, upregulation of HsfA1b expression has been shown to improve heat tolerance in wheat by allowing plants to withstand high temperatures while maintaining growth and yields (Tian et al., 2020). Other innovative approaches involve using CRISPR/Cas to increase the synthesis of osmolytes to protect the plant from heat (Yadav et al., 2023). Combining CRISPR with gene drive technology can provide an opportunity to spread heat tolerance traits in the entire crop rapidly (Mohan et al., 2022). The type and amount of lipids present in the cell determine its fluidity. A well-documented CBF (C-repeat Binding Factor) pathway is involved in temperature regulation through CBF transcription factors, which activate a suite of genes that confer cold tolerance in plants. An innovative approach to enhance cold tolerance could be an increase in fatty acid content in the cell. CRISPR-Cas-mediated upregulation of fatty acid desaturase genes changed the lipid composition in the membrane and provided better support to cellular processes during cold exposure. Improved unsaturated fatty acids content in rice membranes through CRISPR-Cas-mediated OsKASI-2 gene expanded cold tolerance in the crop (Zhang et al., 2024b). Epigenetic modifications by CRISPR-Cas to regulate the cold-related gene expression have also been reported as an innovative and flexible approach to crop improvement in a chilled environment (Chen et al., 2024).

1.6.4 Heavy metals and UV radiation

CRISPR/Cas technology offers a practical approach to negate the adverse effects of heavy metals, UV radiation, and oxidative stress on crop health through precise genetic alterations that increase tolerance to soil contaminated with heavy metals, DNA and oxidative damage, and interrupted photosynthesis. The construction of low heavy metal rice in cadmium-contaminated soil is a challenging task for researchers. Reduction in cadmium uptake and its accumulation in edible parts, as well as increased cadmium tolerance, have been observed by Chen et al. (2023) by editing the OsLCD gene with CRISPR-Cas in rice. Moreover, improved arsenic tolerance in arsenic-contaminated soil and its reduced accumulation have been reported by Xu et al. (2024) after targeting Lsi1 and Lsi2 genes in rice. Modifications in the ZmHMA3 gene resulted in enhanced zinc tolerance in maize by improving compartmentalization and detoxification of this metal (Lv et al., 2023). CRISPR's ability to improve UV tolerance has been proved by modifying the OsCOP1 gene in rice, which made the crop resistant to UV-B radiation and protected against DNA damage under high UV (Hu et al., 2024). The CAT (catalase) gene family, particularly OsCAT3, is crucial for a reduction in oxidative stress in plants through detoxifying hydrogen peroxide and superoxide radicals and was targeted to mitigate oxidative damage in rice (Jiang et al., 2023).

1.7 BIOTIC STRESS RESISTANCE

Biotic stress includes viral, fungal, and bacterial threats that destroy 10–30% of agricultural production per year worldwide. The use of pesticides not only contaminates water and soil but also generates pesticide-resistant microbial species, requiring higher concentrations to control them, which results in bioaccumulation of chemicals in the environment. Genomic editing opens a new vista to improve plant defense mechanisms against pathogens, which reduces disease pressure on plant growth and productivity, which is highly relevant to the lack of food availability in the world due to increasing populations. CRISPR/Cas techniques have been successfully employed for disease resistance against fungal, bacterial, and viral infections. For example, CRISPR-Cas mediated knockout of susceptibility genes such as *TaPDIL5* or *OsDjA2* and *OsERF* provided a broad-spectrum virus resistance in wheat and rice (Kan et al., 2022; Távora et al., 2022). Replication inhibition of the sugarcane mosaic virus (SCMV) in maize was achieved by knocking out ZmPDRP1 (Xie et al., 2024). The *StNRL1* gene knockout in potatoes enhanced resistance to botj late blight and increased susceptibility to early blight induced by Phytophthora infestans and Alternaria alternate, respectively (Norouzi et al., 2024). Engineering plant immune receptors with CRISPR/Cas has enabled a much more effective perception of bacterial pathogens. Enhanced recognition and response against bacterial infections have been reported in rice and wheat by editing *FERONI* and *SlWak1*, and the *FLS* gene, which encodes a receptor involved in pathogen recognition (Huang et al., 2020; Zhang et al., 2020). To reduce fungal infections, two approaches could be adapted (1) knockout susceptibility genes in plants that fungi exploit, and (2) edit the genome of the plant pathogenic fungus. MLO (Mildew Locus O) genes knockout in wheat and soybean confer resistance to powdery mildew normally caused by fungal infections (Li et al., 2022; Bui et al., 2023). Editing of the MeRPPL1 gene showed enhanced fungal resistance in the cassava crop (Ramulifho and Rey, 2024). Overexpression of phenolic compounds production genes showed reduced insect herbivory in soybean and maize (Razzaq et al., 2023; Kumari et al., 2024). Production of high levels of insecticidal proteins in plants through genetic modifications results in an effective natural defense in plants against pests. CRISPR-Cas modification of overexpression of insecticidal encoding genes for Cry and VIP proteins improved pest resistance in crops (Dubovskiy et al., 2024).

1.8 IMPROVED NUTRITION AND ELIMINATING ANTI-NUTRIENTS

Millions of people, particularly in the developing world, suffer from micronutrient malnutrition. Biofortification here refers to the enhancement of the nutritional traits of crops by the use of genome editing. An example would be the application of CRISPR-Cas9 to improve the β-carotene content of rice; the product is commercially known as Golden Rice (Datta et al., 2021). Its enhanced vitamin A content is intended to help combat vitamin A deficiency, which is one of the leading causes of preventable blindness among children. Improving iron and zinc content in staple crops such as wheat and maize has been done in a similar vein by editing the *OsNAS* gene to avoid anemia and impaired immune function. CRISPR-Cas has been used in developing "Golden Maize" by boosting the biosynthesis of pro-vitamin A through modification of the *PSY1*, *Crtl*, and *LCYB* genes in the crop (Sobrino-Mengual et al., 2024). Increased levels of vitamins and essential amino acids in cassava have been achieved with CRISPR-Cas editing, which significantly improved the nutritional quality of this staple crop (Otun et al., 2023). Reduction in anti-nutritional factors can enhance the nutritional quality of the crop. For instance, a reduction in phytic acid production enhanced the bioavailability of zinc and iron in soybean by targeting the *GmIPK1* gene using CRISPR-Cas technology (Song et al., 2022). Knockout of the *ITPK* gene by CRISPR/Cas9 in rapeseeds exhibited a 35% reduction in phytic acid without a change in plant performance (Sashidhar et al., 2020). Targeting of a conserved region of the α-gliadin genes using CRISPR-Cas9 generated transgene-free low-gluten wheat lines for gluten-intolerant individuals (Hischenhuber et al., 2006). In addition to the above-mentioned applications of genome editing, CRISPR-Cas technology has been utilized for enhanced production of fatty acids, oil contents, γ-Aminobutyric acid and carotenoid contents, improving cooking and eating quality and flavor, prolonging shelf life, enhancing nitrogen use efficiency, and reduction in the use of chemical inputs (Wang et al., 2022), indicating the strength of genome editing for sustainable agriculture.

1.9 ETHICAL CONCERNS AND REGULATORY CHALLENGES

Although there is a lot of promise for increasing food security through CRISPR-Cas-mediated genome editing, off-target effects or the disruption of significant genetic circuits, and illegal or irresponsible experimentation are the main worries in the scientific community. Off-target effects of CRISPR-Cas9-based genome editing in functional studies of plant genome have already been reported (Zhang et al., 2015; Zhang et al., 2018). Nonetheless, with the recent advance in the technology, the frequency of off-target mutations in plant genome editing is rare and undetectable (Eş et al., 2019). Anderson et al. (2016) revealed the presence of less genetic variation in transgenic soybean developed through Agrobacterium-mediated transformation compared to radiation mutagenesis. Moreover, a few off-target modifications have been observed by Li, Manghwar et al. (2019) in CRISPER-Cas9-edited cotton plants, indicating the reduction in non-targeted mutations related risks and concerns. Extension of CRISPR-Cas tools with the discovery of novel Cas proteins, Cas 12 and Cas 13, and the combination of prime editing and base editing with CRISPR-Cas 9 presented groundbreaking developments, enabling direct and precise editing of target DNA sequences (Chen and Liu, 2023; Naeem et al., 2023). However, a huge amount of work needs to be done with full dedication to develop highly precise CRISPR-Cas tools to eradicate the risk of off-target mutations completely, which could lead to unexpected and unpredictable consequences in the future.

The big ethical and regulatory concern about genome editing arose in November 2018 after the claim of a Chinese scientist, Jiankui He, about the use of CRISPR-Cas9 to create genetically modified babies immune to the human immunodeficiency virus (HIV) (Zhang et al., 2020). The genome editing in human embryo germ lines for any purpose, including clinical use, is highly prohibited because of ethical and scientific concerns (Li, Walker et al., 2019), but the above-mentioned incident highlighted the probability of illegal and misuse of CRISPR technology.

Besides the tremendous potential of rewriting the genome system of the target organisms, the concept of using genome editing in plant science has raised controversy about environmental and health risks:

1. Genetically modified plants having insecticidal genes may have a detrimental effect on biodiversity by injuring non-target insect species;
2. The natural gene transfer between modified plants and insects may result in the emergence of resistant weeds and insects;
3. Novel gene combinations may result in long-term unpredictable impacts on human health;
4. Construction of herbicide-tolerant plants may encourage farmers to employ broad-spectrum herbicides, which could harm animals in that region;
5. The presence of antibiotic marker genes in plants could result in antibiotic-resistant infectious microorganisms that would need strong antibiotics to kill and would be a serious health risk for humans;
6. The production of new unnatural proteins in modified plants could result in unknown and unwanted allergic reactions in humans and animals;
7. Some believe that genome editing is "tampering with nature" and would have long-term dangerous effects on the environment.

Due to safety concerns, poor communication with and poor engagement with civil society, genome editing in agriculture is perceived negatively in public opinion, creating social, cultural, and spiritual conflicts, thus generating regulatory and governance challenges worldwide (Ahmad et al., 2023). Many countries, including several European countries, have restrictions on the cultivation of modified crops and the consumption of their products because of public concern. However, CRISPR technology does not use a foreign gene in many types and is still evolving, there is still a lot to learn about the long-term effects of CRISPR-edited plants on human health and the environment. Scientists are working on the tool with the aim of introducing minor changes in the native genome through CRISPR-Cas technology to obtain the desired traits instead of changing larger pieces of DNA, in order to address safety concerns and challenges. Therefore, the implementation of a more dependable, inclusive, transparent, and customized policy and regulatory approach by the governing bodies of countries to gene editing can foster public trust in the critical time of food insecurity and climate change.

1.10 CONCLUSION

The CRISPR/Cas9 system has emerged as one of the most utilized and versatile technologies to modulate crop-specific traits in a more precise and efficient manner. Due to its unparalleled capability for gene modulation, this technique has helped develop a number of varieties of crops exhibiting desired performances with respect to agronomic traits. However, most of the work on gene editing for crop improvement is still at a descriptive stage, mostly limited to describing the function and regulation mechanisms of the genome. The technology has shown the potential to revolutionize the improvement in resilience, yield, and nutritional quality of staple crops such as rice and maize by creating stress tolerance, resistance, and productivity improvement through accurate modifications at the genomic level. Nonetheless, the commercialization of gene-edited crops is still a long way off. The delivery of CRISPR cargoes to cells and off-target mutations remains a huge challenge. In addition, public concern and strict government regulation policies on gene-editing technology stand as another obstacle to innovations in the field. Though there are still some challenges that have to be overcome, indeed, based on its current use and future perceptions, the CRISPR/Cas system holds enormous potential to emerge as a powerful DNA and RNA editing technology in modern agriculture. Most of the successful genome-editing experiments predominantly involve gene knockout by CRISPR, while CRISPR/Cas-based knock-in approaches provide a broader canvas for targeted gene insertion, which is essential to obtain beneficial traits such as stress tolerance, disease resistance, and improved nutritional contents. Gene knock-in relies on the HDR repair system, which has lower efficiency than the NHEJ DNA repair system, posing a problem. Several strategies, such as enhancing HDR pathways, developing better CRISPR-Cas variants, and exploring alternative genome-editing approaches like CRISPR nickase and prime editing, are under investigation to improve gene knock-in with minimal off-target effects. CRISPR- Cas genome editing should not be considered a standalone solution to all agriculture issues and food security. Combining CRISPR-Cas, with other techniques like precision agriculture, advanced breeding techniques, artificial intelligence, prime editing, base editing, engineering of Cas proteins, and use of nanoparticle-based delivery systems, can maximize its impact. The continuous improvements in technology are crucial to amplifying its role in advancing agricultural innovation, but to harness the full potential of CRISPR-Cas, serious efforts need to be made to resolve the ethical and regulatory issues. Clear communication between scientists, politicians, policy regulators, and civil society will improve public acceptability for modified crops and will help achieve the zero-hunger goal and maintain food sustainability for the growing human population.

REFERENCES

Abdul Rahim, A., Uzair, M., Rehman, N., Fiaz, S., Attia, K. A., Abushady, A. M., et al. (2024). CRISPR/Cas9 mediated TaRPK1 root architecture gene mutagenesis confers enhanced wheat yield. *Journal of King Saud University – Science* 36, 103063. doi: 10.1016/j.jksus.2023.103063.

Achary, V. M. M., Reddy, M. K. (2021). CRISPR-Cas9 mediated mutation in GRAIN WIDTH and WEIGHT2 (GW2) locus improves aleurone layer and grain nutritional quality in rice. *Scientific Reports* 11, 21941. https://doi.org/10.1038/s41598-021-00828-z.

Ahmad, A., Jamil, A., Munawar, N. (2023). GMOs or non-GMOs? The CRISPR conundrum. *Frontiers in Plant Science* 14, 1232938. doi: 10.3389/fpls.2023.1232938.

Ahmad, A., Munawar, N., Khan, Z., Qusmani, A. T., Khan, S. H., Jamil, A., Ashraf, /S., Ghouri, M. Z., Aslam, S., Mubarik, M. S., Munir, A., Sultan, Q., Abd-Elsalam, K. A., Qari, S. H. (2021). An outlook on global regulatory landscape for genome-edited crops. *International Journal of Molecular Science* 22(21), 11753. doi: 10.3390/ijms222111753.

Anderson, J. E., Michno, J.-M., Kono, T. J., Stec, A. O., Campbell, B. W., Curtin, S. J., Stupar, R. M. (2016). Genomic variation and DNA repair associated with soybean transgenesis: A comparison to cultivars and mutagenized plants. *BMC Biotechnology* 16, 41.

Anderson P. K., Cunningham, A. A., Patel, N. G., Morales, F. J. et al. (2004). Emerging infectious diseases of plants: Pathogen pollution, climate change and agrotechnology drivers. *Trends in Ecology and Evolution* 19, 535–544.

Babu K., Kathiresan, V., Kumari, P., Newsom, S., Parameshwaran, H. P., Chen, X., Liu, J., Qin, P. Z., Rajan, R. (2021). Coordinated actions of Cas9 HNH and RuvC nuclease domains are regulated by the bridge Helix and the target DNA sequence. *Biochemistry* 60(49), 3783–3800.

Barrangou, R., Fremaux, C., Deveau, H., Richards, M., et al. (2007). CRISPR provides acquired resistance against viruses in prokaryotes. *Science* 315, 1709–1712.

Bolotin, A., Quinquis, B., Sorokin, A., Ehrlich, S. D. (2005). Clustered regularly interspaced short palindrome repeats (CRISPRs) have spacers of extrachromosomal origin. *Microbiology* 151, 2551–2561.

Brouns, S. J., Jore, M. M., Lundgren, M., Westra, E. R., et al. (2008). Small CRISPR RNAs guide antiviral defense in prokaryotes. *Science* 321, 960–964.

Bui, T. P., Le, H., Ta, D. T., Nguyen, C. X., Le, N. T., Tran, T. T., et al. (2023). Enhancing powdery mildew resistance in soybean by targeted mutation of MLO genes using the CRISPR/

Cas9 system. *BMC Plant Biology* 23, 533. doi: 10.1186/s12870-023-04549-5

Chen, F., Chen, L., Yan, Z., Xu, J., Feng, L., He, N., Guo, M., Zhao, J., Chen, Z., Chen, H., Yao, G., Liu, C. (2024). Recent advances of CRISPR-based genome editing for enhancing staple crops. *Frontiers in Plant Science* 15, 1478398. doi: 10.3389/fpls.2024.1478398.

Chen, H., Ye, R., Liang, Y., Zhang, S., Liu, X., Sun, C., et al. (2023). Generation of low-cadmium rice germplasms via knockout of OsLCD using CRISPR/Cas9. *Journal of Environmental Sciences* 126, 138–152. doi: 10.1016/j.jes.2022.05.047

Chen, P. J., Liu, D. R. (2023). Prime editing for precise and highly versatile genome manipulation. *Nature Reviews Genetics* 24, 161–177. doi: 10.1038/s41576-022-00541-1.

Christian, M., Cermak, T., Doyle, E. L., Schmidt, C., Zhang, F., Hummel, A., Voytas, D. F. (2010). Targeting DNA double-strand breaks with TAL effector nucleases. *Genetics* 186, 757–761. doi: 10.1534/genetics.110.120717.

Datta, S. K., Majumder, S., Datta, K. (2021). Molecular breeding for improved β-carotene synthesis in Golden Rice: Recent progress and future perspectives. In M. A. Hossain, L. Hassan, K. Md. Ifterkharuddaula, A. Kumar, R. Henry (Eds.), *Molecular Breeding for Rice Abiotic Stress Tolerance and Nutritional Quality*. Hoboken, NJ: Wiley, pp. 287–303. doi: 10.1002/9781119633174.ch15

Delgado L., Schuster M., Torero M. (2017). The reality of food losses: A new measurement methodology. IFPRI Discussion Paper 1686, Washington, DC: International Food Policy Research Institute.

Deltcheva, E., Chylinski, K., Sharma, C., et al. (2011). CRISPR RNA maturation by trans-encoded small RNA and host factor RNase III. *Nature* 471, 602–607. https://doi.org/10.1038/nature09886.

Deveau, H., Barrangou, R., Garneau, J. E., Labonte, J., Fremaux, C., et al. (2008). Phage response to CRISPR-encoded resistance in Streptococcus thermophilus. *Journal of Bacteriology* 190, 1390–1400.

Dubovskiy, I. M., Grizanova, E. V., Gerasimova, S. V. (2024). Plant recombinant gene technology for pest control in the twenty-first century: from simple transgenesis to CRISPR/Cas. In A. Kumar, S. Arora, S. Ogita, Y.-Y. Yau, K. Mukherjee (Eds.), *Gene Editing in Plants: CRISPR-Cas and Its Applications*. Singapore: Springer Nature Singapore, pp. 17–72.

Eş, I., Gavahian, M., Marti-Quijal, F. J., Lorenzo, J. M., Khaneghah, A. M., Tsatsanis, C., Kampranis, S. C., et al. (2019). The application of the CRISPR-Cas9 genome editing machinery in food and agricultural science: Current status, future perspectives, and associated challenges. *Biotechnol. Adv.* 37, 410–421.

FAO. (2009). Global agriculture towards 2050. Paper presented to High-level Expert Forum on "How to feed the world in 2050." *FAO*, Rome, 12–13 October 2009.

Farhat, S., Jain, N., Singh, N., Sreevathsa, R., Dash, P. K., Rai, R., et al. (2019). CRISPR-Cas9 directed genome engineering for enhancing salt stress tolerance in rice. *Seminars in Cell & Developmental Biology* 96, 91–99. doi: 10.1016/j.semcdb.2019.05.003

Feng, C., Gao, H., Zhou, Y., Jing, Y., Li, S., Yan, Z., et al. (2023). Unfolding molecular switches for salt stress resilience in soybean: recent advances and prospects for salt-tolerant smart plant production. *Frontiers in Plant Science* 14, 1162014. doi: 10.3389/fpls.2023.1162014

Garneau, J., Dupuis, M. È., Villion, M., et al. (2010). The CRISPR/Cas bacterial immune system cleaves bacteriophage and plasmid DNA. *Nature* 468, 67–71. https://doi.org/10.1038/nature09523.

Hischenhuber, C., Crevel, R., Jarry, B., Mäki, M., Moneret-Vautrin, D., Romano, A., Troncone, R., Ward, R. (2006). Safe amounts of gluten for patients with wheat allergy or coeliac disease. *Alimentary Pharmacology & Therapeutics* 23, 559–575. doi: 10.1111/j.1365-2036.2006.02768.x

Hu, S., Chen, Y., Qian, C., Ren, H., Liang, X., Tao, W., et al. (2024). Nuclear accumulation of rice UV-B photoreceptors is UV-B- and OsCOP1-independent for UV-B responses. *Nature Communications* 15, 6396. doi: 10.1038/s41467-024-50755-6

Huang, J., Chen, W., Gao, L., et al. (2024). Rapid improvement of grain appearance in three-line hybrid rice via CRISPR/Cas9 editing of grain size genes. *Theoretical and Applied Genetis* 137, 173. https://doi.org/10.1007/s00122-024-04627-8

Huang Y.-Y., Liu X.-X., Xie Y., Lin X.-Y., Hu Z.-J., Wang H., et al. (2020). Identification of FERONIA-like receptor genes involved in rice-Magnaporthe oryzae interaction. *Phytopathology Research* 2, 14. doi: 10.1186/s42483-020-00052-z

Ishino, Y., Shinagawa, H., Makino, K., Amemura, M., Nakata, A. (1987). Nucleotide sequence of the iap gene, responsible for alkaline phosphatase isozyme conversion in *Escherichia coli*, and identification of the gene product. *Journal of Bacteriology* 169, 5429–5433.

Jankele, R., Svoboda, P. (2014). TAL effectors: Tools for DNA targeting. *Briefings in Functional Genomics* 13, 409–419. doi: 10.1093/bfgp/elu013.

Jiang, W., Ye, Q., Wu, Z., Zhang, Q., Wang, L., Liu, J., et al. (2023). Analysis of CAT gene family and functional identification of OsCAT3 in rice. *Genes* 14, 138. doi: 10.3390/genes14010138

Jinek, M., Chylinski, K., Fonfara, I., Hauer, M., Doudna, J. A., Charpentier, E. (2012). A programmable dual-RNA–guided DNA endonuclease in adaptive bacterial immunity. *Science* 337(6096), 816–821.

Kan, J., Cai, Y., Cheng, C., Jiang, C., Jin, Y., Yang, P. (2022). Simultaneous editing of host factor gene TaPDIL5-1 homoeoalleles confers wheat yellow mosaic virus resistance in hexaploid wheat. *New Phytologist* 234, 340–344. doi: 10.1111/nph.v234.2.

Kumar, M., Prusty, M. R., Pandey, M. K., Singh, P. K., Bohra, A., Guo, B., Varshney, R. K. (2023). Application of CRISPR/Cas9-mediated gene editing for abiotic stress management in crop plants. *Frontiers in Plant Sciience* 14, 1157678. doi: 10.3389/fpls.2023.1157678.

Kumari, R., Saha, T., Kumar, P., Singh, A. K. (2024). CRISPR/Cas9-mediated genome editing technique to control fall armyworm (Spodoptera frugiperda) in crop plants with special reference to maize. *Physiology and Molecular Biology of Plants* 30, 1161–1173. doi: 10.1007/s12298-024-01486-x

Li, J., Manghwar, H., Sun, L., Wang, P., Wang, G., Sheng, H., Zhang, J. et al. (2019). Whole genome sequencing reveals rare off-target mutations and considerable inherent genetic or/and somaclonal variations in CRISPR/Cas9-edited cotton plants. *Plant Biotechnology Journal* 17, 858–868.

Li, J. R., Walker, S., Nie, J. B., Zhang, X. Q. (2019). Experiments that led to the first gene-edited babies: The ethical failings and the urgent need for better governance. *Journal of Zhejiang University - Science. B* 20, 32–38.

Li, S., Lin, D., Zhang, Y., Deng, M., Chen, Y., Lv, B., et al. (2022). Genome-edited powdery mildew resistance in wheat without growth penalties. *Nature* 602, 455–460. doi: 10.1038/s41586-022-04395-9

Liu, Q., Yang, F., Zhang, J., Liu, H., Rahman, S., Islam, S., Ma, W., She, M. (2021). Application of CRISPR/Cas9 in crop quality improvement. *International Journal of Molecular Science* 22(8), 4206. doi: 10.3390/ijms22084206. PMID: 33921600; PMCID: PMC8073294.

Lv, G.-H., Li, Y.-Q., Chen, J.-J., Wu, Z.-X., Wu, W.-M., Wu, X.-H., et al. (2023). ZmHMA3 enhances heavy mental Zn stress tolerance and mediates Zn transport in Maize. *Research Square.* doi: 10.21203/rs.3.rs-2970282/v1

Mohan, C., Satish, L., Muthubharathi, B. C., Selvarajan, D., Easterling, M., Yau, Y.-Y. (2022). CRISPR-Cas technology: A genome-editing powerhouse for molecular plant breeding, In S. Arora, A. Kumar, S. Ogita, Y. Y. Yau (Eds.), *Biotechnological Innovations for Environmental Bioremediation.* Singapore: Springer Nature Singapore, pp. 803–879.

Mohr, T., Horstman, J., Gu, Y. Q., Elarabi, N. I., Abdallah, N. A., Thilmony, R. (2022). CRISPR-Cas9 gene editing of the Sal1 gene family in wheat. *Plants* 11, 2259. doi: 10.3390/plants11172259

Mojica, F. J., Diez-Villasenor, C., Garcia-Martinez, J., Soria, E. (2005). Intervening sequences of regularly spaced prokaryotic repeats derive from foreign genetic elements. *Journal of Molecular Evolution* 60, 174–182.

Munawar, N., Ahmad, A. (2022). An introduction to genome editing techniques. In A. Ahmad, S. H. Khan, Z. Khan (Eds.), *The CRISPR/Cas Tool Kit for Genome Editing*. Singapore: *Springer*. https://doi.org/10.1007/978-981-16-6305-5_1.

Naeem, M., Alkhnbashi, O. S. (2023). Current bioinformatics tools to optimize CRISPR/Cas9 experiments to reduce off-target effects. *International Journal of Molecular Science* 24, 6261. doi: 10.3390/ijms24076261.

Norouzi, M., Nazarain-Firouzabadi, F., Ismaili, A., Ahmadvand, R., Poormazaheri, H. (2024). CRISPR/Cas StNRL1 gene knockout increases resistance to late blight and susceptibility to early blight in potato. *Frontiers in Plant Science* 14, 1278127. doi: 10.3389/fpls.2023.1278127

Otun, S., Escrich, A., Achilonu, I., Rauwane, M., Lerma-Escalera, J. A., Morones-Ramírez, J. R., et al. (2023). The future of cassava in the era of biotechnology in Southern Africa. *Critical Reviews in Biotechnology* 43, 594–612. doi: 10.1080/07388551.2022.2048791

Pourcel, C., Salvignol, G., Vergnaud, G. (2005). CRISPR elements in Yersinia pestis acquire new repeats by preferential uptake of bacteriophage DNA, and provide additional tools for evolutionary studies. *Microbiology* 151, 653–663.

Price, L. J. (2022). Investigating barley WRKYs for salinity tolerance by scrutinising CRISPR edited mutants. College of Science, Health, Engineering and Education, Murdoch University, Perth, Australia.

Ramulifho, E., Rey, C. (2024). A coiled-coil nucleotide-binding domain leucine-rich repeat receptor gene MeRPPL1 plays a role in the replication of a geminivirus in cassava. *Viruses* 16, 941. doi: 10.3390/v16060941

Razzaq, M. K., Hina, A., Abbasi, A., Karikari, B., Ashraf, H. J., Mohiuddin, M., et al. (2023). Molecular and genetic insights into secondary metabolic regulation underlying insect-pest resistance in legumes. *Functional and Integrative Genomics* 23, 217. doi: 10.1007/s10142-023-01141-w

Rodríguez-Leal, D., Lemmon, Z. H., Man, J., Bartlett, M. E., Lippman, Z. B. (2017). Engineering quantitative trait variation for crop improvement by genome editing. *Cell* 171, 470–480. doi: 10.1016/j.cell.2017.08.030.

Sashidhar, N., Harloff, H. J., Potgieter, L., Jung, C. (2020). Gene editing of three BnITPK genes in tetraploid oilseed rape leads to significant reduction of phytic acid in seeds. *Plant Biotechnology Journal* 18, 2241–2250. doi: 10.1111/pbi.13380.

Savary, S., Willocquet, L., Pethybridge, S. J., Esker, P., et al. (2019). The global burden of pathogens and pests on major food crops. *Nature Ecology & Evolution* 3, 430–439.

Scafetta, N. (2024). Impacts and risks of "realistic" global warming projections for the 21st century. *Geoscience Frontiers* 2024; 15, 2. https://doi.org/10.1016/j.gsf.2023.101774.

Sobrino-Mengual, G., Alvarez, D., Twyman, R. M., Gerrish, C., Fraser, P. D., Capell, T., et al. (2024). Activation of the native PHYTOENE SYNTHASE 1 promoter by modifying near-miss cis-acting elements induces carotenoid biosynthesis in embryogenic rice callus. *Plant Cell Reports* 43, 118. doi: 10.1007/s00299-024-03199-7

Song, J. H., Shin, G., Kim, H. J., Lee, S. B., Moon, J. Y., Jeong, J. C., et al. (2022). Mutation of GmIPK1 gene using CRISPR/Cas9 reduced phytic acid content in soybean seeds. *International Journal of Molecular Science* 23, 10583. doi: 10.3390/ijms231810583.

Távora, F. T. P. K., Meunier, A. C., Vernet, A., Portefaix, M., Milazzo, J., Adreit, H., et al. (2022). CRISPR/Cas9-targeted knockout of rice susceptibility genes OsDjA2 and OsERF104 reveals alternative sources of resistance to Pyricularia oryzae. *Rice Science* 29, 535–544. doi: 10.1016/j.rsci.2022.04.001

Tian, X., Wang, F., Zhao, Y., Lan, T., Yu, K., Zhang, L., et al. (2020). Heat shock transcription factor A1b regulates heat tolerance in wheat and Arabidopsis through OPR3 and jasmonate signalling pathway. *Plant Biotechnology Journal* 18, 1109–1111. doi: 10.1111/pbi.13268.

Wang, W., Pan, Q., Tian, B., He, F., Chen, Y., Bai, G., Akhunova, A., Trick, H. N., Akhunov, E. (2019). Gene editing of the wheat homologs of TONNEAU1-recruiting motif encoding gene affects grain shape and weight in wheat. *Plant Journal* 100, 251–264. doi: 10.1111/tpj.14440.

Wang, X., Guo, Y., Wang, Y., Peng, Y., Zhang, H., Zheng, J. (2024). ZmHDT103 negatively regulates drought stress tolerance in Maize seedlings. *Agronomy* 14, 134. doi: 10.3390/agronomy14010134

Wang, Y., Zafar, N., Ali, Q., Manghwar, H., Wang, G., Yu, L., Ding, X., Ding, F., Hong, N., Wang, G., Jin, S. (2022). CRISPR/Cas genome editing technologies for plant improvement against biotic and abiotic stresses: Advances, limitations, and future perspectives. *Cells* 11(23), 3928. doi: 10.3390/cells11233928.

Xie, J., Fei, X., Yan, Q., Jiang, T., Li, Z., Chen, H., et al. (2024). The C4 photosynthesis bifunctional enzymes, PDRPs, of maize are co-opted to cytoplasmic viral replication complexes to promote infection of a prevalent potyvirus sugarcane mosaic virus. *Plant Biotechnology Journal* 22, 1812–1832. doi: 10.1111/pbi.14304

Xu, R., Yang, Y., Qin, R., Li, H., Qiu, C., Li, L., Wei, P., Yang, J. (2016). Rapid improvement of grain weight via highly efficient CRISPR/Cas9-mediated multiplex genome editing in rice. *Journal of Genetics and Genomics* 43, 529. doi: 10.1016/j.jgg.2016.07.003.

Xu, Y. (2010). *Molecular Plant Breeding.* Egham: CAB International.

Xu X., Sun S.-K., Zhang W., Tang Z., Zhao F.-J. (2024). Editing silicon transporter genes to reduce arsenic accumulation in rice. *Enviromental Science & Technology* 58, 1976–1985. doi: 10.1021/acs.est.3c10763.

Yadav, R. K., Tripathi, M. K., Tiwari, S., Tripathi, N., Asati, R., Chauhan, S., et al. (2023). Genome editing and improvement of Abiotic stress tolerance in crop plants. *Life* 13, 1456. doi: 10.3390/life13071456.

Yuan, B., Bi, C., Tian, Y., Wang, J., Jin, Y., Alsayegh, K., et al. (2024). Modulation of the microhomology-mediated end joining pathway suppresses large deletions and enhances homology-directed repair following CRISPR-Cas9-induced DNA breaks. *BMC Biology* 22, 101. doi: 10.1186/s12915-024-01896-z.

Yuste-Lisbona, F. J., Fernández-Lozano, A., Pineda, B., Bretones, S., Ortíz-Atienza, A., García-Sogo, B., Müller, N. A., Angosto, T., Capel, J., Moreno, V., et al. (2020). ENO regulates tomato fruit size through the floral meristem development network. *Proceedings of the National Academy of Sciences. USA*. 117, 8187–8195. doi: 10.1073/pnas.1913688117.

Zhang, D., Hussain, A., Manghwar, H., Xie, K., Xie, S., Zhao, S., Larkin, R. M., Qing, P., Jin, S., Ding, F. (2020). Genome editing with the CRISPR-Cas system: An art, ethics and global regulatory perspective. *Plant Biotechnology Journal* 18, 1651–1669.

Zhang, L., Wang, S., Bai, B., Chen, Y., Xiang, Z., Chen, C., et al. (2024b). OsKASI-2 is required for the regulation of unsaturation levels of membrane lipids and chilling tolerance in rice. *Plant Biotechnology Journal* 22, 2157–2172. doi: 10.1111/pbi.14336

Zhang, Q., Xing, H. L., Wang, Z. P., Zhang, H. Y., Yang, F., Wang, X. C., Chen, Q. J. (2018). Potential high-frequency off-target mutagenesis induced by CRISPR/Cas9 in Arabidopsis and its prevention. *Plant Molecular Biology* 96, 445–456.

Zhang, X. H., Tee, L. Y., Wang, X.-G., Huang, Q. S., Yang, S. H. (2015). Off-target effects in CRISPR/Cas9-mediated genome engineering. *Molecular Therapy –. Nucleic Acids* 4, e264.

Zsögön, A., Čermák, T., Naves, E. R., Notini, M. M., Edel, K. H., Weinl, S., Freschi, L., Voytas, D. F., Kudla, J., Peres L. E. P. (2018). De novo domestication of wild tomato using genome editing. *Nature Biotechnology* 36, 1211–1216. doi: 10.1038/nbt.4272.

2 Genome-Editing Technologies

A Comprehensive Overview

Muhammad Ibrar, Hira Khursheed, Alman Fatima, Muhammad Zawwad Raza, and Muhammad Riaz

2.1 INTRODUCTION

Ever since the discovery of deoxyribonucleic acid (DNA) as genetic material in 1944 (Avery et al., 1944), seminal advances have been made to reveal how genetic instructions are stored and disseminated to the next generation of all living beings. The central outcome of those advances is the resolution of the DNA molecule as a double helix structure comprised of two strands held together through hydrogen bonding present between two of four types of nucleotides (Sasisekharan and Pattabiraman, 1978). The genetic instructions are concealed in a specific order of the nucleotides, which constitute functional units called genes. The genes generate short sequences of ribonucleic acid (RNA) that encode proteins to accomplish biological functions in cells. This flow of genetic information is found to be interrupted by external stimuli such as UV irradiation and oxidative stress (Besaratinia and Pfeifer, 2006; Poetsch et al., 2022) as well as internal stimuli, such as replication errors, resulting in the alteration of nucleotide sequences called mutations (Bellelli and Boulton, 2021; Murat et al., 2022). Mutations typically disrupt the normal function of genes in living cells, which can result in diseases such as cancer or diabetes (Karczewski et al., 2020, 2021). For a long time, the functional consequences of the genes were primarily studied through a molecular method called forward genetics, which includes the induction of random mutations in the genome and correlating them with corresponding phenotypic changes in model organisms. This method is robust but limited by both the time and cost constraints (Lawson and Wolfe, 2011; Miyoshi et al., 2019). Reverse genetics then emerged as an alternative approach in which the function of specific genes is studied by inducing genomic change at the desired genomic site (Lawson and Wolfe, 2011). This has paved the basis for the development of a very powerful technology known today as genome editing and, thus, permitting researchers to essentially customize the genome of living beings to cope with diseases and potentially resolve the food calamities that humanity is on the verge of experiencing.

The first and foremost prerequisite of genome editing is the introduction of nicks on both DNA strands, also called double-stranded breaks (DSBs), preferably around the target genomic site with considerably high efficiency and specificity. These DSBs subsequently trigger several cellular DNA repair processes of which non-homologous end joining (NHEJ) and homology-directed repair (HDR) are much more frequent (Figure 2.1).

The NHEJ is an error-prone process that introduces random insertions and deletions (indels) of nucleotides at the repair junction. In contrast, the HDR uses a homologous template, typically a sister chromatid that emerges during cell replication, to carry out seamless DSB repair (Figure 2.1). These two naturally existing methods are intentionally exploited in genome editing for different favorable purposes. The NHEJ is commonly used to disrupt genes as it introduces indels while repairing the DSBs. Alternatively, the HDR is used to perform gene knock-in or replace mutations in the existing genes at the target site in the genome. In nature, the DSBs are introduced by the endonucleases, which work as molecular scissors to scavenge the DNA. These endonucleases are readily fused with DNA-binding proteins to obtain locus specificity in the genome. To date, three types of DNA binding proteins have been exploited for site-specific endonuclease engineering: Meganuclease, Zinc finger proteins (ZFNs), and Transcription Activator-like effectors (TALEs) (Bogdanove and Voytas, 2011; Elrod-Erickson and Pabo, 1999; Pavletich and Pabo, 1991; Shi and Berg, 1995). More recently, the discovery of CRISPR-Cas9 has accelerated the work of introducing desired genomic changes at a specific DNA sequence, paving the way for potentially correcting the disease-associated mutations and genes in healthcare and introducing drought-resistant or pest-resistant genes into crops to enhance productivity and yield for food availability (Jinek et al., 2012; Jinek et al., 2013; Ran, Hsu, Lin, et al., 2013; Ran, Hsu, Wright, et al., 2013).

2.2 MEGANUCLEASES

Meganucleases (MegNs) are naturally occurring homing endonucleases in microbes and various eukaryotes (Belfort and Roberts, 1997; Chevalier and Stoddard, 2001; Jurica and Stoddard, 1999). They can recognize relatively large DNA sequences spanning over 12–45 base pairs to generate accurate DSBs (Jacquier and Dujon, 1985) (Figure 2.2A). MegNs recognize the target DNA sequence through a DNA binding domain and perform DNA digestion with an active site containing conserved divalent cation binding residues. MegNs are broadly grouped into five families based on their protein sequence and structural motifs:

DOI: 10.1201/9781003500933-3

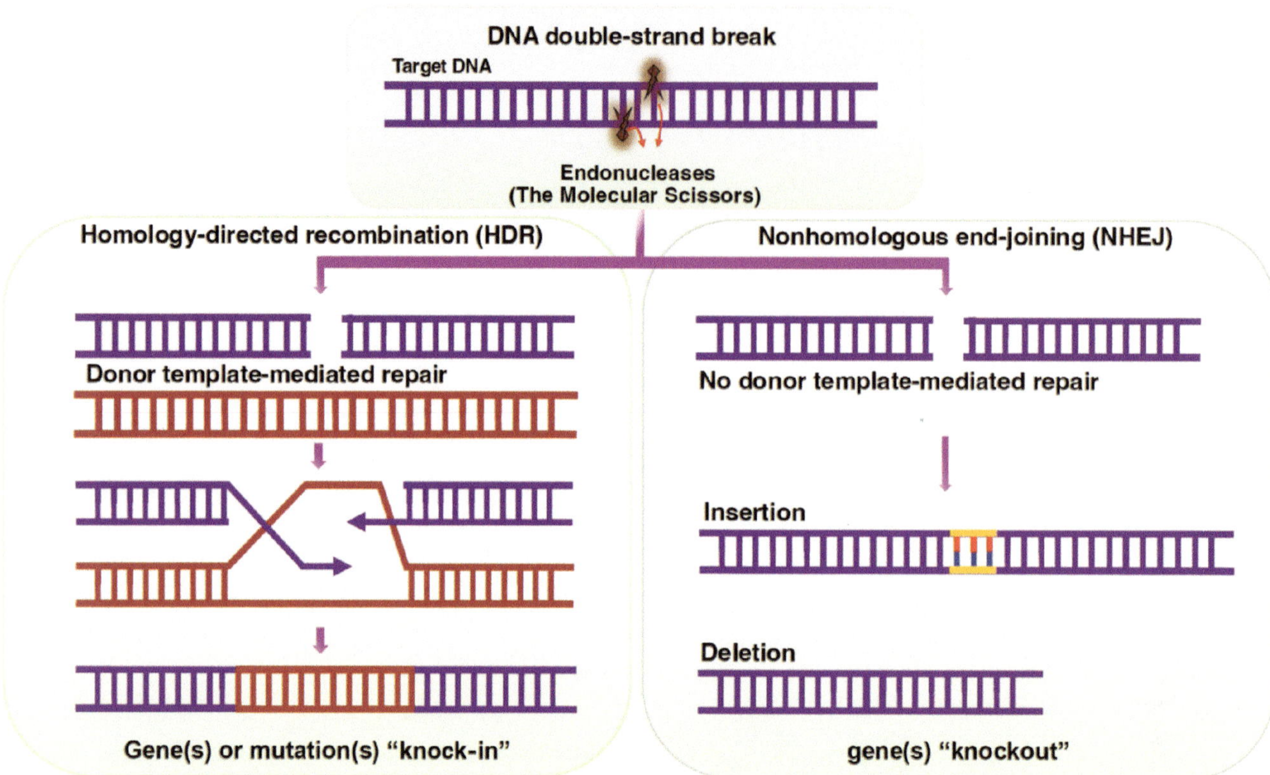

FIGURE 2.1 The mechanism and outcome of two major DSB-induced repair pathways: Homology-directed recombination (HDR) and Nonhomologous end-joining (NHEJ).

1. The HNH (Histidine-Asparagine-Histidine) family of MegNs possesses a unique module of 30–40 amino acids that are involved in making three highly conserved HNH residues capable of approaching the target DNA sequence nonspecifically and generating DSBs (Flick et al., 1998).
2. The PD-(D/E) XK family of MegNs possesses a critical domain of active residues capable of recognizing up to 17 base pairs of target DNA sequence with high specificity (Singleton et al., 2004; Zhao et al., 2009). This family mainly encompasses nucleases but can also exist in a variety of other functional formats in addition to DNA restriction enzymes (Vyverman et al., 2012).
3. The GIY-YIG family of MegNs contains a catalytic motif consisting of a compact domain of ~100 amino acids essential for DNA hydrolysis at the target sequence site. They can generate DSBs in the monomeric form to allow genomic modifications via NHEJ or HDR repair pathways (Zhou et al., 2013).
4. The His-Cys box family of MegNs is relatively small, and the members are identified from the conserved nuclear regions of small and large subunits of the ribosomal prokaryotic DNA (Johansen et al., 1993). These MegNs possess a conserved series of 100 histidine and cysteine residues that constitute a centrally active region involved in generating DSB on the target DNA sequence by coordinating with the divalent cations (Chevalier and Stoddard, 2001). This category of MegNs typically forms homodimers to recognize minimally 14 base pairs in the target DNA sequence for cleavage induction to generate 4–5 nucleotides 3' overhang (Elde et al., 2000; Ellison and Vogt, 1993).
5. The LAGLIDADG family makes up a large, well-described family of MegNs with the enzymatically active domain represented by a well-defined sequence motif, termed as LAGLIDADG (Arnould et al., 2011). LAGLIDADG endonucleases possess one or two of the conserved LAGLIDADG amino acid sequence motifs. Enzymes with one LAGLIDADG motif, such as I-CreI and I-CeuI, typically form homodimers and recognize DNA with palindromic or nearly palindromic symmetry. The enzymes with two motifs, such as I-SceI, act as monomers to recognize symmetric DNA targets. Nevertheless, the range for DNA target of both LAGLIDADG endonucleases varies from 18–22 base pairs (Taylor et al., 2012).

MegNs can be tailored to generate chimeric variants with enhanced specificity for the targeted sequences (Chevalier et al., 2002; Jacquier and Dujon, 1985). Such protein engineering techniques have been adopted for performing targeted genomic modifications to several plants, including consumable crops (Pixley et al., 2022; Shi et al., 2023; Wu et al., 2020). For instance, a custom-designed meganuclease was employed to develop low-nicotine tobacco plants by editing the nicotine-synthesizing gene (USDA-APHIS, 2020). Moreover, a re-engineered meganuclease was designed for specific cleavage of an endogenous target sequence adjacent to a transgenic insect control locus in cotton plants (D'Halluin et al., 2013). Despite such critical applications in the agricultural field, MegNs lack the ability of modular binding and versatile programmability to target potentially more interesting loci in the genome of higher eukaryotes. This has largely prevented the utilization of MegNs as routine tools for targeted genomic modifications.

2.3 ZINC FINGER NUCLEASES (ZFNS)

Zinc-finger nucleases (ZFNs) are purely artificial, chimeric proteins generated by the fusion of two key components, a DNA binding domain (zinc-finger-binding domain proteins) and an endonuclease (a nonspecific DNA cleavage domain found in the *Fok-I* restriction endonuclease) (Urnov et al., 2010) (see Figure 2.2B). Discovered in 1985, zinc-finger

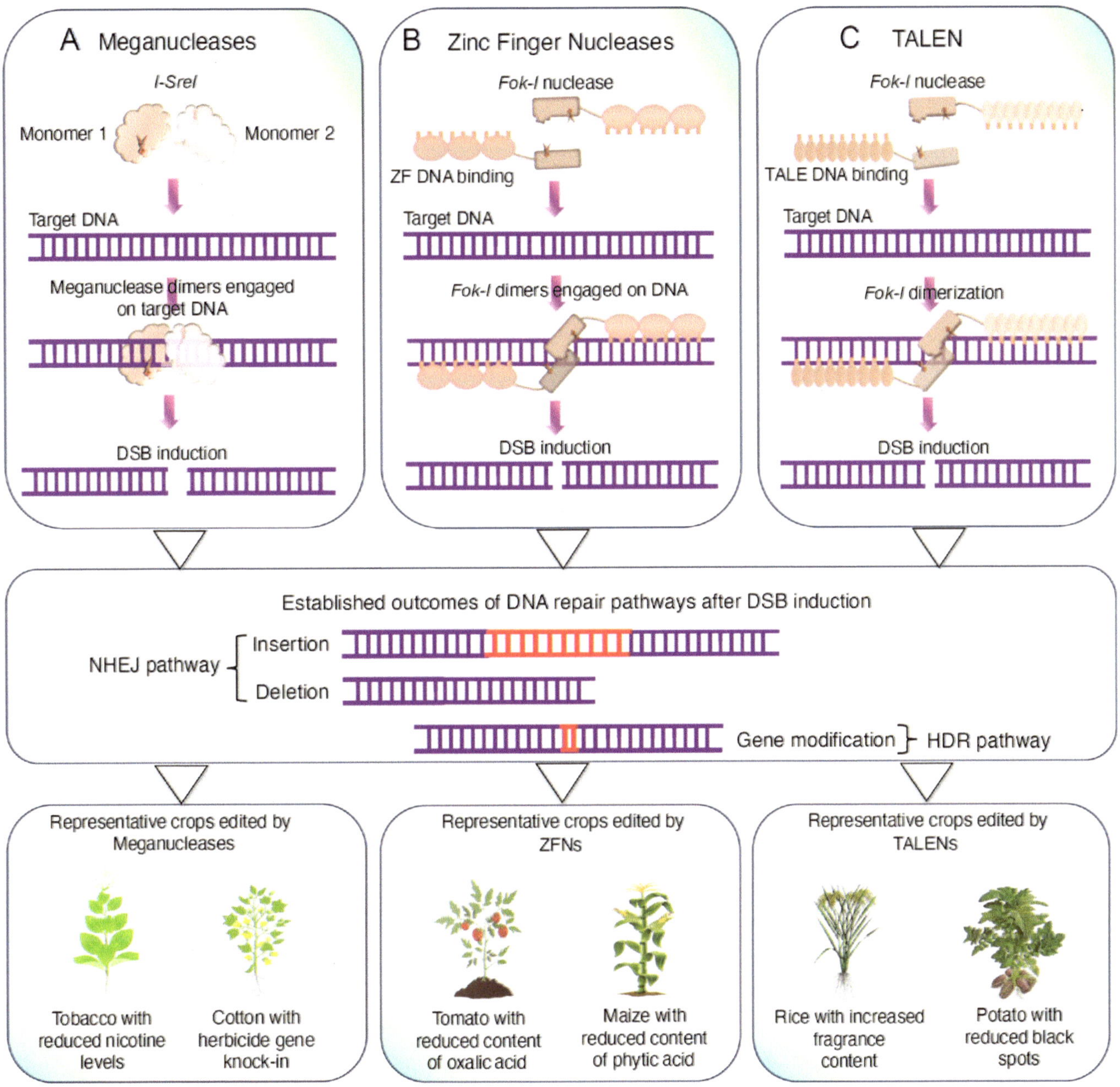

FIGURE 2.2 Application of DNA binding proteins and endonucleases for genome editing: A: Schematic overview and mode of action of a typical meganuclease (I-SreI) to create DSB at the target genomic site. B: Zinc finger nucleases (ZFNs) consist of a DNA-binding (Cys2His2) and a DNA cleavage domain (Fok-I restriction enzyme) which are designed to bind DNA in tail-to-tail orientation. C: Transcription activator-like effector nucleases (TALENs) consist of DNA binding and a cleavage domain.

proteins (ZFPs) belong to one of the most abundant groups of proteins with an extensive array of molecular functions (Klug and Rhodes, 1987). Their name stems from the role of the zinc (Zn) ion in maintaining a finger-like structure of ZNFs. The genetic regulation property of zinc-finger proteins owes to their ability to interact with DNA, RNA, PAR (poly-ADP-ribose), and other proteins. A single zinc finger (ZF) motif comprises ~30 amino acids and has a simple beta-beta alpha (ββα) structure. This structure is stabilized by the chelation of zinc ions. Each ZF binds to three base pairs while a variety of ZF motifs identify between 9 and 18 base pairs of certain DNA sequences. Amino acid residues at positions −1 to +6 relative to the start of the α-helix of each ZF motif are crucial for sequence-specific interaction of the ZF motif with the DNA site (Elrod-Erickson and Pabo, 1999; Pavletich and Pabo, 1991; Shi and Berg, 1995). Therefore, these amino acids are mostly manipulated to generate ZF motifs to target unique DNA sequences within the genome. Additionally, longer DNA sequences can be targeted by linking several ZF motifs to form ZFPs. As such, the ZF has the potential to be a platform for programmable genome targeting because of its modular structure.

The second key component in ZFNs is an endonuclease, for that, a non-specific DNA cleavage domain of the *Fok-I* restriction enzyme is used (see Figure 2.2B). *Fok-I* restriction enzyme is a bacterial type-IIS restriction endonuclease that recognizes the non-palindromic pentadeoxyribonucleotide, 5′-GGATG-3′: 5′-CATCC-3′, in duplex DNA and cleaves 9/13 nucleotides downstream of the recognition site (Sugisaki and Kanazawa, 1981; Szybalski et al., 1991). *Fok-I* restriction enzyme consists of two separate protein domains: a DNA-sequence recognition domain and an endonuclease domain (DNA-cleavage domain). To generate a ZFN, the recognition domain of *Fok-I* is replaced with ZF motifs or ZFPs to achieve site-specific cleavage in the target genome. A targeted genomic location can be directed toward both upstream and downstream regions by two ZFNs, which generally cause the dimerization of *Fok-I* cleavage domains, causing DNA DSBs, which are repaired by HDR and NHEJ pathways. This results in exact sequence substitution or random indels for knockout or knock-in of any desired gene.

The modified ZF arrays can bind to specific genomic loci effectively, but the creation of ZF arrays for most genomic sites is still a challenge (Urnov et al., 2010). This is due to the possibility of crosstalk between nearby ZF motifs, which limits a ZF array's ability to bind to the intended DNA region. As such, screening several ZF arrays is necessary to produce two functioning ZFNs. ZFNs are very tiny and can be included in Adeno-associated viruses (AAV), a common and feasible vector for in vivo gene editing, making them a useful tool for gene therapy. One of the main issues with genome-editing technology is "off-target (OT) editing," which is modifications in the DNA sequence in undesirable genomic locations. OT editing is of particular importance in light of the potential uses in gene therapy of ZFNs. Notably, ZFNs have been extensively investigated for their impact on OT editing effects in vitro, and it has been shown that ZFNs can cause unique cleavage at hundreds of OT sites. ZFNs can modify a small number of OT sites found in human cells in vitro, demonstrating the OT functions of ZFNs in vivo.

The challenge of building ZF arrays has prevented their broad use in non-specialized laboratories, in contrast to TALENs and CRISPR-Cas9. Indeed, it is still difficult to create ZF domains that can identify all DNA triplets, particularly the 5′-CNN-3′ and 5′-TNN-3′ variants. ZFNs do not have the same level of target adaptability as more modern genome-editing techniques. However, ZFNs have the potential to mediate specific and effective genome editing, as demonstrated by several ongoing clinical trials in the healthcare field (Holt et al., 2010; Nathwani, 2019) as well as in agricultural settings (Pixley et al., 2022).

2.4 TRANSCRIPTION ACTIVATOR-LIKE EFFECTOR NUCLEASES (TALENS)

Transcription Activator-Like Effector Nucleases (TALENs) represent a revolutionary class of genome-editing tools that have transformed our ability to precisely modify genetic material. Inspired by the sophisticated mechanisms employed by certain bacteria, particularly the strain of *Xanthomonas*, to manipulate plant genomes, TALENs harness the power of nature to enable targeted modifications of DNA sequences with unprecedented precision. The gram-negative bacterium *Xanthomonas*, one of the plant pathogens, naturally produces TALE proteins as their key virulence factor and injects them into plant cells' cytoplasm using a needle-like structure called a type III secretion system (Gohre and Robatzek, 2008; Kay and Bonas, 2009; Rossier et al., 1999). These effector TALEs trigger a transcriptional reprogramming of the plant cell by mimicking the eukaryotic transcriptional machinery and contribute to the spread of pathogens and disease symptoms in the host (Boch et al., 2009; Bogdanove et al., 2010). It is noteworthy that the whole mechanism of pathogen-plant war is dependent on DNA recognition by TALEs within the host genome. Boch et al. and Moscou and Bogdanove in complementary studies revealed this DNA recognition mechanism of TALEs (Boch et al., 2009; Moscou and Bogdanove, 2009; Voytas and Joung, 2009). This revelation prompted scientists to transform this natural process into a gene-editing instrument by combining the DNA-binding segment of TALEs with a nuclease domain, usually sourced from the *Fok-I* endonuclease. Consequently, the resulting fusion protein, termed TALEN, can identify particular DNA sequences within the genome and instigate double-strand breaks (DSBs) at these sites (Bogdanove and Voytas, 2011). This composite protein is commonly referred to as TALENs (see Figure 2.2C).

The DNA-binding domain of TALENs is made up of several repeat motifs that vary slightly in length (~ 33–35 amino acids) but are nearly identical in structure except for two hypervariable polymorphisms at amino acid positions 12 and 13, known as repeat variable dinucleotides (RVDs). These RVDs have distinct preferences for binding to specific nucleotides. The most prevalent RVDs, namely HD, NG, NI, and NN, correspond to each of the four nucleotides C, T, A, and

G, respectively, and are responsible for nucleotide specificity of TALENs (Bogdanove and Voytas, 2011). The combination and length of repeat motifs and RVD define the target DNA sequence. A customized DNA-binding domain can be constructed to target a specific sequence only by modulating the RVD nucleotides while keeping the same repeat backbone. These DNA-recognition domains can then be fused with a variety of proteins including transcriptional activators, site-specific recombinases, as well as endonucleases. To stimulate the NHEJ and HDR events, TALENs are employed as a tool by connecting the binding domain with the non-specific cleavage domain of the type II restriction endonuclease known as *Fok-I*. Like ZFNs, TALENs are commonly used with obligate heterodimer *Fok-I* domains to address the possibility of OT events.

TALENs offer researchers a precise and versatile tool for genome editing in a wide range of organisms, including animals, microbes, and plants. TALENs are being used to modify the genomes of several model and non-model species, animals, and crops (Joung and Sander, 2013). The application of TALENs in cancer treatment holds promise for targeted gene therapies. Additionally, TALENs' successful introduction of the first genome-edited crop to the market (Menz et al., 2020) showcases the technology's potential to revolutionize agriculture by enabling precise modifications to enhance crop yield, nutritional content, and resistance to environmental stressors, ushering in a new era of sustainable food production (Ku and Ha, 2020; Menz et al., 2020; Pixley et al., 2022).

Key differences exist between TALENs and ZFNs: (1) TALEN repeats are three to four times larger than ZFNs when found for each base pair of the targeted DNA. This might interfere with viral delivery techniques, particularly AVV-based ones in mammalians. (2) TALENs exhibit a higher degree of modularity compared to ZFNs. In TALENs, each repeat unit within the binding domain targets a single DNA base pair, whereas zinc fingers in ZFNs generally recognize 3–4 base pairs each. This inherent modularity of TALENs confers several advantages, including enhanced specificity, simplified design, greater customization capabilities, increased flexibility, and reduced risk of OT effects. To achieve targeted gene editing, ZFN assembly requires an archive of high-quality, well-characterized modules, as finger-to-finger crosstalk may lead to imprecise DNA recognition. Unlike ZFNs, context-dependent effects between repeat units have not yet been discovered for TALENs. To recognize practically any target sequence, TALEN repeats can be assembled in a variety of ways. The thymidine sequence at the start of the TALE binding site needs to be followed, which is the only restriction. When looking for possible target sites within a locus, this must be taken into account.

While TALENs serve as potent instruments for genome editing, they do come with certain limitations. These include their relatively large size, the labor-intensive nature of their design, particularly when multiplexing, and the associated costs for targeting multiple loci or undertaking large-scale genome-editing endeavors. Despite these limitations, TALENs remain valuable tools for precise genome editing and continue to be widely used in various research and biotechnological applications, including genome editing of several crops for their enhanced yield (Menz et al., 2020; Pixley et al., 2022) (see Figure 2.2). Ongoing advances in genome-editing technology, such as the development of CRISPR-Cas systems, offer complementary approaches that address some of the limitations associated with TALENs.

2.5 CRISPR-CAS9 (CLUSTERED REGULARLY INTERSPACED SHORT PALINDROMIC REPEATS ASSOCIATED CAS-9 PROTEIN)

The CRISPR-Cas9 system has rapidly transformed the landscape of targeted genome engineering since its introduction. Offering a potentially simpler and more efficient substitute to earlier technologies like MegNs, ZFNs, and TALENs, CRISPR-Cas9 has emerged as a powerful tool for precise genetic modifications. Just like TALENs, CRISPR-Cas9 also harnesses the power of nature to enable targeted modifications of DNA sequences but with significantly improved precision and ease of design (Cong et al., 2013; Jinek et al., 2012; Jinek et al., 2013). Originally discovered as part of bacteria and archaea's adaptive immune systems (Barrangou et al., 2007; Brouns et al., 2008; Ishino et al., 2018), CRISPR-Cas loci encode RNA-guided machinery capable of targeting and cleaving foreign DNA. Among the CRISPR-Cas systems, the class-II system is the most commonly used for gene editing. This system relies only on a single DNA endonuclease, such as Cas9 or Cas12a (also called Cpf1), guided by CRISPR-RNA (crRNA) to precisely cleave double-stranded DNA in animal and plant species (Figure 2.3A).

As previously outlined, an effective gene-editing tool consists of two essential components: a DNA-binding module, which enables precise targeting of specific genomic sites, and an endonuclease, which is responsible for inducing DSBs in the DNA. In contrast to ZFNs and TALENs, which both depend on protein domains as their DNA-binding module, CRISPR-Cas9 employs non-coding, DNA-targeting complementary RNAs, commonly referred to as guide RNA (gRNA) or single guide RNA (sgRNA). The sgRNA consists of a crRNA sequence fused to the trans CRISPR RNA (tracrRNA), where the crRNA spans 20 nucleotides and matches the target DNA sequence, while the tracrRNA acts as a link between crRNA and the endonuclease (Cas9) to form a ribonucleoprotein (RNP) complex (Cong et al., 2013; Jinek et al., 2013; Mali et al., 2013).

On the other hand, Cas9 serves as the molecular scissors, capable of precisely cutting DNA at specific target sequences. The specificity of the CRISPR/Cas9 is ensured by the "protospacer-adjacent motif" (PAM). PAM is usually a 2–6 base pairs conserved DNA sequence and is generally found 3–4 nucleotides downstream from the cleavage site (Gleditzsch et al., 2019). PAM sequences are often recognized by distinct Cas proteins. For example, the *Lachnospiraceae* bacteria Cas12a (LbCas12a) and *Acidaminococcus* Cas12a (AsCas12a) identify a TTTV PAM sequence at the 5′ end

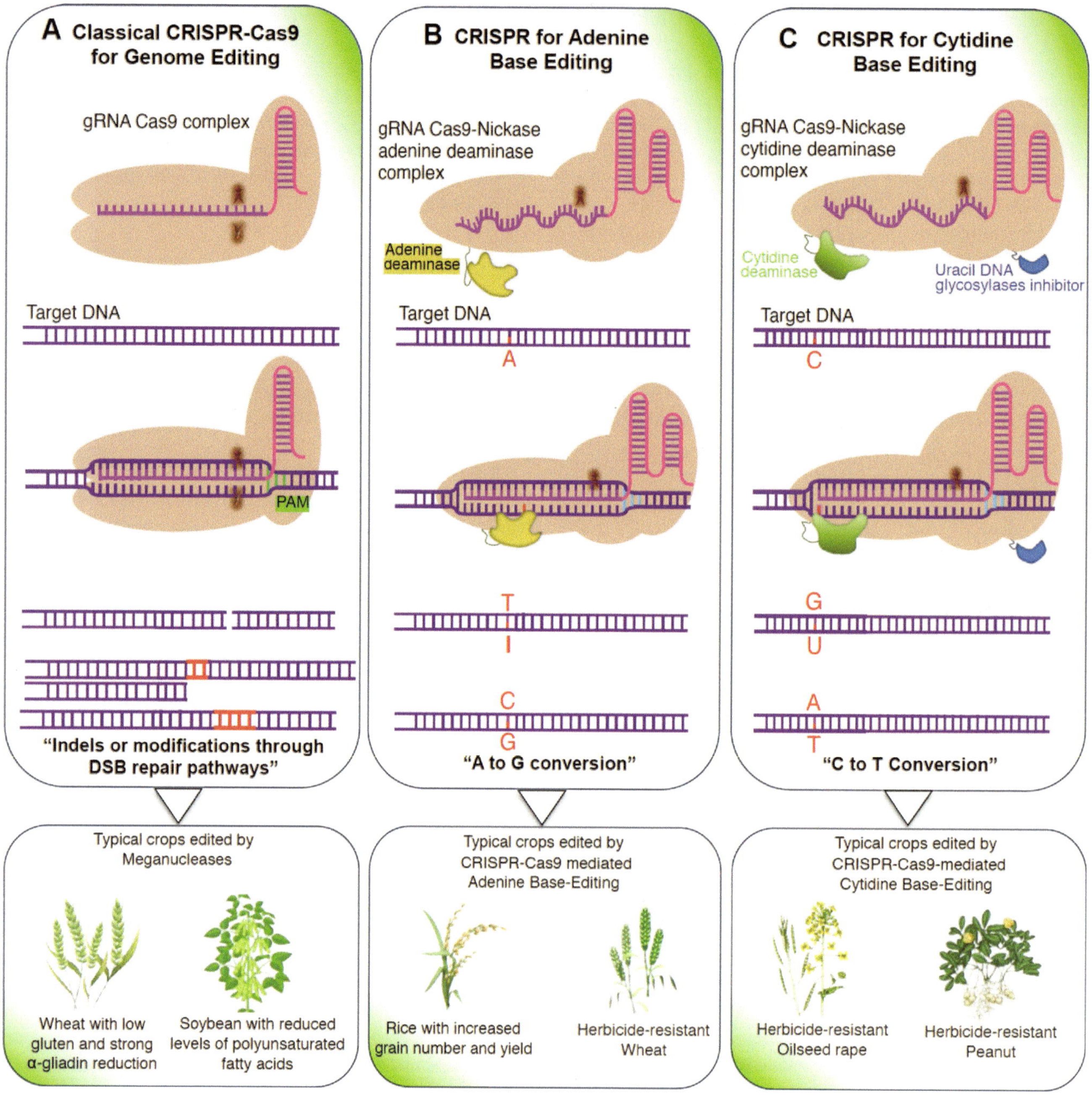

FIGURE 2.3 CRISPR technologies for genome editing: A: Cas9 nuclease is fused with the single-stranded gRNA to form a complex that targets a specific genomic region through RNA-DNA complementarity directly upstream to the protospacer adjacent motif (PAM; NGG) resulting in DSB. B and C: CRISPR-Cas9 is fused to either adenine or cytidine deaminases to precisely induce adenine base editing to guanine or cytidine to thymidine, respectively.

of the proto-spacer region, respectively (Chen et al., 2020). The *Streptococcus pyogenes* Cas9 (SpCas9) is a commonly used strain and recognizes an NGG PAM sequence at the 3′ end of the proto-spacer region. The requirement for a PAM sequence ensures the specificity of the CRISPR-Cas9 system, as it prevents Cas9 from randomly cleaving the host organism's genome. Different Cas9 orthologs have specific PAM sequences they recognize, providing researchers with options for targeting different genomic regions. Understanding the role of PAM sequences is crucial for designing effective CRISPR-Cas9 experiments and optimizing target site selection. However, some natural CRISPR-Cas systems target RNA instead of dsDNA and consequently do not require a PAM to specify the site of the dsDNA unwinding (Marraffini and Sontheimer, 2010).

Mechanistically, PAM serves as an initiation code for the binding of Cas9 to the genomic DNA. As a first step for the CRISPR-Cas9 function, Cas9 scans the host DNA for the presence of PAM. Once PAM is identified by Cas9, it binds with PAM and initiates the unwinding of DNA duplex for subsequent binding of gRNA. Once a matching sequence is found, Cas9 undergoes a conformational change that activates its endonuclease activity, leading to the precise cleavage of the target DNA through the formation of DSBs in the target DNA. These DSBs trigger the cell's DNA repair machinery, which can result in either NHEJ or HDR. The NHEJ pathway often leads to small insertions or deletions (indels) at the cleavage site, introducing gene disruptions or frameshift mutations (gene knockout). The HDR pathway, on the other hand, can be exploited to introduce precise genetic modifications by providing a donor DNA repair template along with the CRISPR-Cas9 system. This intricate architecture enables precise and efficient genomic targeting, underscoring the versatility and utility of the CRISPR-Cas9 system in genome-editing applications.

2.5.1 CRISPRing for Gene Knock-in and Knockout

CRISPR-Cas9-mediated gene knock-in and knockout are pivotal methods in genetic manipulation, each offering distinct capabilities and applications. CRISPR-Cas9 gene knock-in involves the precise insertion of specific DNA sequences or nucleotide bases into targeted genomic locations. It initiates with the design of a gRNA to direct the CRISPR-Cas9 enzyme to the desired site, where it induces a DSB. A donor DNA template, comprising the desired sequence to be inserted and flanking homology arms, is then introduced. HDR then guides the integration of the donor DNA into the genome. Verification methods such as PCR and DNA sequencing ensure insertion accuracy. The CRISPR-Cas9-mediated gene knock-in holds promise for diverse applications, including fundamental research and biotechnological improvements of edible crops for disease resistance and enhanced grain number and yield (Baliou et al., 2018; Im et al., 2016) (see Figure 2.3B). On the other hand, CRISPR gene knockout utilizes the precision of the CRISPR-Cas9 system to selectively disrupt or inactivate target genes within living organisms. It involves the custom-designed gRNA guiding Cas9 to cleave the DNA at the specific target site. Subsequently, cell repair machinery attempts to mend the broken DNA strands. This often results in insertions or deletions (indels) that disrupt the function of the targeted gene, effectively "knocking it out." This technique also shows potential in therapeutic applications, offering a means to treat genetic disorders by disabling disease-causing genes in health and agricultural settings. Despite its promise, CRISPR-Cas9-mediated gene knockout has limitations, such as OT effects and variable efficiency, which require ongoing research efforts for refinement.

Concerning OT effects, PAM-distal regions are usually more vulnerable to mismatches between the single-guide RNA spacer area and the OT site than PAM proximal regions. The OT effects of Cas9 can be reduced by using single-guide RNA that has fewer possible genomic DNA OT sites. As the Cas9 protein has two endonuclease domains, a mutation in one of them can produce a nickase variant of Cas9 (nCas9). When co-expressed with two single-guide RNAs that target opposing DNA strands of an on-target location, nCas9 can be used to minimize unwanted indels (Lee et al., 2023). The two nicks are made at opposing DNA strands by nCas9, simulating a DSB. A DNA single-strand break (SSB) is only produced by nCas9 at a particular OT site, effectively avoiding the OT indels caused by DSBs. Even though endogenously produced cytidine deaminases can transform an SSB into a DSB (Stavnezer and Schrader, 2006), SSBs can still result in certain amounts of indels at particular OT sites. A further reduction in OT editing was achieved by fusing the *Fok-I* cleavage domain with catalytically dead Cas9 (dCas9). At on-target locations where two single-guide RNAs cause the dimerization and activation of *Fok-I* nuclease, the fusion protein can cause DSB. There is no DNA break at OT sites because the *Fok-I* cleavage domain is not active as a monomer.

2.5.2 CRISPRing for Base Editing

CRISPR for base editing involves a precise molecular mechanism that allows for targeted modification of individual nucleotides within the DNA or RNA sequence. It employs a specialized enzyme complex called a base editor, which is created by fusing a catalytically inactive form of the Cas9 enzyme with a deaminase enzyme. Guided by a gRNA, the base editor is directed to a specific target site in the genome through complementary base pairing. Once at the target site, the Cas9 enzyme creates a single-strand nick in the DNA, exposing one of the DNA strands. Subsequently, the deaminase enzyme catalyzes the conversion of a specific nucleotide base within a defined editing window surrounding the target site. For instance, cytosine base editors (CBEs) convert cytosine (C) to uracil (U), while adenine base editors (ABEs) convert adenine (A) to inosine (I) as outlined in Figure 2.3B and C. The cell's natural DNA repair mechanisms recognize the altered base and attempt to repair the DNA, ultimately resulting in a permanent change in the nucleotide sequence. CRISPR-Ca9 base editing offers a highly specific and efficient approach to genetic manipulation, enabling precise, single-nucleotide changes without the need for DSBs or HDR DNA repair pathways to be activated. Both approaches of base editing have been widely employed in manipulating plant genomics to gain increased pest resistance and consequently improved annual yield (Menz et al., 2020; Shi et al., 2023; Wu et al., 2020).

2.6 CONCLUSION

In this chapter, we have given an overview of the advances and potential applications of genome-editing technologies, including meganucleases, ZFNs, TALENs, and the groundbreaking clustered CRISPR-Cas9. Through meticulous examination, we illuminated the principles, advantages, and limitations of each technology, highlighting their contributions to genetic manipulation and their impact across various fields including agriculture biotechnologies. These technologies have revolutionized genetic engineering by providing precise and efficient tools to modify DNA sequences. Meganucleases, with their inherent sequence specificity, laid the foundation for targeted genome editing, followed by ZFNs and TALENs, which offered improved precision and versatility, albeit with challenges in design complexity and off-target effects. However, the emergence of CRISPR-Cas9 has heralded a new era in genome editing, offering unparalleled efficiency, simplicity, and multiplexing capabilities. The prospects of genome-editing technologies are bright, with ongoing efforts focused on enhancing specificity, reducing off-target effects, and expanding the range of applications in agricultural settings. Moreover, the integration of machine learning and synthetic biology approaches holds promise for accelerating the development and optimization of these tools. The continuous refinement of these tools holds great potential for advances in basic research and agriculture, paving the way for transformative breakthroughs in understanding and manipulating the genetic code of living organisms. Finally, there is an utmost need to make new policies and establish regulatory authorities for the controlled use of genome-editing technologies to develop genome-edited crops to help improve the yield of edible crops.

REFERENCES

Arnould, S., Delenda, C., Grizot, S., Desseaux, C., Paques, F., Silva, G. H., and Smith, J. (2011). The I-CreI meganuclease and its engineered derivatives: applications from cell modification to gene therapy. *Protein Engineering, Design and Selection*, 24(1–2), 27–31. https://doi.org/10.1093/protein/gzq083

Avery, O. T., Macleod, C. M., and McCarty, M. (1944). Studies on the chemical nature of the substance inducing transformation of Pneumococcal types: Induction of transformation by a Desoxyribonucleic acid fraction isolated from *Pneumococcus* Type III. *Journal of Experimental Medicine*, 79(2), 137–158. https://doi.org/10.1084/jem.79.2.137

Baliou, S., Adamaki, M., Kyriakopoulos, A. M., Spandidos, D. A., Panayiotidis, M., Christodoulou, I., and Zoumpourlis, V. (2018). CRISPR therapeutic tools for complex genetic disorders and cancer (Review). *International Journal of Oncology*, 53(2), 443–468. https://doi.org/10.3892/ijo.2018.4434

Barrangou, R., Fremaux, C., Deveau, H., Richards, M., Boyaval, P., Moineau, S., Romero, D. A., and Horvath, P. (2007). CRISPR provides acquired resistance against viruses in prokaryotes. *Science*, 315(5819), 1709–1712. https://doi.org/10.1126/science.1138140

Belfort, M., and Roberts, R. J. (1997). Homing endonucleases: keeping the house in order. *Nucleic Acids Research*, 25(17), 3379–3388. https://doi.org/10.1093/nar/25.17.3379

Bellelli, R., and Boulton, S. J. (2021). Spotlight on the Replisome: Aetiology of DNA replication-associated genetic diseases. *Trends in Genetics*, 37(4), 317–336. https://doi.org/10.1016/j.tig.2020.09.008

Besaratinia, A., and Pfeifer, G. P. (2006). Investigating human cancer etiology by DNA lesion footprinting and mutagenicity analysis. *Carcinogenesis*, 27(8), 1526–1537. https://doi.org/10.1093/carcin/bgi311

Boch, J., Scholze, H., Schornack, S., Landgraf, A., Hahn, S., Kay, S., Lahaye, T., Nickstadt, A., and Bonas, U. (2009). Breaking the code of DNA binding specificity of TAL-type III effectors. *Science*, 326(5959), 1509–1512. https://doi.org/10.1126/science.1178811

Bogdanove, A. J., Schornack, S., and Lahaye, T. (2010). TAL effectors: Finding plant genes for disease and defense. *Current Opinion in Plant Biology*, 13(4), 394–401. https://doi.org/10.1016/j.pbi.2010.04.010

Bogdanove, A. J., and Voytas, D. F. (2011). TAL effectors: customizable proteins for DNA targeting. *Science*, 333(6051), 1843–1846. https://doi.org/10.1126/science.1204094

Brouns, S. J., Jore, M. M., Lundgren, M., Westra, E. R., Slijkhuis, R. J., Snijders, A. P., Dickman, M. J., Makarova, K. S., Koonin, E. V., and van der Oost, J. (2008). Small CRISPR RNAs guide antiviral defense in prokaryotes. *Science*, 321(5891), 960–964. https://doi.org/10.1126/science.1159689

Chen, P., Zhou, J., Wan, Y., Liu, H., Li, Y., Liu, Z., Wang, H., Lei, J., Zhao, K., Zhang, Y., Wang, Y., Zhang, X., and Yin, L. (2020). A Cas12a ortholog with stringent PAM recognition followed by low off-target editing rates for genome editing. *Genome Biology*, 21(1), 78. https://doi.org/10.1186/s13059-020-01989-2

Chevalier, B. S., Kortemme, T., Chadsey, M. S., Baker, D., Monnat, R. J., and Stoddard, B. L. (2002). Design, activity, and structure of a highly specific artificial endonuclease. *Molecular Cell*, 10(4), 895–905. https://doi.org/10.1016/s1097-2765(02)00690-1

Chevalier, B. S., and Stoddard, B. L. (2001). Homing endonucleases: Structural and functional insight into the catalysts of intron/intein mobility. *Nucleic Acids Research*, 29(18), 3757–3774. https://doi.org/10.1093/nar/29.18.3757

Cong, L., Ran, F. A., Cox, D., Lin, S., Barretto, R., Habib, N., Hsu, P. D., Wu, X., Jiang, W., Marraffini, L. A., and Zhang, F. (2013). Multiplex genome engineering using CRISPR/Cas systems. *Science*, 339(6121), 819–823. https://doi.org/10.1126/science.1231143

D'Halluin, K., Vanderstraeten, C., van Hulle, J., Rosolowska, J., van den Brande, I., Pennewaert, A., D'Hont, K., Bossut, M., Jantz, D., Ruiter, R., and Broadhvest, J. (2013). Targeted molecular trait stacking in cotton through targeted double-strand break induction. *Plant Biotechnology Journal*, 11(8), 933–941. https://doi.org/10.1111/pbi.12085

Elde, M., Willassen, N. P., and Johansen, S. (2000). Functional characterization of isoschizomeric His-Cys box homing endonucleases from Naegleria. *European Journal of Biochemistry*, 267(24), 7257–7266. https://doi.org/10.1046/j.1432-1327.2000.01862.x

Ellison, E. L., and Vogt, V. M. (1993). Interaction of the intron-encoded mobility endonuclease I-PpoI with its target site. *Molecular and Cellular Biology*, 13(12), 7531–7539. https://doi.org/10.1128/mcb.13.12.7531-7539.1993

Elrod-Erickson, M., and Pabo, C. O. (1999). Binding studies with mutants of Zif268. Contribution of individual side chains to binding affinity and specificity in the Zif268 zinc finger-DNA complex. *Journal of Biological Chemistry*, 274(27), 19281–19285. https://doi.org/10.1074/jbc.274.27.19281

Flick, K. E., Jurica, M. S., Monnat, R. J., Jr., and Stoddard, B. L. (1998). DNA binding and cleavage by the nuclear intron-encoded homing endonuclease I-PpoI. *Nature*, 394(6688), 96–101. https://doi.org/10.1038/27952

Gleditzsch, D., Pausch, P., Muller-Esparza, H., Ozcan, A., Guo, X., Bange, G., and Randau, L. (2019). PAM identification by CRISPR-Cas effector complexes: diversified mechanisms and structures. *RNA Biology*, 16(4), 504–517. https://doi.org/10.1080/15476286.2018.1504546

Gohre, V., and Robatzek, S. (2008). Breaking the barriers: Microbial effector molecules subvert plant immunity. *Annual Review of Phytopathology*, 46, 189–215. https://doi.org/10.1146/annurev.phyto.46.120407.110050

Holt, N., Wang, J., Kim, K., Friedman, G., Wang, X., Taupin, V., Crooks, G. M., Kohn, D. B., Gregory, P. D., Holmes, M. C., and Cannon, P. M. (2010). Human hematopoietic stem/progenitor cells modified by zinc-finger nucleases targeted to CCR5 control HIV-1 in vivo. *Nature Biotechnology*, 28(8), 839–847. https://doi.org/10.1038/nbt.1663

Im, W., Moon, J., and Kim, M. (2016). Applications of CRISPR/Cas9 for gene editing in hereditary movement disorders. *Journal of Movement Disorders*, 9(3), 136–143. https://doi.org/10.14802/jmd.16029

Ishino, Y., Krupovic, M., and Forterre, P. (2018). History of CRISPR-Cas from encounter with a mysterious repeated sequence to genome editing technology. *Journal of Bacteriology*, 200(7). https://doi.org/10.1128/JB.00580-17

Jacquier, A., and Dujon, B. (1985). An intron-encoded protein is active in a gene conversion process that spreads an intron into a mitochondrial gene. *Cell*, 41(2), 383–394. https://doi.org/10.1016/s0092-8674(85)80011-8

Jinek, M., Chylinski, K., Fonfara, I., Hauer, M., Doudna, J. A., and Charpentier, E. (2012). A programmable dual-RNA-guided DNA endonuclease in adaptive bacterial immunity. *Science*, 337(6096), 816–821. https://doi.org/10.1126/science.1225829

Jinek, M., East, A., Cheng, A., Lin, S., Ma, E., and Doudna, J. (2013). RNA-programmed genome editing in human cells. *Elife*, 2, e00471. https://doi.org/10.7554/eLife.00471

Johansen, S., Embley, T. M., and Willassen, N. P. (1993). A family of nuclear homing endonucleases. *Nucleic Acids Research*, 21(18), 4405. https://doi.org/10.1093/nar/21.18.4405

Joung, J. K., & Sander, J. D. (2013). TALENs: a widely applicable technology for targeted genome editing. *Nature Reviews Molecular Cell Biology*, 14(1), 49-55. https://doi.org/10.1038/nrm3486

Jurica, M. S., and Stoddard, B. L. (1999). Homing endonucleases: structure, function and evolution. *Cellular and Molecular Life Sciences*, 55(10), 1304–1326. https://doi.org/10.1007/s000180050372

Karczewski, K. J., Francioli, L. C., Tiao, G., Cummings, B. B., Alfoldi, J., Wang, Q., Collins, R. L., Laricchia, K. M., Ganna, A., Birnbaum, D. P., Gauthier, L. D., … MacArthur, D. G. (2020). The mutational constraint spectrum quantified from variation in 141,456 humans. *Nature*, 581(7809), 434–443. https://doi.org/10.1038/s41586-020-2308-7

Karczewski, K. J., Francioli, L. C., Tiao, G., Cummings, B. B., Alfoldi, J., Wang, Q., Collins, R. L., Laricchia, K. M., Ganna, A., Birnbaum, D. P., Gauthier, L. D., … MacArthur, D. G. (2021). Author Correction: The mutational constraint spectrum quantified from variation in 141,456 humans. *Nature*, 590(7846), E57. https://doi.org/10.1038/s41586-020-03174-8

Kay, S., and Bonas, U. (2009). How Xanthomonas type III effectors manipulate the host plant. *Current Opinions in Microbiology*, 12(1), 37–43. https://doi.org/10.1016/j.mib.2008.12.006

Klug, A., and Rhodes, D. (1987). Zinc fingers: A novel protein fold for nucleic acid recognition. *Cold Spring Harbor Symposia on Quantitativ Biology*, 52, 473–482. https://doi.org/10.1101/sqb.1987.052.01.054

Ku, H. K., and Ha, S. H. (2020). Improving nutritional and functional quality by genome editing of crops: status and perspectives. *Frontiers in Plant Science*, 11, 577313. https://doi.org/10.3389/fpls.2020.577313

Lawson, N. D., and Wolfe, S. A. (2011). Forward and reverse genetic approaches for the analysis of vertebrate development in the zebrafish. *Developmental Cell*, 21(1), 48–64. https://doi.org/10.1016/j.devcel.2011.06.007

Lee, J., Lim, K., Kim, A., Mok, Y. G., Chung, E., Cho, S. I., Lee, J. M., and Kim, J. S. (2023). Prime editing with genuine Cas9 nickases minimizes unwanted indels. *Nature Communications*, 14(1), 1786. https://doi.org/10.1038/s41467-023-37507-8

Mali, P., Yang, L., Esvelt, K. M., Aach, J., Guell, M., DiCarlo, J. E., Norville, J. E., and Church, G. M. (2013). RNA-guided human genome engineering via Cas9. *Science*, 339(6121), 823–826. https://doi.org/10.1126/science.1232033

Marraffini, L. A., and Sontheimer, E. J. (2010). Self versus non-self discrimination during CRISPR RNA-directed immunity. *Nature*, 463(7280), 568–571. https://doi.org/10.1038/nature08703

Menz, J., Modrzejewski, D., Hartung, F., Wilhelm, R., and Sprink, T. (2020). Genome edited crops touch the market: A view on the global development and regulatory environment. *Frontiers in Plant Science*, 11, 586027. https://doi.org/10.3389/fpls.2020.586027

Miyoshi, C., Kim, S. J., Ezaki, T., Ikkyu, A., Hotta-Hirashima, N., Kanno, S., Kakizaki, M., Yamada, M., Wakana, S., Yanagisawa, M., and Funato, H. (2019). Methodology and theoretical basis of forward genetic screening for sleep/wakefulness in mice. *Proceedings of the National Academy of Sciences, U S A*, 116(32), 16062–16067. https://doi.org/10.1073/pnas.1906774116

Moscou, M. J., and Bogdanove, A. J. (2009). A simple cipher governs DNA recognition by TAL effectors. *Science*, 326(5959), 1501. https://doi.org/10.1126/science.1178817

Murat, P., Perez, C., Crisp, A., van Eijk, P., Reed, S. H., Guilbaud, G., and Sale, J. E. (2022). DNA replication initiation shapes the mutational landscape and expression of the human genome. *Science Advances*, 8(45), eadd3686. https://doi.org/10.1126/sciadv.add3686

Nathwani, A. C. (2019). Gene therapy for hemophilia. *Hematology: American Society of Hematology Education Program*, 2019(1), 1–8. https://doi.org/10.1182/hematology.2019000007

Pavletich, N. P., and Pabo, C. O. (1991). Zinc finger-DNA recognition: Crystal structure of a Zif268-DNA complex at 2.1 A. *Science*, 252(5007), 809–817. https://doi.org/10.1126/science.2028256

Pixley, K. V., Falck-Zepeda, J. B., Paarlberg, R. L., Phillips, P. W. B., Slamet-Loedin, I. H., Dhugga, K. S., Campos, H., and Gutterson, N. (2022). Genome-edited crops for improved food security of smallholder farmers. *Nature Genetics*, 54(4), 364–367. https://doi.org/10.1038/s41588-022-01046-7

Poetsch, M., Markwerth, P., Konrad, H., Bajanowski, T., and Helmus, J. (2022). About the influence of environmental factors on the persistence of DNA: A long-term study. *International Journal of Legal Medicine*, 136(3), 687–693. https://doi.org/10.1007/s00414-022-02800-6

Ran, F. A., Hsu, P. D., Lin, C. Y., Gootenberg, J. S., Konermann, S., Trevino, A. E., Scott, D. A., Inoue, A., Matoba, S., Zhang, Y., and Zhang, F. (2013). Double nicking by RNA-guided CRISPR Cas9 for enhanced genome editing specificity. *Cell*, 154(6), 1380–1389. https://doi.org/10.1016/j.cell.2013.08.021

Ran, F. A., Hsu, P. D., Wright, J., Agarwala, V., Scott, D. A., and Zhang, F. (2013). Genome engineering using the CRISPR-Cas9 system. *Nature Protocols*, 8(11), 2281–2308. https://doi.org/10.1038/nprot.2013.143

Rossier, O., Wengelnik, K., Hahn, K., and Bonas, U. (1999). The Xanthomonas Hrp type III system secretes proteins from plant and mammalian bacterial pathogens. *Proceedings of the National Academy of Sciences, U S A*, 96(16), 9368–9373. https://doi.org/10.1073/pnas.96.16.9368

Sasisekharan, V., and Pattabiraman, N. (1978). Structure of DNA predicted from stereochemistry of nucleoside derivatives. *Nature*, 275(5676), 159–162. https://doi.org/10.1038/275159a0

Shi, L., Li, X., Xue, L., Zhang, J., Huang, B., Sun, Z., Zhang, Z., Dai, X., Han, S., Dong, W., and Zhang, X. (2023). Creation of herbicide-resistance in allotetraploid peanut using CRISPR/Cas9-meditated cytosine base-editing. *Plant Biotechnology Journal*, 21(10), 1923–1925. https://doi.org/10.1111/pbi.14114

Shi, Y., and Berg, J. M. (1995). A direct comparison of the properties of natural and designed zinc-finger proteins. *Chemistry & Biology*, 2(2), 83–89. https://doi.org/10.1016/1074-5521(95)90280-5

Singleton, M. R., Dillingham, M. S., Gaudier, M., Kowalczykowski, S. C., and Wigley, D. B. (2004). Crystal structure of RecBCD enzyme reveals a machine for processing DNA breaks. *Nature*, 432(7014), 187–193. https://doi.org/10.1038/nature02988

Stavnezer, J., and Schrader, C. E. (2006). Mismatch repair converts AID-instigated nicks to double-strand breaks for antibody class-switch recombination. *Trends in Genetics*, 22(1), 23–28. https://doi.org/10.1016/j.tig.2005.11.002

Sugisaki, H., and Kanazawa, S. (1981). New restriction endonucleases from Flavobacterium okeanokoites (FokI) and Micrococcus luteus (MluI). *Gene*, 16(1–3), 73–78. https://doi.org/10.1016/0378-1119(81)90062-7

Szybalski, W., Kim, S. C., Hasan, N., and Podhajska, A. J. (1991). Class-IIS restriction enzymes: A review. *Gene*, 100, 13–26. https://doi.org/10.1016/0378-1119(91)90345-c

Taylor, G. K., Petrucci, L. H., Lambert, A. R., Baxter, S. K., Jarjour, J., and Stoddard, B. L. (2012). LAHEDES: The LAGLIDADG homing endonuclease database and engineering server. *Nucleic Acids Research*, 40(Web Server issue), W110–W116. https://doi.org/10.1093/nar/gks365

Urnov, F. D., Rebar, E. J., Holmes, M. C., Zhang, H. S., and Gregory, P. D. (2010). Genome editing with engineered zinc finger nucleases. *Nature Reviews Genetics*, 11(9), 636–646. https://doi.org/10.1038/nrg2842

USDA-APHIS. (2020). USDA Animal and Plant Health Inspection Service 2020 Impact Report. Available at: www.aphis.usda.gov/publications/aphis_general/rpt-aphis-impact-2020.pdf

Voytas, D. F., and Joung, J. K. (2009). Plant science: DNA binding made easy. *Science*, 326(5959), 1491–1492. https://doi.org/10.1126/science.1183604

Vyverman, M., De Baets, B., Fack, V., and Dawyndt, P. (2012). Prospects and limitations of full-text index structures in genome analysis. *Nucleic Acids Reearch*, 40(15), 6993–7015. https://doi.org/10.1093/nar/gks408

Wu, J., Chen, C., Xian, G., Liu, D., Lin, L., Yin, S., Sun, Q., Fang, Y., Zhang, H., and Wang, Y. (2020). Engineering herbicide-resistant oilseed rape by CRISPR/Cas9-mediated cytosine base-editing. *Plant Biotechnology Journal*, 18(9), 1857–1859. https://doi.org/10.1111/pbi.13368

Zhao, L., Pellenz, S., and Stoddard, B. L. (2009). Activity and specificity of the bacterial PD-(D/E)XK homing endonuclease I-Ssp6803I. *Journal of Molecular Biology*, 385(5), 1498–1510. https://doi.org/10.1016/j.jmb.2008.10.096

Zhou, T., Yang, L., Lu, Y., Dror, I., Dantas Machado, A. C., Ghane, T., Di Felice, R., and Rohs, R. (2013). DNAshape: A method for the high-throughput prediction of DNA structural features on a genomic scale. *Nucleic Acids Research*, 41(Web Server issue), W56–W62. https://doi.org/10.1093/nar/gkt437

3 CRISPR-Cas Systems for Genome Editing

Evolution, Classification, and Applications

Nayla Munawar, Rabail Afzal, Aftab Ahmad, and Anam Niamat

3.1 INTRODUCTION

The burgeoning world's population and continuously changing climatic conditions have had adverse effects on crop production and biodiversity. Biodiversity depletion poses an alarming situation for human survival, food availability, and quality of health due to global warming. Food demand is expected to double in the next five decades, surpassing the current availability. In the current scenario, a method is quickly needed to increase crop quality and production to meet emerging food requirements (Maja and Ayano 2021). With the progress in technology, molecular biology has given innovative solutions. Deliberate alteration of the DNA by inserting, removing, or modifying at desired sites holds profound significance in the era of emerging genetic technology and biotechnology. The successful alteration undergoes activation or deactivation of the targeted genes, hence delivering valuable outcomes and providing insights to study the function of the gene and genome diversity. Genome editing plays a vital role not only in elucidating gene functions but also in deciphering the regulatory mechanisms of organisms (Kritzer and Luedtke 2005). The revelation of restriction endonucleases (restriction enzymes) in the late 1970s evolved as a turning point in the history of genetic engineering followed by numerous discoveries to date (Newell-McGloughlin and Re 2007). With the development of CRISPR/Cas (Clustered Regularly Interspaced Short Palindromic Repeats/CRISPR-associated) technology, the current era of genetic engineering and genome modifications has undergone revolutionary changes (Ali et al. 2023).

Prior to the emergence of CRISPR/Cas, researchers mainly relied on two genome-editing tools through the application of restriction enzymes: (1) Zinc finger nucleases (ZFN) and (2) Transcription activator-like effector nucleases (TALENS). The enzymes used in both techniques have a DNA binding domain to bind to the targeted DNA and a restriction domain that cuts the DNA at the targeted site. The key distinction lies in their efficiency in DNA binding, with TALENs exhibiting higher effectiveness. Both techniques suffer from disadvantages, such as being time-consuming, incurring high costs, and difficulties in protein engineering (Liu, Sun, and Yu 2024).

Overall, genome editing, as discussed, was once remembered as a laborious and imprecise technology with numerous flaws, which has finally found a breakthrough in precision, efficiency and reliability. With the low cost of development, less time, and more efficient results, CRISPR technology has deepened its roots in biological research. The accuracy and efficiency in altering genes facilitate unraveling the mysteries of unidentified complex mechanisms of the underlying cell regulatory pathways and have taken biological research into new frontiers (Ali et al. 2023).

CRISPR, a defensive mechanism of bacteria, protects it from foreign invaders on the basis of its recognition and defensive strategy (Ahmed et al. 2018). CRISPR loci are extended by various Cas genes and palindromic sequences, interspaced by spacers, which identify foreign invasive sequences. These palindromic sequences are found in the chromosomal or plasmid DNAs. The Cas gene encodes the nuclease protein and degrades the foreign element, either by cutting the specific genetic part or modifying it by mutation. Scientists have improved the plant's genome by enhancing crop productivity and disease resistance by employing these mechanisms (Ostria-Hernández et al. 2015).

The convergence of genetic engineering with molecular biology and therapeutic applications has been introduced with remarkable precision and accuracy. The transformative power of therapeutic genome editing using CRISPR technology, ushering in the modern epoch of precision medicine, comes with numerous benefits (Li et al. 2020). The two most impactful genetic therapeutic mechanisms introduced so far are gene therapy, which aims to restore impaired gene activity by utilizing viral transgene expression, and RNA interference (RNAi), which suppresses the undesirable or defective genes by knocking down the respective mRNA (Bobbin and Rossi 2016). Ongoing experiments have proved the wide-range adaptability of CRISPR to many genetic conditions, enabling researchers to identify various biomedical applications for personalized treatments.

Considering the importance and need of CRISPR in the current era, this chapter delves into the evolution, classification, and diverse applications of the CRISPR/Cas system, elucidating its transformative power as a genomic revolution in biological and therapeutic interventions.

3.2 EVOLUTION OF THE CRISPR/CAS SYSTEMS

Evolving from microbial genomes, the CRISPR/Cas system has undergone a revolutionary journey in leading genetic engineering into a new era. Bacteria and archaea engage in

DOI: 10.1201/9781003500933-4

an ongoing battle against selfish genetic elements like phages, transposons, and plasmids, employing various defense mechanisms. One well-known defense mechanism involves the restriction-modification enzyme systems, which target invaders based on methylation patterns of DNA. Recently, a novel defense system, termed the CRISPR (Clustered Regularly Interspaced Short Palindromic Repeats) associated system or CRISPR/Cas, was discovered. This system operates on a different principle, utilizing CRISPR-associated proteins (Cas) to combat invaders (Koonin and Makarova 2019). For years, the presence of distinct arrays of short repeats interspersed with unique spacers known as CRISPR has been observed in genomes of both bacteria and archaea, yet their potential functions remained elusive. The initial "prokaryotic RNAi" hypothesis suggested that CRISPR/Cas systems might target the mRNAs of foreign invaders (Koonin and Makarova 2009).

Clustered regularly interspaced short palindromic repeats (CRISPR) were initially recognized within the DNA sequences of *Escherichia coli* (*E. coli*) bacteria and formally documented in 1987 (Toro et al. 2013). During this era, discerning the biological role of these peculiar repetitive sequences proved challenging owing to the limited availability of DNA sequence data, particularly concerning mobile genetic elements. The core function of this distinctive sequence remained mysterious till the mid-2000s. In 1993, CRISPRs were initially identified in archaea, specifically within *Haloferax mediterranei* (Li et al. 2012), and later found in an expanding array of bacterial and archaeal genomes, coinciding with the advent of genomics in life sciences. The conservation of these sequences across two of the three fundamental domains of life was pivotal in recognizing their significance. In the early 2000s, the revelation of sequence resemblance between the spacer regions of CRISPRs and sequences from bacteriophages, archaeal viruses, and plasmids finally illuminated CRISPR's function as an immune system. Concurrently, many genes earlier believed to encode DNA repair proteins specific to hyper-thermophilic archaea were found to be closely linked with CRISPR and were consequently labeled as *cas* (CRISPR-associated) genes (Marshall and Santangelo 2020). Comparative genomic analyses indicated that CRISPR and Cas proteins (the products of *cas* genes) collaborate, forming an acquired immunity system that shields prokaryotic cells from intruding viruses and plasmids. This mechanism bears a resemblance to the eukaryotic RNA interference (RNAi) system (Marshall and Santangelo 2020).

In 2017, the initial experimental insights into the workings of the CRISPR system emerged from the research efforts of two French food scientists, Rodolphe Barrangou and Philippe Horvath. They conducted their studies using bacterial yogurt cultures of *Streptococcus thermophilus* while working for the Danish company Danisco. Presently, CRISPR repeats have been identified in the majority of archaeal genomes and approximately half of the investigated bacterial genomes. However, they have not been detected in sequences of eukaryotic or viral DNA (Barrangou and Horvath 2017).

Initially, various acronyms were used for CRISPR, complicating early research identification. The term "CRISPR" was coined by Jansen et al. in 2002, who also described cas1–4 genes (Jansen et al. 2002). Similar genes were noted by Eugene Koonin's team the same year. Since then, a diverse range of CRISPR-associated genes has been found in prokaryotes, categorized into adaptive (e.g., Cas1, Cas2) and effector modules. Current classification divides CRISPR–Cas systems into two classes and six types, with numerous subtypes identified. The key contrast between the classes is that the effector module of Class I systems consists of a complex of several proteins, while in Class II it is a single multidomain protein (Cas9, Cas12, or Cas13) (Makarova et al. 2019). The most studied Cas proteins are those responsible for the precise cutting of foreign DNA, often referred to as "genetic scissors." One key example is the nuclease Cas9, first linked to CRISPR repeats in a paper by Bolotin. Another important discovery by Bolotin was a specific pattern near CRISPR spacers, later recognized as PAMs (protospacer adjacent motifs), crucial for the system's function. PAM motifs signal foreign DNA for destruction by the prokaryotic immune system, while spacers lacking PAMs are not targeted (Bolotin et al. 2005).

In the CRISPR/Cas9 system, a key player is the crRNA (CRISPR-associated RNA), which guides prokaryotic immune system proteins to extrinsic genetic material. Discovered by John van der Oost's team at Wageningen University, the crRNA originates from the transcription of the CRISPR locus, initially forming a precursor molecule called pre-crRNA, which is later cleaved into individual fragments (Brouns et al. 2008). The last essential component needed to assemble a functional CRISPR/Cas9 system in vitro was discovered in 2011 by Emmanuelle Charpentier's team: a short RNA molecule crucial for nuclease activity (Li et al. 2023). This molecule, identified during the study of crRNA processing, was named tracrRNA (trans-activating CRISPR RNA) (Deltcheva et al. 2011). The identification of essential components of the CRISPR/ Cas9 system initiated a competition to be the first to utilize the technology of genetic editing in human and animal cells. Emmanuelle Charpentier and Jennifer Doudna made groundbreaking discoveries in DNA manipulation with the CRISPR/Cas9 system, earning them the first-ever Nobel Prize awarded to two women in 2020. Their work with the "genetic scissors" technique is of immense significance, allowing manipulation of model organism genomes, modification of important crop and animal traits, and potentially revolutionary changes in medicine, particularly in treating genetic diseases (Yeung 2022).

3.3 EVOLUTIONARY DYNAMICS IN THE CRISPR/CAS SYSTEM

The mechanism of CRISPR/Cas is typically divided into three stages: (1) adaptation, where new 30–84 bp long, unique spacers homologous to proto-spacer sequences in viral genomes or other foreign DNA molecules are integrated into the CRISPR repeat cassettes; (2) expression and processing of pre-crRNA into short guide crRNAs; and (3) interference, wherein the foreign DNA or RNA is targeted by the Cas proteins complex that includes a crRNA guide, resulting in cleavage of the unique target site. The CRISPR/Cas systems represent a universal immune mechanism capable of adapting to defend against various mobile genetic elements (MGEs). Unlike innate immune systems such as restriction-modification modules, which are ubiquitous and abundant, CRISPR/Cas systems exhibit extensive diversification. They are classified into two classes based on the design principles of their effector modules: Class I systems feature multi-subunit effector complexes composed of several Cas proteins, while Class II systems have a single, large, multidomain protein as the effector. Classification of CRISPR/Cas systems is complex due to the absence of universal Cas proteins for phylogenetic markers. Existing classifications rely on a range of criteria, including the organization of cas operons, signature *cas* genes, and phylogenies of conserved Cas proteins (Iranzo et al. 2013). Nonetheless, the latest discovery of a signaling pathway integrated into Type III systems of Class I may provide insight, suggesting that Type III effector modules could have originated from a stress-induced programmed cell death signal transduction system. The later advancement of Class I effector complexes involved serial gene duplication and displacement, particularly of genes encoding proteins with RNA recognition motif domains. Alongside the numerous contributions of mobile genetic elements (MGEs) to CRISPR/Cas evolution, the reverse flow of information is noteworthy: MGEs have been found to recruit minimalist variants of CRISPR/Cas systems for functions yet to be fully understood (Faure et al. 2019).

3.3.1 CRISPR/Cas System Restricts HGT

CRISPR/Cas systems have the capability to integrate segments of invading DNA, such as fragments from bacteriophages or mobile genetic elements, as spacers within CRISPR loci. Active CRISPR/Cas systems require a set of Cas genes to transcribe and process CRISPR arrays into short CRISPR RNAs (crRNAs). These crRNAs, each containing a single spacer, must then bind to the Cas endonuclease. Together, this complex utilizes crRNA base complementarity to identify and degrade DNA from elements containing the spacer sequence upon subsequent reinfection of the cell. Essentially, CRISPR/Cas systems serve as a molecular memory of past infections, providing bacteria and archaea with adaptive immunity against foreign DNA. Horizontal gene transfer (HGT) significantly influences bacterial evolution and serves as a major source of genome expansion (Wheatley and MacLean 2020). While CRISPR/Cas systems primarily target parasitic genetic elements such as lytic phages, their xenogenic immunity may also limit horizontal gene transfer (HGT) more broadly. Increasingly, it's recognized that CRISPR/Cas systems target various other mobile genetic elements, including integrative conjugative elements (ICE), plasmids, and DNA acquired through transformation. Experimental studies across multiple systems have demonstrated that CRISPR/Cas can effectively prevent HGT over short time scales. For instance, in Staphylococcus epidermidis, CRISPR/Cas systems with a spacer targeting a highly conserved nickase on staphylococcal conjugative plasmids have effectively prevented plasmid transformation. However, bioinformatic studies have produced conflicting results regarding the broader impact of CRISPR/Cas on horizontal gene transfer (HGT) over longer time scales (Sorek, Lawrence, and Wiedenheft 2013). While some studies found no correlation between CRISPR/Cas activity and HGT frequency, others reported that the presence of CRISPR/Cas systems only partially limits the assimilation of antibiotic-resistance genes in certain bacterial pathogens. The simplest interpretation of this relationship is that CRISPR/Cas systems limit the acquisition of mobile genetic elements. However, recent experimental evidence suggests that CRISPR/Cas systems can be lost due to selection pressures favoring the elimination of immunity to newly acquired mobile elements that provide a fitness advantage (Butiuc-Keul et al. 2021).

We propose that the CRISPR/Cas system is a crucial constraint to horizontal gene transfer. It is vital to understand the evolutionary mechanisms that ensure its maintenance or drive its loss to decipher how this pathogen adapts to new environments and stressors.

3.4 CLASSIFICATION OF THE CRISPR/CAS SYSTEMS

The CRISPR/Cas system functions as an adaptive immune system in bacteria and archaea, protecting it from foreign invasion owing to its variable features. These systems possess a rich and diverse mechanism across multiple domains, including protein composition, pre-crRNA processing, genome architecture, the mechanism of interference and adaptation, effector complex formation, etc. The composition of the CRISPR/Cas system involves repeated spacers which are transcribed into CRISPR RNA (crRNA) and trans- activating CRISPR RNA (tracrRNA) with a cluster of CRISPR-associated (Cas genes). These genes encode Cas proteins with endonuclease activity. When the prokaryotes encounter foreign invasion, the respective DNA is fragmented by Cas proteins and incorporated as new spacers into the CRISPR array. Upon encountering them again, the crRNA promptly recognizes and binds to the foreign DNA, which directs the Cas protein to cut the target DNA quickly and protects the host (Lino et al. 2018).

The Cas genes serve as the basis of the classification of the CRISPR/Cas system into two classes and six different types. In the CRISPR system, Class I uses multi-subunit protein complexes, and Class II relies on a single, large Cas protein for its function. Further, Class I includes Types I, III and IV, while Class II has Types II, V, and VI. These types are classified by the existence of specific Cas genes, i.e. Cas3 in Type I; Cas9 in Type II; and Cas10 in Type III. Types I, II, and III have unique characteristics, well-developed mechanisms, and a widespread presence in microbial species. Types I and III possess distinct Cas genes that, when combined with crRNA, enable the identification and eradication of foreign DNA.

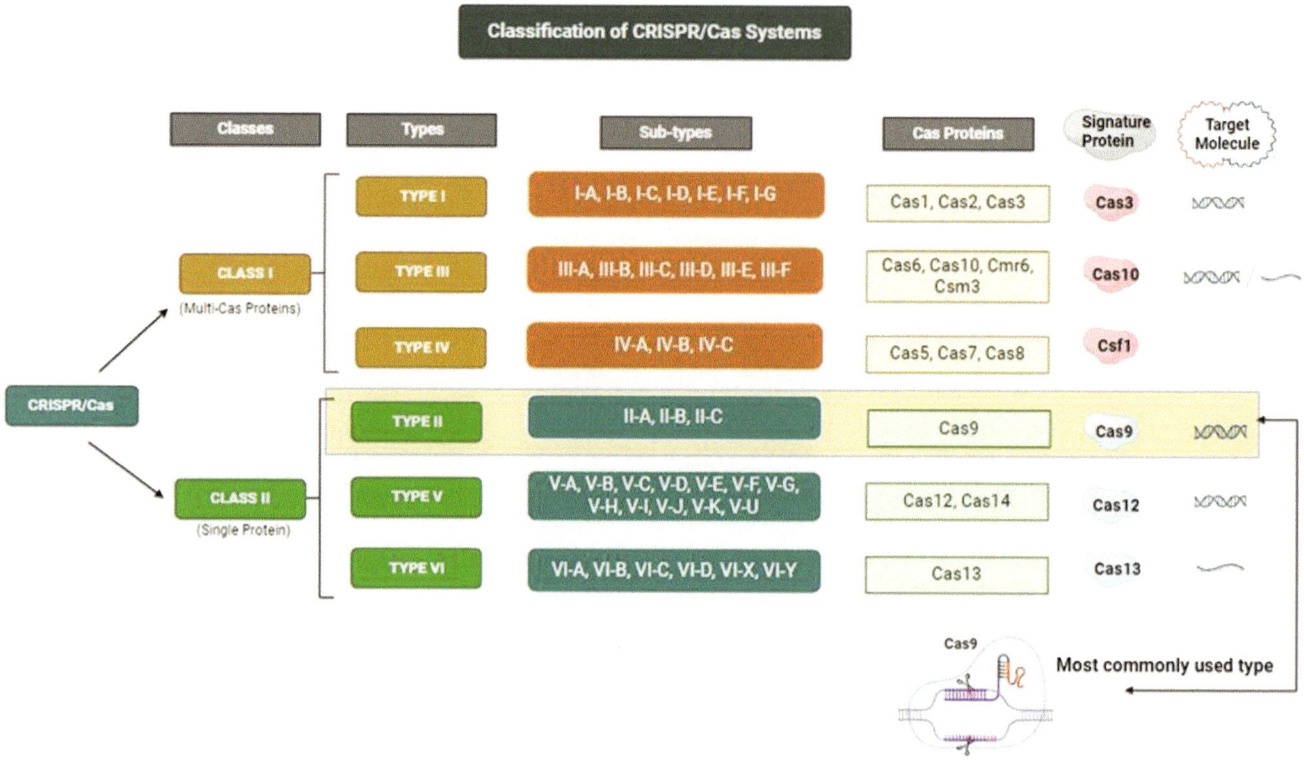

FIGURE 3.1 Classification of CRISPR-Cas systems.

On the other hand, Type II possesses fewer Cas genes, with their biological significance less well understood. However, it is capable of performing the function of cleaving DNA and generating crRNA. It possesses two separate domains, namely, RuvC and HNH. These domains are accountable for cutting the targeted double-stranded DNA (DSBs) by cutting one strand with each domain, resulting in precise and accurate cleavage. This precision makes this type highly accurate and efficient, with low cost in genetic engineering, contributing to its widespread adoption (Makarova, Wolf, and Koonin 2018).

In contrast, Types IV, V, and VI have received less attention due to their infrequent occurrence in archaeal and bacterial species, as well as their unclear complex mechanisms, which have resulted in a lack of data and opportunities for practical investigations. Despite this, the roles of each Cas protein in these types and their interactions with targeted DNA are still being investigated due to their intricate molecular mechanisms. Thus, the limited occurrence, complex mechanisms, and underdevelopedapplications contribute to the restricted understanding and slower adaptation of these types (Liu et al. 2020). See Figure 3.1.

3.4.1 Class I CRISPR Systems

3.4.1.1 Type I CRISPR/Cas

Among the types of Class I CRISPR/Cas system, Type I is the most promising. This system involves seven further subtypes (I-A to I-G) encoding a multi-domain protein Cas3 gene. The presence of Cas3 with helicase and exonuclease activity makes the mechanism easier to induce single-strand breaks and digest DNA using nuclease enzymes. The Cas3 protein contains two main domains, i.e., one to cleave the DNA (N-terminal HD phophohydrolase) and the other to unwind the double-stranded DNA (C-terminal DExH helicase). Both of these domains work side by side to degrade the foreign DNA. This system is present in the majority of bacteria and archaea.

Although Cas3 alone cannot recognize an invader's DNA and protect bacteria from infections, it does so in conjunction with other proteins to form a complex known as the crRNA-guided surveillance complex, or CASCADE. The CASCADE complex, which is composed of multiple Cas proteins, functions to identify and bind target sequences that are complementary to the crRNA spacer. The crRNA-guided surveillance complex was first identified in *E. coli* K12 (Type I-E) and is comprised of five Cas proteins. Cas6e, previously known as Cse3 or CasE, aids in the maturation of crRNA, which remains attached to the CASCADE complex and is responsible for detecting and cleaving invader DNA. Similar complexes have also been reported in *S. solfataricus*, *Pseudomonas aeruginosa* (Type I-F), and *Bacillus halodurans* (Type I-C) (Jackson et al. 2014; Li et al. 2015).

3.4.1.2 Type III CRISPR/Cas Systems

Type III systems are numerous and similar to those of Type I, including the presence of the cas10 gene in their operons. The Cas10 protein carries a Palm domain as an RNA recognition motif and cyclase catalytic domain that have long been known to be involved in RNA recognition and cleavage. Unlike the other types, Type III systems uniquely target both DNA and RNA, which makes them a versatile CRISPR Type. Overall, there are fewer subtypes of Type III systems compared to Type I, and the architecture of their operons is more diverse in terms of functional adaptations (Kolesnik et al. 2021).

3.4.1.3 Type IV CRISPR/Cas Systems

While Type IV systems technically belong to Class I, less is known about them than Types I and III. In contrast to other systems of Class I, the effector complexes in Type IV are much smaller as they only consist of a limited but a diverse set of proteins. Type IV systems do not include domains necessary for target nucleic acid cleavage or spacer acquisition, making it functionally different from the other CRISPR systems. The precise role and mechanism of this system in CRISPR immunity are still unknown (Pinilla-Redondo et al. 2019).

3.4.2 Class II: Type II CRISPR/Cas Systems

Among the various types of CRISPR systems, the Type II CRISPR/Cas system stands out as one of the most widely studied and applied systems, particularly because of its use of the Cas9 protein, which has become a cornerstone in genome-editing technologies.

3.4.2.1 Understanding the Type II CRISPR/Cas System

The Type II CRISPR/Cas systems belong to Class II, known for their simplicity and widespread use in genome editing, are divided into three subtypes: IIA, IIB, and IIC. These subtypes are distinguished by their unique components and mechanisms of action, particularly in how they process the CRISPR RNA (crRNA) and target DNA for cleavage (Makarova et al. 2019).

The IIA subtype is the most studied and commonly used in research, particularly the system from *Streptococcus pyogenes*. In addition to cas1, cas2, and cas 9, the IIA subtype possesses a csn2 gene encoding for Csn2 protein involved in spacer integration but is not required for interference. In this subtype, the Cas9 protein is guided by a dual RNA structure composed of crRNA and a trans-activating crRNA (tracrRNA), which forms a complex that directs the Cas9 to the target DNA sequence. The tracrRNA is essential in processing the pre-crRNA into mature crRNA in conjunction with the host RNase III, enabling the Cas9 protein to recognize and cleave the target DNA at specific sites adjacent to the protospacer adjacent motif (PAM) (Chaudhuri, Halder, and Datta 2022).

On the other hand, the IIB and IIC subtypes show variations in their processing and targeting mechanisms. The IIB subtype, which is less common than IIA, lacks the csn2 gene but possesses another additional gene of the *cas4* family, differentiating it from IIA. Type IIB systems also rely on tracrRNA for the processing of pre-crRNA. Cas9 is the primary effector protein responsible for cleaving the target DNA, guided by crRNA and tracrRNA. However, the role of Cas4 in crRNA maturation, its interaction with Cas9, and the targeting process are still under investigation. The IIC subtype, found in species such as *Neisseria meningitidis*, is unique in that it possesses a minimalistic system with a smaller Cas9 protein and a streamlined crRNA processing mechanism that doesn't rely heavily on accessory proteins. The IIC subtype is particularly noteworthy for its compact nature, containing only the cas1, cas2, and cas9 genes in the operon and lacking csn2 or cas4 genes. The IIC subtype does not always require RNase III for crRNA maturation. Cas9 protein itself plays a dual role in crRNA processing, simplifying the overall mechanism. The simplified architecture of IIC makes it an attractive candidate for gene-editing applications requiring delivery systems with size constraints, such as in viral vectors. These differences in RNA processing, protein composition, and overall system complexity highlight the diversity within the Type II CRISPR/Cas systems, each offering distinct advantages and applications in genome editing (Singh 2020; Arroyo-Olarte, Rodríguez-Hernández, and Morales-Ríos 2023).

3.4.2.2 The Role of Cas9: The Signature Protein of the Type II System

The Cas9 protein is a signature protein of the Type II CRISPR system that has attracted extensive attention because it can induce site-specific double-strand breaks in DNA—one of the most important features that make it an effective genome editor. Cas9 is an endonuclease, thus capable of cleaving DNA strands. This cleavage specificity is provided by the guide RNA, which is a single RNA molecule created by fusing crRNA and tracrRNA. By means of complementary base pairing, this gRNA guides Cas9 to the DNA sequence of interest, hence assuring the cleavage in the correct genomic position (Chylinski et al. 2013).

The process by which Cas9 mediates DSBs consists of a series of steps. First, the gRNA binds to the target DNA sequence, which must be immediately adjacent to a protospacer adjacent motif (PAM) sequence, usually "NGG" in the case of the commonly used *Streptococcus pyogenes* Cas9 (SpCas9). The PAM sequence is very important because it makes the target DNA distinguishable from the rest of the host genome and therefore protects the bacterial genome from its immune system (Jiang and Doudna 2017).

Upon binding of the gRNA to the target DNA sequence, Cas9 undergoes a conformational change that activates its nuclease domains. Cas9 contains two different nuclease domains, RuvC and HNH, each cutting one strand of the DNA, thus creating a DSB. A DSB in the DNA now activates the cellular mechanisms for repair, which can be further used for genome editing. This break can then be repaired by one of two pathways: NHEJ, which frequently introduces small insertions or deletions that can disrupt the gene, or HDR, which allows for the precise insertion of a DNA template into the site of the break (Jiang and Doudna 2017).

3.4.2.3 Types V and VI in the CRISPR systems

Type V CRISPR systems primarily use Cas12 as a signature effector protein (formerly designated Cpf1) guided by gRNA, which can target both strands of the target DNA through its single RuvC nuclease domain, unlike Cas 9 which uses the HNH-nuclease domain and RuvC-like nuclease domain to cut the complementary and non-complementary strands of the targeted DNA. Type V systems also require a guide RNA such as tracrRNA, similar to Type II CRISPR systems, to function effectively. Cas12 proteins of the Type V subtype systems are diverse at N-terminal however, the C-terminus with only one RuvC endonuclease domain is unified and is the defining characteristic of the type C CRISPR system (Makarova et al. 2015).

Type VI CRISPR systems, which are unique within Class II, exclusively target single-stranded RNA instead of DNA. The signature endonuclease protein used by Type VI systems is Cas13, which is crRNA-guided RNases with two distinct and independent catalytic centers, having RNA-specific editing capabilities. One catalytic site processes the pre-crRNA, and the other is composed of two R-X4-H motifs responsible for the cleavage of single-stranded RNA (ssRNA). An active Cas13-crRNA complex not only cleaves the target complementary ssRNA but remains active and also cleaves non-specific RNAs encountered in the cell through accessible uridine (U) or adenosine (A) sites within the secondary structures of the RNA referred to as "collateral" or "bystander" cleavage. Gootenberg et al. (2017) developed a CRISPR-based diagnostic tool known as SHERLOCK (Specific High Sensitivity Enzymatic Reporter unLOCKing) by exploiting the unique RNA-targeting ability of Cas13. This tool has demonstrated the ability to detect the Zika virus and some strains of dengue fever, indicating its potential to expand its applications after further reprogramming to respond to viral and bacterial outbreaks, as it lacks the sensitivity required in its current form for many research and clinical applications (Gootenberg et al. 2017; Perčulija et al. 2021).

3.5 APPLICATIONS OF CRISPR/CAS

CRISPR/Cas has revolutionized the study of gene function by providing precise and versatile tools for gene editing.

The precise and programmable modification of the DNA sequence has provided researchers with an opportunity to manipulate genes for various applications, from research in model organisms to therapeutic applications and agriculture. The applications of this versatile tool are discussed throughout this book; however, a brief overview of successful CRISPR applications with a few examples in different fields is described below and shown in Figure 3.2.

3.5.1 Genome Editing in Model Organisms to Understand Cellular Mechanisms

Model organisms such as mice, zebrafish, and fruit flies are essential for studying human diseases and biological processes. CRISPR/Cas has expanded the scope of gene editing in these organisms, enabling scientists to understand gene functions and pathways for a better understanding of disease mechanisms and the development of potential therapies for humans. Genetically modified, knockout, and knock-in mice models have been developed by scientists worldwide and are now available commercially to the scientific community for research aimed at understanding the genetic basis of diseases such as cancer, diabetes, and neurological disorders (Yamamoto et al. 2023). A dystrophin gene knockout mouse model of Duchenne Muscular Dystrophy reflecting the human disease condition was created using CRISPR. CRISPR's efficiency made it an essential tool for studying genetic interactions related to aging and neurodegeneration in fruit flies. The CRISPR-mediated genome-edited models of Drosophila have been used to target genes linked to neural circuits and brain function to explore neural development and behavior (Zhu et al. 2017). Specific gene disruption employing CRISPR in zebrafish embryos enabled the identification of developmental genetics and embryogenesis. Zebrafish is also a significant model organism to study cardiovascular diseases (Zhu et al. 2017).

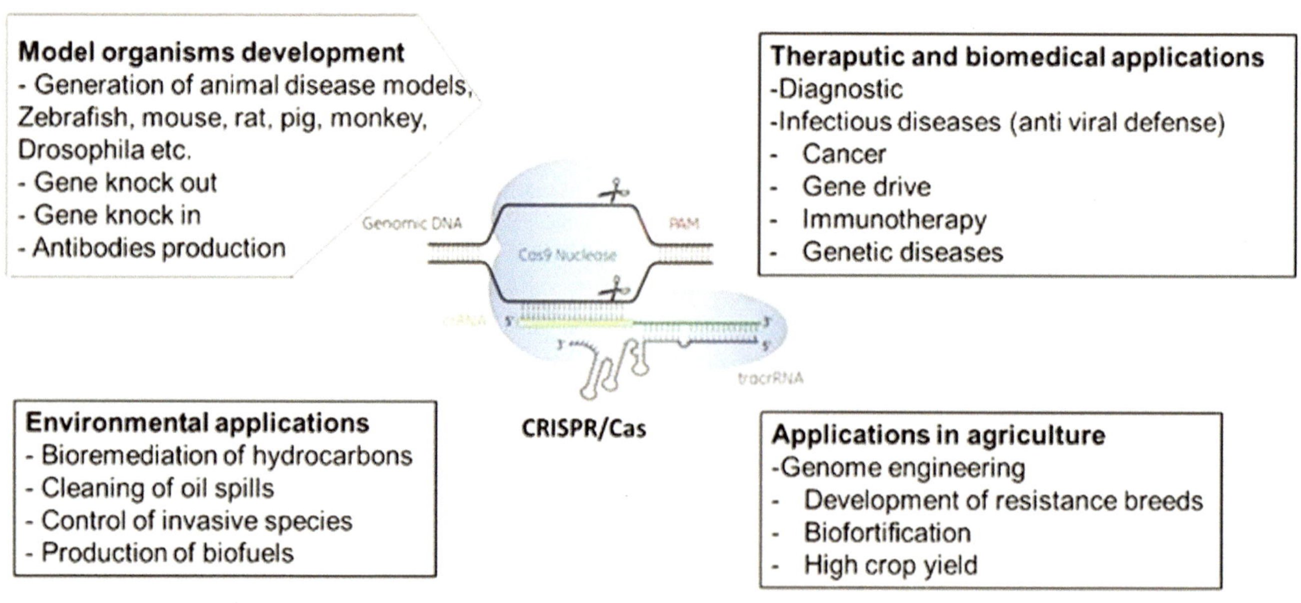

FIGURE 3.2 Diverse applications of CRISPR-Cas in various fields.

3.5.2 Therapeutic Applications

One of the immense promises of the CRISPR/Cas technology lies in gene therapy, offering the potential to correct the DNA mutations and treat previously incurable genetic disorders . For instance, correcting a single gene mutation in hemoglobin by CRISPR has offered a permanent cure for sickle cell anemia (Demirci et al. 2021). Moreover, gene corrections by CRISPR in stem cells for cystic fibrosis, Duchenne muscular dystrophy (DMD), and beta-thalassemia treatment have shown promising results in vitro, displaying the potential for the enduring treatment of these genetic disorders with CRISPR technology (Aslan and Yuka 2023).

CRISPR-based immunotherapy has been employed successfully against cancer by editing T-cells, enhancing their ability to recognize and attack tumor cells . CRISPR-edited CAR-T cells can fight against leukemia and lymphoma (Rezalotfi et al. 2022). Enhanced anti-tumor activity has been observed by CRISPR-engineered T-cells having the potential to remove a gene responsible for immune inhibition in cancer patients (Allemailem et al. 2023). CRISPR/Cas offers specific oncogene targeting to develop new lines of defense against cancer. Furthermore, targeting viral DNA or RNA by the CRISPR/Cas systems is also being developed for the treatment of infectious diseases. CRISPR-mediated genome editing in HIV-infected human cells targets viral DNA integrated into the host genome, disrupts the HIV proviral DNA, and significantly reduces viral load (Kaminski et al. 2016). The development of new CRISPR/Cas systems, which provide resistance to viral infections, potentially leads to a cure for difficult-to-treat viral infectious diseases such as Hepatitis B and the herpes simplex virus.

3.5.3 Agricultural Applications

Crop improvement with CRISPR showed pronounced advantages over traditional plant breeding techniques. The technique allows for precise DNA modifications, offering the ability to create crops with resistance to diseases and environmental stress, enhanced yield, and high nutritional value without the introduction of foreign genes, indicating its transformative potential in agriculture. A higher drought tolerance rice crop by gene knockout with CRISPR has been reported (Wang et al. 2020). Resistance to bacterial blight in rice, fungal infections in wheat, and combating citrus greening disease in citrus plants by targeting specific genes using CRIPSR resulted in enhanced disease resistance in these crops (Rato et al. 2021). The CRISPR/Cas gene modification system has been effectively employed to increase beta-carotene (precursor of vitamin A) content in rice to address vitamin deficiencies in people who depend on rice as a staple food. Additionally, CRISPR has been used to confer biofortification in tomatoes, cassava, and other crops to increase the levels of essential vitamins and minerals (Kiran 2020).

3.5.4 Environmental Applications

In addition to several applications in the above-mentioned fields, CRISPR/Cas systems are being used in environmental sustainability to manage ecosystems and address pollution through bio-remediation, controlling invasive species, and improving the efficiency of biofuel production. The use of CRISPR-driven gene drives to spread malaria resistance in mosquitos is one of the efficacious examples of controlling the populations of invasive species or disease vectors (Subica 2023). Engineering of microorganisms using CRISPR/Cas to degrade hydrocarbons, heavy metals, and other pollutants offers new hope for cleaning up oil spills and contaminated sites. Genetically modified bacteria could provide a more efficient and sustainable way of addressing pollution (Rafeeq et al. 2023). Modified Pseudomonas species using CRISPR to improve their capacity to degrade hydrocarbons in oil spills indicates the potential of CRISPR to address bioremediation of marine environments where pollutant degradation remains a significant challenge (Wu et al. 2021). However, ethical concerns apply, which need to be addressed with the advanced development of the CRISPR systems.

3.6 CONCLUSION

The CRISPR/Cas system has emerged as a transformative technology offering innovative solutions to some of the most pressing issues in medicine, food security, and environmental conservation. On one hand, the precision and accuracy in targeting genes extend CRISPR applications from basic research in model organisms to therapeutic interventions and agricultural improvements, on the other hand, ethical, regulatory, and safety aspects of its use are the most tenacious challenge to be addressed. However, continuously refining the technology toward greater precision is brightening its future, expanding its applications in therapeutics and diagnostics of infectious diseases like COVID-19, thus making CRISPR a powerful tool in global health monitoring (Gootenberg et al. 2017). Constant improvement and innovation in designing the tools will unlock broader possibilities across diverse scientific and industrial sectors in the future.

REFERENCES

Ahmed, W., Hafeez, M.A., Ahmad, R. and Mahmood, S. (2018). CRISPR-Cas system in regulation of immunity and virulence of bacterial pathogens. *Gene Reports*, 13, 151–157. https://doi.org/10.1016/j.genrep.2018.10.004.

Ali, A., Zafar, Z.Z., Farooq, Z., Ahmed, S.R., Ijaz, A., Anwar, Z., Abbas, H., Tariq, M.S., Tariq, H., Mustafa, M., Bajwa, M.H., Shaukat, F., Razzaq, A., and Maozhi, R. (2023). Breakthrough in CRISPR/Cas system: Current and future directions and challenges. *Biotechnology Journal*. https://doi.org/10.1002/biot.202200642.

Allemailem, K.S., Alsahli, M.A., Almatroudi, A., Alrumaihi, F., Al Abdulmonem, W., Moawad, A.A., Alwanian, W.M., Almansour, N.M., Rahmani, A.H., and Khan, A.A. (2023). Innovative strategies of reprogramming immune system cells by targeting CRISPR/Cas9-Based genome-editing tools: A new era of cancer management. *International Journal of Nanomedicine*, 18, 5531–5559. https://doi.org/10.2147/ijn.s424872.

Arroyo-Olarte, R.D., Rodríguez-Hernández, K.D., and Morales-Ríos, E. (2023). Genome engineering in bacteria: Current and prospective applications. In V. Gurtler and M. Calcutt (Eds.), *Methods in Microbiology*, Vol. 52, *Genome Engineering*. Science Direct, pp. 35–76. https://doi.org/10.1016/bs.mim.2023.01.003.

Aslan, A. and Yuka, S.A. (2023). Stem cell-based therapeutic approaches in genetic diseases. *Advances in Experimental Medicine and Biology*, 1436, 19–53. https://doi.org/10.1007/5584_2023_761.

Barrangou, R. and Horvath, P. (2017). A decade of discovery: CRISPR functions and applications. *Nature Microbiology*, 2(7). https://doi.org/10.1038/nmicrobiol.2017.92.

Bobbin, M.L. and Rossi, J.J. (2016). RNA interference (RNAi)-based therapeutics: Delivering on the promise? *Annual Review of Pharmacology and Toxicology*, 56(1), 103–122. https://doi.org/10.1146/annurev-pharmtox-010715-103633.

Bolotin, A., Quinquis, B., Sorokin, A., and Ehrlich, S.D. (2005). Clustered regularly interspaced short palindrome repeats (CRISPRs) have spacers of extrachromosomal origin. *Microbiology*, 151(8), 2551–2561. https://doi.org/10.1099/mic.0.28048-0.

Brouns, S.J.J., Jore, M.M., Lundgren, M., Westra, E.R., Slijkhuis, R.J.H., Snijders, A.P.L., Dickman, M.J., Makarova, K.S., Koonin, E.V., and van der Oost, J. (2008). Small CRISPR RNAs guide antiviral defense in Prokaryotes. *Science*, 321(5891), 960–964. https://doi.org/10.1126/science.1159689.

Butiuc-Keul, A., Farkas, A., Carpa, R., and Iordache, D. (2021). CRISPR-Cas system: The powerful modulator of accessory genomes in Prokaryotes. *Microbial Physiology*, 32(1–2), 1–16. https://doi.org/10.1159/000516643.

Chaudhuri, A., Halder, K., and Datta, A. (2022). Classification of CRISPR/Cas system and its application in tomato breeding. *Theoretical and Applied Genetics*, 135(2), 367–387. https://doi.org/10.1007/s00122-021-03984-y.

Chylinski, K., Le Rhun, A., and Charpentier, E. (2013). The tracrRNA and Cas9 families of type II CRISPR-Cas immunity systems. *RNA Biology*, 10(5), 726–737. https://doi.org/10.4161/rna.24321.

Deltcheva, E., Chylinski, K., Sharma, C.M., Gonzales, K., Chao, Y., Pirzada, Z.A., Eckert, M.R., Vogel, J., and Charpentier, E. (2011). CRISPR RNA maturation by trans-encoded small RNA and host factor RNase III. *Nature*, 471(7340), 602–607. https://doi.org/10.1038/nature09886.

Demirci, S., Leonard, A., Essawi, K., and Tisdale, J.F. (2021). CRISPR/Cas9 to induce fetal hemoglobin for the treatment of sickle cell disease. *Molecular Therapy: Methods & Clinical Development*, 23. https://doi.org/10.1016/j.omtm.2021.09.010.

Faure, G., Shmakov, S.A., Yan, W.X., Cheng, D.R., Scott, D.A., Peters, J.E., Makarova, K.S., and Koonin, E.V. (2019). CRISPR–Cas in mobile genetic elements: Counter-defence and beyond. *Nature Reviews Microbiology*, 17(8), 513–525. https://doi.org/10.1038/s41579-019-0204-7.

Gootenberg, J.S., Abudayyeh, O.O., Lee, J.W., Essletzbichler, P., Dy, A.J., Joung, J., Verdine, V., Donghia, N., Daringer, N.M., Freije, C.A., Myhrvold, C., … Zhang, F. (2017). Nucleic acid detection with CRISPR-Cas13a/C2c2. *Science*, 356(6336), 438–442. https://doi.org/10.1126/science.aam9321.

Iranzo, J., Lobkovsky, A.E., Wolf, Y.I., and Koonin, E.V. (2013). Evolutionary dynamics of the Prokaryotic adaptive immunity system CRISPR-Cas in an explicit ecological context. *Journal of Bacteriology*, 195(17), 3834–3844. https://doi.org/10.1128/jb.00412-13.

Jackson, R.N., Golden, S.M., van Erp, P.B.G., Carter, J., Westra, E.R., Brouns, S.J.J., van der Oost, J., Terwilliger, T.C., Read, R.J., and Wiedenheft, B. (2014). Crystal structure of the CRISPR RNA-guided surveillance complex from Escherichia coli. *Science*, 345(6203), 1473–1479. https://doi.org/10.1126/science.1256328.

Jansen, R., Embden, J.D.V., Gaastra, W., and Schouls, L.M. (2002). Identification of genes that are associated with DNA repeats in prokaryotes. *Molecular Microbiology*, 43(6), 1565–1575. https://doi.org/10.1046/j.1365-2958.2002.02839.x.

Jiang, F. and Doudna, J.A. (2017). CRISPR–Cas9 structures and mechanisms. *Annual Review of Biophysics*, 46(1), 505–529. https://doi.org/10.1146/annurev-biophys-062215-010822

Kaminski, R., Chen, Y., Fischer, T., Tedaldi, E., Napoli, A., Zhang, Y., Karn, J., Hu, W., and Khalili, K. (2016). Elimination of HIV- 1 genomes from human T-lymphoid cells by CRISPR/Cas9 gene editing. *Scientific Reports*, 6(1). https://doi.org/10.1038/ srep22555.

Kiran, K. (2020). Advanced approaches for biofortification. In T.R. Sharma, R. Deshmukh, and H. Sonah (Eds.), *Advances in Agri-Food Biotechnology*. Springer, pp. 29–55. https://doi.org/10.1007/978-981-15-2874-3_2.

Kolesnik, M.V., Fedorova, I., Karneyeva, K.A., Artamonova, D.N., and Severinov, K.V. (2021). Type III CRISPR-Cas systems: Deciphering the most complex Prokaryotic immune system. *Biochemistry (Moscow)*, 86(10), 1301–1314. https://doi.org/10.1134/s0006297921100114.

Koonin, E.V. and Makarova, K.S. (2009). CRISPR-Cas: An adaptive immunity system in prokaryotes. *F1000 Biology Reports*, 1. https://doi.org/10.3410/b1-95.

Koonin, E.V. and Makarova, K.S. (2019). Origins and evolution of CRISPR-Cas systems. *Philosophical Transactions of the Royal Society B: Biological Sciences*, 374(1772), 20180087. https://doi.org/10.1098/rstb.2018.0087.

Kritzer, J.A. and Luedtke, N.W. (2005). Chemical biology: Innovative solutions for diverse challenges. *Chemistry & Biology*, 12(6), 617–620. https://doi.org/10.1016/j.chembiol.2005.06.004.

Li, H., Yang, Y., Hong, W., Huang, M., Wu, M., and Zhao, X. (2020). Applications of genome editing technology in the targeted therapy of human diseases: Mechanisms, advances and prospects. *Signal Transduction and Targeted Therapy*, 5(1), 1–23. https://doi.org/10.1038/s41392-019-0089-y.

Li, M., Liu, H., Han, J., Liu, J., Wang, R., Zhao, D., Zhou, J., and Xiang, H. (2012). Characterization of CRISPR RNA biogenesis and Cas6 cleavage-mediated inhibition of a Provirus in the Haloarchaeon Haloferax mediterranei. *Journal of Bacteriology*, 195(4), 867–875. https://doi.org/10.1128/jb.01688-12.

Li, T., Yang, Y., Qi, H., Cui, W., Zhang, L., Fu, X., He, X., Liu, M., Li, P., and Yu, T. (2023). CRISPR/Cas9 therapeutics: Progress and prospects. *Signal Transduction and Targeted Therapy*, 8(1), 1–23. https://doi.org/10.1038/s41392-023-01309-7.

Li, Y., Pan, S., Zhang, Y., Ren, M., Feng, M., Peng, N., Chen, L., Liang, Y.X., and She, Q. (2015). Harnessing Type I and Type III CRISPR-Cas systems for genome editing. *Nucleic Acids Research*, 44(4), e34–e34. https://doi.org/10.1093/nar/gkv1044.

Lino, C.A., Harper, J.C., Carney, J.P., and Timlin, J.A. (2018). Delivering CRISPR: A review of the challenges and approaches. *Drug Delivery*, 25(1), 1234–1257. https://doi.org/10.1080/10717544.2018.1474964.

Liu, Q., Sun, Q., and Yu, J. (2024). Gene editing's sharp edge: Understanding Zinc Finger Nucleases (ZFN), Transcription Activator-Like Effector Nucleases (TALEN) and Clustered Regularly Interspaced Short Palindromic Repeats (CRISPR). *Transactions on Materials, Biotechnology and Life Sciences*, 3, pp.170–179. https://doi.org/10.62051/e47ayw75.

Liu, Z., Dong, H., Cui, Y., Cong, L., and Zhang, D. (2020). Application of different types of CRISPR/Cas-based systems in bacteria. *Microbial Cell Factories*, 19(1). https://doi.org/10.1186/s12934-020-01431-z.

Maja, M.M. and Ayano, S.F. (2021). The impact of population growth on natural resources and farmers' capacity to adapt to climate change in low-income countries. *Earth Systems and Environment*, 5(2), 271–283. https://doi.org/10.1007/s41748-021-00209-6.

Makarova K. S., Wolf Y. I., Alkhnbashi O. S., Costa F., Shah S. A., Saunders S. J., et al. (2015). An updated evolutionary classification of CRISPR-Cas systems. *Nature Reviews Microbiology*, 13, 722–736. 10.1038/nrmicro3569.

Makarova, K.S., Wolf, Y.I., Iranzo, J., Shmakov, S.A., Alkhnbashi, O.S., Brouns, S.J.J., Charpentier, E., Cheng, D., et al. (2019). Evolutionary classification of CRISPR-Cas systems: A burst of class 2 and derived variants. *Nature Reviews Microbiology*, 18(2), 67–83. https://doi.org/10.1038/s41579-019-0299-x.

Makarova, K.S., Wolf, Y.I., and Koonin, E.V. (2018). Classification and nomenclature of CRISPR-Cas systems: Where from here? *The CRISPR Journal*, 1(5), 325–336. https://doi.org/10.1089/crispr.2018.0033.

Marshall, C.J. and Santangelo, T.J. (2020). Archaeal DNA repair mechanisms. *Biomolecules*, 10(11), p.1472. https://doi.org/10.3390/biom10111472.

Newell-McGloughlin, M. and Re, E. (2007). The dawning of the age of Biotechnology 1970–1990. In M. Newell-McGloughlin and E. Re (Eds.), *The Evolution of Biotechnology*. Springer, pp. 45–91. https://doi.org/10.1007/1-4020-5149-2_3.

Ostria-Hernández, M.L., Sánchez-Vallejo, C.J., Ibarra, J.A., and Castro-Escarpulli, G. (2015). Survey of clustered regularly interspaced short palindromic repeats and their associated Cas proteins (CRISPR/Cas) systems in multiple sequenced strains of Klebsiella pneumoniae. *BMC Research Notes*, 8(1). https://doi.org/10.1186/s13104-015-1285-7.

Perčulija, V., Lin, J., Zhang, B., and Ouyang, S. (2021). Functional features and current applications of the RNA-Targeting Type VI CRISPR-Cas systems. *Advanced Science*, 2004685. https://doi.org/10.1002/advs.202004685.

Pinilla-Redondo, R., Mayo-Muñoz, D., Russel, J., Garrett, R.A., Randau, L., Sørensen, S.J., and Shah, S.A. (2019). Type IV CRISPR–Cas systems are highly diverse and involved in competition between plasmids. *Nucleic Acids Research*, 48(4), 2000–2012. https://doi.org/10.1093/nar/gkz1197.

Rafeeq, H., Afsheen, N., Rafique, S., Arshad, A., Intisar, M., Hussain, A., Bilal, M., and Iqbal, H.M.N. (2023). Genetically engineered microorganisms for environmental remediation. *Chemosphere*, 310, 136751. https://doi.org/10.1016/j.chemosphere.2022.136751.

Rato, C., Carvalho, M.F., Azevedo, C., and Oblessuc, P.R. (2021). Genome editing for resistance against plant pests and pathogens. *Transgenic Research*, 30(4), 427–459. https://doi.org/10.1007/s11248-021-00262-x.

Rezalotfi, A., Fritz, L., Förster, R., and Bošnjak, B. (2022). Challenges of CRISPR-Based gene editing in primary T cells. *International Journal of Molecular Sciences*, 23(3), 1689. https://doi.org/10.3390/ijms23031689.

Singh, V. (2020). An introduction to genome editing CRISPR-Cas systems. In V. Singh and P.K. Dhar (Eds.), *Genome Engineering via CRISPR-Cas9 System*. Academic Press. Available at: www.sciencedirect.com/science/article/pii/B9780128181409000015.

Sorek, R., Lawrence, C.M., and Wiedenheft, B. (2013). CRISPR-mediated adaptive immune systems in bacteria and archaea. *Annual Review of Biochemistry*, 82(1), 237–266. https://doi.org/10.1146/annurev-biochem-072911-172315.

Subica, A.M. (2023). CRISPR in public health: The health equity implications and role of community in gene-editing research and applications. *American Journal of Public Health*, 113(8), 874–882. https://doi.org/10.2105/AJPH.2023.307315.

Toro, M., Cao, G., Ju, W., Allard, M., Barrangou, R., Zhao, S., Brown, E., and Meng, J. (2013). Association of Clustered Regularly Interspaced Short Palindromic Repeat (CRISPR) elements with specific serotypes and virulence potential of Shiga Toxin-Producing Escherichia coli. *Applied and Environmental Microbiology*, 80(4), 1411–1420. https://doi.org/10.1128/aem.03018-13.

Wang, B., Zhong, Z., Wang, X., Han, X., Yu, D., Wang, C., Song, W., Zheng, X., Chen, C., and Zhang, Y. (2020). Knockout of the OsNAC006 transcription factor causes drought and heat sensitivity in rice. *International Journal of Molecular Sciences*, 21(7), 2288. https://doi.org/10.3390/ijms21072288.

Wheatley, R.M. and MacLean, R.C. (2020). CRISPR-Cas systems restrict horizontal gene transfer in Pseudomonas aeruginosa. *The ISME Journal*. https://doi.org/10.1038/s41396-020-00860-3.

Wu, C., Li, F., Yi, S., and Ge, F. (2021). Genetically engineered microbial remediation of soils co-contaminated by heavy metals and polycyclic aromatic hydrocarbons: Advances and ecological risk assessment. *Journal of Environmental Management*, 296, 113185. https://doi.org/10.1016/j.jenvman.2021.113185.

Yamamoto, S., Kanca, O., Wangler, M.F., and Bellen, H.J. (2023). Integrating non-mammalian model organisms in the diagnosis of rare genetic diseases in humans. *Nature Reviews Genetics*. https://doi.org/10.1038/s41576-023-00633-6.

Yeung, T. (2022). Embryo editing technology and its effect on conceptions of motherhood. In *Proceedings of the World Congress on New Technologies*. https://doi.org/10.11159/icbb22.005.

Zhu, P., Wu, F., Mosenson, J., Zhang, H., He, T.-C., and Wu, W.-S. (2017). CRISPR/Cas9-mediated genome editing corrects dystrophin mutation in skeletal muscle stem cells in a mouse model of muscle dystrophy. *Molecular Therapy – Nucleic Acids*, 7, 31–41. https://doi.org/10.1016/j.omtn.2017.02.007.

4 CRISPR-Design Technologies

Enhancing Accuracy and Precision in Gene Editing

Aftab Ahmad, Sameer H. Qari, Rana Ali Haider, Annena Ijaz, Muhammad Faheem, Anam Niamat, Noor ul Huda, and Muhammad Kashif Zahoor

4.1 INTRODUCTION

The CRISPR-Cas system is an adaptive immune system of bacteria and archaea that uses ~20 bp crRNA to guide Cas nucleases to identify and remove the DNA of invading phages. This method has evolved into a potent genome-editing tool for introducing specific mutations at targeted genomic sites in a variety of organisms, such as plants and animals. Since then, various CRISPR/Cas genome-editing tools have been used for DNA, RNA, and protein editing and have many applications in health, medicine, and agriculture (Chuang and Lin, 2021). CRISPR-Cas9 is considered the first and most widely used genome-editing CRISPR system, but several others exist (Li et al., 2023). Researchers have also developed Cas nucleases, such as dCas9 and nCas9, for applications beyond genome editing. Researchers are currently using a range of CRISPR/Cas applications, such as base editing, prime editing, gene expression control, and precise genetic modifications.

CRISPR-Cas-based genome editing has become a revolutionary and powerful gene-editing tool in research and practice due to its simplicity, accuracy, cost-effectiveness, and adaptability. The rapid development of CRISPR-Cas technologies and methods in the last decade has revolutionized the fields of precision medicine, gene therapy, and crop genetic improvement. In nature, the CRISPR-Cas9 mechanism functions like an immune defense against viruses in bacteria and archaea, storing details about viruses in the genomes of prokaryotes (Garcia-Robledo et al., 2020), including CRISPR arrays made up of segments (regions) separated by palindromic repeats. The process of transcribing these sequences and further processing the resulting transcripts leads to the creation of CRISPR-RNA (crRNA), also referred to as guide RNA in active systems. A unique aspect of the crRNA guide is its strict match with the original virus DNA (protospacer). Within the CRISPR-Cas9 mechanism, the transcribed transactivating-RNA (tracrRNA) and crRNA also need to come together to form a ribonucleoprotein complex (RNP complex) with the Cas nuclease (Amirkhanov and Stepanov, 2019). The crRNA facilitates the connection of the RNP complex with the trinucleotide protospacer adjacent motifs (PAM, NGG in the case of SpCas9) present in the phage or DNA element's genome.

In CRISPR-Cas9, the PAM is located right after the protospacer and is crucial for self-protection rather than discrimination, thus avoiding an immune response. When the crRNA matches its target with an adjacent PAM, the Cas9 nuclease causes a double-stranded break (DSB) 3 bp before the PAM, leading to the degradation of the phage or invading DNA element. For the creation of CRISPR-Cas9 systems for gene editing, crRNA and tracrRNA are converted into guide RNAs (sgRNAs) to direct Cas9 to specific locations in the genome (Jiang and Doudna, 2017). Thus, the fundamental structure of this system requires just two components: sgRNA and Cas9 nuclease, both of which can be produced by a plasmid within the cell to be edited. The DSB is then repaired by nonhomologous end joining (NHEJ) or homology-directed repair (HDR). The NHEJ method often results in small insertions or deletions (indels) that were previously believed to be random but recent studies suggest they are somewhat predictable (Kosicki et al., 2022). This method is frequently used to induce mutations, leading to knockout genes (KOs). The HDR method necessitates homologous repair templates, which can be endogenous such as sister chromatids or exogenous in the form of plasmids or single-stranded oligo-deoxy-ribonucleotides (ssODNs), and can result in precise repairs or modifications, such as mutations or gene breaks or splicing.

4.2 THE CRISPR-CAS MECHANISM

The CRISPR/Cas9 genome-editing mechanism can be broadly divided into three steps: adaptation, expression, and interferences. The designed sgRNA directs Cas9 to the target sequence in the genome by binding to the target site. In the absence of the sgRNA, the Cas9 protein remains inactive. The Cas9 nuclease generates a DSB at the target site 3 bp upstream of the PAM. The PAM sequence for Cas9-based cleavage is a short (2–5 base pairs long) conserved DNA sequence located adjacent to the target site. The widely used spCas9 effector recognizes 5Ê1-NGG-3Ê1 as a PAM sequence (N can be any nucleotide base). When the gRNA binds to the target with an appropriate PAM, it causes the local DNA to melt and subsequently forms an RNA-DNA hybrid, but the extent to which the Cas9 enzyme melts the target DNA is unclear. The Cas9 protein is then activated to perform DNA cleavage. The HNH domain cleaves the complementary strand, while the RuvC domain cleaves the non-complementary strand of the target DNA, producing blunt-ended DSBs. Finally, the DSB is repaired by the cellular machinery (McGinn and Marraffini, 2019).

DOI: 10.1201/9781003500933-5

4.2.1 gRNA and Its Importance for Gene Editing

Between 2012 and 2013, the CRISPR/Cas system was first adapted to eukaryotic gene-editing tool. The generation of gRNAs is important for site-specific gene editing by CRISPR-Cas systems designed in eukaryotes, as gRNAs act as molecular GPS, guiding Cas9 to target sequences in the genome. Several bioinformatics tools, such as CHOPCHOP, zinc finger targeting (ZiFiT), and CRISPR-direct, were developed to efficiently use gRNA for CRISPR-mediated gene targeting in various diseases. All of these platforms are used for gRNA design in model and commercial crops. ZiFiT is also used to produce zinc fingers and TALEN modules, but Cas9 guide RNA constructs and CRISPR are not currently available. gRNA generation tools play an important role in gene editing by enabling efficient and accurate gRNA generation by targeting genome editing with CRISPR-Cas (Pallarès-Masmitjà et al., 2019). This tool uses computational techniques to predict and optimize gRNA based on specific criteria such as target accessibility, target cleavage, off-target effect, and Cas protein binding efficiency. Many well-known gRNA design tools have been developed, each offering unique features and the potential to support various applications in CRISPR-based gene editing. Specialized tools such as beditor and pegFinder for base editing and primer editing, respectively, have also been developed for specific applications of CRISPR-Cas.

Over time, numerous design tools have been created to enhance the precision and effectiveness of gRNAs in reducing the activity of targets. Many of these tools include various competitive methods to measure on-target and off-target effects, providing users with alternative tasks. Some of these tools use their own sophisticated algorithms to identify potential off-target sgRNAs in the host genome to assist users in gRNA selection. For example, CHOPCHOP provides users with additional capabilities for gRNA design, including off-target prediction and specificity measurement based on sequence and secondary structure. Similarly, E-CRISP uses its own SAE (Specificity, Annotation, Efficiency) scores, as well as scores not used by CHOPCHOP, to determine the quality of each sgRNA. Another tool developed, CCTop, empirically assigns an off-target score to each segment. Additionally, CRISPOR is a versatile platform that lists gRNAs by differential score to assess the potential output of targets in the genome of interest and predict the activity of the target (Gupta and Shukla, 2017).

The CRISPR design tool (Table 4.1) is one of the most widely used platforms developed by Zhang's lab at the Broad Institute of MIT, USA. Users can enter a target sequence or genomic region to create a gRNA suitable for Cas9. The tool identifies potential gRNAs in a genome, based on sequence specificity, GC content, and resistance. Benchling's CRISPR design tool is a well-known platform that provides users with the ability to create suitable gRNAs. It combines features such as efficiency, specificity, and target yield to produce high-quality gRNA to meet specific applications. Overall, this gRNA design tool provides researchers working on CRISPR-Cas-based genome editing with significant resources to create precise, efficient gRNAs for specific applications. Recently, AI-based prediction models (such as DeepCRISPR, CRISTA, and DeepHF) have been used in gRNA design, thus contributing to the advancement of CRISPR-based gene therapy in agriculture, healthcare, and medicine.

4.2.2 Selection of an Appropriate Cas Nuclease

Based on evidence from different studies, it has been shown that the CRISPR locus and the Cas9 endonuclease work together. The gene encoding the Cas9 protein is located near the CRISPR site. Selection of the appropriate Cas nuclease is critical to the success and safety of CRISPR-Cas-based gene-editing techniques. Different Cas nucleases have been identified for processing dsDNA, ssDNA, RNA, and proteins. For example, Cas9 and Cas12 can cut double-stranded DNA, while Cas14 can cut single-stranded DNA. In contrast, Cas13 is characterized by its ability to modify RNA (Anzalone et al., 2020). Additionally, Cas effectors have been designed to recognize alternative PAM binding that results in single-strand breaks (nCas9) or to ligate DNA only at the target site without inducing DSBs (dCas9). Similarly, different Cas nucleases have different properties that affect the efficiency, specificity, and output of gene editing. In Section 4.2.3, we discuss important features to consider when choosing the right Cas nuclease (Lee et al., 2016).

4.2.3 Compatibility of the PAM Sequence

Different Cas proteins recognize different PAM sequences to bind to the target DNA or RNA. A PAM sequence that is suitable for the selected Cas nuclease must be present adjacent to the target. For example, the widely used Cas9 from *Streptococcus pyogenes* (SpCas9) recognizes the 5'-NGG-3'-PAM sequence on the target, thereby creating DSBs in the genome. However, SpCas9 can still recognize the 5-NAG-3- and 5-NGA-3-PAM sequences and does not cause any changes in the genome (Hu et al., 2018). To overcome this problem, researchers have developed SpCas9s, such as SpCas9-NG and X-Cas9, that can use various PAM sequences, thus enabling targeted targeting in the genome (Liu et al., 2021).

4.2.4 Specificity

Some Cas nucleases are more effective than others, so selecting specific Cas effectors is important, especially for clinical applications, to reduce targets. For example, low-target Cas nucleases such as Cas12a and engineered Cas9 nucleases are preferred for clinical applications. Engineered Cas9 variants such as eSpCas9, HypaCas9, and Hi-FiCas9 interact more strongly with the DNA backbone, resulting in higher specificity and reduced targeting. Similarly, Cas12b and CasX may compete in terms of specificity and small size and therefore may be suitable for translational and clinical applications (Suresh, 2021).

TABLE 4.1
CRISPR Tools for gRNA design and off-target prediction

Tool	Organism	Language	Cas nuclease	Description	Database/ web server	Website (accessed 10 November 2022)
CHOPCHOP	More than 100	Python	Cas9, Cas12, Cas13	A pioneering online instrument developed by Harvard University for the creation of gRNA sequences based on both matches and mismatches	Web server	http://crispor.tefor.net/
Cas-Designer	More than 100	Python	More than 20 Cas nucleases	Forecasts a range of off-target and on-target effects using the Cas-OFFinder model	Web server	www.rgenome.net/cas-designer/
Guidescan	Mouse and human	Python	Cpf1 and Cas9	Predicts off-target effects based on sequence and structural features	Web server	https://guidescan.com/
sgRNAcas9	Mouse	Pearl	spCas9	A web-based tool to predict off-target effects	Dataset	www.biootools.com/software.html
EupaDGT	Eukaryotic pathogen	Python	More than 10 Cas nucleases	Machine-learning-based tool to predict on- and off-target effects simultaneously	Web-based	http://grna.ctegd.uga.edu/
WU-CRISPR	Human and mouse	Pearl	SpCas9	Machine-learning-algorithm-based tool that can predict off-target effects by providing sequences between 20 and 30,000 bp	Web-based	http://crispr.wustl.edu/
CRISPR-P	49 plant species	Python	More than 14 Cas nucleases orthologs	Web-based off-target and on-target prediction tools for a wide range of plant species	Web-based	http://crispr.hzau.edu.cn/CRISPR2
CRISPRz	Zebrafish, human, and mouse	Python	spCas9	Trained on large datasets from zebrafish, humans, and mice to generate a gRNA dataset	Web-based	https://research.nhgri.nih.gov/CRISPRz/
CRISPOR	More than 100 species	Python	More than 30 Cas orthologs	Design gRNA dataset based on match/match in seed regions	Web tool	http://crispor.tefor.net/
PnB Designer	6 species	Python	Cas9	A newly developed tool for creating guide RNA for both base editing and prime editing	Web tool	www.crisprindelphi.design/

4.2.5 Importance of the PAM in CRISPR-Cas

Posterior adjacent motifs (PAMs) are short DNA sequences located near DNA target sites in the genome. The PAM sequence is important for CRISPR-Cas systems because it helps the CRISPR-Cas complex recognize and bind to target sites in the genome. The exact sequence of the PAM varies depending on the type of Cas nuclease used, as different Cas nucleases recognize different PAMs to create DSBs in the target (Gleditzsch et al., 2019). For example, the widely used SpCRISPR-Cas9 system recognizes 5-NGG-3™, just as the PAM recognizes CRISPR-Cas complexes. If the PAM is not near the target, the CRISPR-Cas complex cannot recognize and remove the target (Hu et al., 2018).

4.3 SPECIFICITY AND PRECISION OF THE CRISPR-CAS SYSTEM

The PAM sequence plays a crucial role in locating the correct CRISPR-Cas target site within the genome. It helps distinguish the target sequence from similar sequences in the host genome, thus confirming the specificity of the CRISPR system. Consequently, the PAM sequence is essential for ensuring CRISPR-Cas specificity and minimizing off-target effects. The specificity provided by the PAM is also important for the safety of CRISPR-Cas systems in clinical applications, where it should be minimized (Strobel and Dervan, 1990).

4.3.1 Site-Specific Cleavage of the Target Site

PAM is required for CRISPR-Cas-based cleavage of genomic targets. When the CRISPR-Cas complex scans the DNA for the presence of the target, the PAM sequence is the first to appear. Once the PAM is recognized, the CRISPR-Cas method searches for adjacent sequences that match the corresponding crRNA (Gleditzsch et al., 2019). If the sequence matches the crRNA, the Cas nuclease creates a DSB at that site. Therefore, the presence of PAM adds an additional layer of specific and precise therapy by ensure precise cleavage of the target site.

4.3.2 Engineered PAM to Expand Applications of CRISPR-Cas

Characterizing PAM sequences and their important roles in CRISPR-Cas allows researchers to design CRISPR-Cas systems with specific editing and knockdown targets. Using CRISPR-Cas technology in conjunction with PAM rules, researchers are able to target previously inaccessible genomic sequences, thus this advancement enhances the use of CRISPR-Cas-based gene editing for treating various diseases and enables multiplexed gene editing for different genetic modifications in seed crops (Ma et al., 2019). Hence, the PAM sequence is an important part of the CRISPR-Cas system that improves specificity, mode of action, and reduced targeting.

4.4 PRINCIPLES OF GRNA DESIGN FOR DIFFERENT APPLICATIONS OF CRISPR

gRNA (guide-RNA) is a key component of the CRISPR-Cas system that enables the Cas nuclease to recognize targets in the genome. sgRNA (synthetic guide RNA) is an RNA molecule with a stem-loop structure that facilitates the binding of the CRISPR-Cas complex to the target site by adding nucleotides to the target sequence, creating DSBs. The first 10–12 nucleotides of the 20 bp gRNA (adjacent to the PAM sequence) are characterized as the seed sequence, and a mismatch in this region may abolish the cleavage activity of the Cas9.

It is recommended to fully sequence the target DNA before designing the gRNA to ensure that there are no SNPs that could reduce the function of the target (Manghwar et al., 2020). In addition, reducing the length of the sgRNA to 17 nucleotides (truncated sgRNA) does not affect the functionality and specificity of the CRISPR-Cas system.

The efficiency and specificity of CRISPR-Cas-based applications depend on the careful design of gRNAs suitable for specific applications. In Section 4.4.1, we discuss the principles of designing gRNAs for CRISPR applications. It is important to track the target site before starting gRNA design. First, the appropriate PAM corresponding to the selected Cas nuclease should be adjacent to the target in the DNA. This is important because Cas recognizes different PAMs depending on its location. More importantly, the efficiency of the CRISPR-Cas system increases as the PAM length increases but may reduce

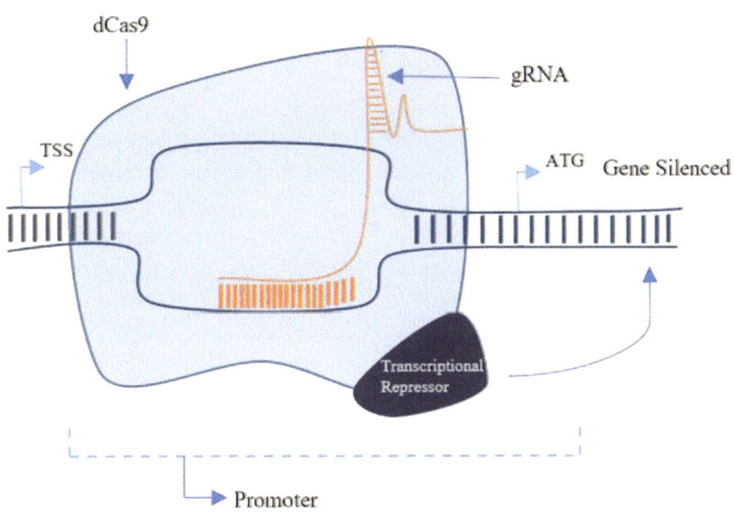

FIGURE 4.1 CRISPRi: the CRISPR system comprises a guide RNA (gRNA), target DNA, and a catalytically inactive Cas9 enzyme (dCas9), thus making dCas9-gRNA complex (upstream within the 300 bp window) fused with transcriptional inhibitory factors, resulting in reduced gene expression and knockdown.

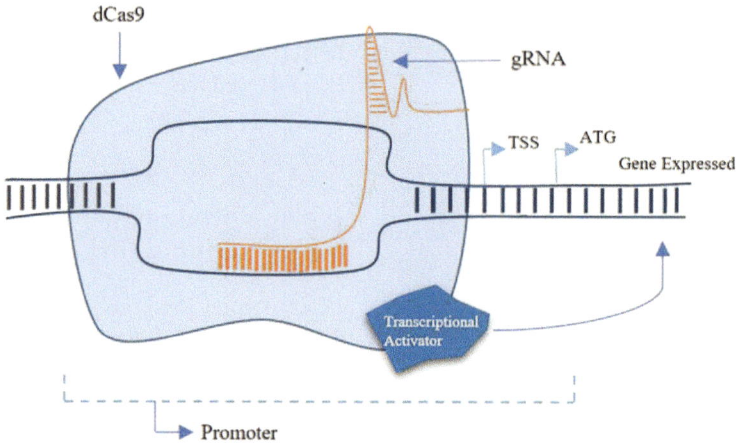

FIGURE 4.2 CRISPRa: the CRISPR system comprising a guide RNA (gRNA), target DNA and "dead" endonuclease (dCas9), without its catalytic domain, thus making dCas9-gRNA complex (downstream within the 400 bp window) fused with transcriptional activating factors, resulting in increased expression and favoring upregulation of a particular gene.

the number of targets. Second, gRNAs are usually 20 bp long, which provides equal binding to the target and specificity of the CRISPR-Cas system. The GC content of the gRNA varies between 40% and 60%, which makes the gRNA-DNA hybrids stable. Increasing the GC content above 60% may lead to a very stable gRNA-DNA complex, which may affect the specificity of the CRISPR-Cas system . Finally, it is important to ensure that the gRNA does not form a secondary structure that would prevent the gRNA from binding to the target. In Section 4.4.1, we discuss the specific guidelines for gRNA design in different CRISPR applications (Jamal et al., 2018).

4.4.1 CRISPR-I

Unlike CRISPR-Cas-mediated gene knockout, CRISPRi relies on dCas9 (deadCas9), which binds to the target site in DNA without creating DSB in the genome. dCas9 is a modified version of Cas9 that contains two mutations in the RuvC and HNH domains (D10A and H841A) to eliminate the endonuclease activity of Cas9. Therefore, the CRISPRi complex (dCas9-KRAB and gRNA) binds only to the target site in DNA without inducing DSBs. CRISPRi has been successfully shown to suppress gene expression in different organisms, including bacteria, archaea, and eukaryotic cells. The CRISPRi system works by combining guide gRNA with dCas9 fused with inhibitory domains (KRAB, MeCP2, and HP1) to suppress gene expression without creating permanent changes in the genome. In CRISPRi, guide RNAs (gRNAs) typically target regions about 100–300 base pairs downstream of the transcription start site (TSS) or within the first 100–200 base pairs of the open reading frame (ORF). They inhibit transcription by physically obstructing the RNA polymerase's ability to bind to the promoter region, thereby blocking the elongation process (see Figure 4.1).

4.4.2 CRISPR-A

The development of tools to unlock gene expression is revolutionary for biotechnology and biological sciences. For example, in metabolic engineering, programmed spatiotemporal expression of complex multigene pathways is important to reduce cellular load and increase production of valuable metabolites. CRISPRa is a powerful tool for regulating gene expression. CRISPRa also uses dCas9 to activate gene mutations without creating permanent changes in the host genome (Manghwar et al., 2020). In CRISPRa, dCas9 is used to insert transcriptional activation domains (VP64, VP16, VPR, and SAM systems) into the promoter region of a gene, thereby activating transcription of the target gene. But the biggest challenge is understanding the rules that define the correct targets for each promoter in the genome. CRISPRi and CRISPRa have been used to manipulate and amplify multiple genes simultaneously. For CRISPRa, gRNA located approximately 200–300 bp upstream of the TSS (transcription start site) of the target gene can cause gene activation (Figure 4.2).

4.4.3 Gene Knockout Using CRISPR-Cas

Gene knockout using CRISPR-Cas has revolutionized molecular biology by providing an efficient and effective way to disrupt specific genes. Compared to T-DNA-based mutagenesis, CRISPR-Cas has the advantages of specificity, high efficiency, and versatility. Using CRISPR-Cas-based gene knockouts, researchers can gain insights into functional genomics, disease modeling, metabolic engineering, therapeutic development, self-healing and screening pathways, synthetic biology, and agricultural biotechnology. Gene knockout using CRISPR-Cas has been successfully demonstrated in bacteria, plants, and animals through scientific evidence of research and in-use techniques. Designing gRNA for gene knockout using CRISPR-Cas is an important step toward achieving efficient and effective gene editing (Jamal et al., 2018).

The following are important considerations for designing an efficient gRNA for knock-out applications of CRISPR-Cas.:

1. Select a target site within an exon, preferably near the 5' end of the gene. It causes permanent mutation of genes and eventually produces non-functional proteins.

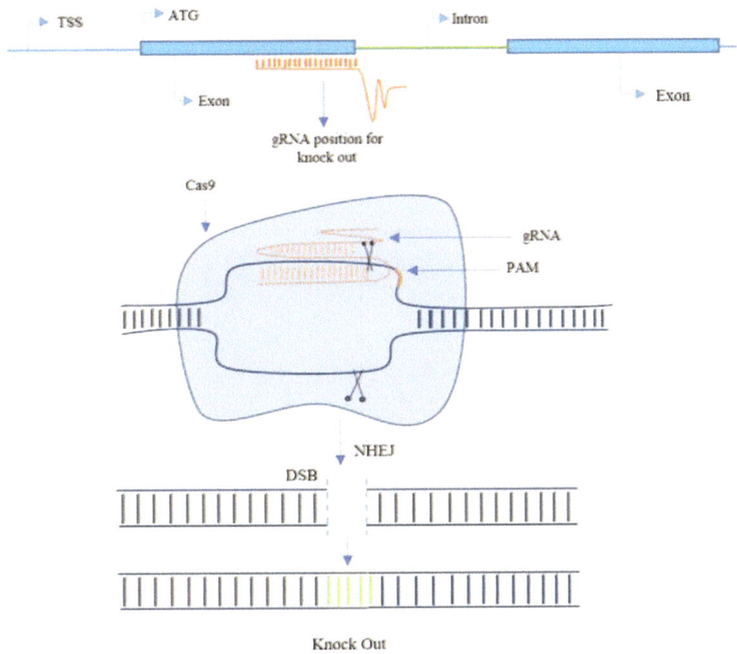

FIGURE 4.3 Schematic picture of genome editing mediated by CRISPR-associated nuclease 9 (Cas9), and DNA repairing. The CRISPR/Cas9 complex aligns with the complementary DNA sequence in the host cells. The Cas9 protein, which is guided by a desired single-strand guide RNA (gRNA), cuts the double-stranded DNA, leading to the gene knockout. Subsequently, DNA repair occurs through the nonhomologous end joining (NHEJ) pathway.

2. Functional domains and storage domains are still important targets in ensuring efficient gene knockout applications of CRISPR-Cas systems.
3. Multiple gRNAs targeting different regions could be another option to increase the likelihood of an effective gene knockout (Figure 4.3).

4.4.4 CRISPR Base Editing

Base editing using CRISPR-Cas is a powerful tool for introducing precise changes to specific regions in the genome without perpetuating DSBs. It allows scientists to make changes in the genome as small as a single nucleotide. Unlike CRISPR/Cas9, which uses Cas9 to create a DSB at the target site, base editing relies on nCas9 to create a single-stranded break or dCas9, which binds only to the target site (Yili et al., 2021). The basic mechanism consists of gRNA, Cas (dCas9 or nCas9) that binds to the target site, and enzymes that convert single DNA into target-specific nucleotides (cytidine deaminase or adenine deaminase). There are two types of base editors: ABE (Adenine Base Editor) and CBE (Cytosine Base Editor), which allow for all four types of base changes (A to G, G to A, C to T, and T for C). Cytidine deaminase converts cytosine (C) to uracil (U) by removing the amine group during replication; uracil (U) is read as thymine (T), effectively converting C-G to T-A. In contrast, adenine deaminase converts adenine (A) to inosine (I), which is recognized as guanine (G) when repeated, thus converting the A-T base pair to a G-C base pair. gRNA directs dCas9 or nCas9 to the target site, while the base editor (ABE or CBE) binds to dCas9 or nCas9 to make the desired change in the target. Base editing through CRISPR-Cas is a powerful tool for introducing precise genetic modification with many applications, such as disease treatment, cancer research, genetics, crop improvement, and drug research. Although base editing systems have been successfully applied in different studies, many challenges such as targeting, off-target identification, and delivery efficiency need to be addressed regarding the general usability and safety of the mechanistic basis.

Designing useful gRNAs on a mechanistic basis requires careful attention to the specificity and efficiency of target modification. Here we discuss important points to consider when designing gRNAs for base modification:

1. Ensure the presence of an appropriate PAM sequence near the target site. The editable (target) base must be present within the editing window of the base editing system (Table 4.2).
2. Confirm the editing window and make sure that the target base is located within it. For example, for CBEs, the target base must be present at positions 4–8, counting from the PAM side. Similarly, for ABEs, the target base must be located within positions 4–7 from the PAM side.
3. Local sequence context is also important for target site as it may affect the efficiency of the base editing system. For example, repetitive sequences and high GC contents may hinder the binding of the gRNA (Figure 4.4).

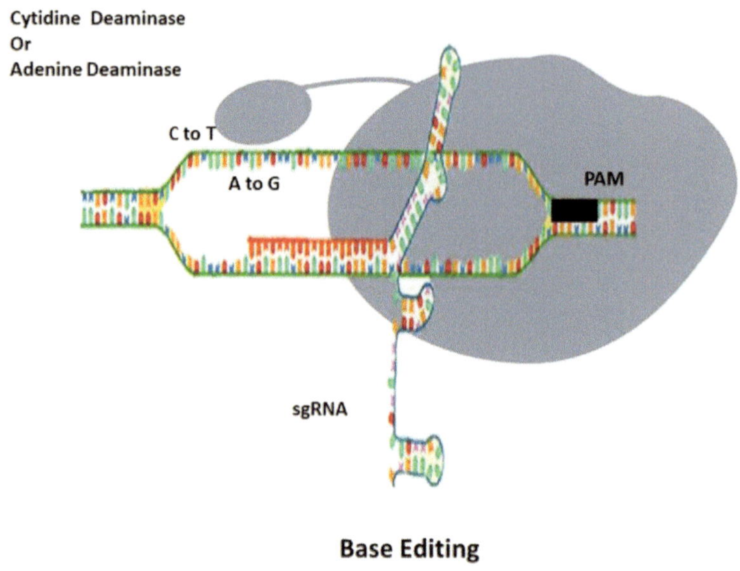

FIGURE 4.4 gRNA design for base editing systems.

4.4.5 CRISPR Prime Editing

Like base editing, prime editing is a CRISPR-based system to introduce precise changes to the DNA without creating DSBs. Prime editing combines nCas9 with reverse transcriptase and a modified gRNA called Prime editing gRNA (pegRNA). PegRNA recruits nCas9 to create a single-stranded break (SSB) at the target site and encodes a modification pattern that includes the desired modification (Lau et al., 2020). Reverse transcriptase then uses this modified template to synthesize new DNA at the SSB site, allowing for all modifications, including deletions, insertions, and substitutions. Compared to base editing, prime editing has minimal off-target risks and higher precision, making it a valuable tool for translational and health applications. The application of prime editing has been successfully demonstrated with good results in plant and animal sciences.

pegRNA is more complex than the standard gRNA used in traditional CRISPR-Cas9 systems. A pegRNA consists of a primer binding site (PBS) and an sgRNA (spacer) that contains a template for introducing precise modifications at the target site. The design of pegRNA includes several factors to improve its targeting and efficiency (Hu and Yu, 2024). The main steps in pegRNA production are:

1. Select the target site (spacer) with the shortest distance to the desired editing site (nucleotides). The spacer must be as close as possible (typically 3–10 nucleotides) to the desired editing site.
2. Select the target site (spacer) and ensure the presence of an appropriate PAM sequence for the Cas nuclease being used in the study.
3. The 20bp sequence of the spacer (gRNA) should be selected to create an optimal nick for the desired edit.
4. Design the PBS and RT template (with desired edits) at the 3'-end of the gRNA. PBS must be 8–15 nucleotides in length and should be complementary to the sequence upstream of the nick site. The RT template should be 10–30 nucleotides in length and the first base of the 3'-extension should not be cytosine (C).
5. For PE3 systems, an additional nicking sgRNA is also required to target the complementary strand at approximately 60–100 bp downstream of the pegRNA cutting site.
6. It is important to note that the pegRNA's strong secondary structure may interfere with its stability and functions.
7. GC contents of the pegRNA should be 40–60% for stable binding and efficient transcription (Figure 4.5).

4.4.6 CRISPR RNA Editing (Cas-13)

CRISPR-Cas13 is a multifunctional RNA editing system that relies on an effector nuclease to edit RNA. It is a type II CRISPR-Cas system that works similarly to the widely used CRISPR-Cas9, but specifically targets RNA instead of DNA. CRISPR-Cas13 is a powerful tool that controls gene expression at the transcriptional level without creating permanent changes in the genome. Like CRISPR-Cas9, crRNA directs the Cas13 nuclease to find and remove complementary RNA sequences, enabling gene expression to be regulated at the RNA level. CRISPR-Cas13 is also highly sensitive to splicing and can remove nonspecific RNA targets once activated. RNA editing using CRISPR-Cas has the potential for many applications in research and applications, including RNA imaging, diagnostics, targeting RNA diseases in plants and animals, and functional labeling (Pickar-Oliver and Gersbach, 2019). These techniques provide a variety of ways to study cellular processes without creating permanent changes in DNA. To date, four Cas13 proteins have been identified, including Cas13a

TABLE 4.2
CRISPR base editing tools for gRNA design and off-target prediction

Tool name	Description	Features	Website
BE-Designer	A web-based application for designing base editing sites for CRISPR base editors, including cytosine and adenine base editors	Predicts off-target areas, offers multiple genomes, provides PAM flexibility, and allows viewing of target sequences	www.rgenome.net/be-designer
BE-Hive	A comprehensive base editing design tool that predicts both on-target and off-target editing results	Integrated with AI algorithms, user-friendly interface, supports both cytosine and adenine base editors	www.crisprbehive.design
BE-Analyzer	An online tool to assess the outcomes of base editing experiments, focusing on efficiency and specificity	Provides detailed analysis of results, supports high-throughput data processing, and evaluates editing efficiency	www.rgenome.net/be-analyzer/#!
DeepBaseEditor	A deep learning tool that predicts the effects of base editing	High accuracy, supports various base editors, and configurable for experimental purposes	https://deepcrispr.info/DeepBaseEditor
ABEmax	Specifically designed for adenine base editing, targeting A-to-G conversions	Off-target prediction, PAM flexibility, detailed representations of target sites	https://bio.tools/t?page=1&q=ABEmax&sort=score
BEable-GPS	A genome-wide tool for predicting base editing sites	Genome-wide search, detailed PAM sequence information, supports both cytosine and adenine base editors	https://bio.tools/BEable-GPS
beditor	A tool for developing and optimizing gRNAs for base editing, focusing on accuracy and off-target rates	Customizable criteria for site selection, comprehensive off-target analysis, supports various genomic assemblies	https://pypi.org/project/beditor/
CRISPR-BE2	An upgraded tool for generating gRNAs for base editing	Extensive off-target prediction, integrates with other CRISPR tools, offers visualizations and reports.	https://crispr-be2.com/
EditR	An R package for analyzing and planning base editing studies	Extensive analysis, customizable experimental design options, integrates with R-based bioinformatics tools	http://baseeditr.com/
BEtarget	An open-source tool for creating gRNAs with a focus on minimizing off-target effects	User-friendly interface, supports custom genomic assemblies, integrates with other bioinformatics tools	https://skl.scau.edu.cn/betarget/
CRISPR-BETS	A program offering thorough analysis and design capabilities for optimizing gRNAs and reducing off-target effects	Genome-wide search, extensive PAM sequence and target location information, supports various base editors	https://bioinfor.yzu.edu.cn/crisprbets/
CRISPR-DBE	A tool for optimizing gRNAs for adenine base editors, focusing on high accuracy and low off-target rates	Off-target prediction, integrates with other CRISPR tools, customizable target site selection	https://crispr-dbe.com/

(also known as C2c2), Cas13b, Cas13c, and Cas13d. The ability of Cas13a has been successfully demonstrated for many applications in animals and plants, such as RNA degradation, transcriptional tagging, splicing editing, and virus detection. In addition, several CRISPR-Cas13-mediated RNA editing platforms have also been developed, including editing catalytically inactive Cas13 (dCas13) with adenine/cytidine in ADAR2 to achieve A to I (G) conversion (2 RNA), combined to use the RESCUE system for C to U transfer.

Designing an effective gRNA for RNA editing with CRISPR-Cas13 involves several important steps. Here are the key guidelines for designing a gRNA for Cas13:

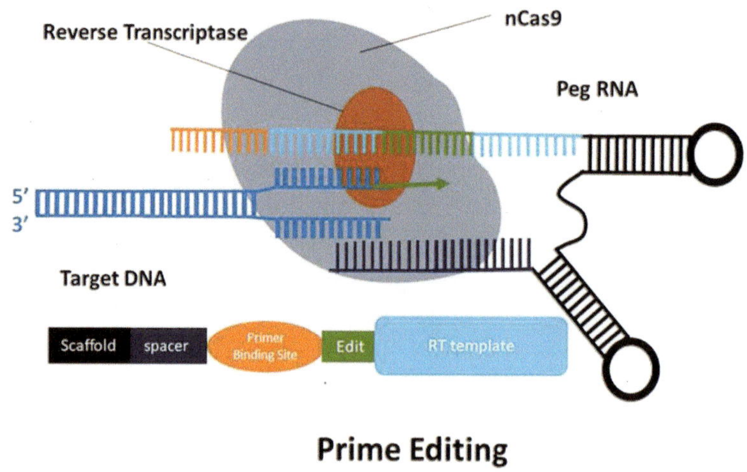

FIGURE 4.5 PegRNA for prime editing.

1. Choose an accessible target in the RNA. Avoid highly structured and bound regions as a target site.
2. Select a unique target site in the RNA to avoid off-targets.
3. Select an appropriate PFS sequence for the Cas13 being used in the study, as different Cas13 variants prefer different nucleotides adjacent to the target site.
4- Select a 28-nucleotide long crRNA (gRNA) with an adjacent PFS sequence. Cas13 variants typically require a 28-nucleotide long crRNA to cleave the target RNA.

4.4.7 CRISPR-Cas14: Single-Strand DNA Cleavage System

CRISPR-Cas14 is a type of CRISPR-Cas that emerged from resistant bacteria. Cas14 is considered a small nuclease with a unique ability to target ssDNA. Compared with CRISPR-Cas9 and CRISPR-Cas12 proteins, CRISPR-Cas14 does not require a PAM for target recognition and cleavage, which can span multiple locations in the genome. The unique capabilities of CRISPR-Cas14 open up new avenues in medicine, disease therapy, diagnostics, therapeutics, and synthetic products. The small size of Cas14 is particularly useful for the efficient delivery of CRISPR agents in gene therapy and other delivery vectors that can limit selection. Below are instructions for creating a good gRNA for Cas14.

1. Select a unique target site in the ssDNA.
2. Select a gRNA with 20 nucleotides in length.
3. Avoid regions for designing gRNA that are likely to be double-stranded.

4.4.8 CRISPR-Cas12

Like Cas9, CRISPR-Cas12 is a powerful and versatile gene-editing tool that cuts double-stranded DNA. However, unlike Cas9, which produces blunt ends at the DSB, Cas12 produces staggered ends that can facilitate insertion or deletion of the target. Additionally, Cas9 recognizes a G-rich PAM, while Cas12 recognizes a T-rich PAM (5-TTTN-3). Compared to Cas9, Cas12 also has a cleavage product (which can nonspecifically cleave a DNA sample after activation), making it ideal for diagnostic applications. CRISPR-Cas12 has been used in the development of diagnostic tools such as DETECTOR (Wei et al., 2021). DETECTOR detects nucleic acids efficiently and sensitively. Together, these capabilities make CRISPR-Cas12 a significant advancement in targeting T-rich regions of DNA due to its small size, disease detection, and efficient host delivery.

4.5 FACTORS AFFECTING GRNA

In the context of CRISPR-Cas9 genome editing, gRNA efficacy is determined by its ability to match with the target site and the creation of DSB by the Cas nuclease. The efficiency of gRNA is influenced by various parameters such as GC contents of the gRNA, the secondary structure, chromatin access, the expression level of the gRNA, specificity, gRNA length, and the presence of a suitable PAM site adjacent to the target site. For example, high specificity increases efficiency and reduces off-target effects. In addition, gRNAs with balanced GC contents (usually around 40–60%) are more effective, as GC content can affect the stability of the gRNA. Moreover, recognition and binding of Cas9 at the target site depend on the presence of an appropriate PAM adjacent to the target site (Javaid and Choi, 2021). gRNA binds to the target site in a sequence-dependent manner and an ideal gRNA should be unique, with no off-target sites in the genome. The accessibility of the target site also affects the binding ability and efficiency of the gRNA. Accessibility of the target site can be affected by DNA methylation and chromatin conformation. It is important to note that binding of gRNA to non-target sites in the genome may have adverse effects on the organism. Gene-editing efficiency can also be affected by the type of cells, the format of Cas reagents and their delivery method, and the exposure time of the target site to the Cas nuclease. For example, the efficiency of gRNA delivery and gene editing

TABLE 4.3
Key factors influencing CRISPR-Cas9 gene editing

Factor	Impact	Reference
Sequence specificity	Precision of the gRNA sequence matching the target DNA	(Liang et al., 2017)
GC content	Optimal GC content (~40–60%) enhances binding stability	(Filippova et al., 2019)
PAM sequence	Required for Cas9 to recognize and bind to the target region	(Cencic et al., 2014)
Secondary sequences	Strong secondary structures in gRNA may inhibit target binding	(Riesenberg et al., 2022)
Chromatin accessibility	Tightly packed chromatin can block Cas9 from accessing the target site	(Verkuijl and Rots, 2019)
Truncation of 3' end	Trimming the 3′ end of sgRNA can improve specificity	(Fu et al., 2014)
Shortening of 5' end	Shortening the 5′ complementary region (by up to 3 nt) enhances specificity	(Fu et al., 2014)
Expression level	High and stable levels of gRNA and Cas9 improve editing efficiency	(Yuen et al., 2017)
Addition of guanine	Adding 2 guanines at the 5′ end of gRNA improves specificity	(Mullally et al., 2020)

varies between cell types. All these factors need optimization to increase gRNA specificity and efficiency in CRISPR-based gene-editing applications (Javaid and Choi, 2021).

The gRNA segment influences the Cas9-gRNA complex in multiple ways. For example, the scaffold part plays an important role in the activity of the Cas9-gRNA complex. The scaffold helps to maintain the structure and stability of gRNA, ensuring it can effectively guide Cas9 to the target site. Without the scaffold, gRNA is unable to form the correct structure, leading to the reduced activity of the Cas9. Similarly, gRNAs expressed from the U6 promoter must begin with G for efficient gene editing in mammalian cells. While the Cas9-gRNA complex scans the genome, the spacer sequence of the gRNA determines how long Cas9 spends scanning target and non-target sites. Various factors that influence the efficiency of the gRNA in CRISPR-Cas have been summarized in Table 4.3.

4.6 GRNA DESIGN AND OFF-TARGETING

gRNA and PAM provide specificity to the Cas nuclease. Although Cas9 and other Cas nucleases, such as Cas12, are considered specific, these nucleases can be recruited to unintended sites in the genome due to the non-specific binding of the gRNA to the non-target sites. Cleavage or binding of the Cas nuclease at the unintended sites in the genome is termed off targets in the CRISPR-Cas. Off-targets in CRISPR-Cas may occur for many reasons, such as single nucleotide mismatch between gRNA and non-target site, multiple mismatches between gRNA and non-target site, and non-canonical PAM sites. Off-targets may result in unwanted genetic changes in the genome, leading to cytotoxicity, genotoxicity, chromosomal rearrangements, abnormal phenotypes, and serious consequences, especially in animal studies and human health. Therefore, off-targets are a significant concern in gene-editing applications of CRISPR, especially in health and translational applications. Minimizing off-target effects is critical to the accuracy and efficiency of genome-editing applications in clinical studies and human health. In contrast, off-targets are not a significant concern in gene-editing applications of CRISPR-Cas in plants. Off-target effects in plants are usually considered less critical than in applications in human health. For example, plants can better tolerate off-targets as they usually have multiple copies of the genome. In addition, regenerating a plant from the single cell allows the screening and selection of plants with minimal off targets. Similarly, minor off-targets in gene-edited plants might not significantly impact plant performance. While in human health applications and clinical applications, minor off-targets may lead to serious health problems. Nonetheless, off-targets are still a major concern in the regulation and biosafety of gene-edited plants, and researchers are continuously improving the precision of CRISPR-Cas systems to make the gene-edited plants safe (Naeem et al., 2020). Genome-wide characterization of off-targets in gene-edited plants is essential for regulatory compliance and consumer acceptance (Haeussler, 2020).

4.7 METHODS TO MINIMIZE OFF-TARGETS

Various techniques have been developed to decrease the expected frequency of off-target occurrences in gene-edited plants. These techniques include selecting a high-fidelity Cas nuclease, optimally selecting gRNA, and selecting an effective delivery system for CRISPR reagents. Additionally, employing nCas9 with two gRNAs may further enhance the specificity of the Cas9 nuclease. Many of these techniques can be coupled to improve the efficiency and specificity of the CRISPR-Cas systems to enhance the safety and specificity of the gene-edited plants.

4.7.1 Optimal gRNA Selection

Off targets can be reduced by selecting a unique gRNA that exhibits minimal homology with other genome regions. Although some CRISPR techniques, such as base editing and prime editing, exhibit little flexibility in the target site, a flexible target site allows researchers to design an optimal gRNA with reduced off-targets, such as knockout of the gene with CRISPR-Cas. Various online prediction programs, such as CRISPR, Cas-OFFinder, and CCTop, enable researchers to design gRNA while scoring the likelihood of their off-targets. Although these software programs are user-friendly and effectively narrow down the target site, they typically consider DNA sequences, ignoring chromatin accessibility and other genomic contexts (Ong et al., 2017).

4.7.2 Optimal Cas Selection

Once an optimal gRNA with minimal or no off-targets in the genome has been selected, choosing and considering high-fidelity Cas nuclease is essential. Despite similar sequences to the gRNA in the genome, some Cas nucleases show resilience to the off-target effects caused by the sequence inconsistencies. Therefore, high-fidelity Cas nucleases are recommended to minimize the off-targets. For example, spCas9s, including evoCas9, sSpCas9(1.1), HypaCas9, and SpCas9HF1, have been developed for effective binding with gRNA. However, it is essential to note that most of the modified Cas enzymes inhibit cleavage at these off-target sites; they are not effective for all off-target sites (Pieniążek et al., 2012). For example, for CRISPRi and CRISPRa, where DSB is not required but Cas localization is only required for activating or suppressing a gene, Cas modifications do not reduce off-targets.

4.7.3 Closely Spaced Dual gRNA

To avoid off-targets, researchers also use closely spaced dual gRNAs and nCas9. gRNAs drive the nCas9 to the target site in the genome. A double-stranded break (DSB) occurs when two closely spaced gRNAs bind their respective target sites, enabling efficient gene edits without any off-targets (Pallarès-Masmitjà et al., 2019). It is important to note that off-targets with nCas9 at unintended sites cause only single-strand breaks, which the cell quickly repairs, minimizing the mutations. In the dual gRNA approach, off-target cleavage is less likely to occur because the chances of off-target regions for two closely spaced gRNAs are lower than for a single gRNA.

4.8 COMPUTATIONAL TOOLS TO EVALUATE OFF-TARGETS

CRISPR-Cas systems have become transformative tools in agricultural research. For example, CRISPR/Cas9 is a powerful tool capable of modifying genes across various species to achieve precise modifications in the genome. The rapid progress in CRISPR-Cas technology has also resulted in the development of bioinformatics tools tailored for CRISPR-Cas research and product development. Researchers have developed various tools to help researchers with the selection of an effective target site, predicting gene-editing outcomes, experimental validation, and the design of effective RNA to minimize off-target effects (Naeem and Alkhnbashi, 2023).

Several tools are available for gRNA design, including E-CRISP, CHOPCHOP, CRISPR-ERA, CRISPOR, and GuideScan. These tools work on a similar principle: they start by identifying candidate target sites in the gene of interest, considering the specific applications such as knockout, inhibition, or activation, and recognizing the respective PAM sequence for the Cas nuclease being used (Gerashchenkov et al., 2020). The target sites are then evaluated and categorized into high-performance and low-performance targets. Low-performance targets are evaluated based on competing factors and generally offer substantial predictive accuracy (up to 98% confidence). In contrast, objective performance scores are less effective at distinguishing between targets. CHOPCHOP is a computational tool that generates gRNA based on a simple BLAST match or mismatch between the gRNA and the target site. It is a versatile tool capable of designing gRNAs for over 100 species. Another approach involves a CRISPR-RGE-based website integrated with Cas-OFFinder, a machine learning-driven search engine, which identifies both on-target and off-target effects to generate an effective gRNA. Similarly, CRISTA is a web-based application that also generates gRNAs by evaluating on-target and off-target effects, and it supports over 100 species, though it is limited to spCas9. In addition, GuideScan is a web tool developed to create custom gRNAs specifically for editing human and mouse genomes. CRISPRdo is another tool that can generate gRNA for Cas9 and Cpf1 nucleases specifically. It can simultaneously assess both on-target and off-target events, making it particularly useful for animal models such as zebrafish, mice, humans, and certain worms. Moreover, sgRNACas9 is an important tool that can predict off-target outcomes based on specific patterns and sequences, with applications including the creation of spCas9 gRNAs for mouse genome modifications.

EupaDGT is a new network-based method for designing gRNAs to modify the genomes of eukaryotic organisms, compatible with various Cas variants. WU-CRISPR is a machine learning-based tool specifically for spCas9, used to generate gRNAs for modifying mouse and human genomes. CRISPR-P is a web-based design tool aimed at editing crop and model plant genomes (Gerashchenkov et al., 2020). A recently developed CRISPRz tool is trained on a large database tailored for spCas9. PhytoCRISPR is another tool designed to generate gRNAs for modifying plant genomes, supporting both SpCas9 and Cas12. After evaluating results with these tools, CRISPOR, a popular web-based program, is used to generate gRNA sequences and supports over 30 different Cas variants. Following the recent developments of base editing and prime editing in CRISPR, new web-based tools PE-Designer and BE-Designer have been developed to generate gRNAs specifically for prime and base editing, respectively.

The reliability of target prediction is influenced not only by the score but also by the algorithm used to identify off-target candidates across the genome. Off-target sites, with 1–3 mismatches in the gRNA, can be detected using read-through tools (Table 4.4). For example, E-CRISP, CRISPR-ERA, and CHOPCHOP use Bowtie/Bowtie2 for this purpose, though they may detect overlapping sites. In contrast, CRISPOR and GuideScan often employ alternative methods such as BWA and custom test data structures, offering more intuitive and versatile off-target predictions. It is important to note that all prediction methods primarily provide information about target activity based on the similarity between the gRNA and

TABLE 4.4
Tools for off-target analysis

Tools	Analysis basis	Input	Output	Supported experiment	Supported Cas9 nucleases
CRISPResso2	NGS	FASTQ	Sequence alignment, NHEJ/HDR events	For CRISPR/Cas9, base editors	Cas9, Cpf1
CasAnalyzer	NGS	FASTQ	Sequence alignment, HDR/NHEJ events	CRISPR/Cas9	spCas9, StCas9, HFCas9, SaCas9, Cpf1, CjCas9
CRISPR-GA	NGS	FASTQ	INDEL frequency, recombination due to HRD events	Only Cas9	Cas9
TIDE/TIDER	Sanger	ABI	INDELS/HDR frequency	Only CRISPR	spCas9, SaCas9, FnCas9, AsCpf1, stCas9
ICE	Sanger	ABI	INDELS/HDR frequency	Only CRISPR/Cas9	Cas9, SaCas9

the target. These methods do not account for experimental variables, such as Cas9 concentration, which can also affect target activity (Bradford and Perrin, 2019).

4.9 ROLE OF AI IN MINIMIZING OFF-TARGETS

Significant advances have been made in developing algorithms for predicting off-target effects. For example, DeepCRISPR, developed by MIT researchers, can predict off-targets to minimize errors. Similarly, Cas-OFFinder can predict both on-target and off-target to design an effective gRNA. Recently, AI-based tools such as CRISTA, TIGER, and CRISPA-MCA have been developed to predict off-targets and enabling gRNA design for several species. Using AI-driven tools, researchers can significantly reduce off-targets, thus making CRISPR more precise, especially for therapeutic and translational applications. Off-target predicting tools are generally trained on data from in vitro cleavage experiments, including those involving immortalized cell lines. Such cell lines offer a controlled environment for gene-editing studies, which is beneficial for creating predictive models. Typically, these prediction methods focus on sequence-based data relative to the CRISPR/Cas9 sequences or similar genes. However, physical genome properties like chromatin accessibility and DNA methylation patterns are often underutilized in AI models, despite their importance in understanding the three-dimensional structure and packaging of DNA in cells (Lin et al., 2023). These factors can influence the accessibility of specific genomic regions for gene therapy. A new system called piCRISPR addresses this gap by integrating structural features with physiological properties such as chromatin accessibility and DNA methylation. Extensive testing with large datasets of CRISPR/Cas9 editing samples has shown that piCRISPR outperforms other methods in predicting off-target effects. Among its models, the CNN S5E2 has achieved the highest accuracy (AU-ROC = 0.998). Additionally, piCRISPR can identify new off-target cleavage sites, facilitating CRISPR/Cas9 study design and reducing the likelihood of resistance (Rao et al., 2023).

4.10 ON TARGET EFFICIENCY

Genome edits are typically verified using short-read or Sanger sequencing methods. While these techniques are effective at detecting small disruptions and deletions—key outcomes of CRISPR-Cas9 editing—they may miss larger genomic alterations. Long-read sequencing offers a solution to this limitation. A study by Kosicki et al. (2022) demonstrated that combining long-PCR with long-read sequencing could identify large deletions and rearrangements at target sites in genome-edited cells. Additionally, Cas9-induced structural variations (SVs) were detected in vivo. Recent reports have also noted other types of genomic modifications at target loci, such as segmental or complete chromosomal deletions and chromosomal breakage (Table 4.5).

Recent experiments in human embryos have revealed that this effect also occurs at off-target sites. Genomic abnormalities in chromosomal regions or in unplanned or probe-edited genes can have unpredictable functional consequences and are therefore more worrisome than unexpected changes at target loci. To investigate whether off-target SVs are involved, their genomic location must first be determined. The off-target site can be estimated by computational tools, but a more reliable method is to test for off-target activity of Cas9 in vitro using hybridization assays.

4.11 BIOINFORMATICS TOOLS TO EVALUATE GRNA EFFICIENCY

There are an increasing number of online CRISPR tools available, provided by both academic and commercial institutions, with a list of the most widely used resources provided in Table 4.6.

The primary purpose of these CRISPR resources is to provide the information needed to identify gRNAs and select gRNAs for experimental use. While some tools may "recommend" gRNAs for a specific purpose, it is the end user's responsibility to evaluate whether gRNAs are suitable for the application. For most applications, however, the CRISPR tool

TABLE 4.5
Tools for gRNA on-target prediction

On-target software	Model	PAM	URL
DeepCRISPR	CNN1	NGG	www.deepcrispr.net/
DeepCpf1	CNN	TTTN	http://deepcrispr.info/
DeepSpCas9	CNN	NGG	https://deepcrispr.info/DeepSpCas9/
CRISPRater	L1-reg2	NGG	https://crispr.cos.uni-heidelberg.de/
WU-CRISPR	SVM3	NGG	https://crisprdb.org/index.html
SgRNAScorer	SVM (C)6	NGG, NAG,NNAGAAW,NNNNGMTT	https://frederick.cancer.gov/resources/repositories/sgrnascorer
TUSCAN	RF4	NGG	https://github.com/BauerLab/TUSCAN
SSC	Elastic Net	NGG	http://crispr.dfci.harvard.edu/SSC/
CRISPRScan	Linear reg	NGG	www.crisprscan.org/?page=welcome
SAM	GBRT5+Â SVM	NGG	https://bio.tools/samtools
Azimuth1.0	Logistic reg	NGG	https://github.com/MicrosoftResearch/Azimuth
Azimuth2.0	GBRT	NGG	https://github.com/MicrosoftResearch/Azimuth/releases/tag/v2.0
CRISPRpred	SVM	NGG	https://bio.tools/crisprpredseq
ge-CRISPR	SVM	NGG	http://bioinfo.imtech.res.in/manojk/gecrispr/index.php

TABLE 4.6
List of common CRISPR resources

Name	Features	URL
RGEN Tools	Analyzes CRISPR off-target effects and predicts guide RNA specificity	www.rgenome.net/
Benchling	Cloud platform for designing CRISPR studies and sharing genomic data	https://benchling.com/crispr
IDT	Offers gRNA synthesis and CRISPR products	https://sg.idtdna.com/site/order/designtool/index/CRISPR_CUSTOM
Synthego	Offers synthetic guide RNA products and CRISPR kits for genome editing	http://design.synthego.com
CHOPCHOP	Tool for choosing CRISPR targets	http://chopchop.cbu.uib.no/
Addgene	Offers genetic resources for CRISPR research	www.addgene.org/crispr/
Crisflash	Off-target site prediction based on gRNA sequences	https://github.com/crisflash/crisflash
WTSI Genome Editing (WGE)	Helps choose gRNAs and target sites	https://www.sanger.ac.uk/tool/wge-crispr-design-tool/
Breaking-Cas	Tool for gRNA design	http://bioinfogp.cnb.csic.es/tools/breakingcas/
Horizon Discovery	Gene editing using CRISPR tools	https://dharmacon.horizondiscovery.com/gene-editing/crispr-cas9/crispr-design-tool/
Broad Institute GPP	Provide tools for guide RNA design and CRISPR screening	https://portals.broadinstitute.org/gpp/public/
CRISPR-ML	Tool for optimizing gRNAs using machine learning	https://www.microsoft.com/en-us/research/project/crispr/
CRISPOR	Tool for designing guide RNA and checking off-target sites	https://crispor.gi.ucsc.edu/
Deskgen	Gene editing tool used for designing CRISPR experiments and gRNAs	www.deskgen.com
E-CRISP	Creates and evaluates gRNAs for genome editing	www.e-crisp.org/E-CRISP/

is only part of the overall design process, which begins with identifying the target location.

4.12 AI TOOLS FOR GRNA DESIGN

Artificial intelligence is a specialized field in computer science dedicated to creating machines or systems that can perform tasks that normally require human intelligence. Consisting of machine learning and deep learning, AI has become powerful and effective, changing the way diagnosis, treatment, and even gene therapy are performed.

AI models play a crucial role in gene therapy and gRNA design by analyzing complex genomic data, identifying events, and predicting potential outcomes (Table 4.7). Developing

TABLE 4.7
Computational methods to assess RNA expression efficiency

Tool	Enzymes	Data source
E-CRISP	Cas9	–
CRISPRscan	Cas9, Cpf1	Zebrafish
sgRNAScorer	Cas9, Cpf1	Human
SSC	Cas9	Human, Mouse
WU-CRISPR	Cas9	Human, Mouse
Azimuth	Cas9	Human, Mouse
CRISPRater	Cas9	Human
CRISPRpred	Cas9	Human, Mouse
CASPER	Cas9, Cpf1	–
DeepCpf1	Cpf1	Human
SAM	Cas9	Human, Mouse, Zebrafish
TUSCAN	Cas9	Human, Mouse, Zebrafish
uCRISPR	Cas9	–

these models involves advanced machine learning and deep learning techniques. AI is instrumental in guiding the genome-editing process by helping to pinpoint the most effective editing options, including the creation of gRNAs that target and suppress unwanted gene expression while enhancing the efficacy of gene therapy. The development of gene-editing capabilities begins with the collection of genomic data, which can include genetic information, patient medical records, and test results. This data is then meticulously prepared to ensure completeness and accuracy, filtering out noise and irrelevant information so that the AI model can focus on meaningful patterns (Lin et al., 2023). Choosing the right model is crucial for the success of gene therapy applications. While traditional machine learning methods, such as decision trees or support vector machines, can be useful for specific tasks, deep learning models, particularly deep neural networks, are more adept at handling the complex interactions within genetic data (Table 4.8).

TABLE 4.8
AI tools for gRNA design

Name	Features	Website
CRISPR Design Tool MIT	Suitable for designing gRNA for various species Perdicts gRNA efficiency and off-target effects	http://genetargeter.mit.edu/
CRISPR- ERA	Provides tools for target site selection	http://crispr-era.stanford.edu/
CHOPCHOP	Provides off-target analysis User-friendly interphase for the designing of gRNA	https://chopchop.cbu.uib.no/
E-CRISP	Offers detailed examination of possible off-target regions Enables design of CRISPR experiments	www.e-crisp.org/
CRISPOR	Predicts gRNA specificity and efficiency	http://crispor.gi.ucsc.edu/
CCTOP	Prioritize gRNA design based on off-target site predictions and various scoring methods	https://cctop.cos.uni-heidelberg.de/
CRISPRdirect	Provides quick and straightforward gRNA design with off-target predictions for various species	https://crispr.dbcls.jp/
Benchling	It is a cloud-based tool that offers a comprehensive gRNA design, including off-target predictions and cloning capabilities	www.benchling.com/crispr
GUIDE	User-friendly interphase for designing gRNAs and with a focus on minimizing off-target effects	https://www.stemcell.com/crispr-guide-design-algorithm
FlashFry	Flexible and fast tool for gRNA design Specifically suitable for large-scale screens	https://github.com/mckennalab/FlashFry
Cas Designer	Provides comprehensive design for CRISPR/Cas systems	www.rgenome.net/cas-designer/
CRISPRseek	R Package for recodnizing off-target sites and designing efficient gRNAs	https://bioconductor.org/packages/release/bioc/html/CRISPRseek.html
CRISPR-P	Focuses on plant genomes and provides a comprehensive platfrom for plant specific gRNA	http://crispr.hzau.edu.cn/
DESKGEN	Offers gRNA design with integrated lab workflows and detailed off-target analysis	www.deskgen.com/lander
CrispRGold	Prioritizes gRNA based on off-target and on-target activity using advanced scoring methods	https://crisprgold.mdc-berlin.de/
CRISPR-PLANT	Provides efficient gRNA and analysis in agricultural biotechnology	http://omap.org/crispr/
WGE (Wellcome Sanger Institute)	Provides a genome editing database and gRNA design for humans and mouse genomes	https://www.sanger.ac.uk/science/tools/#
CHOPCHOP V3	Latest version with improved algorithms for selecting optimal gRNA targets with detailed off-target predictions	https://chopchop.cbu.uib.no/

(*continued*)

TABLE 4.8 (Continued)
AI tools for gRNA design

Name	Features	Website
Cas Analyzer	Analysis of the results of CRISPR experiments	www.rgenome.net/cas-analyzer/
Off-Spotter	Specifically designed for off-target predictions and analysis. Enhancing the specifity of gRNA design	https://cm.jefferson.edu/Off-Spotter/
Geneious Prime	Comprehensive molecular biology suite with gRNA design tools integrated with cloning and sequence analysis	https://www.graphstats.net/geneious-prime
CLC Genomics Workbench	Offers advanced CRISPR tools for designing gRNA Analyzing off-targets Visualizing genomic data	https://digitalinsights.qiagen.com/trial-request/
CRISPRminer	Database and tool for mining CRISPR arrays and designing gRNA for various microorganism	www.microbiome-bigdata.com/CRISPRminer/
CRISPR Genome Analyzer	Analyzes CRISPR genome editing outcomes Provides insight into gRNA efficiency and specificity	www.cropbioengineering.iastate.edu/cgat/
GT-Scan	Online tool for rapid gRNA target site identification across multiple species and genomes	https://gt-scan.csiro.au/gt-scan
DeepCRISPR	Utilizes deep learning to predict gRNA efficiency and minimize off-target effects, offering high accuracy in gRNA design	https://deepcrispr.info/

4.13 EMERGING TRENDS AND PROSPECTS

Computational biology, AI, and data-driven approaches are increasingly becoming essential in gRNA design tools to improve efficiency, specificity, and versatility and minimize off-targets in CRISPR-Cas. As AI and machine learning tools can analyze large datasets, they are being integrated with design tools to predict on-target efficiency, enabling efficient design of gR NA. AI tools trained on large datasets can better predict the on-target efficiency with minimum off-targets specifically for organisms with complex and poorly characterized genomes. In addition, deep learning methods can better predict off-targets, considering sequence contexts, chromatin accessibility, and genome structure, thus improving CRISPR-Cas for gene therapies and clinical trials. Multiplex gene editing, targeting multiple genes simultaneously, is a developing application of CRISPR-Cas. Computational tools for designing multiple gRNAs, enabling multiple gene modifications in a single experiment, are essential for improving complex and multiple traits in animals and plants. gRNA design tools are the latest applications of CRISPR-Cas, including prime editing and base editing, which are emerging. These applications require optimization of gRNAs, especially in prime editing, to drive the Cas9 to introduce precise base editing without introducing DSB at the target site. Structural genomics, such as chromatin accessibility and sequence contexts, are being incorporated into the gRNA design to improve the efficiency of gRNA. Next-generation gRNA design tools are focused on reducing off-targets, enhancing on-target efficiency and gene-editing outcomes. The automated design tools for synthetic biology and high throughput screens can rapidly design thousands of gRNA to speed up functional genomics and characterize new genes and therapeutic targets. Multi-omics data such as genomics, proteomics, epigenomics, and transcriptomics will likely be incorporated into the future gRNA design tools. These trends highlight the evolution of gRNA design tools to improve the specificity of gene editing, thus paving the way for precise and customizable CRISPR applications in agriculture and human health.

REFERENCES

Amirkhanov, R.N. and G.A. Stepanov. 2019. Systems of delivery of CRISPR/Cas9 ribonucleoprotein complexes for genome editing. *Russian Journal of Bioorganic Chemistry* 45: 431–437.

Anzalone, A.V., L.W. Koblan, and D.R. Liu. 2020. Genome editing with CRISPR–Cas nucleases, base editors, transposases and prime editors. *Nature Biotechnology* 38: 824–844.

Bradford, J. and D. Perrin. 2019. A benchmark of computational CRISPR-Cas9 guide design methods. *PLoS Computational Biology* 15: e1007274.

Cencic, R., Miura, H., Malina, A., Robert, F., Ethier, S., Schmeing, T.M., ... and Pelletier, J. 2014. Protospacer adjacent motif (PAM)-distal sequences engage CRISPR Cas9 DNA target cleavage. *PloS One*, 9(10): e10921.

Chuang, C.-K. and W.-M. Lin. 2021. Points of view on the tools for genome/gene editing. *International Journal of Molecular Science* 22: 9872.

Filippova, J.A., Matveeva, A.M., Zhuravlev, E.S., Balakhonova, E.A., Prokhorova, D.V., Malanin, S.J., ... and Stepanov, G.A. 2019. Are small nucleolar RNAs "CRISPRable"? A report on box C/D small nucleolar RNA editing in human cells. *Frontiers in Pharmacology*, 10: 1246.

Fu, Y., Sander, J.D., Reyon, D., Cascio, V.M., and Joung, J.K. 2014. Improving CRISPR-Cas nuclease specificity using truncated guide RNAs. *Nature Biotechnology*, 32(3): 279–284.

Garcia-Robledo, J.E., M.C. Barrera, and G.J. Tobón. 2020. CRISPR/Cas: From adaptive immune system in prokaryotes to therapeutic weapon against immune-related diseases: CRISPR/Cas9 offers a simple and inexpensive method for disease modeling, genetic screening, and potentially for disease therapy. *International Review of Immunology* 39: 11–20.

Gerashchenkov, G.A., N.A. Rozhnova, B.R. Kuluev, O.Y. Kiryanova, G.R. Gumerova, A.V. Knyazev, Z.R. Vershinina, E.V Mikhailova, D.A. Chemeris, and R.T. Matniyazov. 2020. Design of guide RNA for CRISPR/Cas plant genome editing. *Molecular Biology* 54: 24–42.

Gleditzsch, D., P. Pausch, H. Müller-Esparza, A. Özcan, X. Guo, G. Bange, and L. Randau. 2019a. PAM identification by CRISPR-Cas effector complexes: diversified mechanisms and structures. *RNA Biology* 16: 504–517.

Gupta, S.K. and P. Shukla. 2017. Gene editing for cell engineering: trends and applications. *Critical Reviews in Biotechnology* 37: 672–684.

Haeussler, M. 2020. CRISPR off-targets: A question of context. *Cell Biology and Toxicology* 36: 5–9.

Hu, H. and F. Yu. 2024. Advances in CRISPR/Cas-based genome editing: Break new ground for plant functional genomics. In *CRISPR and Plant Functional Genomics*. Boca Raton, FL: CRC Press, pp. 1–19.

Hu, J.H., S.M. Miller, M.H. Geurts, W. Tang, L. Chen, N. Sun, C.M. Zeina, X. Gao, H.A. Rees, and Z. Lin. 2018. Evolved Cas9 variants with broad PAM compatibility and high DNA specificity. *Nature* 556: 57–63.

Jamal, M., A. Ullah, M. Ahsan, R. Tyagi, Z. Habib, and K. Rehman. 2018. Improving CRISPR-Cas9 on-target specificity. *Current Issues in Molecular Biology* 26: 65–80.

Javaid, N. and S. Choi. 2021a. CRISPR/Cas system and factors affecting its precision and efficiency. *Frontiers in Cell and Developmental Biology* 9: 761709.

Jiang, F. and J.A. Doudna. 2017. CRISPR–Cas9 structures and mechanisms. *Annual Review of Biophysics* 46: 505–529.

Kosicki, M., F. Allen, F. Steward, K. Tomberg, Y. Pan, and A. Bradley. 2022. Cas9-induced large deletions and small indels are controlled in a convergent fashion. *Nature Communications* 13: 3422.

Lau, C.-H., C. Tin, and Y. Suh. 2020. CRISPR-based strategies for targeted transgene knock-in and gene correction. *Faculty Reviews* 9: 20.

Lee, C.M., T.J. Cradick, E.J. Fine, and G. Bao. 2016. Nuclease target site selection for maximizing on-target activity and minimizing off-target effects in genome editing. *Molecular Therapy* 24: 475–487.

Li, Z.-H., J. Wang, J.-P. Xu, J. Wang, and X. Yang. 2023. Recent advances in CRISPR-based genome editing technology and its applications in cardiovascular research. *Military Medical Research* 10: 12.

Liang, Z., Chen, K., Li, T., Zhang, Y., Wang, Y., Zhao, Q., ... and Gao, C. 2017. Efficient DNA-free genome editing of bread wheat using CRISPR/Cas9 ribonucleoprotein complexes. *Nature Communications*, 8(1): 14261.

Lin, J., X. Chen, and K.-C. Wong. 2023. An artificial intelligence approach for gene editing off-target quantification: Convolutional self-attention neural network designs and considerations. *Statistics in Biosciences* 15: 657–668.

Liu, Z., S. Chen, Y. Jia, H. Shan, M. Chen, Y. Song, L. Lai, and Z. Li. 2021. Efficient and high-fidelity base editor with expanded PAM compatibility for cytidine dinucleotide. *Science China Life Sciences* 64(8): 1185.

Ma, D., Z. Xu, Z. Zhang, X. Chen, X. Zeng, Y. Zhang, T. Deng, M. Ren, Z. Sun, and R. Jiang. 2019. Engineer chimeric Cas9 to expand PAM recognition based on evolutionary information. *Nature Communications* 10: 560.

Manghwar, H., B. Li, X. Ding, A. Hussain, K. Lindsey, X. Zhang, and S. Jin. 2020. CRISPR/Cas systems in genome editing: methodologies and tools for sgRNA design, off-target evaluation, and strategies to mitigate off-target effects. *Advanced Science* 7: 1902312.

McGinn, J. and L.A. Marraffini. 2019. Molecular mechanisms of CRISPR–Cas spacer acquisition. *Nature Reviews Microbiology* 17: 7–12.

Mullally, G., Van Aelst, K., Naqvi, M.M., Diffin, F.M., Karvelis, T., Gasiunas, G., ... and Szczelkun, M.D. 2020. 5′ modifications to CRISPR–Cas9 gRNA can change the dynamics and size of R-loops and inhibit DNA cleavage. *Nucleic Acids Research*, 48(12): 6811–6823.

Naeem, M. and O.S. Alkhnbashi. 2023. Current bioinformatics tools to optimize CRISPR/Cas9 experiments to reduce off-target effects. *International Journal of Molecular Science* 24: 6261.

Naeem, M., S. Majeed, M.Z. Hoque, and I. Ahmad. 2020. Latest developed strategies to minimize the off-target effects in CRISPR-Cas-mediated genome editing. *Cells* 9(7): 1608.

Ong, S.H., Y. Li, H. Koike-Yusa, and K. Yusa. 2017. Optimised metrics for CRISPR-KO screens with second-generation gRNA libraries. *Scientific Reports* 7: 7384.

Pallarès Masmitjà, M., N. Knödlseder and M. Güell. 2019. CRISPR-gRNA design. In Y. Luo (Ed.,), *CRISPR Gene Editing: Methods and Protocols*. New York: Humana, pp. 3–11.

Pickar-Oliver, A. and C.A. Gersbach. 2019. The next generation of CRISPR–Cas technologies and applications. *Nature Reviews Molecular Cell Biology* 20: 490–507.

Pieniążek, P., Ł. Tekieli, P. Musiałek, A. Kabłak, T. Przewłocki, R. Motyl, K. Dzierwa, P. Paluszek, M. Hlawaty, and K. Żmudka. 2012. Carotid artery stenting according to the tailored- CAS algorithm is associated with a low complication rate at 30 days: data from the TARGET- CAS study. *Polish Heart Journal (Kardiologia Polska)* 70: 378–386.

Rao, M., E. McDuffie and C. Sachs. 2023. Artificial intelligence/machine learning-driven small molecule repurposing via off-target prediction and transcriptomics. *Toxics* 11: 875.

Riesenberg, S., Helmbrecht, N., Kanis, P., Maricic, T., and Pääbo, S. 2022. Improved gRNA secondary structures allow editing of target sites resistant to CRISPR-Cas9 cleavage. *Nature Communications*, 13(1): 489.

Strobel, S.A. and P.B. Dervan. 1990. Site-specific cleavage of a yeast chromosome by oligonucleotide-directed triple-helix formation. *Science* 249 : 73–75.

Suresh, S. 2021. Beginner's guide to CRISPR-Cas9-based gene editing. *Biochemistry (London)* 43: 36–40.

Verkuijl, S.A., and Rots, M.G. 2019. The influence of eukaryotic chromatin state on CRISPR–Cas9 editing efficiencies. *Current Opinion in Biotechnology*, 55, 68–73.

Wei, Y., Z. Yang, C. Zong, B. Wang, X. Ge, X. Tan, X. Liu, Z. Tao, P. Wang, and C. Ma. 2021. Trans single-stranded DNA cleavage via CRISPR/Cas14a1 activated by target RNA without destruction. *Angewandte Chemie* 133: 24443–24449.

Yili, F., L.I.U. Sicheng, C. Ruodan,and X.I.E. Anyong. 2021. Target binding and residence: a new determinant of DNA double-strand break repair pathway choice in CRISPR/Cas9 genome editing. *Journal of Zhejiang University Science B* 22: 73.

Yuen, G., Khan, F.J., Gao, S., Stommel, J.M., Batchelor, E., Wu, X., and Luo, J. 2017. CRISPR/Cas9-mediated gene knockout is insensitive to target copy number but is dependent on guide RNA potency and Cas9/sgRNA threshold expression level. *Nucleic Acids Research*, *45*(20): 12039–12053.

5 Available Cargoes and Their Delivery Tools in Genome Editing

Hassan Zafar, Muhammad Faheem, Anam Niamat, Aftab Ahmad, Ahmad Munir, Muhammad Naeem Khan, and Muhammad Naeem

5.1 INTRODUCTION

The fields of molecular biology, genetics, and biotechnology have been transformed by the advances in genome-editing technologies. Leading the industry are three prominent technologies: Zinc Finger Nucleases (ZFNs), Transcription Activator-Like Effector Nucleases (TALENs), and CRISPR-Cas9 (Clustered Regularly Interspaced Short Palindromic Repeats-associated protein 9). Each of these tools possesses distinct mechanisms and uses, propelling significant progress in genome editing.

ZFNs, one of the pioneering engineered nucleases in genome editing, are comprised of a DNA-binding domain with zinc finger motifs and a DNA-cleaving domain deriving from the *FokI* restriction enzyme. By customizing the zinc fingers to target specific DNA sequences, ZFNs facilitate precise double-strand breaks (DSBs) at designated genomic sites. The subsequent repair of these breaks by cellular mechanisms can result in gene disruption or correction through homologous recombination or non-homologous end joining methods (Carroll, 2011). ZFNs offer the advantage of targeting a broad spectrum of DNA sequences with great precision. Nonetheless, creating and putting together functional ZFNs can pose a significant technical and time-related challenge, which hampers their extensive use (Urnov et al., 2010).

TALENs have emerged as a more adaptable option to ZFNs. TALENs (transcription activator-like effectors nucleases) are derived from proteins produced by plant pathogenic bacteria, and are linked to the FokI nuclease domain. The DNA-binding portion of TALEN is extremely flexible and can easily be adjusted to identify particular DNA sequences. Unlike ZFNs, the TALEN structure simplifies the design process by linking each repeat to a single DNA base pair (Bogdanove and Voytas, 2011). TALENs have proved to be highly effective and specific in editing genomes, their flexibility and ease of customization have led to their popularity for accurate genetic modifications across a range of organisms (Mussolino and Cathomen, 2012).

Due to its simplicity, versatility, and efficiency, CRISPR-Cas9 has quickly emerged as the most commonly utilized genome-editing tool. The CRISPR-Cas9 system, originating from a bacterial immune response, is composed of the Cas9 nuclease and a single-guide RNA (sgRNA). This sgRNA guides Cas9 to a specific DNA sequence by complementary base pairing, leading to Cas9 triggering a double-strand break in the DNA (Jinek et al., 2012; Ahmad et al., 2021; Aslam et al., 2022). The simplicity in creating sgRNAs to target nearly any sequence, along with the precision in making edits through homologous directed repair (HDR) or causing gene disruptions through non-homologous end joining (NHEJ), have established CRISPR-Cas9 as a valuable tool for genetic manipulation, used in various fields such as functional genomics, disease modeling, and therapeutic research (Doudna and Charpentier, 2014; Sultan et al., 2022). Among the genome-editing technologies—including ZFNs, TALENs, and CRISPR-Cas9—CRISPR-Cas9 stands out for its user-friendly nature and scalability compared to the other two options. In Table 5.1, concise information about the delivery methods for all three (ZFNs, TALENs, and CRISPR-Cas9) are given.

5.2 DELIVERY OF ZINC FINGER NUCLEASES (ZFNS) AND TRANSCRIPTION ACTIVATOR-LIKE EFFECTOR NUCLEASES (TALENS)

The delivery of ZFNs and TALENs into target cells is a critical step in genome-editing applications. Effective delivery systems ensure that ZFNs reach the nucleus of target cells to induce DNA DSBs. Delivery methods for both techniques have their own unique characteristic advantages and limitations, making their use more specific according to needs. Factors that contribute to these characteristics include desired efficiency, type of target cell, and acceptable risk level. Recent advances in these techniques have tried to increase the specificity, efficiency as well as the safety of TALEN and ZFN delivery methods, increasing their potential in therapeutic and plant research. Various methods have been developed to deliver ZFNs (Urnov et al., 2010; Shamshirgaran et al., 2022). These methods include viral vectors, non-viral vectors, and physical methods, increasing their potential in therapeutic research.

5.2.1 Viral Vectors

Viral vectors are one of the most efficient and commonly used delivery systems for ZFNs and TAELNs. They leverage the natural ability of viruses to infect cells and deliver genetic material, facilitating efficient delivery of genome-editing cargoes in the nucleus of the cell (Nayerossadat et al., 2012). Several types of viral vectors are used, including Adeno-Viruses (AV), Adeno Associated Viruses (AAVs), and Lenti-Viruses (LV). Each of them has its own specific characteristic features, such as AVV has the ability of transduction in both reproducing and non-reproducing cells, adenovirus transfer

DOI: 10.1201/9781003500933-6

TABLE 5.1
Comparison of delivery cargoes of genome-editing tools

Feature	TALENs	ZFNs	CRISPR-Cas9
Components	TALENs (Transcription Activator-Like Effector Nucleases)	ZFNs (Zinc Finger Nucleases)	Cas9 protein, gRNA
DNA recognition	TAL effector repeats	Zinc finger domains	gRNA
Nuclease	FokI	FokI	Cas9
Guide design	Modular repeats specific to target sequence	Modular zinc fingers specific to target sequence	Customizable gRNA matching target sequence
Size	Large proteins	Large proteins	Cas9 (relatively large), gRNA (small)
Delivery methods			
Viral vectors	AAV, Lentivirus	AAV, Lentivirus	AAV, Lentivirus, Adenovirus
Non-viral vectors	Plasmid DNA, mRNA	Plasmid DNA, mRNA	Plasmid DNA, mRNA, RNP complexes
Physical methods	Electroporation, Microinjection	Electroporation, Microinjection	Electroporation, Microinjection, LNPs
Editing efficiency	High	High	High
Off-target effects	Moderate to low	Moderate to high	Variable, depending on gRNA and Cas9 variant
Ease of design	Moderate	Complex	Relatively simple
Cost	Moderate to high	High	Lower relative to TALENs and ZFNs
Applications			
Gene knockout	Yes	Yes	Yes
Gene knock-in	Yes	Yes	Yes
Gene activation	No	No	Yes (with CRISPRa)
Base editing	No	No	Yes (with base editors)
Prime editing	No	No	Yes
Scalability	Moderate	Low to moderate	High

has high transduction ability, and lentiviruses have a broad spectrum of target cells. All these features make effective DSBs, ultimately increasing the efficiency of genome-editing tools (Lundstrom, 2018).

5.2.2 Adenoviral Vectors (AV)

Adenoviral vectors (AV) transduce a wide range of dividing and non-dividing cells and can carry large genetic payloads in animal cells. In animal therapeutics, AV has been used to deliver ZFNs and TALENs to correct genetic mutations in models of diseases such as Duchenne Muscular Dystrophy (DMD) (Mendell et al., 2010). DMD is caused by a mutation in the dystrophin gene which normally expresses dystrophin protein, which helps in the normal functioning of muscles. ZFN has been successfully designed to target the mutated dystrophin gene in the dystrophin-diseased mice model (Mendell et al., 2010). ZFN was delivered using adenoviral vectors and was injected in mice. This study revealed that ZFN produced a double-stranded break at the target site, correcting the muted gene into a normal healthy gene, ultimately restoring the normal expression of dystrophin in mice models. This indicates the therapeutic potential of adenoviral vectors as delivery cargoes in the treatment of genetic diseases resulting from mutations in genes (Aslesh et al., 2021).

In comparison to ZFN delivery, TALEN constructs have also been delivered using AV and helped to treat genetic disorders. In animal therapeutics, AVs have been successfully applied to deliver TALEN constructs in the cells of mice liver to treat Hereditary Tyrosinemia type I (HTI). HTI is caused by a mutation in the Fah gene and was successfully restored using AV vectors, delivering the TALEN construct in mice models. This study confirmed the restored function of enzyme proteins via the successful restoration of the Fah gene, showing the potential of AV-mediated delivery of the TALEN construct in the treatment of genetic diseases (Trevisan et al., 2020).

5.2.3 Adeno Associated Viral Vectors (AAV)

Due to their less immunogenicity, AAV-mediated delivery of ZFNs has shown promise in targeting liver cells to treat metabolic liver diseases. These diseases are caused by defects in metabolic genes that express metabolic enzymes, that are directly involved in the metabolism of substances in the liver (Lisowski et al., 2014). The main limitation of AAV vectors is their small packaging capacity, which can constrain the delivery of large constructs of ZFN and TALEN.

5.2.4 Lenti-Viral Vectors (LV)

Lentiviruses can integrate their genetic material into the host genome, providing stable and long-term expression of nucleases. This feature makes lenti-viral vectors suitable for applications requiring sustained ZFN and TALEN expression, such as in hematopoietic stem cells for gene therapy (Lombardo et al., 2007). LVs have an efficient ability to transfer ZFN as well as TALEN in animal and plant cells to deal with genetic dilemmas.

In a study by Martin et al. (2016), LVs were used to target CCR5 gene in human T cells using ZFNs as a genome-editing tool. CCR-5 is responsible for the expression of the Co receptor which is directly linked with the entry of HIV in humans. In this study, LVs successfully transferred ZFNs construct in the T cell lines and disrupted the CCR-5 gene, increasing resistance against HIV. This study aimed to highlight the potential of LVs in gene therapy in human cell lines, offering a strategic approach to deal with various viral infections.

In living organisms, a lenti-viral vector system was developed to enhance therapeutic applications by combining TALEN-mediated excision of HIV proviral DNA with the simple selection of gene-transduced and excised cells in T cell lines. TALENs targeting the HIV long terminal repeat (LTR) site were delivered via a lenti-viral vector system, efficiently removing more than 80% of the lenti-viral DNA from T cells, demonstrating its potential for HIV therapy (Ebina et al., 2015). However, risk of immunogenicity remains a major concern in using viral vectors as a vehicle which could be overcome using non-viral vectors. Details of these vectors are given below.

5.2.5 Non-Viral Vectors

Non-viral delivery systems offer a safer alternative to viral vectors, minimizing the risks associated with immunogenicity and insertional mutagenesis. Non-viral methods include plasmid DNA, mRNA, and nanoparticle-based delivery. Plasmid DNA delivery involves the use of plasmids encoding ZFN and TALENs constructs. Electroporation and chemical transfection agents, such as liposomes, facilitate the entry of plasmids into cells. While plasmid-based delivery is relatively simple and cost-effective, it often results in low transfection efficiency and transient expression (Wang et al., 2015). mRNA-based delivery of ZFN-encoding mRNA can overcome some limitations of plasmid DNA, such as the risk of genomic integration. mRNA delivery leads to transient expression, reducing the risk of off-target effects and insertional mutagenesis. Lipid nanoparticles (LNPs) are commonly used to encapsulate and deliver mRNA to target cells efficiently (Kormann et al., 2011). Various nanoparticle-based systems, including gold nanoparticles, lipid nanoparticles, and polymer-based nanoparticles, have been explored for ZFN delivery. These systems can enhance the stability and cellular uptake of ZFNs, offering targeted and efficient delivery (Rosi et al., 2006). Some notable practical examples in both animals and plant genome editing of these delivery systems are discussed in the following paragraphs.

A significant example of using non-viral vectors in plants is improved dough contents in wheat (*Triticum aestivum*) via Polyethylene Glycole (PEG) mediated delivery of ZFN. In this study, ZFN was delivered to target the glutenin gene, which is responsible for determination of dough contents in wheat. PEG-mediated delivery of ZFN allowed the targeted DSB in the glutenin gene, ultimately improving the dough content in wheat (Koeppel et al., 2019). In living organisms, Lipid nanoparticles (LNPs) have been used for delivering ZFNs to specifically target and edit genes in immune cells, such as CD4+ T cells. For example, LNP-conjugated CD4 antibodies enabled targeted delivery of ZFN mRNA to CD4+ T cells, achieving gene editing in T cell populations within the spleen and lymph nodes. This method holds promise for immunotherapy and gene-editing applications, including potential treatments for HIV (Tombacz et al., 2021).

Just like ZFN, TALEN has also been delivered in plants via PEG-mediated transformation in maize. TALEN was prepared to target the ARGOS8 gene which is associated with drought tolerance in the protoplast of maize. PEG-mediated transformation of TALEN in the protoplast of maize produced DSB in the ARGOS8 gene locus, increasing the drought tolerance in maize (Varroto., 2023). This application shown the potential of non-viral vectors to cope with environmental stress in crops. Gold nanoparticles (AuNPs) have been used for therapeutic applications, including the delivery of TALENs to target specific genes in cancer treatment. The functionalization of AuNPs with nucleic acids, such as TALEN constructs, enables precise modulation of oncogenes or tumor suppressor genes, demonstrating the potential of AuNP-mediated delivery systems for efficient gene editing both in vitro and in vivo cancer models (Ferreira et al., 2020).

5.2.6 Physical Methods

Physical methods of ZFN and TALENs delivery include microinjection, electroporation, and hydrodynamic delivery. These techniques physically introduce ZFNs and TALENs into cells, often achieving high transfection efficiency. Microinjection involves directly injecting ZFNs and TALENs into individual cells using a fine needle. This method is highly precise and is commonly used in embryos for germline editing (Carlson et al., 2012). However, it is labor-intensive and not suitable for large-scale applications. Electroporation uses electrical pulses to create transient pores in the cell membrane, allowing ZFN constructs to enter the cells. This method is widely used for its efficiency and ease of use. It has been successfully applied to deliver ZFNs into various cell types, including stem cells (Gaj et al., 2013). Hydrodynamic delivery involves the rapid injection of a large volume of DNA solution into the bloodstream, leading to the uptake of genetic material by target cells, particularly in the liver. This method has been used to deliver ZFNs and TALENs for in vivo genome editing in animal models (Liu et al., 1999).

In plants, biolistic delivery (particle bombardment) has also proved significant practical examples. One of examples was the plant study in which TALENs were delivered via

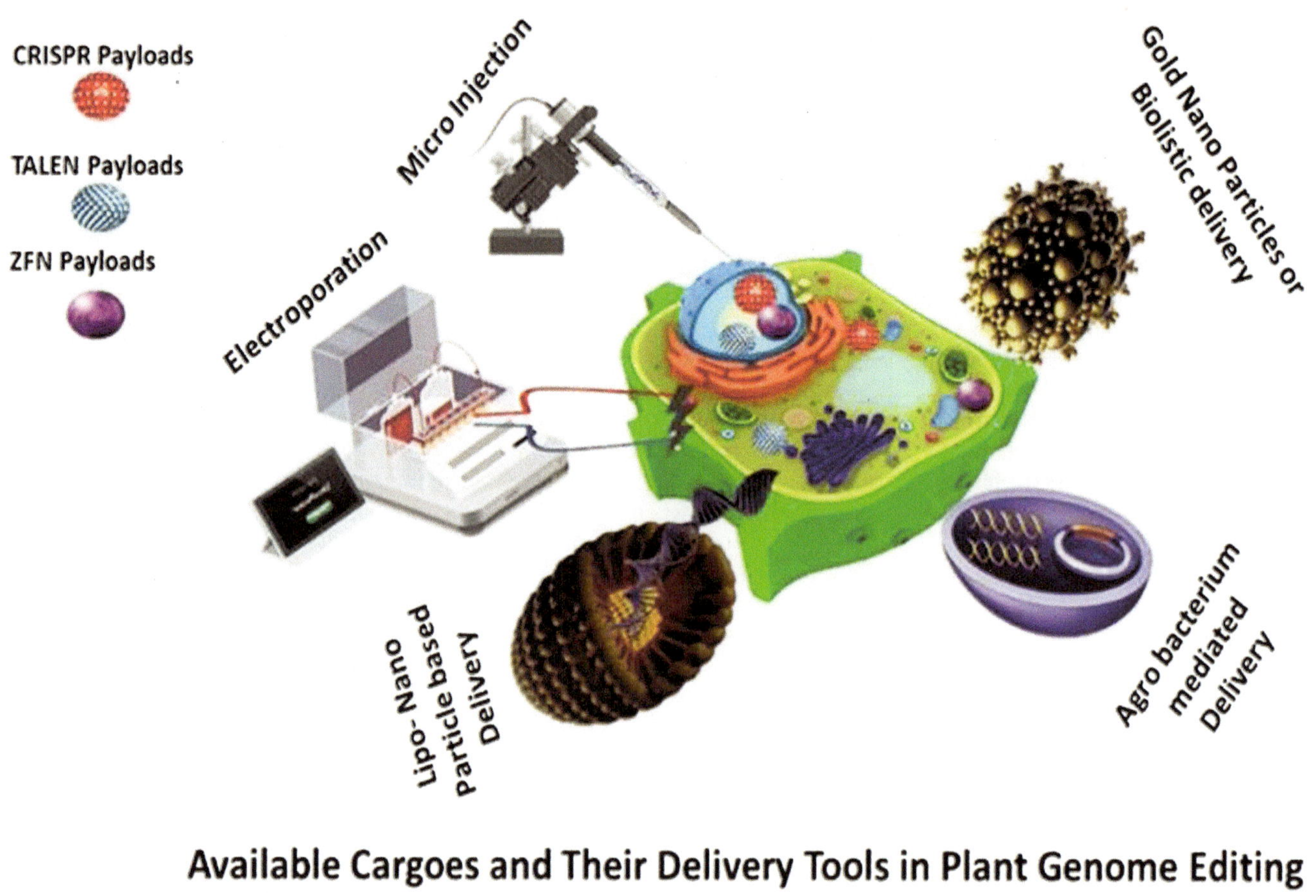

FIGURE 5.1 Available delivery cargoes in plants.

particle bombardment into rice calli, enabling homologous recombination (HR)-mediated gene replacement in the acetolactate synthase gene (OsALS). This resulted in herbicide-resistant rice lines, demonstrating the feasibility of TALEN-mediated genome editing in rice with successful heritable gene edits in the T1 generation (Li et al., 2016).

In therapeutic applications, TALEN has been delivered using the electroporation method in living organisms. In animal models, TALENs have been successfully delivered using this method. For example, electroporation of TALEN mRNA into rat embryos resulted in efficient genome editing at targeted loci, producing knockout rats with a survival rate of 91%. This method offers a simpler and more effective alternative to microinjection for generating knockout animals (Kaneko et al., 2014). Figure 5.1 shows the available delivery cargoes in plants.

5.3 DELIVERY OF CRISPR/CAS9

Clustered Regularly Interspaced Short Palindromic Repeats (CRISPR)-Cas9 technology has revolutionized the field of genome editing via its unique and precise targeting ability in both plants and animals. It is easy to apply as compared with ZFN and TALEN. It delivered via a number of ways including viral, non-viral as well as physical methods (Wani et al., 2022; Munir et al., 2024).

5.3.1 Adeno-Associated Virus (AAV) Vectors

Viral vectors, particularly Adeno-Associated Virus (AAV) vectors, have become a popular choice for delivering CRISPR-Cas9 components due to their high transduction efficiency, broad host range, and relatively low immunogenicity. AAV vectors can package and deliver the CRISPR-Cas9 system into target cells, ensuring stable expression and facilitating targeted gene editing. Their ability to transduce both dividing and non-dividing cells makes AAV vectors highly versatile for applications in animals (Yang et al., 2022).

Adeno-associated virus (AAV) vectors are one of the most popular viral delivery systems for CRISPR-Cas9 due to their ability to infect dividing and non-dividing cells and their relatively low immunogenicity but with a small AAV loading capacity that is about 4.7 kb (Pei, 2022). The Cas9 protein gRNA is larger and it can also inhibit its distribution. This limitation can be addressed by using small Cas9 orthologs such as SaCas9 from Staphylococcus aureus (Mohan et al., 2022). AAV vectors efficiently infect tissues and can target both dividing and non-dividing cells, providing versatility in

CRISPR-mediated genome editing. Despite these advantages, AAV use may be challenged by pre-existing immunity in some individuals, which may reduce their efficacy (Hinderer et al., 2018). Recent advances include the development of cleaved protein structures to replicate Cas9 in the cell, thus overcoming the limitations of packaging materials, and the use of tissue-specific promoters for targeted delivery has also improved, making the applications of the CRISPR-Cas9 system broader in both plant and animal research (Zheng et al., 2018). Following are some important examples of the successful delivery of CRISPR using AAV.

> In therapeutics engineered AAV2g9 vector has been utilized to treat neurological disorders such as schizophrenia disease using CRISPR/Cas 9 as genome editing tool. A single injection consisting of AAV2g9 vector containing CRISPR/Cas9 payloads was given via intracranial route to target MIR-137 gene in rats. This gene was responsible for schizophrenia and after successful knock out of MIR-137 gene using CRISPR, resulted in improved health condition with no adverse effects on liver. This indicated the use of engineered AAV in neurological research (Murlidharan et al., 2016).

5.3.2 Lenti-Viral Vectors

Compared to AAVs, lenti-viral vectors have superior packaging capacity, making them suitable for delivery of full-length Cas9 proteins and gRNAs. Lenti-viral vectors are particularly effective in modifying cell division, providing significant advantages for some applications (Sakuma et al., 2014). However, the risk of insertion mutagenesis due to genomic integration and greater immunogenicity are notable limitations compared to AAV. To mitigate these risks, auto-inactivating lenti-viral vectors have been developed to reduce the chances of insertion mutations, and nonhomologous lenti-viral vectors are being sought for transient expression (Zufferey et al., 1998).

In human therapeutics, lenti-viral vectors have shown their potential to target various animal genes using the CRISPR-Cas9 system. One of the most suitable examples of CRISPR-Cas9 delivery using lenti-viral as a vector for delivery is the targeting of the CCR5 gene in human T lymphocytes to increase resistance against HIV infection. The CCR5 gene expresses a co-receptor that is used by the HIV virus to gain entry into the T lymphocytes. A lenti-viral vector containing CRISPR consisting of Cas9 enzyme and gRNA for CCR5 gene payload has been used to target T-cells ex vivo. The gRNA provides complementary nucleotide sequences for target site in the CCR5 gene to the Cas9 endonuclease enzyme. After binding, Cas9 produced DSB and the outcome was increased resistance against HIV. After modification, T cells were then transformed back into patients, providing functional treatment of HIV (Xiao et al., 2019).

5.3.3 Delivery of CRISPR-Cas9 by Adenoviral Vectors

Adenoviral vectors are widely used for delivering the CRISPR-Cas9 system due to their high transduction efficiency and ability to infect a broad range of cell types. These vectors can deliver large genetic payloads and are particularly useful for transient expression of the CRISPR-Cas9 components, making them ideal for applications in both plants and animals (Ehrke-Schulz et al., 2017). Adenoviral vectors provide high packaging capacity, enabling delivery of large Cas9 variants and donor DNA. Multiple cell lines can be obtained, and high transduction efficiency achieved. However, because they do not integrate into the host genome, their high immunogenicity and transient expression have substantial drawbacks. To overcome these issues, "gutless" adenoviral vectors have been designed to reduce immunogenicity and delay expression (Mitani and Kubo, 2002). In addition, the use of tissue-specific targeting ligands increases the specificity of adenovirus delivery (Kreppel and Kochanek, 2008).

Adenoviral vectors efficiently work in living organisms. In a study by Stephens et al. (2019), the mice model was used for the treatment of the Hemophilia B (HB), which is caused by the defective FIX gene. In this study, the healthy FIX gene was knocked in using the adenoviral vector-mediated delivery of the CRISPR-Cas9 system. The phenotypic correction in the juvenile mice demonstrated that adenoviral vectors efficiently delivered the CRISPR-Cas9 in the mice model. This study proved that the adenoviral vectors can be used for therapeutic purposes (Stephens et al., 2019). Both examples illustrate the versatility and effectiveness of adenoviral vectors in delivering CRISPR-Cas9 across different biological systems.

5.3.4 Non-Viral Delivery Methods

The non-viral delivery methods for the CRISPR-Cas9 offer significant benefits including, avoidance of insertional mutagenesis as well as lower immunogenicity. These strategies encompass various ranges of techniques like biological, chemical as well as physical approaches (Shalaby et al., 2020). In the biological method, cell-penetrating peptides are implicated that exploit the natural cellular uptake mechanism to enhance the intracellular delivery of CRISPR-Cas9 components (Shalaby et al., 2020). In the chemical methods, the CRISPR-Cas9 components are delivered with the help of encapsulation in the target cells, in such chemical strategies nanoparticles and polymer-based techniques are included. Encapsulation ensures the efficient uptake and release of the CRISPR components (Carboni et al., 2019). An example of encapsulation is the lipid nanoparticles (LNPs) that have emerged as an important delivery method for nucleic acids, including mRNA and CRISPR RNP complexes. They hold the material in place, protect it from degradation, and improve cellular permeability. LNPs are not immunogenic and can deliver large payloads, making them suitable for in vivo applications (Kaczmarek et al., 2017). However, transfection efficiency may vary among cell types, and potential toxicity at high doses is a concern. Recent progress includes the development of ionizable lipids to increase delivery efficiency and reduce toxicity, and the use of PEGylation to improve stability and circulation time in vivo (Cullis and Hope, 2017).

Various physical methods of delivery of CRISPR components have been investigated, including microinjection, electroporation, nanoparticle-mediated delivery, biolistic-delivery (gene gun), and hydrodynamic injections.

Electroporation works on the principle of electrolysis that involves the application of a strong electric field to temporarily permeate the cell membrane, allowing the CRISPR components to be absorbed. This approach is particularly effective for ex vivo applications. It boasts high transfection efficiency and can deliver DNA, RNA, or proteins that are components of CRISPR (Smith et al., 2015). However, its use is limited to ex vivo applications, and the potential to damage cells due to electroshock is a major drawback. Advances in this field have included optimizing electrophoretic parameters to reduce cell damage and improve efficiency, as well as microfluidic devices designed for controllable and reproducible electro filtration (Chang et al., 2016).

Microinjection allows CRISPR components to be delivered directly into cells via a microneedle. This method is very successful but labor-intensive and technically demanding. Despite these difficulties, microinjection provides direct delivery to the nucleus, resulting in increased efficacy (Mashiko et al., 2013). Recent developments include automated microinjection systems to increase throughput and stability, and the use of microneedle arrays for high-throughput microinjection in small organisms or tissues (Hashimoto and Takemoto, 2015).

Nanoparticle-mediated delivery consists of nanoparticles, such as gold or silica, that can be functionalized to efficiently deliver CRISPR components. These nanoparticles provide tunable properties for targeting and payload release. However, the potential cytotoxicity and complex synthesis functionalization process are challenges that need to be addressed (Xu et al., 2014). Recent advances include the development of biodegradable nanoparticles to reduce toxicity and the use of magnetic nanoparticles for targeted delivery using external magnetic fields (Sun et al., 2015). A recent study demonstrated the use of gold nanoparticles functionalized with CRISPR-Cas9 components to target the PLK-1 gene in cancer cells. The gold nanoparticles were conjugated with Cas9 protein and guide RNA, facilitating efficient delivery into the target cells. Upon entering the cells, the CRISPR-Cas9 system induced precise gene editing at the PLK-1 locus, leading to significant inhibition of cancer cell proliferation. This method showcased the potential of gold nanoparticles for delivering CRISPR-/Cas9 components, highlighting their tunable properties for targeted gene editing. The study also addressed cytotoxicity concerns by optimizing the nanoparticle synthesis process, ensuring minimal adverse effects on the cells. This example illustrates the promising application of nanoparticle-mediated delivery in therapeutic gene editing (Xu et al., 2014; Sun et al., 2015).

Biolistic delivery, or gene guns, use small, high-speed particles to deliver CRISPR components into cells. This method is particularly suitable for plant cells and other cells that are difficult to culture. Although bio-delivery is efficient for plant genome editing and eliminates the need for vectors, it can cause physical damage to cells and is usually limited to surface or tissue cells (Altpeter et al., 2005;Taylor and Fauquet, 2002).

Hydrodynamic injection involves rapid injection of multiple CRISPR components into the bloodstream, resulting in a transient increase in pressure that improves liver cell uptake of this pathway for liver-targeted gene exchange and production is dense and cost-effective. However, it is only hepatocytes that can cause tissue damage due to high stress (Liu et al., 1999). Recent modifications in injection protocols aim to improve safety and efficacy, and targeted delivery strategies are being developed to extend use beyond the liver (Zhang et al., 2001). These methods enable precise genetic modification without the potential risk which is linked to the viral methods, so these non-viral approaches will provide efficient as well as versatile alternatives for genome editing in the various living organisms. Figure 5.2 shows the available cargoes for genome editing in animals.

5.4 ADVANCES AND CHALLENGES IN DELIVERY TECHNOLOGIES

Achieving tissue-specific delivery of CRISPR-Cas9 components is crucial for therapeutic applications. Advances in targeting ligands and receptor-mediated endocytosis have improved the precision of delivery systems, allowing for selective editing in target tissues while minimizing off-target effects (Wang et al., 2018; Ahmad et al., 2024). Off-target effects remain a significant challenge in CRISPR-Cas9 genome editing. The development of high-fidelity Cas9 variants and improved gRNA design algorithms has mitigated some of these issues (Kleinstiver et al., 2016). Additionally, transient expression systems, such as delivering Cas9 as mRNA or RNP complexes, reduce the duration of Cas9 activity, thereby decreasing the likelihood of off-target cleavage (Kim et al., 2014). Scalability and manufacturing of CRISPR-Cas9 delivery systems are critical for clinical translation. Viral vector production, in particular, requires stringent quality control and large-scale manufacturing capabilities. Non-viral delivery systems, such as LNPs, offer potential advantages in scalability and cost-effectiveness (Zhang et al., 2018). The immunogenicity and safety of CRISPR-Cas9 delivery systems are paramount for in vivo applications. Strategies to reduce immune responses, such as using less immunogenic Cas9 orthologs and optimizing delivery vectors, are essential for the successful implementation of CRISPR-based therapies (Crudele and Chamberlain, 2018; Ahmad et al., 2024)..

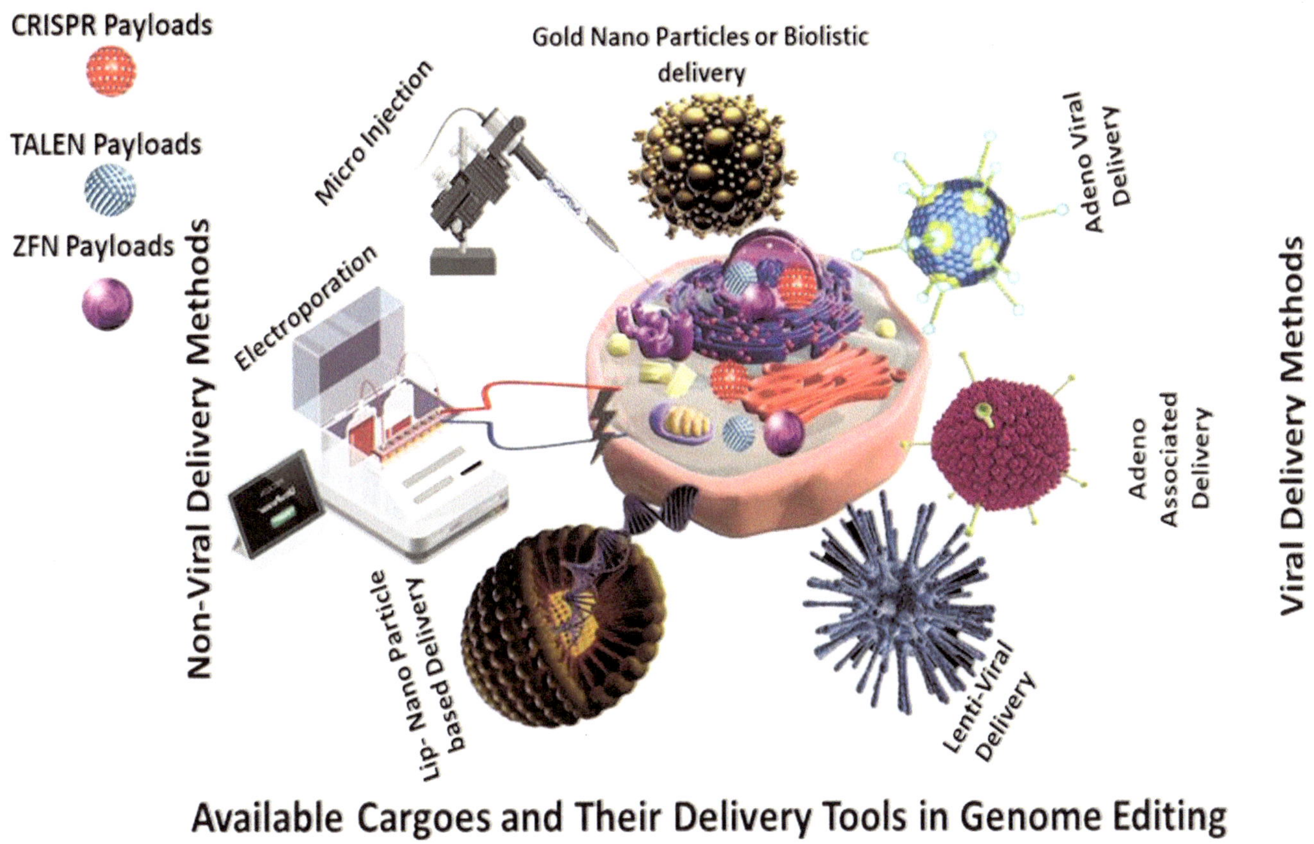

FIGURE 5.2 Available cargoes for genome editing in animals.

5.5 CONCLUSION

The fields of molecular genetics and biotechnology have been revolutionized by genome-editing technologies. There are three prominent technologies: Zinc Finger Nucleases (ZFNs), Transcription Activator-Like Effector Nucleases (TALENs), and CRISPR-Cas9 (Clustered Regularly Interspaced Short Palindromic Repeats-associated protein 9). Each of these tools possesses distinct mechanisms and uses, propelling significant progress in genome editing. All of these technologies are delivered using viral, non-viral and physical methods. The delivery of CRISPR-Cas9 components is a crucial factor in the success of genome editing. Various delivery methods, including viral and non-viral systems, have been developed and refined to enhance efficiency, specificity, and safety of genome editing. Each method has its own advantages and limitations, making the choice of delivery tool dependent on the specific application and target cells or tissues. Ongoing research and advances in delivery technologies are aimed at addressing current challenges, such as tissue-specific targeting, reducing off-target effects, and improving scalability and safety. These developments will be key to unlocking the full potential of CRISPR-Cas9 for precise and targeted genome modifications.

REFERENCES

Ahmad, A., Abd-Elsalam, K. A., Ali, M., Majeed, H. N., Arif, A., Munir, A., Khan, Z., and Munawar, N. (2024). Regulatory triggers of CRISPR-edited crops. In K.A. Abd-Elsalam and A. Ahmad (Eds.), *Global Regulatory Outlook for CRISPRized Plants* (pp. 91–112). Academic Press. https://doi.org/10.1016/b978–0-443-18444-4.00016-8

Ahmad, A., Munawar, N., Khan, Z., Qusmani, A. T., Khan, S. H., Jamil, A., Ashraf, S., Ghouri, M. Z., Aslam, S., Mubarik, M. S., Munir, A., Sultan, Q., Abd-Elsalam, K. A., and Qari, S. H. (2021). An outlook on global regulatory landscape for genome-edited crops. *International Journal of Molecular Sciences*, 22(21), 11753. https://doi.org/10.3390/ijms222111753

Altpeter, F., et al. (2005). Particle bombardment and the genetic enhancement of crops: myths and realities. *Molecular Breeding*, 15(3), 305–327.

Aslam, S., Munir, A., Aslam, H. M., Khan, S. H., and Ahmad, A. (2022). Genome editing advances in soybean improvement against biotic and abiotic stresses. In S. H. Wani, N. ul Rehman Sofi, M. A. Bhat, and F. Lin (Eds.), *Soybean Improvement* (pp. 241–274). Springer. https://doi.org/10.1007/978-3-031-12232-3_13

Aslesh, T., Erkut, E., and Yokota, T. (2021). Restoration of dystrophin expression and correction of Duchenne muscular dystrophy by genome editing. *Expert Opinion on Biological Therapy*, 21(8), 1049–1061. https://doi.org/10.1080/14712598.2021.1872539

Bogdanove, A. J., and Voytas, D. F. (2011). TAL effectors: customizable proteins for DNA targeting. *Science*, 333(6051), 1843–1846. https://doi.org/10.1126/science.1204094

Carboni, V., Maaliki, C., Alyami, M., Alsaiari, S., and Khashab, N. (2019). Synthetic vehicles for encapsulation and delivery of CRISPR/Cas9 gene editing machinery. *Advanced Therapeutics*, 2(4), 1800085. https://doi.org/10.1002/adtp.201800085

Carlson, D. F., Tan, W., Lillico, S. G., Stverakova, D., Proudfoot, C., Christian, M., ... and Fahrenkrug, S. C. (2012). Efficient TALEN-mediated gene knockout in livestock. *Proceedings of the National Academy of Sciences*, 109(43), 17382–17387.

Carroll, D. (2011). Genome engineering with zinc-finger nucleases. *Genetics*, 188(4), 773–782. https://doi.org/10.1534/genetics.111.131433

Chang, L., et al. (2016). Microfluidic devices for manipulation and analysis of biological cells. *Annual Review of Biomedical Engineering*, 18, 315–340.

Crudele, J. M., and Chamberlain, J. S. (2018). Cas9 immunity creates challenges for CRISPR gene editing therapies. *Nature Communications*, 9(1), 3497.

Cullis, P. R., and Hope, M. J. (2017). Lipid nanoparticle systems for enabling gene therapies. *Molecular Therapy*, 25(7), 1467–1475.

Doudna, J. A., and Charpentier, E. (2014). Genome editing: The new frontier of genome engineering with CRISPR-Cas9. *Science*, 346(6213), 1258096. https://doi.org/10.1126/science.1258096

Ebina, H., Kanemura, Y., Misawa, N., Sakuma, T., Kobayashi, T., Yamamoto, T., and Koyanagi, Y. (2015). A high excision potential of TALENs for integrated DNA of HIV-based lentiviral vector. *PLOS ONE*, 10(3), e0120047. https://doi.org/10.1371/journal.pone.0120047

Ehrke-Schulz, E., Schiwon, M., Leitner, T., Dávid, S., Bergmann, T., Liu, J., and Ehrhardt, A. (2017). CRISPR/Cas9 delivery with one single adenoviral vector devoid of all viral genes. *Scientific Reports*, 7(1). https://doi.org/10.1038/s41598-017-17180-w

Ferreira, D., Fontinha, D., Martins, C., Pires, D., Fernandes, A. R., and Baptista, P. V. (2020). Gold nanoparticles for vectorization of nucleic acids for cancer therapeutics. *Molecules*, 25(15), 3489. https://doi.org/10.3390/molecules25153489

Gaj, T., Guo, J., Kato, Y., Sirk, S. J., and Barbas, C. F. (2013). Targeted gene knockout by direct delivery of zinc-finger nuclease proteins. *Nature Methods*, 10(5), 424–426.

Hashimoto, M., and Takemoto, T. (2015). Kormann, M. S., Hasenpusch, G., Aneja, M. K., Nica, G., Flemmer, A. W., Herber-Jonat, S., ... and Rosenecker, J. (2011). Expression of therapeutic proteins after delivery of chemically modified mRNA in mice. *Nature Biotechnology*, 29(2), 154–157.

Hinderer, C., Katz, N., Buza, E. L., Dyer, C., Goode, T., Bell, P., Richman, L. K., and Wilson, J. M. (2018). Severe toxicity in nonhuman primates and piglets following high-dose intravenous administration of an adeno-associated virus vector expressing human SMN. *Human Gene Therapy*, 29(3), 285–298. https://doi.org/10.1089/hum.2018.015

Jinek, M., Chylinski, K., Fonfara, I., Hauer, M., Doudna, J. A., and Charpentier, E. (2012). A programmable dual-RNA-guided DNA endonuclease in adaptive bacterial immunity. *Science*, 337(6096), 816–821. https://doi.org/10.1126/science.1225829

Kaczmarek, J. C., Kowalski, P. S., and Anderson, D. G. (2017). Advances in the delivery of RNA therapeutics: From concept to clinical reality. *Genome Medicine*, 9(1), 60. https://doi.org/10.1186/s13073-017-0450-0

Kaneko, T., Sakuma, T., Yamamoto, T., and Mashimo, T. (2014). Simple knockout by electroporation of engineered endonucleases into intact rat embryos. *Scientific Reports*, 4(1). https://doi.org/10.1038/srep06382

Kim, S., Kim, D., Cho, S. W., Kim, J., and Kim, J. S. (2014). Highly efficient RNA-guided genome editing in human cells via delivery of purified Cas9 ribonucleoproteins. *Genome Research*, 24(6), 1012–1019. https://doi.org/10.1101/gr.171322.113

Kleinstiver, B. P., Pattanayak, V., Prew, M. S., Tsai, S. Q., Nguyen, N. T., Zheng, Z., and Joung, J. K. (2016). High-fidelity CRISPR–Cas9 nucleases with no detectable genome-wide off-target effects. *Nature*, 529, 490–495. https://doi.org/10.1038/nature16526

Koeppel, I., Hertig, C., Hoffie, R., and Kumlehn, J. (2019). Cas Endonuclease technology—A quantum leap in the advancement of barley and wheat genetic engineering. *International Journal of Molecular Sciences*, 20(11), 2647. https://doi.org/10.3390/ijms20112647

Kormann, M. S. D., Hasenpusch, G., Aneja, M. K., Nica, G., Flemmer, A. W., Herber-Jonat, S., Huppmann, M., Mays, L. E., Illenyi, M., Schams, A., Griese, M., Bittmann, I., Handgretinger, R., Hartl, D., Rosenecker, J., and Rudolph, C. (2011). Expression of therapeutic proteins after delivery of chemically modified mRNA in mice. *Nature Biotechnology*, 29(2), 154–157. https://doi.org/10.1038/nbt.1733

Kreppel, F., and Kochanek, S. (2008). Modification of adenovirus gene transfer vectors with synthetic polymers: A scientific review and technical guide. *Molecular Therapy*, 16(1), 16–29. https://doi.org/10.1038/sj.mt.6300321

Li, T., Liu, B., Chen, C. Y., and Yang, B. (2016). TALEN-mediated homologous recombination produces site-directed DNA base change and herbicide-resistant rice. *Journal of Genetics and Genomics*, 43(5), 297–305. https://doi.org/10.1016/j.jgg.2016.03.005

Lisowski, L., Dane, A. P., Chu, K., Zhang, Y., Cunningham, S. C., Wilson, E. M., ... and Kay, M. A. (2014). Selection and evaluation of clinically relevant AAV variants in a xenograft liver model. *Nature*, 506(7488), 382–386.

Liu, J., Li, C., Yu, Z., Huang, P., Wu, H., Wei, C., Zhu, N., Shen, Y., Chen, Y., Zhang, B., Deng, W.-M., and Jiao, R. (n.d.). Efficient and specific modifications of the Drosophila genome by means of an easy TALEN strategy. State Key Laboratory of Brain and Cognitive Science, Institute of Biophysics, The Chinese Academy of Sciences.

Liu, X., LeCluyse, E. L., Brouwer, K. R., and Gan, L. S. (1999). Biliary excretion in primary rat hepatocytes cultured in a collagen-sandwich configuration. *American Journal of Physiology-Gastrointestinal and Liver Physiology*, 277(1), G12–G21.

Lombardo, A., Genovese, P., Beausejour, C. M., Colleoni, S., Lee, Y. L., Kim, K. A., ... and Naldini, L. (2007). Gene editing in human stem cells using zinc finger nucleases and integrase-defective Lenti-Viral vector delivery. *Nature Biotechnology*, 25(11), 1298–1306.

Lundstrom, K. (2018). Viral vectors in gene therapy. *Diseases*, 6(2), 42. https://doi.org/10.3390/diseases6020042

Martin, F., Sánchez-Hernández, S., Gutierrez-Guerrero, A., and Benabdellah, K. (2016). Gene delivery technologies for efficient genome editing: Applications in gene therapy. In M. S. D. Korman (Ed.). *Modern Tools for Genetic Engineering*. IntechOpen.

Mashiko, D., Fujihara, Y., Satouh, Y., Miyata, H., Isotani, A., and Ikawa, M. (2013). Generation of mutant mice by pronuclear injection of circular plasmid expressing Cas9 and single guided RNA. *Scientific Reports*, 3, 3355. https://doi.org/10.1038/srep03355

Mendell, J. R., Campbell, K., Rodino-Klapac, L. R., Sahenk, Z., Shilling, C., Lewis, S., ... and Walker, C. M. (2010). Dystrophin immunity in Duchenne's muscular dystrophy. *New England Journal of Medicine*, 363(15), 1429–1437.

Mitani, K., and Kubo, S. (2002). Adenovirus as an integrating vector. *Current Gene Therapy*, 2(2), 135–144. https://doi.org/10.2174/1566523024605591

Mohan, C., Satish, L., Muthubharathi, B. C., Selvarajan, D., Easterling, M., and Yau, Y.-Y. (2022). CRISPR-Cas technology: A genome-editing powerhouse for molecular plant breeding. In S. Arora, A. Kumar, S. Ogita, and Y-Y. Yau (Eds.), *Biotechnological Innovations for Environmental Remediation* (pp. 803–879). Springer. https://doi.org/10.1007/978-981-16-9001-3_32

Munir, A., Amin, I., Zahoor, M. K., Majeed, H. N., Almoammar, H., Ghaffar, A., and Ahmad, A. (2024). Multiplex genome editing in plants through CRISPR-Cas. In K. A. Abd-Elsalam, A. Ahmad, and B. Zhang (Eds.), *CRISPRized Horticulture Crops* (pp. 127–142). Springer. https://doi.org/10.1016/b978-0-443-13229-2.00012-0

Murlidharan, G., Sakamoto, K., Rao, L., Corriher, T., Wang, D., Gao, G., ... and Asokan, A. (2016). CNS-restricted transduction and CRISPR/Cas9-mediated gene deletion with an engineered AAV vector. *Molecular Therapy-Nucleic Acids*, 5.

Mussolino, C., and Cathomen, T. (2012). TALE nucleases: Tailored genome engineering made easy. *Current Opinion in Biotechnology*, 23(5), 644–650.

Nayerossadat, N., Ali, P., and Maedeh, T. (2012). Viral and nonviral delivery systems for gene delivery. *Advanced Biomedical Research*, 1(1), 27. https://doi.org/10.4103/2277-9175.98152

Pei, G. (2022). Research status of CRISPR/Cas9 delivery system. *Highlights in Science, Engineering and Technology*, 21, 96–100. https://doi.org/10.54097/hset.v21i.3143

Rosi, N. L., Giljohann, D. A., Thaxton, C. S., Lytton-Jean, A. K., Han, M. S., and Mirkin, C. A. (2006). Oligonucleotide-modified gold nanoparticles for intracellular gene regulation. *Science*, 312(5776), 1027–1030.

Sakuma, T., Barry, M. A., and Ikeda, Y. (2012). Lentiviral vectors: Basic to translational. *Biochemical Journal*, 443(3), 603–618. https://doi.org/10.1042/BJ20120146

Shalaby, K., Aouida, M., and El-Agnaf, O. (2020). Tissue-specific delivery of CRISPR therapeutics: Strategies and mechanisms of non-viral vectors. *International Journal of Molecular Sciences*, 21(19), 7353. https://doi.org/10.3390/ijms21197353

Shamshirgaran, Y., Liu, J., Sumer, H., Verma, P. J., and Taheri-Ghahfarokhi, A. (2022). Tools for efficient genome editing; ZFN, TALEN, and CRISPR. *Methods in Molecular Biology*, 2495, 29–46. https://doi.org/10.1007/978-1-0716-2301-5_2

Smith, C., Gore, A., Yan, W., Abalde-Atristain, L., Li, Z., He, C., Wang, Y., Brodsky, R. A., Zhang, K., Cheng, L., and Ye, Z. (2014). Whole-genome sequencing analysis reveals high specificity of CRISPR/Cas9 and TALEN-based genome editing in human iPSCs. *Cell Stem Cell*, 15(1), 12–13.

Stephens, C. J., Lauron, E. J., Kashentseva, E., Lu, Z. H., Yokoyama, W. M., and Curiel, D. T. (2019). Long-term correction of hemophilia B using adenoviral delivery of CRISPR/Cas9. *Journal of Controlled Release,* 298, 128–141. https://doi.org/10.1016/j.jconrel.2019.02.009

Sultan, Q., Ashraf, S., Munir, A., Khan, S. H., Munawar, N., Abd-Elsalam, K. A., and Ahmad, A. (2022). Beyond genome editing: CRISPR approaches. In A. Ahmad, S. H. Khan, and Z. Khan (Eds.), *The CRISPR/Cas Tool Kit for Genome Editing*, pp.187–218. Springer Nature. https://doi.org/10.1007/978-981-16-6305-5_6

Sun, D., Hussain, H. I., Yi, Z., Rookes, J. E., Kong, L., and Cahill, D. M. (2016). Mesoporous silica nanoparticles enhance seedling growth and photosynthesis in wheat and lupin. *Plant Physiology and Biochemistry*, 108, 459–470. https://doi.org/10.1016/j.plaphy.2016.08.012

Taylor, N.J. and Fauquet, C.M. (2002) Microparticle bombardment as a tool in plant science and agricultural biotechnology. *DNA and Cell Biology*, 21, 963–977. nhttp://dx.doi.org/10.1089/104454902762053891

Tombácz, I., Laczkó, D., Shahnawaz, H., Muramatsu, H., Natesan, A., Yadegari, A., ... and Parhiz, H. (2021). Highly efficient CD4+ T cell targeting and genetic recombination using engineered CD4+ cell-homing mRNA-LNPs. *Molecular Therapy*, 29(11), 3293–3304.

Trevisan, M., Masi, G., and Palù, G. (2020). Genome editing technologies to treat rare liver diseases. *Translational Gastroenterology and Hepatology*, 5, 23–23. https://doi.org/10.21037/tgh.2019.10.10

Urnov, F. D., Rebar, E. J., Holmes, M. C., Zhang, H. S., and Gregory, P. D. (2010). Genome editing with engineered zinc finger nucleases. *Nature Reviews Genetics*, 11(9), 636–646. https://doi.org/10.1038/nrg2842

Varotto, S. (2023). Current status and future prospective of genome editing application in maize. In A. Richroch et al. (Eds.), *A Roadmap for Plant Genome Editing* (pp. 165–182). Springer Nature.

Wang, H., Park, H., Liu, J., and Sternberg, P. W. (2018). An efficient genome editing strategy to generate putative null mutants in Caenorhabditis elegans using CRISPR/Cas9. *G3: Genes, Genomes, Genetics*, 8(11), 3607–3616. https://doi.org/10.1534/g3.118.200662

Wang, W., Li, W., Ma, N., Steinhoff, G., and Lv, C. (2015). Efficient genetic modification of mouse embryonic stem cells by plasmid transfection. *Journal of Biotechnology,* 208, 32–40.

Wani, A. K., Akhtar, N., Singh, R., Prakash, A., Raza, S. H. A., Cavalu, S., Chopra, C., Madkour, M., Elolimy, A., and Hashem, N. M. (2022). Genome centric engineering using ZFNs, TALENs and CRISPR-Cas9 systems for trait improvement and disease control in Animals. *Veterinary Research Communications*, 47(1), 1–16. https://doi.org/10.1007/s11259-022-09967-8

Xiao, Q., Chen, S., Wang, Q., Liu, Z., Liu, S., Deng, H., Hou, W., Wu, D., Xiong, Y., Li, J., and Guo, D. (2019). CCR5 editing by Staphylococcus aureus Cas9 in human primary CD4+ T cells and hematopoietic stem/progenitor cells promotes HIV-1 resistance and CD4+ T cell enrichment in humanized mice. *Retrovirology*, 16(1). https://doi.org/10.1186/s12977-019-0477-y

Xu, H., Xiao, T., Chen, C. H., Li, W., Meyer, C. A., Wu, Q., Wu, D., Cong, L., Zhang, F., Liu, J. S., Brown, M., and Liu, X. S. (2015). Sequence determinants of improved CRISPR sgRNA design. *Genome Research*, 25(8), 1147–1157. https://doi.org/10.1101/gr.191452.115

Yang, W., Yan, J., Zhuang, P., Ding, T., Chen, Y., Zhang, Y., Zhang, H., and Cui, W. (2022). Progress of delivery methods for

CRISPR-Cas9. *Expert Opinion on Drug Delivery*, 19(8), 913–926. https://doi.org/10.1080/17425247.2022.2100342

Zhang, S., Shen, J., Li, D., and Cheng, Y. (2021). Strategies in the delivery of Cas9 ribonucleoprotein for CRISPR/Cas9 genome editing. *Theranostics*, 11(2), 614–648. https://doi.org/10.7150/thno.47007

Zhang, X., Dong, X., Sawyer, G. J., Collins, L., and Fabre, J. W. (2004). Regional hydrodynamic gene delivery to the rat liver with physiological volumes of DNA solution. *The Journal of Gene Medicine*, 6(6), 626–632. https://doi.org/10.1002/jgm.595

Zhang, X. R., He, J. B., Wang, Y. Z., and Du, L. L. (2018). A cloning-free method for CRISPR/Cas9-mediated genome editing in fission yeast. *G3: Genes, Genomes, Genetics*, 8(6), 2067–2077. https://doi.org/10.1534/g3.118.200164

Zufferey, R., Dull, T., Mandel, R. J., Bukovsky, A., Quiroz, D., Naldini, L., and Trono, D. (1998). Self-inactivating lentivirus vector for safe and efficient in vivo gene delivery. *Journal of Virology*, 72(12), 9873–9880. https://doi.org/10.1128/JVI.72.12.9873-9880.1998

Part II

CRISPR-Cas Systems and Their Applications

6 CRISPR/Cas9

The Most Widely Adopted Tool for Genome Editing

Ruben D. Arroyo-Olarte, Karla Daniela Rodríguez-Hernández, Karen Betsabe Miranda Ortega, and Ricardo Reyes-Chilpa

6.1 CRISPR-CAS: AN ADAPTIVE IMMUNE SYSTEM AGAINST FOREIGN GENETIC ELEMENTS

CRISPR/Cas systems are the new generation of genome-editing tools, recognized with the Nobel Prize in Chemistry in 2020. CRISPRs are regions in the bacterial genome that help defend against plasmids and other non-native genetic materials (viruses). They are present in about 50% of bacteria and 90% of archaea and have two components: a protein called CAS for cutting DNA and a guide RNA molecule (gRNA) which together form a complex that recognizes and cuts specific segments of DNA. The editing process involves several stages to insertion or deletion of DNA, or to induce changes and replace a native gene with specificity and efficiency (Arroyo-Olarte et al., 2023). The CRISPR immune response in prokaryotes consists of three basic steps (Kumar, Shumayla, & Upadhyay, 2024):

Step 1 Adaptation: This is the process where an acquiring foreign DNA fragment from an invading virus is processed into short segments that are inserted into the CRISPR sequence as new spacers (protospacers) and then are used to produce RNAs guide that specifically target the related foreign nucleic acid.

Step 2 Interference: Repeats and spacers are produced into the bacterial DNA undergoing transcription, copying DNA to the RNA chain, then it is cut into short pieces called CRISPR-RNAs (crRNAs) and used as guides to recognize the target DNA or RNA, which is then cleaved by the nuclease moiety of the CRISPR effector complex.

Step 3 Biogenesis, expression and maturation: In this step, the long transcript of the pre-CRISPR crRNA matrix is processed to produce mature and functional crRNA, it guides the bacterial molecular machinery to destroy the viral material and serves as an excellent guide.

6.2 THE CRISPR-CAS SYSTEMS

CRISPR-Cas systems' stages are mediated by a distinct set of Cas proteins and several accessory genes that comprise functional modules. According to the structure, the number of subunits and the function of the Cas protein, the CRISPR-Cas systems can be categorized into two main classes depending on the number of effector proteins or nucleases (Arroyo-Olarte et al., 2023).

The Class I system has a multi-subunit effector protein and thus is more complex; the Class II system has a single-subunit effector protein. The classification, based on the type of effector Cas proteins and/or modules, is further divided into types and subtypes. Class-I CRISPR system, types I, III, and IV use Cas3, Cas1, and CsfI as effector proteins, respectively (Zakrzewska & Burmistrz, 2023).

Class II, types II, V, and VI use Cas9, Cas12, and Cas13 proteins as effector modules, respectively. Another category is about their functional mechanisms, divided into their subtypes. CRISPR/Cas systems, type I and subtypes III-A and III-B are called CRISPR-associated complex for antiviral defense (Cascade) complex including Cas3, Casm, and Cmr, respectively. Types II, III, and V, single multidomain proteins including Cas9, Cas10, and Cpf1, respectively, perform the expression, maturation, and interference stages (Arroyo-Olarte et al., 2023).

An example is the very conserved CRISPR-Cas systems of Xanthomonas species, an important genus of plant-associated bacteria that causes significant yield losses of economically important crops worldwide. The CRISPR-Cas loci and spacers are viruses (phages), a few are plasmids, and are the main mechanism to control the invasion of foreign DNA. The regulation of CRISPR-Cas is by the diffusible signal factor (DSF) and the quorum sensing signal molecule (QS), mainly when the cell density increases, during phage infection or due to environmental stresses like starvation, avoiding a possible reduction in bacterial fitness (Prajapati & Tyagi 2024). Cas proteins in different species of Xanthomonas are the subtype I-C and I-F CRISPR-Cas systems, an ancient acquisition throughout evolution, according to the cost-benefit of protecting against biological threats and fitness maintenance in challenging conditions and might be the specific key to understand the plant-pathogen interactions (Martins et al., 2024).

6.3 MOLECULAR MECHANISM OF CRISPR-CAS9

The CRISPR-Cas9 system as a genome-editing tool, can generally be divided into three steps: recognition, cleavage, and repair. They consist of several conformational dynamic motions of Cas apoproteins for binding and accommodating

DOI: 10.1201/9781003500933-8

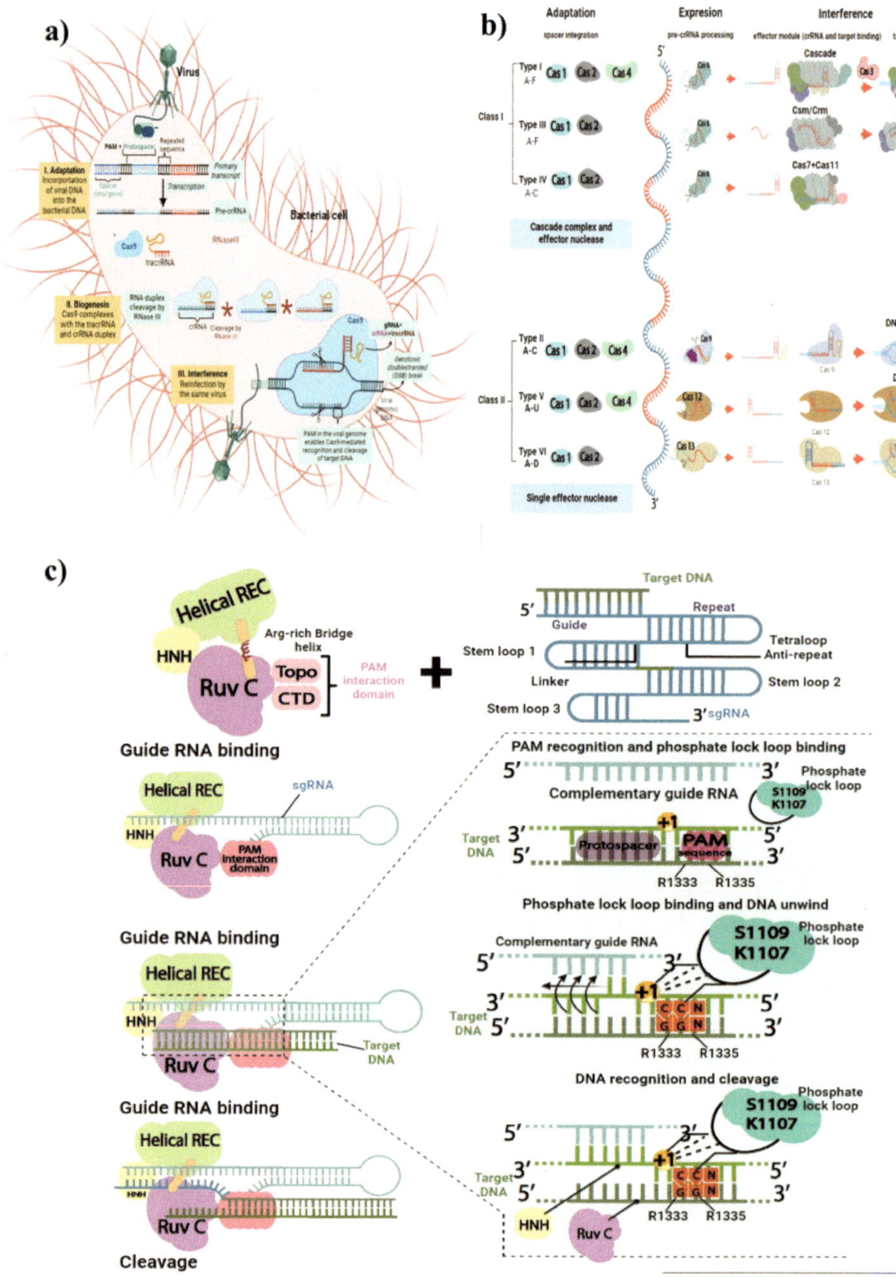

FIGURE 6.1 (a) The three stages of the CRISPR-Cas. *Adaptation*: A viral DNA is injected into the bacterial genome through the phage to bacterial cell activates the adaptation module proteins (PAM-protospacer), these proteins excise spacer-sized fragments of phage DNA (protospacers that are adjacent to a PAM in the viral genome) for incorporation into CRISPR loci. *Interference:* individual gRNAs are bound by Cas protein effectors (e.g., Cas9). Cas-crRNA complex is identified as a complementary sequence adjacent to the protospacer adjacent motif (PAM). *Biogenesis:* CRISPR loci are transcribed, and pre-crRNA is processed by a Cas9/RNase III complex at repeat sequences to generate mature crRNAs that couple to tracrRNA (gRNA). The invading DNA is cleaved by Cas-nuclease to prevent subsequent infection, and after a new phage infection with protospacer (and PAM) sequences matching a crRNA appears in the cell, specific Cas/gRNA complexes bind to viral DNA and cleave it. (b) Classification and functional modules of CRISPR-Cas systems. Class I is composed of multiple Cas proteins and Class II a single multidomain protein forms a crRNA-binding complex. (c) Molecular mechanism of target DNA recognition and cleaving by Cas9.

RNA and DNA nucleic acids. The structure of the Cas9 endonuclease is composed of two lobes: The recognition (REC) lobe accommodates target nucleic acid sequences. The nuclease lobe performs DNA cleavages through the catalytic HNH and RuvC domains, and the HDR domain repairs and executes the precise gene insertion or replacement by adding a donor DNA template with sequence homology at the predicted double-stranded break (DSB) site by the host cellular machinery (Asmamaw & Zawdie 2021).

The guide RNA binding recognition and cleavage of dsDNA strictly require the presence of a protospacer adjacent motif (PAM). Cas9 results in the formation of a PAM binding site (PAM-interaction domain), conformed by TOPO and CTD subdomains. Cas9-sgRNA engages the PAM GG dinucleotide using Arg1333 and Arg1335, and positions the target DNA duplex such that the +1 phosphate interacts with the phosphate lock loop, resulting in local strand separation immediately upstream of the PAM. The base-pairing between the displaced target DNA strands and the seed region of the guide RNA promotes further stepwise strand displacement and propagation of the guide-target heteroduplex. The endonuclease domains of Cas9 (HNH and RuvC) generate site-specific breaks in the double-stranded DNA (dsDNA) target. The non-target DNA strand depends on the base-pair complementarity of the target DNA strand (tDNA) to the RNA guide template, SpCas9 adopts an overall bilobed architecture, in which the sgRNA: target DNA heteroduplex resides within a central channel between the α-helical recognition (REC). The REC lobe englobes two metal-ion-dependent nuclease domains, termed HNH and RuvC, which are responsible for cleaving the Tdna (Arroyo-Olarte, Rodríguez-Hernández, & Morales-Ríos, 2023) (see Figure 6.1).

6.4 GENOME-EDITING VIA CRISPR-CAS9

6.4.1 DNA Repair Mechanisms and Their Applications for CRISPR-Cas9 Genome Editing

The Cas9 protein from *Streptococcus pyogenes* (SpCas9) is the most widely employed CRISPR-based gene-editing tool. This is mostly due to its relatively common PAM sequence requirement: NGG (where N can be any nucleotide), with a theoretical frequency of once every 8 bp for a random double-strand DNA sequence. The actual frequency of the PAM motif will vary across genomes and is expected to be much rarer in AT-rich genomes. For example, NGG occurs once every 42 bases in the human genome (Scherer, 2008). New Cas9 variants with less stringent or even absent PAM requirements have been generated, to expand the target site recognition of CRISPR-Cas9 methods(Walton et al., 2020). The crRNA and tracrRNA are usually fused into a single-guide RNA molecule (sgRNA) with the same activity. An optional element of the CRISPR-Cas systems is the recombination template (RT), composed of 5' and 3' homology arms, the desired edit (insertion, deletion, or specific mutation) and an internal sequence that disrupts the PAM target site, to prevent re-targeting of the successfully edited genome. Cleavage by CRISPR nucleases in the absence of successful recombination is often lethal in bacteria due to the formation of a double-strand break (DSB), and it serves as a powerful counterselection avoiding the necessity of inserting a large antibiotic-resistance cassette marker into the genome. In this way the DSB drives editing through homologous recombination (HR) or via non-homologous end joining (NHEJ) (Shuman & Glickman, 2007). It is, therefore, the DNA repair systems of the host species/strain which actually perform the desired editing. In Figure 6.2B, we show the DNA repair mechanisms and their function during CRISPR-Cas9 genome editing. It is important to note that multiple sgRNAs with different specificities can direct Cas9 to the corresponding sites in the same cell or individual, this is known as multiplex editing (Cong et al., 2013). Multiplex editing holds many applications, such as engineering of functionally related genes to control metabolic pathways or complex features (Ma et al., 2015; Xie et al., 2015).

6.4.2 CRISPR-Cas9 Delivery Methods: Advantages and Disadvantages

Traditionally CRISPR-Cas9 machinery is encoded in plasmids, following either single- or dual-vector strategy for Cas9 nuclease, guide RNA and/or, in the case of homology-directed repair, a DNA template encoding desired mutations.

Plasmid transfection and genome-editing efficiency in plants are difficult due to several factors such as: large plasmid size (the Cas9 gene alone is over 4 Kb), thick cellulose cell walls and polyploidy. Therefore, traditional chemical transfection (e.g., via polyethylene glycol, PEG) and electroporation require protoplast generation.

A more efficient physical alternative to traditional electroporation and much more commonly used for genome engineering in plants is biolistics (particle bombardment).

However, one of the most commonly used transgene delivery methods in plants and that is currently being applied for CRISPR-Cas9 machinery delivery is transformation with *Agrobacterium tumefaciens* in susceptible plant species (e.g., leguminous plants). Similarly viral transfection can also be performed. In Figure 6.2B we show the main CRISPR-Cas9 delivery methods with particular emphasis in plants.

6.4.3 Drawbacks of CRISPR-Cas9-mediated Genome Editing and Current Alternatives: Not All CRISPRs Are the Same

The delivery of CRISPR machinery in plants is usually performed via *Agrobacterium*- or particle bombardment-mediated transformation (Figure 6.2B). Although these are effective and well-characterized approaches, the random integration of Cas9 and gRNA cassettes into plant genomes raises several biosafety concerns such as, but not limited to, regulatory bans against transgenic plants. CRISPR transgenes can be eliminated through breeding, however, this process is time-consuming and labor-consuming, and is not viable for species with long juvenile phases (e.g., trees) and for those which are only vegetatively propagated (e.g., bananas). Another alternative is plasmid-based transformation, which can avoid genome integration, but requires efficient and species-specific promoters and expression systems, particularly for multiplexing which is a must for polyploid plants where more than one gRNA is needed for complete target gene(s) inactivation. On the other hand, continuous overexpression of SpCas9 can be cytotoxic in several species across the realms of life, such as bacteria, fungi, plants, and mammals (Foster

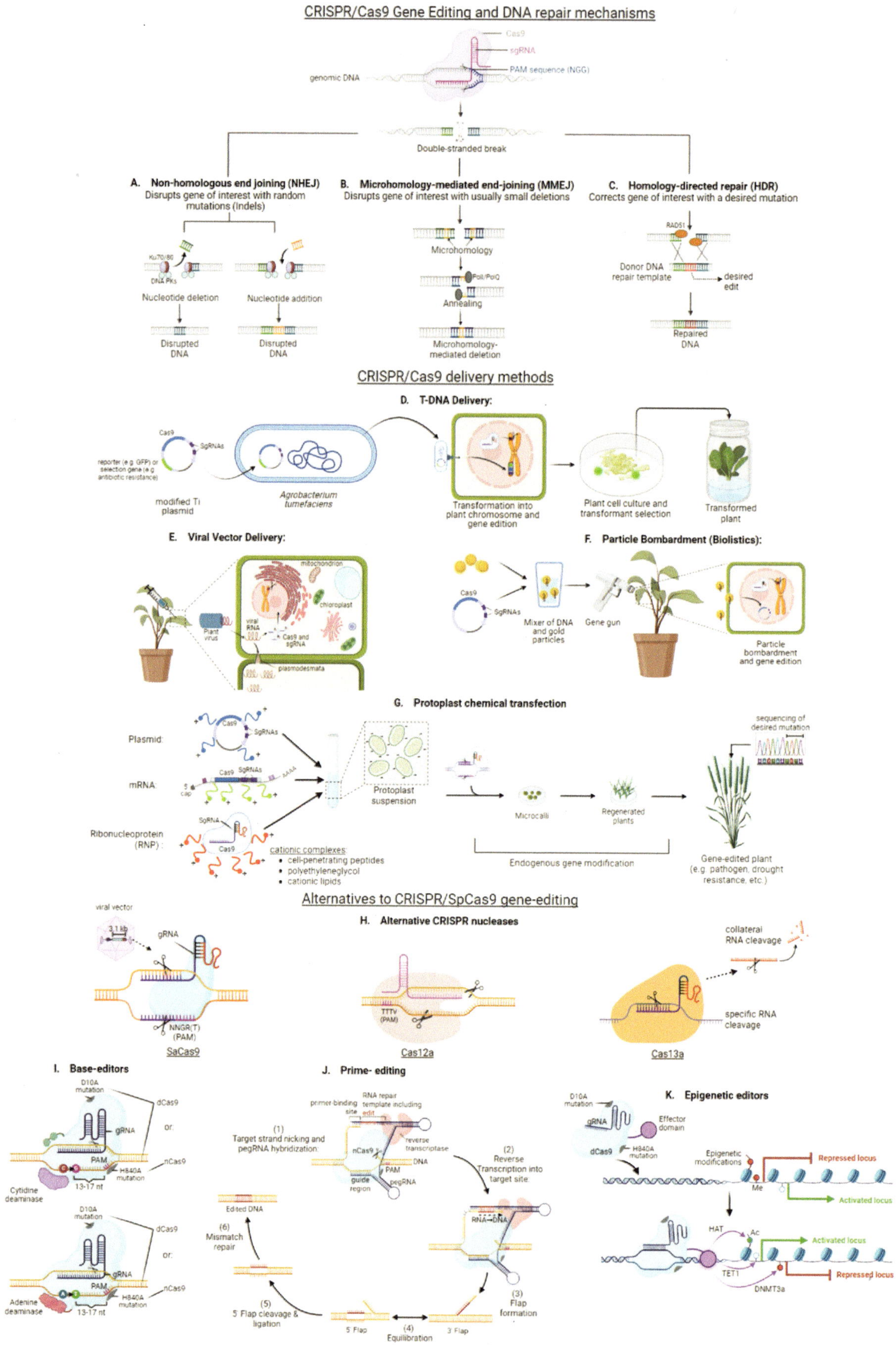

FIGURE 6.2 (A–C) CRISPR/Cas9 genome editing and DNA repair mechanisms. (D–G) Main CRISPR/Cas9 delivery methods, with particular emphasis in plants. (H–K) Current alternative approaches to CRISPR/SpCas9-mediated genome editing.

et al., 2018; W. Jiang et al., 2014; Morgens et al., 2017). Particularly for higher organisms such as mammals, and to a lesser extent in plants, there is an additional risk associated with the pre-existing immunological response against the bacterial components of the CRISPR machinery, especially during the continuous expression of SpCas9 associated with plasmids and genome integration (Charlesworth et al., 2019).

To cope with the risks associated with SpCas9 expression and/or genome integration, Cas9/gRNA ribonucleoprotein (RNP) transfection offers multiple advantages. The first advantage is that we can screen in vitro multiple guide RNAs for their efficiency prior to transfection and in vivo genome editing. Additionally, the RNPs remain only for a couple of days within the cell to perform the genome edition before being degraded by the proteasomal machinery, avoiding most cytotoxic effects, and limiting to a great extent the so-called off-target events of genome editions beyond the intended target gene, via an incomplete complementarity of the guide RNA to other genome regions.

There are still other disadvantages posed by SpCas9 that cannot be resolved with the RNP approach, including its great size (4.2 kb) which makes difficult its insertion into viral vectors. In this regard smaller Cas9 orthologs from other species have been characterized to be used with the same efficiency as SpCas9. Cas9 form *Staphylococcus aureus* (SaCas9, 3.1 Kb) is currently the most used and it has been tested in multiple organisms such as protozoa (Medeiros et al., 2017), plants (Jia et al., 2017), and mammals (Hunker et al., 2020; Yin et al., 2017).

Another issue with Cas9 in general is its dependence on the host cell DNA repair machinery for a correct and efficient genome edition. It has been shown that these mechanisms can be optimized with single-strand breaks rather than with the typical blunt double-strand break. The SpCas9-RuvC domain has been mutagenized (D10A) to function as a DNA nickase to produce single-strand breaks instead of the potentially lethal DSB. The resulting nicking SpCas9 (nCas9) has been useful as a genome-editing tool in cases where transformation with SpCas9 plasmids leads to no colonies, especially for large-scale genome deletions (Standage-Beier et al., 2015). Due to the non-lethal nature of single-strand breaks, nCas9 is not a counter-selection tool, often leading to low efficient genome editing (McAllister et al., 2017; Song et al., 2017). Alternatively, two gRNAs targeting adjacent genome regions at opposing DNA strands can be used with nCas9 to generate a staggered double-strand break (Ran et al., 2013). This approach would be more specific and less susceptible to off-target edits, but also has more requirements including the presence of the PAM in both strands and in close proximity and finding two gRNAs with high activity targeting a short sequence of base pairs. In *Corynebacterium glutamicum*, where neither exogenous SpCas9 nor nCas9 expression is possible due to toxicity and consequent plasmid loss, this issue has been solved by inserting the SpCas9 into the genome under the control of a native promoter. The resulting strain showed a low rate of escape colonies and a high gene-editing efficiency when transformed with a plasmid encoding a specific gRNA and recombination template (Cameron Coates et al., 2019).

The continuous research in the CRISPR gene-editing systems has created new CRISPR tools such as natural double-nicking CRISPR nucleases like the Type V-A Cas12a (also known as Cpf1) from *Francisella novicida* (FnCas12a) and they have also been characterized and studied as an alternative for genetic engineering. Cas12a orthologs require one single gRNA and are usually smaller than SpCas9, recognizing a T-rich PAM and introducing a 5′ 5-nt overhang upon DNA cleavage (Zetsche et al., 2015) (Figure 6.2H). As an alternative to Cas9, in bacteria such as *Corynebacterium glutamicum*, a FnCas12a-encoding plasmid could be successfully transformed and used for genome editing (Y. Jiang et al., 2017). Cas12a has since then been applied for genome editing in other bacteria, for example, *Yersinia pestis* and *Mycobacterium smegmatis* (Yan et al., 2017). In plants, grapevine genome has been successfully edited with the Cas12a ortholog from Lachnospiraceae bacterium (LbCas12a) (Ren et al., 2023). Rice and citrus protoplasts have also been gene-edited via Cas12 orthologs (Zhang et al., 2022).

Another CRISPR nuclease is Cas13a, which has no need for a PAM but targets RNA, having potential applications as a mRNA downregulator. However, Cas13a, at least in eukaryotic cells, has a non-specific activity once it gets activated by specific gRNA binding to target sequence (Figure 6.2H). This unspecific activity has recently had applications as an RNA virus detection tool with the so-called SHERLOCK system (Kellner et al., 2019).

One of the most recent advances in CRISPR-based genome editing is the base-editor (Figure 6.2I). Base-editors can perform specific single-nucleotide edits without a double-strand break or recombination template. The most extended base-editor, BE3, is composed of a chimera of nCas9 or dCas9 to provide strong, specific gRNA-programmable DNA binding, and cytidine-deaminases, for example, APOBEC1, to conduct C to T editing in the target gene (Komor et al., 2016) (Figure 6.2I). Other variants of the system include an adenine-deaminase (A to G conversion) instead of a cytidine-deaminase (Gaudelli et al., 2017) or Cas12a instead of nCas9, which has recently been applied in rice to perform efficient multiplex genome editing (Cheng et al., 2023). The advantages of this system include its relative innocuity compared to Cas9-induced DSBs, and its independence from recombination machinery to introduce specific single-nucleotide mutations in a target gene. However, transcriptome-wide off-target RNA editing activity has been demonstrated during the continuous expression of base-editors, particularly those based on cytidine-deaminases in mammalian and plant cells (Xin et al., 2019). Similarly, embryonic cells expressing base-editors show a higher-than-normal frequency of single-nucleotide polymorphisms (Zuo et al., 2019). These reports are consistent with the well-documented anti-DNA and anti-RNA virus replication activity of cytidine-deaminases like APOBEC1 and APOBEC3G, mainly through hypermutating viral genomes (Gee et al., 2011; Ikeda et al., 2008; Noguchi et al., 2005; Sawyer et al., 2004).

Although base-editors are optimized for single-base edits, replacing larger stretches of genomic DNA by inserting sequences such as an epitope tag or a deletion usually requires a foreign DNA donor to repair a Cas9-induced DSB. Prime editing (Figure 6.2J) avoids these requirements by combining an nCas9 and a reverse transcriptase (RT) that uses a pegylated guide RNA (pegRNA) that works as both a guide RNA and as a template for RT to generate a desired DNA sequence to be integrated in the target locus. Notably, the PAM requirement (NGG) in prime-editing is rather relaxed as a target site as far as 29 bp downstream of a PAM can be efficiently edited (Anzalone et al., 2019). Prime editing also shows higher or similar efficiency and fewer byproducts than homology-directed repair and induces much lower off-target editing than Cas9 nuclease at known Cas9 off-target sites in human cells (Anzalone et al., 2019). Prime editing has also been successfully applied in mice (Liu et al., 2020) and plants (Lin et al., 2020), although its extreme large size (about 7000 bp) may be a limiting factor for its regular use as a genome editor.

Finally, sometimes it is desirable to go beyond genome editing and instead manipulate the natural gene-expression control mechanisms of the cell, such as DNA methylation and histone acetylation. In this regard, epigenetic CRISPR editors are more likely to be approved by heavy transgenic legislation and still have significant effects turning on and off relevant genes for plant growth, pathogen immunity, climate resistance, etc. In this case, dCas9 is combined with a guide RNA complementary to the target regulatory region (e.g., a gene promoter) and an effector protein domain with either DNA methylase (e.g., DNMT3a), demethylase (e.g., TET1) or histone acetyltransferase (e.g., HAT p300) activity, to perform inheritable epigenetic editing (Figure 6.2K). The proof-of concept for epigenetic editors in plants has been demonstrated by recent works in the plant model *Arabidopsis thaliana* (de Melo et al., 2020; Gallego-Bartolomé et al. 2018; Papikian et al., 2019) (see Figure 6.2).

6.5 CURRENT APPLICATIONS OF CRISPR-CAS9 GENOME EDITING

Genome editing via CRISPR-Cas9 is still the main application in plants and other eukaryotic and prokaryotic organisms. However as shown in Figure 6.2H, the CRISPR toolbox has expanded beyond traditional SpCas9. In Table 6,1 we show some of the main applications of CRISPR-Cas-mediated genome editing in some major biological groups, including animals, protozoa, bacteria, and particularly in plants.

TABLE 6.1
Strategies for CRISPR-mediated genome-editing

Strategies for editing	Organism	Results	Efficiency (%)	References
Genome editing by using CRISPR-Cas9 system	*Plasmodium falciparum*	Transgenic parasites constitutively expressing Cas9 were transfected with a plasmid containing repair DNA sequence and sgRNA. Experimental validation of this method was done by site-directed mutagenesis of kelch13, a gene involved in artemisinin resistance.	100	(Kirti et al., 2021)
	Trypanosoma cruzi	CRISPR-Cas9 system was used for the genetic engineering of *T. cruzi*, demonstrating rapid and efficient knockout of multiple endogenous genes, including essential genes		(Peng et al., 2015) (Kirti et al., 2021)
	Trypanosoma cruzi	They used either vectors containing single guide RNA (sgRNA) and Cas9 or one vector containing sgRNA and Cas9 plus donor DNA for homologous recombination to rapidly generate mutant cell lines in which the PFR1, PFR2, and GP72 genes have been disrupted key components of the parasite flagellum		(Lander et al., 2015) (Kirti et al., 2021)
	Trypanosoma cruzi *Trypanosoma brucei*	They show for the first time that RNP complexes of Cas9/gRNA can be delivered into epimastigotes and trypomastigotes of *T. cruzi*, with or without single-stranded repair oligonucleotides, to knock out or tag target genes with very high efficiency		(Soares, Medeiros et al., 2017) (Kirti et al., 2021)

TABLE 6.1 (Continued)
Strategies for CRISPR-mediated genome-editing

Strategies for editing	Organism	Results	Efficiency (%)	References
	Trypanosoma cruzi (Dm28c)	With assessment by western blot and fluorescence assays, they demostrated that the Cas9 was indeed being expressed and translocated into the cell nucleus and that showed that is possible to reduce costs and to maintain a highly efficient CRISPR/Cas9 system to disrupt genes in *T. cruzi*		(Romagnoli et al., 2018) (Kirti et al., 2021)
	Trypanosoma cruzi	They demonstrate that extracellular amastigotes (EA) can replicate free of host cells and can be utilized in a variety of assays, including exogenous gene expression and CRISPR/Cas9-mediated gene knockout		(Takagi et al., 2019) (Kirti et al., 2021)
	Leishmania major *Leishmania mexicana* *Trypanosoma brucei*	They developed a new protocol for single-guide RNA (sgRNA) delivery using PCR-generated DNA templates which are transcribed in vivo by T7 RNA polymerase and an online resource (LeishGEdit.net) for automated primer design for functional studies of proteins and its localization screening after its transfection in promastigotes and bloodstream forms	100	(Beneke et al., 2017) (Kirti et al., 2021)
	Trypanosoma brucei.	They use a new system for homology-directed, precision base editing; a single-stranded oligo deoxyribonucleotide repair template, delivered to introduce a single (aquaglyceroporin) AQP2 – T791G/L264R mutation in *T. brucei* that would benefit from temporal control, to aid high efficiency and precision genome editing	~ 100	(Rico et al., 2018) (Kirti et al., 2021)
	Babesia bovis	They designed a basic plasmid expressing Cas9 and human dihydrofolate reductase (hDHFR) for *B. bovis* to evaluate its usefulness and efficiency of epitope tagging, introduction of a point mutation, and gene replacement. This advance will aid functional analysis of the genome and underpin the characterization of candidate drug and vaccine targets.		(Hakimi et al., 2019) (Kirti et al., 2021)
	Trichomonas vaginalis	Is the first report on the use of the CRISPR/Cas9 system in *T. vaginalis* demostrating Cas9-gRNA-directed modification of a nanoluciferase reporter gene using homology-directed repair and expands the ability to manipulate the genome of this pathogen		(Janssen et al., 2018) (Kirti et al., 2021)
The native type I-E CRISPR-Cas system	*Lactobacillus crispatus*	They detail how the native type I-E CRISPR-Cas system, with a 5′-AAA-3′ protospacer adjacent motif and a 61-nucleotide guide crRNA, can be repurposed for efficient chromosomal targeting and genome editing in L. crispatus, with potential for probiotic enhancement, biotherapeutic engineering, and mucosal vaccine delivery	100*, 36**, 19***	(Hidalgo-Cantabrana et al., 2019) (Sen & Mukhopadhya& 2024)

(*continued*)

TABLE 6.1 (Continued)
Strategies for CRISPR-mediated genome-editing

Strategies for editing	Organism	Results	Efficiency (%)	References
CRISPR-Cas and PRESA (phage-delivered resistance eradication with subsequent antibiotic)	*Escherichia coli*	The CRISPR-Cas9 system was recombined with a newly isolated E.coli lysogenic phage and coupled with a conventional antibiotic. Results were the transfer of antibiotic-resistant plasmids were blocked - no resistant mutants were generated in the process.		(Liu et al., 2020) (Sen & Mukhopadhyay 2024)
CRISPR-based antimicrobials	*Enterococcus faecalis*	This study showed the presence of two CRISPR-Cas systems in E. faecalis in agricultural., Results provided that this system contributes to transfer antibiotic resistance genes via plasmids and acted as a barrier to the spread of antibiotic-resistant genes in the manure environment.		(Upreti et al., 2023) (Sen & Mukhopadhyay 2024)
Type I-F CRISPRi (CSYi) gene silencing system	*Pseudomonas aeruginosa*	They optimized a gene silencing system based on Type I-F CRISPRi (CSYi) for the characterization of antibiotic-resistant genes in clinical strains of P. aeruginosa		(S. Chen et al., 2023) (Sen & Mukhopadhyay 2024)
The endogenous type I-B CRISPR-Cas system	*Clostridioides difficile*	They show that the endogenous type I-B CRISPR-Cas system in C. difficile can be repurposed as an antimicrobial agent by the expression of a self-targeting CRISPR that redirects endogenous CRISPR-Cas3 activity against the bacterial chromosome		(Selle et al., 2020) (Sen & Mukhopadhyay 2024)
Nonviral Genome Editing Based on a Polymer-Derivatized CRISPR Nanocomplex	*Staphylococcus aureus*	They show that nanosized CRISPR complexes (Cr-Nanocomplex) were successfully formed while maintaining the functional activity of Cas9 endonuclease to induce double-strand DNA cleavage, and demonstrated this Cr-Nanocomplex can be efficiently delivered into Methicillin-resistant S. aureus (MRSA), and allow the editing of the bacterial genome		(Kang et al., 2017) (Sen & Mukhopadhyay 2024)
CRISPR-Cas9 by engineered probiotics	*Escherichia coli Citrobacter rodentium*	They generated a potent conjugative delivery vehicle for CRISPR-cas9 that can eliminate > 99.9% of targeted antibiotic-resistant E. coli in the mouse gut microbiota using a single dose and then applied this system to a C. rodentium infection model, achieving full clearance		(Neil et al., 2021) (Sen & Mukhopadhyay 2024)
toxin-antitoxin (TA) and CRISPR-Cas	*Acinetobacter*	Their study demonstrates a proof-of-concept antimicrobial strategy named ATTACK, which AssociaTes TA and CRISPR-Cas to Kill multidrug-resistant (MDR) pathogens		(R. Wang et al., 2023) (Sen & Mukhopadhyay 2024)
CRISPR/Cas9 System for Plasmid Elimination and Bacterial Killing	*Bacillus anthracis Bacillus cereus*	The CRISPR/Cas9 system was used to eliminate native plasmids from *B. anthracis* and *B. cereus*, and specifically killed *B. anthracis*. They found that the plasmid elimination percentage was slightly higher when the sgRNA targeted the replication initiation region (*), rather than the non-replication initiation region (**).	96–100* 88–92**	(X. Wang et al., 2021) (Sen & Mukhopadhyay 2024)

TABLE 6.1 (Continued)
Strategies for CRISPR-mediated genome-editing

Strategies for editing	Organism	Results	Efficiency (%)	References
Genome editing by using CRISPR-Cas9 system	*Klebsiella pneumoniae*	They examined the drug-resistant characteristics of K. pneumoniae clinical isolates using cas1 gene. In addition, the study also discovered differences in resistance between two groups of strains that the I-Eb group had significantly lower resistance rates to several antibiotics, than the I-Ea group.		(Li, 2025)
CRISPR-Cas9-based vectors generated single and simultaneous multiple gene knockouts	*Oryza glaberrima* *Oryza sativa*	They developed protocols for genetically transforming African landraces (*O. glaberrima* y *O. sativa*) to allow the use of CRISPR-Cas mediated breeding approaches. The system CRISPR-Cas9-based vectors generated single and simultaneous multiple gene knockouts. This resulted in mutants with significantly improved seed yield.		(Lacchini et al., 2020) (Kumar, Shumayla, & Upadhyay 2024)
base editor base on CRISPR/Cas9	*winter wheat variety Kenong199* and *the Japonica rice variety Zhonghua11*	Here, they describe a new plant adenine base-editor based on an evolved tRNA adenosine deaminase fused to the nickase CRISPR/Cas9, enabling A•T to G•C conversion at frequencies up to 7.5% in protoplasts and 59.1% in regenerated rice and wheat plants. With this new adenine base-editing system, it is now possible to precisely edit all base pairs.		(Chen et al., 2018) (Kumar, Shumayla, & Upadhyay 2024)
CRISPR/Cas9 Mediated Genome Editing	*the bread wheat variety Kenong 199 and YZ814*	They demonstrated that CRISPR/Cas9 RNPs delivered into immature embryo cells of bread wheat by particle bombardment were effective in performing targeted genome editing		(Z. Liang et al., 2017) (Kumar, Shumayla, & Upadhyay 2024)
	Japonica rice (*Oryza sativa*) variety Zhonghua11 and the winter wheat variety Kenong199	Here, they adapted prime-editors for use in plants through codon, promoter, and editing-condition optimization. Regenerated prime-edited rice plants were obtained at frequencies of up to 21.8%.		(Lin et al., 2020) (Kumar, Shumayla, & Upadhyay 2024)
	Nicotiana benthamiana, Arabidopsis thaliana and Oryza sativa	In this study, developed and tested a multifaceted multiplex CRISPR/Cas9 toolbox that consists of 37 Golden Gate- and Gateway-compatible vectors. They successfully activated and repressed transcription of both protein-coding and noncoding genes and imprinted genes in Arabidopsis.		(Lowder et al., 2015) (Kumar, Shumayla, & Upadhyay 2024)
	Rice ZH11	Here they used chemically modified donor DNA and CRISPR-Cas9 to insert sequences of up to 2,049 base pairs (bp), including enhancers and promoters, into the rice genome	25	(Lu et al., 2020) (Kumar, Shumayla, & Upadhyay 2024)
CRISPR–Act3.0	*Arabidopsis thaliana* Rice *japónica* Kitaake Tomato M82	Here, their method a highly robust CRISPRa system working in rice, *Arabidopsis* and tomato: CRISPR-Act3.0. This system results in fourfold to sixfold higher activation than the state-of-the-art CRISPRa systems.		(Pan et al., 2021) (Kumar, Shumayla, & Upadhyay 2024)

(continued)

TABLE 6.1 (Continued)
Strategies for CRISPR-mediated genome-editing

Strategies for editing	Organism	Results	Efficiency (%)	References
CRISPR/Cas9 Mediated Genome Editing	*Penicillium subrubescens*	In this study, they developed an efficient genetic transformation method for the filamentous ascomycete *P. subrubescens* and demonstrated the establishment of CRISPR/Cas9 genome editing for this fungus		(Salazar-Cerezo et al., 2020) (de Castro Leitão et al., 2024)
	Candida albicans	They describe a transient CRISPR-Cas9 system for efficient gene deletion in C. albicans, that suggest that two mechanisms lead to homozygous deletions. They find that CRISPR elements can be expressed from genes that are present only transiently, and they develop a transient CRISPR system that further accelerates C. albicans genetic manipulation.		(Min et al., 2016) (de Castro Leitão et al., 2024)
	Mucor circinelloides	A a plasmid-free CRISPR-Cas9 system was first developed for the genetic modification of a Mucoral fungus. It does not require plasmid construction and its expression in the recipient organism. Instead, it involves the direct introduction of the guide RNA and the Cas9 enzyme and, in the case of homology-directed repair (HDR), the template DNA into the recipient strain.		(Nagy et al., 2017) (de Castro Leitão et al., 2024)
CRISPR/Cas9 Mediated Genome Editing	*Bombyx mori*	Here they reported highly efficient and multiplex genome editing in B. mori cell line and heritable site-directed mutagenesis of Bmku70, which is required for NHEJ pathway and they also related to antigen diversity, telomere length maintenance and subtelomeric gene silencing, using CRISPR/Cas9 system		(Ma et al., 2014) (Kumar, Shumayla, & Upadhyay 2024)
	Human embryonic kidney cells (HEK293T)	Here, they engineer the type II bacterial CRISPR system to function with custom guide RNA (gRNA) in human cells. Their results establish an RNA-guided editing tool for facile, robust, and multiplexable human genome engineering.		(Mali et al., 2013) (Kumar, Shumayla, & Upadhyay 2024)
CRISPR-Cas3 system	293T human cells	Here they demonstrate that type I-E CRISPR mediates distinct DNA cleavage activity in human cells. Notably, Cas3, with helicase and nuclease activity. This Cas3-mediated directional and broad DNA degradation can be used to introduce functional gene knockouts and knock-ins, this have potential therapeutic applications.		(Morisaka et al., 2019) (Kumar, Shumayla, & Upadhyay 2024)
CRISPR/Cas9 Mediated Genome Editing	*Danio rerio Caenorhabditis elegans*	Here studied the nuclease activity of SpG and SpRY by targeting 40 sites in zebrafish and C. elegans. They induced mutations in vivo, at a lower rate than SpCas9 in equivalent formulations. They found that the CRISPRscan algorithm could help to predict SpG and SpRY targets with high activity in vivo. Finally, they applied SpG and SpRY to generate knock-ins by homology-directed repair.		(Vicencio et al., 2022) (Kumar, Shumayla, & Upadhyay 2024)

TABLE 6.1 (Continued)
Strategies for CRISPR-mediated genome-editing

Strategies for editing	Organism	Results	Efficiency (%)	References
CRISPR-Cas12a	*HEK293T cells (CRL-3216), HeLa cells (CRM-CCL-2), N2A cells (CCL131) MCF7 cells (HTB-22)*	Here, they use the smaller Cas12a protein to develop four circular RNA-mediated prime editor (CPE) systems: niCPE**, nuCPE*, sniCPE*** and snuCPE. CPE systems preferentially recognize T-rich genomic regions and possess a potential multiplexing capacity in comparison to corresponding Cas9-based systems.	10.42* 24.89** 40.75***	(R. Liang et al., 2024)
CRISPR-Cas12b	HEK293T cells	Here they explore the diversity of the Cas12b family and identify a promising candidate for human gene editing from *Bacillus hisashii*, BhCas12b. This work establishes a third RNA-guided nuclease platform, in addition to Cas9 and Cpf1/Cas12a, for genome editing in human cells.		(Strecker et al., 2019) (Duan et al., 2024)
Cas12i and the developed Improving Editing Activity by Synergistic Engineering (MIDAS)	Human cells (HEK293T)	Here, they have developed a method termed Improving Editing Activity by Synergistic Engineering (MIDAS). MIDAS robustly and significantly increased the gene-editing efficiency of Cas12i, Cas12b, and CasX in human cells.		(Y. Chen et al., 2022) (Duan et al., 2024)
In vivo CRISPR base editing	*Macaca fascicularis*	Here they demonstrated that CRISPR base editors that are delivered in vivo using lipid nanoparticles can efficiently and precisely modify disease-related genes in living cynomolgus *M. fascicularis*. In their results, CRISPR base-editors can make precise single nucleotide changes in therapeutic target genes in the liver, and potentially in other organs.		(Musunuru et al., 2021) (Leclerc, Siroky, & Miller 2024)
CRISPR-Cas9 Mediated Genome Editing	Human cells: U-2 OS HeLa DLD-1	Here they investigated the use of baculoviral vectors as a delivery vehicle for CRISPR/Cas9-based genome-editing tools. They demonstrated the transduction of a panel of cell lines with Cas9 and a sgRNA sequence, which results in the efficient knockout of all four targeted subunits of the chromosomal passenger complex (CPC).		(Hindriksen et al., 2017) (Leclerc, Siroky, & Miller 2024)
Engineered CRISPR-Cas12a variants	*Human cells:* *S170R* *E174R* *S542R* *K548R*	They used structure-guided protein engineering to attempt to expand the PAM recognition of Cas12a nucleases, focusing on AsCas12a. Four of the 10 variants they tested in human cells displayed higher gene-editing activities on sites with canonical and non-canonical PAMs relative to wild-type AsCas12a.		(Kleinstiver et al., 2019) (Duan et al., 2024)

(continued)

TABLE 6.1 (Continued)
Strategies for CRISPR-mediated genome-editing

Strategies for editing	Organism	Results	Efficiency (%)	References
CRISPR-dCas9 mediated epigenome editing	*Arabidopsis thaliana*	sgRNA targeting the *AREB-1* promoter region targets dCas9 fusion with *A. thaliana* histone acetyl-transferase.	~200x (control) to 400X (under drought stress) enhanced *AREB-1* gene transcription and consequent protein accumulation. Potent drought tolerance was induced for up to 30 days.	(De Melo et al., 2020)
		Targeted DNA demethylation of *A. thaliana* genome using a dCas9-SunTag system fused to human TET1 catalytic domain.	~6 to 12X enhanced *FWA* gene transcription. Global genome-wide CG methylation was undisturbed outside the *FWA* promoter.	(Gallego-Bartolomé et al., 2018)
		CRISPR-based methylation targeting system, combining SunTag with the catalytic domain of the *Nicotiana tabacum* DRM methyltransferase	Targeted methylation at the *FWA* promoter and ~100% *FWA* gene silencing. Inheritable methylation patterns up to 4 generations, with early-flowering phenotype.	(Papikian et al., 2019)

REFERENCES

Anzalone, A. V., Randolph, P. B., Davis, J. R., Sousa, A. A., Koblan, L. W., Levy, J. M., Chen, P. J., Wilson, C., Newby, G. A., Raguram, A., & Liu, D. R. (2019). Search-and-replace genome editing without double-strand breaks or donor DNA. *Nature*, 576(7785), 149–157. https://doi.org/10.1038/s41586-019-1711-4

Arroyo-Olarte, R. D., Rodríguez-Hernández, K. D., & Morales-Ríos, E. (2023). Genome engineering in bacteria: Current and prospective applications. In V. Gurtler & M. Calcutt (Eds.), *Methods in Microbiology* , Vol. 52, *Genome Engineering* (pp. 35–76). https://doi.org/10.1016/bs.mim.2023.01.003

Asmamaw, M., & Zawdie, B. (2021). Mechanism and applications of CRISPR/Cas-9-mediated genome editing. *Biologics: Targets & Therapy,* 15, 353–361. https://doi.org/10.2147/BTT.S326422

Beneke, T., Madden, R., Makin, L., Valli, J., Sunter, J., & Gluenz, E. (2017). A CRISPR Cas9 high-throughput genome editing toolkit for kinetoplastids. *Royal Society Open Science*, 4(5), 170095. https://doi.org/10.1098/rsos.170095.

Cameron Coates, R., Blaskowski, S., Szyjka, S., van Rossum, H. M., Vallandingham, J., Patel, K., Serber, Z., & Dean, J. (2019). Systematic investigation of CRISPR–Cas9 configurations for flexible and efficient genome editing in Corynebacterium glutamicum NRRL-B11474. *Journal of Industrial Microbiology and Biotechnology,* 46(2), 187–201. https://doi.org/10.1007/s10295-018-2112-7

Charlesworth, C. T., Deshpande, P. S., Dever, D. P., Camarena, J., Lemgart, V. T., Cromer, M. K., Vakulskas, C. A., Collingwood, M. A., Zhang, L., Bode, N. M., Behlke, M. A., Dejene, B., Cieniewicz, B., Romano, R., Lesch, B. J., Gomez-Ospina, N., Mantri, S., Pavel-Dinu, M., Weinberg, K. I., & Porteus, M. H. (2019). Identification of preexisting adaptive immunity to Cas9 proteins in humans. *Nature Medicine*, 25(2). https://doi.org/10.1038/s41591-018-0326-x

Chen, L., Li, W., Katin-Grazzini, L., Ding, J., Gu, X., Li, Y., ... & Li, Y. (2018). A method for the production and expedient screening of CRISPR/Cas9-mediated non-transgenic mutant plants. *Horticulture Research*, 5.

Chen, S. Gong B., Cong Z., Lei C., & Nie Z. (2023). Nucleic acid-assisted CRISPR-Cas systems for advanced biosensing and bioimaging. *TrAC Trends in Analytical Chemistry,* 159, 116931.

Cheng, Y., Zhang, Y., Li, G., Fang, H., Sretenovic, S., Fan, A., Li, J., Xu, J., Que, Q., & Qi, Y. (2023). CRISPR–Cas12a base editors confer efficient multiplexed genome editing in rice. *Plant Communications*, 4(4), 100601. https://doi.org/10.1016/j.xplc.2023.100601

Cong, L., Ran, F. A., Cox, D., Lin, S., Barretto, R., Habib, N., Hsu, P. D., Wu, X., Jiang, W., Marraffini, L. A., & Zhang, F. (2013). Multiplex genome engineering using CRISPR/Cas systems. *Science (New York, N.Y.),* 339(6121), 819–823. https://doi.org/10.1126/science.1231143

de Castro Leitão, M., de Albuquerque Vieira, H. L., Torres, F. A. G., Perez, A. L. A., Piva, L. C., Reis, V. C. B., & Coelho, C. M. (2024). Biotechnological applications of CRISPR-Cas systems in fungi. *CRISPR-Cas System in Translational Biotechnology,* 237–263. doi: 10.1016/B978-0-323-91808-4.00011-0 de Melo, B. P., Lourenço-Tessutti, I. T., Paixão, J. F. R., Noriega, D. D., Silva, M. C. M., de Almeida-Engler, J., Fontes, E. P. B., & Grossi-de-Sa, M. F. (2020). Transcriptional modulation of AREB-1 by CRISPRa improves plant physiological performance under severe water deficit. *Scientific Reports,* 10(1), 16231. https://doi.org/10.1038/s41598-020-72464-y

Duan, Z., Liang, Y., Sun, J., Zheng, H., Lin, T., Luo, P., Wang, M., Liu, R., Chen, Y., Guo, S., Jia, N., Xie, H., Zhou, M., Xia, M., Zhao, K., Wang, S., Liu, N., Jia, Y., Si, W., Chen, Q., … Zhu, J. K. (2024). An engineered Cas12i nuclease that is an efficient genome editing tool in animals and plants. *Innovation Cambridge (Mass.)*, 5(2), 100564. https://doi.org/10.1016/j.xinn.2024.100564

Foster, A. J., Martin-Urdiroz, M., Yan, X., Wright, H. S., Soanes, D. M., & Talbot, N. J. (2018). CRISPR-Cas9 ribonucleoprotein-mediated co-editing and counterselection in the rice blast fungus. Scientific Reports, 8(1). https://doi.org/10.1038/s41598-018-32702-w

Gaudelli, N. M., Komor, A. C., Rees, H. A., Packer, M. S., Badran, A. H., Bryson, D. I., & Liu, D. R. (2017). Programmable base editing of A•T to G•C in genomic DNA without DNA cleavage. *Nature*, 551(7681), 464–471. https://doi.org/10.1038/nature24644

Gallego-Bartolomé, J., Gardiner, J., Liu, W., Papikian, A., Ghoshal, B., Kuo, H. Y., Zhao, J. M., Segal, D. J., & Jacobsen, S. E. (2018). Targeted DNA demethylation of the Arabidopsis genome using the human TET1 catalytic domain. *Proceedings of the National Academy of Sciences of the United States of America*, 115(9), E2125–E2134. https://doi.org/10.1073/pnas.1716945115.

Gee, P., Ando, Y., Kitayama, H., Yamamoto, S. P., Kanemura, Y., Ebina, H., Kawaguchi, Y., & Koyanagi, Y. (2011). APOBEC1-mediated editing and attenuation of Herpes Simplex Virus 1 DNA indicate that neurons have an antiviral role during Herpes Simplex Encephalitis. *Journal of Virology*, 85(19), 9726–9736. https://doi.org/10.1128/JVI.05288-11

Hakimi, H., Ishizaki, T., Kegawa, Y., Kaneko, O., Kawazu, S. I., & Asada, M. (2019). Genome Editing of *Babesia bovis* Using the CRISPR/Cas9 System. *mSphere*, 4(3), e00109-19. https://doi.org/10.1128/mSphere.00109-19

Hidalgo-Cantabrana, C., Goh, Y. J., Pan, M., Sanozky-Dawes, R., & Barrangou, R. (2019). Genome editing using the endogenous type I CRISPR-Cas system in *Lactobacillus crispatus*. *Proceedings of the National Academy of Sciences of the United States of America*, 116(32), 15774–15783. https://doi.org/10.1073/pnas.1905421116

Hindriksen, S., Bramer, A. J., Truong, M. A., Vromans, M. J. M., Post, J. B., Verlaan-Klink, I., Snippert, H. J., Lens, S. M. A., & Hadders, M. A. (2017). Baculoviral delivery of CRISPR/Cas9 facilitates efficient genome editing in human cells. *PloS One*, 12(6), e0179514. https://doi.org/10.1371/journal.pone.0179514

Hunker, A. C., Soden, M. E., Krayushkina, D., Heymann, G., Awatramani, R., & Zweifel, L. S. (2020). Conditional single vector CRISPR/SaCas9 viruses for efficient mutagenesis in the adult mouse nervous system. *Cell Reports*, 30(12). https://doi.org/10.1016/j.celrep.2020.02.092

Ikeda, T., Ohsugi, T., Kimura, T., Matsushita, S., Maeda, Y., Harada, S., & Koito, A. (2008). The antiretroviral potency of APOBEC1 deaminase from small animal species. *Nucleic Acids Research*, 36(21), 6859–6871. https://doi.org/10.1093/nar/gkn802

Janssen, B. D., Chen, Y.-P., Molgora, B. M., Wang, S. E., Simoes-Barbosa, A., & Johnson, P. J. (2018). CRISPR/Cas9-mediated gene modification and gene knock out in the human-infective parasite *Trichomonas vaginalis*. *Scientific Reports*, 8(1), 270. https://doi.org/10.1038/s41598-017-18442-3

Jia, H., Xu, J., Orbović, V., Zhang, Y., & Wang, N. (2017). Editing citrus genome via SaCas9/sgRNA system. *Frontiers in Plant Science*, 8. https://doi.org/10.3389/fpls.2017.02135

Jiang, W., Brueggeman, A. J., Horken, K. M., Plucinak, T. M., & Weeks, D. P. (2014). Successful transient expression of Cas9 and single guide RNA genes in Chlamydomonas reinhardtii. *Eukaryotic Cell*, 13(11). https://doi.org/10.1128/EC.00213-14

Jiang, Y., Qian, F., Yang, J., Liu, Y., Dong, F., Xu, C., Sun, B., Chen, B., Xu, X., Li, Y., Wang, R., & Yang, S. (2017). CRISPR-Cpf1 assisted genome editing of Corynebacterium glutamicum. *Nature Communications,* 8, 15179. https://doi.org/10.1038/ncomms15179

Kang, Y. K., Kwon, K., Ryu, J. S., Lee, H. N., Park, C., & Chung, H. J. (2017). Nonviral genome editing based on a polymer-derivatized CRISPR nanocomplex for targeting bacterial pathogens and antibiotic resistance. *Bioconjugate Chemistry*, 28(4), 957–967. https://doi.org/10.1021/acs.bioconjchem.6b00676

Kellner, M. J., Koob, J. G., Gootenberg, J. S., Abudayyeh, O. O., & Zhang, F. (2019). SHERLOCK: nucleic acid detection with CRISPR nucleases. *Nature Protocols*, 14(10), 2986–3012. https://doi.org/10.1038/s41596-019-0210-2

Kirti, A., Sharma, M., Rani, K., & Bansal, A. (2021). CRISPRing protozoan parasites to better understand the biology of diseases. *Progress in Molecular Biology and Translational Science,* 180, 21–68. https://doi.org/10.1016/bs.pmbts.2021.01.004

Komor, A. C., Kim, Y. B., Packer, M. S., Zuris, J. A., & Liu, D. R. (2016). Programmable editing of a target base in genomic DNA without double-stranded DNA cleavage. *Nature*, 533(7603), 420–424. https://doi.org/10.1038/nature17946

Kumar, A., Shumayla, & Upadhyay, S. K. (2023). CRISPR-Cas genome editing in plants. In S. K. Upadhyay (Ed.), *Applications of genome engineering in plants*. (pp. 1–19). Wiley. https://doi.org/10.1002/9781394183913

Lacchini, E., Kiegle, E., Castellani, M., Adam, H., Jouannic, S., Gregis, V., & Kater, M. M. (2020). CRISPR-mediated accelerated domestication of African rice landraces. *PloS One*, 15(3), e0229782. https://doi.org/10.1371/journal.pone.0229782

Lander, N., Chiurillo, M. A., Vercesi, A. E., & Docampo, R. (2017). Endogenous C-terminal Tagging by CRISPR/Cas9 in *Trypanosoma cruzi*. *Bio-protocol*, 7(10), e2299. https://doi.org/10.21769/BioProtoc.2299

Leclerc, D., Siroky, M. D., & Miller, S. M. (2024). Next-generation biological vector platforms for in vivo delivery of genome editing agents. *Current Opinion in Biotechnology,* 85, 103040. https://doi.org/10.1016/j.copbio.2023.103040

Li, X., Wang, L., Lin, J., Gu, Y., Liu, Z., & Hu, J. (2025). Detection of CRISPR–Cas and type I R–M systems in Klebsiella pneumoniae of human and animal origins and their relationship to antibiotic resistance and virulence. *Microbiology Spectrum*, 13:e00009-24.https://doi.org/10.1128/spectrum.00009-24

Liang, R., He, Z., Zhao, K. T., Zhu, H., Hu, J., Liu, G., ... & Gao, C. (2024). Prime editing using CRISPR-Cas12a and circular RNAs in human cells. *Nature Biotechnology,* 1–9.

Liang, Z., Chen, K., Li, T., Zhang, Y., Wang, Y., Zhao, Q., ... & Gao, C. (2017). Efficient DNA-free genome editing of bread wheat using CRISPR/Cas9 ribonucleoprotein complexes. *Nature Communications*, 8(1), 14261.

Lin, Q., Zong, Y., Xue, C., Wang, S., Jin, S., Zhu, Z., Wang, Y., Anzalone, A. V., Raguram, A., Doman, J. L., Liu, D. R., & Gao, C. (2020). Prime genome editing in rice and wheat. *Nature Biotechnology*, 38(5), 582–585. https://doi.org/10.1038/s41587-020-0455-x

Liu, Y., Li, X., He, S., Huang, S., Li, C., Chen, Y., Liu, Z., Huang, X., & Wang, X. (2020). Efficient generation of mouse models with the prime editing system. *Cell Discovery*, 6(1), 27. https://doi.org/10.1038/s41421-020-0165-z

Lowder, L. G., Zhang, D., Baltes, N. J., Paul III, J. W., Tang, X., Zheng, X., ... & Qi, Y. (2015). A CRISPR/Cas9 toolbox for multiplexed plant genome editing and transcriptional regulation. *Plant Physiology*, 169(2), 971-985.

Lu, Y., Tian, Y., Shen, R. *et al.* Targeted, efficient sequence insertion and replacement in rice. *Nature Biotechnology* 38, 1402–1407 (2020). https://doi.org/10.1038/s41587-020-0581-5

Ma, X., Zhang, Q., Zhu, Q., Liu, W., Chen, Y., Qiu, R., Wang, B., Yang, Z., Li, H., Lin, Y., Xie, Y., Shen, R., Chen, S., Wang, Z., Chen, Y., Guo, J., Chen, L., Zhao, X., Dong, Z., & Liu, Y. G. (2015). A robust CRISPR/Cas9 system for convenient, high-efficiency multiplex genome editing in monocot and dicot plants. *Molecular Plant*, 8(8), 1274–1284. https://doi.org/10.1016/j.molp.2015.04.007

Ma, Y., Zhang, L., & Huang, X. (2014). Genome modification by CRISPR/Cas9. *The FEBS Journal*, 281(23), 5186–5193. https://doi.org/10.1111/febs.13110

Mali, P., Yang, L., Esvelt, K. M., Aach, J., Guell, M., DiCarlo, J. E., Norville, J. E., & Church, G. M. (2013). RNA-guided human genome engineering via Cas9. *Science (New York, N.Y.)*, 339(6121), 823–826. https://doi.org/10.1126/science.1232033

Martins, P. M. M., Granato, L. M., Morgan, T., Nalin, J. L., Takita, M. A., Alfenas-Zerbini, P., & de Souza, A. A. (2024). Analysis of CRISPR-Cas loci distribution in Xanthomonas citri and its possible control by the quorum sensing system. *FEMS Microbiology Letters,* 371, fnae005. https://doi.org/10.1093/femsle/fnae005

McAllister, K. N., Bouillaut, L., Kahn, J. N., Self, W. T., & Sorg, J. A. (2017). Using CRISPR-Cas9-mediated genome editing to generate C. difficile mutants defective in selenoproteins synthesis. Scientific Reports. https://doi.org/10.1038/s41598-017-15236-5

Medeiros, L. C. S., South, L., Peng, D., Bustamante, J. M., Wang, W., Bunkofske, M., Perumal, N., Sanchez-Valdez, F., & Tarleton, R. L. (2017). Rapid, selection-free, high-efficiency genome editing in protozoan parasites using CRISPR-cas9 ribonucleoproteins. *MBio*, 8(6), e01788–17. https://doi.org/10.1128/mBio.01788-17

Min, K., Ichikawa, Y., Woolford, C. A., & Mitchell, A. P. (2016). Candida albicans gene deletion with a transient CRISPR-Cas9 system. *mSphere,* 1(3), e00130-16. https://doi.org/10.1128/mSphere.00130-16

Morgens, D. W., Wainberg, M., Boyle, E. A., Ursu, O., Araya, C. L., Kimberly Tsui, C., Haney, M. S., Hess, G. T., Han, K., Jeng, E. E., Li, A., Snyder, M. P., Greenleaf, W. J., Kundaje, A., & Bassik, M. C. (2017). Genome-scale measurement of off-target activity using Cas9 toxicity in high-throughput screens. *Nature Communications*, 8. https://doi.org/10.1038/ncomms15178

Morisaka, H., Yoshimi, K., Okuzaki, Y., Gee, P., Kunihiro, Y., Sonpho, E., Xu, H., Sasakawa, N., Naito, Y., Nakada, S., Yamamoto, T., Sano, S., Hotta, A., Takeda, J., & Mashimo, T. (2019). CRISPR-Cas3 induces broad and unidirectional genome editing in human cells. *Nature Communications*, 10(1), 5302. https://doi.org/10.1038/s41467-019-13226-x

Musunuru, K., Chadwick, A. C., Mizoguchi, T., Garcia, S. P., DeNizio, J. E., Reiss, C. W., Wang, K., Iyer, S., Dutta, C., Clendaniel, V., Amaonye, M., Beach, A., Berth, K., Biswas, S., Braun, M. C., Chen, H. M., Colace, T. V., Ganey, J. D., Gangopadhyay, S. A., Garrity, R., … Kathiresan, S. (2021). In vivo CRISPR base editing of PCSK9 durably lowers cholesterol in primates. *Nature*, 593(7859), 429–434. https://doi.org/10.1038/s41586-021-03534-y

Nagy, G., Szebenyi, C., Csernetics, Á., Vaz, A. G., Tóth, E. J., Vágvölgyi, C., & Papp, T. (2017). Development of a plasmid-free CRISPR-Cas9 system for the genetic modification of Mucor circinelloides. *Scientific Reports,* 7(1), 16800. https://doi.org/10.1038/s41598-017-17118-2

Neil, K., Allard, N., Roy, P., Grenier, F., Menendez, A., Burrus, V., & Rodrigue, S. (2021). High-efficiency delivery of CRISPR-Cas9 by engineered probiotics enables precise microbiome editing. *Molecular Systems Biology*, 17(10), e10335. https://doi.org/10.15252/msb.202110335

Noguchi, C., Ishino, H., Tsuge, M., Fujimoto, Y., Imamura, M., Takahashi, S., & Chayama, K. (2005). G to A hypermutation of hepatitis B virus. *Hepatology*, 41(3), 626–633. https://doi.org/10.1002/hep.20580

Pan, C., Sretenovic, S., & Qi, Y. (2021). CRISPR/dCas-mediated transcriptional and epigenetic regulation in plants. *Current Opinion in Plant Biology,* 60, 101980. https://doi.org/10.1016/j.pbi.2020.101980

Papikian, A., Liu, W., Gallego-Bartolomé, J., & Jacobsen, S. E. (2019). Site-specific manipulation of Arabidopsis loci using CRISPR-Cas9 SunTag systems. Nature Communications, 10(1). https://doi.org/10.1038/s41467-019-08736-7

Peng, D.., Kurup, S.P., Yao, P.Y., Minning, T.A, Tarleton, R. L. (2015).CRISPR-Cas9-mediated single-gene and gene family disruption in Trypanosoma cruzi. *mBio* 6:10.1128/mbio.02097-14. https://doi.org/10.1128/mbio.02097-14

Prajapati, R., & Tyagi, K. (2024). CRISPR/Cas-mediated multiplex genome editing in plants and applications. In S. K. Upadhyay (Ed.), *Applications of genome engineering in plants* (pp. 20–39). Wiley.

Ran, F. A., Hsu, P. D., Lin, C.-Y., Gootenberg, J. S., Konermann, S., Trevino, A. E., Scott, D. A., Inoue, A., Matoba, S., Zhang, Y., & Zhang, F. (2013). Double nicking by RNA-guided CRISPR Cas9 for enhanced genome editing specificity. *Cell*, 154(6), 1380–1389. https://doi.org/10.1016/j.cell.2013.08.021

Ren, C., Gathunga, E. K., Li, X., Li, H., Kong, J., Dai, Z., & Liang, Z. (2023). Efficient genome editing in grapevine using CRISPR/LbCas12a system. *Molecular Horticulture*, 3(1), 21. https://doi.org/10.1186/s43897-023-00069-w

Rico, E., Jeacock, L., Kovářová, J., & Horn, D. (2018). Inducible high-efficiency CRISPR-Cas9-targeted gene editing and precision base editing in African trypanosomes. *Scientific Reports*, 8(1), 7960. https://doi.org/10.1038/s41598-018-26303-w

Romagnoli, B. A. A., Picchi, G. F. A., Hiraiwa, P. M., Borges, B. S., Alves, L. R., & Goldenberg, S. (2018). Improvements in the CRISPR/Cas9 system for high efficiency gene disruption in Trypanosoma cruzi. *Acta Tropica,* 178, 190–195. https://doi.org/10.1016/j.actatropica.2017.11.013

Salazar-Cerezo, S., Kun, R. S., de Vries, R. P., & Garrigues, S. (2020). CRISPR/Cas9 technology enables the development of the filamentous ascomycete fungus Penicillium subrubescens as a new industrial enzyme producer. *Enzyme and Microbial Technology,* 133, 109463. https://doi.org/10.1016/j.enzmictec.2019.109463

Sawyer, S. L., Emerman, M., & Malik, H. S. (2004). Ancient adaptive evolution of the primate antiviral DNA-editing enzyme APOBEC3G. *PLoS Biology*, 2(9), e275. https://doi.org/10.1371/journal.pbio.0020275

Scherer, S. (2008). *A Short Guide to the Human Genome*. Cold Spring Harbor Laboratory Press.

Selle, K., Fletcher, J. R., Tuson, H., Schmitt, D. S., McMillan, L., Vridhambal, G. S., Rivera, A. J., Montgomery, S. A., Fortier, L. C., Barrangou, R., Theriot, C. M., & Ousterout, D. G. (2020). In vivo targeting of Clostridioides difficile using phage-delivered CRISPR-Cas3 antimicrobials. *mBio,* 11(2), e00019-20. https://doi.org/10.1128/mBio.00019-20

Sen, D., & Mukhopadhyay, P. (2024). Antimicrobial resistance (AMR) management using CRISPR-Cas based genome editing. *Gene and Genome Editing,* 7, 100031. https://doi.org/10.1016/j.ggedit.2024.100031

Shuman, S., & Glickman, M. S. (2007). Bacterial DNA repair by non-homologous end joining. *Nature Reviews Microbiology*, 5(11), 852–861. https://doi.org/10.1038/nrmicro1768

Song, X., Huang, H., Xiong, Z., Ai, L., & Yang, S. (2017). CRISPR-Cas9D10A nickase-assisted genome editing in Lactobacillus casei. Applied and Environmental Microbiology. https://doi.org/10.1128/AEM.01259-17

Standage-Beier, K., Zhang, Q., & Wang, X. (2015). Targeted large-scale deletion of bacterial genomes using CRISPR-Nickases. *ACS Synthetic Biology*, 4(11), 1217–1225. https://doi.org/10.1021/acssynbio.5b00132

Strecker, J., Jones, S., Koopal, B., Schmid-Burgk, J., Zetsche, B., Gao, L., Makarova, K. S., Koonin, E. V., & Zhang, F. (2019). Engineering of CRISPR-Cas12b for human genome editing. *Nature Communications*, 10(1), 212. https://doi.org/10.1038/s41467-018-08224-4

Soares Medeiros, L. C., South, L., Peng, D., Bustamante, J. M., Wang, W., Bunkofske, M., Perumal, N., Sanchez-Valdez, F., & Tarleton, R. L. (2017). Rapid, selection-free, high-efficiency genome editing in protozoan parasites Using CRISPR-Cas9 ribonucleoproteins. *mBio*, 8(6), e01788-17. https://doi.org/10.1128/mBio.01788-17

Takagi, Y., Akutsu ,Y., Doi, M., & Furukawa, K. (2019). Utilization of proliferable extracellular amastigotes for transient gene expression, drug sensitivity assay, and CRISPR/Cas9-mediated gene knockout in Trypanosoma cruzi. *PLOS Neglected Tropical Diseases* 13(1): e0007088. https://doi.org/10.1371/journal.pntd.0007088

Upreti C., Kumar P., Durso L. M. & Palmer K.L. (2023). CRISPR-Cas inhibits plasmid transfer and immunizes bacteria against antibiotic resistance acquisition in manure. bioRxiv, https://doi.org/10.1101/2023.09.26.559507

Vicencio, J., Sánchez-Bolaños, C., Moreno-Sánchez, I., Brena, D., Vejnar, C.E., Kukhtar, D., Ruiz-López, M., Cots-Ponjoan, M., Rubio, A., Melero, N.R., Crespo-Cuadrado, J., Carolis, C., Pérez-Pulido, A.J., Giráldez, A.J., Kleinstiver, B.P., Cerón, J., & Moreno-Mateos, M.A. (2022) Genome editing in animals with minimal PAM CRISPR-Cas9 enzymes. *Nature Communications*. 13, 2601.

Walton, R. T., Christie, K. A., Whittaker, M. N., & Kleinstiver, B. P. (2020). Unconstrained genome targeting with near-PAMless engineered CRISPR-Cas9 variants. *Science*, 368(6488), 290–296. https://doi.org/10.1126/science.aba8853

Wang, R., Shu, X., Zhao, H., Xue, Q., Liu, C., Wu, A., Cheng, F., Wang, L., Zhang, Y., Feng, J., Wu, N., & Li, M. (2023). Associate toxin-antitoxin with CRISPR-Cas to kill multidrug-resistant pathogens. *Nature Communications,* 14(1), 2078. https://doi.org/10.1038/s41467-023-37789-y

Wang, X., Lyu, Y., Wang, S., Zheng, Q., Feng, E,. Zhu, L., Pan, C, Wang, S., Wang, D., Liu, X., & Wang, H. (2021) Application of CRISPR/Cas9 system for plasmid elimination and bacterial killing of Bacillus cereus group strains. *Frontiers in Microbiology,* 12: 536357. doi: 10.3389/fmicb.2021.536357

Xie, K., Minkenberg, B., & Yang, Y. (2015). Boosting CRISPR/Cas9 multiplex editing capability with the endogenous tRNA-processing system. *Proceedings of the National Academy of Sciences of the United States of America*, 112(11), 3570–3575. https://doi.org/10.1073/pnas.1420294112

Xin, H., Wan, T., & Ping, Y. (2019). Off-targeting of base editors: BE3 but not ABE induces substantial off-target single nucleotide variants. *Signal Transduction and Targeted Therapy,* 4(1), 9. https://doi.org/10.1038/s41392-019-0044-y

Yan, M. Y., Yan, H. Q., Ren, G. X., Zhao, J. P., Guo, X. P., & Sun, Y. C. (2017). CRISPR-Cas12a-assisted recombineering in bacteria. *Applied and Environmental Microbiology*, 83(17), e00947–17. https://doi.org/10.1128/AEM.00947-17

Yin, C., Zhang, T., Qu, X., Zhang, Y., Putatunda, R., Xiao, X., Li, F., Xiao, W., Zhao, H., Dai, S., & Qin, X. (2017). In vivo excision of HIV-1 provirus by saCas9 and multiplex single-guide RNAs in animal models. Molecular Therapy, 25(5). https://doi.org/10.1016/j.ymthe.2017.03.012

Zakrzewska, M., & Burmistrz, M. (2023). Mechanisms regulating the CRISPR-Cas systems. *Frontiers in Microbiology,* 14, 1060337. https://doi.org/10.3389/fmicb.2023.106033

Zetsche, B., Gootenberg, J. S., Abudayyeh, O. O., Slaymaker, I. M., Makarova, K. S., Essletzbichler, P., Volz, S. E., Joung, J., van der Oost, J., Regev, A., Koonin, E. V., & Zhang, F. (2015). Cpf1 is a single RNA-guided endonuclease of a Class 2 CRISPR-Cas system. *Cell,* 163(3), 759–771. https://doi.org/10.1016/j.cell.2015.09.038

Zhang, Y., Cheng, Y., Fang, H., Roberts, N., Zhang, L., Vakulskas, C. A., Niedz, R. P., Culver, J. N., & Qi, Y. (2022). Highly efficient genome editing in plant protoplasts by Ribonucleoprotein delivery of CRISPR-Cas12a nucleases. Frontiers in Genome Editing, 4. https://doi.org/10.3389/fgeed.2022.780238

Zuo, E., Sun, Y., Wei, W., Yuan, T., Ying, W., Sun, H., Yuan, L., Steinmetz, L. M., Li, Y., & Yang, H. (2019). Cytosine base editor generates substantial off-target single-nucleotide variants in mouse embryos. Science (New York, N.Y.) , 364(6437), 289–292. https://doi.org/10.1126/science.aav9973

7 CRISPR/Cas12

An Emerging CRISPR/Cas System for Precise DNA Manipulation

Neslihan Yaşar, Doğa Selin Kayıhan, Ceyhun Kayıhan, and Emre Aksoy

7.1 HISTORY AND BASIC MECHANISMS OF GENOME EDITING AND CRISPR-CAS SYSTEMS

"Genetic engineering" or "genome editing" requires the usage of some technologies including meganucleases (MegNs)/homing endonucleases, transcription activator-like effector nucleases (TALENs), zinc-finger nucleases (ZFNs), or clustered regularly interspaced short palindromic repeats CRISPR/Cas (CRISPR-associated) system (Gaj et al., 2016). MegNs are known to be naturally occurring endodeoxyribonucleases with their high DNA specificity whereas TALENs and ZFNs are artificially engineered proteins and they can target specific regions in the genome. TALENs are known to be transcriptional activators enabling DNA recognition by a central repeat domain. These enzymes use the *FokI* nuclease domain which is a type IIS restriction enzyme for site-specific manipulation of the genetic information (Miller et al., 2011; Kim et al., 2013; Khalil, 2020). Similarly, ZFNs provide cleavage of specific target sites via the cleavage activity of *FokI* (van der Oost and Patinios, 2023). For both the TALENs and ZFNs, some insertions or deletions at specific sites are achieved by repairing broken DNA with error-prone non-homologous end joining (NHEJ) (Kim et al., 2013). These enzymes recognize the DNA by specific DNA-protein contacts (van der Oost and Patinios, 2023). However, the CRISPR-Cas system uses DNA-RNA interactions to recognize the DNA (Khalil, 2020). This system is known to be one of the most powerful applications of genome editing and also is a more efficient and recent alternative to other genome-editing tools. It aims to alter specifically targeted sequences in the DNA (Rojo et al., 2018).

Initially, this system is described as the adaptive immunity of prokaryotic cells to protect the organism against foreign genetic material (Makarova et al., 2015). In 1987, the CRISPR loci were discovered in *Escherichia coli* and thought to be the adaptive immune system of bacteria since there were unusual DNA repeats and some other DNA fragments. Later, these DNA fragments are called "spacers" which come from viral genetic information. These spacers are separated by short palindromic repeats and all of these motifs are found in intergenic sequences (Gostimskaya, 2022).

The CRISPR-Cas system contains three phases: adaptation, expression, and interference. During the adaptation, the foreign DNA, known as "protospacers," are incorporated into the CRISPR locus of bacteria as a new spacer. When the cell encounters the same invader later, these spacers provide an adaptive immunity to defend against this infection (Makarova et al., 2015). In the expression phase, this CRISPR locus is transcribed into the precursor CRISPR RNA (pre-crRNA), which is complementary to the spacer sequence, to produce CRISPR RNA (crRNA) eventually, after processing and maturation of the pre-crRNA (Makarova et al., 2015; Rojo et al., 2018). After the formation of this complex, it binds to the Cas protein which cuts the long pre-crRNA to separate it into crRNA/transactivating crRNA (tracrRNA) complexes by Rnase III. These complexes guide the Cas protein for its target DNA (Rojo et al., 2018; Tröder and Zevnik, 2022). The interference phase begins with the guidance of the crRNA/tracrRNA complex to the Cas protein to target a 3-5 bp-long seed sequence which is called protospacer adjacent motif (PAM) found in the upstream of the crRNA binding site (Gaj et al., 2016; Rojo et al., 2018). After recognition of the PAM sequence, the complex cleaves the cognate nucleic acids (Makarova et al., 2015). At that point, the target sequence is unwound and cleaved by the nuclease domain of the Cas protein, resulting in a double-strand break (DSB). At the end of the interference stage, the desired DNA repair template is inserted and the cleaved DNA strands are repaired by the homology-directed repair (HDR) mechanism. This repaired spacer sequence is now adapted into the genome of bacteria (Rojo et al., 2018). The acquisition of the spacer to the bacterial genome provides a chronological record of the previous infections from foreign nucleic acids, similar to the adaptive immunity in the higher complex organisms (van der Oost et al., 2014; Tröder and Zevnik, 2022). Using the CRISPR-Cas system for targeted genome modifications in a variety of species including plants, mice, pigs, rabbits, monkeys, insects, and human cells, has been rapidly and effectively achieved (Hryhorowicz et al., 2017).

Recent findings and the rapid evolution of Cas genes require a proper annotation and classification of CRISPR-Cas systems (Makarova et al., 2015). According to the Cas genes and locus arrangements, the CRISPR-Cas systems are divided into two main classes, six different types, and 33 subtypes (Khalil, 2020; Hryhorowicz et al., 2023). The Cas proteins and crRNAs are necessary for all of these systems. Types I, III, and IV belong to the Class I system which is associated with multiprotein crRNA effector complex whereas Types II, V, and VI belong to the Class II system associated with a single multidomain protein crRNA effector complex, such

DOI: 10.1201/9781003500933-9

as Cas9. Types I, II, and V target and cleave the DNA, Type VI recognizes RNA, and the Type III system cleaves both the DNA and RNA. The signature proteins: for the Type I system is Cas3, for the Type III system it is Cas10, for the Type IV system it is DinG (Csf4), for the Type II system it is Cas9, for the Type V system it is Cas12, and for the Type VI system it is Cas13 (Makarova and Koonin, 2015). This chapter focuses on the features and mechanisms of the CRISPR-Cas12 system, its advantages and disadvantages in usage, its applications in different organisms, and also its roles in the detection and diagnosis of diseases at a molecular level.

7.2 THE CRISPR-CAS12 SYSTEM AND ITS UNIQUENESS

The Cas9 protein, approximately 1,000–1,600 amino acids long, is the best-known and most widely used for genome editing (Oakes et al., 2014). The CRISPR-Cas9 system provides precise genome modifications for various plant and animal species. A more recent CRISPR-Cas12 system, promising to be preferable to the CRISPR-Cas9 system, also can be used for precise genome editing due to its distinct features and characteristics. Currently, both the Cas9 and Cas12 systems are widely used in numerous fields due to their high specificity, high sensitivity, and ease of use (Yang et al., 2023) and both of these proteins are probably evolved from TnpB-like proteins (Sasnauskas et al., 2023). Here, the question that needs to be answered is: What are the uniqueness and specific characteristics of the CRISPR-Cas12 system among other CRISPR systems during the defending against foreign genetic materials and also in its applications? To answer this question; the structure of the effector protein, PAM sequence motifs, DNA cleavage, target sequence, off-target effects, and applications should be assessed in detail with their strengths and weaknesses.

As mentioned above, the Cas12 protein is the signature protein of the Type V system of CRISPR. The greatest number of subtypes discovered so far are found in Type V CRISPR systems (Xiao et al., 2019). The subtypes a, b, c, d, e, f, g, h, i, k, l, m, n, and u are found in the Type V CRISPR system with different functions. The effector proteins in Type V systems are divided into three main branches in their phylogenetic tree according to their homologies. The first class involves Cas12-a, -c, -d, -e; the second one includes Cas12-b, -h, -i; and the last one involves smaller effectors such as Cas12-f, -g, -m, -n, -u2. All of the Type V effector proteins identified to date are listed in Table 7.1 (Zetsche et al., 2015; Koonin et al., 2017; Yan et al., 2019; Tong et al., 2021; Wu et al., 2022; Chen et al., 2023; Sun et al., 2023).

Researchers discovered and characterized the CRISPR-Cas12 system in 2015, and concluded that the effector protein in this system is Cpf1 endonuclease (for theCRISPR-Cas12a system), guided by a single crRNA, lacking tracrRNA (Zetsche et al., 2015). Among the CRISPR-Cas12 systems, the ones that have been studied and detailed the most by researchers are Cas12a and Cas12b. To target the DNA and generate DSBs, the Cpf1 (Cas12a) proteins recognize crRNA-DNA heteroduplexes, without the need for a tracrRNA. The Cpf1 (CRISPR from *Prevotella* and *Francisella* 1) is a large, approximately 1,300 amino acid-long protein found in diverse bacteria and two archaea (Zetsche et al., 2015; Shmakov et al., 2017; Hryhorowicz et al., 2023). The Cas12b effector protein also known as C2c1, in contrast to the Cas12a, requires a dual RNA (both crRNA and tracrRNA) for its targeting activity. The Cas12b protein is relatively smaller than the Cas12a protein with its approximately 1,100 amino acids-long structure. So, it is considered to be useful for genome editing in vivo due to its smaller size and the ability to perform intracellular delivery by Strecker et al. (2019). However, they also stated that since this protein shows its most efficient double-strand cleavage activity at 48°C, it may require some modifications to make it active at lower temperatures for use in mammalian genome editing. Both the Cas12a and Cas12b effector proteins need a T-rich PAM sequence to cleave double strands and create a staggered end. When compared to Cas12a, Cas12b protein has a higher targeting specificity due to its single base pair mismatch tolerating ability in the protospacer (Ming et al., 2020). As a result, both the Cas12a and Cas12b systems have advantages and disadvantages but are promising genome-engineering tools.

Generally, the common unique property of Type V CRISPR-Cas systems includes only a single RNA-guided effector protein, containing a RuvC nuclease domain and a nuclease lobe domain (NUC) (Senthilnathan et al., 2023). This single RuvC nuclease domain cleaves single-stranded DNA (ssDNA) nonspecifically and/or also double-stranded DNA (dsDNA) adjacent to the PAM sequence (Yan et al., 2018). Depending on the type of effector Cas protein, only a single crRNA may be sufficient for guidance; or both the tracrRNA and crRNA, a dual RNA, may be used for targeting. Some of the Cas proteins cleave the strands of their targets, such as Cas12a, whereas some of them create nicks by cleaving one of the target strands, such as Cas12i. This nicking feature of Cas12i may provide high-fidelity gene editing. Also, the collateral (trans) cleavage activity of Cas12g for its RNA targets might be useful for the sensitive detection of nucleic acids (Swarts, 2019). Some Cas12 proteins do bind to their target but do not show target cleavage activity called a "binding-only" mechanism, such as Cas12c, Cas12k, and Cas12m (Wu et al., 2022). The binding-only mechanism provides transcriptional silencing in these effector proteins having a non-canonical form. This mechanism was first discovered in the CRISPR-Cas12c system, providing immunity to the bacteria against a viral attack by binding to the target to block the replication and expression of invaders instead of degrading the target (Huang et al., 2022; Badon et al., 2024).

The PAM sequence should be 5' T-rich upstream of the target sequence for recognition by the Cas12 protein for its editing activity (Senthilnathan et al., 2023). This recognition initiates the unwinding of target DNA with the help of a guide sequence of the crRNA (Tong et al., 2021). In comparison with the Cas9, the Cas12 protein can process the precrRNA itself, without a need for tracrRNA or RNase III activity, due to its own intrinsic RNase activity. It makes

TABLE 7.1
Proper PAM sequences, requirements, and targets of the effector proteins in the type V CRISPR systems are shown. A short-complementarity untranslated RNA (scoutRNA) may be required together with crRNA for the targeting and cleaving activity of the effector protein

Subtype	Effector protein(s)	PAM sequence	Requirements	Target	References
Cas12a	Cpf1	5'-TTV[a]	crRNA	ssDNA and dsDNA	Zetsche et al., 2015
Cas12b	C2c1	5'AT-rich	tracrRNA and crRNA	ssDNA and dsDNA	Koonin et al., 2017
Cas12c	Cas12c1 Cas12c2	5'TG 5'TN	crRNA and tracrRNA/ scoutRNA	ssDNA and dsDNA (no cleavage, binding only)	Tong et al., 2021; Harrington et al., 2020; Yan et al., 2019
Cas12d	CasY	TA or TG	crRNA and scoutRNA	ssDNA and dsDNA	Tong et al., 2021
Cas12e	CasX	TTCN	tracrRNA and crRNA	dsDNA	Badon et al., 2024; Liu et al., 2019
Cas12f (Cas14)	Cas12f1:Cas14a Cas12f2:Cas14b Cas12f3:Cas14c	No need for PAM sequence	tracrRNA and crRNA	ssDNA	Makarova et al., 2019
Cas12g	Cas12g1	No need for PAM sequence	tracrRNA and crRNA	ssRNA and ssDNA	Yan et al., 2019; Swarts, 2019
Cas12h1		5'-RTR[b]	crRNA	ssDNA and dsDNA	Yan et al., 2019
Cas12i	Cas12i1 Cas12i2	5'-TTN	crRNA	ssDNA and dsDNA	Yan et al., 2019
Cas12j	CasΦ	TBN[c]	crRNA	ssDNA and dsDNA	Tong et al., 2021
Cas12k	C2c5 (Cas12u5)	5'-GTN	tracrRNA and crRNA	dsDNA (no cleavage, binding only)	Strecker et al., 2019
Cas12l	Casπ-1 Casπ-2 Casπ-3 Casπ-4	5' C-rich	tracrRNA and crRNA	dsDNA	Sun et al., 2023
Cas12m	C2c4 (Cas12u1)	5V'-TTN	crRNA	dsDNA (no cleavage, binding only)	Wu et al., 2022; Omura et al., 2023
Cas12n	C2c9 (Cas12u4)	5'-AAN	tracrRNA and crRNA	dsDNA	Chen et al., 2023; Badon et al., 2024
Cas12u	Cas12u1:C2c4 (Cas12m) Cas12u2:C2c8 Cas12u3:C2c10 (Cas12f1) Cas12u4:C2c9 (Cas12n) Cas12u5:C2c5 (Cas12k)				Chen et al., 2023

[a] V represents A, C, and G.
[b] R represents A and G.
[c] B represents C, G, and T/.

the CRISPR-Cas12 system more practical for multiplex genome editing (Hillary and Ceasar, 2023; Senthilnathan et al., 2023). As mentioned above, depending on the protein, the Type V CRISPR-Cas systems can be associated with a single crRNA or both crRNA and tracrRNA for an efficient cut (Omura et al., 2023). Additionally, the occurrence of off-target effects is lower in Type V CRISPR systems which makes it more efficient and reliable (Senthilnathan et al., 2023).

When the Cas12 protein encounters a possible target site next to a PAM sequence, the non-target strand (NTS) is displaced, it creates an R-loop, forming base-pair hybridization between the target DNA strand and the crRNA. At that point, the R-loop is formed as a result of Cas12 matching the <17 base pair of the target sequence (Pickar-Oliver and Gersbach, 2019; Tong et al., 2021). Following the formation of the R-loop, the active RuvC domain of the Cas12 cleaves the non-target strand with the aid of the PAM sequence. The nuclease activity of the Cas12 protein produces staggered or sticky ends. In contrast to the Cas9 protein which creates blunt ends, the staggered ends generated by the Cas12 protein might be advantageous in some cases such as precisely combining the DNA sequences in a right orientation. Nevertheless, how the RuvC nuclease domain of Cas12 cleaves the targeted DNA strand is not yet clearly understood (Hillary and Ceasar, 2023; Senthilnathan et al., 2023). The mechanism of the CRISPR-Cas12 system can be seen in Figure 7.1.

In the Cpf1, Cas12a protein, three nuclease activities are found: (1) targeted nuclease activity known as canonical (cis) cleavage activity; (2) non-targeted nuclease activity known as collateral (trans) cleavage activity; and (3) endoribonuclease activity responsible for crRNA processing (Feng et al., 2023). Just as most of the Type V CRISPR effector proteins target dsDNA, they also target ssDNA. This ssDNA targeting and cleaving are achieved not only by PAM sequence, but also the complementarity of the target sequence to the crRNA provides cleavage. This targeted nuclease activity is called canonical or cis cleavage activity. In the meantime, the Type V CRISPR effector proteins also have a distinct non-specific or non-targeted ssDNA nuclease activity. Generally, among different Type V effector protein variants, the RuvC domain is responsible for catalyzing targeted cleavage or nicking activity. However, most of these variants also have an ambiguous collateral cleavage activity. The nuclease domain of the effector protein cleaves the targeted dsDNA and then the PAM-distal product is released from the targeted dsDNA material, whereas the effector protein remains attached to the PAM-proximal strand. In the current situation, the RuvC domain of the protein remains capable of accessing ssDNA and performing non-targeted nuclease activity which is called nonselective cleavage of collateral ssDNA or trans cleavage activity (Tong et al., 2021; Wu et al., 2022).

7.3 APPLICATIONS OF THE CRISPR-CAS12 SYSTEM

By using the CRISPR-Cas systems, it is possible to achieve genomic alterations for different purposes in various species and cell lines including bacteria, fungi, plants, human cell lines, animals, and other organisms such as *Drosophila*, sponges, zebrafish, and insects (Senthilnathan et al., 2023). The purpose of use of the CRISPR-Cas systems might be modifying the genome for gene knockout, inversion, deletion, addition, correction; regulation of transcription, epigenetic engineering, and several other purposes in the fields of biology, medicine (e.g., drug development), and agriculture (e.g., crop improvement, breeding). In the scope of this chapter, current applications of the Type V CRISPR systems in different organisms for many purposes are discussed. The Cas12 effector proteins have been currently used for lots of purposes such as genome editing, molecular diagnosis, and nucleic acid detection, due to their ease of application, higher specificity, and higher sensitivity. The CRISPR-Cas12 system is preferable for these purposes since it has fewer steps during the amplification of target nucleic acids and collateral cleavage of ssDNAs. To date, genome editing has been achieved by using the CRISPR-Cas12 system in microorganisms, fungi, plants, animals, and humans. This chapter summarizes the applications of CRISPR-Cas12 systems which can be useful for further applications and aspects (Senthilnathan et al., 2023; Yang et al., 2023).

For genome-editing applications, the Cas12a and Cas12b effector proteins have been precisely used today. When the CRISPR-Cas12 system is used for genome-editing applications, how is the repair of the DBS that occurs after cleavage activity performed by the effector protein? The process involves induction of a DSB in the presence of a plasmid expressing the Cas protein and an artificial gRNA, and then repair of this break via NHEJ or HDR mechanisms. If there is no homologous DNA template, the NHEJ mechanism takes place which introduces random indels to the cleavage

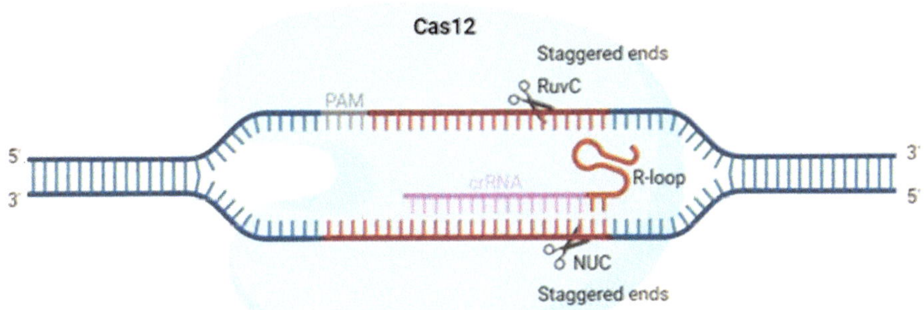

FIGURE 7.1 Schematic representation of the Type V CRISPR-Cas system mechanism. The Cas12 protein induces a double-strand break by creating a staggered end. The base-pair hybridization between the target DNA strand and the crRNA and the R-loop formation are shown. Following the formation of the R-loop, the active RuvC domain of the Cas12 cleaves the non-target strand and the nuclease lobe domain (NUC) cleaves the target strand.

site. Most of the time this mechanism results in inactivation of the gene. If there is a homologous DNA template present; replacement, deletion, or insertion of genes/nucleotides can be achieved by HDR mechanism (Tong et al., 2021). NHEJ-mediated repair is mostly used by eukaryotes which are prone to mistakes, whereas the error-free HDR mechanism results in less efficient repair than NHEJ (Pickar-Oliver and Gersbach, 2019).

7.4 GENOME EDITING IN MICROORGANISMS WITH THE CRISPR-CAS12 SYSTEMS

The CRISPR-Cas12 system might be used for genome editing of microorganisms such as fungi, bacteria, and cyanobacteria in different areas including biotechnology, pharmaceutical production, cosmetic compounds, and bio-based materials (Adiego-Pérez et al., 2019). Some products such as antibiotics and enzymes produced as a result of metabolic activities of microorganisms have been used by industry. Most of the time, the genome of the organism of interest needs to be modified for strain improvement. For this purpose, an efficient and sensitive tool for gene editing is necessary (Senthilnathan et al., 2023; Yang et al., 2023). There are some applications of CRISPR-Cas12a system in microorganisms such as *cyanobacteria* (Hesp et al., 2020), *Amycolatopsis mediterranei* (Zhou et al., 2020), *Corynebacterium glutamicum* (Jiang et al., 2017; Dong et al., 2020), *Rhodobacter capsulatus* (Zhang and Yuan, 2020), *Zymomonas mobilis* (Shen et al., 2019), *Clostridium beijerinckii* (Zhang et al., 2018), *E. coli* MG1655 (Ao et al., 2018), *Clostridium difcile* (Hong et al., 2018).

7.5 CRISPR-CAS12-MEDIATED GENOME EDITING IN HUMAN CELL LINES AND ANIMALS

For therapeutic and clinical purposes, some Cas12 effector proteins and their orthologs including Cas12a, Cas12e, and Cas12f have been used and tested in human cell lines and mice. Successful genome editing was performed by using Cas12a by inducing pluripotent stem cells (IPSCs) from Duchenne muscular dystrophy (DMD) patients to fix DMD target genes in mouse embryos (Zhang et al., 2017). In another study, silencing of the CCR5 receptor on the T cells of HIV-infected patients showed that the Cas12a protein provides safer and more specific editing than the Cas9 protein (Liu et al., 2020). In addition to these, many cell lines such as HEK293T, HEPG2, A375, Huh-7, human cancer HCT116, U2OS cell lines, and T cells have been modified with CRISPR-Cas12 system (Senthilnathan et al., 2023). For editing, different transfection methods are required for different cell lines, including the use of a commercial transfection agent, electroporation, adeno-associated virus (AAV) transduction, lenti-viral transduction, or ribonucleoprotein transfection (Senthilnathan et al., 2023).

7.6 GENOME EDITING IN PLANTS USING THE CRISPR-CAS12 SYSTEM

In the scope of agricultural production, sustainability for producing more crops with higher quantity and quality, while enduring biotic and climatic challenges is necessary. Safe and genetically modified organism (GMO)-free crops should be produced with valuable traits by genome-editing applications. For genome editing in plants, Cas12a and Cas12b have been used in maize, rice, soybean, cotton, and *Arabidopsis thaliana* (Senthilnathan et al., 2023; Yang et al., 2023). To perform the targeted mutagenesis, Cpf1 from *Francisella novicida* (FnCpf1) was expressed with its targeting crRNA in tobacco and rice and the results showed that transgenic tobacco and rice had mostly deletions on targeted sites (Endo et al., 2016). They concluded that FnCpf1 can be used for targeted mutagenesis in plants. In *Physcomitrella patens*, the efficiency of Cas12 effector protein is mostly unknown. For this reason, Pu et al. (2019) studied the targeting of multiple loci in *Physcomitrella patens* by using LbCas12a (Cas12a from *Lachnospiraceae bacterium*) and a mature crRNA-containing cassette. Similar to the study described below, their results revealed that the frequency of mutation, which was mostly a deletion, reached up to 100% in a single locus. They demonstrated that LbCas12a can be used to create loss of function transgenics in *P. patens*.

In rice, to provide resistance to rice leaf blight disease, Yu et al. (2021) performed a targeted mutagenesis via LbCas12a for the rice *Xa13* gene which had a crucial role in this disease. The results revealed that the transgenic rice seeds had improved resistance to the disease due to the deletion of two core nucleotides in the promoter of *Xa13*.

Hot and sweet cultivars of *Capsicum annuum* (pepper) were investigated in terms of efficiency of designed crRNA for LbCas12a effector protein to edit the target *CaMLO2* gene (Kim et al., 2020). The isolated protoplasts of soil-grown leaf and callus are transfected via PEG-mediated transformation and both the resulting transgenic cultivars of pepper carry edited *CaMLO2* gene with several deletions.

Lee et al. (2018) compared the activities of Cas9 and LbCas12a effector proteins by targeting the coding region of the same gene, *glossy2* in maize. Their results revealed that 90–100% of the transgenic maize were edited by Cas9 causing InDel mutations whereas 0–60% of the LbCas12a were edited in transgenic maize having on-target mutations on the *glossy2* gene. They suggested that the LbCas12a effector protein requires further research and optimization for higher editing efficiency.

LbCas12a protein and its target specifically designed crRNA were delivered into isolated protoplasts of soybean and wild tobacco to edit the *FAD2* paralogues in soybean and *AOC* in wild tobacco by Kim et al. (2017). In this study, LbCas12a induced several nucleotide deletions on targeted sites. Additionally, there were not any significant off-target mutations in the soybean genome. As a result, they suggested that LbCas12a and crRNA complex may be used as an effective plant genome-editing tool.

In *Arabidopsis thaliana*, targeted mutagenesis was tested for 11 targeted sites across four loci via two Cas12b (C2c1) endonucleases: BvCas12b (from *Bacillus* sp. V3-13) and BhCas12b V4 (from *Bacillus hisashii*). Both Cas12b effector proteins caused mutations through large deletions in multiple loci. The editing pattern was determined mostly as 5–13 nucleotide deletions. Also, no notable alterations were found on off-target sites in transgenic *Arabidopsis* (Wu et al., 2020).

The Cas12b effector protein was used for editing of tetraploid cotton (*Gossypium hirsutum*). The Cas12b protein is known for its cleavage ability at high temperatures and temperature inducibility in mammals (Wang et al., 2020). For this reason, temperature-resistant cotton was used to evaluate the editing capability of Cas12b effector protein at higher temperatures (Wang et al., 2020). The results suggest that in the transgenic cotton grown for 4 days at 45°C, the Cas12b protein performed the most efficient cleavage activity and there were no off-target mutations observed. So, the Cas12b protein was considered a precise genome-editing tool for cotton (Wang et al., 2020). As a result, of the applications of both the Cas12a and the Cas12b effector proteins in various plant species, it has been determined that the most common mutations are deletions in plants. Accordingly, this editing activity of Cas12a and Cas12b proteins causes loss of function in the targeted genes.

7.7 DETECTION AND DIAGNOSIS VIA THE CRISPR-CAS12 SYSTEM

The CRISPR-Cas12 system can also be used for detection and diagnostic purposes. In this scope, two important diagnostic tools exist: DNA endonuclease-targeted CRISPR trans-reporter (DETECTR) (Chen et al., 2018) and one HOur low-cost Multipurpose highly Efficient System (HOLMES) (Leung et al., 2022). In the DETECTR, the signal is amplified by recombinase polymerase amplification (RPA). This tool is based on the Type V effector enzyme, LbCas12a, for cleaving ssDNA in three steps. The first step involves the complementation of the effector Cas12 protein with crRNA to the reporter probes of DNA. After recognition of the target DNA, the Cas12 protein performs collateral cleavage activity for dsDNA or ssDNA. In the second step, the probes of target DNA attach to quencher molecules and fluorophores; and in the last step, these DNA probes are degraded, resulting in the release of quencher molecules and fluorophores. This provides a fluorescent signal emitted from the cleavage of target dsDNA or ssDNA (Hillary and Ceasar, 2023). With the help of this tool, human papillomavirus (HPV) is detected by Chen et al. (2018). The HOLMES nucleic acid detection system also uses collateral cleavage of ssDNA via LbCas12a effector protein. Within one hour, this tool can detect DNA or RNA viruses, and single nucleotide polymorphisms (SNPs) (Li et al., 2019). The effector protein and crRNA complex cleaves its target and the cleaved DNA probe releases a fluorescence signal which can be detected by fluorescence spectrophotometer. The signal amplification is achieved by PCR or RT-PCR. Both the DETECTR and HOLMES tools require the capturing of fluorescent signals so they are not useful in field studies (Bai et al., 2019). The newer version of this tool, called HOLMESv2, provides quantitative detection of nucleic acids, unlike DETECTR and HOLMES. The HOLMESv2 uses thermophilic AacCas12b effector protein, providing a wide range of working temperatures (Li et al., 2019). Loop-mediated isothermal amplification (LAMP) or reverse transcription LAMP (RT-LAMP) can be used for the amplification of target sequences (Leung et al., 2022). The optimum working temperature for this tool is determined as 55°C. In this system, DNA or RNA templates are amplified by using a 5' to 3' DNA polymerase for detection. For RNA detection, it is the most straightforward method of CRISPR-based biosensing (Li et al., 2019).

7.8 ADVANTAGES AND DISADVANTAGES OF USING CRISPR-CAS12 FOR GENOME EDITING

Both the Cas9 and Cas12 proteins are known as efficient and sensitive genome-editing tools for various organisms. However, the CRISPR-Cas12 system may be preferable to the CRISPR-Cas9 system since its cleavage activity results in the generation of staggered or sticky ends in the double strand whereas the cleavage activity of Cas9 protein generates blunt ends. This situation may be advantageous in some cases such as precisely combining the DNA sequences in the right orientation (Hillary and Ceasar, 2023). The detection and diagnosis purposes can be achieved in a time-saving manner via the CRISPR-Cas12 system rather than the CRISPR-Cas9 system (Li et al., 2019; Hillary and Ceasar, 2023). However, if the Cas12 protein promotes a homology-directed repair mechanism, this situation causes the Cas12 system to depend on a host DNA repair mechanism. Additionally, the repair of DNA via a homology-directed repair mechanism is related to cell division, which causes the Cas12 system in some cells to be ineffective during an active cell division (e.g., neuron cells) (Al-Shayeb et al., 2020).

REFERENCES

Adiego-Pérez, B., Paola, R., Daran, J. M., Verwaal, R., Roubos, J. A., Daran-Lapujade, P., and van der Oost, P. 2019. Multiplex genome editing of microorganisms using CRISPR-Cas. *FEMS Microbiology Letters,* 366(8), fnz086. https://doi.org/10.1093/femsle/fnz086.

Al-Shayeb, B., Sachdeva, R., Chen, L. X., Ward, F., Munk, P., Devoto, A., and Amano, Y. 2020. Clades of huge phages from across Earth's ecosystems. *Nature*, 578(7795), 425–431.

Ao, X., Yao, Y., Li, T., Yang, T.-T., Dong, X., Zheng, Z.-T., Chen, G-Q., Wu, Q., and Guo, Y. 2018. A multiplex genome editing method for Escherichia coli based on CRISPR-Cas12a. *Frontiers in Microbiology,* 9, 2307.

Badon, I. W., Oh, Y., Kim, H. J., and Lee, S. H. 2024. Recent application of CRISPR-Cas12 and OMEGA system for genome editing. *Molecular Therapy*, 32(1), 32–43.

Bai, J., Mao, L., Chen, H., Sun, Y., Li, Q. and Zhang, R., 2019. *A new automatic detection method for bundle branch block using ecgs*. In *Health Information Science: 8th International Conference, HIS 2019, Xi'an, China, October 18–20, 2019, Proceeings 8* (pp. 168–180). Springer International Publishing.

Chen, J. S., Ma, E., Harrington, L. B., Da Costa, M., Tian, X., Palefsky, J. M., and Doudna, J. A. 2018. CRISPR-Cas12a target binding unleashes indiscriminate single-stranded DNase activity. *Science,* 360(6387), 436–439.

Chen, W., Ma, J., Wu, Z., Wang, Z., Zhang, H., Fu, W., Pan, D., Shi, J., and Ji, Q. 2023. Cas12n nucleases, early evolutionary intermediates of type V CRISPR, comprise a distinct family of miniature genome editors. *Molecular Cell,* 83(15), 2768–2780.e6. https://doi.org/10.1016/j.molcel.2023.06.014

Dong, J., Kan, B., Liu, H., Zhan, M., Wang, S., Xu, G., Han, R., and Ni, Y. 2020. CRISPR-Cpf1-assisted engineering of Corynebacterium glutamicum SNK118 for enhanced L-ornithine production by NADP-dependent glyceraldehyde-3-phosphate dehydrogenase and NADH-dependent glutamate dehydrogenase. *Applied Biochemistry and Biotechnology,* 191, 955–967. https://doi.org/10.1007/s12010-020-03231-y

Endo, A., Masafumi, M., Kaya, H., and Toki, S. 2016. Efficient targeted mutagenesis of rice and tobacco genomes using Cpf1 from Francisella novicida. *Scientific Reports*, 6(1), 38169.

Feng, W., Zhang, H., and Le, X. C. 2023. Signal amplification by the trans-cleavage activity of CRISPR-Cas systems: Kinetics and performance. *Analytical Chemistry*, 95(1), 206–217.

Gaj, T., Sirk, S. J., Shui, S. L., and Liu, J. 2016. Genome-editing technologies: Principles and applications. *Cold Spring Harbor Perspectives in Biology*, 8(12), a023754. https://doi.org/10.1101/cshperspect.a023754

Gostimskaya, I. 2022. CRISPR-Cas9: A history of its discovery and ethical considerations of its use in genome editing. *Biochemistry. Biokhimiia*, 87(8), 777–788. https://doi.org/10.1134/S0006297922080090

Harrington, L. B., Ma, E., Chen, J. S., Witte, I. P., Gertz, D., Paez-Espino, D., Al-Shayeb, B., Kyrpides, N. C., Burstein, D., Banfield, J. F., and Doudna, J. A. 2020. A scoutRNA is required for some Type V CRISPR-Cas systems. *Molecular Cell*, 79(3), 416–424.e5. https://doi.org/10.1016/j.molcel.2020.06.022

Hesp, K., Flores Alvarez, J. L., Alexandru, A.-M, van der Linden, J., Martens, D. E., Wijfels, R. H., and Pomponi, S. A. 2020. CRISPR/Cas12amediated gene editing in Geodia barretti sponge cell culture. *Frontiers in Marine Science,* 7, 1134.

Hillary, V. E., and Ceasar, S. A. 2023. A review on the mechanism and applications of CRISPR/Cas9/Cas12/Cas13/Cas14 proteins utilized for genome engineering. *Molecular Biotechnology*, 65(3), 311–325.

Hong, W., Zhang, J., Cui, G., Wang, L., and Wang, Y. 2018. Multiplexed CRISPR-Cpf1-mediated genome editing in Clostridium difficile toward the understanding of pathogenesis of C. difficile infection. *ACS Synthetic Biology,* 7, 1588–1600.

Hryhorowicz, M., Lipiński, D., and Zeyland, J. 2023. Evolution of CRISPR/Cas systems for precise genome editing. *International Journal of Molecular Sciences,* 24(18), 14233. https://doi.org/10.3390/ijms241814233

Hryhorowicz, M., Lipiński, D., Zeyland, J. et al. 2017. CRISPR/Cas9 immune system as a tool for genome engineering. *Archivum Immunologiae Therapiae Experimentalis,* 65, 233–240. https://doi.org/10.1007/s00005-016-0427-5

Huang, C. J., Adler, B. A., and Doudna, J. A. 2022. A naturally DNase-free CRISPR-Cas12c enzyme silences gene expression. *Molecular Cell,* 82(11), 2148–2160.

Jiang, Y., Qian, F., Yang, J., Liu, Y., Dong, F., Xu, C., Sun, B., Chen, B., Xu, X., and Li, Y. 2017. CRISPR-Cpf1 assisted genome editing of Corynebacterium glutamicum. *Nature Communications,* 8, 1–11.

Khalil, A. M. 2020. The genome editing revolution: Review. *Journal of Genetic Engineering and Biotechnology,* 18, 68. https://doi.org/10.1186/s43141-020-00078-y

Kim, H., Choi, J., and Won, K. H. 2020. A stable DNA-free screening system for CRISPR/RNPs-mediated gene editing in hot and sweet cultivars of Capsicum annuum. *BMC Plant Biology,* 20, 1–12.

Kim, H., Kim, S. T., Ryu, J., Kang, B. C., Kim, J. S., and Kim, S. G. 2017. CRISPR/Cpf1-mediated DNA-free plant genome editing. *Nature Communications*, 8(1), 14406.

Kim, Y., Kweon, J., and Kim, JS. 2013. TALENs and ZFNs are associated with different mutation signatures. *Nature Methods,* 10, 185. https://doi.org/10.1038/nmeth.2364

Koonin, E. V., Makarova, K. S., and Zhang, F. 2017. Diversity, classification and evolution of CRISPR-Cas systems. *Current Opinion in Microbiology,* 37, 67–78.

Lee, K., Zhang, Y., Kleinstiver, B. P., Guo, J. A., Aryee, M. J., Miller, J., ... and Wang, K. 2019. Activities and specificities of CRISPR/Cas9 and Cas12a nucleases for targeted mutagenesis in maize. *Plant Biotechnology Journal*, 17(2), 362–372.

Leung, R. K. K., Cheng, Q. X., Wu, Z. L., Khan, G., Liu, Y., Xia, H. Y., and Wang, J. 2022. CRISPR-Cas12-based nucleic acids detection systems. *Methods,* 203, 276–281.

Li, Y., Li, S., Wang, J., and Liu, G. 2019. CRISPR/Cas systems towards next-generation biosensing. *Trends in Biotechnology*, 37(7), 730–743.

Liu, J. J., Orlova, N., Oakes, B. L. et al. 2019. CasX enzymes comprise a distinct family of RNA-guided genome editors. *Nature,* 566, 218–223. https://doi.org/10.1038/s41586-019-0908-x

Liu, Z., Liang, J., Chen, S., Wang, K., Liu, X., Liu, B., Xia, Y., Guo, M., Zhang, X., and Sun, G. 2020. Genome editing of CCR5 by AsCpf1 renders CD4+ T cells resistance to HIV-1 infection. *Cell & Bioscience,* 10, 1–13.

Makarova, K. S., and Koonin, E. V. 2015. Annotation and classification of CRISPR-Cas systems. *Methods in Molecular Biology (Clifton, N.J.),* 1311, 47–75. https://doi.org/10.1007/978-1-4939-2687-9_4

Makarova, K. S., Wolf, Y. I., Alkhnbashi, O. S., Costa, F., Shah, S. A., Saunders, S. J., Barrangou, R., Brouns, S. J., Charpentier, E., Haft, D. H., Horvath, P., Moineau, S., Mojica, F. J., Terns, R. M., Terns, M. P., White, M. F., Yakunin, A. F., Garrett, R. A., van der Oost, J., Backofen, R., … Koonin, E. V. 2015. An updated evolutionary classification of CRISPR-Cas systems. *Nature Reviews Microbiology,* 13(11), 722–736. https://doi.org/10.1038/nrmicro3569

Makarova, K. S., Wolf, Y. I., Iranzo, J., Shmakov, S. A., Alkhnbashi, O. S., Brouns, S. J. J., … Koonin, E. V. 2019. Evolutionary classification of CRISPR–Cas systems: A burst of class 2 and derived variants. *Nature Reviews Microbiology. doi:10.1038/s41579-019-0299-x*

Miller, J., Tan, S., Qiao, G. et al. 2011. A TALE nuclease architecture for efficient genome editing. *Nature Biotechnology,* 29, 143–148. https://doi.org/10.1038/nbt.1755

Ming, M., Ren, Q., Pan, C. et al. 2020. CRISPR–Cas12b enables efficient plant genome engineering. *Nature Plants,* 6, 202–208. https://doi.org/10.1038/s41477-020-0614-6

Oakes, B. L., Nadler, D. C., and Savage, D. F. 2014. Protein engineering of Cas9 for enhanced function. *Methods in Enzymology,* 546, 491–511. https://doi.org/10.1016/B978-0-12-801185-0.00024-6

Omura, S. N., Nakagawa, R., Südfeld, C. et al. 2023. Mechanistic and evolutionary insights into a type V-M CRISPR–Cas effector enzyme. *Nature Structural Molecular Biology,* 30, 1172–1182. https://doi.org/10.1038/s41594-023-01042-3

Perez Rojo, F., Nyman, R. K. M., Johnson, A. A. T., Navarro, M. P., Ryan, M. H., Erskine, W., and Kaur, P. 2018. CRISPR-Cas systems: Ushering in the new genome editing era. *Bioengineered,* 9(1), 214–221. https://doi.org/10.1080/21655979.2018.1470720

Pickar-Oliver, A., and Gersbach, C.A. 2019. The next generation of CRISPR–Cas technologies and applications. *Nature Reviews Molecular Cell Biology,* 20, 490–507. https://doi.org/10.1038/s41580-019-0131-5

Pu, X., Liu, L., Li, P., Huo, H., Dong, X., Xie, K., ... and Liu, L. 2019. A CRISPR/LbCas12a-based method for highly efficient multiplex gene editing in Physcomitrella patens. *The Plant Journal*, 101(4), 863–872.

Sasnauskas, G., Tamulaitiene, G., Druteika, G. et al. 2023. TnpB structure reveals minimal functional core of Cas12 nuclease family. *Nature,* 616, 384–389. https://doi.org/10.1038/s41586-023-05826-x

Senthilnathan, R., Ilangovan, I., Kunale, M., Easwaran, N., Ramamoorthy, S., Veeramuthu, A., and Kodiveri Muthukaliannan, G. 2023. An update on CRISPR-Cas12 as a versatile tool in genome editing. *Molecular Biology Reports*, 50(3), 2865–2881.

Shen, W., Zhang, J., Geng, B., Qiu, M., Hu, M., Yang, Q., Bao, W., Xiao, Y., Zheng, Y., and Peng, W. 2019. Establishment and application of a CRISPR–Cas12a assisted genome-editing system in Zymomonas mobilis. *Microbial Cell Factories,* 18, 1–11.

Shmakov, S., Smargon, A., Scott, D. et al. 2017. Diversity and evolution of class 2 CRISPR–Cas systems. *Nature Reviews Microbiology,* 15, 169–182. https://doi.org/10.1038/nrmicro.2016.184

Strecker, J., Jones, S., Koopal, B. et al. 2019. Engineering of CRISPR-Cas12b for human genome editing. *Nature Communications,* 10, 212. https://doi.org/10.1038/s41467-018-08224-4

Strecker, J., Ladha, A., Gardner, Z., Schmid-Burgk, J. L., Makarova, K. S., Koonin, E. V., and Zhang, F. 2019. RNA-guided DNA insertion with CRISPR-associated transposases. *Science (New York, N.Y.),* 365(6448), 48–53. https://doi.org/10.1126/science.aax9181

Sun, A., Li, CP., Chen, Z. et al. 2023. The compact Casπ (Cas12l) 'bracelet' provides a unique structural platform for DNA manipulation. *Cell Research,* 33, 229–244. https://doi.org/10.1038/s41422-022-00771-2

Swarts, D. C. 2019. Stirring up the Type V alphabet soup. *The CRISPR Journal,* 2(1), 14–16. doi:10.1089/crispr.2019.29044.dcs

Tong, B., Dong, H., Cui, Y., Jiang, P., Jin, Z., and Zhang, D. 2021. The versatile type V CRISPR effectors and their application [illegible] 8,

Tröder, S. E., and Zevnik, B. 2022. History of genome editing: From meganucleases to CRISPR. *Laboratory Animals*, 56(1), 60–68. https://doi.org/10.1177/0023677221994613

van der Oost, J., and Patinios, C. 2023. The genome editing revolution. *Trends in Biotechnology,* 41(3), 396–409. https://doi.org/10.1016/j.tibtech.2022.12.022

van der Oost, J., Westra, E., Jackson, R. et al. 2014. Unravelling the structural and mechanistic basis of CRISPR–Cas systems. *Nature Reviews Microbiology,* 12, 479–492. https://doi.org/10.1038/nrmicro3279

Wang, Q., Alariqi, M., Wang, F., Li, B., Ding, X., Rui, H., ... and Jin, S. 2020. The application of a heat-inducible CRISPR/Cas12b (C2c1) genome editing system in tetraploid cotton (G. hirsutum) plants. *Plant Biotechnology Journal*, 18(12), 2436–2443.

Wu, F., Qiao, X., Zhao, Y., Zhang, Z., Gao, Y., Shi, L., ... and Kong, D. 2020. Targeted mutagenesis in Arabidopsis thaliana using CRISPR-Cas12b/C2c1. *Journal of Integrative Plant Biology*, 62(11), 1653–1658.

Wu, W. Y., Mohanraju, P., Liao, C., Adiego-Pérez, B., Creutzburg, S. C. A., Makarova, K. S., Keessen, K., Lindeboom, T. A., Khan, T. S., Prinsen, S., Joosten, R., Yan, W. X., Migur, A., Laffeber, C., Scott, D. A., Lebbink, J. H. G., Koonin, E. V., Beisel, C. L., and van der Oost, J. 2022. The miniature CRISPR-Cas12m effector binds DNA to block transcription. *Molecular Cell*, 82(23), 4487–4502.e7. https://doi.org/10.1016/j.molcel.2022.11.003

Xiao, R., Li, Z., Wang, S., Han, R., and Chang, L. 2019. Structural basis for substrate recognition and cleavage by the dimerization-dependent CRISPR–Cas12f nuclease. *Nucleic Acids Research*, 49(7), 4120–4128. https://doi.org/10.1093/nar/gkab179

Yan, F., Wang, W,. and Zhang, J., 2019. CRISPR-Cas12 and Cas13: the lesser known siblings of CRISPR-Cas9. *Cell Biology and Toxicology*, 35, 489–492. https://doi.org/10.1007/s10565-019-09489-1

Yan, W. X., Hunnewell, P., Alfonse, L. E., Carte, J. M., Keston-Smith, E., Sothiselvam, S., … Scott, D. A. 2018. Functionally diverse type V CRISPR-Cas systems. Science, eaav7271. doi:10.1126/science.aav7271

Yang, Y., Wang, D., Lü, P., Ma, S., and Chen, K. 2023. Research progress on nucleic acid detection and genome editing of CRISPR/Cas12 system. *Molecular Biology Reports*, 50(4), 3723–3738.

Yu, K., Liu, Z., Gui, H., Geng, L., Wei, J., Liang, D., ... and Chen, X. 2021. Highly efficient generation of bacterial leaf blight-resistant and transgene-free rice using a genome editing and multiplexed selection system. *BMC Plant Biology*, 21(1), 1–10.

Zetsche, B., Gootenberg, J. S., Abudayyeh, O. O., Slaymaker, I. M., Makarova, K. S., Essletzbichler, P., Volz, S. E., Joung, J., van der Oost, J., Regev, A., Koonin, E. V., and Zhang, F. 2015. Cpf1 is a single RNA-guided endonuclease of a class 2 CRISPR-Cas system. *Cell,* 163(3), 759–771. https://doi.org/10.1016/j.cell.2015.09.038

Zhang, J., Hong, W., Zong, W., Wang, P., and Wang, Y. 2018. Markerless genome editing in Clostridium beijerinckii using the CRISPRCpf1 system. *Journal of Biotechnology,* 284, 27–30.

Zhang, Y., Long, C., Li, H., McAnally, J. R., Baskin, K. K., Shelton, J. M., Bassel-Duby, R., and Olson, E. N. 2017. CRISPR-Cpf1 correction of muscular dystrophy mutations in human cardiomyocytes and mice. *Science Advanced,* 3, e1602814.

Zhang, Y., and Yuan, J. 2020. *CRISPR/Cas12a mediated genome engineering in photosynthetic bacteria.* bioRxiv.

Zhou, Y., Liu, X., Wu, J., Zhao, G., and Wang, J. 2020. CRISPR-Cas12aassisted genome editing in Amycolatopsis mediterranei. *Frontiers in Bioengineering and Biotechnology,* 8, 698.

Zischewski, J., Fischer, R., and Bortesi, L. 2017. Detection of on-target and of-target mutations generated by CRISPR/Cas9 and other sequence-specifc nucleases. *Biotechnology Advances,* 35, 95–104.

8 Beyond Traditional Gene Editing

Expanding Applications of CRISPR-Cas

Aftab Ahmad, Muhammad Rasheed, Sidra Ashraf, Ayesha Mobeen, Hira Yousaf, Noor ul Huda, Ahmed Al-Harrasi, and Sameer H. Qari

8.1 INTRODUCTION

Eukaryotes possess vast and diverse genomes, consisting of billions of base pairs. Moreover, regulating gene expression in eukaryotes is more complex than in prokaryotes. Genetic engineering enabled researchers to modify eukaryotic genomes by introducing desired traits. However, it always involved random insertions of the transgenes in the genome. Recently, site- directed nucleases (SDNs), a notable breakthrough in genome editing, allow precise and accurate modifications at a target site within the genome (Adli, 2018). The success of gene editing with programmable nucleases relies on the cellular repair process of an induced double-strand break (DSB) in the genome (Kim et al., 2014). Programmable nucleases, such as zinc-finger nucleases (ZFNs) (Handel et al., 2011), transcription activator-like effector nucleases (TALENs) (LaFountaine et al., 2015), and the CRISPR/Cas system (Barrangou et al., 2016), have been successfully demonstrated for precise gene-editing applications in bacteria, animals, and plants.

The CRISPR/Cas systems are highly diverse and classified into two classes, which encompass six types and 33 subtypes, according to the latest taxonomy (Makarova et al., 2020). The class 1 CRISPR/Cas systems comprise types I, III, and IV, which rely on multiple Cas proteins for gene editing applications. In contrast, the class 2 systems, including Types II, V, and VI, require a single Cas nuclease for gene-editing purposes (Makarova et al., 2020). Class 2 Cas effectors such as Cas9 (Type II), Cas12a (Type V), and Cas13 (Type VI) have been extensively used for gene-editing studies using CRISPR-Cas (Tong et al., 2021). As all these systems use a CRISPR-RNA (crRNA) to drive Cas nuclease to the target site in the genome, it is simple to "program" these proteins to the desired site in the genome by designing a spacer complementary to that target site (Lino et al., 2018). CRISPR technology is continuously evolving with new gene-editing approaches and CRISPR-Cas systems. For example, CRISPR/Cas13 is a powerful RNA editing system with diverse applications in the RNA world (Nidhi et al., 2021). The successful implementation of this groundbreaking system has greatly expanded the potential applications of CRISPR technology, specifically in manipulating gene regulation at the RNA level (Akinci et al., 2021). In addition, dCas9 enables researchers to manipulate gene regulation at the transcriptional dCas9 fused with a transcription repressor, such as the KRAB domain, to the promoter region of a gene, it is possible to inhibit transcription, effectively suppressing the gene's activity (Cai et al., 2023). In contrast, recruiting an activator domain, such as VP64 with dCas9, to the promoter region of a gene may enhance transcription, leading to the activation of the gene (Chen et al., 2017). In addition, dCas9 allows scientists to observe chromatin's organization and dynamic changes surrounding a particular gene by combining dCas9 with fluorescent proteins like GFP. For example, CRISPR Rainbow is an advanced system to visualize multiple genes simultaneously in living cells (Wang et al., 2018). The versatility of CRISPR-Cas can be enhanced through various modifications of the dCas9, enabling it to perform tasks, such as base editing and epigenetic modifications (Goell et al., 2021). Prime editing is a recent development of CRISPR technology that uses nCas9 fused with reverse transcriptase to achieve precise base modifications in the genome without the provision of a donor DNA template (Kantor et al., 2020). The CRISPR/Cas systems such as dCas9, Cas12, and Cas13 hold immense potential beyond the genome-editing application of the CRISPR-Cas, particularly in agriculture and human health. Scientists are particularly interested in applications of CRISPR-Cas to improve agricultural productivity, leading to food security in developing countries, especially using multiplex gene editing (Bhatia et al., 2023). This chapter highlights the extensive potential of CRISPR/Cas systems for beyond-gene-editing applications of CRISPR-Cas.

8.2 DEVELOPMENTS IN CRISPR-CAS TECHNOLOGY

The CRISPR technology has been recognized as a powerful tool for introducing precise gene modifications in the genome. The technique mainly relies on introducing a double-strand break (DSB) in the DNA at the target site, subsequently repaired by either the non-homologous end joining (NHEJ) mechanism or the homology-directed repair (HDR) mechanism (Yang et al., 2020). NHEJ is a preferred DSB repair system in the cell that results in indels at the DSB site, leading to gene knockouts. This process is template-independent and can insert or delete a few

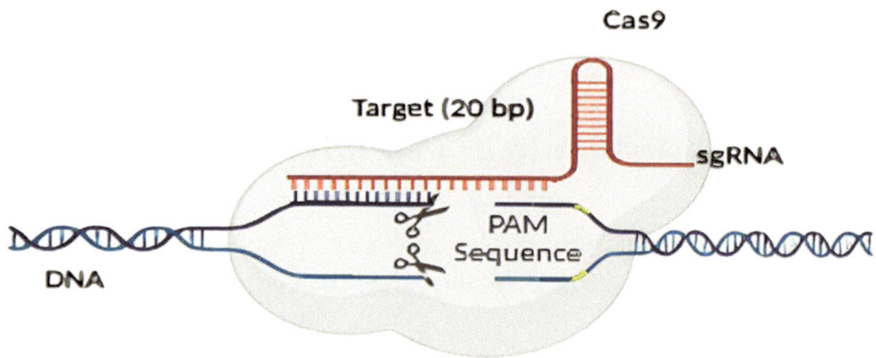

FIGURE 8.1 Schematic diagram of CRISPR-Cas9.

animals (Weiss et al., 2024). In contrast, HDR is a precise and error-free process that involves using a donor DNA template to precisely insert the desired DNA fragments at the target site in the genome, leading to gene insertions or replacements. Nevertheless, the HDR mechanism exhibits lower efficiency compared to NHEJ and operates exclusively during specific cell cycle phases, primarily G2 and S. Therefore, CRISPR holds immense potential to introduce either random mutations or make precise edits at the desired location in the genome in diverse organisms (Chen et al., 2024).

CRISPR-Cas9 has been recognized as one of the most widely used CRISPR systems for genome-editing applications. The spCas9 isolated from *S. pyrogens, was* one of the first CRISPR/Cas systems that was engineered for eukaryotic gene editing (Serajian et al., 2021). Engineered CRISPR/Cas9 system is composed of two components: a gRNA (sgRNA) and a Cas nuclease. The sgRNA is a short RNA molecule, typically consisting of 20 nucleotides that are complementary to the target region. The RNA-guided Cas9 nuclease contains two functional HNH and RuvC nuclease domains to create DSB at the target site. Once sgRNA binds its complementary sequence in the genome with an adjacent PAM site, Cas9 creates a DSB three bp upstream to the PAM. Multiplex gene editing is a significant advantage of CRISPR-Cas9 with emerging applications in metabolic engineering, synthetic biology, and studying cellular pathways (Sander et al., 2014) (Figure 8.1).

However, Cas9 has its limitations, such as off-target effects, specific PAM requirements, and the large size (984–1629 amino acids) of Cas9. Novel members of CRISPR systems, such as Cas12, Cas13, Cas14, and CasX, have been characterized and demonstrated for gene editing in various organisms. These systems offer unique advantages that help overcome the limitations of the conventional CRISPR/Cas9 system. For example, compared with Cas9, Cas12 is a smaller protein that requires a T-rich PAM and produces staggered ends of the DNA at the DSB site (Hillary et al., 2023) (Figure 8.2).

CRISPR-Cas13 is an RNA editing system that has been demonstrated for RNA editing in animals and plants without introducing any permanent mutations in the genome. Cas13 has been successfully demonstrated for earlier detection of pathogenic agents, including bacteria and viruses, in animals and plants (Singh, Bindal, et al., 2022) (Figure 8.3).

Cas14 is a recently discovered member of the CRISPR family that specifically targets single-stranded DNA (ssDNA). It is significantly smaller than Cas9, approximately half its size, and it can cleave ssDNA at certain locations without requiring a PAM site. The system is categorized into three subtypes: Cas14a, Cas14b, and Cas14c (He et al., 2023) (Figure 8.4).

CRISPR/CasX is also a newly identified member of the CRISPR/Cas family. It is small, has a unique PAM requirement, and creates staggered cuts at the DSB. Recent biochemical studies have shown that CasX has characteristics of both Cas9 and Cas12, functioning as a hybrid nuclease. This approach has been successfully used for genome editing in bacterial and human cells (Sultan et al., 2022) (Figure 8.5).

Several factors influence the precision and specificity of CRISPR/Cas systems in gene editing, including the frequency of off-target effects. Studies have demonstrated that the accuracy of the Cas9 enzyme is significantly affected by the size of the guide RNA (sgRNA). Extending the length of sgRNA can improve the precision of Cas9, while the optimal length for accurate editing is 20 base pairs. Off-target effects at unintended sites within the genome represent significant challenges in CRISPR/Cas-mediated genome editing (Guo et al., 2023). Recent research has shown that CRISPR/Cas9 has the potential to edit a target site, with a maximum of five mismatches between the sgRNA and the target DNA sequence (Lee et al., 2020). Hence, it is crucial to accurately predict potential off-targets to ensure the safe utilization of CRISPR/Cas9, particularly in research projects focused on developing novel therapies (Sullivan et al., 2020). To assist with this, numerous AI tools have been developed to precisely predict off-target effects (Cheng et al., 2023). Similarly, techniques such as truncated sgRNA and a Cas9 variant with nicking activity have been used to minimize off-target effects. To overcome the limitations of PAM, alternate Cas variants with unique PAM sequences have been characterized from other bacteria (Arroyo-Olarte et al., 2021).

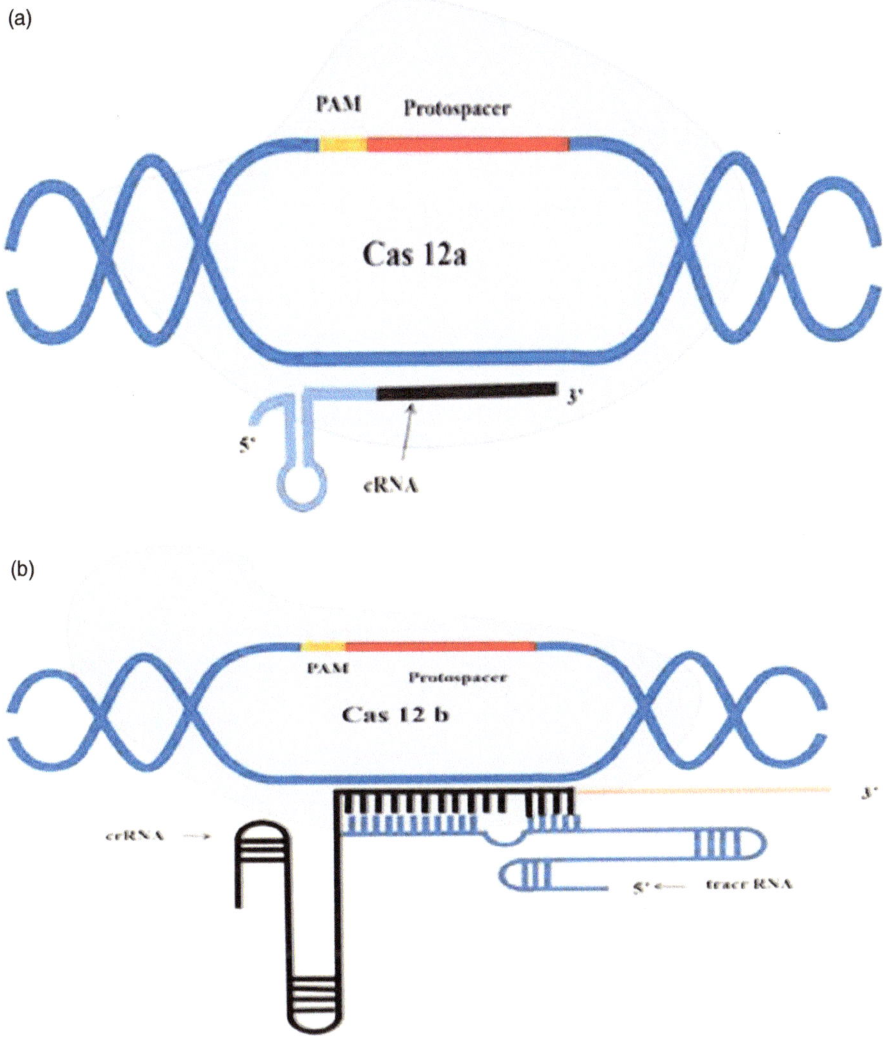

FIGURE 8.2 Schematic diagram of CRISPR-Cas12a and 12b.

8.3 CRISPR POTENTIAL BEYOND TRADITIONAL GENE-EDITING APPLICATIONS

CRISPR/Cas systems have been successfully demonstrated for precise gene-editing applications in plants, animals, and bacteria. Initially, applications of the CRISPR-Cas9 system were mainly focused on gene knockout using the NHEJ repair pathway. However, today, beyond-gene editing applications of CRISPR-Cas9 are exceeding its conventional gene-editing applications such as CRISPRi, CRISPRa, base editing, prime editing, and CRISPR-based diagnostics. All these beyond-gene-editing applications of CRISPR-Cas depend on the sgRNAs to drive the dCas9 or nCas9 to the target site in the genome to perform actions such as binding, nicking, or recruiting other molecules such as activators, repressors, fluorescent proteins, base editors or reverse transcriptase to the target site (Tavakoli et al., 2021). In addition, the CRISPR/Cas system has been used in laboratories to develop gene novel proteins with new functions (Wang et al., 2021). In this chapter, we investigate the various uses of CRISPR/Cas that extend beyond the scope of genome editing.

8.4 CRISPR-BASED DIAGNOSTIC METHODS IN AGRICULTURE AND HUMAN HEALTH

8.4.1 CRISPR-Based Diagnosis of Plant Pathogens

CRISPR/Cas technology has been widely employed in various fields, including disease detection, pharmaceutical advancement, cell imaging, and gene expression regulation (Serajian et al., 2021). Precise and early diagnosis of diseases is essential for proper treatment. Existing diagnostic methods such as PCR, real-time PCR, and ELISA are very precise and sensitive but require expensive equipment and professional expertise (Yang et al., 2022). Recently, CRISPR-based diagnostics methods have enabled precise and accurate

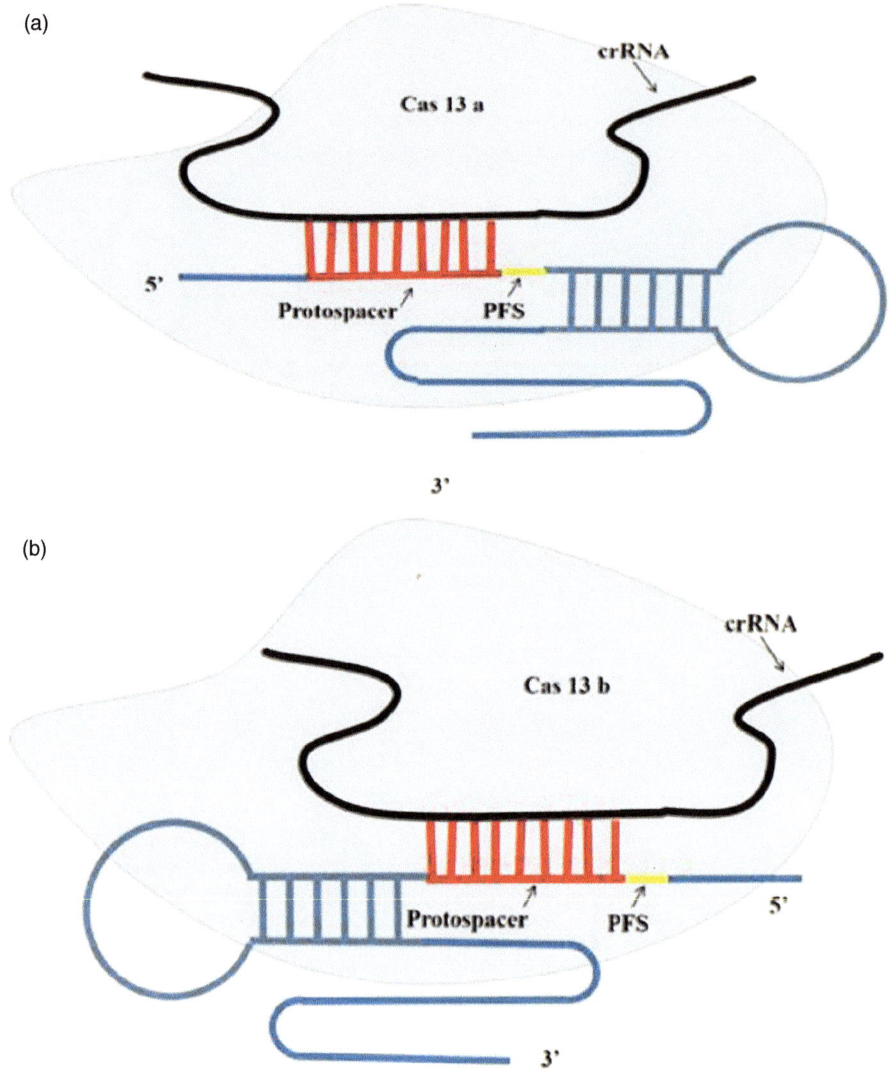

FIGURE 8.3 Schematic diagram of CRISPR-Cas13a and 13b.

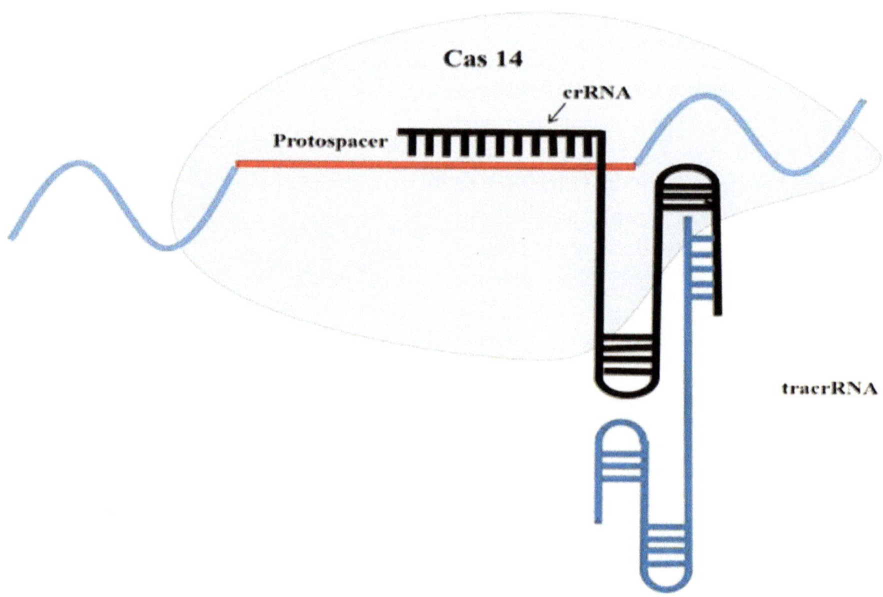

FIGURE 8.4 Schematic diagram of CRISPR-Cas14. CRISPR-Cas14 together with crRNA and tracRNA binds and cleaves single-stranded DNA without PAM recognition.

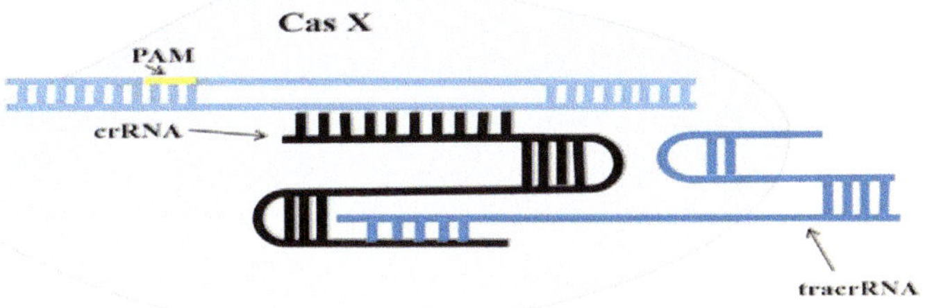

FIGURE 8.5 Schematic diagram of CRISPR-CasX. RNA-dependent plasmid with two natural RNAs, including crRNA and tracRNA, binds and cleaves double-stranded DNA.

importance of CRISPR/Cas-based diagnostic tools, leading to the development of advanced and precise diagnostic methods that can be used directly at the point of care (Kostyusheva et al., 2022). CRISPR technology, combined with AI and NGS, holds immense potential for rapid and earlier diagnosis and management of diseases and precision medicine. CRISPR-based methodologies are currently under investigation for the complex challenge of identifying and forecasting cancer. Pathogens can be detected by programming Cas12 and Cas13 proteins to target specific DNA or RNA sequences. Both CRISPR systems have been extensively employed for the diagnosis of viral infections in humans and animals and, to a lesser degree, in plants. The designed sgRNA will specifically bind to the distinct DNA regions of the pathogen's genome, allowing the Cas enzyme to detect the presence of the disease-causing agents in the plant sample (Singh et al., 2022).

The applications of Cas12 and Cas13-based disease diagnosis in plants offer a significant advantage for on-site testing. It enables farmers to make well-informed decisions regarding disease management in real time, resulting in enhanced efficiency and sustainability in agricultural practices (Yuan et al., 2022). However, CRISPR-based disease diagnosis faces challenges and researchers need to overcome these limitations to fully realize the potential of CRISPR in developing disease diagnostics and tracking of new infections. For instance, one significant limitation of most existing CRISPR-based diagnostics is that they require pre-amplification to detect targets below the femtomolar range. While the primers used for pre-amplification enhance specificity, they also add complexity to the assay, increase expenses, and lengthen reaction time. Potential improvements include integrating non-PCR-based signal-amplification techniques and modifying the Cas enzyme, crRNA, and reporter molecule (Rahman et al., 2022). It has been suggested that future applications of CRISPR-based diagnostics will likely incorporate technologies that combine synthetic biology and materials science with detecting nucleic acids on solid surfaces, wearable devices, and medical equipment (Pandya et al., 2023).

application of CRISPR/Cas12a, among the several variants of Cas12, is being studied for its potential in diagnostic applications using its unique collateral cleavage properties. CRISPR/Cas12a is a molecular scissor that cleaves single-stranded DNA within its vicinity once activated by gRNA and the presence of its target DNA template. HOLMES (One-Hour Low- Cost Multipurpose Highly Efficient System), a system used in detecting nucleic acid and DETECTR (DNA endonuclease-targeted CRISPR trans reporter) engage with the Cas-12a system for their function. These diagnostic tools have been successfully used to identify the Japanese encephalitis virus (JAV) and the SARS-CoV-2 virus (Sajal, 2024). CRISPR/Cas9 as a diagnostic molecular tool commenced in 2016, when Cas9 was combined with a DNA amplification technique to detect the Zika virus (Kostyusheva et al., 2022). Similarly, CRISPR-Cas9 triggered exponential amplification reaction (CAS-EXPAR) was used to identify miRNAs. This method involves a site-specific nuclease that targets a specific DNA sequence, which is then detected using a fluorescence-based method (Huang et al., 2024). In addition, the Finding Low Abundance Sequences by Hybridization (FLASH) technique was employed to identify drug-resistant bacteria such as *Staphylococcus aureus* and *Plasmodium falciparum* (Tram et al., 2023). Moreover, CRISPR-Cas9-mediated lateral flow nucleic acid assay (CASLFA) was used to diagnose the African swine flu pathogen (Kostyusheva et al., 2022) and FnCas9 editor linked uniform detection assay (FELUDA) was used to identify SARC-CoV-2 (Azhar et al., 2021). These studies highlight the beyond-gene-editing potential of CRISPR, and accurate diagnosis for therapeutic purposes in the rapidly advancing world. It is important to note that the pandemic of SARC-CoV-19 has highlighted the importance of rapid, efficient, and cost-effective molecular diagnostics in controlling the transmission of diseases and future pandemics. Molecular diagnostics using Cas9 is a time-consuming process due to its complex, multi-step gene-editing process, whereas CRISPR/Cas12 and CRISPR/Cas13 have enabled highly efficient diagnostic techniques that are nearly optimal in terms of time and cost-effectiveness, rapid response, and high sensitivity (Weng et al., 2023).

enzyme. CRISPR-Cas13a system has been characterized for its efficiency, robustness, high levels of accuracy and precision, and versatility in diagnostic techniques. The accuracy of this system during diagnostic tests is 99% and the sensitivity is 100% with an extremely low level of detection limit (Zhao et al., 2022). In addition, the improved version of this system does not require a pre-amplification step if complementary amplification is performed and it may detect pathogen nucleic acid at as low as the attomolar level. This system has been successfully employed for the detection of both contagious and non-contagious diseases (Wang et al., 2021). The CRISPR/Cas13a system can also detect single nucleotide polymorphisms (SNPs) which could be used to diagnose point mutations. CRISPR/Cas13a has been demonstrated to detect pathogenic microbes and predict the prognosis of oncogene mutations. For instance, CRISPR-Cas13 based Specific High-sensitivity Enzymatic Reporter Unlocking (SHERLOCK) and Heating Un-extracted Diagnostic Samples to Obliterate Nucleases (HUDSON) have been developed for diagnostic purposes (Fapohunda et al., 2022). These methods have been used in detecting DNA-based viruses such as Epstein-Barr virus (EBV) and cytomegalovirus (CMV). SHERLOCK has been specifically designed to detect RNA, focusing on RNA viruses such as Dengue, Ebola, and COVID-19. It garnered significant attention throughout the pandemic for operating under stringent and highly restrictive conditions (Kumaran et al., 2023). In conclusion, these methods represent the potential of CRISPR-based diagnostic methods for highly sensitive and precise detection of pathogens in agriculture, human health, and therapeutics.

8.5 FINE TUNING TRANSCRIPTION THROUGH THE CRISPR/DCAS SYSTEM

RNA interference (RNAi) has long been a commonly used approach in laboratories to suppress gene expression at the post-transcriptional level. Despite its applications to all organisms, RNAi has encountered significant challenges, including off-target effects, reduced efficacy over time, and restricted to transcriptional level (Miki et al., 2004). The developments in CRISPR/Cas9 and its associated systems, such as CRISPR/dCas9 (used for gene activation known as CRISPRa, and interference known as CRISPRi), as well as CRISPR/nCas9 (base editing and prime editing), have entirely revolutionized the landscape of functional genomics in all organisms (Munawar et al., 2021). CRISPR/Cas9 is noteworthy for its straightforwardness, accuracy, and low off-target effects, making it a potent tool for biological research. Shortly after the discovery, scientists developed CRISPR/dCas9 to precisely control gene expression using a catalytically inactivated (dCas9) version of Cas9. dCas9, the inactive form of Cas9, cannot create DSB in the genome but retains its ability to bind at the target site, and has been successfully employed to selectively recruit activators, repressors, or fluorescent markers at the targeted location in genome (Boettcher et al., 2015; Kampmann, 2018; Schuster et al., 2019). CRISPR/dCas9 has demonstrated significant prospects due to its extensive range of applications in gene control, chromatin studies, genome imaging, and epigenomics. It can actively control the expression of numerous genes using multiple gRNAs, exerting simultaneous repressing and activating effects at different genes (Goell & Hilton, 2021). CRISPRi is a reliable technique for reducing gene expression at the DNA level. It directly attaches to the transcription start sites or TATA boxes, stopping RNA polymerase from initiating gene transcription. For example, Gilbert et al. (2013) have demonstrated that gene expression could be programmed by fusing dCas9 with transcriptional repressors, such as Krüppel-associated box (KRAB), in human cell lines (Figure 8.6).

CRISPRa is a highly effective technique for targeted activation of gene expression in a genome. CRISPRa is employed by combining CRISPR/dCas9 with transcriptional activators such as VP64 to enhance the expression of specific genes. These systems have a wide range of applications, including genetic screens and the overexpression of target genes. The first-generation CRISPRa system dCas9- VP64 was created by combining dCas9 with VP64, a potent transcriptional activator. Similarly, adding

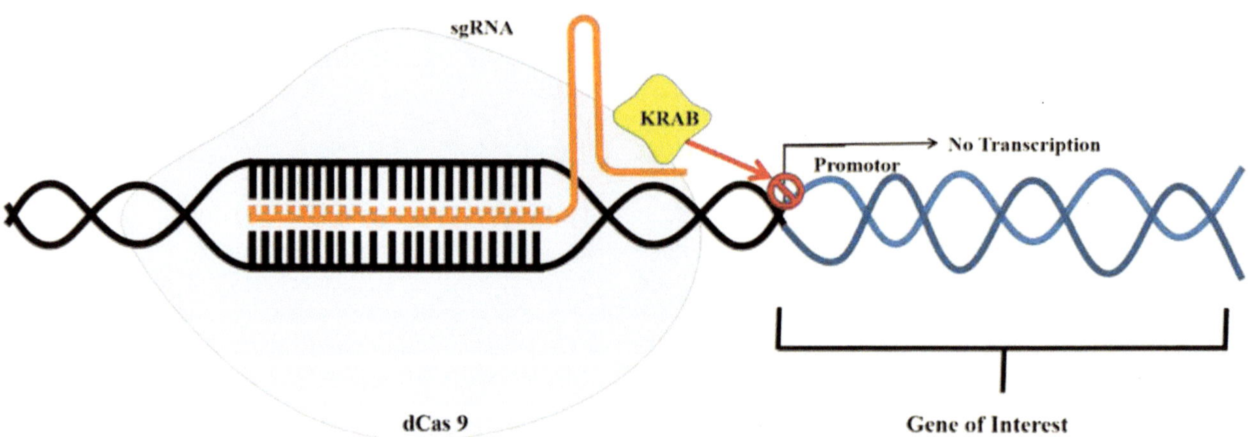

FIGURE 8.6 CRISPRi mechanism. dCas9 is fused with a transcriptional repressor, typically the Krüppel-associated box (KRAB) domain, to inhibit gene expression.

one VP64 to the N-terminus of dCas9-VP64, known as VP64-dCas9-VP64, significantly enhanced gene transcription (Omachi et al., 2022). In addition, using dCas9, protein tagging systems including SunTag and MS2-MCP were designed to amplify the gene expression of the targeted gene to improve the effectiveness of activation. Using an alternative approach to transcriptional inducers, dCas9 has been combined with epigenetic modifiers such as Tet1, histone acetylase, p300 and a CpG DNA demethylase to stimulate the expression of native genes. Moreover, several transcriptional activators, such as HSF1, Rta and p65 have recently been discovered (Kazi et al., 2021). Numerous CRSIPRa systems have been developed by combining dCas9 with transcription activation domains, such as CRISPRa-VPR, CRISPR-TREE, CRISPR-SPH, and CRISPR-SAM systems. These approaches have demonstrated the efficacy of CRISPR-dCas9 for precise gene regulation in diverse organisms, such as mice, humans, and fly cell lines (Zalatan et al., 2015).

8.6 DIRECTED EPIGENOME MODULATION WITH THE CRISPR-DCAS9 SYSTEM

Epigenetic regulation of genes refers to controlling gene expression by modifying the DNA bases while keeping the nucleotide sequences unchanged in the genome. Several mechanisms, such as histone modification, DNA methylation, and the influence of non-coding RNA, are involved in the epigenetic regulation of genes. These mechanisms are important in controlling cellular activities and organizing genome structure. For example, histone proteins can undergo several modifications, including acetylation, phosphorylation and methylation, which often enhance gene expression (Chinnusamy et al., 2009). Conversely, modifications such as sumoylation and biotinylation generally suppress gene expression. In addition, short, non-coding RNAs can modify the chromatin structure and affect gene expression. DNA methylation patterns, which differ significantly across plants and animals, are vital in controlling gene expression in both kinds of organisms (Bure et al., 2022) (Figure 8.7).

Epigenetic modifiers, such as readers, erasers, and writers, are essential for modifying the epigenetic signatures in the genome. A project of the encyclopedia of DNA elements and the roadmap epigenome mapping consortium has been focusing on the importance of these epigenetic alterations in controlling gene(s) expression in a cell-specific way by modifying histone structures and chromatin remodeling. Despite being at the developing stage, CRISPR/dCas9 has enabled researchers to redesign the epigenome landscape of eukaryotic genomes. Researchers have successfully integrated dCas9 with several epigenetic modifiers to precisely introduce or remove epigenetic signatures (Holtzman et al., 2018). This approach has enabled scientists to precisely evaluate the impact of epigenetic modifications on gene expression and functions. Alterations in the DNA methylation patterns have been identified as a significant contributor to the development of cancers. Although medicines like 5-azacytidine, approved by the FDA, can hinder these aberrant epigenetic modifications, their non-selective impact on DNA methylation in the genome is still a challenge (Singh, Kumar, et al., 2022).

Furthermore, CRISPR-dCas9 is a versatile approach to changing additional epigenetic markers. It is capable of restoring genes that have been silenced or have a modified chromatin architecture. In addition, combining dCas9 with the acetyltransferase P300 or LSD1 results in a substantial influence on gene expression through the modification of histone marks that regulate enhancer activity and local chromatin structure (Chiarella et al., 2020).

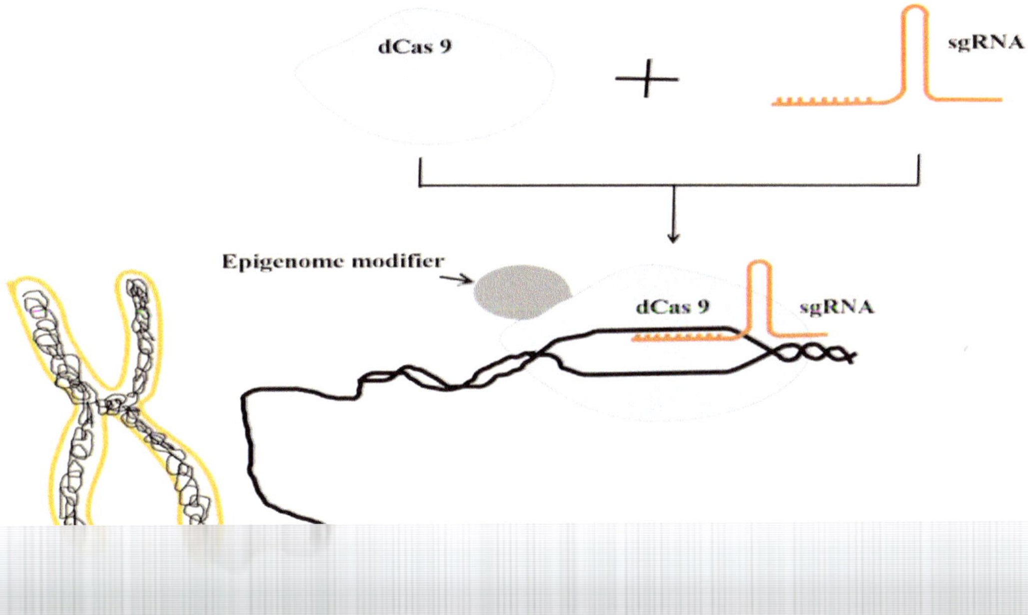

8.7 CRISPR-DRIVEN SITE-SPECIFIC BASE EDITING

The Base Editors (BE) are gene-editing tools that combine the CRISPR/Cas system with an individual deaminase that enables the precise replacement of a single base at the target site in DNA or RNA without causing a DSB or provision of a donor DNA template in living cells. Two groups of DNA base editors have been described: adenine base editors (ABEs) which convert an A•T base pair to a G•C base pair, while cytosine base editors (CBEs) convert a C•G base pair into a T•A base pair. ABEs and CBEs may collectively facilitate all four potential transition mutations (G to A, A to G, C to T and T to C) at a target site in the genome (Huang et al., 2021).

In 2016, Liu and his coworkers designed the CBE1 (1st generation of the base- editor). In this study, a rat-derived cytosine deaminase Apolipoprotein B mRNA editing enzyme catalytic subunit 1 (APOBEC1) was fused to the N-terminus of the dCas9 to achieve site-specific base editing. CBE1 catalyzes the conversion of cytosine to uracil in a specific region of the non-targeted strand. Uracil is then identified by the cellular replication machinery as thymine, leading to a shift from C-G to T-A. Although, CBE1 may perform an efficient conversion of targeted base-editing in laboratory conditions (achieving up to 37% editing with a 1.1% rate of indel creation), it was not effective in human cells because of the cellular-mediated repair of the U-G intermediate in DNA through the base excision repair (BER) pathway (Kantor et al., 2020). Base excision repair (BER) of uracil-guanine (U-G) mismatches in DNA is catalyzed by an enzyme called uracil N-glycosylase (UNG). UNG recognizes the U-G mismatch and cleaves the glycosidic bond between the uracil and the deoxyribose backbone of DNA. This cleavage results in the reversal of the U-G intermediate generated by the base editor, restoring the original cytosine-guanine (C-G) base pair. Due to the low editing efficiency and limitations of CBE1, researchers have developed a series of base editors (CBE2) with enhanced efficiencies. This was achieved by combining a uracil DNA glycosylase inhibitor (UGI) with the C-terminus of BE1, effectively blocking the activity of UNG. Suppression of BER in BE2 systems led to a threefold increase in editing efficiency in human cells (Kantor, 2022).

Further improvements led to the development of third-generation CBE3 by combining CBE2 with an improved version of nCas9 for even better activity. Conversely, SpCas9-based CBEs encountered difficulties because of their reliance on G/C-rich PAM sequences. Recent advances have used dCas12-based CBEs and Cas variants with different PAM criteria to overcome these constraints. Applications of CBEs were limited to converting C-G to T-A, prompting researchers to broaden the scope of correctable genetic mutations (Wong et al., 2023). By using ABEs, genetic abnormalities caused by mutations may be able to be repaired. Like CBEs, ABEs were developed by fusing CRISPR/dCas9 with adenosine deaminase, enabling the conversion of adenine to inosine,

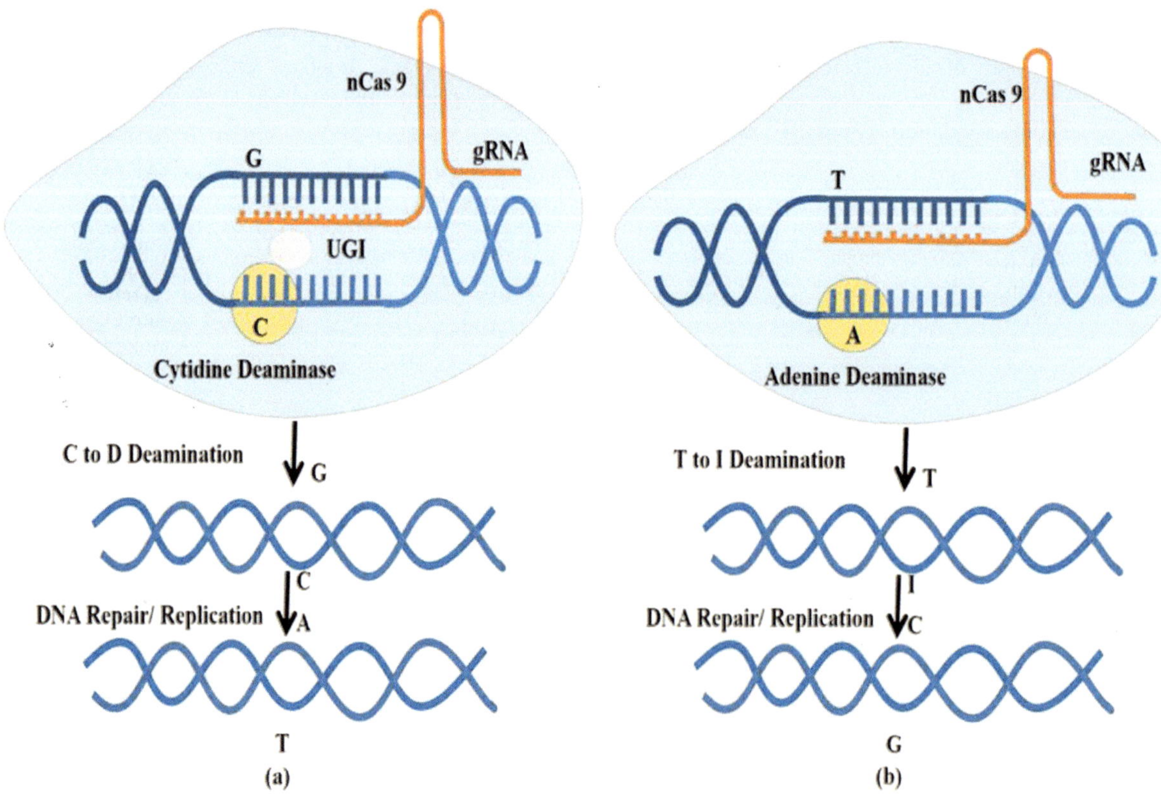

FIGURE 8.8 The catalytically impaired nickase Cas9 (nCas9) used for base editing without creating DSB. (a) Cytosine Base Editor (CBE) and (b) Adenine Base Editor (ABE).

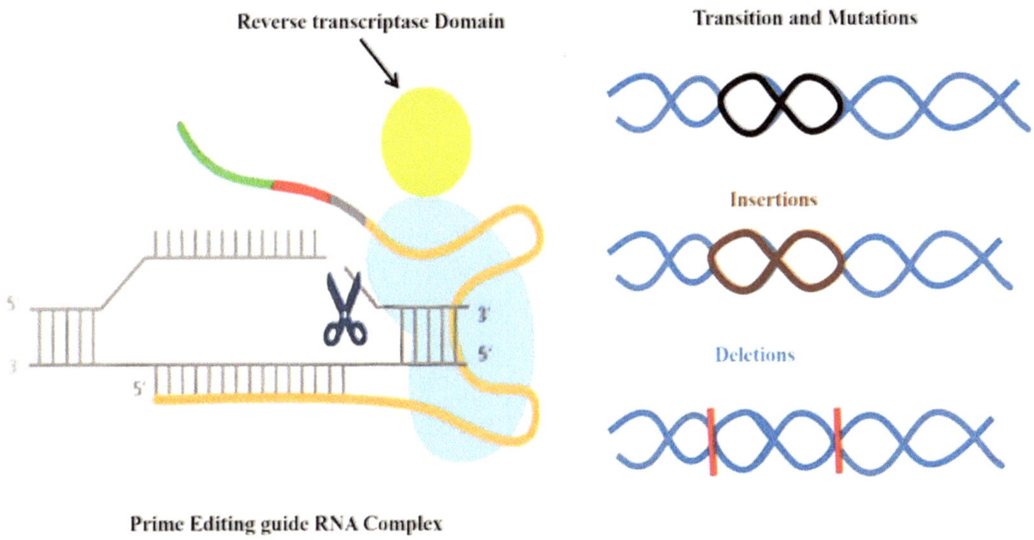

FIGURE 8.9 Prime editing.

which is then read as guanine by the cellular machinery, thus substituting an A-T pair with a G-C pair at the target site (Figure 8.8). In contrast to base editing in DNA, CRISPR-Cas13 based REPAIR system has been demonstrated for base editing in RNA. REPAIR allows for temporary gene modifications without making permanent changes to the DNA, offering a reversible solution for gene editing (Lo et al., 2022).

Compared with HDR-based gene correction, base editing is simple and highly efficient as it doesn't require a donor DNA template or a DSB to install precise modifications at the target site. Multiple studies have showcased the potential of CRISPR-based base editing for accurate modification in the genome. For instance, Zong et al. (2017) successfully changed cytosine to thymine in maize, rice, and wheat, with a success rate of 43.48%. In a similar study, Shimatani et al. (2018) engineered several herbicide-resistant mutations in plants. In addition, Barrero et al. (2022) showed that editing a base in the Sub1A-2 gene, which is sensitive to flooding, could change a vulnerable allele into a resistant one in rice.

8.8 PRIME EDITING (PE): A TRANSFORMATIVE TOOL FOR GENE EDITING

Prime editing is an innovative CRISPR/Cas tool that allows for precise and inheritable modifications at the target site without providing a donor template. While base editing has effectively introduced single base modifications in the past, it has several limitations. These limitations include the presence of a PAM sequence close to the editing site, a narrow base-editing window, potential off-target effects, and enabling only four types of base transitions. To address these limitations, the Liu group at Harvard MIT, developed the prime editing (PE) system to introduce precise base editing at the target without donor template. This method involves a CRISPR/nCas9 in the genome. PE uses a longer form of single-guide RNA called pegRNA and a nickase version of Cas9 (nCas9) fused with an engineered reverse transcriptase (RT) (Doman et al., 2022). In this system, the pegRNA guides the nCas9 to the target sequence and carries an extension encoding the desired genetic changes. The reverse transcriptase reads this extension and synthesizes a complementary DNA (cDNA) directly from it, bypassing the need for a donor template at the target site. When the pegRNA directs nCas9 to make a single-strand break (SSB) at the target site, it exposes a 3'-hydroxyl group, priming the DNA for modification. The reverse transcriptase then extends this nicked strand using the blueprint provided by the pegRNA. This action creates two flaps of DNA: a 5' flap of the original, unedited strand and a 3' flap that includes the desired genetic modifications. Although the 5' flap naturally aligns with the unedited DNA strand, cellular enzymes like FEN1 are used to remove the 5' flap, thus integrating the new, intended changes into the genome (Alok et al., 2021). This makes prime editing a versatile and powerful gene-editing tool, especially for base editing and gene therapy in human health and therapeutic applications (Figure 8.9).

The first generation of primer editors, PE1, combined Moloney murine leukaemia virus reverse transcriptase (MMLV RT) with nCas9 and pegRNA. Nevertheless, their effectiveness was somewhat limited. In the second-generation PEs, Anzalone and colleagues introduced different M-MLV RT variants with improved DNA binding ability and that were more stable at high temperatures, significantly boosting their efficiency. The third generation, PE3, introduces a secondary cut in the unedited strand, enabling the cell's repair mechanisms to preferably use the newly edited strand as a blueprint for correction. This innovation has greatly enhanced the precision and efficiency of base editing with reduced off-target effects. Primer editors have been successfully employed to make

The future of PE technology looks promising as researchers continue to refine these storage systems, recorders, and DNA writers. PE-based DNA recording technology stands out as a potential transformative tool, allowing the insertion of arbitrary sequences at the target site in the genome that could revolutionize how we monitor and interact with biological and environmental systems (Cong et al., 2022).

8.9 COMPREHENSIVE GENE DISCOVERY THROUGH CRISPR-BASED GENOME-WIDE SCREENING

Genome-wide screening is a method to generate and analyze a collection of mutants to identify the set of genes that regulate specific functions. CRISPR/Cas technology has significantly enhanced the efficiency of genome-wide screening studies. Researchers have developed genome-wide CRISPR/Cas guide RNA (gRNA) libraries for numerous organisms, including humans, mice, *E. coli*, and flies, targeting every gene in their genomes for functional genomic studies (Sander & Joung, 2014). However, designing and managing these extensive libraries is more complex than editing individual genes. Researchers have developed multiple softwares, webtools and AI algorithms to aid in the efficient designing and screening of these gRNAs. For example, AI tools including CRISPR Direct, CHOPCHOP, CRISPR-P, and E-CRISPR, assist in designing the gRNAs with minimum off-target effect (Sultan et al., 2022). Furthermore, E-CRISP and Cas-OFFinder enable users to evaluate off-target effects. While CHOPCHOP software employs empirical data to assess the off-target effects, each tool offers distinct features. For instance, CRISPR-ERA is focused on transcriptional regulation, while BE-Designer is designed explicitly for base- editing studies. Similarly, CRISPR-PLANT has been tailored for targeting plant genomes (Liu et al., 2015) and FlyCRISPR is specifically aimed at designing gRNAs in Drosophila (Gratz et al., 2013). A study by (Ma et al., 2021) used CRISPR-based genome-wide screening to investigate the mutant responses to abiotic stresses.

8.10 DYNAMIC GENOME VISUALIZATION THROUGH CRISPR/CAS-MEDIATED GENOME IMAGING

Several studies have demonstrated that chromatin's arrangement and dynamic organization play a vital role in important genomic activities, such as DNA replication, gene expression and DNA repair (Nair et al., 2017). Studying these essential processes requires live cell imaging to visualize the spatio-temporal organization of the chromatin and genome. Nevertheless, genome imaging of specific sites within living cells offers considerable difficulty. In past decades, researchers employed fluorescently labeled DNA-binding proteins to identify precise locations in the genome. However, this method was primarily effective for distinct regions such as centromeres and telomeres, while imaging a variety of sequences using these methods remained a complicated and challenging task. Alternatively, fluorescent in situ hybridization (FISH) is a technique that provides greater adaptability and is essential for precisely identifying different genomic locations. However, due to the fixation and denaturation requirement, it was not suitable for live cell imaging (Poon et al., 2001). On the other hand, CRISPR/Cas systems such as CRISPR-dCas9 and CRISPR-dCas12 have enabled precise chromatin modeling in live cells. CRISPR technology provides numerous benefits compared to conventional imaging techniques, such as its simple and flexible design, the capacity to target multiple genetic locations simultaneously, and its compatibility with live cell imaging. For this, dCas9 can be fused with fluorescent proteins such as GFP to provide customizable DNA markers that enable the visualization of chromatin in live cells. In addition, the dCas9 technique also enables multicolor labeling, allowing the simultaneous observation of multiple genomic sites. This can be achieved through two approaches: one uses various dCas9 proteins fused with different fluorescent proteins, and the other uses specially engineered sgRNAs that attract distinct RNA-binding proteins tagged with multiple fluorescent proteins (Tian et al., 2023). Studies have shown that sgRNAs containing eight aptamers provide more precise imaging of genomic sites. Impressively, this system has enabled the simultaneous visualization of more than three genomic sites using three different dCas9 orthologs, recognizing specific PAM sequences and sgRNA scaffolds (Zhang, 2019). Besides chromatin imaging, the CRISPR/dCas9 system has been used for chromosomal painting. Chromosomal painting requires introducing hundreds of sgRNAs across the chromosome. Researchers have also explored using the CRISPR/dCas system with synthetic dyes, fluorescent proteins, or luminescent nanocrystal particles for imaging genomic sites. Organic dyes are notably smaller, sensitive, and more photostable than fluorescent proteins. So far, three dye systems—Halo-tag, RNA-aptamers-based, and molecular beacon-based—have been used to image genomic loci in living cells (Sultan et al., 2022). Unlike fluorescent proteins and organic dyes, quantum dots are exceptionally effective for genome imaging due to their smaller size and higher sensitivity. Despite these advances, challenges remain in chromatin imaging with CRISPR/dCas9, such as issues with target accessibility, off-target effects, target availability, and background fluorescence. These challenges must be addressed to harness the full potential of this innovative technology (Wu et al., 2019) (Figure 8.10).

8.11 DECODING CELLULAR LINEAGE WITH THE CRISPR/CAS TECHNOLOGY

Understanding how cells develop and differentiate from a single zygote into a complete multicellular organism is fundamental to biology. This process, seen in all sexually reproducing eukaryotes, involves an intricate sequence of developmental and differentiation events (McKenna et al., 2019). Scientists can uncover the mechanisms that drive tissue and organism development by studying these early cell division patterns and lineage across different cell types. Tracking

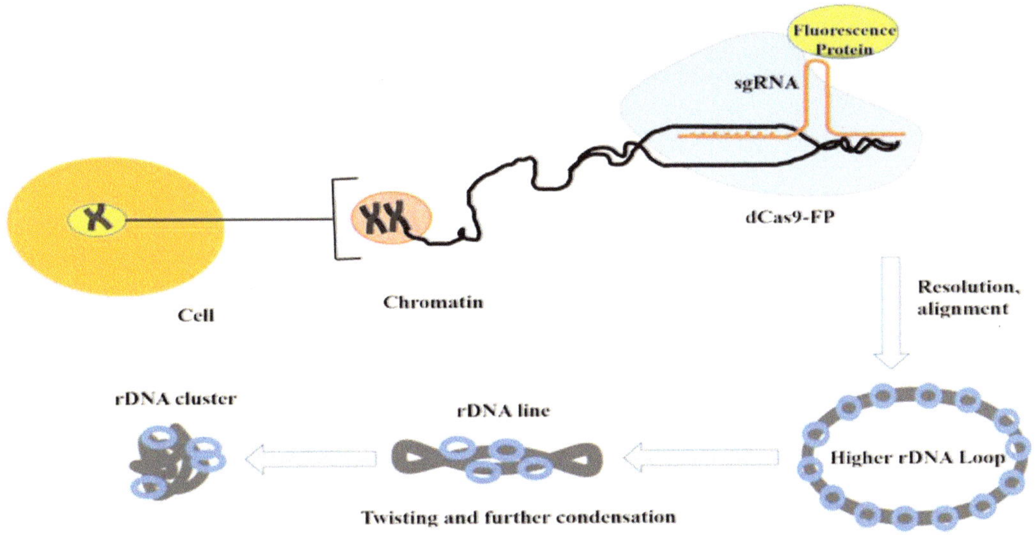

FIGURE 8.10 Live-cell imaging of chromatin by CRISPR.

cell lineage relies on a heritable genetic marker that can be transmitted via cell divisions and the capacity to readily trace this marker (Woodworth et al., 2017). In the past, researchers employed a range of procedures, such as the conventional practice of staining cells with dyes, despite the dye becoming less concentrated with each cell cycle (Nair et al., 2021). In recent times, durable heritable markers such as transposable elements, viral DNA barcodes and fluorescent proteins have been employed for tracking. Scientists have also used naturally existing indicators, such as epigenetic markers and microsatellite repeats, to examine the connections between lineages (Laghi et al., 2008). Although these strategies have proven beneficial, they frequently encounter constraints due to the scarcity of markers and challenges in establishing their connection with gene expression. The tremendous progress in RNA sequencing, specifically in single-cell transcriptomic sequencing (scRNA-seq), has fundamentally transformed our ability to analyze the transcriptomes of numerous individual cells, facilitating the recognition and description of various cell types in different tissues (Jovic et al., 2022). Nevertheless, there are still obstacles to overcome, such as variations in gene expression at various stages of development, which might make lineage tracking more complex. However, integrating single- cell RNA sequencing (scRNA-seq) with dCas9 barcode significantly advances this domain, greatly enhancing the ability to investigate cell lineage. Researchers begin by incorporating precise DNA fragments (barcodes) into the genome using CRISPR-based DNA barcodes. CRISPR/Cas specifically targets these barcodes to create alterations (indels) that can be monitored over time, offering a dynamic method to investigate cell lineage during development (Schultzhaus et al., 2021).

For example, Bowling et al. (2020) developed a mouse line The study generated more than 44,000 transcribed barcodes to investigate the behavior of fetal hepatic hematopoietic cell clones. Similarly, Kalhor et al. (2018) demonstrated a novel approach to track the developmental lineages in mammalian models using CRISPR-based vivo barcoding. The researchers employed a self-targeting CRISPR/Cas9 method that used homing guide RNAs (hgRNAs) to generate distinct barcodes. These barcodes were then used to track the mouse brain's location and timing. Furthermore, CRISPR barcoding has profoundly influenced cancer biology, allowing scientists to investigate the genetic diversity within tumors and gain insights into cancer cells' evolutionary and adaptive processes. This technology introduces a revolutionary way of cell lineage monitoring, providing unparalleled insights into cellular growth and the progression of diseases.

8.12 CRISPR-DRIVEN GENE DRIVES: REDEFINING INHERITANCE AND POPULATION CONTROL

Gene drives are autonomous systems that rapidly transmit desirable genetic traits throughout a population, surpassing the speed of ordinary Mendelian inheritance. Although the potential of gene drives was widely acknowledged, the first efforts to use them were hindered by the difficulty of accurately controlling their insertion into the genome (Hammond et al., 2017). CRISPR/Cas9 technology has revolutionized this field by simplifying, speeding up, and refining the process. CRISPR/Cas9 allows for the design of customized gene drives that include the CRISPR/Cas system and a targeted genetic trait to be introduced within the population. The CRISPR components specifically cut the DNA at a target site on the wild-type chromosome. The cell then uses a high-fidelity to Mendelian inheritance, a genetically altered allele would have a 50% chance of being transmitted to the next generation, meaning it wouldn't necessarily spread throughout the entire population. However, when linked to a drive allele, the altered trait is forced to be inherited by nearly all offspring, effectively propagating it throughout the population. Incorporating a CRISPR-based gene drive into a wild population begins by engineering a gene that replaces the normal sequence on the chromosome with one that includes the drive allele, composed of Cas9, gRNA and the modified allele (Noble et al., 2018). This innovative approach holds tremendous potential for addressing ecological and health-related challenges. CRISPR-based gene drives have shown effectiveness in addressing vector-borne diseases, such as dengue, zika and malaria, in the laboratory (Wang et al., 2022). Moreover, engineered synthetic CRISPR-based gene drives show immense promise for altering or eliminating populations that pose environmental threats. Furthermore, CRISPR-based gene drives have the ability to aid in the management of agricultural pests, the control of diseases in animals and humans, the eradication of invasive species, the protection of endangered species, and the tagging of populations for research purposes (Llanos 2019). However, the use of CRISPR-based gene drives needs strict biosafety precautions because of potential risks, including off-target effects, unintentional transmission to other species, and environmental consequences.

8.13 CRISPR-DRIVEN DIRECTED EVOLUTION: ENGINEERING GENES FOR DESIRED TRAITS

Directed evolution is a method used for designing novel protein variants, often resulting in significantly improved properties. Phage-assisted continuous evolution (PACE) and other traditional methods have effectively modified essential proteins used in medications (Popa et al., 2020). Nevertheless, earlier directed evolution techniques were restricted in their ability to alter each nucleotide within targeted genomic regions. The integration of CRISPR technology with directed evolution has revolutionized the field by enabling the development of tailor-made proteins that are significant in medicine, cell biology, and agriculture. The EvolvR platform is an interesting CRISPR-based system integrating nCas9 with a mutation-inducing DNA polymerase (PolI3M). In this system, nCas9 creates a nick in the DNA strand and detaches from it. Then, the PolI3M enzyme takes over, modifying the DNA at the nick. EvolvR has been employed in mammalian studies to achieve accurate mutations in important genes (Lyzenga et al., 2021). In plant science, a new variant of the OsSF3B1 protein present in rice was modified to make resistance against the herbicide GEX1A, showcasing the ability of CRISPR technology to improve plant durability against stresses (Khan et al., 2021). The prospects of EvolvR are significant, with the potential to completely transform the fields of agriculture, medicine, and health. Researchers can use this tool to investigate genomic regions, explore protein interactions, identify novel proteins, and customize bacterial genomes.

8.14 SYNTHETIC BIOLOGY MEETS CRISPR: CUTTING-EDGE SOLUTIONS FOR BIOTECHNOLOGY

Biomolecules are essential organic components, such as carbohydrates, proteins, lipids, and nucleic acids, that play an essential role in maintaining metabolic functions and survival of living organisms. Certain valuable biomolecules are in high demand, yet their renewable resources are unable to meet this need (Carbajal-Valenzuela et al., 2022). Microbes have emerged as a preferred method of producing these useful macromolecules through alternative methods. Researchers employ many techniques such as stimulating gene expression and deleting some genes, to manipulate bacteria in order to enhance their production of the desired biomolecule (Bueso et al., 2018). The application of the CRISPR/Cas system has made it easier and less time-consuming to manipulate bacterial genes in order to boost the production of biomolecules. Developing metabolic pathways has several challenges, such as optimizing metabolic flux, managing undesired enzymes, and addressing genes that exhibit unexpected behavior (Chen et al., 2018). In order to address these issues, scientists collect the most efficient biological elements discovered in nature. These components are subsequently assembled to provide a novel pathway for synthesizing the desired biomolecule, without interfering with the natural functioning of the host organism. Using this method, scientists have the power to customize genomes in order to optimize microorganisms, increasing their capacity to generate targeted biomolecules while eliminating superfluous functionalities (Kohman et al., 2018). However, researchers have discovered an alternative application for the dCas9. Rather than modifying genes, it may now be employed to deactivate them, a process known as CRISPRi. dCas9 inhibits the conversion of genes into proteins by disrupting the transcription pathway. Researchers have figured out how to control when this gene-blocking action happens by using special switches called inducible promoters. Nevertheless, because of dCas9's high efficacy in gene suppression, even a minor level of its activity inappropriately can result in a substantial decline in the targeted gene's function. To provide enhanced control over dCas9-mediated silence by implementing further regulation at the translational level, it is necessary to introduce a modification to the dCas9 gene by inserting a TAG stop codon (Koopal et al., 2019). Researchers are able to include a synthetic amino acid known as L-biphenylalanine (BipA) into a precise location within the dCas9 protein. Without BipA, the dCas9 gene produces a dCas9 protein that's incomplete and cannot function properly. But when BipA is around, the TAG codon is translated, and we end up with a fully functional, full-length dCas9 protein. This new version of dCas9, with BipA in it (dCas9-BipA), can still effectively shut down gene activity (Schultenkämper et al., 2020). Comparison of regular dCas9 worked to dCas9-BipA found that the latter caused a lot less unintended gene shutdown—about 14 times less, in fact. This new system has developed the reduced background levels of unwanted gene silencing, giving us more precise control over when and how

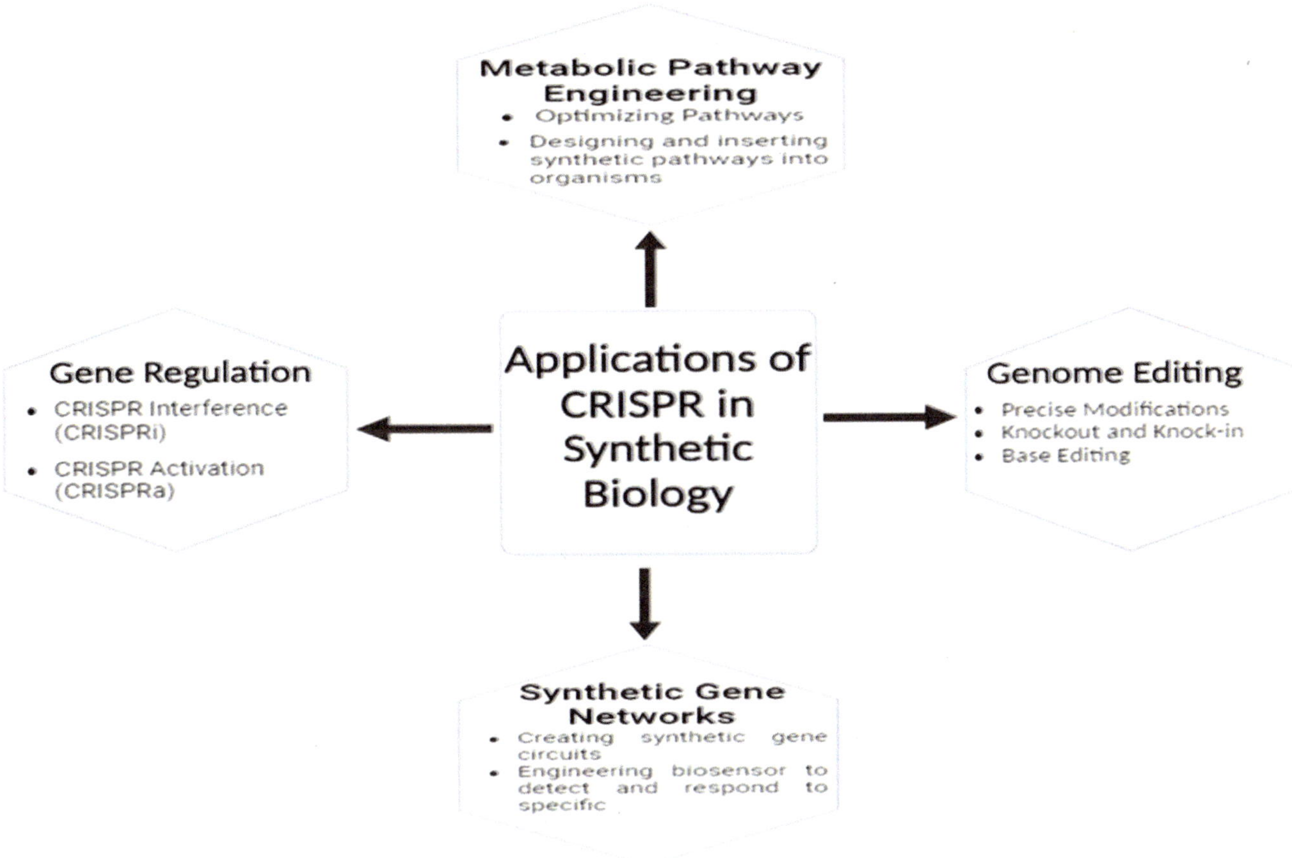

FIGURE 8.11 Schematic diagram of applications of CRISPR in synthetic biology.

much our target genes are turned on or off (Tiwari et al., 2014) (Figure 8.11).

8.15 CRISPR/CAS13-MEDIATED RNA EDITING: A NEW ERA IN GENE REGULATION

Editing of DNA can sometimes lead to unintended changes in the genetic code, known as off-targets, which can pose potential problems. On the other hand, RNA editing offers multiple benefits compared to genome editing as it does not cause permanent modifications to DNA and can be reversed. RNA editing doesn't rely on mechanisms like HDR, making it applicable even to cells that aren't actively dividing. CRISPR/Cas13 is rapidly emerging as a significant tool for precisely editing single-stranded RNA. Cas13 is equipped with two domains called HEPN RNase, found in both higher eukaryotes and prokaryotes, enabling precise cleavage of RNA, especially with a preference for specific sequences flanking the target site (O'Connell, 2019). Four families of Cas13 have been identified: Cas13a, Cas13b, Cas13c, and Cas13d. Each variant acts as a single effector RNase responsible for the processing of RNA and programmed cleavage. CRISPR/Cas13 has several benefits over other RNA editing technologies such as

Unlike RNAi, Cas13 can specifically target pre-mRNA, non-coding nuclear RNAs and nuclear RNAs by incorporating a nuclear localization signal. Cas13 and its variants also open avenues for various RNA manipulations, including accurately editing RNA sequences, combating RNA viruses, visualizing RNA, tracking RNA, targeting specific pathogenic splicing isoforms, and regulating gene expression (post-transcriptionally and reversibly). Building upon Cas13, systems like REPAIR and RESCUE have emerged for precise RNA base editing. However, the potential for multiplexing with REPAIR and RESCUE remains to be confirmed. Similarly, advanced diagnostic tools such as PACMAN and SHERLOCK have been developed, leveraging Cas13's collateral cleavage activity for ultra-sensitive detection (Chan et al., 2021). For a deeper exploration of the applications of CRISPR/Cas13, refer to Chapter 7.

8.16 FUTURE PROSPECTS

Advances in science often stem from innovative techniques and groundbreaking discoveries. Science gets inspiration from nature, particularly in biology, where researchers harness the potency of natural products to develop powerful and restriction enzymes. The emergence of the CRISPR/Cas toolkit for genome editing is a recent development in scientists' ability. The successful outcome of CRISPR/Cas9 also highlights another fundamental premise of science: that research can lead to revolutionary breakthroughs. The advances and applications of CRISPR/Cas technology during the past decade have been really remarkable. This progress includes several advances, such as using CRISPRi/a with dCas9, genome imaging, directed evolution, base editing, cell lineage tracing prime editing, epigenetic modulation, and DNA data storage. Moreover, recent progress in guide RNA (gRNA) has broadened the application range of CRISPR. These breakthroughs have greatly expanded our knowledge of gene regulation, epigenetics, genome organization, and functional genomics on a comprehensive level.

However, CRISPR faces technical, ethical, and regulatory challenges despite its potential innovations. Although there is great potential for therapy, successfully administering these tools to living organisms remains a huge challenge. Although viral delivery vectors are commonly used for in vivo CRISPR administration, concerns persist over their potential to induce cancer and trigger immunological responses. Furthermore, there is a growing interest in the ethical and regulatory concerns related to CRISPR/Cas. Global discussions vary on the regulatory framework for CRISPR products. In the United States, SDN1 and SDN2 CRISPR crops have been chosen for deregulation. However, in European Union countries, all CRISPR-edited crops are classified as GM crops and subjected to thorough regulatory laws before commercialization. Similarly, the scientists are concerned about the environmental dissemination of gene drives. While gene drives can potentially address vector-borne diseases, it is necessary to build additional regulatory frameworks before they can be practically implemented.

To conclude, CRISPR has revolutionized the field of study by enabling scientists to change genomes. This technology is not only revolutionizing the fields of food and medicine but also fundamentally altering our perception and interaction with the natural environment. In the future, researchers will further investigate and extend the boundaries of CRISPR and exploit it to tackle challenges like as enhancing human well-being, advancing agriculture, and eliminating infectious diseases.

REFERENCES

Adli, M. (2018). The CRISPR tool kit for genome editing and beyond. *Nature Communications,* 9(1), 1911.

Akinci, E., Hamilton, M. C., Khowpinitchai, B., & Sherwood, R. I. (2021). Using CRISPR to understand and manipulate gene regulation. *Development*, 148(9), dev182667.

Alok, A., Chauhan, H., Upadhyay, S. K., Pandey, A., Kumar, J., & Singh, K. (2021). Compendium of plant-specific CRISPR vectors and their technical advantages. *Life*, 11(10), 1021.

Arroyo-Olarte, R. D., Bravo Rodriguez, R., & Morales-Ríos, E. (2021). Genome editing in bacteria: CRISPR-Cas and beyond. *Microorganisms*, 9(4), 844.

Azhar, M., Phutela, R., Kumar, M., Ansari, A. H., Rauthan, R., Gulati, S., … Singh, S. (2021). Rapid and accurate nucleobase detection using FnCas9 and its application in COVID-19 diagnosis. *Biosensors and Bioelectronics,* 183, 113207.

Barrangou, R., & Doudna, J. A. (2016). Applications of CRISPR technologies in research and beyond. *Nature Biotechnology*, 34(9), 933–941.

Barrero, L. S., Willmann, M. R., Craft, E. J., Akther, K. M., Harrington, S. E., Garzon-Martinez, G. A., … McCouch, S. R. (2022). Identifying genes associated with abiotic stress tolerance suitable for CRISPR/Cas9 editing in upland rice cultivars adapted to acid soils. *Plant Direct,* 6(12), e469.

Bhatia, S., & Yadav, S. K. (2023). CRISPR-Cas for genome editing: classification, mechanism, designing and applications. *International Journal of Biological Macromolecules,* 238, 124054.

Boettcher, M., & McManus, M. T. (2015). Choosing the right tool for the job: RNAi, TALEN, or CRISPR. *Molecular Cell*, 58(4), 575–585.

Bowling, S., Sritharan, D., Osorio, F. G., Nguyen, M., Cheung, P., Rodriguez-Fraticelli, A., … Li, B. E. (2020). An engineered CRISPR-Cas9 mouse line for simultaneous readout of lineage histories and gene expression profiles in single cells. *Cell,* 181(6), 1410–1422. e1427.

Bueso, Y. F., Lehouritis, P., & Tangney, M. (2018). In situ biomolecule production by bacteria; a synthetic biology approach to medicine. *Journal of Controlled Release,* 275, 217–228.

Bure, I. V., Nemtsova, M. V., & Kuznetsova, E. B. (2022). Histone modifications and non-coding RNAs: mutual epigenetic regulation and role in pathogenesis. *International Journal of Molecular Sciences*, 23(10), 5801.

Cai, R., Lv, R., Shi, X. e., Yang, G., & Jin, J. (2023). CRISPR/dCas9 tools: epigenetic mechanism and application in gene transcriptional regulation. *International Journal of Molecular Sciences*, 24(19), 14865.

Carbajal-Valenzuela, I. A., Apolonio–Hernandez, N. M., Gutierrez-Chavez, D. V., González-Arias, B., Jimenez-Hernandez, A., Rico-García, E., … Guevara-González, R. G. (2022). Biological macromolecules as nutraceuticals. In A. K. Nayak, A. K. Dhara, & D. Pal (Eds.), *Biological macromolecules* (pp. 97–138). Elsevier.

Chan, K. G., Ang, G. Y., Yu, C. Y., & Yean, C. Y. (2021). Harnessing CRISPR-Cas to combat COVID-19: From diagnostics to therapeutics. *Life,* 11(11), 1210.

Chen, M., & Qi, L. S. (2017). Repurposing CRISPR system for transcriptional activation. In L. C. Li (Ed.), *RNA Activation* (vol. 983, pp. 147–157). Springer.

Chen, X., Du, J., Yun, S., Xue, C., Yao, Y., & Rao, S. (2024). Recent advances in CRISPR/Cas-based genome insertion technologies. *Molecular Therapy-Nucleic Acids*, 35.

Chen, X., Gao, C., Guo, L., Hu, G., Luo, Q., Liu, J., … Liu, L. (2018). DCEO biotechnology: Tools to design, construct, evaluate, and optimize the metabolic pathway for biosynthesis of chemicals. *Chemical Reviews*, 118(1), 4–72.

Cheng, X., Li, Z., Shan, R., Li, Z., Wang, S., Zhao, W., … Fei, T. (2023). Modeling CRISPR-Cas13d on-target and off-target effects using machine learning approaches. *Nature Communications*, 14(1), 752.

Chiarella, A. M., Lu, D., & Hathaway, N. A. (2020). Epigenetic control of a local chromatin landscape. *International Journal of Molecular Sciences*, 21(3), 943.

Chinnusamy, V., & Zhu, J.-K. (2009). Epigenetic regulation of stress responses in plants. *Current Opinion in Plant Biology*, 12(2), 133–139.

Cong, J., Lau, J., Liu, G., Neuendorffer, S., Pan, P., Vissers, K., & Zhang, Z. (2022). FPGA HLS today: Successes, challenges, and opportunities. *ACM Transactions on Reconfigurable Technology and Systems (TRETS)*, 15(4), 1–42.

Doman, J. L., Sousa, A. A., Randolph, P. B., Chen, P. J., & Liu, D. R. (2022). Designing and executing prime editing experiments in mammalian cells. *Nature Protocols*, 17(11), 2431–2468.

Fapohunda, F. O., Qiao, S., Pan, Y., Wang, H., Liu, Y., Chen, Q., & Lü, P. (2022). CRISPR Cas system: a strategic approach in detection of nucleic acids. *Microbiological Research,* 259, 127000.

Gilbert, L. A., Larson, M. H., Morsut, L., Liu, Z., Brar, G. A., Torres, S. E., ... Doudna, J. A. (2013). CRISPR-mediated modular RNA-guided regulation of transcription in eukaryotes. *Cell,* 154(2), 442–451.

Goell, J. H., & Hilton, I. B. (2021). CRISPR/Cas-based epigenome editing: Advances, applications, and clinical utility. *Trends in Biotechnology*, 39(7), 678–691.

Gratz, S. J., Wildonger, J., Harrison, M. M., & O'Connor-Giles, K. M. (2013). CRISPR/Cas9-mediated genome engineering and the promise of designer flies on demand. *Fly*, 7(4), 249–255.

Guo, C., Ma, X., Gao, F., & Guo, Y. (2023). Off-target effects in CRISPR/Cas9 gene editing. *Frontiers in Bioengineering and Biotechnology,* 11, 1143157.

Hammond, A. M., & Galizi, R. (2017). Gene drives to fight malaria: Current state and future directions. *Pathogens and Global Health*, 111(8), 412–423.

Handel, E.-M., & Cathomen, T. (2011). Zinc-finger nuclease based genome surgery: It's all about specificity. *Current Gene Therapy*, 11(1), 28–37.

He, Y., Yan, W., Long, L., Dong, L., Ma, Y., Li, C., ... Xia, W. (2023). The CRISPR/Cas system: A customizable toolbox for molecular detection. *Genes*, 14(4), 850.

Hillary, V. E., & Ceasar, S. A. (2023). A review on the mechanism and applications of CRISPR/Cas9/Cas12/Cas13/Cas14 proteins utilized for genome engineering. *Molecular Biotechnology*, 65(3), 311–325.

Holtzman, L., & Gersbach, C. A. (2018). Editing the epigenome: Reshaping the genomic landscape. *Annual Review of Genomics and Human Genetics*, 19(1), 43–71.

Huang, D., Xu, C., Jiang, C., Chen, Q., Xu, Z., & Fang, X. (2024). Advances and challenges of signal readout systems in CRISPR-based biosensors for point-of-care testing of nucleic acid. *TrAC Trends in Analytical Chemistry,* 178, 117856.

Huang, T. P., Newby, G. A., & Liu, D. R. (2021). Precision genome editing using cytosine and adenine base editors in mammalian cells. *Nature Protocols*, 16(2), 1089–1128.

Jovic, D., Liang, X., Zeng, H., Lin, L., Xu, F., & Luo, Y. (2022). Single-cell RNA sequencing technologies and applications: A brief overview. *Clinical and Translational Medicine*, 12(3), e694.

Kalhor, R., Kalhor, K., Leeper, K., Graveline, A., Mali, P., & Church, G. M. (2018). A homing CRISPR mouse resource for barcoding and lineage tracing. *bioRxiv*, 280289.

Kampmann, M. (2018). CRISPRi and CRISPRa screens in mammalian cells for precision biology and medicine. *ACS Chemical Biology*, 13(2), 406–416.

Kantor, A., McClements, M. E., & MacLaren, R. E. (2020). CRISPR-Cas9 DNA base-editing and prime-editing. *International Journal of Molecular Sciences*, 21(17), 6240.

Karlson, C. K. S., Mohd-Noor, S. N., Nolte, N., & Tan, B. C. (2021). CRISPR/dCas9-based systems: Mechanisms and applications in plant sciences. *Plants,* 10(10), 2055.

Kazi, T. A., & Biswas, S. R. (2021). CRISPR/dCas system as the modulator of gene expression. *Progress in Molecular Biology and Translational Science,* 178, 99–122.

Khan, Z., Saboor, T., Ashfaq, M., Saddique, A., & Khanum, P. (2021). Disease resistance in crops through CRISPR/Cas. In A. Ahmad, S. H. Khan, & Z. Khan (Eds.). *CRISPR crops: The future of food security* (pp. 151–175). Springer.

Kim, H., & Kim, J.-S. (2014). A guide to genome engineering with programmable nucleases. *Nature Reviews Genetics*, 15(5), 321–334.

Kohman, R. E., Kunjapur, A. M., Hysolli, E., Wang, Y., & Church, G. M. (2018). From designing the molecules of life to designing life: future applications derived from advances in DNA technologies. *Angewandte Chemie International Edition*, 57(16), 4313–4328.

Koopal, B., Kruis, A. J., Claassens, N. J., Nobrega, F. L., & Van Der Oost, J. (2019). Incorporation of a synthetic amino acid into dCas9 improves control of gene silencing. *ACS Synthetic Biology,* 8(2), 216–222.

Kostyusheva, A., Brezgin, S., Babin, Y., Vasilyeva, I., Glebe, D., Kostyushev, D., & Chulanov, V. (2022). CRISPR-Cas systems for diagnosing infectious diseases. *Methods,* 203, 431–446.

Kumaran, A., Jude Serpes, N., Gupta, T., James, A., Sharma, A., Kumar, D., ... Pandey, S. (2023). Advancements in CRISPR-based biosensing for next-gen point of care diagnostic application. *Biosensors*, 13(2), 202.

LaFountaine, J. S., Fathe, K., & Smyth, H. D. (2015). Delivery and therapeutic applications of gene editing technologies ZFNs, TALENs, and CRISPR/Cas9. *International Journal of Pharmaceutics*, 494(1), 180–194.

Laghi, L., Bianchi, P., & Malesci, A. (2008). Differences and evolution of the methods for the assessment of microsatellite instability. *Oncogene*, 27(49), 6313–6321.

Lee, H. J., Kim, H. J., & Lee, S. J. (2020). CRISPR-Cas9-mediated pinpoint microbial genome editing aided by target-mismatched sgRNAs. *Genome Research*, 30(5), 768–775.

Li, Y., Li, W., & Li, J. (2021). The CRISPR/Cas9 revolution continues: From base editing to prime editing in plant science. *Journal of Genetics and Genomics*, 48(8), 661–670.

Lino, C. A., Harper, J. C., Carney, J. P., & Timlin, J. A. (2018). Delivering CRISPR: A review of the challenges and approaches. *Drug Delivery*, 25(1), 1234–1257.

Liu, H., Wei, Z., Dominguez, A., Li, Y., Wang, X., & Qi, L. S. (2015). CRISPR-ERA: A comprehensive design tool for CRISPR-mediated gene editing, repression and activation. *Bioinformatics*, 31(22), 3676–3678.

Llanos, R. L. (2019). CRISPR-based gene drive as a management strategy for invasive species control. Online opinion piece. www.researchgate.net/profile/Richard-Llanos-2/publication/330913097_Gene_Drive_for_Conservation_CRISPR-Based_Gene_Drive_as_a_Management_Strategy_for_Invasive_Species_Control_Opinion_piece/links/5c5b4642299bf1d14cb18977/Gene-Drive-for-Conservation

Lyzenga, W. J., Pozniak, C. J., & Kagale, S. (2021). Advanced domestication: harnessing the precision of gene editing in crop breeding. *Plant Biotechnology Journal*, 19(4), 660–670.

Ma, L., & Liang, Z. (2021). CRISPR technology for abiotic stress resistant crop breeding. *Plant Growth Regulation*, 94(2), 115–129.

Makarova, K. S., Wolf, Y. I., Iranzo, J., Shmakov, S. A., Alkhnbashi, O. S., Brouns, S. J., … Horvath, P. (2020). Evolutionary classification of CRISPR–Cas systems: A burst of class 2 and derived variants. *Nature Reviews Microbiology,* 18(2), 67–83.

McKenna, A., & Gagnon, J. A. (2019). Recording development with single cell dynamic lineage tracing. *Development*, 146(12), dev169730.

Miki, D., & Shimamoto, K. (2004). Simple RNAi vectors for stable and transient suppression of gene function in rice. *Plant and Cell Physiology*, 45(4), 490–495.

Munawar, N., & Ahmad, A. (2021). CRISPR/Cas system: an introduction. In A. Ahmad, S. H. Khan, & Z. Khan (Eds.), *CRISPR crops: The future of food security* (pp. 1–35). Springer.

Nair, A., & Manohar, S. M. (2021). A flow cytometric journey into cell cycle analysis. *Bioanalysis*, 13(21), 1627–1644.

Nair, N., Shoaib, M., & Sørensen, C. S. (2017). Chromatin dynamics in genome stability: roles in suppressing endogenous DNA damage and facilitating DNA repair. *International Journal of Molecular Sciences*, 18(7), 1486.

Nidhi, S., Anand, U., Oleksak, P., Tripathi, P., Lal, J. A., Thomas, G., … Tripathi, V. (2021). Novel CRISPR–Cas systems: an updated review of the current achievements, applications, and future research perspectives. *International Journal of Molecular Sciences,* 22(7), 3327.

Noble, C., Adlam, B., Church, G. M., Esvelt, K. M., & Nowak, M. A. (2018). Current CRISPR gene drive systems are likely to be highly invasive in wild populations. *Elife,* 7, e33423.

O'Connell, M. R. (2019). Molecular mechanisms of RNA targeting by Cas13-containing type VI CRISPR–Cas systems. *Journal of Molecular Biology*, 431(1), 66–87.

Omachi, K., & Miner, J. H. (2022). Comparative analysis of dCas9-VP64 variants and multiplexed guide RNAs mediating CRISPR activation. *PloS One*, 17(6), e0270008.

Pandya, K., Jagani, D., & Singh, N. (2023). CRISPR-cas systems: Programmable nuclease revolutionizing the molecular diagnosis. *Molecular Biotechnology*, 66(8), 1739–1753,

Phelps, M. P., Seeb, L. W., & Seeb, J. E. (2020). Transforming ecology and conservation biology through genome editing. *Conservation Biology*, 34(1), 54–65.

Poon, S. S., & Lansdorp, P. M. (2001). Quantitative fluorescence in situ hybridization (Q-FISH). *Current Protocols in Cell Biology*, 12(1), 18.14. 11–18.14. 21.

Popa, S. C., Inamoto, I., Thuronyi, B. W., & Shin, J. A. (2020). Phage-assisted continuous evolution (PACE): A guide focused on evolving protein–DNA interactions. *ACS Omega,* 5(42), 26957–26966.

Rahman, M.-u., Zulfiqar, S., Raza, M. A., Ahmad, N., & Zhang, B. (2022). Engineering abiotic stress tolerance in crop plants through CRISPR genome editing. *Cells*, 11(22), 3590.

Sajal, S. (2024). CRISPR-Cas system and its role in the development of viral disease diagnostics. In S. Joshi, D. Verma, & R. K. Gupta (Eds.), *CRISPR-Cas system in translational biotechnology* (pp. 131–149). Elsevier.

Sander, J. D., & Joung, J. K. (2014). CRISPR-Cas systems for editing, regulating and targeting genomes. *Nature Biotechnology*, 32(4), 347–355.

Schultenkämper, K., Brito, L. F., & Wendisch, V. F. (2020). Impact of CRISPR interference on strain development in biotechnology. *Biotechnology and Applied Biochemistry*, 67(1), 7–21.

Schultzhaus, Z., Wang, Z., & Stenger, D. (2021). CRISPR-based enrichment strategies for targeted sequencing. *Biotechnology Advances,* 46, 107672.

Schuster, A., Erasimus, H., Fritah, S., Nazarov, P. V., van Dyck, E., Niclou, S. P., & Golebiewska, A. (2019). RNAi/CRISPR screens: From a pool to a valid hit. *Trends in Biotechnology*, 37(1), 38–55.

Serajian, S., Ahmadpour, E., Oliveira, S. M. R., Pereira, M. d. L., & Heidarzadeh, S. (2021). CRISPR-Cas technology: emerging applications in clinical microbiology and infectious diseases. *Pharmaceuticals*, 14(11), 1171.

Shimatani, Z., Fujikura, U., Ishii, H., Matsui, Y., Suzuki, M., Ueke, Y., … Kondo, A. (2018). Inheritance of co-edited genes by CRISPR-based targeted nucleotide substitutions in rice. *Plant Physiology and Biochemistry,* 131, 78–83.

Singh, M., Bindal, G., Misra, C. S., & Rath, D. (2022). The era of Cas12 and Cas13 CRISPR-based disease diagnosis. *Critical Reviews in Microbiology*, 48(6), 714–729.

Singh, M., Kumar, V., Sehrawat, N., Yadav, M., Chaudhary, M., Upadhyay, S. K., … Dilbaghi, N. (2022). *Current paradigms in epigenetic anticancer therapeutics and future challenges. Paper presented at the Seminars in Cancer Biology.*

Sullivan, N. T., Allen, A. G., Atkins, A. J., Chung, C.-H., Dampier, W., Nonnemacher, M. R., & Wigdahl, B. (2020). Designing safer CRISPR/Cas9 therapeutics for HIV: Defining factors that regulate and technologies used to detect off-target editing. *Frontiers in Microbiology,* 11, 1872.

Sultan, Q., Ashraf, S., Munir, A., Khan, S. H., Munawar, N., Abd-Elsalam, K. A., & Ahmad, A. (2022). Beyond genome editing: CRISPR approaches. In A. Ahmad, S. H. Khan, & Z. Khan (Eds.), *The CRISPR/Cas tool kit for genome editing* (pp. 187–218). Springer.

Tavakoli, K., Pour-Aboughadareh, A., Kianersi, F., Poczai, P., Etminan, A., & Shooshtari, L. (2021). Applications of CRISPR-Cas9 as an advanced genome editing system in life sciences. *BioTech*, 10(3), 14.

Tian, M., Zhang, R., & Li, J. (2023). Emergence of CRISPR/Cas9-mediated bioimaging: a new dawn of in-situ detection. *Biosensors and Bioelectronics,* 232, 115302.

Tiwari, M., Sharma, D., & Trivedi, P. K. (2014). Artificial microRNA mediated gene silencing in plants: Progress and perspectives. *Plant Molecular Biology*, 86, 1–18.

Tong, B., Dong, H., Cui, Y., Jiang, P., Jin, Z., & Zhang, D. (2021). The versatile type V CRISPR effectors and their application prospects. *Frontiers in Cell and Developmental Biology,* 8, 622103.

Tram, T. T. B., Ha, V. T. N., Trieu, L. P. T., Ashton, P. M., Crawford, E. D., Thu, D. D. A., … Anscombe, C. (2023). FLASH-TB: an application of next-generation CRISPR to detect drug resistant tuberculosis from direct sputum. *Journal of Clinical Microbiology,* 61(4), e01634-01622.

Wang, G.-H., Du, J., Chu, C. Y., Madhav, M., Hughes, G. L., & Champer, J. (2022). Symbionts and gene drive: two strategies to combat vector-borne disease. *Trends in Genetics*, 38(7), 708–723.

Wang, H., Xu, X., Nguyen, C. M., Liu, Y., Gao, Y., Lin, X., … Qi, L. S. (2018). CRISPR-mediated programmable 3D genome positioning and nuclear organization. *Cell,* 175(5), 1405–1417. e1414.

Wang, I. E., Cooper, G., & Mousa, S. A. (2021). Diagnostic approaches for COVID-19 and its associated complications. *Diagnostics,* 11(11), 2071.

Wang, Y., Xue, P., Cao, M., Yu, T., Lane, S. T., & Zhao, H. (2021). Directed evolution: methodologies and applications. *Chemical Reviews*, 121(20), 12384–12444.

Weiss, T., Kumar, J., Chen, C., Guo, S., Schlegel, O., Lutterman, J., ... Zhang, F. (2024). Dual activities of an X-family DNA polymerase regulate CRISPR-induced insertional mutagenesis across species. *Nature Communications*, 15(1), 6293.

Weng, Z., You, Z., Yang, J., Mohammad, N., Lin, M., Wei, Q., ... Zhang, Y. (2023). CRISPR-Cas biochemistry and CRISPR-based molecular diagnostics. *Angewandte Chemie International Edition*, 62(17), e202214987.

Wong, P. K., Mohamad Zamberi, N. N., Syafruddin, S. E., Cheah, F. C., Azmi, N., Law, J. X., & Chua, E. W. (2023). The promises and pitfalls of CRISPR-mediated base editing in stem cells. *The CRISPR Journal*, 6(3), 196–215.

Woodworth, M. B., Girskis, K. M., & Walsh, C. A. (2017). Building a lineage from single cells: genetic techniques for cell lineage tracking. *Nature Reviews Genetics*, 18(4), 230–244.

Wu, X., Mao, S., Ying, Y., Krueger, C. J., & Chen, A. K. (2019). Progress and challenges for live-cell imaging of genomic loci using CRISPR-based platforms. *Genomics, Proteomics and Bioinformatics*, 17(2), 119–128.

Yang, H., Ren, S., Yu, S., Pan, H., Li, T., Ge, S., ... Xia, N. (2020). Methods favoring homology-directed repair choice in response to CRISPR/Cas9 induced-double strand breaks. *International Journal of Molecular Sciences*, 21(18), 6461.

Yang, S.-M., Lv, S., Zhang, W., & Cui, Y. (2022). Microfluidic point-of-care (POC) devices in early diagnosis: A review of opportunities and challenges. *Sensors,* 22(4), 1620.

Yuan, B., Yuan, C., Li, L., Long, M., & Chen, Z. (2022). Application of the CRISPR/Cas System in Pathogen detection: A review. *Molecules*, 27(20), 6999.

Zalatan, J. G., Lee, M. E., Almeida, R., Gilbert, L. A., Whitehead, E. H., La Russa, M., ... Qi, L. S. (2015). Engineering complex synthetic transcriptional programs with CRISPR RNA scaffolds. *Cell,* 160(1), 339–350.

Zhang, F. (2019). Development of CRISPR-Cas systems for genome editing and beyond. *Quarterly Reviews of Biophysics,* 52, e6.

Zhao, L., Qiu, M., Li, X., Yang, J., & Li, J. (2022). CRISPR-Cas13a system: A novel tool for molecular diagnostics. *Frontiers in Microbiology,* 13, 1060947.

Zong, Y., Wang, Y., Li, C., Zhang, R., Chen, K., Ran, Y., ... Gao, C. (2017). Precise base editing in rice, wheat and maize with a Cas9-cytidine deaminase fusion. *Nature Biotechnology*, 35(5), 438–440.

9 CRISPR-Based RNA Editing Systems (Cas13 and Cas7-11) in Plants

Vijay Sheri and Baohong Zhang

9.1 INTRODUCTION

A nucleic acid-targeted defense mechanism, the clustered regularly interspaced short palindromic repeats (CRISPR), and its associated system (Cas), an adaptive immunity system are present in many species of archaea and bacteria. These systems are made up of two components, the primary component is the CRISPR genomic locus that comprises short repeating elements (repeats) separated through distinct sequences (spacers), which can arise from mobile genetic components (MGEs), such as bacteriophages, transposons, and plasmids. The second component is Cas protein, which is encoded by Cas genes, that are typically found next to the CRISPR array, the Cas proteins are involved in CRISPR/Cas systems as effectors (Hille & Charpentier, 2016; Stern et al., 2010; Vercoe et al., 2013). The early identification of this mechanism in procaryotes inspired researchers to employ the CRISPR/Cas genome-editing approach as a more powerful tool for altering RNA and DNA in a diversity of species.

CRISPR/Cas technologies enable targeted mutagenesis in agricultural plants while maintaining the native plant phenotype, making them more effective and user-friendly tools. One of the first Cas proteins to be thoroughly examined was Cas9 nuclease of *Streptococcus pyogenes* which remains the most extensively employed Type II CRISPR/Cas9 system. This ultimately contributed to its broad application in DNA manipulation. in bacteria, animals, and plants (Li et al., 2013; Sretenovic & Qi, 2021; Xie & Yang, 2013; Zhang et al., 2017; Zhou et al., 2021). CRISPR/Cas9 genome editing consists of three steps: (1) recognition, (2) cleavage, as well as (3) repair. These sgRNA are intended to recognize the targeted region in a gene of interest with a complementary base pairing. The Cas9 nuclease creates double-strand breaks (DSBs) at three base pairs upstream of the protospacer adjacent motifs (PAM), these breaks are either repaired by the processes of homology-directed repair (HDR) or non-homologous end join (NHEJ) (Mengstie & Wondimu, 2021). Considering the capability to identify polynucleotides, like RNA or DNA, the structure of the Cas genes, which targets identification, and cleavage techniques, each type and subtype posses unique properties (Makarova et al., 2020).

The CRISPR system is primarily separated into two classes Class 1 and Class 2, based on the type of effector protein. The Cas proteins are further categorized into six kinds (I–VI) and various subtypes according on their modifications and actions (Jiang & Doudna, 2017; Koonin et al., 2017; Makarova et al., 2018). The Class 1 system (Types I, III, and IV) employs several Cas protein activator complexes, whereas Class 2 systems (Types II, V, and VI) use a single protein to change the target (Makarova et al., 2015). Cas proteins of Types II and V both modify DNA, whereas those of Type VI alter RNA (Cao et al., 2020). RNA editing provides several benefits over more conventional DNA editing technologies. However, there are certain drawbacks to CRISPR-based DNA editing, for instance, the need for a Protospacer Adjacent Motif (PAM) location, off-target alterations, and minimal viral proficiency. A few of these restrictions can possibly be overcome with recently discovered CRISPR-based RNA editing (Types VI and III) systems, which use the Cas13 and Cas7-11 proteins to allow sequence-specific degradation of single-stranded RNA molecules (Goswami et al., 2022; Wolter & Puchta, 2018). This chapter covers the most recent achievements in RNA editing with CRISPR/Cas13 and Cas7-11, with a focus on their potential uses in plants.

9.2 CRISPR/CAS13

Cas13 are single RNA-guided effectors from the Class II, Type VI editing systems based on CRISPR-Cas, which were more versatile than Cas9 and Cpf1, as they do not need a PAM sequence at the target region. Cas13 effector proteins generally contain two HEPN (Higher Eukaryotic and Prokaryotic Nucleotide-binding) RNase modules (Makarova et al., 2020; Shmakov et al., 2017). An individual catalytic site is shaped by the two HEPN domains, which cleaves RNA when activated through base pairing among their guide and a matching target RNA (Abudayyeh et al., 2016; Cox et al., 2017; Konermann et al., 2018; Smargon et al., 2017; Xu et al., 2021; Yan et al., 2018). The higher eukaryotic and prokaryotic nucleotide (HEPN)-binding domains have distinct RNase activity, one of them processes their own CRISPR arrays as well as another for RNA targeting cleavage (Anantharaman et al., 2013; East-Seletsky et al., 2016). The locations of these HEPN domains vary across the Cas13 proteins; they are found at the N-terminus and C-terminus of Cas13b, Cas13X, and Y (Smargon et al., 2017; Xu et al., 2021) as well as in the center and C terminus of Cas13a, 13c, and 13d (Abudayyeh et al., 2016) (Figure 9.1).

The Cas13 effector proteins have two lobes: a nuclease (NUC) lobe and a recognition (REC) lobe. The NUC lobe is made up of two functionally connected higher eukaryotic and prokaryotic nucleotide-binding domains, HEPN-1 and

 DOI: 10.1201/9781003500933-11

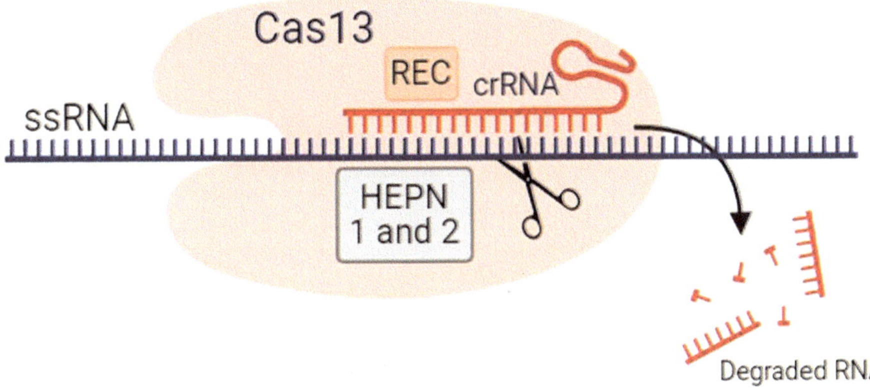

FIGURE 9.1 The structure of Class II Type VI CRISPR/Cas13; A single protein effector with two lobes HEPN, REC along with crRNA forms a complex, and cleaves ssRNA at precise target region. Cas13 is an RNA-directed and RNA-specific nuclease system, that identifies and degrades single-stranded RNA.

TABLE 9.1
CRISPR/Cas13 summary

Type VI Cas13 subtypes	Pre-crRNA processing site	Protospacer flanking site (PFS), direct repeat lengths	Application	Reference
VI-A/Cas13a (1389 aa)	HEPN Domains were located at the C terminus and in the center, while Helical-1 and HEPN-2 Domains are engaged in pre-crRNA processing	A, U and C 35–39 nt	RNA knockdown, Virus resistance, Nucleic acid detection	(Abudayyeh et al., 2016; East-Seletsky et al., 2016; Liu et al., 2017)
VI-B/Cas13b (1127 aa)	RRI-2 domain, HEPN Domains in N-terminus and C-terminus	A, U and G 30 nt	Virus resistance, base editing, and RNA knockdown	(Smargon et al., 2017)
VI-C/Cas13c (1121 aa)	HEPN domain are located in the middle and C-terminus	28–30 nt	RNA knockdown	(Huynh et al., 2020; Shmakov et al., 2017)
VI-D/Cas13d (954 aa)	Helical-1 domain in REC lobe, HEPN 1 and 2 in NUC lobe are in center and C terminus	no PFS constraints 23–30 nt	RNA editing, virus interference, virus resistance	(Konermann et al., 2018; Yan et al., 2018; Zhan et al., 2019)
VI-X/Cas13x (775 aa)	The Cas13X protein takes on a bilobed configuration with REC and NUC lobes	no PFS constraints 36 nt	RNA interference	(Xu et al., 2021)
VI-Y/Cas13y (803 aa)	No detailed structural information for Cas13Y is given	–	RNA editing	(Xu et al., 2021)

HEPN-2 (Zhu & Huang, 2019). The REC lobe consists of the (NTD) N-terminal domain and the Helical-1 domain (Liu & Pei, 2022). Cas13 is further classified into six subtypes according to phylogeny, characteristics, and functional classification: subtype VI-A (Cas13a, formerly C2c2), VI-B (Cas13b, formerly C2c4), VI-C (Cas13c, formerly C2c7), VI-D (Cas13d), VI-X (Cas13X) and VI-Y (Cas13Y) (Shmakov et al., 2015, 2017; Smargon et al., 2017; Xu et al., 2021; Yan et al., 2018). CRISPR/Cas13a targets the single-stranded RNA constructed using a computational sequencing database-mining method (Zhang et al., 2017). Cas13c (C2c7), have a similar prototypical location and crRNA structure similar to Cas13a (Zhang et al., 2017). Cas13d, the smallest subfamily of single effector Cas proteins, offers more interference from RNA viruses than other variants (Konermann et al., 2018; Zhang et al., 2018; Zhang et al., 2019). The Cas13X and Cas13Y are compact CRISPR/Cas13 systems, and have been identified as valuable tools for therapeutic applications

9.3 CRISPR/CAS13 SYSTEM IN PLANT GENOME EDITING

CRISPR/Cas13 systems can be used in multiple organisms for RNA interference, detection, editing, and targeting (Burmistrz et al., 2020; Kordyś et al., 2022; Wang et al., 2019). CRISPR/Cas13 is able to target endogenous RNAs and viral RNAs in human cells and plant (Gunitseva et al., 2023). CRISPR/Cas13 systems provide several benefits for agriculture due to their simplicity, resilience, and ability to target multiple targets simultaneously (Aman, Ali, et al., 2018). The first evidence was of *Leptrotrichia wadei* Cas13a (LwCas13a) being able to knock down targeted RNA more selectively in human cells than RNAi (Abudayyeh et al., 2017). A number of potential orthologs of Cas13, besides LshCas13a and LwCas13a, have been discovered and used in RNA alterations (Cox et al., 2017; Konermann et al., 2018; Mahas & Mahfouz, 2018; Shmakov et al., 2015; Xu et al., 2021; Yan et al., 2018). For various purposes, such as base editing, nucleic acid detection, and live cell imaging of RNA, Cas13 proteins were also altered by deactivating HEPN domains also combining effector spheres (Abudayyeh et al., 2017; Cox et al., 2017; Gootenberg et al., 2017).

Current research has shown that the Cas13 has less off-target impacts and higher accuracy in genome-editing applications (Konermann et al., 2018). CRISPR-Cas13 RNA targeting is an encouraging method for plant research (Wolter & Puchta, 2018). The most prevalent type of plant virus is an RNA virus, and many plant DNA viruses have an RNA intermediary at some stage of their life cycle (Nicaise, 2014; Roossinck, 2003). Using LwaCas13a, Abudayyeh et al. (2017) effectively targeted the PDS, HCT, and EPSPS genes in rice protoplasts. It was demonstrated that LwaCas13a may effectively disrupt the cytoplasmic RNA pool in plants when seven out of nine gRNAs produced more than 50% knockdown. Furthermore, a method for fighting plant RNA viruses has been made possible by CRISPR-Cas13 technology (Table 9.2). Turnip mosaic virus (TuMV) in *Arabidopsis* and *N. benthamiana* can be inhibited by CRISPR/Cas13a, as shown by (Aman, Ali, et al., 2018; Aman, Mahas, et al., 2018) using codon-optimized LshCas13a. LshCas13a from *Leptotrichia shahii* has demonstrated a modest level of efficacy in engineering immunity against RNA viruses in both monocot and dicot plants, including potato virus Y (PVY) and tobacco turnip mosaic RNA virus (TuMV) (Khan et al., 2018; Sharma et al., 2021; Zhan et al., 2019).

Another application of the dCas13a technology is to observe the transport and localization of specific endogenic RNA in live cells using a florescent protein. Also, this method was used to track the migration of virus RNA inside infested cells. Likewise, modified PUF proteins were used to sense the location of tobacco mosaic virus RNA in leaves of *N. benthamiana* (Peña & Heinlein, 2012; Tilsner et al., 2009). Furthermore, target transcript binding activates Cas13a's collateral RNA degrade activity, causing programmed cell death or stasis in certain cells and prevents viral infection from spreading to nearby cells (Abudayyeh et al., 2016). The creation of diagnostic techniques like SHERLOCK (Specific High sensitivity Enzymatic Reporter Unlocking) enables in vitro identification of specific DNA or RNA components in a pool of non-target nucleic acids using attomolar specificity, and has benefited from the philandering RNase action of Cas13a (Gootenberg et al., 2017). Additionally, by combining dCas13a, that restores deleterious modifications without affecting the surrounding bases, similar Cas13b, with RNA-editing enzymes like cytidine deaminase (which converts cytosine to uracil) and ADAR (Adenosine Deaminase Acting on RNA, converts adenosine to Inosine), the system may also be used to alter bases (Cox et al., 2017). All those characteristics generate the CRISPR/Cas13a method an adaptable tool in investigating RNA biology and RNA editing to understand the roles of RNAs.

9.4 CRISPR/CAS7-11 SYSTEMS

Genome-editing methods using CRISPR/Cas are constantly evolving because of innovative applications and extra considerations. Cas7-11 is an establishing class 1 single-effector protease that targets RNA without causing toxicity to the cell and without collateral cleavage (Catchpole & Terns, 2021; Özcan et al., 2021; Goswami et al., 2022) (Figure 9.2). Its main advantage is that it has substantially less off-target

TABLE 9.2
Applications of CRISPR/Cas13 in plants

CRISPR/Cas13	Plant species	Application	Reference
LshpCas13a	*N. benthamiana*	Interference of RNA contrary to the TuMV virus.	(Aman, Ali, et al., 2018)
LshpCas13a	*A. thaliana*	RNA interference against TuMV virus	(Aman, Mahas, et al., 2018)
LshCas13a	*O. Sativa*, *N. benthamiana*	Inhibition of viral infection in both dicot and monocot plants.	(Zhang et al., 2019)
LwaCas13a	*O. Sativa*	EPSP, HCT	(Abudayyeh et al., 2017)
CasRx13d	*N. benthamiana*	CasRx demonstrated substantial interference efficiency and high specificity.	(Mahas & Mahfouz, 2018)
Cas13a	*Solanum tuberosum* L.	Generating virus-resistant potato cultivars through RNA genome target.	(Zhan et al., 2019)

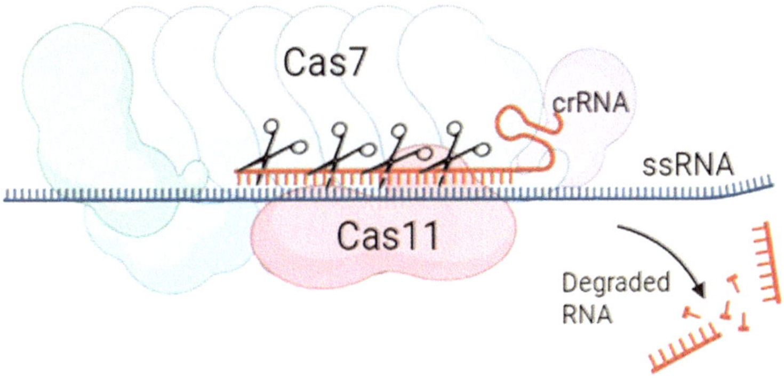

FIGURE 9.2 The structure of Class I, Type III CRISPR/Cas7-11; a multi-protein effector complex with Cas7-11 and crRNA that allows RNA-directed RNA degradation without any toxicity or collateral activities and has potential uses in RNA interference.

toxicity than other RNA editing technologies, allowing for more efficient experimental design and clinical applications. Like class 2 nucleases, Cas7-11 converts pre-crRNA into mature crRNA and then cleaves RNA at a site indicated by gRNA. Using a single multi-protein effector called Cas7-11 (gRAMP), the Type III CRISPR/Cas system cleaves RNA by engaging with a caspase such as protease Csx29 (Yu et al., 2022).

Cas7-11 is a Type III CRISPR/Cas system made up of seven domains (Cas7, Cas11, CTE, and INS) and four interdomain linkers: The Cas7 is fusion of four domains Cas7.1, Cas7.2, Cas7.3, and Cas7.4, these four Cas7 domains combine to form a helical filament for holding crRNA, and a Cas11 domain facilitates the formation of crRNA-target RNA duplexes. The Cas7.1 domain is used to recognize and process crRNA maturation and the Cas7.2 and Cas7.3 domains to cleave target RNAs at specific sites using the conserved catalytic residues (Kato et al., 2022; Yu et al., 2022). Cas7-11 with dual RNase capabilities converts pre-CRISPR RNA into mature CRISPR RNA and cleaves the target RNA without collateral action and cell toxicity, offering an alternative to RNA knockdown (Özcan et al., 2021). Cas7-11 is feasible to specifically target RNA to cause interference such as against diverse plant viruses, RNA editing has yet to be investigated, however, it has the potential to become a far more efficient technique.

9.5 CONCLUSION

Plant genome research can considerably benefit from the application of CRISPR/Cas13 and Cas7-11 systems for RNA targeting and editing plants. The major advantage of CRISPR-based RNA editing in plants is to confer virus resistance, viral diagnostics, RNA base editing and RNA knockdown with minimal off-target effects. However, it is becoming increasingly used for studying mRNA, circRNA, and other non-coding RNA. Furthermore, due to its great specificity, CRISPR/Cas13 nucleic acid detection has gained popularity in recent research, particularly for the diagnosis of

by using orthologs such as LwaCas13a, PspCas13b, and RfxCas13d. Additionally, RNA-guided RNA degradation is made possible by Cas7-11, which has no side effects and less toxicity to cells. Thus, CRISPR-based editing of RNA technologies has the ability to become an effective tool for crop improvement, providing new opportunities for diverse applications in agriculture.

REFERENCES

Abudayyeh, O. O., Gootenberg, J. S., Essletzbichler, P., Han, S., Joung, J., Belanto, J. J., Verdine, V., Cox, D. B. T., Kellner, M. J., Regev, A., Lander, E. S., Voytas, D. F., Ting, A. Y., & Zhang, F. (2017). RNA targeting with CRISPR-Cas13. *Nature*, 550(7675), 280–284. https://doi.org/10.1038/nature24049

Abudayyeh, O. O., Gootenberg, J. S., Konermann, S., Joung, J., Slaymaker, I. M., Cox, D. B. T., Shmakov, S., Makarova, K. S., Semenova, E., Minakhin, L., Severinov, K., Regev, A., Lander, E. S., Koonin, E. V., & Zhang, F. (2016). C2c2 is a single-component programmable RNA-guided RNA-targeting CRISPR effector. *Science*, 353(6299). https://doi.org/10.1126/science.aaf5573

Aman, R., Ali, Z., Butt, H., Mahas, A., Aljedaani, F., Khan, M. Z., Ding, S., & Mahfouz, M. (2018). RNA virus interference via CRISPR/Cas13a system in plants. *Genome Biology*, 19(1), 1. https://doi.org/10.1186/s13059-017-1381-1

Aman, R., Mahas, A., Butt, H., Aljedaani, F., & Mahfouz, M. (2018). Engineering RNA virus interference via the CRISPR/Cas13 machinery in Arabidopsis. *Viruses,* 10(12), 732. https://doi.org/10.3390/v10120732

Anantharaman, V., Makarova, K. S., Burroughs, A. M., Koonin, E. V, & Aravind, L. (2013). Comprehensive analysis of the HEPN superfamily: Identification of novel roles in intra-genomic conflicts, defense, pathogenesis and RNA processing. *Biology Direct*, 8(1), 15. https://doi.org/10.1186/1745-6150-8-15

Burmistrz, M., Krakowski, K., & Krawczyk-Balska, A. (2020). RNA-Targeting CRISPR-Cas systems and their applications. *International Journal of Molecular Sciences*, 21(3), 1122. https://doi.org/10.3390/ijms21031122

Cao, Y., Zhou, H., Zhou, X., & Li, F. (2020). Control of plant

Catchpole, R. J., & Terns, M. P. (2021). New Type III CRISPR variant and programmable RNA targeting tool: Oh, thank heaven for Cas7-11. *Molecular Cell*, 81(21), 4354–4356.

Cox, D. B. T., Gootenberg, J. S., Abudayyeh, O. O., Franklin, B., Kellner, M. J., Joung, J., & Zhang, F. (2017). RNA editing with CRISPR-Cas13. *Science*, 358(6366), 1019–1027. https://doi.org/10.1126/science.aaq0180

East-Seletsky, A., O'Connell, M. R., Knight, S. C., Burstein, D., Cate, J. H. D., Tjian, R., & Doudna, J. A. (2016). Two distinct RNase activities of CRISPR-C2c2 enable guide-RNA processing and RNA detection. *Nature*, 538(7624), 270–273. https://doi.org/10.1038/nature19802

Gootenberg, J. S., Abudayyeh, O. O., Lee, J. W., Essletzbichler, P., Dy, A. J., Joung, J., Verdine, V., Donghia, N., Daringer, N. M., Freije, C. A., Myhrvold, C., Bhattacharyya, R. P., Livny, J., Regev, A., Koonin, E. V., Hung, D. T., Sabeti, P. C., Collins, J. J., & Zhang, F. (2017). Nucleic acid detection with CRISPR-Cas13a/C2c2. *Science*, 356(6336), 438–442. https://doi.org/10.1126/science.aam9321

Goswami, H. N., Rai, J., Das, A., & Li, H. (2022). Molecular mechanism of active Cas7-11 in processing CRISPR RNA and interfering target RNA. *ELife*, 11. https://doi.org/10.7554/eLife.81678

Gunitseva, N., Evteeva, M., Borisova, A., Patrushev, M., & Subach, F. (2023). RNA-dependent RNA Targeting by CRISPR-Cas systems: Characterizations and applications. *International Journal of Molecular Sciences*, 24(8), 6894. https://doi.org/10.3390/ijms24086894

Hille, F., & Charpentier, E. (2016). CRISPR-Cas: Biology, mechanisms and relevance. *Philosophical Transactions of the Royal Society B: Biological Sciences*, 371(1707), 20150496. https://doi.org/10.1098/rstb.2015.0496

Huynh, N., Depner, N., Larson, R., & King-Jones, K. (2020). A versatile toolkit for CRISPR-Cas13-based RNA manipulation in Drosophila. *Genome Biology*, 21(1), 279. https://doi.org/10.1186/s13059-020-02193-y

Jiang, F., & Doudna, J. A. (2017). CRISPR-Cas9 structures and mechanisms. *Annual Review of Biophysics*, 46(1), 505–529. https://doi.org/10.1146/annurev-biophys-062215-010822

Kato, K., Zhou, W., Okazaki, S., Isayama, Y., Nishizawa, T., Gootenberg, J. S., Abudayyeh, O. O., & Nishimasu, H. (2022). Structure and engineering of the type III-E CRISPR-Cas7-11 effector complex. *Cell*, 185(13), 2324–2337.e16. https://doi.org/10.1016/j.cell.2022.05.003

Khan, M. Z., Amin, I., Hameed, A., & Mansoor, S. (2018). CRISPR-Cas13a: Prospects for plant virus resistance. *Trends in Biotechnology*, 36(12), 1207–1210. https://doi.org/10.1016/j.tibtech.2018.05.005

Konermann, S., Lotfy, P., Brideau, N. J., Oki, J., Shokhirev, M. N., & Hsu, P. D. (2018). Transcriptome engineering with RNA-targeting type VI-D CRISPR effectors. *Cell*, 173(3), 665–676.e14. https://doi.org/10.1016/j.cell.2018.02.033

Koonin, E. V, Makarova, K. S., & Zhang, F. (2017). Diversity, classification and evolution of CRISPR-Cas systems. *Current Opinion in Microbiology,* 37, 67–78. https://doi.org/10.1016/j.mib.2017.05.008

Kordyś, M., Sen, R., & Warkocki, Z. (2022). Applications of the versatile CRISPR-Cas13 RNA targeting system. *WIREs RNA*, 13(3). https://doi.org/10.1002/wrna.1694

Li, D., Qiu, Z., Shao, Y., Chen, Y., Guan, Y., Liu, M., Li, Y., Gao, N., Wang, L., Lu, X., Zhao, Y., & Liu, M. (2013). Heritable gene targeting in the mouse and rat using a CRISPR-Cas system. *Nature Biotechnology*, 31(8), 681–683. https://doi.org/10.1038/nbt.2661

Liu, L., Li, X., Ma, J., Li, Z., You, L., Wang, J., Wang, M., Zhang, X., & Wang, Y. (2017). The molecular architecture for RNA-guided RNA cleavage by Cas13a. *Cell*, 170(4), 714–726.e10. https://doi.org/10.1016/j.cell.2017.06.050

Liu, L., & Pei, D.-S. (2022). Insights gained from RNA editing targeted by the CRISPR-Cas13 family. *International Journal of Molecular Sciences,* 23(19), 11400. https://doi.org/10.3390/ijms231911400

Mahas, A., & Mahfouz, M. (2018). Engineering virus resistance via CRISPR-Cas systems. *Current Opinion in Virology,* 32, 1–8. https://doi.org/10.1016/j.coviro.2018.06.002

Makarova, K. S., Wolf, Y. I., Iranzo, J., Shmakov, S. A., Alkhnbashi, O. S., Brouns, S. J. J., Charpentier, E., Cheng, D., Haft, D. H., Horvath, P., Moineau, S., Mojica, F. J. M., Scott, D., Shah, S. A., Siksnys, V., Terns, M. P., Venclovas, Č., White, M. F., Yakunin, A. F., … Koonin, E. V. (2020). Evolutionary classification of CRISPR–Cas systems: A burst of class 2 and derived variants. *Nature Reviews Microbiology*, 18(2), 67–83. https://doi.org/10.1038/s41579-019-0299-x

Makarova, K. S., Wolf, Y. I., & Koonin, E. V. (2018). Classification and nomenclature of CRISPR-Cas systems: Where from here? *The CRISPR Journal*, 1(5), 325–336. https://doi.org/10.1089/crispr.2018.0033

Mengstie, M. A., & Wondimu, B. Z. (2021). Mechanism and applications of CRISPR/Cas-9-Mediated genome editing. *Biologics: Targets and Therapy,* 15, 353–361. https://doi.org/10.2147/BTT.S326422

Nicaise, V. (2014). Crop immunity against viruses: Outcomes and future challenges. *Frontiers in Plant Science*, 5. https://doi.org/10.3389/fpls.2014.00660

Özcan, A., Krajeski, R., Ioannidi, E., Lee, B., Gardner, A., Makarova, K. S., Koonin, E. V., Abudayyeh, O. O., & Gootenberg, J. S. (2021). Programmable RNA targeting with the single-protein CRISPR effector Cas7-11. *Nature*, 597(7878), 720–725. https://doi.org/10.1038/s41586-021-03886-5

Peña, E. J., & Heinlein, M. (2012). RNA transport during TMV cell-to-cell movement. *Frontiers in Plant Science*, 3. https://doi.org/10.3389/fpls.2012.00193

Roossinck, M. J. (2003). Plant RNA virus evolution. *Current Opinion in Microbiology*, 6(4), 406–409. https://doi.org/10.1016/S1369-5274(03)00087-0

Sharma, S. K., Gupta, O. P., Pathaw, N., Sharma, D., Maibam, A., Sharma, P., Sanasam, J., Karkute, S. G., Kumar, S., & Bhattacharjee, B. (2021). CRISPR-Cas-led revolution in diagnosis and management of emerging plant viruses: New avenues toward food and nutritional security. *Frontiers in Nutrition*, 8. https://doi.org/10.3389/fnut.2021.751512

Shmakov, S., Abudayyeh, O. O., Makarova, K. S., Wolf, Y. I., Gootenberg, J. S., Semenova, E., Minakhin, L., Joung, J., Konermann, S., Severinov, K., Zhang, F., & Koonin, E. V. (2015). Discovery and functional characterization of diverse Class 2 CRISPR-Cas systems. *Molecular Cell*, 60(3), 385–397. https://doi.org/10.1016/j.molcel.2015.10.008

Shmakov, S., Smargon, A., Scott, D., Cox, D., Pyzocha, N., Yan, W., Abudayyeh, O. O., Gootenberg, J. S., Makarova, K. S., Wolf, Y. I., Severinov, K., Zhang, F., & Koonin, E. V. (2017). Diversity and evolution of class 2 CRISPR-Cas systems. *Nature Reviews Microbiology*, 15(3), 169–182. https://doi.org/10.1038/nrmicro.2016.184

Smargon, A. A., Cox, D. B. T., Pyzocha, N. K., Zheng, K., Slaymaker, I. M., Gootenberg, J. S., Abudayyeh, O. A., Essletzbichler, P., Shmakov, S., Makarova, K. S., Koonin, E. V., & Zhang, F. (2017). Cas13b is a type VI-B CRISPR-associated RNA-guided RNase differentially regulated by accessory proteins Csx27 and Csx28. *Molecular Cell*, 65(4), 618–630.e7. https://doi.org/10.1016/j.molcel.2016.12.023

Sretenovic, S., & Qi, Y. (2021). Plant prime editing goes prime. *Nature Plants*, 8(1), 20–22. https://doi.org/10.1038/s41477-021-01047-0

Stern, A., Keren, L., Wurtzel, O., Amitai, G., & Sorek, R. (2010). Self-targeting by CRISPR: Gene regulation or autoimmunity? *Trends in Genetics*, 26(8), 335–340. https://doi.org/10.1016/j.tig.2010.05.008

Tilsner, J., Linnik, O., Christensen, N. M., Bell, K., Roberts, I. M., Lacomme, C., & Oparka, K. J. (2009). Live-cell imaging of viral RNA genomes using a Pumilio-based reporter. *The Plant Journal*, 57(4), 758–770. https://doi.org/10.1111/j.1365-313X.2008.03720.x

Vercoe, R. B., Chang, J. T., Dy, R. L., Taylor, C., Gristwood, T., Clulow, J. S., Richter, C., Przybilski, R., Pitman, A. R., & Fineran, P. C. (2013). Cytotoxic chromosomal targeting by CRISPR/Cas systems can reshape bacterial genomes and expel or remodel pathogenicity islands. *PLoS Genetics*, 9(4), e1003454. https://doi.org/10.1371/journal.pgen.1003454

Wang, F., Wang, L., Zou, X., Duan, S., Li, Z., Deng, Z., Luo, J., Lee, S. Y., & Chen, S. (2019). Advances in CRISPR-Cas systems for RNA targeting, tracking and editing. *Biotechnology Advances*, 37(5), 708–729. https://doi.org/10.1016/j.biotechadv.2019.03.016

Wolter, F., & Puchta, H. (2018). The CRISPR/Cas revolution reaches the RNA world: Cas13, a new Swiss Army knife for plant biologists. *The Plant Journal*, 94(5), 767–775. https://doi.org/10.1111/tpj.13899

Xie, K., & Yang, Y. (2013). RNA-Guided genome editing in plants using a CRISPR-Cas system. *Molecular Plant*, 6(6), 1975–1983. https://doi.org/10.1093/mp/sst119

Xu, C., Zhou, Y., Xiao, Q., He, B., Geng, G., Wang, Z., Cao, B., Dong, X., Bai, W., Wang, Y., Wang, X., Zhou, D., Yuan, T., Huo, X., Lai, J., & Yang, H. (2021). Programmable RNA editing with compact CRISPR-Cas13 systems from uncultivated microbes. *Nature Methods*, 18(5), 499–506. https://doi.org/10.1038/s41592-021-01124-4

Yan, W. X., Chong, S., Zhang, H., Makarova, K. S., Koonin, E. V., Cheng, D. R., & Scott, D. A. (2018). Cas13d is a compact RNA-targeting type VI CRISPR effector positively modulated by a WYL-domain-containing accessory protein. *Molecular Cell*, 70(2), 327–339.e5. https://doi.org/10.1016/j.molcel.2018.02.028

Yu, G., Wang, X., Zhang, Y., An, Q., Wen, Y., Li, X., Yin, H., Deng, Z., & Zhang, H. (2022). Structure and function of a bacterial type III-E CRISPR-Cas7-11 complex. *Nature Microbiology*, 7(12), 2078–2088. https://doi.org/10.1038/s41564-022-01256-z

Zhan, X., Zhang, F., Zhong, Z., Chen, R., Wang, Y., Chang, L., Bock, R., Nie, B., & Zhang, J. (2019). Generation of virus-resistant potato plants by RNA genome targeting. *Plant Biotechnology Journal*, 17(9), 1814–1822. https://doi.org/10.1111/pbi.13102

Zhang, C., Konermann, S., Brideau, N. J., Lotfy, P., Wu, X., Novick, S. J., Strutzenberg, T., Griffin, P. R., Hsu, P. D., & Lyumkis, D. (2018). Structural basis for the RNA-guided ribonuclease activity of CRISPR-Cas13d. *Cell*, 175(1), 212–223.e17. https://doi.org/10.1016/j.cell.2018.09.001

Zhang, H., Cheng, Q.-X., Liu, A.-M., Zhao, G.-P., & Wang, J. (2017). A novel and efficient method for bacteria genome editing employing both CRISPR/Cas9 and an antibiotic resistance Cassette. *Frontiers in Microbiology*, 8. https://doi.org/10.3389/fmicb.2017.00812

Zhang, T., Zhao, Y., Ye, J., Cao, X., Xu, C., Chen, B., An, H., Jiao, Y., Zhang, F., Yang, X., & Zhou, G. (2019). Establishing CRISPR/Cas13a immune system conferring RNA virus resistance in both dicot and monocot plants. *Plant Biotechnology Journal*, 17(7), 1185–1187. https://doi.org/10.1111/pbi.13095

Zhou, J., Yuan, M., Zhao, Y., Quan, Q., Yu, D., Yang, H., Tang, X., Xin, X., Cai, G., Qian, Q., Qi, Y., & Zhang, Y. (2021). Efficient deletion of multiple circle RNA loci by CRISPR-Cas9 reveals Os06circ02797 as a putative sponge for OsMIR408 in rice. *Plant Biotechnology Journal*, 19(6), 1240–1252. https://doi.org/10.1111/pbi.13544

Zhu, Y., & Huang, Z. (2019). Recent advances in structural studies of the CRISPR-Cas-mediated genome editing tools. *National Science Review*, 6(3), 438–451. https://doi.org/10.1093/nsr/nwy150

10 CRISPR/Cas-Mediated Base Editing in Plants

Aman Prakash, Raj Bahadur Singh, Nidhi Kumari, Pragalbh Tiwari, Sanjana Singh, Shashi Ranjan, and Sayanti Mandal

10.1 INTRODUCTION

CRISPR stands for Clustered Regularly Interspaced Short-Palindromic Repeats and Cas is a CRISPR-associated protein. It is a prokaryotic adaptive phage immunity mechanism that is used to cleave the invading nucleic acids. An array of CRISPR-Cas systems is present across varied species of bacteria and archaea. They have differences in their components and action mechanism. For example, multi-protein effector complexes are found in Class I CRISPR-Cas system while single effector protein appears in Class 2 system. The target specificity of CRISPR-Cas9 depends on the base pairing of nucleic acid rather than protein-DNA recognition. CRISPR-Cas relies on DNA-RNA recognition and binding for sequence-specific nucleic acid cleavage, they can be employed to introduce double-strand breaks (DSB) at any desired target site at a minimal cost (Chatterjee Saha et al., 2023; Khare et al., 2023; Saini et al., 2023). The CRISPR/Cas9 genome-editing mechanism originates from a bacterial defense system against viruses, known for its role in immunity. This technology has become highly valued for its precise and efficient genome-editing capabilities. CRISPR/Cas9 and its variations have been used to edit a wide array of genes. Found in certain bacteria, the CRISPR locus, along with its proteins, serves as a defense against phages. Transcription of the CRISPR locus results in the production of CRISPR RNA (crRNA) and trans-activating CRISPR RNA (tracr RNA), which, together with Cas9, target and interact with phage DNA. The guide RNA (gRNA) is an engineered molecule that combines crRNA and tracr RNA, guiding the system to the specific DNA sequence to be edited (Hsu et al., 2014).

It is estimated that by 2050, the world population will reach 9.6 billion, which will lead to an increase in the demand for staple crops by 60%. Worldwide agriculture is encountering a list of unrivalled challenges. It is documented that the increase rate of the yields in agriculture by the green revolution is dropping. There are certain expectations that climate change will restrict plant production and will affect agriculture practices. Providing food and nourishment to such a huge population with decreased arable land, and the shortage of availability of water resources, is a gigantic challenge (Kumar et al., 2020).

There is a need for the generation and improvement of cultivars with strengthened resilience to adverse environments and with high and enhanced quality. The prominent drawbacks of conventional crop breeding methodologies are that they are time-consuming, complicated, and very laborious. The more efficient and time-saving strategies save the need for hours (Saini et al., 2023).

First in 2013, three independent groups implemented the CRISPR/Cas 9 system in rice (*Oryza sativa*), *Nicotiana benthamiana*, wheat (*Triticum aestivum*), and *Arabidopsis thaliana.* Since then, CRISPR/Cas 9 has been employed for gene editing in a list of crop species by introducing into them the values of agriculture traits. It was the first time in history that the plant breeders have extensive control over the specific introduction of a sequence variation (targeted). It was a groundbreaking breakthrough for the expeditious improvement of agricultural crops. Even since then, the constant enhancement of CRISPR/Cas systems, such as CRISPR/Cpf1 and nucleotide substitution tools for base editing, has led to the wide adaptation, efficient cost, and ease of use of targeted genetic manipulation assets being put into various crops (Ali et al., 2018; Jiang et al., 2013; Saini et al., 2023; Selma et al., 2022; Wu et al., 2018; Yin et al., 2019; Khare et al., 2023) (Figure 10.1).

Over the past 20 years, the genomes of numerous crops have been enhanced by incorporating genes with significant breeding value, addressing challenges inherent to traditional breeding methods. Moving forward, innovative gene-editing techniques are essential to advance crop productivity, quality, and resilience to diverse environmental stresses (Aldemita et al., 2015; Araki & Ishii, 2015; Chen & Lin, 2013).

However, CRISPR-Cas has the potential of doing more than just editing specific loci for the improvement of crops. An array of novel plant biotechnologies has come to light which are proficient in promoting gene regulation and protein engineering also.

10.2 HISTORY AND CLASSIFICATION OF CRISPR

CRISPR-Cas systems are organized into two main classes, six types, and 33 subtypes (Janik et al., 2020; Makarova et al., 2020; Pyzocha & Chen, 2018; Rojo et al., 2018; Shmakov et al.,

DOI: 10.1201/9781003500933-12

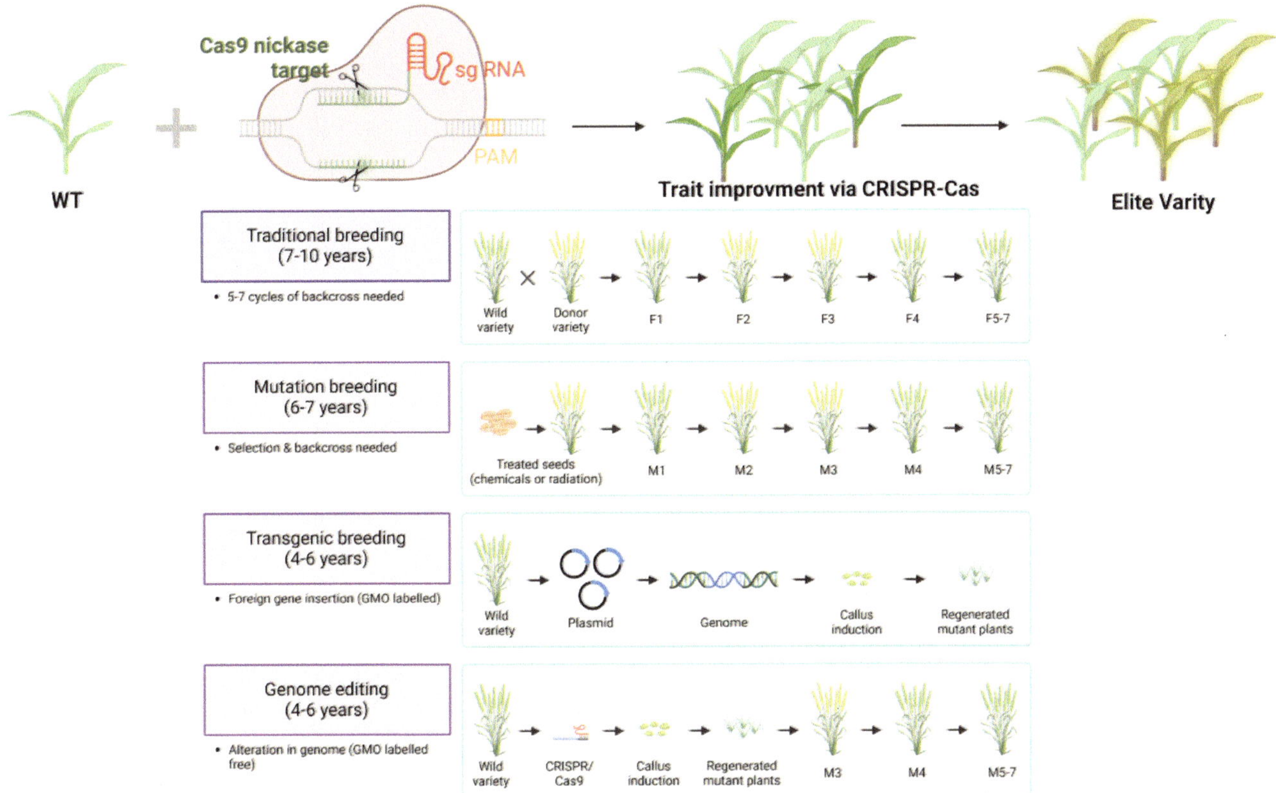

FIGURE 10.1 Potential advantages of CRISPR-Cas9 for generations of elite plant variety. A comparison of the most popular methods for improving crops via plant breeding that save both time and money. (Image modified from Ahmar et al., 2021; image created with BioRender.com.)

2017). Class 1 of the CRISPR-Cas system is characterized by multiprotein effector complexes, while Class 2 is defined by a single, multifunctional effector protein with multiple domains. Each class is further divided into three types. Class 1 includes Types I, III, and IV, while Class 2 comprises Types II, V, and VI. Each type is distinguished by the unique architecture of its effector modules, which contain some specific Cas proteins (Haft et al., 2005; Koonin et al., 2017; Shmakov et al., 2017). Furthermore, each type is divided into several subtypes, each characterized by slight differences in locus organization and encoding subtype-specific Cas proteins. The defining features of the type and subtype of CRISPR-Cas systems are the Cas genes and the proteins they encode, which are both genetically and functionally diverse. This diversity reflects the range of biochemical functions they perform at various stages of CRISPR-mediated immunity. The RNA recognition motif (RRM) is common in many Cas proteins, and numerous Cas protein families contain functional domains that interact with nucleic acids, helicase, and nuclease motifs. Genetically, cas1 and cas2 are universally present across different types and subtypes, while signature genes such as cas3, cas9, and cas10 have been identified for Types I, II, and III, respectively (Barrangou, 2013; Makarova et al., 2011, 2013). Figure 10.2)

10.3 THE CRISPR-MEDIATED BASE EDITING MECHANISM

Base editors, such as cytosine base editors (CBEs) and adenine base editors (ABEs), that have been developed from CRISPR/Cas9 combine base deaminases with nickase Cas9 (nCas9) or dead Cas9(dCas9) to change C to T or A to G. These base editors have shown themselves to be effective instruments for engineering the trait deletion or insertion (indel) in crop breeding that confer desired agronomic features. An ABE has been developed by some research perspectives that allows for exact A-to-G conversion in wheat and rice. Additionally, in maize, wheat, and rice, the CBE causes C-to-T conversion. The base excision repair (BER) process is triggered by CBEs and ABEs through uracil glycosylation, which frequently results in C-to-G replacements. This pathway provides the foundation for the development of C-to-G base editors (CGBEs), which have emerged recently (Lee et al., 2023). For *in vivo* DNA editing investigations, base editing (BE) techniques that combine a fusion of deaminase with partial (nickase, nCas9) or entirely (dead form, dCas9) nuclease deficient Cas9 are excellent because they can diversify specific DNA locations without requiring donor DNA templates or causing double-

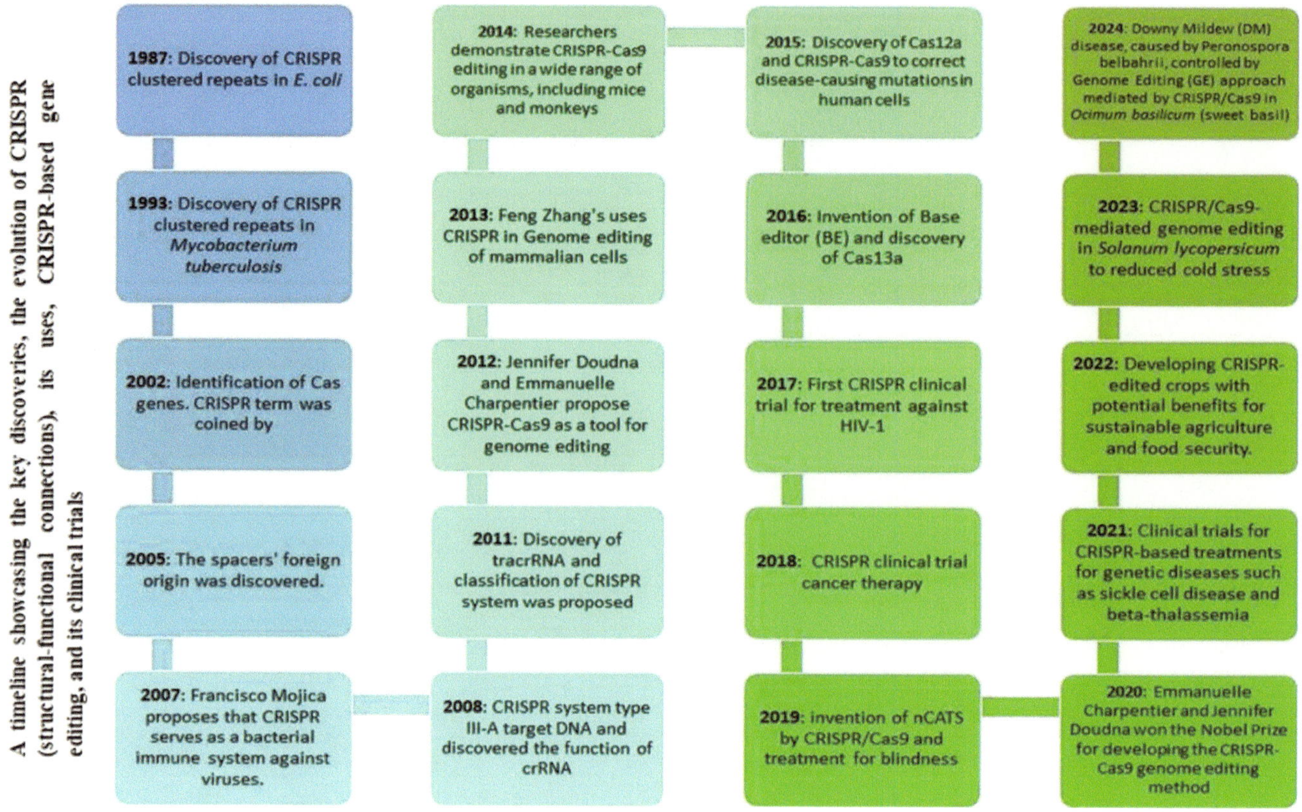

FIGURE 10.2 History of CRISPR. (Image adapted from Kumar et al. (2020); Nidhi et al. (2021); Zhang et al. (2021).)

nucleobase substitutions. The latter methods typically have poorer efficiency and a higher rate of insertion-deletions than exact base alterations. Due to their inability to alter more than one type of nucleotide and their small editable window (EW, DNA length accessible for nucleotide substitution), BEs can only slightly diversify the targeted loci (Shelake et al., 2024).

The two types of base editors are ABE and CBE. For CBE, a cytidine deaminase that catalyzes the deamination of cytosine (C) in a small window of the non-targeted strand is coupled with the Cas9 nickase (nCas9) or catalytically dead Cas protein (dCas9). The process of deamination changes the original C to uracil (U), which is identified as thymine (T) during DNA replication. This leads to a single-base substitution that occurs from CG to TA. Whereas for ABE, adenosine deaminase, which catalyzes the deamination of adenine (A), is united with either nCas9 or dCas9 in ABE. During DNA replication, deamination changes the original A to inosine (I), which is known as guanine (G). Therefore, ABE allows base replacements from AT to GC (Bharat et al., 2020). The Glycosylate Base Editors (GBEs) consist of an nCas9, a cytidine deaminase, and a uracil-DNA glycosylase (Ung). Dead Cas9-Activation-induced cytidine deaminase (dCas9-AID) cleaves one DNA strand at a nearby position while also removing the amine group of C bases on only one strand within the target editing area. Base editing was carried out using dCas9-AID in place of nCas9-AID in order to ascertain whether the pattern of DNA damage was the primary cause of C-to-A changes. There were no detectable C-to-A mutations. from which the specific DNA damage induced by nCas9-AID was postulated to be produced by Ung-initiated C-to-A conversions. After being demonstrated, this repair mechanism could be a fruitful and significant area of study that could be applied to plant or animal cells for targeted C-to-A base editing (Zhao et al., 2020). By creating Cas9 enzymes that can identify non-NGG PAMs and hence extend the editing window, the base editing system has also been enhanced. Where the enzyme cannot cause a double-strand break (DSB), the single guide RNA (sgRNA) directs dCas9 to the target site in the genome. However, cytidine deaminase, for example, can alter a single base in the target area since dCas9 is still attached to it (Rengasamy et al., 2024). Although the base editing pathway uses an inactive CRISPR-Cas9 nuclease (which cannot create double-strand breaks) linked to a deaminase enzyme (cytosine or adenosine) component that modifies nucleotide conversion, it still depends on guideRNA-guided target discovery in the genome. Acetolactate synthase (ALS), a gene involved in herbicide tolerance in wheat, is a prime candidate for base editing. Point mutations in ALS can provide sufficient herbicide tolerance while having little effect on plant productivity (M. Kumar et al., 2023).

Base substitutions were effectively introduced in a few transgene-free herbicide-resistant soybeans that carried a homozygous P180S mutation in those genes. Agronomic properties are not adversely affected by this P180S mutation. In addition, scientists were able to create allele-specific markers to specifically identify P180S mutations in that gene, which will aid in the breeding of soybeans resistant to herbicides (Wei et al., 2023).

A new type of base editor called AYBE has been developed through genetic engineering for A-to-Y (Y = C/T) editing in rice. The process involves isolating an endogenous rice N-methylpurine DNA glycosylase (OsMPG) and engineering two plant rice A-to-K (K = G or T) base editors, such as rAKBE into four types. Their performances were examined in both rice protoplast and stable lines. Of these, rAKBE with a mutant OsMPG performed the best with an expanded editing window, enabling increased A-to-G transition editing, along with decreased A-to-T transversion and the occurrences of fewer insertion-deletions within the editing window of targets (Y. Li et al., 2024). To select target sites by accompanying sgRNAs and cause targeted C to T (or G to A) conversions within a window of around 5 nucleotides, a fusion construct consisting of Cas9 and cytidine deaminase was created in the base editing approach (BE1). Intermediate wobble G:U base pairs were produced by cytidine deaminase's C-to-T conversions during the first stage of base editing. Uracil DNA glycosylase (UDG) activity has a detrimental effect on base editing efficiency, according to an analysis of G:U pair resolution results in cells. The G:U to G:C pair is reverted when UDG detects the sporadic G:U pairs and initiates base-excision repair (BER) to catalytically remove the uracil. According to the BER mechanism, base editing efficiencies could rise if UDG is inhibited. Relentlessly, combining BE1 with a Uracil DNA glycosylase inhibitor (UGI) produced the second-generation base editor (BE2). The basic editing efficiencies of BE2 were almost three times higher than those of BE1, as expected. The base editing procedures do not depend on NHEJ or HDR, which is why low indel rates (less than 0.1%) were found for BE1 and BE2 (Lee et al., 2021)

10.4 FUTURE PROSPECTS OF BASE EDITING IN PLANTS

CRISPR–Cas systems have been a game-changer in plant science, enabling precise genetic alterations that enhance the agricultural traits of various crops. While significant advances have been made to boost the efficiency of these gene-editing tools, there remains a gap in fully meeting the demands of plant genome engineering. Consequently, further enhancements are necessary, particularly in the realm of delivery methods. Streamlining and innovating new delivery techniques are crucial for the cost-effective application of gene editing in plants. Currently, the focus is shifting toward direct delivery into reproductive cells such as sperm, eggs, and zygotes, which could circumvent limitations related to species-specificity and conventional breeding methods. Pollen-based transformation is one such method that could bypass these limitations. Additionally, cutting-edge delivery systems using nanotechnology and viral-like particles are being explored. For instance, carbon nanotubes have successfully facilitated the in plant biotechnology (Azameti & Dauda, 2021; Demirer et al., 2019). Additional nanomaterials include layered double hydroxides (Mitter et al., 2017), mesoporous silica nanoparticles, and polyethyleneimine (Cunningham et al., 2018). These substances are used in various applications due to their unique properties at the nanoscale. While base editing systems have significantly propelled the progress of genome editing with their straightforwardness and accuracy, there is still an area of this technology that requires further investigation and numerous hurdles to surmount before these systems can be deemed dependable for genome editing. Presently, the base editors in use are limited to converting purines to other purines and pyrimidines to other pyrimidines. To broaden the scope of base editing applications in plant genomics, it is imperative to innovate base editors capable of interchanging purines with pyrimidines and vice versa (Gao, 2018; Marx, 2018; Monsur et al., 2020). In recent times a new approach has been introduced, a novel genome-editing approach also known as prime editing. This method combines a Cas9 nickase with reverse transcriptase, enabling the execution of all types of base conversions within human cells. The prime editing technique has demonstrated its effectiveness by achieving precise base transversions and small insertions or deletions (indels) in both human cells and mouse neurons (Anzalone et al., 2019; B. Yang et al., 2019). Lin and colleagues (Lin et al., 2020) have recently made strides in refining prime editing for plant genomes, focusing on the optimization of promoters, codons, and editing conditions. While prime editing may not be as efficient as base editors (BEs) for creating transition mutations, it has the unique ability to produce a wider array of genetic alterations, including substitutions, insertions, deletions, and transversions, thereby broadening the possibilities of genome editing. Cpf1-based cytosine base editors (CBEs), known for catalyzing the conversion of cytosine to thymine, face limitations due to their requirement for T-rich protospacer-adjacent motif (PAM) sequences. Despite this, they have been effectively used in human cells (Li et al., 2018), and certain CBEs have also demonstrated the ability to facilitate cytosine-to-guanine and cytosine-to-adenine substitutions (Shimatani et al., 2017; Veillet et al., 2019). Base editing's versatility is further enhanced by the use of various modified or engineered Cas9 proteins, which offer greater PAM sequence flexibility. To enhance the precision, reliability, and longevity of base editing tools, it is essential to continue developing their PAM sequence compatibility (Hua et al., 2019; Molla & Yang, 2019; B. Yang et al., 2019).

RNA base editing technologies, such as RESCUE and REPAIR, offer a higher level of precision in the genetic enhancement of agricultural crops compared to DNA base editors. Consequently, it is advisable for future research to focus on these systems for their potential applications in crop improvement (Azameti & Dauda, 2021).

More than half of the genes recently identified in plants have sequences that resemble those already known in other developmental roles can be obtained through the analysis of mRNA and protein expression patterns across different cell types, developmental stages, or in response to stress factors. The impact of transcription and post-translational modifications on gene function is also significant. Functional genomics approaches, particularly those involving RNA, such as gene knockouts or overexpression, are crucial for linking gene function to observable traits or agricultural characteristics. While the methods for studying gene function in plants and crops offer several advantages, they are not without their limitations (Kumar et al., 2020).

Base editors (BEs) have the capability to modify either cytosine (C) or adenine (A) within their target window, yet it's not always predictable which specific base will be altered. While BEs like A3A-PBE with extensive editing windows are favored for their ability to edit a broader range of bases, precision-focused applications necessitate BEs with narrower windows due to the reduced specificity found in those with larger windows (Zong et al., 2018). Tan and colleagues (Tan et al., 2019) have developed a highly precise CBE that excels in single-nucleotide substitutions. Prime base editors (PBEs) designed with a singular nucleotide editing window may hold the key to achieving both high efficiency and target specificity in genome editing. ABEs have been instrumental in inducing effective mutations and modifying traits in Arabidopsis, such as altering flowering time, plant height, disease resistance, and the albino phenotype through the editing of specific genes like AtFT, AtPDS3, and AteIF4E. Similarly, BEs have enhanced plant structure and nitrogen absorption in rice. Both ABEs and CBEs are pivotal in advancing plant structure, resilience, and nutrient assimilation. Using genes that confer resistance is a potent strategy for crafting genome-editing tools in research settings. Nonetheless, the generation of herbicide-resistant crops for market use must be approached with consideration for sustainable agricultural practices and societal acceptance of genetic modification. Expanding the scope of base editing is crucial for its dependability. While existing RNA base editors (RBEs) can change adenine to inosine, and recent developments like the RESCUE system offer cytosine to uracil editing in RNA, further research is essential to refine RBE systems for cytosine to uracil conversions in DNA (Monsur et al., 2020).

10.5 APPLICATIONS OF BASE-EDITED CROPS

Genome editing using the CRISPR/Cas9 system has blossomed as a successful tool, radically altering the fields of agriculture and plant science due to its simplicity, versatility, and exceptional precision. In the CRISPR/Cas9 architecture, the Cas9-sgRNA complex navigates the DNA, producing DSBs at specified places identified, as the Protospacer Adjacent Motif (PAM) serves as a marker, indicating specific sites where the Cas9-sgRNA complex acts, while the sgRNA aligns with the desired DNA sequence (Jinek et al., 2014). These DSBs are then repaired by natural DNA repair mechanisms known as NHEJ and homology-directed repair (HDR) which represent two distinct pathways for repairing DNA breaks. Error-prone NHEJ causes often result in random insertions and deletions, whereas HDR is characterized by its precision technique that results in gene insertions or replacements (Voytas & Gao, 2014). The CRISPR/Cas9 system is a very successful and durable method for genome editing, with wide applicability in crop and model plant modification due to its adaptability and precision (Liang et al., 2017; Shao et al., 2017; Shi et al., 2017; Tomlinson et al., 2019; Zsogon et al., 2018). Nonetheless, obstacles remain particularly accidental mutations in off-target regions and issues with PAM specificity, which are major concerns with this approach (Mishra et al., 2020).

Several key agronomic features depend on point mutations or modest changes in gene sequences (Doebley et al., 2006; Ma et al., 2016; Li et al., 2017). Employing CRISPR/Cas9 for gene replacement via HDR is a promising route for correcting these mutations and speeding up crop development (Li et al., 2018; Sun et al., 2016; Wang et al., 2017). Nonetheless, the irregular nature of HDR events, as well as the difficulties involved with supplying template DNA efficiently, have long hampered plant growth (Ran et al., 2017). Furthermore, although the CRISPR/Cas9 system is proficient in gene knockout/knock-in, it cannot directly convert one base to another, emphasizing the need for other approaches that assure stable and accurate genome editing in crops (Mishra et al., 2020). Single nucleotide polymorphisms (SNPs) in the genome affect many agriculturally important properties, and base editing has become an important method for correcting these mutations and accelerating crop improvement. Targeted editing of individual genes impacted by SNPs is made easier by the effectiveness of adenine as well as cytosine base editors, which have been demonstrated in several important crops and model plants (Hua et al., 2018; Li et al., 2018; Lu & Zhu 2017; Ren et al., 2018,).

In addition to bacteria, several plant species have adopted base editing, as newer versions of prime editors tailored for plants have recently been created, aiming to perform precise and efficient editing tasks. Plants have been treated with DSB-induced HDR editing techniques to improve crop quality. However, because single-nucleotide alterations are credited with most agricultural advances, their usefulness has been limited. Because of their gene-repair system, plants seldom experience HDR (Que et al., 2019), because the DSB-induced Cas9 system's generation of unwanted byproducts (Zhang et al., 2015) renders this editing strategy unfeasible. Incorporating BEs and PEs into the fundamental DNA sequences has streamlined the exact manipulation of crops, allowing for the introduction of novel features with greater ease (Abdullah et al., 2020).

10.5.1 Application of Base Editing in Plants

Agriculture has made extensive use of base editing technologies, especially as crops can be improved by single nucleotide alterations. Crops have the potential to significantly increase in qualities directly relevant to bioproduction, such as growth rates, yield, and resistance to environmental stress, by incorporating functional SNPs (Thomas, 2014; see Table 10.1).

TABLE 10.1
Some common examples of CRISPR CAS technology used in the plant-based system

Target plant	Gene target	Mechanism	Benefit	Reference
Papaver somniferum L.	4'OMT2 gesne	Gene knockout mechanism involved. Synthetic binary plasmids viral-based TRV were developed for transcription of sgRNA and by means of Agrobacterium-mediated transformation; the gene was loaded into plant cells using Cas9	Gene involved in regulation of benzylisoquinoline alkaloid (BIAs) biosynthesis. Improving the metabolism and production of secondary metabolites using CRISPR/Cas9 system for high value medicinal plants.	Alagoz et al. (2016)
Salvia miltiorrhiza	Rosmarinic acid synthase (SmRAS) gene	gRNA was given along with U6 promoter of *Arabidopsis*.	CRISPR/Cas9 technique editing to modify the pathway of phenolics production. S. miltiorrhiza is a medicinal plant which is used in cardiovascular diseases, such as atherosclerosis, thrombosis, and angina pectoris	(Alok et al., 2020; Wang et al., 2017; Zhou et al., 2018
Salvia miltiorrhiza	diterpene synthase gene (SmCPS1)	*A. rhizogenes*- mediated transformation using CRISPR/Cas9 is a simple and efficient genome editing tool in *S. miltiorrhiza*, thus paving the way for large-scale genome editing	The tanshinone biosynthesis is important for pathway elucidation of secondary metabolites, quality improvement, and yield increases for this valuable traditional Chinese medicinal herb.	Li et al. (2017)
Solanum lycopersicum	Mildew resistant LOCUS O (Mlo) gene	The sixteen genes of Mlo, SlMlo 1 is known to be the main grantor of disease. Transformants were detected by Illumina sequencing.	Non-transgenic tomato having resistance against powdery mildew pathogen *Oidium neolycopersici* was developed using CRISPR/Cas9 technique.	(Nekrasov et al., 2017; Zheng et al., 2016).
Solanum lycopersicum	Phytoene desaturase gene	Hindering the conversion to beta and alpha-carotene with the help of CRISPR/Cas9.	Enhancing the production of lycopene in tomato.	(Li et al., 2018)
O. sativa	Carotenoid cleavage dioxygenase 7 (CCD7) gene encodes for (SLs)	Branching of shoot in plants can be inhibited by Strignolactones (SLs). CRISPR/Cas9 technique was employed for site specific disruption of CCD7.	CRISPR/Cas9 for site-specific modification in plants	(Butt et al., 2018)
Watermelon, Tomato, Rice, Maize and Wheat	Acetolactate synthase (ALS) gene	Cytidine base editors (CBE) in CRISPR/Cas9 system by combining adenine or cytidine deaminase with Cas9 nickase (Cas9n)	Cytidine bases were edited successfully generating plants with chlorsulfuran resistance and efficiency of around 70% in tomato	Bhalothia et al., 2020; Shimatani et al., 2017; Tian et al., 2018; Veillet et al., 2019; Zong et al., 2017;)

(*continued*)

TABLE 10.1 (Continued)
Some common examples of CRISPR CAS technology used in the plant-based system

Target plant	Gene target	Mechanism	Benefit	Reference
Musa balbisiana	eBSV sequences are mainly integrated in the B genome	Inactivate the eBSV by editing the virus sequences by CRISPR/ Cas9	Editing of endogenous banana streak virus in banana leads to genetic improvement of banana by conventional breeding is key for the development of improved varieties with disease and pest resistance and higher yields.	(Tripathi et al., 2019)
Cotton (Gossypium hirsutum)	LbCpf1 gene	Temperature-sensitive CRISPR/LbCpf1 was successfully applied in allotetraploid cotton, enhancing genome editing efficiency. CRISPR/LbCpf1 (LbCas12a) is a temperature-sensitive system for plant genome editing.	Creation of nontransgenic, gossypol-free cotton for breeding. Determination of optimum temperature for cotton genome editing.	(Li et al., 2021; Malzahn et al., 2019)
Oilseed rape	BnALS1	The BnALS1 gene was precisely edited at position P197by CBE system, conferring tribenuron-methyl resistance to oilseed rape	Provide tolerance from heribicides	(Azameti & Dauda, 2021; Wu et al., 2020)
Nicotiana benthamiana	Replication associated protein (Rep) gene and the βC1 gene of the betasatellite.	Multiplexing approach with CRISPR/Cas9 showed efficient resistance in plants.	The CRISPR/Cas9 system was successfully used to develop resistance against Cotton leaf curl virus, showing delayed symptoms and lower virus titres, indicating potential for resistance in model plants. Can be used in editing cotton plants for virus resistance.	(Khan et al., 2020)
Arabidopsis thaliana	WRKY3 and WRKY4 genes	CRISPR/Cas9-mediated mutagenesis	WRKY3 and WRKY4 genes regulate plant defense response to stresses. Reduces salt and Me-JA stress tolerance.	(J. Li et al., 2021)
Arabidopsis thaliana	5-oxoprolinase gene	CRISPR/Cas9 edited *OXP1* gene confers resistance to sulfonamide compounds. O Gene editing in (oxp1/ CRISPR), leds to inhibition of the folic acid biosynthetic pathway.	Resistance to sulfonamide compounds, demonstrated by reduced sensitivity to sulfamethoxazole and tolerance to other sulfonamide compounds and heavy metals.	(Baeg et al., 2021C)
Solanum lycopersicum	miR482b and miR482c gene	Multiplex CRISPR/Cas9 system	CRISPR/Cas9 is an efficient strategy for generating pathogen-resistant tomatoes, and cross-regulation between miRNAs may reveal the novel mechanism in tomato-P. infestans interactions.	(Hong et al., 2021)

TABLE 10.1 (Continued)
Some common examples of CRISPR CAS technology used in the plant-based system

Target plant	Gene target	Mechanism	Benefit	Reference
Strawberry	*FvICE1* gene	CRISPR/cas9-mediated gene edited strawberries mutant showed lower cold and drought tolerance	Tolerance to cold and drought at the phenotypic and physiological levels. FvICE1 functioned as a positively regulator of cold and drought	(Han et al., 2023; Kumar et al., 2023)
Cassava (Manihot esculenta)	CYP79D1 and CYP79D2 gene	Knockout of using CRISPR in cassava	Reduces toxic cyanogen production, enhancing food safety and reducing processing needs in various cultivars	(Gomez et al., 2023)
Alfalfa (Medicago sativa L.)	*Flowering locus Ta1 (MsFTa1)*	CRISPR/Cas9-mediated mutagenesis using *Agrobacterium*-mediated transformation	Reduced lignin, acid detergent fibre (ADF) and neutral detergent fibre (NDF) content and significantly higher crude protein (CP) and mineral contents especially in the stems	(Wolabu et al., 2023)
Barley	Protein Disulfide Isomerase (PDIL5-1)	CRISPR/Cas9 genome editing to accelerate breeding for disease resistance in barley	The mutated plants were more resistant to yellow mosaic virus disease than wild-type plants, with no adverse effects on growth or yield	(Hamdan et al., 2023)
Soybean (*Glycine max* (L.) Merr.)	base substitutions in *GmAHAS3* and *GmAHAS4*	Cytosine base editor (BE3) by using CRISPR/Cas9-mediated base editing	To generate herbicide-resistant soybean	(Wei et al., 2023)
Solanum lycopersicum	*SlNPR1* expression and protein content	CRISPR/Cas9-mediated genome editing of SlNPR1	Symptoms of chilling injury in plant can be reduced by accumulation of hydrogen peroxide (H_2O_2), superoxide anion (O_2^-), and malonic dialdehyde (MDA), and reduction in proline content, antioxidant enzymatic activities, and soluble protein content.	(Kumar et al., 2023; Shu et al., 2023).
Ocimum basilicum (sweet basil)	DMR6 (Downy Mildew Resistant 6)	*O. basilicum* "Italiko," the elite cultivar used to produce "Pesto Genovese D.O.P," was developed	Downy Mildew (DM) disease, caused by *Peronospora belbahrii*, leading to huge crop losses can be controlled by genome editing (GE) approach mediated by CRISPR/Cas9.	(Laura et al., 2024)

BEs provide a way to add desired SNPs and inculcate desired features without worrying about linkage drag. Additionally, numerous studies demonstrate that BEs have been instrumental in the advancement of herbicide-resistant crop varieties (Chen et al., 2017; Li et al., 2018; Shimatani et al., 2017; Tian et al., 2018; Zhang et al., 2019; Zhong et al., 2017; Zong et al., 2018). Beyond herbicide resistance, there is a rising interest in generating disease-resistant crops by rectifying

Caranta, 2006). Studies recommend using resistant alleles capable of encoding intact translational initiation components rather than loss-of-function alleles, which have a narrower resistance spectrum (Bastet et al., 2017; Ruffel et al., 2006).

Making numerous amino acid modifications employing genome-editing tools reliant on DSB mechanisms and mutagenesis procedures can be difficult. Fortunately, base editing provides a preferable solution for such alterations.

or off-target editing using BE/PE systems show enormous promise in enabling crop bio-production or enhancement due to their high accuracy and effectiveness (Abdullah et al., 2020).

10.6 ADVANTAGES AND CHALLENGES OF BASE EDITING

It is not feasible to increase crop productivity by altering metabolic pathways using a traditional breeding approach. However, using a transgenic technique to overexpress the genes can help achieve it to some extent. It takes a lot of time and effort to introduce desirable traits into crops with polyploid genomes, like oil palm, using traditional breeding techniques. With the advent of genome-editing technologies like base editing, these obstacles can now be overcome to achieve productive advances. Using Cas9-mediated gene targeting, BE is a powerful genome-editing technique that may cause targeted point mutations in DNA or RNA. SNPs in plants' genomes are primarily responsible for several significant features, and base editing tools are crucial for repairing point mutations that lead to desirable phenotypic changes (Yara et al., 2020)

By altering the target genome areas, the base editing technique can be used to create plants that are resistant to a variety of diseases. Climate change, degrading soil, and many other limitations make it very difficult to meet the increasing need for higher food yields. Traditional breeding has focused on raising food yields in the past few decades. BEs may rapidly alter a species' characteristics and enable carefully targeted genetic alterations in plants. Using different techniques, base modification can also be used to create crops with higher yields. The economic value and edibility of a crop are intimately correlated with the quality of crop products. Humans can obtain a substantial amount of lycopene from tomato-based foods and tomatoes. Beta- and lycopene-containing carotenoids are the main pigments found in tomatoes. Because of their potent antioxidant properties, carotenoids are vital functional components; SlDDB1, SlDET1, and SlCYC-B are important players in the accumulation of carotenoid pigments (Li et al., 2022).

Furthermore, BE can be used to control nitrogen intake, improve crop shape, and increase the nutritional content of crop plants. Research has shown that a C→T mutation (Thr327Met) in this gene boosts nitrogen use efficiency in rice. NRT1.1B encodes a nitrogen transporter. Thus, germplasm development and crop enhancement can benefit from base editing. To increase crop quality, we can choose useful mutant loci by building sgRNA libraries using base editing to conduct screening for mutant loci that are not found in nature. To maintain agricultural output, crop diseases can be controlled economically and environmentally by breeding stable, disease-resistant plants. For instance, rice blast is a serious rice disease. Breeders need to think of ways to mitigate or eliminate rice blight. Pi-d2 is a rice blast resistance gene of significant agricultural importance. We may currently more readily investigate gene function thanks to advances in bioinformatics and sequencing technologies. As more and more disease-resistant genes are found, precision editing of these genes using BEs combining bioinformatics and BE technology may be a potential strategy to shield plants from biotic stressors ((Li et al., 2022).

In addition to interfering with crop plants' access to nutrients, sunshine, and living space, weeds can obstruct airflow, block light, and harbor pests and diseases in agricultural areas. Furthermore, the productivity and quality of crops are reduced when parasitic weeds take up nutrients from the crop plants. It follows that implementing weed management techniques is crucial. BEs provide a quick and accurate method for creating new plant lines resistant to herbicides. Building sgRNA libraries for gene screening allows base editing to identify more herbicide-resistant loci for agricultural production through precise and targeted replacement (Li et al., 2022).

In the field of biological research, base editing is a revolutionary technique that can transform one base into another. The simple accessibility of functional genomic resources for building gRNAs for precision editing in targeted areas is one of the essential requirements for base editing technology. It was only lately that tools that enable accurate, effective, and DSB-free editing down to a single base have been made available. With the development of BE and, subsequently, PE, researchers now have access to potent gene-editing instruments. These powerful instruments have been applied in many domains, including agriculture, medicine, and synthetic metabolic engineering. These days, as life sciences research continues to grow, animal, plant, and bacterial models have swiftly embraced these techniques. To advance the bio-production of bacteria and plants, these tools must be used. Since it can be challenging to keep up with research due to the discipline's rapid breakthroughs, this study outlines the developments in BE and PE technology in bacteria and plants (Abdullah et al., 2020).

Even though interest in BE and PE editing tools is growing, there are still issues that need more attention. The restricted target locations of BE and PE resulting from the PAM specificity of Cas proteins present a significant hurdle. Though they have been developed, BE fused to various Cas proteins such as xCas9 (PAM recognized, NG and more), Cas12a (PAM recognized, TTTV and more), SaCas9(D10A) (PAM recognized, NNGRRT), VQRCas9 (D10A) (PAM recognized, NGA, AGTG, AGCG), VRERCas9(D10A) (PAM recognized, NGCG), SaKKH-Cas9 (D10A) (PAM recognized, NNNRRT), and Cas9 (D10A) (PAM recognized, NG and more) have all been developed, yet they are still restricted to narrow PAM accessibility. The efficiency of editing is the next restriction related to BE/PE.

Several Cas proteins have been fused to BE, but the resulting proteins' effectiveness is still less than that of the original BE. The Liu group has concentrated on creating advanced base editors, such as ABE8e, which claims to offer higher editing efficiency than the original ABE7.10, and fourth-generation CBE BE4, with higher Cas compatibility and increased editing efficiencies. Hua and colleagues improved efficiency by decomposing the rice adenine deaminase. PEs, on the other hand, have received less research attention because they are still relatively young (Hua et al., 2018; Hua et al., 2019).

Off-target editing is yet another challenge for BE/PE. CRISPR-mediated base editing is a considerably more accurate method that allows base conversions without disrupting genes. Off-target editing persists as an immense worry, though. However, according to a recent study, rice crops are susceptible to unanticipated and unforeseen genome-wide off-target mutations caused by CBEs BE3 and high-fidelity BE3 (HF1-BE3). To reduce off-target edits in crops, YEE-BE3, one of the more effective CBE variants, could potentially be used (Mishra et al., 2020).

A serious restriction in base editors lowers editing efficiency and specificity. In cytosine deaminase base editors can modify any C that occurs within the broad activity window of about 4–5 nucleotides, or up to 9 nt., eliminating unnecessary sequences from the deaminase and experimenting with various proline-rich linkers of various lengths that can increase precision and reduce the catalytic window. Therefore, these very effective and exact base editors can be useful tools for precise crop breeding (Mishra et al., 2020).

Although base editing has been effectively used to boost the yields of wheat and rice, its application to other crops is still in the early stages. This method can be applied to major crops like oil palm by using the various Cas9 variations and base editors that have been developed thus far. Base editing is a relatively new editing technique that has the potential to be vital in synthetic biology, crop improvement, metabolic engineering, fundamental research, and evolutionary engineering. To realize their full potential, both editing tools, PE in particular, and its implementation in bacteria and plants will be essential for the improvement of their bio-production capabilities, and need significant study (Yara et al., 2020).

10.7 CONCLUSION

The CRISPR-Cas9 system, through its intensive development, is yielding significant benefits in the realm of science. It enables transcriptional regulation, genomic alterations, and epigenetic editing with remarkable simplicity and efficiency.

Base editing represents a cutting-edge advance that enables the precise and efficient alteration of individual bases within the genomes of both plants and animals. Functioning programmatically, it stands out for not inducing DSBs, making it applicable to a wide range of cell types, whether they are in a state of division or not. In recent years, this technology has been successfully implemented in various species, offering a level of accuracy superior to other genome-editing methods due to its non-reliance on DSB.

Despite its successes, there is still room for improvement in base editing to broaden its effectiveness and range of application. This involves refining the technology to minimize unintended off-target effects and the occurrence of bystander [illegible]

crops, which is crucial for maintaining sustainable food production in the face of ongoing global changes.

CRISPR-Cas9 holds immense potential in the agricultural sector, where it is used to accelerate the breeding of livestock and crops. However, there are several controversies associated with CRISPR, including off-target effects, the immunogenicity of Cas9 nucleases, and the potential carcinogenic effects of CRISPR components. These issues necessitate thorough investigation and scientific clarification.

REFERENCES

Ahmar, S., Mahmood, T., Fiaz, S., Mora-Poblete, F., Shafique, M. S., Chattha, M. S., & Jung, H. (2021). Advantage of nanotechnology-based genome editing system and its application in crop improvement. Frontiers in Plant Science, 12. https://doi.org/10.3389/fpls.2021.663849

Alagoz, Y., Gurkok, T., Zhang, B., & Unver, T. (2016). Manipulating the biosynthesis of bioactive compound alkaloids for next-generation metabolic engineering in opium poppy using CRISPR-Cas 9 genome editing technology. *Scientific Reports*, 6(1), 1–9. https://doi.org/10.1038/srep30910

Aldemita, R. R., Reaño, I. M. ar. E., Solis, R. O., & Hautea, R. A. (2015). Trends in global approvals of biotech crops (1992–2014). *GM Crops & Food,* 6(3), 150–166. https://doi.org/10.1080/21645698.2015.1056972

Ali, Z., Eid, A., Ali, S., & Mahfouz, M. M. (2018). Pea early-browning virus-mediated genome editing via the CRISPR/Cas9 system in Nicotiana benthamiana and Arabidopsis. *Virus Research,* 244, 333–337. https://doi.org/https://doi.org/10.1016/j.virusres.2017.10.009

Alok, A., Sandhya, D., Jogam, P., Rodrigues, V., Bhati, K. K., Sharma, H., & Kumar, J. (2020). The rise of the CRISPR/Cpf1 system for efficient genome editing in plants. *Frontiers in Plant Science,* 11, 494295. https://doi.org/10.3389/fpls.2020.00264

Anzalone, A. V, Randolph, P. B., Davis, J. R., Sousa, A. A., Koblan, L. W., Levy, J. M., Chen, P. J., Wilson, C., Newby, G. A., Raguram, A., & Liu, D. R. (2019). Search-and-replace genome editing without double-strand breaks or donor DNA. *Nature*, 576(7785), 149–157. https://doi.org/10.1038/s41586-019-1711-4

Araki, M., & Ishii, T. (2015). Towards social acceptance of plant breeding by genome editing. *Trends in Plant Science*, 20(3), 145–149. https://doi.org/10.1016/j.tplants.2015.01.010

Azameti, M. K., & Dauda, W. P. (2021). Base editing in plants: Applications, challenges, and future prospects. *Frontiers in Plant Science*, 12(July). https://doi.org/10.3389/fpls.2021.664997

Baeg, G. J., Kim, S. H., Choi, D.M. et al .(2021). CRISPR/Cas9-mediated mutation of 5-oxoprolinase gene confers resistance to sulfonamide compounds in *Arabidopsis. Plant Biotechnology Reports,* 15, 753–764. https://doi.org/10.1007/s11816-021-00718-w

Barrangou, R. (2013). CRISPR-Cas systems and RNA-guided interference. *Wiley Interdisciplinary Reviews: RNA*, 4(3), 267–278. https://doi.org/10.1002/wrna.1159

Bastet, A., Robaglia, C., & Gallois, J. L. (2017). eIF4E [illegible]

polymorphism in eIF4E by CRISPR-Cas9 base editing is associated with resistance to potyviruses. *Plant Biotechnology Journal,* 17, 1736–1750.

Bhalothia, P., Yajnik, K., Alok, A., & Upadhyay, S. K. (2020). The current progress of CRISPR/Cas9 development in plants. *In Genome Engineering via CRISPR-Cas9 System* (pp. 123-129). Academic Press.

Bharat, S. S., Li, S., Li, J., Yan, L., & Xia, L. (2020). Base editing in plants: Current status and challenges. *The Crop Journal*, 8(3), 384–395. https://doi.org/10.1016/j.cj.2019.10.002

Butt, H., Rao, G. S., Sedeek, K., Aman, R., Kamel, R., Mahfouz, M. (2020). Engineering herbicide resistance via prime editing in rice. *Plant Biotechnology Journal,* 18, 2370–2372.

Chatterjee Saha, S., Mishra, T., Mandal, S., Prasanth, A., Mohture, V. M., & Mane, A. B. (2023). CRISPR applications in medicinal and aromatic plants. CRISPRized Horticulture Crops, 365–380. https://doi.org/10.1016/B978-0-443-13229-2.00015-6

Chen, H., & Lin, Y. (2013). Promise and issues of genetically modified crops. *Current Opinion in Plant Biology*, 16(2), 255–260. https://doi.org/https://doi.org/10.1016/j.pbi.2013.03.007

Chen, Y., Wang, Z., Ni, H., Xu, Y., Chen, Q., & Jiang, L. (2017). CRISPR/Cas9-mediated base-editing system efficiently generates gain-of-function mutations in Arabidopsis. *Science China Life Science,* 60, 520–523.

Cunningham, F. J., Goh, N. S., Demirer, G. S., Matos, J. L., & Landry, M. P. (2018). Nanoparticle-mediated delivery towards advancing plant genetic engineering. *Trends in Biotechnology*, 36(9), 882–897. https://doi.org/10.1016/j.tibtech.2018.03.009

Demirer, G. S., Zhang, H., Matos, J. L., Goh, N. S., Cunningham, F. J., Sung, Y., Chang, R., Aditham, A. J., Chio, L., Cho, M.-J., Staskawicz, B., & Landry, M. P. (2019). High aspect ratio nanomaterials enable delivery of functional genetic material without DNA integration in mature plants. *Nature Nanotechnology*, 14(5), 456–464. https://doi.org/10.1038/s41565-019-0382-5

Gao, C. (2018). The future of CRISPR technologies in agriculture. *Nature Reviews Molecular Cell Biology*, 19(5), 275–276. https://doi.org/10.1038/nrm.2018.2

Gomez, M. A., Berkoff, K. C., Gill, B. K., Iavarone, A. T., Lieberman, S. E., Ma, J. M., Schultink, A., Karavolias, N. G., Wyman, S. K., Chauhan, R. D., Taylor, N. J., Staskawicz, B. J., Cho, M., Rokhsar, D. S., & Lyons, J. B. (2023). CRISPR-Cas9-mediated knockout of CYP79D1 and CYP79D2 in cassava attenuates toxic cyanogen production. *Frontiers in Plant Science,* 13, 1079254. https://doi.org/10.3389/fpls.2022.1079254

Haft, D. H., Selengut, J., Mongodin, E. F., & Nelson, K. E. (2005). A guild of 45 CRISPR-associated (Cas) protein families and multiple CRISPR/cas subtypes exist in prokaryotic genomes. *PLoS Computational Biology*, 1(6), 0474–0483. https://doi.org/10.1371/journal.pcbi.0010060

Hamdan, M. F., & Tan, B. C. (2024). Genetic modification techniques in plant breeding: A comparative review of CRISPR/Cas and GM technologies. *Horticultural Plant Journal.* https://doi.org/10.1016/j.hpj.2024.02.012

Han, J., Li, X., Li, W., Yang, Q., Li, Z., Cheng, Z., Lv, L., Zhang, L., & Han, D. (2023). Isolation and preliminary functional analysis of FvICE1, involved in cold and drought tolerance in *Fragaria vesca* through overexpression and CRISPR/Cas9 technologies. *Plant Physiology and Biochemistry,* 196, 270–280. https://doi.org/10.1016/j.plaphy.2023.01.048

Hong, Y., Meng, J., He, X., Zhang, Y., Liu, Y., Zhang, C., Qi, H., & Luan, Y. (2021). Editing miR482b and miR482c Simultaneously by CRISPR/Cas9 enhanced tomato resistance to *Phytophthora infestans. Phytopathology*, 111(6), 1008–1016. https://doi.org/10.1094/PHYTO-08-20-0360-R

Hsu, P. D., Lander, E. S., & Zhang, F. (2014). Development and applications of CRISPR-Cas9 for genome engineering. *Cell,* 157(6), 1262–1278. https://doi.org/10.1016/j.cell.2014.05.010

Hua, K., Tao, X., Yuan, F., Wang, D., & Zhu, J. K. (2018). Precise AT to GC base editing in the rice genome. *Molecular Plant,* 11, 627–30.

Hua, K., Tao, X., & Zhu, J. K. (2019). Expanding the base editing scope in rice by using Cas9 variants. *Plant Biotechnology Journal*, 17(2), 499–504. https://doi.org/10.1111/pbi.12993

Janik, E., Niemcewicz, M., Ceremuga, M., Krzowski, L., Saluk-Bijak, J., & Bijak, M. (2020). Various aspects of a gene editing system: CRISPR-Cas9. *International Journal of Molecular Sciences,* 21(24), 1–20. https://doi.org/10.3390/ijms21249604

Jiang, W., Zhou, H., Bi, H., Fromm, M., Yang, B., & Weeks, D. P. (2013). Demonstration of CRISPR/Cas9/sgRNA-mediated targeted gene modification in Arabidopsis, tobacco, sorghum and rice. *Nucleic Acids Research*, 41(20), e188–e188. https://doi.org/10.1093/nar/gkt780

Jiang, Z., Hong, X., Zhang, S., Yao, R., Xiao, Y. (2020). CRISPR base editing and prime editing: DSB and template-free editing systems for bacteria and plants. *Synthetic and Systems Biotechnology*, 5(4), 277–292.

Jinek, M., Jiang, F., Taylor, D. W., Sternberg, S. H., Kaya, E., & Ma, E. (2014). Structures of Cas9 endonucleases reveal RNA-mediated conformational activation. *Science*, 343(6176), 1247997.

Khan, S., Mahmood, M. S., Rahman, S. U., Rizvi, F., & Ahmad, A. (2020). Evaluation of the CRISPR/Cas9 system for the development of resistance against *Cotton leaf curl virus* in model plants. *Plant Protection Science*, 56(3), 154–162. doi: 10.17221/105/2019-PPS

Khare, P., Mandal, S., Verma, A. K., Parashar, D., Jadhav, K., & Gupta, S. (2023). CRISPR: The Janus god of modern science. CRISPR-Cas System in Translational Biotechnology, 23–56. https://doi.org/10.1016/B978-0-323-91808-4.00006-7

Koonin, E. V., Makarova, K. S., & Zhang, F. (2017). Diversity, classification and evolution of CRISPR-Cas systems. *Current Opinion in Microbiology,* 37, 67–78. https://doi.org/https://doi.org/10.1016/j.mib.2017.05.008

Kumar, M., Prusty, M. R., Pandey, M. K., Singh, P. K., Bohra, A., Guo, B., & Varshney, R. K. (2023). Application of CRISPR/Cas9-mediated gene editing for abiotic stress management in crop plants. *Frontiers in Plant Science,* 14, 1157678. https://doi.org/10.3389/fpls.2023.1157678

Kumar, P., Alok, A., & Kumar, J. (2020). Expanding the potential of CRISPR-Cas9 technology for crops improvement. *Advances in Synthetic Biology*, 251–274. https://doi.org/10.1007/978-981-15-0081-7_15

Kumar, V., Singh, D. and Wesley, C. J. (2023). Effect of plant growth regulator (GA3 and NAA) on growth, yield and quality of tomato (Solanum Lycopersicum L.). *International Journal of Environment and Climate Change*, 13(10), 2636–43. https://doi.org/10.9734/ijecc/2023/v13i102928

Laura, M., Forti, C., Barberini, S., Ciorba, R., Mascarello, C., Giovannini, A., Pistelli, L., Pieracci, Y., Lanteri, A. P., Ronca, A., Minuto, A., Ruffoni, B., Cardi, T., & Savona, M. (2022). Highly efficient CRISPR/Cas9 mediated gene editing in Ocimum basilicum 'FT Italiko' to induce resistance to Peronospora belbahrii. *Plants*, 12(13), 2395.

Lee, J., Oh, N., Yun, J., Choi, H. S., Seo, J., Kang, J., & Jung, C. (2023). Application of CRISPR-based C-to-G base editing in rice protoplasts. *Applied Biological Chemistry, 66(*1), 1–5. https://doi.org/10.1186/s13765-023-00775-5"

Lee, J. H., Won, H. J., Hoang Nguyen Tran, P., Lee, S. M., Kim, H. Y., Jung, J. H. (2021). Improving lignocellulosic biofuel production by CRISPR-Cas9-mediated lignin modification in barley. *GCB Bioenergy*, 13(4), 742–752. doi: 10.1111/gcbb.12808"

Li, B., Liang, S., Alariqi, M., Wang, F., Wang, G., Wang, Q., Xu, Z., Yu, L., Zafar, M. N., Sun, L., Si, H., Yuan, D., Guo, W., Wang, Y., Lindsey, K., Zhang, X., & Jin, S. (2020). The application of temperature sensitivity CRISPR/LbCpf1 (LbCas12a) mediated genome editing in allotetraploid cotton (*G. Hirsutum*) and creation of nontransgenic, gossypol-free cotton. *Plant Biotechnology Journal*, 19(2), 221. https://doi.org/10.1111/pbi.13470

Li, C., Zong, Y., Wang, Y., Jin, S., Zhang, D., Song, Q., Zhang, R., & Gao, C. (2018). Expanded base editing in rice and wheat using a Cas9-adenosine deaminase fusion. *Genome Biology,* 19, 1–9.

Li, J., Sun, Y., Du, J., Zhao, Y., & Xia, L. (2017). Generation of targeted point mutations in rice by a modified CRISPR/Cas9 system. *Molecular Plant,* 10, 526–529.

Li, J., Wei, Q., Cheng, Y. et al. (2024). Cas12a-mediated gene targeting by sequential transformation strategy in *Arabidopsis thaliana. BMC Plant Biology,* 24, 665 https://doi.org/10.1186/s12870-024-05375-z

Li, X., Wang, Y., Liu, Y., Yang, B., Wang, X., Wei, J., Lu, Z., Zhang, Y., Wu, J., Huang, X., Yang, L., & Chen, J. (2018). Base editing with a Cpf1-cytidine deaminase fusion. *Nature Biotechnology*, 36(4), 324–327. https://doi.org/10.1038/nbt.4102

Li, Y., Li, S., Li, C., Zhang, C., Yan, L., Li, J., He, Y., Guo, Y., & Xia, L. (2024). Fusion of a rice endogenous N-methylpurine DNA glycosylase to a plant adenine base transition editor ABE8e enables A-to-K base editing in rice plants. *ABIOTECH*, 5(2), 127. https://doi.org/10.1007/s42994-024-00138-8

Li, Y., Liang, J., Deng, B., Jiang, Y., Zhu, J., Chen, L., Li, M., & Li, J. (2022). Applications and prospects of CRISPR/Cas9-mediated base editing in plant breeding: Functional genomics and comparative genomics analysis in plants. *Current Issues in Molecular Biology*, 45(2), 918–935.

Liang, Z., Chen, K., Li, T., Zhang, Y., Wang, Y., & Zhao, Q. (2017). Efficient DNA -free genome editing of bread wheat using CRISPR/Cas9ribonucleoprotein complexes. *Nature Communication,* 8, 14261.

Lin, Q., Zong, Y., Xue, C., Wang, S., Jin, S., Zhu, Z., Wang, Y., Anzalone, A. V, Raguram, A., Doman, J. L., Liu, D. R., & Gao, C. (2020). Prime genome editing in rice and wheat. *Nature Biotechnology*, 38(5), 582–585. https://doi.org/10.1038/s41587-020-0455-x

Lu, Y., & Zhu, J. K. (2017). Precise editing of a target base in the rice genome using a modified CRISPR/Cas9 system. *Molecular Plant,* 10, 523–525.

Ma, Y., Zhang, J., Yin, W., Zhang, Z., Song, Y., & Chang, X. (2016). Targeted AID-mediated mutagenesis (TAM) enables efficient genomic diversification in mammalian cells. *Nature Methods,*

V. (2011). Evolution and classification of the CRISPR-Cas systems. *Nature Reviews Microbiology,* 9(6), 467–477. https://doi.org/10.1038/nrmicro2577

Makarova, K. S., Wolf, Y. I., Iranzo, J., Shmakov, S. A., Alkhnbashi, O. S., Brouns, S. J. J., Charpentier, E., Cheng, D., Haft, D. H., Horvath, P., Moineau, S., Mojica, F. J. M., Scott, D., Shah, S. A., Siksnys, V., Terns, M. P., Venclovas, Č., White, M. F., Yakunin, A. F., … Koonin, E. V. (2020). Evolutionary classification of CRISPR–Cas systems: A burst of class 2 and derived variants. *Nature Reviews Microbiology*, 18(2), 67–83. https://doi.org/10.1038/s41579-019-0299-x

Makarova, K. S., Wolf, Y. I., & Koonin, E. V. (2013). The basic building blocks and evolution of CRISPR–Cas systems. *Biochemical Society Transactions*, 41(6), 1392–1400. https://doi.org/10.1042/BST20130038

Malzahn, A. A., Tang, X., Lee, K., Ren, Q., Sretenovic, S., Zhang, Y., Chen, H. *et al.* (2019). Application of CRISPR-Cas12a temperature sensitivity for improved genome editing in rice, maize, and Arabidopsis. *BMC Biology,* 17, 9.

Marx, V. (2018). Base editing a CRISPR way. *Nature Methods*, 15(10), 767–770. https://doi.org/10.1038/s41592-018-0146-4

Mishra, R., Joshi, R. K., & Zhao, K. (2020). Base editing in crops: Current advances, limitations and future implications. *Plant Biotechnology,* 1, 20–31.

Mitter, N., Worrall, E. A., Robinson, K. E., Li, P., Jain, R. G., Taochy, C., Fletcher, S. J., Carroll, B. J., Lu, G. Q., & Xu, Z. P. (2017). Clay nanosheets for topical delivery of RNAi for sustained protection against plant viruses. *Nature Plants*, 3(January). https://doi.org/10.1038/nplants.2016.207

Molla, K. A., & Yang, Y. (2019). CRISPR/Cas-mediated base editing: Technical considerations and practical applications. *Trends in Biotechnology*, 37(10), 1121–1142. https://doi.org/10.1016/j.tibtech.2019.03.008

Monsur, M. B., Shao, G., Lv, Y., Ahmad, S., Wei, X., Hu, P., & Tang, S. (2020). Base editing: The ever expanding clustered regularly interspaced short palindromic repeats (CRISPR) tool kit for precise genome editing in plants. Genes, 11(4). https://doi.org/10.3390/genes11040466

Nekrasov, V., Wang, C., Win, J., Lanz, C., Weigel, D., & Kamoun, S. (2017). Rapid generation of a transgene-free powdery mildew resistant tomato by genome deletion. *Scientific Reports,* 7, 482. https://doi.org/10.1038/s41598-017-00578-x

Nidhi, S., Anand, U., Oleksak, P., Tripathi, P., Lal, J. A., Thomas, G., Kuca, K., & Tripathi, V. (2021). Novel crispr–cas systems: An updated review of the current achievements, applications, and future research perspectives. *International Journal of Molecular Sciences*, 22(7), 1–42. https://doi.org/10.3390/ijms22073327

Pyzocha, N. K., & Chen, S. (2018). Diverse Class 2 CRISPR-Cas effector proteins for genome engineering applications. *ACS Chemical Biology*, 13(2), 347–356. https://doi.org/10.1021/acschembio.7b00800

Que, Q., Chen, Z., Kelliher, T., Skibbe, D., Dong, S., & Chilton, M. D. (2019). Plant DNA repair pathways and their applications in genome engineering. *Methods in Molecular Biology*, 1917, 3–24.

Ran, Y., Liang, Z., & Gao, C. (2017). Current and future editing reagent delivery systems for plant genome editing. *Science China Life Science,* 60, 490–505.

Ren, B., Yan, F., Kuang, Y., Li, N., Zhang, D., & Zhou, Y. (2018).

Rengasamy, B., Manna, M., Thajuddin, N. B. et al.(2024). Breeding rice for yield improvement through CRISPR/Cas9 genome editing method: Current technologies and examples. *Physiology and Molecular Biology of Plants*, 30, 185–198. https://doi.org/10.1007/s12298-024-01423-y

Robaglia, C., & Caranta, C. (2006). Translation initiation factors: A weak link in plant RNA virus infection. *Trends in Plant Science*, 11, 40–5.

Rojo, F. P., Nyman, R. K. M., Johnson, A. A. T., Navarro, M. P., Ryan, M. H., Erskine, W., & Kaur, P. (2018). Crispr-Cas systems: Ushering in the new genome editing era. *Bioengineered*, 9(1), 214–221. https://doi.org/10.1080/21655979.2018.1470720

Ruffel, S., Gallois, J. L., Moury, B., Robaglia, C., Palloix, A., & Caranta, C. (2006). Simultaneous mutations in translation initiation factors eIF4E and eIF(iso)4E are required to prevent pepper veinal mottle virus infection of pepper. *Journal of General Virology*, 87, 2089–2098.

Saini, H., Thakur, R., Gill, R., Tyagi, K., & Goswami, M. (2023). CRISPR/Cas9-gene editing approaches in plant breeding. *GM Crops and Food*, 14(1), 1–17. https://doi.org/10.1080/21645698.2023.2256930

Selma, S., Gianoglio, S., Uranga, M., Vázquez-Vilar, M., Espinosa-Ruiz, A., Drapal, M., Fraser, P. D., Daròs, J.-A., & Orzáez, D. (2022). Potato virus X-delivered CRISPR activation programs lead to strong endogenous gene induction and transient metabolic reprogramming in Nicotiana benthamiana. *The Plant Journal*, 111(6), 1550–1564. https://doi.org/https://doi.org/10.1111/tpj.15906

Shao, G., Xie, L., Jiao, G., Wei, X., Sheng, Z., & Tang, S. (2017). CRISPR/CAS9 – mediated editing of the fragrant Gene Badh2 in rice. *Chinese Journal of Rice Science*, 31(2), 216–222.

Shelake, R. M., Pramanik, D., & Kim, Y. (2024). CRISPR base editor-based targeted random mutagenesis (BE-TRM) toolbox for directed evolution. *BMB Reports*, 57(1), 30. https://doi.org/10.5483/BMBRep.2023-0086

Shi, J., Gao, H., Wang, H., Laftte, H. R., & Archibald, R. L. (2017). ARGOS8 variants generated by CRISPRCas9 improve maize grain yield under field drought stress conditions. *Plant Biotechnology Journal*, 15, 207–216.

Shimatani, Z., Kashojiya, S., Takayama, M., Terada, R., Arazoe, T., Ishii, H., Teramura, H., Yamamoto, T., Komatsu, H., Miura, K., Ezura, H., Nishida, K., Ariizumi, T., & Kondo, A. (2017). Targeted base editing in rice and tomato using a CRISPR-Cas9 cytidine deaminase fusion. *Nature Biotechnology*, 35(5), 441–443. https://doi.org/10.1038/nbt.3833

Shmakov, S., Smargon, A., Scott, D., Cox, D., Pyzocha, N., Yan, W., Abudayyeh, O. O., Gootenberg, J. S., Makarova, K. S., Wolf, Y. I., Severinov, K., Zhang, F., & Koonin, E. V. (2017). Diversity and evolution of class 2 CRISPR–Cas systems. *Nature Reviews Microbiology*, 15(3), 169–182. https://doi.org/10.1038/nrmicro.2016.184

Shu, P., Li, Y., Xiang, L., Sheng, J., & Shen, L. (2023). SlNPR1 modulates chilling stress resistance in tomato plant by alleviating oxidative damage and affecting the synthesis of ferulic acid. *Scientia Horticulturae, 307*, 111486. https://doi.org/10.1016/j.scienta.2022.111486

Sun, Y., Zhang, X., Wu, C., He, Y., Ma, Y., & Hou, H. (2016). Engineering herbicide-resistant rice plants through CRISPR/Cas9-mediated homologous recombination of Acetolactate synthase. *Molecular Plant*, 9, 628–631.

Tan, J., Zhang, F., Karcher, D., & Bock, R. (n.d.). Engineering of high-precision base editors for site-specific single nucleotide replacement. *Nature Communications*, 2019. https://doi.org/10.1038/s41467-018-08034-8

Thomson, M. J. (2014). High-throughput SNP genotyping to accelerate crop improvement. *Plant Breeding Biotechnology*, 2, 195–212.

Tian, S., Jiang, L., Cui, X., Zhang, J., Guo, S., Li, M., Zhang, H., Ren, Y., Gong, G., Zong, M., Liu, F., & Chen, Q. (2018). Engineering herbicide-resistant watermelon variety through CRISPR/Cas9-mediated base-editing. *Plant Cell Report*, 37, 1353–1356.

Tomlinson, L., Yang, Y., Emenecker, R., Smoker, M., Taylor, J., & Perkins, S. (2019). Using CRISPR/Cas9 genome editing in tomato to create a gibberellin -responsive dominant dwarf DELLA allele. *Plant Biotechnology Journal*, 17, 132–140.

Tripathi, J. N., Ntui, V. O., Ron, M., Muiruri, S. K., Britt, A., & Tripathi, L. (2019). CRISPR/Cas9 editing of endogenous banana streak virus in the B genome of Musa spp. Overcomes a major challenge in banana breeding. *Communications Biology*, 2(1), 1–11. https://doi.org/10.1038/s42003-019-0288-7

Veillet, F., Perrot, L., Chauvin, L., Kermarrec, M. P., Guyon-Debast, A., Chauvin, J. E., Nogué, F., & Mazier, M. (2019). Transgene-free genome editing in tomato and potato plants using Agrobacterium-mediated delivery of a CRISPR/Cas9 cytidine base editor. International Journal of Molecular Sciences, 20(2). https://doi.org/10.3390/ijms20020402

Voytas, D. F., & Gao, C. (2014). Precision genome engineering and agriculture: Opportunities and regulatory challenges. *PLoS Biology*, 12, 1–6.

Wang, M., Lu, Y., Botella, J. R., Mao, Y., Hua, K., & Zhu, J. K. (2017). Gene targeting by homology-directed repair in rice using a gemini-virus -based CRISPR/Cas9 system. *Molecular Plant*, 10, 1007–1010.

Wei, T., Jiang, L., You, X., Ma, P., Xi, Z., & Wang, N. N. (2023). Generation of herbicide-resistant soybean by base editing. *Biology, 12*(5), 741. https://doi.org/10.3390/biology12050741"

Wolabu, T. W., Mahmood, K., Jerez, I. T., Cong, L., Yun, J., Udvardi, M., Tadege, M., Wang, Z., & Wen, J. (2023). Multiplex CRISPR/Cas9-mediated mutagenesis of alfalfa FLOWERING LOCUS Ta1 (MsFTa1) leads to delayed flowering time with improved forage biomass yield and quality. *Plant Biotechnology Journal*, 21(7), 1383–1392. https://doi.org/10.1111/pbi.14042

Wu, J., Chen, C., Xian, G., Liu, D., Lin, L., Yin, S., Sun, Q., Fang, Y., Zhang, H., & Wang, Y. (2020). Engineering herbicide-resistant oilseed rape by CRISPR/Cas9-mediated cytosine base-editing. *Plant Biotechnology Journal*, 18(9), 1857. https://doi.org/10.1111/pbi.13368

Wu, R., Lucke, M., Jang, Y., Zhu, W., Symeonidi, E., Wang, C., Fitz, J., Xi, W., Schwab, R., & Weigel, D. (2018). An efficient CRISPR vector toolbox for engineering large deletions in Arabidopsis thaliana. *Plant Methods*, 14(1), 65. https://doi.org/10.1186/s13007-018-0330-7

Yang, B., Yang, L., & Chen, J. (2019). Development and application of base editors. *The CRISPR Journal*, 2(2), 91–104. https://doi.org/10.1089/crispr.2019.0001

Yang, L., Yang, B., & Chen, J. (2019). One prime for all editing. *Cell*, 179(7), 1448–1450. https://doi.org/10.1016/j.cell.2019.11.030

Yara, R., Cao, H., Jin, L., Mengdi, Y., & Zhou, L. (2020). CRISPR/ Cas mediated base editing: A practical approach for genome editing in oil palm. *3 Biotech,* 10, 1–7.

Yin, K., Han, T., Xie, K., Zhao, J., Song, J., & Liu, Y. (2019). Engineer complete resistance to Cotton Leaf Curl Multan virus by the CRISPR/Cas9 system in Nicotiana benthamiana. *Phytopathology Research,* 1(1), 9. https://doi.org/10.1186/s42483-019-0017-7

Zhang, H., Qin, C., An, C., Zheng, X., Wen, S., Chen, W., Liu, X., Lv, Z., Yang, P., Xu, W., Gao, W., & Wu, Y. (2021). Application of the CRISPR/Cas9-based gene editing technique in basic research, diagnosis, and therapy of cancer. *Molecular Cancer*, 20(1), 1–22. https://doi.org/10.1186/s12943-021-01431-6

Zhang, R., Liu, J., Chai, Z., Chen, S., Bai, Y., Zong, Y., Chen, K., Li, J., Jiang, L., & Gao, C. (2019). Generation of herbicide tolerance traits and a new selectable marker in wheat using base editing. *Native Plants,* 5, 480–485.

Zhang, X. H., Tee, L. Y., Wang, X. G., Huang, Q. S., Yang, S. H. (2015). Off-target effects in CRISPR/ Cas9-mediated genome engineering. *Molecular Therapy Nucleic Acids,* 4, e264.

Zhao. D., Li. J., Li. S., Xin. X., Hu. M., Price. M. A., et al. (2020). Glycosylase base editors enable C-to-A and C-to-G base changes. *Nature Biotechnology,* 39, 35–40. 10.1038/s41587-020-0592-2

Zong, Y., Song, Q., Li, C., Jin, S., Zhang, D., Wang, Y., Qiu, J. L., & Gao, C. (2018). Efficient c-to-t base editing in plants using a fusion of ncas9 and human apobec3a. *Nature Biotechnology*, 36(10), 950. https://doi.org/10.1038/nbt.4261

Zong, Y., Wang, Y., Li, C., Zhang, R., Chen, K., Ran, Y., Qiu, J. L., Wang, D., & Gao, C. (2017). Precise base editing in rice, wheat and maize with a Cas9-cytidine deaminase fusion. *Nature Biotechnology,* 35, 438–440.

Zsögön, A., Cermák, T., Naves, E. R., Notini, M. M., & Edel, K. H. (2018). De novo domestication of wild tomato using genome editing. *Nature Biotechnology,* 36, 1211–1216.

11 Prime Editing through CRISPR-Cas and Its Applications in Plants

Revati Wanikar, Priyanjali Dixit, Maya Khater, Medha Dange, Rohini Yevale, and Sayanti Mandal

11.1 INTRODUCTION

In the last few years remarkable research has been done for the establishment of precise modification of genes by using CRISPR-Cas9-mediated genome-editing tools. Prime editing (PE) is a novel method of genome editing that combines a reverse transcriptase enzyme with the CRISPR-Cas9 system (Huang et al., 2023). It was created as an improvement to the original CRISPR-Cas9 technology and provides a more accurate and adaptable way to make specific changes to the DNA sequence. In conventional CRISPR-Cas9 editing, a short RNA molecule directs the Cas9 protein to a specific region of the genome where it breaks the DNA. Through the introduction of a modified Cas9 protein linked to a reverse transcriptase enzyme, prime editing increases the capabilities of CRISPR-Cas9 (Huang et al., 2021; Khare et al., 2023). Furthermore, prime editing improves the safety and precision of CRISPR-Cas9 by decreasing the possibility of off-target effects in comparison to the traditional method (Vu et al., 2024; Zhao et al., 2021). Anzalone et al. (2019) have announced prime editing, a method for overcoming the constraints of conventional genome-editing tools. This unique prime editing technique uses a longer gRNA (longer than conventional CRISPR gRNA) termed prime editing gRNA (pegRNA) and a Cas9 H840 nickase-Reverse Transcriptase (RT) fusion protein to target genomic loci without HDR or donor DNA templates. The main benefit of a prime editing method over previous CRISPR-Cas systems is that it avoids double-strand breaks (DSBs), which may damage target genes by inducing insertions or deletions at non-target regions (Anzalone et al., 2019). Prime editing was swiftly accepted by scientists as a flexible genome-editing tool for modifying a small gene because of these flexibilities (Table 11.1).

CRISPR-Cas9 technology has revolutionized the field of plant breeding and genetic engineering by offering precise and efficient tools for modifying plant genomes. Some of the applications of CRISPR-Cas9 in plants include crop improvement, disease resistance, herbicide tolerance, and modifications to the plant metabolism (Liu et al., 2023). CRISPR-Cas9 permits precise alterations in crop plants to introduce beneficial features, including greater yield, improved nutritional value, and improved resistance to diseases, pests, and abiotic stresses and optimized plant design (Shan et al., 2013; Erdoğan et al., 2023). Plants can be made more resistant to bacterial, fungal, and viral infections by using CRISPR-Cas9. Plants can be made more resistant to particular diseases, lowering the need for chemical pesticides, by changing the susceptibility genes that are targeted for modifications (Erdoğan et al., 2023; Khare et al., 2023). It is possible to create crops that are resistant to a particular type of herbicide which enables selective weed removal while reducing the negative effects on agricultural plants. Researchers can examine the function of particular genes in plant growth, development, and responses to environmental cues by making targeted mutations or gene knockouts. CRISPR-Cas9 can be used to change plant metabolic pathways to enhance the synthesis of particular substances, such as medicines, biofuels, or industrial chemicals. CRISPR-Cas9 allows for precise stacking of multiple genetic modifications in a single plant. This enables the simultaneous introduction of multiple traits into crops, streamlining the breeding process and accelerating the development of improved varieties (Hillary and Ceasar, 2022). This chapter includes the recent approach of CRISPR-Cas9, i.e., prime editing and its mechanism, and also focuses on applications of original CRISPR-Cas9 with some PE technology for crop improvement.

11.2 THE WORKING MECHANISM OF PRIME EDITING

Traditional methods of generating improved new varieties of plants always have been done by breeding and selection. These classical methods are laborious and time-consuming. Anzalone et al. (2019) have recently demonstrated a novel genome-editing technology called prime editing in mammalian and yeast cells. This new technology is basically a "Search and Replace" tool for genome editing that can mediate targeted insertions and deletion. Precise genome editing was achieved without the generation of DSBs or requiring a donor DNA template, which occurs in the case of genome editing via HDR.

Prime editing (PE) is the first technology to adopt precise genome-editing. This technique allows all possible base-to-base conversions, insertions as well as deletions, which further do not require DSBs or donor DNA (Anzalone et al., 2019). The wide editing spectrum makes possible the correction of up to 89% of different human genetic diseases (Anzalone et al., 2019). PEs are comprised of a reverse transcriptase

DOI: 10.1201/9781003500933-13

TABLE 11.1
Comparison of different classes of CRISPR/Cas-derived genome editing agents, such as base editor and prime editor

Property	CRISPR-Cas	Base editing	Prime editing
Origin	Diverse bacteria (*Streptococcus pyogenes*)	Nature	Nature
Construction	Easy	Easy	Easy
Components	Donor DNA, sgRNA	Base editor (fusion Cas9n + deaminase) and sgRNA	Prime editor (fusion Cas9n + RT) and pegRNA
Endonuclease	Cas9/Cas12/Cas13/ Cas14	dCas	pegRNA
Target	DNA/RNA	DNA/RNA	DNA
Length of target sequence (bp)	19–22	15–20	8–15
Edit type	Insertions, deletions, replacements, and inversions	PAM distal transition/point mutation	PAM proximal transitions/point mutations/small insertions and deletions
Mutation rate	Low	High	Very high
Multiplexing	Easier	Easier	Not yet tested
Cost-effectiveness	High	Very high	Very high
Mutation rate	Low	High	Very high
Pros	Highly efficient	Precise and single base editing	Target flexibility

Source: Adapted from Hillary and Ceasar (2022).

(RT) that incorporates Cas9 nickase (H840A mutation) which can produce only ss breaks and a prime-editing guide RNA (pegRNA) which is 3'extended sgRNA with a primer binding site (PBS) and a RT template. Both these specify the target site and encode the desired edit; details are shown in Figure 11.1 (Anzalone et al., 2019; Lu et al., 2022; Zhao et al., 2023). In Figure 11.1, the PE is a fusion protein that combines a reverse transcriptase (RT) with a nickase Cas9 (nCas9). On the non-target strand, the nCas9 forms an SSB. The prime-editing guide RNA (pegRNA) 3′ end hybridiszes with the released 3′ end, and the RT domain subsequently reverse transcribes the signal. The newly synthesized DNA strand is modified by the reverse transcription, which integrates the changes contained in the pegRNA. By achieving equilibrium between the edited 3′ flap and the unedited 5′ flap, endogenous cleavage and ligation of the 5′ flap, and DNA repair, the intended edit is incorporated into the genome in a stable manner.

The nCas9 of PE protein nicks the non-target strand thereby exposing a 3′-ss DNA after binding the PE system to the target DNA under the guidance of pegRNA. Then the cas9 nickase cleaves the strand containing the PAM sequences, ultimately exposing its 3'hydroxyl group. This 3'-ss DNA later hybridizes to the PBS. This binding primes the associated RT enzyme to extend the nicked 3′-ssDNA using the RT template. At the end of this step, the edited DNA strand has two redundant ssDNA flaps: a flap on the 5′ end with unmodified sequence and a flap at the 3′ end with the desired edited sequence. Of these two, the fully complementary 5′ flap is thermodynamically preferred to hybridize with the unedited strand. However, the inherent vulnerability of 5′ flaps to removal by endogenous structure-specific endonucleases typically leads to hybridization of the edited 3′ flap. This results in the formation of heteroduplex. Ultimately, ligation and a DNA mismatch repair result in the permanent incorporation of the chosen medications (Anzalone et al., 2019; Lu et al., 2022; Zhao et al., 2023). The prime editing process can perform various types of edits, including insertions, deletions, and point mutations of small DNA fragments and offers greater targeting flexibility.

11.2.1 Improvements in Prime Editing Techniques

Although PE has overcome the shortcomings of earlier gene-editing techniques, this technique is still in its initial phases and needs to overcome multiple limitations in order to understand and use with the maximum potential (Lu et al., 2022; Zhao et al., 2023). A few of the shortcomings of PE are described here. First is the PAM sequence-dependent recognition of Cas9 and, second, is the invariable efficiency of this PE system among various cell types. This may be because of differences in the target sequences across various cell types. Third, a major issue is the original PE enzyme is very large and therefore, that makes it difficult to load using the classical delivery tools. To make the PE system fully efficient, the following improvements have been made in the system (Anzalone et al., 2019; Zhao et al., 2023).

11.2.2 Relaxation in PAM Restriction for PE

The current PE system completely relies on the SpCas9 enzyme that requires NGG PAM sequences. These occur on average once in every 16 randomly selected genomic loci, greatly limiting the range of target sequences. In this regard, researchers have been working to look for and modify Cas9 enzymes which can expand their compatibility with various PAMs' range of targets for gene editing (Lu et al., 2022). Several researchers have developed various SpCas9 variants expanding the targetable PAM sites (Keithstiver et al., 2015; Hu et al., 2018; Nishmasu et al., 2018; Miller et al., 2020).

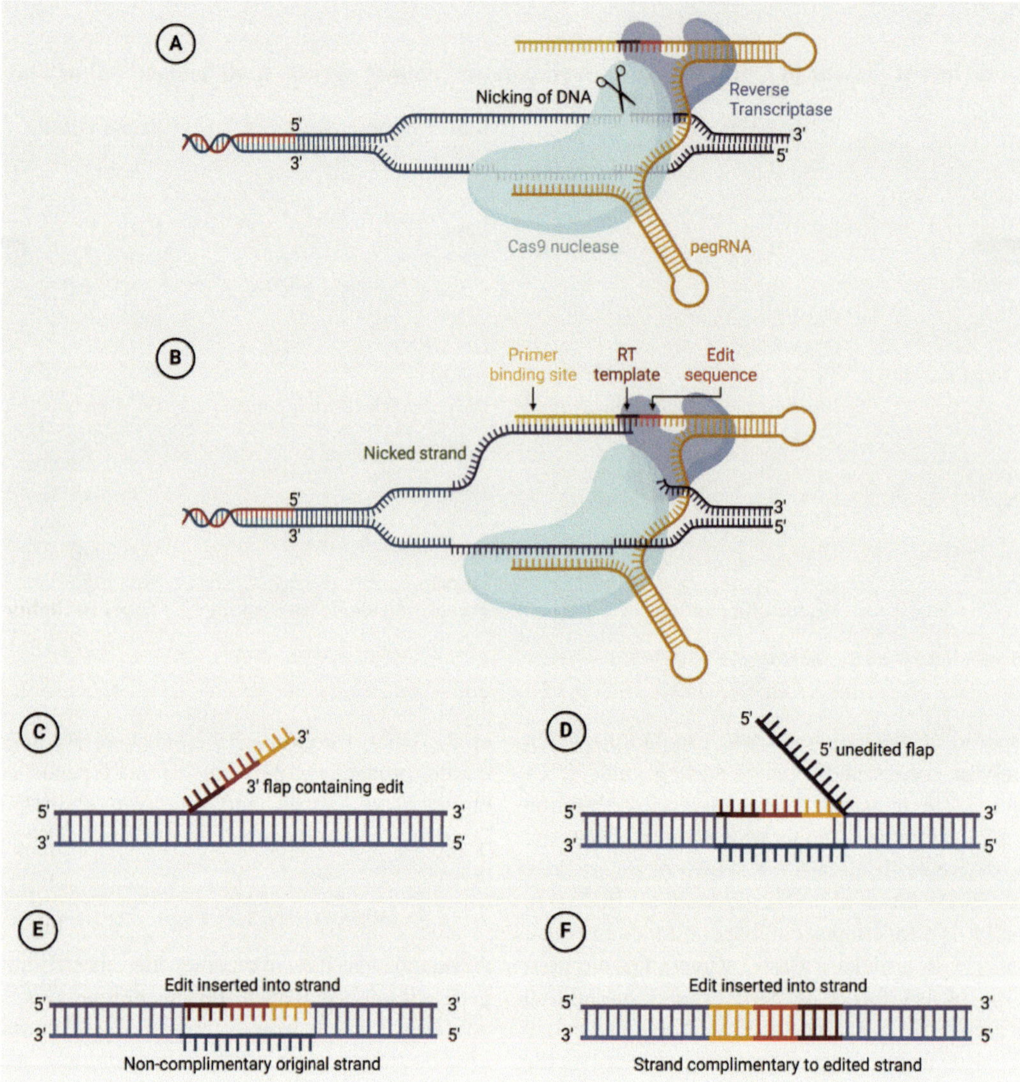

FIGURE 11.1 Schematic diagram representing the mechanism of prime editing (PE). (Adapted and modified from Matsoukas (2020). Image was created with BioRender.com.)

11.2.3 Improvement of Editing Efficiency

Low editing efficiency is a major challenge of PE. There are several reports stating improvement in efficiency. During refinement different generations of PE system were produced. The maximum editing efficiency that can be extended by the PE system reached approximately 50%, which is almost a 10-fold improvement compared to the first generation of PE systems (Anzalone et al., 2020). Also engineering of pegRNA remains the main method to improve PE efficiency in the future.

11.2.4 Delivery of PE Agents to the Targets

The success of genome editing depends upon the efficient delivery of the prime editing machinery into the target cells. The full-length PE system is very bulky and precludes the efficient delivery to target cells using simple viral vectors.

11.2.5 Development of Algorithms for the Design of pegRNA

To explore the diversity of PE applicability in various species, tools for PE system design and the prediction of editing effects were merged to facilitate the experiments. Several algorithmic models are debugged to select the target sequence that obtains the highest efficiency based on the predicted results. To quickly achieve the picture of the target region editing, Hwanget al. (2021) invented a PE-Analyzer that can easily analyze the PE high-throughput sequencing results of cells that have undergone PE, which can easily provide information, such as the distribution and specific values of each type of mutation (Hwang et al., 2021).

11.3 RECENT PROGRESS IN PRIME EDITING

11.3.1 Prime Editing System Using Single pegRNA

Anzalone et al. first published a report on PE in 2019, and then within a year the first report on PE performance in precise editing of different types of single base substitution and short insertions/deletions at multiple loci in rice and wheat was also published by Lin et al. (2020). Later many researchers published applications of PE in plants with remarkable efficiency which recently were compiled by Vu et al. (2024).

11.3.2 Paired Prime Editing Approaches for Short DNA Editing

Anzalone et al. (2019) and Lin et al. (2020) have reported that high PE efficiency is seen with short sequence deletions followed by base substitution. It was also shown that efficiency drops as insertion length increases (Lin et al., 2020). However, these limitations of PE efficiency and length limitations for sequence insertions can be overcome by using paired PE systems (Liang et al., 2023; Lin et al., 2021). Moreover, when sequence insertion and short sequence deletions were done simultaneously, the efficiency improved even more (Choi et al., 2022; Vu et al. 2024).

11.3.3 Paired Prime Editing Approaches for Long DNA Editing

In plants, the precise insertion of large DNA fragments into a desired genomic site is made possible with the use of the paired PE-approach which is also called PrimeRoot (Sun et al., 2023). This excellent technology has made possible the precise insertion of genes or alleles of interest into safe sites present in plant genomes. This has noticeably accelerated precise gene or allele pyramiding (Hillary and Ceasar, 2022; Vu et al., 2024).

11.3.4 Dual Peg-RNA

Use of dual peg-RNA was developed in order to optimize the prime editing technique, which uses pegRNAs in trans to encode the same edits, one for forward and other for reverse of each DNA strand simultaneously (NGG-pegRNA and CCN-pegRNA). This approach in plants can improve efficiency by optimizing the melting temperature (Tm) of PBS (Lin et al., 2021)

11.3.5 Enhanced Plant Prime Editor 2 (enpPE2) Systems

The enpPE2 system includes various optimization strategies such as updating the PE architecture to PEmax architecture and the expression of epegRNA under the control of composite promoter (Chen et al., 2022).

11.3.6 Twin Prime Editor (TwinPE)

This system relies on the use of two partially complementary pegRNAs generated at similar sites to be edited. This strategy directly allows the replacement of double strands with the desirable edits and no involvement of single strand invasion as well as DNA mismatch repair. Twin PE shows more efficiency compared to PE, also exhibits fewer unintended indels. But it slightly extends the strand with edits due to pegRNA partial complement (Anzalone et al., 2022).

11.4 APPLICATIONS OF PRIME EDITING

11.4.1 Improving Traits in Agronomically Important Plants

The quality of crops can be improved by using the CRISPR/Cas9 technology and the agronomic traits can be enhanced by modifying the target genes (Chen et al., 2019). This may reduce the wastage of food and increase the consumers' acceptance. A wide spectrum of research with respect to gene editing has been done on a range of plant species, including A. thaliana, wild cabbage (*B. oleracea* L.), cucumber (*Cucumis sativus* L.), soybean (*Glycine max*), barley (*Hordeum vulgare* L.), lettuce (*Lactuca sativa* L.), rice (*Oryza Sativa*), tomato (*Solanum lycopersicum* L.), potato (*S. tuberosum*), sorghum (*Sorghum bicolor*), wheat (*T. aestivum*), grape (*Vitis vinifera* L.), maize (*Zea mays*), and many other crops (Li et al., 2023; Tingting et al., 2023). This chapter mainy highlights the effect on rice (*O. sativa*), maize (*Z. mays*) and wheat (*T. aestivum*) crops which contribute to the world's food security because they are part of a staple diet (Table 11.2).

11.4.2 Rice (Oryza Sativa)

Rice is the major staple food consumed across the globe, however, climate change, growing populations, and increasing demand are affecting the rice production. CRISPR-Cas technology aims to produce observable effects in rice either by gene overexpression or gene knockout. Several genes which encode for traits like grain quality, disease resistance, and anthocynin content have been identified and successfully manipulated in rice. CRISPR-Cas technology improved multiple agronomic traits in rice plants. CRISPR-Cas based mutation in *OsGA20ox2* increased grain yield and successfully altered the plant architecture by editing/knockout of genes (Tabassum et al., 2021). Guo et.al. (2020) reported overexpression and knockout of *OsProDH* gene in rice which encodes for proline dehydrogenase, an enzyme which degrades proline in rice. Proline has a significant role in managing biotic as well as abiotic stresses. Manipulation in *OsProDH* by using CRISPR-Cas technology has resulted in proline accumulation which is reported to reduce Reactive oxygen species (ROS) (Guo et al., 2020). Various approaches have been used in recent years to study and identify regulatory genes which are involved in grain quality. CRISPR-Cas9 is found to be highly efficient at improving rice grain quality with more precision (Jain et al., 2015).

In order to conduct a methodical evaluation in rice plants, Jin et al. (2021) extensively researched and reported that the number and location of mismatches in the primer binding site and spacer of the prime editing guide RNA (pegRNA) affected the editing frequency. Upon evaluating the activity of 12 pegRNAs at 179 locations that were anticipated to be off-target,

TABLE 11.2
CRISPR-Cas9 enables prime editing in plants

S. No.	PEs	Target gene	Mutation type	Editing efficiency	Features of the PE (RT template length (nt) and PBS length (nt))	Trait improvement	References
I	PE2						
1.	PE2	*OsALS, OsIPA1, OsTB1*	2–4 bp Subs, 2 bp Ins	0.00–2.04	13, 15-20	Herbicide resistance, High yield	Butt et al. (2020)
2.	PE2	*OsALS, OsACC*	1–3 bp Subs	1.00–7.60	13,16; 13, 14–23	Herbicide resistance	Jiang, Chai, et al. (2022)
3.	pPE2	*OsPDS, OsACC, OsALS, OsCDC48*	1 bp Ins, 1–2 bp Subs	0.00–29.17	10–13, 11–18	Herbicide resistance	J. Li et al. (2022)
4.	pPE2 (an engineered pegRNA with mpknot)	*OsPDS, OsALS, OsCDC48*	1 bp Ins, 1–2 bp Subs	10.42–25.00	10–13, 13–18	Herbicide resistance	J. Li et al. (2022)
5.	pPE2 (an engineered pegRNA with evopreQ$_1$)	*OsPDS, OsACC, OsALS, OsCDC48*	1 bp Ins, 1–2 bp Subs	2.08–50.00	10–13, 11–18	Herbicide resistance	J. Li et al. (2022)
6.	ePPE(replacing M-MLV-RT with M- MLV-RT-ΔRNaseH)	*OsALS*	2 bp Subs	11.30	8, 17	Herbicide resistance	Zong et al. (2022)
II	PE3						
7.	Sp-PE3	*APO1, GFP, OsALS*	1–2 bp Subs	0.00–17.10	13, 13–16	Herbicide resistance	K. Hua et al. (2020)
8.	PPE3	*OsCDC48, OsALS*	1–3 bp Subs, 6 bp Del	2.60–21.80	10–12, 9–17	Herbicide resistance	Lin et al. (2020)
9.	pPE3	*OsWx, OsACC*	1 bp Subs	0.00–16.70	13, 10	Herbicide resistance	Xu, Li, et al. (2020)
10.	PE3	*OsALS, OsACC, OsEPSPS*	1–3 bp Subs	1.30–70.30	13,16; 13, 14–23	Herbicide resistance	Jiang, Chai, et al. (2022)
11.	pPE3b	*OsACC*	1 bp Subs	6.25	13, 10	Herbicide resistance	Xu, Li, et al. (2020)
12.	PE-P2(nCas9 (H840A)-M-MLV-T2A-hpt)	*OsDEP1, OsALS, OsACC*	1–4 bp Subs	1.70–26.00	11–14, 14–23	Herbicide resistance	Xu, Zhang, et al. (2020)
13.	pCXPE03 (derived by the RPS5A promoter)	*SlGAI, SlALS, SlPDS*	2 bp Subs, 2 bp Ins	0.00–6.70	14, 17–18	Herbicide resistance	Lu et al. (2021)
14.	PPE3-evopreQ1	*OsROC, OsALS, OsCDC48, OsDEP1*	1–3 bp Subs, 3 bp Ins	2.60–47.50 (Normal); 5.00–60.50 (HT)	11–14, 11–18	Herbicide resistance	Zou et al. (2022)
15.	PE-P3-RT-M (C- terminal M-MLV+ multiple synonymous base mutations in RTT)	*OsGS3, OsALS, OsACC, OsChalk5, OsDEP1, OsWaxy,*	3–4 bp Subs	0.00–82.60	8–14, 13–23	-	Xu et al. (2022)
III	PE4						
16.	PE4	*OsALS, OsACC*	1–3 bp Subs	5.20–27.10	13,16; 13, 14–23	Herbicide resistance	Jiang, Chai, et al. (2022)
IV	PE5						

TABLE 11.2 (Continued)
CRISPR-Cas9 enables prime editing in plants

S. No.	PEs	Target gene	Mutation type	Editing efficiency	Features of the PE (RT template length (nt) and PBS length (nt))	Trait improvement	References
17.	PE5	*OsALS, OsACC, OsEPSPS*	1–3 bp Subs	1.60–64.10	13,16; 13, 14–23	Herbicide resistance	Jiang, Chai, et al. (2022)
18.	PEmax ePE3max (the PEmax protein + an engineered pegRNA with evopreQ1)	*OsEPSPS*	3 bp Subs	37.20–39.80	13, 22	Herbicide resistance	Jiang, Chai, et al. (2022)

Source: Adapted from Li et al. (2023).

they found very few off-target alterations (0.00~0.23%). Whole-genome sequencing of 29 rice plants treated with PEs verified that PEs do not cause minor insertions/deletions or genome-wide pegRNA-independent off-target single-nucleotide variations (Jin et al., 2021). Using the Cas9 variant SpCas9, researchers have constructed the prime editors Sp-PE2 and Sp-PE3, in rice. Sp-PE2 was injected with an expression cassette containing pegRNA, the ZmUbi promoter, OsU6, and an inactive EGFP controlled by the CaMV35 promoter. Sp-PE3 was then used to detect mutations. Compared to Sp-PE3, Sp-PE2 is more efficient and exhibits a stronger GFP signal (Hua, Tao, et al., 2020). Xiong et al. (2021) have reported that multiplex coactivation by dCas9-TADDs could simultaneously upregulate *OsGW7* and *OsER1* genes in rice varieties. In another study, polygalacturonate's function in controlling the immunological response of the cell wall was investigated through the *OsPG1* gene which encodes for polygalacturonate in rice varieties (Cao et al., 2021). *OsPG1*'s incorrect splicing was shown to be the cause of the necrotic leaf tips by functional complementation and a knockout test, which demonstrates that *OsPG1* directly controls leaf growth and development. These results demonstrate the effectiveness of CRISPR-Cas9 technology in modifying rice characteristics, and it is anticipated that these techniques will be used to produce more robust, resilient, and nutrient-dense rice, which will contribute to global sustainability. Using CRISPR-Cas9, four yield-related genes—*Gn1a* (which controls grain number), *DEP1* (DENSE AND ERECT PANICLE1; controls panicle architecture), *GS3* (which controls grain size), and *IPA1* (which controls plant architecture)—were effectively altered in rice (Li M. et al., 2016). To create an effective PE system and to achieve herbicide-tolerant varieties in rice, numerous attempts have been made in recent years (Li H. et al., 2020; W. Xu et al., 2020). Xu et al successfully targeted nucleotide substitution of *acetolactate synthase* (*ALS*) gene with frequencies of up to 26% in rice.

11.4.3 Maize (*Zea mays*)

Maize is a major cereal crop grown worldwide for food as well as animal feed. Global yields of maize and the majority of other crops have been greatly impacted by temperature variations, precipitation, and other climatic conditions. CRISPR-Cas technology provides revolutionary changes by introducing new strategies for maize improvement (Yilin et al., 2022). The majority of maize genome-editing research has used biolistic delivery techniques or genetic transformation protocols based on *Agrobacterium tumefaciens* in a few temperate genotypes suitable for Agrobacterium infection and regeneration (Kausch et al., 2021a; Yassitepe et al., 2021). However, the widespread use of genome editing in maize is hampered by the lack of a transformation methodology that can be applied to a larger range of genotypes and low transformation efficiency. The CRISPR-Cas9 technology has been effectively used to modify the genome of maize since 2014. Its uses in maize include verifying the function of important genes, increasing yield, strengthening stress tolerance, enhancing quality, producing male-sterility mutants, and inducing haploidy (Xing et al., 2014).

Using prime editors in maize, Jiang et al. (2020) produced double mutations in *ZmALS1* and *ZmALS2* more efficiently than previously documented. The impact of inflorescence architecture on maize yield is evident in various dimensions, including meristem size, kernel weight, and kernel depth. The CRISPR-Cas9 technology has found several genes that control inflorescence development, making them particularly appealing targets for increasing maize production (Ning et al., 2021).

Li et al. (2020) reported successful gene editing in maize for *ZmPHYC1* and *ZmPHYC2* genes which encode for flowering time and plant height respectively through gene knockout and the gene overexpression method. The important characteristics of high-quality maize are its nutrition content, texture, taste, and flavor. CRISPR/Cas technology improved traits by editing responsible genes such as *ZmWAX1*, *ZmSH2*, *ZmbZIP22*, *ZmBADH2a*, and *ZmBADH2b* (Li et al., 2018; Dong et al., 2019; Gao et al., 2020; Wang et al., 2021).

Ning et al. (2021) demonstrated *ZmACO2*'s ability to stimulate meristem and flower development. It is documented that the knockout alleles of *ZmACO2* and promoter-edited alleles showed considerable increase in ear length, ear weight, and kernel counts in comparison to their wild-type siblings (Ning et al., 2021). Sweet and waxy compound corns (SWCs) have been created using the CRISPR-Cas9 system by concurrently altering *ZmSH2* and *ZmWAXY1* (Dong et al., 2019), notably, the first commercially available genome-editing maize will be CRISPR-Cas9 waxy maize. Prime editing has been shown possible to edit several exogeneous and endogeneous genes in rice and wheat successfully with low editing efficiency whereas PE applications in maize require more optimizations.

11.4.4 Wheat (*Triticum aestivum*)

Wheat is an extensively cultivated crop consumed as a main source of food and also used for production of cereal-based products worldwide. CRISPR technologies have been used to modify wheat genome which involves small random insertions or deletions (indels) as well as deletions or insertion of larger fragments (Li et al., 2021; Mandal and Verma, 2021; Hernandes et al., 2023; Zhou et al., 2023).

Common wheat is now being edited using CRISPR-Cas-mediated genome-editing techniques that have recently been refined and enhanced. These techniques include base editing, multiplex genome editing, PE, and targeted gene mutagenesis (Zhou et al., 2023). PE guide RNA (pegRNA) is used by PE systems primarily to combine a catalytically impaired Cas9 (H840A) (nCas9) with a reverse transcriptase M-MLV-RT (Moloney murine leukaemia virus reverse transcriptase) and encode the appropriate genetic information at the target site. The protein complex in the PE system binds to the target DNA and nicks the non-target strand. Using the RT template in the pegRNA, the resulting 3' DNA terminal primes reverse transcription of fresh DNA containing the desired edit after hybridizing to the PBS in the pegRNA (Anzalone et al., 2019; Li et al., 2021). The main goal of improving wheat is yield, which is mostly determined by the weight of the grains, the number of grains per spike, and the number of spikes per unit area (Liu et al., 2022). *TaGW2* is a gene that encodes RING-type E3 ubiquitin ligase which inhibits cell division. By targeting two or three *TaGW2* homologs, increases in wheat grain yield, length, and width were achieved (Wang et al., 2018; Wang et al., 2022; Zhang et al., 2018).

Wheat quality is also determined by starch content and ratio of amylose and amylopectin. High amylose starch contributes in resistant starch (RS) formation. RS has wide applications in varieties of food products due to its beneficial health effects; it also reduces the risk of diabetes (Huang et al., 2021; Wanikar et al., 2021). By using CRISPR/Cas9 to modify the Starch branching enzyme IIa (TaSBEIIa) gene, highamylose wheat lines with elevated RS levels were produced (Li et al., 2021). In another study, by deleting the Granule-bound starch synthase gene (*TaWaxy*), glutinous wheat with a lower amylose concentration was produced (Zhang et al., 2021). In a recent field trial, shorter plants with larger spikes and a much greater grain weight were produced by editing the wheat auxin response factor gene, *TaARF12* which showed yield improvements of up to 11.1% (Kong et al., 2023).

Several genes can be synergistically manipulated to increase yields as well. For example, simultaneous deletion of the RING-type E3 ligase gene (ZnF-B) and Reduced height 1 (Rht-B1) balanced the levels of the phytohormones gibberellin and brassinolide, leading to semi-dwarf wheat plants with high yield and efficient nitrogen uptake (Song et al., 2023). Future research could potentially lead to the development of high-yielding wheat varieties by simultaneously editing various genes that regulate yield or by modifying genes linked to trade-off attributes like grain weight and number. The most recent improved engineered PPE, known as ePPEplus, was created by stacking different optimization strategies, such as adding the V223A mutation into MMLV RT, updating the PE architecture to PEmax, and expressing an engineered pegRNA with a structured motif, based on ePPE architecture (combining PPE-DRNase H and PPE-NC strategies).Compared to the original PPE and ePPE, respectively, ePPEplus significantly increased editing frequency, reaching up to 21.2% in wheat protoplasts (Ni et al., 2023). ePPEplus has demonstrated remarkable success in producing heritable mutations in wheat plants, with editing rates reaching 86.9% at the *TaGASR7* locus. This significantly broadens the possible applications of PE in wheat (Ni et al., 2023).

11.5 ENVIRONMENTAL STRESS TOLERANCE IN CRISPR PLANTS

Several environmental stresses such as drought, heat, cold, salt, and metal stress have a significant impact on crop production worldwide. Plants have evolved different cellular and transcriptional regulations to cope up and respond to these stresses through morphological, physio-biochemical, as well as cellular changes. Owing to the complex mechanisms underlying abiotic stress tolerance, traditional breeding methods are inefficient to meet the challenge of manipulating multiple genes to generate crop varieties with enhanced resilience to abiotic stresses (Wahab et al., 2022; Nawaz et al., 2023). The advent of genome-editing technologies has facilitated the dissection of tolerance mechanisms by focusing on either "Tolerance (T) genes" that exhibit positive regulatory responses to stress or "Sensitive (S) genes" that are involved in stress susceptibility (Wahab et al., 2022; Nawaz et al., 2023). The use of CRISPR technology enables precise and targeted modifications of specific genes, i.e., manipulating T-genes or disrupting S-genes for developing crop varieties tolerant to abiotic stresses (Erdoğan et al., 2023; Khare et al., 2023). Figure 11.2 illustrates the CRISPR-Cas system in the generation of biotic and abiotic resilient crop plants. Integrating various "OMICS" techniques, genes responsible

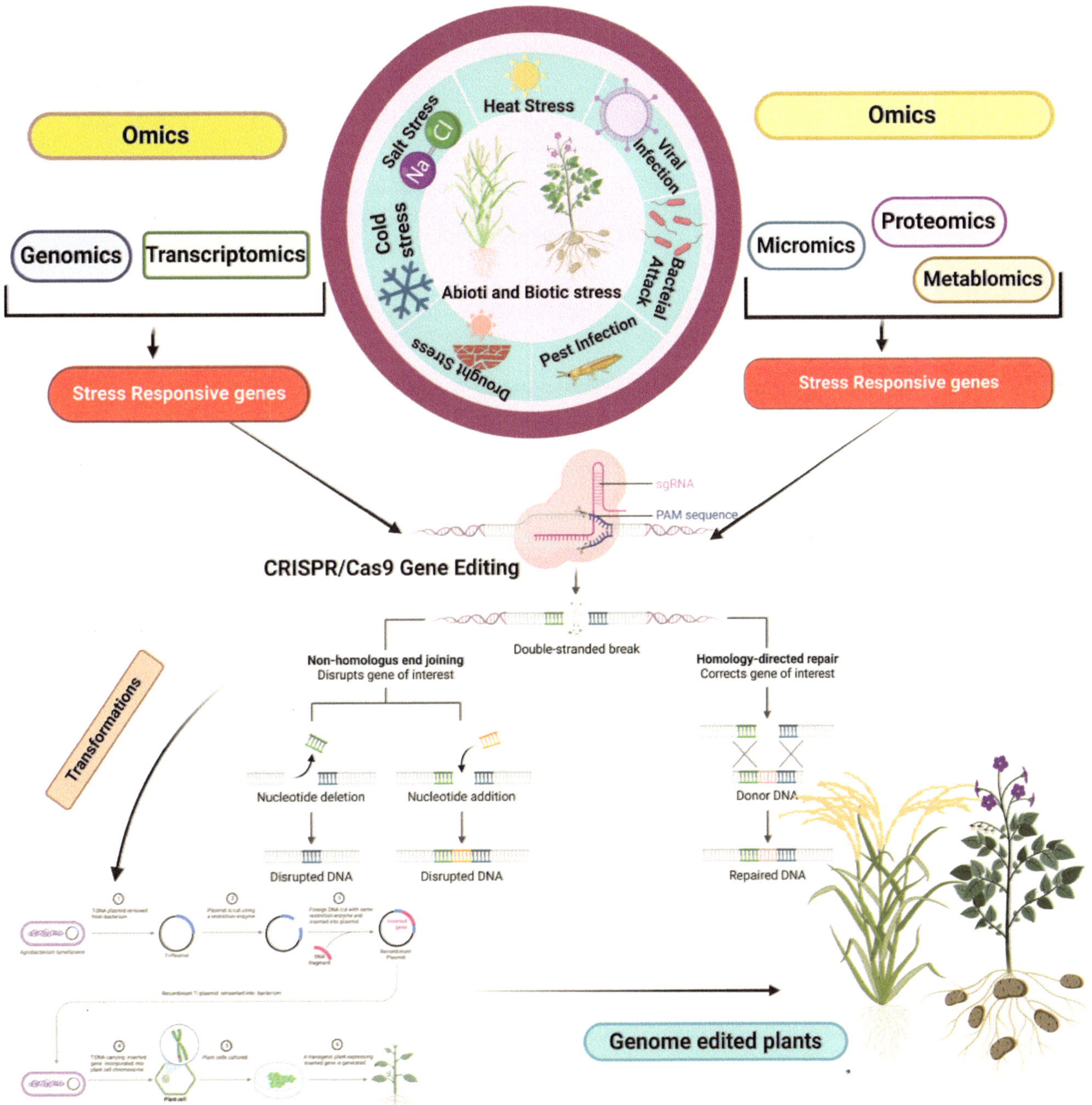

FIGURE 11.2 Schematic illustration of CRISPR-Cas system in the generation of abiotic- and biotic-resilient crop plants. (Image was created with BioRender.com.)

for abiotic and biotic stress tolerance can be identified. These genes can be edited using CRISPR and transformed in the crop plants using *Agrobacterium*-mediated transformation to generate climate resilient plants.

11.5.1 Drought Tolerance

With the increasing impact of global warming, intensified water evaporation contributes to water shortage and drought stress in plants. Drought has become one of the major constraints for crop production all over the world as it adversely affects plant morphology and biochemistry. Drought typically prevents plants from absorbing nutrients and hence they exhibit reduced plant height, withering leaves, and disturbance in buds during the plant developmental phases. As depicted in Table 11.3 CRISPR-Cas9 has been employed in different plants to modulate several structural genes and transcriptional regulators with the aim of increasing agricultural output (Erdoğan et al., 2023; Kumar et al., 2023; Rai et al., 2023). Specifically, studies in maize, tomato, rice, and wheat elucidated many transcription factors required for drought tolerance as indicated by knock-in or knockout mechanisms of their manipulations. Overexpression of several drought-responsive genes and transcription factors enhanced

TABLE 11.3
Applications of CRISPRised plants for improvement of drought stress conditions

Plant species	Target gene	Gene function	Trait	Method of delivery	Reference
Maize	*ARGOS8*	Transcription factor	Enhanced grain yield in field (under stress conditions)	Biolistic gene transfer	J. Shi et al. (2017)
	ZmGA20ox3	GA biosynthesis enzyme	Improved plant height and enhancing drought resistance	Agrobacterium-mediated	Liu et al. (2023)
Tomato	*SlNPR1*	Transcriptional coactivator	Reduced drought tolerance in the mutants	Agrobacterium-mediated	R. Li et al. (2019)
	SlLBD40	Transcription factor	Drought tolerance in the knockout lines	Agrobacterium-mediated	Lun Liu et al. (2020)
	SlMAPK3	Signal transduction	Reduced drought tolerance in the mutants in greenhouse conditions	Agrobacterium-mediated	L. Wang et al. (2017)
Rice	*SAPK2*	Signal transduction	Mutants more sensitive to drought stress	Agrobacterium-mediated	Lou et al. (2017)
	OsSRL1, OsSRL2	Transcription factor	Drought tolerance (higher grain filling under stress)	Agrobacterium-mediated	Liao et al. (2019)
	OsPYL9	Transcription factor	Higher yield under normal and drought conditions (in growth chamber)	Agrobacterium-mediated	Usman et al. (2020)
	OsDSL	Transcription factor	High tolerance to NaCl moderate tolerance to osmotic stress at seedling stage	Agrobacterium-mediated	Santosh et al. (2020)
	OsRR22	Transcriptional regulator/Signal	Salinity tolerance at seedling stage	Agrobacterium-mediated	A. Zhang et al. (2019)
	OsERA1	Transcriptional regulator/Signal	Enhanced response to drought stress	Agrobacterium-mediated	Ogata et al. (2020)
	OsOLP1	Pathogenesis-related protein 5 (PR5)	enhanced drought tolerance	Agrobacterium-mediated	Yan et al. (2023)
	OsPUB7	Plant U-box (PUB) family members	Enhanced Drought and salinity tolerance	Agrobacterium-mediated	Kim et al. (2023)
Wheat	*TaDREB2*	Transcription factor	Enhanced drought tolerance	Protoplast transformation	Kim. et al. (2018)
	TaERF3	Transcription factor	Enhanced drought tolerance	Protoplast transformation	Kim et al. (2018)
Sugar-cane	*ScNLTP*	Structural gene	Alteration of MeJA-induced pathways	Agrobacterium-mediated	Chen et al. (2017)
Soybean	*GmHdz4*	Transcription factor	Higher drought tolerance	Agrobacterium-mediated	Zhong et al. (2022)
Strawberry	*FvICE1*	Transcription factor	Enhanced drought tolerance	Agrobacterium-mediated	Han et al. (2023)

drought tolerance in plants by accumulating signaling molecules and metabolic compounds (Li et al., 2019; Liu et al., 2020; Liu et al., 2023).Tthe expressions of (S) genes enhance susceptibility in plants to drought through hormonal disproportion, declined antioxidant activities, and increasing reactive oxygen species (ROS) production (Erdoğan et al., 2023; Rai et al., 2023).

11.5.2 Cold Tolerance

Low temperature is another abiotic stress that has detrimental effects on plant growth and productivity. Tolerance to cold stress has been linked to different metabolic pathways. In rice, cold stress is known to affect germination rate, growth of seedlings, and grain yield. Annexin gene, which encodes a calcium-dependent phospholipid binding protein, has been identified as a contributor of cold stress resistance using CRISPR-Cas9 editing. Additionally, transcription factors like OsMYB30 have been identified as a negative regulator of cold tolerance in rice (Nawaz et al., 2019; Zeng et al., 2020; Han et al., 2023). In other plants like tomato, strawberry, and Arabidopsis, transcription factors and structural genes have been identified as important contributors to cold tolerance

TABLE 11.4
Applications of CRISPRised plants for improvement of cold stress conditions

Plant species	Target gene	Gene function	Trait	Method of delivery	Reference
Arabidopsis	*AtCBF1, AtCBF2*	Cystein like proteases	Cold tolerance	Agrobacterium-mediated	Chen et al. (2010)
	UGT79B2, UGT79B3	UDP-glycosyltransferases (knock out)	Decreased cold tolerance	Agrobacterium-mediated	Li et al. (2017)
Rice	*OsAnn3, OsAnn5*	annexin (knockout)	Decreased cold tolerane	Agrobacterium-mediated	Shen et al. (2017)
	OsPIN5b, GS3, OsMYB30	Structural genes	Enhanced cold tolerance	Agrobacterium-mediated	Zeng et al. (2020)
	OsPRP1	Structural genes	Decreased cold tolerance	Agrobacterium-mediated	Nawaz et al. (2019)
Tomato	*SlCBF1*	C-repeat binding factors (CBFs)	Decreased cold tolerance	Agrobacterium-mediated	Li, Yue. et al. (2018)
Strawberry	*FvICE1*	Transcription factor	Enhanced cold tolerance	Agrobacterium-mediated	Han et al. (2023)

TABLE 11.5
Applications of CRISPRised plants for improvement of heat stress conditions

Plant species	Target gene	Gene function	Trait	Method of delivery	Reference
Gossypium hirsutum	*GhPGF, GhCLA1*	Structural gene	Heat tolerance	Agrobacterium-mediated	Hu et al. (2021)
Lactuca sativa	*LsNCED4*	Structural gene	Enhanced Heat tolerance	Agrobacterium mediated	Bertier et al. (2018)
Rice	*OsPDS*	Structural gene	Heat tolerance	Gene gun	Nandy et al. (2019)
	OsHSA1	Structural gene	Heat tolerance	Agrobacterium-mediated	Qiu et al. (2018)
	OsNAC006	Transcription factor	Heat tolerance	Agrobacterium-mediated	Yang et al. (2020)
	OsPYL1/4/6		Heat tolerance	Agrobacterium-mediated	Miao et al. (2013)
S. lycopersicum	*SlAGL6*	Structural gene	Heat tolerance	Agrobacterium-mediated	Klap et al. (2017)
	SlCPK28	Protein kinase	Heat tolerance	Agrobacterium-mediated	Hu. et al. (2021)
	SlMAPK3	Protein kinase	Heat tolerance	Agrobacterium-mediated	Yu et al. (2019)
	SlBZR1	Transcription factor	Heat tolerance	Agrobacterium-mediated	Yin et al. (2018)
Zea mays	*ZmTMS5*	Structural gene	Heat tolerance	Particle bombardment	P.Li et al. (2017)
Soybean	*GmHsp90A2*	Heat shock protein	Heat tolerance	Agrobacterium-mediated	Y. Huang. et al. (2019)

(Erdoğan et al., 2023; Kumar et al., 2023) Table 11.4 presents the effects of cold tolerance.

11.5.3 Heat Tolerance

High temperature is a crucial abiotic factor that has affected crop plants' productivity worldwide. Plants react to heat stress by activating complex molecular networks that include heat stress-responsive gene expression, signal transduction, and metabolite production. As shown in Table 11.5, in studies on plants like tomato, rice, maize, and soybean, various heat stress-associated genes that have been modified using CRISPR-Cas9-based genome editing that have enhanced heat stress tolerance in them. Many of them code for heat shock proteins (HSPs) and heat shock transcription factors (HSFs) (Nandy et al., 2019; Wang et al., 2020; Li et al., 2021).

11.5.4 Salt Tolerance

Salinity causes various morphological and physiological changes in plants that are governed by the expression of genes and signaling pathways. Key harmful effects of salinity stress are ionic disturbances and necrosis. The CRISPR-Cas gene-editing system has been employed to enhance salt stress tolerance in various plants targeting several genes. Table 11.6 depicts that in rice, certain genes identified as positive regulators of salt stress (e.g., *OsBBS1*, *OsMIR528*, *OsRAV2*) improved salt tolerance upon CRISPR-Cas9-based genome editing (Duan et al., 2016; Zhao et al., 2017; Bouzroud et al., 2020). However, genes like *SnRK2*, *SAPK-1* and *SAPK-2* when edited using loss of function mutations with CRISPR-Cas9 resulted in increased salinity resistance (Duan et al., 2016). Similar studies in plants like tomato, Arabidopsis, maize, and soybean confirmed many transcription factors,

TABLE 11.6
Applications of CRISPRised plants for improvement of salinity stress conditions

Plant species	Target gene	Gene function	Trait	Method of delivery	Reference
O. sativa	*OsSAPK1, OsSAPK2*	Transcription Factor	Salt stress tolerance	Agrobacterium-mediated	Bouzroud et al. (2020)
	OsBBS1	receptor-like kinase	Salt stress tolerance	Agrobacterium-mediated	Zeng et al. (2018)
	OsMIR408, OsMIR528	Microsatellites	Salt stress tolerance	Agrobacterium-mediated	Zhao et al. (2017)
	OsRAV2	Transcription Factor	Salt stress tolerance	Agrobacterium-mediated	Duan et al. (2016)
	OsGTg-2 (Loss of function)	Transcription Factor	Salt stress tolerance	Agrobacterium-mediated	X. Liu. et al. (2020)
	PIL14	Phytochrome-Interacting Factor	Salt stress tolerance	Agrobacterium-mediated	Weiping et al. (2020)
	OstPQT3	E3 ubiquitin ligase	Salt stress tolerance	Agrobacterium-mediated	Alfatih et al. (2020)
	OsAGO2	RNA induced silencing complex proteins	Salt stress tolerance	Agrobacterium-mediated	Yin et al. (2020)
	OsDST	Transcription Factor	Salt stress tolerance	Agrobacterium-mediated	Santosh. et al. (2020)
	FLN2	fructokinase-like protein2	Salt stress tolerance	Agrobacterium-mediated	Chen et al. (2020)
	OsRR9 and *OsRR10* (Loss of function)	Transcription factor	Salt stress tolerance	Agrobacterium-mediated	C. Zhang. et al. (2019)
	OsDOF15	Transcription factor	Salt stress tolerance	Agrobacterium-mediated	Qin et al. (2019)
	OsSPL10	Transcription factor	Salt stress tolerance	Agrobacterium-mediated	Tao et al. (2019)
	NCA1a, NCA1b	chaperone protein	Salt stress tolerance	Agrobacterium-mediated	Liu et al. (2019)
	RR22	Transcription factor	Salt stress tolerance	Agrobacterium-mediated	C. Zhang et al. (2019)
	OsNAC041	Transcription factor	Salt stress tolerance	Agrobacterium-mediated	Bo et al. (2019)
	OsOTS1 (Loss of function)	Transcription factor	Salt stress tolerance	Agrobacterium-mediated	
Z. mays	*HKT1*	High-affinity potassium transporter	Salt stress tolerance	Agrobacterium-mediated	M. Zhang, Cao et al. (2018)
	miR408	miRNA	Salt stress tolerance	Agrobacterium-mediated	Ruidong et al. (2023)
A. thaliana	*AtSAUR41*	Auxin response gene	Salt stress tolerance	Agrobacterium-mediated	Ting et al. (2020)
Cucurbita moschata	*RBOHD*	Structural gene	Salt stress tolerance	Agrobacterium-mediated	Huang et al. (2019)
A. thaliana	*AtC/VIF1* (Loss of function)	Transcription factor	Salt stress tolerance	Agrobacterium-mediated	Weiping et al. (2020)
Hordeum vulgare	*HvITPK5/6*	Inositol trisphosphate 5/6 kinases	Salt stress tolerance	Agrobacterium-mediated	Vlčko et al. (2020)
G. max L	*GmAITR*	Transcription factor	Salt stress tolerance	Agrobacterium-mediated	Y. Wang et al. (2021)
S. lycopersicum L	*SlARF4* (Loss of function)	Auxin response factor	Salt stress tolerance	Agrobacterium-mediated	Bouzroud et al. (2020)
A. thaliana	*ArathEULS3*	Transcription factor	Salt stress tolerance	Agrobacterium-mediated	Dubiel et al. (2020)
	AtWRKY66	Transcription factor	Salt stress tolerance	Agrobacterium-mediated	M. Zhang et al. (2023)
Rice	*OsRR22* (Loss of function)	Transcription factor	Salt stress tolerance	Agrobacterium-mediated	Sheng et al. (2023)

TABLE 11.6 (Continued)
Applications of CRISPRised plants for improvement of salinity stress conditions

Plant species	Target gene	Gene function	Trait	Method of delivery	Reference
	OsPIP1	Aquaporins	Salt stress tolerance	Agrobacterium-mediated	Tao et al (2023)
	HyPRP1	Hybrid proline-rich proteins	Salt stress tolerance	Agrobacterium-mediated	Tran et al. (2023)
	14-3-3 proteins	Regulatory molecules	Salt stress tolerance	Agrobacterium-mediated	Wang et al. (2023)
	OsWRKY28	Transcription factor	Salt stress tolerance	Agrobacterium-mediated	Y. Zhang et al. (2023)
	OsPMS1	Mismatch repair gene	Salt stress tolerance	Agrobacterium-mediated	J. Li et al. (2023)
	ATS1 (loss of function)	Glycerolipids biosynthetic pathway enzyme	Salt stress tolerance	Agrobacterium-mediated	Y. Liu et al. (2023)
	qSOR1	Quantitative trait locus	Salt stress tolerance	Agrobacterium-mediated	Kitomi et al. (2020)
	osxlg1/osxlg4 (double mutants)	GTP-binding protein	Salt stress tolerance	Agrobacterium-mediated	Biswal et al. (2022)
	gs3, and dep1	The heterotrimeric G protein complex	Salt stress tolerance	Agrobacterium-mediated	Cui et al. (2020)
	OsbHLH024	Transcription factor	Salt stress tolerance	Agrobacterium-mediated	Alam et al. (2022)
	Osbhlh044 (loss of function)	Transcription factor	Salt stress tolerance	Agrobacterium-mediated	Alam et al. (2022)
	SERK2	Transcription factor	Salt stress tolerance	Agrobacterium-mediated	Dong et al. (2020)
	OsVDE	Enzyme in chloroplasts	Salt stress tolerance	Agrobacterium-mediated	Y. Wang et al. (2021)
	OsBadh2	Enzyme in aroma compound synthesis	Salt stress tolerance	Agrobacterium-mediated	Prodhan et al. (2022)
	OsCaM1	Calmodulin gene	Salt stress tolerance	Agrobacterium-mediated	Yang et al. (2021)
	OsGLYI3	Aglyoxalase gene	Salt stress tolerance	Agrobacterium-mediated	C-X. Liu et al (2022)
	OsMPT3	Transporters	Salt stress tolerance	Agrobacterium-mediated	Shengcai et al. (2020)
	BEAR1	Transcription factor	Salt stress tolerance	Agrobacterium-mediated	Teng et al. (2022)
Z. mays	*ZmSTG1*	Transporter	Salt stress tolerance	Agrobacterium-mediated	Mei et al. (2023)

structural genes and transporters as crucial determinants of plants' ability to manage salt stress (Bouzroud et al., 2020; X Liu. et al., 2020; Q Liu. et al., 2020).

11.5.5 Metal Tolerance

Heavy metal exposure is another important stress that adversely affects crop productivity. Accumulation of metal ions not only disturbs ion homeostasis but also creates oxidative stress, culminating in cellular injuries. Plants have developed different detoxification mechanisms to counteract these deleterious effects and various genes direct these mechanisms. The CRISPR-Cas gene-editing system has been successfully used to amend key genes in rice, *Arabidopsis*, tomato, and soybean (Zhang et al., 2018; Baeg et al., 2021; Li et al., 2022; Chen et al., 2023; Yue et al., 2023) (Table 11.7). Increased tolerance to metals like arsenic, cadmium, caesium, copper, etc. has been seen by editing genes that code for transporters, transcription factors, and detoxifying enzymes, etc. (Kumar et al., 2023).

11.6 CONCLUSION

The emergence of the CRISPR-Cas9 system to propel beneficial traits in crops has revitalized the field of plant biotechnology. This genome-editing tool has the potential to modify the

TABLE 11.7
Applications of CRISPRised plants for improvement of metal stress conditions

Plant species	Target gene	Gene function	Trait	Method of delivery	Reference
A. thaliana	Atoxp1	Enzyme in glutathione recycling pathway	Resistance to sulfonamide compounds	Agrobacterium-mediated	Baeg et al. (2021)
O. sativa	OsARM1	f	Response to arsenic stress	Agrobacterium-mediated	F. Wang et al. (2017)
	OsNramp5	Cd transporter	Cadmium uptake and transport	Agrobacterium-mediated	Tang et al. (2017)
	OsLCT1	Cd transporter	Cadmium uptake and transport	Agrobacterium-mediated	Lu et al. (2017)
	OsHAK1	Cs transporter	Caesium uptake	Agrobacterium-mediated	Nieves-Cordones et al. (2017)
	OsPRX2	Thiol-based peroxidases	Potassium deficiency tolerance	Agrobacterium-mediated	Xiaohui et al. (2018)
	OsATX	Cu chaperone	Copper transport and distribution	Agrobacterium-mediated	Zhang et al. (2018)
	OsPMEI12	Pectin methylesterase inhibitors	Resistance to cadmium stress	Agrobacterium-mediated	Z. Li et al. (2022)
	OsLCD	Cd distribution	Resistance to cadmium stress	Agrobacterium-mediated	Chen et al. (2023)
	OsmiR535	MicroRNAs (miRNAs)	Resistance to cadmium stress	Agrobacterium-mediated	Yue et al. (2023)
	ABCG36	The ABC transporter	Resistance to cadmium stress	Agrobacterium-mediated	Fu et al. (2019)
Soybean	*ALS1*	Key enzyme in branched chain amino acid biosynthesis	Resistant to chlorsulfuron	Agrobacterium-mediated	Li et al. (2015)
Tomato	*ALS*	Key enzyme in branched chain amino acid biosynthesis	Resistance to chlorsulfuron	Agrobacterium-mediated	Veillet et al. (2019)
	Sl1	E3 Ubiquitin Ligase	Resistance to cadmium stress	Agrobacterium-mediated	C-X. Liu. et al. (2022)

target genes to improve quality, yield, and provide resistance to manage biotic as well as abiotic stresses. In the near future, there may be more ways to take advantage of CRISPR-Cas9. Given the rapid advancement of science in the current day, this is something to be expected. It is anticipated that agricultural and nutritional demands will be satisfied through crop improvement, contributing to an overall improvement in the standard of living on a worldwide scale. By introducing minor alterations like deletions, insertions, and targeted mutations using CRISPR-Cas, GE plants change the characteristics of their plants. Their agronomic qualities have significantly improved as a result of these GE plants. The CRISPR-Cas9-based genome-editing technology has developed into a potent and popular genetic tool. PE has recently expanded the capabilities of genome editing. It is anticipated that PE will have a significant impact on fundamental research, precision molecular breeding, and disease treatments as this technology continues to be optimized and updated in future years. The PE system still has many shortcomings even with its advancements. First, there is still work to be done to increase PE's editing efficiency. When compared to the original PE system, the PE system's efficiency has significantly been boosted through ongoing optimization. It's important to note that the use of CRISPR-Cas technology in agriculture, especially in genetically modified organisms (GMOs), raises ethical, regulatory, and environmental concerns. Regulations and public perception vary by region, so the adoption of these technologies depends on local policies and societal acceptance. Additionally, thorough risk assessments are necessary to ensure that unintended consequences are minimized and ecological balance is maintained.

REFERENCES

Alam, M.S., Kong, J., Tao, R., Ahmed, T., Alamin, M., Alotaibi, S.S., Abdelsalam, N.R., Xu, J. H. (2022). CRISPR/Cas9 mediated knockout of the *OsbHLH024* transcription factor improves salt stress resistance in rice (*Oryza sativa* L.). *Plants (Basel)* 11(9), 1184. doi: 10.3390/plants11091184.

Alam, M.S., Yang, Z.K., Li, C., Yan, Y., Liu, Z., Nazir, M.M., Xu, J.H. (2022). Loss-of-function mutations of OsbHLH044 transcription factor lead to salinity sensitivity and a greater chalkiness in rice (*Oryza sativa* L.). *Plant Physiology and Biochemistry*, 193, 110–123. doi: 10.1016/j.plaphy.2022.10.033.

Alfatih, A., Wu, J., Jan, S.U., Zhang, Z.-S., Xia, J.-Q., Xiang, C.-B. (2020). Loss of rice *PARAQUAT TOLERANCE 3* confers enhanced resistance to abiotic stresses and increases grain yield in field. *Plant Cell Environment*, 43, 2743–2754. https://doi.org/10.1111/pce.13856.

Anzalone, A.V., Gao, X.D., Podracky, C.J., Nelson, A.T., Koblan, L.W., Raguram, A., Levy, J.M., Mercer, J.A.M., Liu, D.R. (2022). Programmable deletion, replacement, integration and inversion of large DNA sequences with twin prime editing. *Nature Biotechnology*, 40(5), 731–740. https://doi.org/10.1038/s41587-021-01133-w.

Anzalone, A.V., Koblan, L., Liu, D.R. (2020). Genome editing with CRISPR–Cas nucleases, base editors, transposases and prime editors. *Nature Biotechnology*, 38, 824–844. https://doi.org/10.1038/s41587-020-0561-9.

Anzalone, A.V., Randolph, P.B., Davis, J.R., Sousa, A.A., Koblan, L.W., Levy, J.M., Chen, P.J., Wilson, C., Newby, G.A., Raguram, A., Liu, D.R. (2019). Search-and-replace genome editing without double-strand breaks or donor DNA. *Nature*, 576(7785), 149–157. https://doi.org/10.1038/s41586-019-1711-4.

Baeg, G.J., Kim, S.H., Choi, D.M. (2021). CRISPR/Cas9-mediated mutation of 5-oxoprolinase gene confers resistance to sulfonamide compounds in *Arabidopsis*. *Plant Biotechnology Reports*, 15, 753–764. https://doi.org/10.1007/s11816-021-00718-w.

Bertier, L.D., Ron, M., Huo, H., Bradford, K.J., Britt, A.B., Michelmore, R.W. (2018). High-resolution analysis of the efficiency, heritability, and editing outcomes of CRISPR/Cas9-induced modifications of *NCED4* in Lettuce (*Lactuca sativa*). *G3 (Bethesda)*, 8(5), 1513–1521. doi: 10.1534/g3.117.300396.

Biswal, A.K., Wu, T.Y., Urano, D., Pelissier, R., Morel, J.B., Jones, A.M., Biswal, A.K. (2022). Novel mutant alleles reveal a role of the Extra-Large G protein in rice grain filling, panicle architecture, plant growth, and disease resistance. *Frontiers in Plant Science*, 3(12), 782960. https://doi.org/10.3389/fpls.2021.782960

Bo, W., Zhaohui, Z., Huanhuan, Z., Xia, W., Binglin, L., Lijia, Y., Xiangyan, H., Deshui, Y., Xuelian, Z., Chunguo, W., Wenqin, S., Chengbin, C., Yong Z. (2019). Targeted mutagenesis of NAC transcription factor gene, OsNAC041, leading to salt sensitivity in rice. *Rice Science*, 26, 98–108. https://doi.org/10.1016/j.rsci.2018.12.005.

Bouzroud, S., Gasparini, K., Hu, G., Barbosa, M.A.M., Rosa, B.L., Fahr, M., Bendaou, N., Bouzayen, M., Zsögön, A., Smouni, A., Zouine, M. (2020). Down regulation and loss of Auxin Response Factor 4 function using CRISPR/Cas9 alters plant growth, stomatal function and improves tomato tolerance to salinity and osmotic stress. *Genes (Basel)*, 11(3), 272. doi: 10.3390/genes11030272.

Butt, H., Rao, G.S., Sedeek, K., Aman, R., Kamel, R., Mahfouz, M. (2020). Engineering herbicide resistance via prime editing in rice. *Plant Biotechnology Journal*, 18, 2370–2372.

Cao, Y., Zhang, Y., Chen, Y., Yu, N., Liaqat, S., Wu, W., Chen, D., Cheng, S., Wei, X., Cao, L., Zhang, Y., Liu, Q. (2021). OsPG1 encodes a Polygalacturonase that determines cell wall architecture and affects resistance to bacterial blight pathogen in rice. *Rice (N Y)*, 14(1), 36. doi: 10.1186/s12284-021-00478-9.

Chen, E. , Lin-Shiao, E. , Trinidad, M. , Saffari Doost, M. , Colognori, D., Doudna, J.A. (2022). Decorating chromatin for enhanced genome editing using CRISPR-Cas9. *Proceedings of the National Academy of Sciences of the USA*, 119, e2204259119,

Chen, G., Hu, J., Dong, L., Zeng, D., Guo, L., Zhang, G., Zhu, L., Qian, Q. (2020). The tolerance of Salinity in rice requires the presence of a functional copy of *FLN2*. *Biomolecules*, 10(1), 17. https://doi.org/10.3390/biom10010017.

Chen, H., Ye, R., Liang, Y., Zhang, S., Liu, X., Sun, C., Li, F., Yi, J. (2023). Generation of low-cadmium rice germplasms via knockout of OsLCD using CRISPR/Cas9. *Journal of Environmental Sciences*, 126, 138–152. https://doi.org/10.1016/j.jes.2022.05.047.

Chen, H.J., Su, C.T., Lin, C.H., Huang, G.J., Lin, Y.H. (2010). Expression of sweet potato cysteine protease SPCP2 altered developmental characteristics and stress responses in transgenic Arabidopsis plants. *Journal of Plant Physiology*, 167(10), 838–847. doi: 10.1016/j.jplph.2010.01.005.

Chen, K., Wang, Y., Zhang, R., Zhang, H., Gao, C. (2019). CRISPR/Cas genome editing and precision plant breeding in agriculture. *Annual Review of Plant Biology*, 70, 667–697. https://doi.org/10.1146/annurev-arplant-050718-100049.

Chen, Y., Ma, J., Zhang, X. (2017). A novel non-specific Lipid Transfer Protein Gene from Sugarcane (*NsLTPs*), obviously responded to abiotic stresses and signaling molecules of SA and MeJA. *Sugar Technology*, 19, 17–25. https://doi.org/10.1007/s12355-016-0431-4.

Choi, J., Chen, W., Suiter, C.C., Lee, C., Chardon, F.M., Yang, W., Leith, A. Daza, R.M., Martin, B., Shendure, J. (2022). Precise genomic deletions using paired prime editing. *Nature Biotechnology*, 40, 218–226. doi: 10.1038/s41587-021-01025-z.

Cui, Y., Jiang, N., Xu, Z. (2020). Heterotrimeric G protein are involved in the regulation of multiple agronomic traits and stress tolerance in rice. *BMC Plant Biology*, 20, 90. https://doi.org/10.1186/s12870-020-2289-6.

Dong, L., Qi, X.,; Zhu, J., Liu, C., Zhang, X., Cheng, B., Mao, L. Xie, C. (2019). Supersweet and waxy: Meeting the diverse demands for specialty maize by genome editing. *Plant Biotechnology Journal*, 17, 1853–1855. https://doi.org/10.1111/pbi.13144.

Dong, N., Yin, W., Liu, D., Zhang, X., Yu, Z., Huang, W., Liu, J., Yang, Y., Meng, W., Niu, M., Tong, H. (2020). Regulation of Brassinosteroid signaling and salt resistance by SERK2 and potential utilization for crop improvement in rice. *Frontiers in Plant Science*, 11, 621859. doi: 10.3389/fpls.2020.621859.

Duan, Y.B, Li, J., Qin, R.Y., Xu, R.F., Li, H., Yang, Y.C., Ma, H., Li, L., Wei, P.C., Yang, J.B. (2016). Jan identification of a regulatory element responsible for salt induction of rice OsRAV2 through ex situ and in situ promoter analysis. *Plant Molecular Biology*, 90(1–2), 49–62. doi: 10.1007/s11103-015-0393-z.

Dubiel, M., Beeckman, T., Smagghe, G., Van Damme, E.J.M. (2020). Arabidopsis Lectin EULS3 is involved in ABA signaling in roots. *Frontiers in Plant Science*. doi: 10.3389/fpls.2020.00437.

Erdoğan, İ., Cevher-Keskin, B., Bilir, Ö., Hong, Y., Tör, M. (2023). Recent developments in CRISPR/Cas9 genome-editing technology related to plant disease resistance and abiotic stress tolerance. *Biology (Basel)*, 12(7), 1037. doi: 10.3390/biology12071037.

Fu, S., Lu, Y., Zhang, X., Yang, G., Chao, D., Wang, Z., Shi, M., Chen, J., Chao, D.-Y., Li, R., Ma, J.F. Xia, J. (2019). The ABC transporter ABCG36 is required for cadmium tolerance in rice. *Journal of Experimental Botany*, 70(20), 5909–5918. https://doi.org/10.1093/jxb/erz335.

Gao, H., Gadlage, M.J., Lafitte, H.R., Lenderts, B., Yang, M., Schroder, M., Farrell, J., Snopek, K., Peterson, D., Feigenbutz,

L. (2020). Superior field performance of waxy corn engineered using CRISPR-Cas9. *Nature Biotechnology*, 38, 579–581. https://doi.org/10.1038/s41587-020-0444-0.

Guo, M., Zhang, X., Liu, J., Hou, L., Liu, H., Zhao, X. (2020). OsProDH Negatively regulates thermotolerance in rice by modulating proline metabolism and reactive oxygen species scavenging. *Rice (N Y)*, 13(1), 61. doi: 10.1186/s12284-020-00422-3. P

Han, J., Li, X., Li, W., Yang, Q., Li, Z., Cheng, Z., Lv, L., Zhang, L., Han D. (2023). Isolation and preliminary functional analysis of FvICE1, involved in cold and drought tolerance in *Fragaria vesca* through overexpression and CRISPR/Cas9 technologies. *Plant Physiology and Biochemistry*, 196, 270–280. https://doi.org/10.1016/j.plaphy.01.048.

Hernandes-Lopes, J., Yassitepe, J.E.C.T., Koltun, A., Pauwels, L., Silva, V.C.H.D., Dante, R.A., Gerhardt, I.R., Arruda, P. (2023). Genome editing in maize: Toward improving complex traits in a global crop. *Genetic Molecular Biology*, 46, e20220217. doi: 10.1590/1678-4685-GMB-2022-0217.

Hillary, V.E., Ceasar, S.A. (2022). Prime editing in plants and mammalian cells: Mechanism, achievements, limitations, and future prospects. *BioEssays*, 44(9), 2200032. https://doi.org/10.1002/bies.202200032

Hu, J.H., Miller, S.M., Geurts, M.H., Tang, W., Chen, L., Sun, N., Zeina, C.M., Gao, X., Rees, H.A., Lin, Z., Liu, D.R. (2018). Evolved Cas9variants with broad PAM compatibility and high DNA specificity. *Nature*, 556(7699), 57–63. doi: 10.1038/nature26155.

Hu, Z., Li, J., Ding, S., Cheng, F., Li, X., Jiang, Y., Yu, J., Foyer, C.H., Shi, K. (2021). The protein kinase CPK28 phosphorylates ascorbate peroxidase and enhances thermotolerance in tomato. *Plant Physiology*, 186(2), 1302–1317. doi: 10.1093/plphys/kiab120.

Hua, K., Jiang, Y., Tao, X., Zhu, J.K. (2020). Precision genome engineering in rice using prime editing system. *Plant Biotechnology Journal*, 18, 2167–2169.

Hua, K., Tao, X., Liang, W., Zhang, Z., Gou, R., Zhu, J.-K. (2020). Simplified adenine base editors improve adenine base editing efficiency in rice. *Plant Biotechnology Journal*, 18, 770–778. doi: 10.1111/pbi.13244.

Huang, L., Tan, H., Zhang, C., Li, Q., Liu, Q. (2021). Starch biosynthesis in cereal endosperms: An updated review over the last decade. *Plant Communications*, 2, 100237. https://doi.org/10.1016/j.xplc.2021.100237.

Huang, S., Xin, S., Xie, G., Han, J., Liu, Z., Wang, B., Zhang, S., Wu, Q., Cheng, X. (2020). Mutagenesis reveals that the rice OsMPT3 gene is an important osmotic regulatory factor. *The Crop Journal*, 8(3), 465–479. https://doi.org/10.1016/j.cj.2020.02.001.

Huang, Y., Cao, H., Yang, L., Chen, C., Shabala, L., Xiong, M., Niu, M., Liu, J., Zheng, Z., Zhou, L., Peng, Z., Bie, Z., Shabala, S. (2019). Tissue-specific respiratory burst oxidase homolog-dependent H_2O_2 signaling to the plasma membrane H+-ATPase confers potassium uptake and salinity tolerance in *Cucurbitaceae. Journal of Experimental Botany*, 70(20), 5879–5893. https://doi.org/10.1093/jxb/erz328.

Huang, Y., Xuan, H., Yang, C., Guo, N., Wang, H., Zhao, J., Xing, H. (2019). GmHsp90A2 is involved in soybean heat stress as a positive regulator. *Plant Science*, 285, 26–33. doi: 10.1016/j.plantsci.2019.04.016.

Huang, Z., Liu, G. (2023). Current advancement in the applications of prime editing. *Frontier Bioengineering Biotechnology*, 11, 1039315. doi: 10.3389/fbioe.2023.1039315.

Hwang, G.H., Jeong, Y.K., Habib, O., Hong, S.A., Lim, K., Kim, J.S., Bae, S. (2021). PE-Designer and PE-Analyzer: Web-based designand analysis tools for CRISPR prime editing. *Nucleic Acids Research*, 49(W1), W499–W504. https://doi.org/10.1093/nar/gkab319.

Jain, M. (2015). Function genomics of abiotic stress tolerance in plants: A CRISPR approach. *Frontiers in Plant Science*, 6, 375. https://doi.org/10.3389/fpls.2015.00375.

Jiang, Y., Chai, Y., Qiao, D., Wang, J., Xin, C., Sun, W., Cao, Z., Zhang, Y., Zhou, Y., Wang, X.C., Chen, Q.J. (2022). Optimized prime editing efficiently generates glyphosate-resistant rice plants carrying homozygous TAP-IVS mutation in EPSPS. *Molecular Plant*, 15, 1646–1649.

Jiang, Y., Sun, K., An, X. (2022). CRISPR/Cas system: Applications and prospects for maize improvement. *ACS Agricultural Science & Technology*, 2, 174–183. http://dx.doi.org/10.1021/acsagscitech.1c00253.

Jiang, Y.-Y., Chai, Y.-P., Lu, M.-H., Han, X.-L., Lin, Q., Zhang, Y., Zhang, Q., Zhou, Y., Wang, X.-C., Gao, C. (2020). Prime editing efficiently generates W542L and S621I double mutations in two ALS genes in maize. *Genome Biology*, 21, 257. https://doi.org/10.1186/s13059-020-02170-5

Jin, S., Lin, Q., Luo, Y., Zhu, Z., Liu, G., Li, Y., Chen, K., Qiu, J.L., Gao, C. (2021). Genome-wide specificity of prime editors in plants. *Nature Biotechnology*, 39(10), 1292–1299. doi: 10.1038/s41587-021-00891-x.

Kausch, A.P., Nelson-Vasilchik, K., Tilelli, M., Hague, J.P. (2021a). Maize tissue culture, transformation, and genome editing. *In Vitro Cellular and Developmental Biology: Plant*, 57, 653–671. https://doi.org/10.1007/s11627-021-10196-y

Khare, P., Mandal, S., Verma, A.K., Parashar, D., Jadhav, K., Gupta, S. (2023). CRISPR: The Janus god of modern science. *CRISPR-Cas System in Translational Biotechnology*, 23–56. https://doi.org/10.1016/B978-0-323-91808-4.00006-7

Kim, D., Alptekin, B., Budak, H. (2018). CRISPR/Cas9 genome editing in wheat. *Functional & Integrative Genomics*, 1831–1841. https://doi.org/10.1007/s10142-017-0572-x.

Kim, M.-S., Ko, S.-R., Jung, Y.J., Kang, K.-K., Lee, Y.-J., Cho, Y.-G. (2023). Knockout mutants of *OsPUB7* generated using CRISPR/Cas9 revealed abiotic stress tolerance in rice. *International Journal of Molecular Science*, 24, 5338. https://doi.org/10.3390/ijms24065338.

Kitomi, Y., Hanzawa, E., Kuya, N., Inoue, H., Hara, N., Kawai, S., Uga, Y. (2020). Root angle modifications by the DRO1 homolog improve rice yields in saline paddy fields. *Proceedings of the National Academy of Sciences*, 117(35), 21242–21250. doi: 10.1073/pnas.2005911117.

Klap, C., Yeshayahou, E., Bolger, A.M., Arazi, T., Gupta, S.K., Shabtai, S., Usadel, B., Salts, Y., Barg, R. (2017). Tomato facultative parthenocarpy results from Sl*AGAMOUS-LIKE 6* loss of function. *Plant Biotechnology Journal*, 15, 634–647. https://doi.org/10.1111/pbi.12662.

Kleinstiver, B., Prew, M.S., Tsai, S.Q., Topkar, V.V., Nguyen, N.T., Zheng, Z., Gonzales, A.P.W., Li, Z., Peterson, R.T., Yeh, J.-R.J., et al. (2015). Engineered CRISPR-Cas9 nucleases with altered PAM specificities. *Nature*, 523, 481–485. https://doi.org/10.1038/nature14592.

Kong, X., Wang, F., Wang, Z., Gao, X., Geng, S., Deng, Z., Zhang, S., Fu, M., Cui, D., Liu, S., et al. (2023). Grain yield improvement by genome editing ofTaARF12 that decoupled peduncle and rachis development trajectories via differential regulation of gibberellin signalling in wheat. *Plant Biotechnology Journal*, 21(10), 1990–2001. https://doi.org/10.1111/pbi.14107.

Kumar, M.R., Prusty, R.M., Pandey, M., Singh, P.K., Bohra, A., Guo, P., Varshney, K.R. (2023). Application of CRISPR/Cas9-mediated gene editing for abiotic stress management in crop plants. *Frontiers in Plant Science*, 14. https://doi.org/10.3389/fpls.2023.1157678

Lan, T., Zheng, Y., Su, Z., Yu, S., Song, H., Zheng, X., Lin, G., Wu, W. (2019). *OsSPL10*, a SBP-Box gene, plays a dual role in salt tolerance and trichome formation in rice (*Oryza sativa* L.). *G3: Genes, Genomes, Genetics*, 9(12), 4107–4114, https://doi.org/10.1534/g3.119.400700.

Li, B., Liang, S., Alariqi, M., Wang, F., Wang, G., Wang, Q., Xu, Z., Yu, L., Naeem Zafar, M., Sun, L., Si, H., Yuan, D., Guo, W., Wang, Y., Lindsey, K., Zhang, X., Jin, S. (2021). The application of temperature sensitivity CRISPR/LbCpf1 (LbCas12a) mediated genome editing in allotetraploid cotton (*G. hirsutum*) and creation of non-transgenic, gossypol-free cotton. *Plant Biotechnology Journal*, 221–223. doi: 10.1111/pbi.13470.

Li, C., Yue, Y., Chen, H., Qi, W., Song, R. (2018). The ZmbZIP22 transcription factor regulates 27-kD γ-zein gene transcription during maize endosperm development. *Plant Cell*, 30, 2402–2424. https://doi.org/10.1105/tpc.18.00422

Li, H., Li, J., Chen, J., Yan, L., Xia, L. (2020). Precise modifications of both exogenous and endogenous genes in rice by prime editing. *Molecular Plant*, 13, 671–674. https://doi.org/10.1016/j.molp.2020.03.011

Li, J., Chen, L., Liang, J., Xu, R., Jiang, Y., Li, Y., Ding, J., Li, M., Qin, R., Wei, P. (2022). Development of a highly efficient prime editor 2 system in plants. *Genome Biology*, 23, 161.

Li, J., Zhang, C., He, Y., Li, S., Yan, L., Li, Y., Zhu, Z., Xia, L. (2023). Plant base editing and prime editing: The current status and future perspectives. *Journal of Integrative Plant Biology*, 65(2), 444–467. https://doi.org/10.1111/jipb.13425

Li, J., Zhang, H., Si, X., Tian, Y., Chen, K., Liu, J., Chen, H., Gao, C. (2017). Generation of thermosensitive male-sterile maize by targeted knockout of the ZmTMS5 gene. *Journal of Genetic Genomics*, 44(9), 465–468. doi: 10.1016/j.jgg.2017.02.002.

Li, M., Li, X., Zhou, Z., Wu, P., Fang, M., Pan, X., et al. (2016). Reassessment of the four yield-related genes Gn1a, DEP1, GS3, and IPA1 in rice using a CRISPR/Cas9 system. *Frontiers in Plant Science*, 7, 377. doi: 10.3389/fpls.2016.00377

Li, P., Li, Y.-J., Zhang, F.-J., Zhang, G.-Z., Jiang, X.-Y., Yu, H.-M., Hou, B.-K. (2017). The Arabidopsis UDP-glycosyltransferases UGT79B2 and UGT79B3, contribute to cold, salt and drought stress tolerance via modulating anthocyanin accumulation. *Plant Journal*, 89, 85–103. https://doi.org/10.1111/tpj.13324.

Li, R., Liu, C., Zhao, R., et al. (2019). CRISPR/Cas9-Mediated *SlNPR1* mutagenesis reduces tomato plant drought tolerance. *BMC Plant Biology*, 19, 38. https://doi.org/10.1186/s12870-018-1627-4.

Li, R., Zhang, L., Wang, L., Chen, L., Zhao, R., Sheng, J., Shen, L. (2018). Reduction of tomato-plant chilling tolerance by CRISPR-Cas9-mediated SlCBF1 mutagenesis. *Journal of Agricultural and Food Chemistry*, 66(34), 9042–9051. doi: 10.1021/acs.jafc.8b02177.

Li, S., Zhang, C., Li, J., Yan, L., Wang, N., Xia, L. (2021). Present and future prospects for wheat improvement through genome editing and advanced technologies. *Plant Communications*, 2(4), 100211. doi: 10.1016/j.xplc.2021.100211.

Li, W.-Q., Zheng, W.-J., Peng, Y., Shao, Y., Liu, C.-T., Li, J., Hu, Y.-Y., Zhao, B.-R., Mao, B.-G. (2023). OsPMS1 mutation enhances salt tolerance by suppressing ROS accumulation, maintaining Na^+/K^+ homeostasis, and promoting ABA biosynthesis. *Genes*, 14(8), 1621. https://doi.org/10.3390/genes14081621.

Li, Z., Liu, Z.-B., Xing, A., Moon, B.P., Koellhoffer, J.P., Huang, L., Ward, R.T., Clifton, E., Falco, S.C., Cigan, A.M. (2015). Cas9-Guide RNA directed genome editing in soybean. *Plant Physiology*, 169(2), 960–970. https://doi.org/10.1104/pp.15.00783.

Li, Z., Rao, M.J., Li, J., Wang, Y., Chen, P., Yu, H., Ma, C., Wang, L. (2022). CRISPR/Cas9 mutant rice *Ospmei12* involved in growth, cell wall development, and response to phytohormone and heavy metal stress. *International Journal of Molecular Sciences*, 23(24), 16082. https://doi.org/10.3390/ijms232416082.

Liang, Z., Wu, Y., Guo, Y., Wei, S. (2023). Addition of the T5 exonuclease increases the prime editing efficiency in plants. *Journal of Genetics and Genomics*, 50, 582–588. https://doi.org/10.1016/j.jgg.2023.03.008

Liao, S., Qin, X., Luo, L., Han, Y., Wang, X., Usman, B., Nawaz, G., Zhao, N., Liu, Y., Li, R. (2019). CRISPR/Cas9-induced mutagenesis of *Semi-Rolled Leaf1,2* confers curled leaf phenotype and drought tolerance by influencing protein expression patterns and ROS scavenging in rice (*Oryza sativa* L.). *Agronomy*, 9, 728. https://doi.org/10.3390/agronomy9110728

Lin, Q., Jin, S., Zong, Y., Yu, H., Zhu, Z., Liu, G., Kou, L., et al. (2021). High efficiency prime editing with optimized, paired pegRNAs in plants. *Nature Biotechnology*, 39, 923–927. https://doi.org/10.1038/s41587-021-00868-w.

Lin, Q., Zong, Y., Xue, C., Wang, S., Jin, S., Zhu, Z., Wang, Y., Anzalone, A.V., Raguram, A., Doman, J.L., Liu, D.R., Gao, C. (2020). Prime genome editing in rice and wheat. *Nature Biotechnology*, 38, 582–585. https://doi.org/10.1038/s41587-020-0455-x.

Liu, C.-X., Yang, T., Zhou, H., Ahammed, G.J., Qi, Z.-Y., Zhou, J. (2022). The E3 Ubiquitin Ligase Gene *Sl1* is critical for Cadmium tolerance in *Solanum lycopersicum* L. *Antioxidants*, 11(3), 456. https://doi.org/10.3390/antiox11030456.

Liu, J., Cui, L., Xie, Z., Zhang, Z., Liu, E., Peng, X. (2019). Two NCA1 isoforms interact with catalase in a mutually exclusive manner to redundantly regulate its activity in rice. *BMC Plant Biology*, 19(1), 105. doi: 10.1186/s12870-019-1707-0.

Liu, L., Zhang, J., Xu, J., Li, Y., Guo, L., Wang, Z., Zhang, X., Zhao, B., Guo, Y.-D., Zhang N. (2020). CRISPR/Cas9 targeted mutagenesis of SlLBD40, a lateral organ boundaries domain transcription factor, enhances drought tolerance in tomato, *Plant Science*, 301, 110–683. https://doi.org/10.1016/j.plantsci.2020.110683.

Liu, S., Liu, W., Lai, J., Liu, Q., Zhang, W., Chen, Z., Gao, J., Song, S., Liu, J., Xiao, Y. (2022). OsGLYI3, a glyoxalase gene expressed in rice seed, contributes to seed longevity and salt stress tolerance. *Plant Physiology and Biochemistry*, 183, 85–95. doi: 10.1016/j.plaphy.2022.04.028.

Liu, X., Liu, Y., Chen, Z., et al. (2023). Gene editing of ZmGA20ox3 improves plant architecture and drought tolerance in maize. PREPRINT (Version 1). 26 July. https://doi.org/10.21203/rs.3.rs-3176812/v1.

Liu, X., Wu, D., Shan, T. (2020). The trihelix transcription factor *OsGTγ-2* is involved adaption to salt stress in rice. *Plant Molecular Biology*, 103, 545–560. https://doi.org/10.1007/s11103-020-01010-1.

Liu, Y., Wu, G., Ke, X., Zheng, Z., Zheng, Y. (2023). Loss-of-function of *ATS1* enhances Arabidopsis salt tolerance. *Plants*, 12(14), 2646. https://doi.org/10.3390/plants12142646.

Lou, D., Wang, H., Liang, G., Yu, D. (2017). OsSAPK2 confers Abscisic acid sensitivity and tolerance to drought stress in rice. *Frontiers in Plant Science*, 8. https://www.frontiersin.org/articles/10.3389/fpls.2017.00993.

Lu, C., Kuang, J., Shao, T., Xie, S., Li, M., Zhu, L., Zhu, L. (2022). Prime editing: An all-rounder for genome editing. *International Journal of Molecular Sciences*, 23(17), 9862. https://doi.org/10.3390/ijms23179862.

Lu, H.P., Liu, S.M., Xu, S.L., Chen, W.Y., Zhou, X., Tan, Y.Y., Huang, J.Z., Shu, Q.Y. (2017). CRISPR-S: An active interference element for a rapid and inexpensive selection of genome-edited, transgene-free rice plants. *Plant Biotechnology Journal*, 15(11), 1371–1373. doi: 10.1111/pbi.12788.

Lu, Y., Tian, Y., Shen, R., Yao, Q., Zhong, D., Zhang, X., Zhu, J.K. (2021). Precise genome modification in tomato using an improved prime editing system. *Plant Biotechnology Journal*, 19, 415–417.

Mandal, S., Verma, A.K. (2021). Wheat breeding, fertilizers, and pesticides: Do they contribute to the increasing immunogenic properties of modern wheat? *Gastrointestinal Disorders*, 3(4), 247–264. https://doi.org/10.3390/gidisord3040023

Mao, X., Zheng, Y., Xiao, K., Wei, Y., Zhu, Y., Cai, Q., Chen, L., Xie, H., Zhang, J. (2018). OsPRX2 contributes to stomatal closure and improves potassium deficiency tolerance in rice. *Biochemical and Biophysical Research Communications*, 495(1), 461–467. https://doi.org/10.1016/j.bbrc.2017.11.045.

Matsoukas, I.G. (2020). Prime editing: Genome editing for rare genetic diseases without double-strand breaks or donor DNA. *Frontiers in Genetics*, 11, 527211. https://doi.org/10.3389/fgene.2020.00528.

Mei, X., Zhao, Z., Bai, Y., Yang, Q., Gan, Y., Wang, W., Cai, Y. (2023). Salt tolerant Gene 1 contributes to salt tolerance by maintaining photosystem II activity in maize. *Plant, Cell & Environment*, 46(6), 1833–1848. https://doi.org/10.1111/pce.14578.

Miao, J., Guo, D., Zhang, J., Huang, Q., Qin, G., Zhang, X., Wan, J., Gu, H., Qu, L.J. (2013). Targeted mutagenesis in rice using CRISPR-Cas system. *Cell Research*, 1233–1236. doi: 10.1038/cr.2013.123.

Miller, S.M., Wang, T., Randolph, P.B., Arbab, M., Shen, M.W., Huang, T.P., Matuszek, Z., Newby, G.A., Rees, H.A., Liu, D.R. (2020). Continuous evolution of SpCas9 variants compatible with non-G PAMs. *Nature Biotechnology*, 38(4), 471–481. doi: 10.1038/s41587-020-0412-8.

Mo, W., Tang, W., Du, Y., Jing, Y., Bu, Q., Lin, R. (2020). Phytochrome-interacting factor-like14 and slender rice1: Interaction controls seedling growth under salt stress. *Plant Physiology*, 184(1), 506–517. https://doi.org/10.1104/pp.20.00024.

Nandy, S, Pathak, B, Zhao, S, Srivastava, V. (2019). Heat-shock-inducible CRISPR/Cas9 system generates heritable mutations in rice. *Plant Direct*, 3, 1–14. https://doi.org/10.1002/pld3.145.

Nawaz, G., Han, Y., Usman, B., Liu, F., Qin, B., Li, R. (2019). Knockout of OsPRP1, a gene encoding proline-rich protein, confers enhanced cold sensitivity in rice (*Oryza sativa* L.) at the seedling stage 3. *Biotech*, 254. doi: 10.1007/s13205-019-1787-4.

Nawaz, M., Sun, J., Shabbir, S., Khattak, W. A., Ren, G., Nie, X., Bo, Y., Javed, Q., Du, D., Sonne, C. (2023). A review of plants strategies to resist biotic and abiotic environmental stressors. *Science of the Total Environment*, 900, 165832. https://doi.org/10.1016/j.scitotenv.2023.165832

Ni, P., Zhao, Y., Zhou, X., Liu, Z., Huang, Z., Ni, Z., Sun, Q., Zong, Y. (2023). Efficient and versatile multiplex prime editing in hexaploid wheat. *Genome Biology*, 24, 156. https://doi.org/10.1186/s13059-023-02990-1.

Nieves-Cordones, M., Mohamed, S., Tanoi, K., Kobayashi, N.I., Takagi, K., Vernet, A., Guiderdoni, E., Périn, C., Sentenac, H., Véry, A.-A. (2017). Production of low-Cs+ rice plants by inactivation of the K+ transporter OsHAK1 with the CRISPR-Cas system. *Plant Journal*, 92, 43–56. https://doi.org/10.1111/tpj.13632.

Ning, Q., Jian, Y., Du, Y., Li, Y., Shen, X., Jia, H., Zhao, R., Zhan, J., Yang, F., Jackson, D., Liu, L., Zhang, Z. (2021). An ethylene biosynthesis enzyme controls quantitative variation in maize ear length and kernel yield. *Nature Communications*, 12, 1–10. doi: 10.1038/s41467-021-26123-z.

Nishimasu, H., Shi, X., Ishiguro, S., Gao, L., Hirano, S., Okazaki, S., Noda, T., Abudayyeh, O.O., Gootenberg, J.S., Mori, H., Oura, S., Holmes, B., Tanaka, M., Seki, M., Hirano, H., Aburatani, H., Ishitani, R., Ikawa, M., … Nureki, O. (2018). Engineered CRISPR-Cas9 nuclease with expanded targeting space. *Science*, 361(6408), 1259–1262. doi: 10.1126/science.aas9129.

Ogata, T., Ishizaki, T., Fujita, M., Fujita, Y. (2020). CRISPR/Cas9-targeted mutagenesis of *OsERA1* confers enhanced responses to abscisic acid and drought stress and increased primary root growth under nonstressed conditions in rice. *PLoS ONE*, 15(12). https://doi.org/10.1371/journal.pone.0243376.

Prodhan, Z.H., Islam, S.A., Alam, M.S., Li, S., Jiang, M., Tan, Y., Shu, Q. (2022). Impact of OsBadh2 mutations on salt stress response in rice. *Plants (Basel)*, 11(21), 2829. doi: 10.3390/plants11212829.

Qin, H., Wang, J., Chen, X., Wang, F., Peng, P., Zhou, Y., Miao, Y., Zhang, Y., Gao, Y., Qi, Y., Zhou, J., Huang, R. (2019). Rice OsDOF15 contributes to ethylene-inhibited primary root elongation under salt stress. *New Phytology*, 223, 798–813. https://doi.org/10.1111/nph.15824.

Qin, R., Hu, Y., Chen, H., Du, Q., Yang, J., Li, W.-X. (2023). MicroRNA408 negatively regulates salt tolerance by affecting secondary cell wall development in maize. *Plant Physiology*, 192(2), 1569–1583. https://doi.org/10.1093/plphys/kiad135.

Qiu, T., Qi, M., Ding, X., Zheng, Y., Zhou, T., Chen, Y., Han, N., Zhu, M., Bian, H., Wang, J. (2020). The *SAUR41* subfamily of *SMALL AUXIN UP RNA* genes is abscisic acid inducible to modulate cell expansion and salt tolerance in *Arabidopsis thaliana* seedlings. *Annals of Botany*, 125(5), 805–819. https://doi.org/10.1093/aob/mcz160.

Qiu, Z., Kang, S., He, L., Zhao, J., Zhang, S., Hu, J., Zeng, D., Zhang, G., Dong, G., Gao, Z., Ren, D., Chen, G., Guo, L., Qian, Q., Zhu, L. (2018). The newly identified heat-stress sensitive albino 1 gene affects chloroplast development in rice. *Plant Science*, 267, 168–179. doi: 10.1016/j.plantsci.2017.11.015.

Rai, G.K., Khanday, D.M., Kumar, P., Magotra, I., Choudhary, S.M., Kosser, R., Kalunke, R., Giordano, M., Corrado, G., Rouphael, Y., & Pandey, S. (2022). Enhancing crop resilience to drought stress through CRISPR-Cas9 genome editing. *Plants*, 12(12), 2306. https://doi.org/10.3390/plants12122306

Santosh Kumar, V.V., Verma, R.K., Yadav, S.K. (2020). CRISPR-Cas9 mediated genome editing of drought and salt tolerance (*OsDST*) gene in *indica* mega rice cultivar MTU1010.

Physiology and Molecular Biology of Plants, 26, 1099–1110. https://doi.org/10.1007/s12298-020-00819-w.

Shan, Q., Wang, Y., Li, J., Zhang, Y., Chen, K., Liang, Z., Gao, C. (2013). Targeted genome modification of crop plants using a CRISPR-Cas system. *Nature Biotechnology*, 31(8), 686–688. https://doi.org/10.1038/nbt.2650.

Shen, C., Que, Z., Xia, Y., et al. (2017). Knock out of the annexin gene *OsAnn3* via CRISPR/Cas9-mediated genome editing decreased cold tolerance in rice. *Journal of Plant Biology*, 60, 539–547. https://doi.org/10.1007/s12374-016-0400-1.

Sheng, X., Ai, Z., Tan, Y., Hu, Y., Guo, X., Liu, X., Sun, Z., Yu, D., Chen, J., Tang, N. (2023). Novel Salinity-tolerant third-generation hybrid rice developed via CRISPR/Cas9-mediated gene editing. *International Journal of Molecular Sciences*, 24(9), 8025. https://doi.org/10.3390/ijms24098025.

Shi, J., Gao, H., Wang, H., Lafitte, H.R., Archibald, R.L., Yang, M., Hakimi, S.M., Mo, H., Habben, J.E. (2017). ARGOS8 variants generated by CRISPR-Cas9 improve maize grain yield under field drought stress conditions. *Plant Biotechnology Journal*, 15, 207–216. https://doi.org/10.1111/pbi.12603.

Song, L., Liu, J., Cao, B., Liu, B., Zhang, X., Chen, Z., Dong, C., Liu, X., Zhang, Z., Wang, W., et al. (2023). Reducing brassinosteroid signalling enhances grain yield in semi-dwarf wheat. *Nature*, 617, 118–124. https://doi.org/10.1038/s41586-023-06023-6.

Sun, C., Lei, Y., Li, B., Gao, Q., Li, Y., Cao, W., Yang, C., Li, H., Wang, Z., Liu, J., Zhao, K.T., Gao, C. (2023). Precise integration of large DNA sequences in plant genomes using PrimeRoot editors. *Nature Biotechnology.* doi: 10.1038/s41587-023-01769-w.

Tabassum, J., Ahmad, S., Hussain, B., Mawia, A.M., Zeb, A., Ju, L. (2021). Applications and potential of genome-editing systems in rice improvement: Current and future perspectives. *Agronomy*, 11, 1359. https://doi.org/10.3390/agronomy11071359.

Tang, L., Mao, B., Li, Y., Lv, Q., Zhang, L., Chen, C., He, H., Wang, W., Zeng, X., Shao, Y., Pan, Y., Hu, Y., Peng, Y., Fu, X., Li, H., Xia, S., Zhao, B. (2017). Knockout of OsNramp5 using the CRISPR/Cas9 system produces low Cd-accumulating indica rice without compromising yield. *Scientific Reports*, 7(1), 14438. doi: 10.1038/s41598-017-14832-9.

Tao, L., Wang, B., Xin, S., Li, W., Huang, S., Liu, L., Cui, J., Zhang, Q., Cheng, X. (2023). A cluster of mutagenesis revealed an osmotic regulatory role of the OsPIP1 genes in enhancing rice salt tolerance. *The Crop Journal*, 11(4), 1204–1217. https://doi.org/10.1016/j.cj.2023.04.010.

Teng, Y., Lv, M., Zhang, X. (2022). BEAR1 bHLH transcription factor, controls salt response genes to regulate rice salt response. *Journal of Plant Biology*, 65, 217–230. https://doi.org/10.1007/s12374-022-09347-4.

Tingting, L., Jinpeng, Z., Xi, Y., Kejian, W., Yuchun, R., Chun, W. (2023). Development and application of prime editing in plants. *Rice Science*, 30(6), 509–522. https://doi.org/10.1016/j.rsci.2023.07.005

Tran, M.T., Son, G.H., Song, Y.J., Nguyen, N.T., Park, S., Thach, T. V., Kim, J., Sung, Y.W., Das, S., Pramanik, D., Lee, J., Son, K.-H., Kim, S.H., Vu, T.V., Kim, J.-Y. (2023). CRISPR-Cas9-based precise engineering of SlHyPRP1 protein towards multi-stress tolerance in tomato. *Frontiers in Plant Science*, 14. www.frontiersin.org/articles/10.3389/fpls.2023.1186932.

Usman, B., Nawaz, G., Zhao, N., Liao, S., Liu, Y., Li, R. (2020). Precise editing of the *OsPYL9* gene by RNA-guided Cas9 nuclease confers enhanced drought tolerance and grain yield in rice (*Oryza sativa* L.) by regulating Circadian Rhythm and abiotic stress responsive proteins. *International Journal of Molecular Science*, 21, 7854. https://doi.org/10.3390/ijms21217854.

Veillet, F., Perrot, L., Chauvin, L., Kermarrec, M.-P., Guyon-Debast, A., Chauvin, J.-E., Nogué, F., Mazier, M. (2019). Transgene-free genome editing in tomato and potato plants using *Agrobacterium*-mediated delivery of a CRISPR/Cas9 cytidine base editor. *International Journal of Molecular Sciences*, 20(2), 402. https://doi.org/10.3390/ijms20020402.

Vlčko, T., Ohnoutková, L. (2020). Allelic variants of CRISPR/Cas9 induced mutation in an inositol trisphosphate 5/6 Kinase gene manifest different phenotypes in arley. *Plants (Basel)*, 9(2), 195. doi: 10.3390/plants9020195.

Vu, V.T., Nguyen, N.T., Kim, J., Hong, J.C., Kim, J. (2024). Prime editing: Mechanism insight and recent applications in plants. *Plant Biotechnology Journal*, 22, 19–36. doi: 10.1111/pbi.14188.

Wahab, A., Abdi, G., Saleem, M. H., Ali, B., Ullah, S., Shah, W., Mumtaz, S., Yasin, G., Muresan, C. C., Marc, R. A. (2022). Plants' physio-biochemical and phyto-hormonal responses to alleviate the adverse effects of drought stress: A comprehensive review. *Plants*, 11(13). https://doi.org/10.3390/plants11131620

Wang, B., Zhong, Z., Wang, X., Han, X., Yu, D., Wang, C., Song, W., Zheng, X., Chen, C., Zhang, Y. (2020). Knockout of the *OsNAC006* transcription factor causes drought and heat sensitivity in rice. *International Journal of Molecular Science*, 21(7), 2288. doi: 10.3390/ijms21072288.

Wang, F., Chen, M., Yu, L., Xie, L., Yuan, L., Qi, H., Chen, Q. (2017). Osarm1, an r2r3 myb transcription factor, is involved in regulation of the response to arsenic stress in rice. *Frontiers in Plant Science*, 8. https://doi.org/10.3389/fpls.2017.01868.

Wang, L., Chen, L., Li, R., Zhao, R., Yang, M., Sheng, J., Shen, L. (2017). Reduced drought tolerance by CRISPR/Cas9-mediated SlMAPK3 mutagenesis in tomato plants. *Journal of Agricultural and Food Chemistry*, 65(39), 8674–8682. https://doi.org/10.1021/acs.jafc.7b02745.

Wang, N., Shi, Y., Jiang, Q., Li, H., Fan, W., Feng, Y., et al. (2023). A 14-3-3 protein positively regulates rice salt tolerance by stabilizing phospholipase C1. *Plant, Cell & Environment,* 46, 1232–1248. https://doi.org/10.1111/pce.14520.

Wang, T., Xun, H., Wang, W., Ding, X., Tian, H., Hussain, S., Dong, Q., Li, Y., Cheng, Y., Wang, C., Lin, R., Li, G., Qian, X., Pang, J., Feng, X., Dong, Y., Liu, B., Wang, S. (2021). Mutation of GmAITR genes by CRISPR/Cas9 genome editing results in enhanced salinity stress tolerance in soybean. *Frontiers in Plant Science*, 12. www.frontiersin.org/articles/10.3389/fpls.2021.779598.

Wang, W., Pan, Q., He, F., Akhunova, A., Chao, S., Trick, H., Akhunov, E. (2018). Transgenerational CRISPR-Cas9 activity facilitates multiplex gene editing in allopolyploid wheat. *CRISPR Journal*, 1, 65e74. https://doi.org/10.1089%2Fcrispr.2017.0010.

Wang, W., Yu, Z., He, F., Bai, G., Trick, H.N., Akhunova, A., Akhunov, E. (2022). Multiplexed promoter and gene editing in wheat using a virus-based guide RNA delivery system. *Plant Biotechnology Journal*, 20, 2332e2341. https://doi.org/10.1111/pbi.13910.

Wang, Y., Liu, X., Zheng, X., Wang, W., Yin, X., Liu, H., Ma, C., Niu, X., Zhu, J., Wang, F. (2021). Creation of aromatic maize by CRISPR/Cas. *Journal of Integrative Plant Biology*, 63, 1664–1670. https://doi.org/10.1111/jipb.13105

Wanikar, R., Kotwal, S. (2021). Resistant starch: A promising functional food ingredient. Book-Starch evolution and recent advances. *InTechOpen*, 101558. http://dx.doi.org/10.5772/intechopen.

Xing, H.-L., Dong, L., Wang, Z.-P., Zhang, H.-Y., Han, C.-Y., Liu, B., Wang, X.-C., Chen, Q.-J. (2014). A CRISPR/Cas9 toolkit for multiplex genome editing in plants. *BMC Biology*, 14, 1–12. doi:10.1186/s12870-014-0327-y.

Xiong, X., Liang, J., Li, Z., Gong, B.Q., Li, J.F. (2021) Multiplex and optimization of dCas9-TV-mediated gene activation in plants. *Journal of Integrative Plant Biology*, 63(4), 634–645. doi: 10.1111/jipb.13023.

Xu, R., Li, J., Liu, X., Shan, T., Qin, R., Wei, P. (2020). Development of plant prime-editing systems for precise genome editing. *Plant Communications*, 1, 100043.

Xu, W., Yang, Y., Yang, B., Krueger, C.J., Xiao, Q., Zhao, S., Zhang, L., Kang, G., Wang, F., Yi, H., Ren, W., Li, L., He, X., Zhang, C., Zhang, B., Zhao, J., Yang, J. (2022). A design optimized prime editor with expanded scope and capability in plants. *Nature Plants*, 8, 45–52.

Xu, W., Zhang, C., Yang, Y., Zhao, S., Kang, G., He, X., et al. (2020). Versatile nucleotides substitution in plant using an improved prime editing system. *Molecular Plant*, 13, 675–678. doi: 10.1016/j.molp.2020.03.012.

Yan, J., Ninkuu, V., Fu, Z., Yang, T., Ren, J., Li, G., Zeng, H. (2023). OsOLP1 contributes to drought tolerance in rice by regulating ABA biosynthesis and lignin accumulation. *Frontiers in Plant Science,* 14, 1163939. https://doi.org/10.3389/fpls.2023.1163939.

Yang, J., Ji, L., Liu, S., Jing, P., Hu, J., Jin, D., Wang, L., Xie, G. (2021). The CaM1-associated CCaMK-MKK1/6 cascade positively affects lateral root growth via auxin signaling under salt stress in rice. *Journal of Experimental Botany*, 72(18), 6611–6627. doi: 10.1093/jxb/erab287.

Yang, W., Chen, S., Cheng, Y., Zhang, N., Ma, Y., Wang, W., Wang, S. (2020). Cell wall/vacuolar inhibitor of fructosidase 1 regulates ABA response and salt tolerance in Arabidopsis. *Plant Signaling & Behavior*, 15(4), 1744293. https://doi.org/10.1080/15592324.2020.1744293.

Yassitepe, J.E.D.C.T., da Silva, V.C.H., Hernandes-Lopes, J., Dante, R.A., Gerhardt, I.R., Fernandes, F.R. , da Silva, P.A., Vieira, L.R. , Bonatti, V., Arruda, P. (2021). Maize transformation: From plant material to the release of genetically modified and edited varieties. *Frontiers in Plant Science*, 12, 766702. https://doi.org/10.3389/fpls.2021.766702.

Yin, W., Xiao, Y., Niu, M., Meng, W., Li, L., Zhang, X., Liu, D., Zhang, G., Qian, Y., Sun, Z., Huang, R., Wang, S., Liu, C.M., Chu, C., Tong, H. (2020). ARGONAUTE2 enhances grain length and salt tolerance by activating *BIG GRAIN3* to modulate cytokinin distribution in rice. *Plant Cell*, 32(7), 2292–2306. doi: 10.1105/tpc.19.00542.

Yin, Y., Qin, K., Song, X., Zhang, Q., Zhou, Y., Xia, X., Yu, J. (2018). BZR1 transcription factor regulates heat stress tolerance through FERONIA receptor-like kinase-mediated reactive oxygen species signaling in tomato. *Plant and Cell Physiology*, 59(11), 2239–2254. doi: 10.1093/pcp/pcy146.

Yu, W., Wang, L., Zhao, R., et al. (2019). Knockout of *SlMAPK3* enhances tolerance to heat stress involving ROS homeostasis in tomato plants. *BMC Plant Biology*, 19, 354. https://doi.org/10.1186/s12870-019-1939-z.

Yue, E., Rong, F., Liu, Z., Ruan, S., Lu, T., Qian, H. (2023). Cadmium induced a non-coding RNA microRNA535 mediates Cd accumulation in rice. *Journal of Environmental Sciences*, 130, 149–162. https://doi.org/10.1016/j.jes.2022.10.005.

Zeng, D.D., Yang, C.C., Qin, R., Alamin, M., Yue, E.K., Jin, X.L., Shi, C.H. (2018). A guanine insert in OsBBS1 leads to early leaf senescence and salt stress sensitivity in rice (*Oryza sativa L.*). *Plant Cell Reports*, 7(6), 933–946. doi: 10.1007/s00299-018-2280-y.

Zeng, Y., Wen, J., Zhao, W., Wang, Q., Huang, W. (2020). Rational improvement of rice yield and cold tolerance by editing the three genes OsPIN5b, GS3, and OsMYB30 with the CRISPR–Cas9 system. *Frontiers in Plant Science*, 10. doi: 10.3389/fpls.2019.0166320.

Zhang, A., Liu, Y., Wang, F., Li, T., Chen, Z., Kong, D., Bi, J., Zhang, F., Luo, X., Wang, J., Tang, J., Yu, X., Liu, G., Luo, L. (2019). Enhanced rice salinity tolerance via CRISPR/Cas9-targeted mutagenesis of the *OsRR22* gene. *Molecular Breeding*, 39, 47. doi: 10.1007/s11032-019-0954.

Zhang, C., Srivastava, A.K., Sadanandom, A. (2019). Targeted mutagenesis of the SUMO protease, overly tolerant to Salt1 in rice through CRISPR/Cas9-mediated genome editing reveals a major role of this SUMO protease in salt tolerance. *BioRxiv*, 555706. doi: https://doi.org/10.1101/555706.

Zhang, M., Cao, Y., Wang, Z., Wang, Z.Q., Shi, J., Liang, X., Song, W., Chen, Q., Lai, J., Jiang, C. (2018). A retrotransposon in an HKT1 family sodium transporter causes variation of leaf Na+ exclusion and salt tolerance in maize. *New Phytologist*, 217(3), 1161–1176. doi: 10.1111/nph.14882.

Zhang, M., Zhao, R., Wang, H. (2023). OsWRKY28 positively regulates salinity tolerance by directly activating *OsDREB1B* expression in rice. *Plant Cell Reports*, 42, 223–234. https://doi.org/10.1007/s00299-022-02950-2.

Zhang, S., Zhang, R., Gao, J., Song, G., Li, J., Li, W., Qi, Y., Li, Y., Li, G. (2021). CRISPR/Cas9-mediated genome editing for wheat grain quality improvement. *Plant Biotechnology Journal*, 19, 1684e1686.

Zhang, Y., Chen, K., Zhao, F.-J., Sun, C., Jin, C., Shi, Y., Sun, Y., Li, Y., Yang, M., Jing, X., Luo, J., Lian, X. (2018). OsATX1 interacts with heavy metal P1B-Type ATPases and affects copper transport and distribution. *Plant Physiology*, 178(1), 329–344. https://doi.org/10.1104/pp.18.00425.

Zhang, Y., Li, D., Zhang, D., Zhao, X., Cao, X., Dong, L., Liu, J., Chen, K., Zhang, H., Gao, C., et al. (2018b). Analysis of the functions of TaGW2 homoeologs in wheat grain weight and protein content traits. *Plant Journal*, 94, 857e866. https://doi.org/10.1111/tpj.13903.

Zhang, Y., Li, P., Niu, Y., Zhang, Y., Wen, G., Zhao, C., Jiang, M. (2023). Evolution of the *WRKY66* gene family and its mutations generated by the CRISPR/Cas9 system increase the sensitivity to salt stress in *Arabidopsis*. *International Journal of Molecular Sciences*, 24(4), 3071. https://doi.org/10.3390/ijms24043071.

Zhao, C., Qiu, J., Agarwal, G., Wang, J., Ren, X., Xia, H., Guo, B., Ma, C., Wan, S., Bertioli, D.J., Varshney, R.K., Pandey, M.K., Wang, X. (2017). Genome-wide discovery of microsatellite markers from diploid progenitor species, Arachis duranensis and A. ipaensis, and their application in cultivated Peanut (A. hypogaea). *Frontiers in Plant Science*, 8. www.frontiersin.org/articles/10.3389/fpls.2017.01209.

Zhao, Z., Shang, P., Mohanraju, P., Geijsen, N. (2023). Prime editing: Advances and therapeutic applications. *Trends in Biotechnology*, 41(8), 1000–1012. doi: 10.1016/j.tibtech.2023.03.004.

Zhong, X., Hong, W., Shu, Y., Li, J., Liu, L., Chen, X., Islam, F., Zhou, W., Tang, G. (2022). CRISPR/Cas9 mediated gene-editing of GmHdz4 transcription factor enhances drought tolerance in soybean (Glycine max [L.] Merr.). *Frontiers in Plant Science*, 13, 988505. doi: 10.3389/fpls.2022.988505.

Zhou, X., Zhao, Y., Ni, P., Ni, Z., Sun, Q., Zong, Y. (2023). CRISPR-mediated acceleration of wheat improvement: Advances and perspectives. *Journal of Genetics and Genomics*, 50(11), 815–834. doi: 10.1016/j.jgg.2023.09.007.

Zong, Y., Liu, Y., Xue, C., Li, B., Li, X., Wang, Y., Li, J., Liu, G., Huang, X., Cao, X., Gao, C. (2022). An engineered prime editor with enhanced editing efficiency in plants. *Nature Biotechnology*, 40, 1394–1402.

Zou, J., Meng, X., Liu, Q., Shang, M., Wang, K., Li, J., Yu, H., Wang, C. (2022). Improving the efficiency of prime editing with epegRNAs and high-temperature treatment in rice. *Science China, Life Science,* 65, 2328–2331.

12 From Mobile Genetic Elements to Cutting-Edge Genome Editors

Exploring Retrons, CAST, CRASPase, OMEGA, and Fanzor

Ahmad Munir, Sehrish Sarwar, Paulo Zaini, Aftab Ahmad, Nayla Munawar, and Baohong Zhang

12.1 INTRODUCTION

Just imagine a world where we can genetically modify our babies and eliminate most of the genetic diseases right at conception level: this is no longer science fiction. Some of the most significant changes in our field have emerged over the last few years, evolving from a fundamental scientific technique into a hallmark technology of the modern era. Although at its beginning it was focused on less efficient methods like meganucleases, zinc finger nucleases (ZFN) (Ahmad et al., 2024), transcription activator-like effector nucleases (TALEN) (Munir et al., 2024; Widner, 2021), today one of the most efficient systems – CRISPR-Cas is best known (Hwang et al., 2024). However, significant challenges remain, including off-target effects and the imprecise identification of target sites, which can lead to unintended biological consequences. These limitations have driven the quest for new genome-editing systems that can increase the accuracy of the process and expand the range of applications of the technique. The latest advancements in the field of mobile genetic elements (MGEs) – natural DNA pieces that can spread in or across genomes – are opening up new approaches to create novelties. MGEs including transposons, retroelements, plasmids, and phages play a role in genome architecture and facilitate horizontal gene transfer. These improvements are the retron-based systems, CRISPR-associated transposons (CAST), CRASPases, OMEGA systems, and Fanzor proteins which use different characteristics of MGEs to overcome the existing difficulty in genome modifications. One of them is the retron-based gene editing: a relatively recent addition to the field but one that draws a lot of interest. The retrons described here are bacterial retroelements capable of synthesizing single-stranded DNA (ssDNA) in vivo (Hu et al., 2022). This makes it possible for the development of retrons library recombineering (RLR) that allows for the integration of several mutations and distinct genetic sequences into bacterial genomes. As such, retrons can be seen as a significant advancement over traditional off-target CRISPR-Cas approaches, which regularly depend on error-prone DNA repair mechanisms. This capability not only accelerates advance the pace of genetic research; it also has high transformative potential in synthetic biology and biotechnology applications. These advances extend beyond accelerating research, there are special CRISPR-associated transposons (CAST) that integrates the specific DNA targeting ability of CRISPR with the transposition function of transposons. The CAST systems are somewhat distinct from other conventional CRISPR-Cas system that relies on host cell repair mechanisms for DNA integration; they can efficiently deliver large DNA cargos to target genomic sites without causing double-strand breaks. CAST also greatly improves the accuracy of the edit, and reduces off-target mutagenesis that are common with conventional gene-editing tools (Gelsinger et al., 2023; Hu et al., 2022).

Another breakthrough is the development of Craspases – the hybrids of CRISPR DNA-targeting ability and proteolytic activity, enabling programmable protein degradation alongside genome editing. Craspase has a protein called gRAMP which has all the regulatory domains to manipulate genes efficiently. This belongs to Class III-E of the CRISPR system that directly edits RNA and protein sequences in addition to the DNA sequences. These two features give a two-in-one system that can be used to investigate gene regulation and protein-DNA interactions in a controlled manner, or mediate the transition from genome to proteome engineering (Liu et al., 2022).

New advances have resulted in the development of OMEGA systems (Obligate Mobile Element Guided Activity). These newly discovered systems emerge from prokaryotic mobile elements and serve the same purpose as CRISPR-associated endonucleases despite their distinct evolutionary origins. OMEGA proteins can modify specific DNA sequences precisely due to small RNA molecules known as ωRNAs. The small size allows for easy transport through viral vectors and can translate gene editing into clinical therapies. The identification of OMEGA systems reveals the considerable unexplored possibilities of MGEs for developing advanced genome engineering methods. Finally, we present Fanzor proteins as another promising class of RNA-guided nucleases that attract attention as functional equivalents to CRISPR-Cas systems. Unlike bacterial CRISPR-Cas systems, Fanzor's eukaryotic origins confer native compatibility with chromatin structure and nuclear import machinery, it functions well within the eukaryotic genome systems. Their reduced size and broad substrate specificity dramatically reduce off-target activities compared to classical Cas proteins and these systems

DOI: 10.1201/9781003500933-14

are now available as tailored solutions for both therapeutic and agricultural applications. The continuous evolution of these technologies demonstrates nature's remarkable potential as a source of sophisticated genome-editing tools (Tae, 2024). Here, these novel technologies include retrons, CAST systems, Craspases, OMEGA systems, and Fanzor proteins, described in this chapter, and provide the conceptual and technical framework for transformative advances across various sectors. Orchestrating their unique features through gifted tools, researchers are revolutionizing the potential for manipulating the genome again that once had been thought impossible to do.

The potential of these advances extends beyond simple interpretation as genetic changes; they are a fundamental change in the way problems can be addressed in medicine, agriculture, and environmental interactions. We believe that by using these advanced tools researchers are not only playing an innovative game and striving to discover higher potential – they are modifying the fundamental laws of life.

12.2 MOBILE GENETIC ELEMENTS: NATURAL GENETIC MOVERS

Mobile genetic elements (MGEs) are described as mobile DNA sequences which are prone to movement within or between genome of organisms, driving genetic diversity. These are DNA transposons, retrotransposons, retrons, CRISPR associated transposons (CASTs) (Frost et al., 2005; Hwang et al., 2024). These MGEs have greatly advanced genome engineering, eradicating several challenges which are found in conventional techniques. Another important feature of MGE-based technologies is that the technologies can be applied to a wide variety of host organisms, and do not depend upon specific host factors. This flexibility allows the application of MGE-based techniques in prokaryotic and eukaryotic systems. Furthermore, MGEs promote genome-wide mutagenesis in which genetic changes occur throughout the entire genome of the host organism. This capability is critical for functional genomics, therapeutic target identification, and understanding genotype-to-phenotype relationships. Although CRISPR is a versatile method of genome-wide genetic screening, it necessitates the use of individual sgRNAs to target all the genes in the genome and is usually accomplished using oligonucleotide pools and plasmid libraries (Liu et al., 2023). On the other hand, DNA transposons are advantageous in that they integrate randomly and therefore, the generation of gene knockout insertional mutants across the genome is much easier. They are also very efficient in the process of joining long DNA fragments into the host genome, especially if over 10 kilobases in size. This capability permits the integration of several genes or regulatory components, by which the design of intricate genetic networks and metabolic pathways and the development of new characteristics or the enhancement of biological procedures becomes possible. One of the major benefits of MGE-based genome engineering is that multiple loci can be targeted using multiple gRNAs. While the other platforms face challenges in multiplexing, CRISPR can program multiple sgRNAs for simultaneous editing. Nevertheless, target site selection in CRISPR is based on double-strand breaks, which may lower its multiplexed editing capacity. While base editors and prime editors do not employ double-strand breaks, they are also not devoid of multiplexing issues, because multiple gRNAs cannot be processed by the Cas9 from a single array. This often necessitates expression constructs that are complicated, making delivery difficult. On the other hand, CAST systems employ the Cas6 protein that can selectively cleave and process the polycistronic gRNA arrays into individual gRNAs, which makes the CAST systems optimal for multiple targets. MGE-based technologies also enable in situ ssDNA production, which is useful for applications such as HDR, for which ssDNA is necessary for accurate DNA repair. However, MGE-based genome engineering offers multiple advantages: These are: DNA transposons, retrons, CASTs, and DNA transposons and retrons have a broad host range, DNA transposons could generate genome-wide mutations, retrotransposons mutate continuously within defined regions, DNA transposons can integrate large DNA fragments, retrons can multiplex and in situ generate ssDNA. These characteristics make MGE-based approaches rather flexible and highly effective in gene- and genome-scale engineering applications (Lim, 2015).

12.3 CLASSIFICATION OF MOBILE GENETIC ELEMENTS (MGE)

Mobile genetic elements (MGEs) are mobile DNA sequences, which participate in the process of genome evolution, rearrangements, and adaptation. MGEs are involved in various aspects of life and serve crucial functions in HGT (Horizontal Gene Transfer), genome rearrangement, and the transmission of some traits such as antibiotic resistance. Their mechanisms of movement have not only contributed to the natural evolutionary processes but have equally been used to create revolutionary genome-editing processes. The following presents a brief review of the important types of MGEs, and the research that has contributed to the knowledge about their participation in movement of genes and in biotechnological innovations (Hou et al., 2019).

12.3.1 Transposons: Key Elements in Genetic Diversity

Transposons are mobile genetic elements also known as "jumping genes" which can alter their location in a chromosome. Initially identified in maize by Barbara McClintock, transposons were found to control gene expression and genomic organization that paved the way for studying genetic variation (Kidwell, 2005). These elements are divided into two main classes: Class I transposons or retrotransposons, where the transposon first makes an RNA copy of their DNA sequence and then through reverse, it is inserted at a new genomic location; Class II transposons or DNA transposons where the element is cut from the initial location and is then inserted into another location by the aid of the enzyme transposase. Retrotransposons represent a large part of the eukaryotic genomes. For instance, about 40% of human genome consists of retrotransposon-derived sequences. Research on other model organisms such as Drosophila melanogaster has revealed how transposons contribute to genomic plasticity and consequently to fast adaptation to environmental changes, thus increasing evolutionary fitness (Mestre et al., 2020).

12.3.2 Plasmids: Key Facilitators of Horizontal Gene Transfer (HGT) and Tools for Genome Editing

Plasmids can be defined as small, circular, double-stranded, and independent DNA molecules that replicate autonomously and are present in bacteria, and archaea. In contrast to chromosomal DNA, plasmids are not required for the minimal maintenance of the host organism but may contain genes that confer benefits to the organism, for example, resistance to antibiotics, toxins production, or new ways of catabolism. Due to their high motility within bacterial cells and sometimes across species, they are involved in the movement of genes by a process known as horizontal gene transfer (HGT), which is one of the leading promoters of genetic variation and evolution among prokaryotes. Plasmids in this case helps in the transfer of these genes between bacterial cells, thus enhancing the fast adaptation of the bacteria to change in environment. Carattoli et al. (2015) described the involvement of plasmids in the process of antibiotic resistance, indicating that they had the capability to transfer genes between different species and spread resistance factors. This principle of gene mobility has been used in synthetic biology and genome editing where plasmids are commonly employed as vectors to deliver the CRISPR-Cas components into the targeted cells for gene editing. Due to this ability to provide selective advantage, plasmids have been essential in both natural and artificial evolutionary processes and their management forms the basis of modern genetic engineering and genome editing (Harrison & Brockhurst, 2012).

12.3.3 Retrotransposons

Retrotransposons use an RNA intermediate to move within the genome. These are transcribed into RNA, reverse transcribed back into DNA and inserted at new genomic locations. They can be divided into two categories: The two main types are the LTR (Long Terminal Repeat) retrotransposons that have long terminal repeats in their sequence and integrate into the genome by a mechanism like that of a retrovirus and non-LTR retrotransposons, which have no terminal repeats but require reverse transcriptase for reintegration into the genome. Brouha et al. (2003) have shown that LINE-1 retrotransposons are still mobilizable in the human genome and may be involved in the generation of genetic variability and evolution, as well as in disease and cancer progression. The ability of retrotransposons to be involved in insertional mutagenesis has therefore placed them as major players in genetic variation in most organisms (Brouha et al., 2003). The study pointed out the evidence of retrotransposition events in human cells and how such elements can alter the genomic structure and possibly cause diseases. It has been found that LINE-1 activity has been linked to different genetic disorders and cancers because of their ability to integrate within or close to the critical genes, thus possibly causing mutations or gene alterations. For example, even though CRISPR-Cas9 system is not based on the retrotransposons, it has been further developed with the knowledge that was gained from mobile elements. CRISPR-Cas9 enables targeted modification of the genome by creating site-specific double-strand breaks and then using cellular repair pathways to incorporate new genetic material. Some scientists compared the work of the CRISPR-Cas9 system with the integration of retrotransposons – DNA sequences that target specific DNA sequences and integrate new genetic material. Retrotransposons also introduced the development of new methods of genome engineering that follow the same principles of RNA targeting. For instance, the advancement of the retroviral-based gene delivery systems employs elements of retrovirus to transduce CRISPR components into target cells leveraging on the retrotransposon's capacity to integrate into the host genome. These systems are being studied for their ability to achieve a steady and effective editing of genes with a special focus on applications that need sustained expression of the therapeutic genes (Luqman-Fatah & Miyoshi, 2023).

12.3.4 Applications of Retrotransposons in Genome Editing

Retrotransposons offer significant potential as molecular tools in genome engineering, with their utility expanding beyond traditional applications. Recent studies have covered the use of retrotransposons, including their role as molecular markers and gene delivery vectors (Arvas et al., 2023). Long Terminal Repeat (LTR) retrotransposons are emerging as promising alternatives to retroviruses. While retroviruses are effective for gene delivery in animals and insects, their application in plants is limited. LTR retrotransposons provide a viable solution for plant genome engineering, offering a broader range of applications than retroviruses (Ramakrishnan et al., 2023).

This section explores the advanced uses of retrotransposons in genome engineering, focusing on studies that highlight their potential beyond basic insertional mutations. Despite the relatively limited use of retrotransposons in this field, their unique properties have been harnessed to develop innovative approaches for genome modification.

A notable example is the use of the native yeast retroelement Ty1 for in vivo-directed mutagenesis of synthetic DNA within eukaryotic cells. Crook et al. (2016) demonstrated that Ty1 can facilitate targeted mutagenesis, overcoming some limitations of traditional in vitro mutagenesis methods. In vitro approaches often require extensive design-build-test cycles, substantial financial investments, and time-consuming processes, resulting in reduced DNA diversity due to inefficiencies in host strain transformation (Crook et al., 2016). In contrast, in vivo mutagenesis with retrotransposons offers a more streamlined alternative, potentially increasing diversity and efficiency. Traditional methods like PACE (Phage-Assisted Continuous Evolution) and MAGE (Multiplex Automated Genome Engineering) have shown in vivo evolution capabilities but come with their own challenges. PACE and MAGE require sophisticated automated equipment and can be labor-intensive. PANCE (Phage-Assisted Non-Continuous Evolution) (Miller et al., 2020) attempts to address these limitations but still involves manual handling and can result in the loss of variants during selection. Manual MAGE also suffers from similar drawbacks, including time-consuming processes and potential inefficiencies (Wannier et al., 2021).

To address these challenges, a novel kilobase-scale directed evolution tool has been developed, leveraging the unique features of retrotransposons. This innovative system enables continuous in vivo evolution of target genes through iterative cycles of mutagenesis and selection within the retrotransposon cargo region. This approach minimizes the need for additional researcher intervention and offers a more efficient and sustained method for improving gene function. Overall, retrotransposons represent a powerful tool in genome engineering, providing novel solutions for in vivo mutagenesis and directed evolution. Their application in both basic research and practical biotechnology continues to evolve, driven by advances in our understanding of their mechanisms and potential.

12.3.5 Retrons: Harnessing Bacterial Retroelements for Advanced Genome Engineering

Retron, a new type of bacterial retro transposable element, has become one of the most prospective instruments for genome editing. These elements are characterized by their capacity to synthesize intracellular ssDNA via reverse transcription process. Even though the exact biological function of retrons is still unclear, their potential for targeted genome editing is gradually emerging. Retrons are composed of three critical components: it consists of a guide sequence (msr), a targeting sequence (msd), and a reverse transcriptase (RT). The msr sequence is essential in the scenario to start the reverse transcription process while the msd sequence serves as the template. An inverted repeat sequence present in between msr and msd helps in the formation of a second structure of RNA. This structure is important for the proper functioning of the retron since it enables the branching guanosine residue of msr to act as a primer for reverse transcription by the retron RT (Lopez et al., 2022; Simon et al., 2018). In the process of reverse transcription, the msr sequence generates RNA to activate the RT and the msd sequence, which creates a hairpin secondary structure, is reverse transcribed into ssDNA. The branching guanosine at the base of the folded mRNA starts the synthesis of the complementary DNA by forming a 2'-5' phosphodiester linkage with internal guanosine residue of msr. During this process most of the msd sequence is removed by the RNase H activity present in the RT. This leads to the formation of multicopy ssDNA (msDNA) hybrid, containing newly synthesized ssDNA and msrRNA.

The size of retrons is comparatively small, which ranges from 1 to 2 kb, and this has made the process of transferring them into cells easier through transduction, transformation, and conjugation. Due to their small size and high reverse transcription efficiency, retrons can be a useful tool in genetic engineering experiments. As mentioned earlier, retrons have been reviewed in the recent past regarding their biochemical characteristics; for instance, the structure and function of msr, msd, and msDNA as well as the different types of RTs that exist (Simon et al., 2019). Taken together, these studies highlight the future development of retrons for engineering as a promising field for the improvement and expansion of modern techniques for genetic manipulation and functional characterization. The present study shows that retrons represent a new avenue for genome editing through the reverse transcription process that yields specific ssDNA. With further exploration of their capabilities, retrons are set to become standard members of the genetic engineer's arsenals of tools to open the next generation of possibilities in genetic engineering.

12.3.6 Retron Library Recombineering (RLR) for Prokaryotic Genome Editing

The idea that retrons may be used as DNA factory and in vivo genome-editing tool was first investigated in bacterial systems in 2018. A better approach was later devised called retron library recombineering (RLR) to use retrons' genome-editing ability in bacterial models. In this approach, the retron sequence was altered to generate ssDNA harboring a particular mutation between the upstream and downstream sequences that are homologous to genomic target site. The recombinant retron library was electroporated into the target cells, the msr/msd region of the retron was then transcribed and reverse-transcribed by retron RT. This process produced several copies of the mutant DNA that was used as a recombineering donor that annealed to the lagging strand during DNA replication with the help of single-stranded annealing protein (SSAP). SSAPs are significant in enhancing homology-directed repair (HDR) during genome editing since they ensure the pairing of complementary single-stranded DNA sequences. During RLR, SSAP coordinated the binding of the RT-DNA donor sequences to replicate DNA at the target site, by the homologous upstream and downstream sequences which resulted in the incorporation of the desired mutation in the daughter strand during replication.

Initially, the desired mutation was present in a few cells, i.e., 0.1% of the *Escherichia coli* cells containing the retron recombineering system. To increase this efficiency, the researchers altered the bacterial DNA, knocking out the mismatch repair mechanism that would normally fix replication errors and possibly undo the intended mutations. They also inactivated two genes that code for exonucleases that degrade ssDNA. After these changes, more than 90% of the bacterial population incorporated the retron sequence into its genome. Recent studies, however, showed that increasing levels of RT-DNA in target cells might increase the accuracy and efficacy of retron-mediated editing without modifying the repair process. Lopez et al. (2022) gave compelling evidence that RT-DNA synthesis constrains RLR-mediated editing, and upon augmenting the similarity between the 5′ and the 3′ ends of retron ncRNA (a1/a2), the levels of RT-DNA were raised in bacterial cells (Kaur and Pati, 2024; Ellington and Reisch, 2022).

Moreover, some improvements were made; optimization of the bacterial retron cassette in a systematic and modular manner to enhance the multiplex genome-editing capability by using the retron-mediated genome-editing system (REGES). For instance, in REGES, the in vivo-generated ssDNA donor determines the region to be edited, and this can be done with a high degree of flexibility. Ribosome-binding sites, promoters, replicons, donor design, and editing conditions of the retron expression system were fine-tuned and several host factors were altered to achieve better stability and increased availability of ssDNA within the cells. These refinements have made it possible to achieve 100% efficiency for different mutations such as insertions, substitutions and deletions in bacteria. Editing efficiency based on one, two, three, and four genomic loci was 100%, 85 ± 3%, 69 ± 14%, and 25 ± 14%, respectively. Furthermore, REGES allowed the generation of barcoded and pooled variant libraries with degenerate ribosome-binding site (RBS) sequences to fine-tune the endogenous and exogenous gene expression in bacteria. Thus, RLR has become a versatile tool for multiplex genome editing in various applications of metabolic engineering and synthetic biology in prokaryotic species (Kaur & Pati, 2024).

12.3.7 Retron-Based Genome Editing Using CRISPR/Cas in Eukaryotes

In eukaryotic cells, retrons have been used together with the CRISPR/Cas system for the purpose of genome editing as illustrated in Figure 12.1. In *S. cerevisiae*, the expression of the Eco1 retron was improved by the inversion of the original operon arrangement. This was necessary because in its native state the ncRNA is in the 5′UTR of the reverse transcriptase (RT) transcript and for translation to occur, the ribosome must enter internally from a ribosome binding site situated within or in the vicinity of the a2 region of the ncRNA. This positioning situates the entire ncRNA between the 5-mRNA cap and the RT start codon in eukaryotic cells. This increased distance together with the presence of out of frame ATG codons and the ncRNA structure hindered th translation of RT. Also, the possibility of altering the native a1/a2 region could have unknown consequences on the RT translation. As a result, the ncRNA is placed in the 3′ UTR to avoid hindering RT, while the RT is controlled by the RNA Pol II promoter with the start codon close to the 5′ end of the mRNA. In yeast, retron-mediated genome editing was done by co-expressing SpCas9 gRNA and ncRNA from a single transcript along with RT. The gRNA cut at the genomic site with high specificity and the break was then repaired through homology directed repair (HDR) with the use of RT-DNA as a template. Further, the change in the length of a1/a2 complementarity in the ncRNA enhanced the editing efficiency in both bacterial and eukaryotic systems (Lopez et al., 2022).

In human HEK293T cells, levels of RT-DNA production were determined with retrons derived from different bacterial species. The Eco2 retron was found to increase synthesis of RT-DNA in both the WT and the extended a1/a2 regions. A single-promoter system of Cas9 and RT linked with a P2A linker sequence was co-transfected with an H1 promoter-driven ncRNA/gRNA construct. Subsequently, retron-based editing was validated in HEK293T cells using Illumina sequencing. However, the efficiency of precise-to-imprecise editing was slightly more than 1:15 and in yeast it was 74:1. This low efficiency in human cells was explained by either abundant levels of RT-DNA or different DNA repair pathways in yeast and humans. In another study, it was also possible to perform retron-mediated editing in HEK293T cells using the same strategy as above, but gRNA was fused to the ncRNA retron at the 5’ end and Cas9 was linked with bacterial RT. Among the four retrons that are tested, only Ec73 can induce up to 10% HDR events at endogenous loci in human cells. More recently, retrons have been applied to overcome difficulties in delivering external donor DNA containing the desired mutations in human cells. When msDNA was linked to gRNA in HEK293T and K562 cells, the HDR efficiency was also improved.

In plants, retrons in combination with CRISPR/Cas system have been applied for targeted genome modification. To HDR, the Ec86 retron system in Nicotiana benthamiana was used to amplify ssDNA containing the desired mutation. Ec86 ncRNA and RT were produced employing the tobacco rattle virus (TRV)-RNA2 and the pK2GW7 gateway vector with a 35S promoter, respectively. Different retron ncRNA architectures were used, of which the extended msd stem and the a1/a2 region of Ec86 were the most effective in overproducing ssDNA in plant cell. Also, the co-expression of the VirE2 gene from Agrobacterium tumefaciens with the retron editors increased the ssDNA output by 10.7fold, thus, improving HDR efficiency by CRISPR/Cas for genome editing in plants (Buffington et al., 2024).

12.3.8 The RLR Advantage over CRISPR

Retrons are useful for synthesizing single-stranded DNA (ssDNA) in the cells which are being edited and this is a distinct advantage. The Retron Library Recombineering (RLR) system integrates mutations without disrupting the inherent DNA, which is an advantage. Whereas CRISPR-based strategies depend on guide RNAs to generate targeted DNA breaks at unique PAMs and then use donor DNA for repair, no other method is as unique and well distinguished as RLR. It does not require PAM sequences which work only as a "donor-only" system, thus, it is more flexible and effective. When retrons are used with the CRISPR/Cas systems, they appear to enhance the basic "guide + donor" practice. The

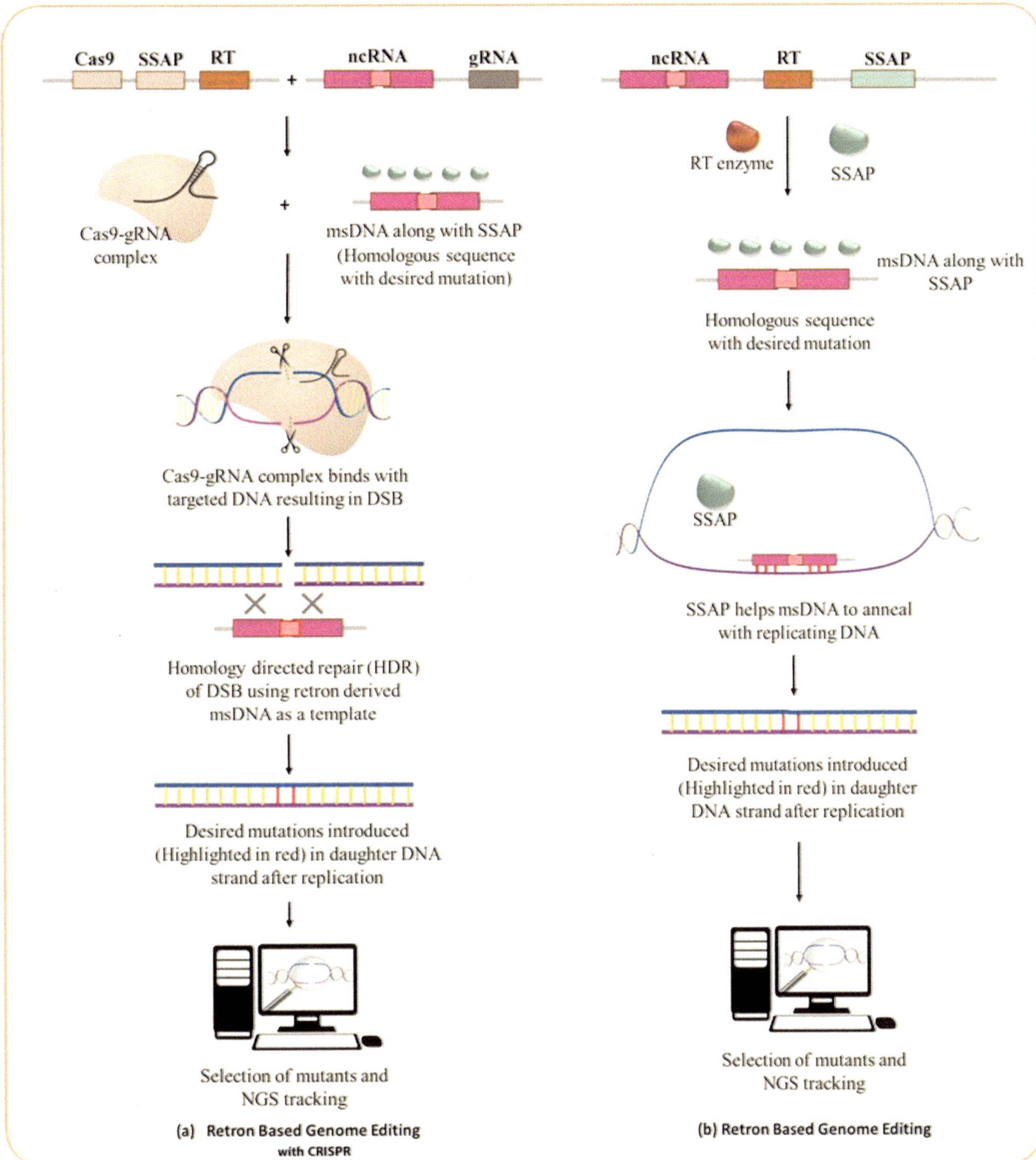

FIGURE 12.1 Comparison of (a) Retron based genome editing combined with CRISPR and (b) Retron based genome editing.

retron-generated ssDNA enhances the number of available donor templates, which enhances the rate of HDR. Further, retrons can be used as natural "barcode" which is convenient for the use of next-generation sequencing (NGS) for mutant screening. Still, another unique characteristic of RLR is that the proportion of cells containing mutation rises with successive bacterial division cycles. This is very different from the CRISPR's approach which works in the "one shot, one kill" fashion, which can be successful at the first attempt only (Simon et al., 2019).

Even though RLR provides a novel and rather effective genome-editing tool with prospects for further expansion in the spheres of biology, there are certain difficulties and limitations associated with it. RLR is one of the few methods that can be performed continuously, which is replication-dependent and can edit across generations. However, this characteristic may also reduce its versatility in some species. Moreover, to increase editing rates, the machinery responsible for mismatch repair in target cells has to be inactivated, which might result in unwanted mutations. Therefore, more modifications and optimization of RLR are needed for the expansion of its application in genome editing. Nevertheless, RLR has several advantages for future genetic investigations because it can enhance the amount of template DNA in cells. This method could be used as an alternative to CRISPR/Cas systems when it results in unwanted effects or could be used in conjunction with the CRISPR/Cas to increase the rate and accuracy of editing. Thus, RLR has the potential for further development opportunities in a broad range of genetic systems that have not yet been explored (Kaur & Pati, 2024).

12.3.9 Diversity of Retrons as Clades

This study demonstrated that retrons are remarkably diverse to create the RT-DNA and genome editing in different organisms, and their efficiency varies among the different clades of retrons. Out of 163 synthesized retrons, only 62 were found to produce measurable levels of RT-DNA; the clades 2, 9, and 10 were found to produce higher levels of RT-DNA while clades 3 and 4 did not seem to produce any RT-DNA at all. This implies that these clades may need other factors or have structural characteristics that prevent the synthesis of RT-DNA. In addition, the study established that while the two retron systems had similar levels of RT-DNA production, retron-Kva1 had a ten-fold better editing efficiency than retron-Eco1, thus implying that while the levels of RT-DNA may be a crucial factor in the recombineering process, it does not explain all the outcomes (Mestre et al., 2020).

Furthermore, the findings indicated that retrons are very efficient in recombineering bacterial and phage genomes; eight retrons outcompeted retron-Eco1 in bacterial editing rates. Nevertheless, the relationship between the levels of RT-DNA production and the efficiency of recombineering seemed to be reduced in the context of phage genome engineering and pointed to the different regulatory mechanisms that exist between bacterial and phage systems. For human genome editing, the study showed that 58 retrons are more efficient than retron-Eco1 and this has been added to the list of tools for precise human genome engineering. Notably, bacterial RT-DNA production was not associated with human editing efficiency, which demonstrates species-related factors that potentially affect retron functionality. This result therefore implies that more research should be conducted to determine the characteristics of retrons within different cellular contexts. The outcomes of the study highlight the possibilities of retron-based systems for genome editing across various species; nonetheless, the research also points to the lack of knowledge of the factors that may affect the effectiveness of retromer. Thus, the findings on retron-Kva1 and other retrons with enhanced editing capabilities are encouraging, yet the study could have delved further into the molecular basis of these differences, especially regarding ncRNA architecture and its interactions with host proteins (Mestre et al., 2020).

12.4 GENOME EDITING WITH CAST: A TRANSPOSON-BASED CRISPR INNOVATION

It is sometimes important to integrate large double-stranded DNA (dsDNA) into the genome of the host organism to make it capable of performing intricate functions or even correcting genetic defects. Different strategies have been employed in this regard such as: the incorporation of adjustable synthetic gene circuits, biosensing modules and functional metabolic pathways for the heterologous expression of desired gene. Although λ-Red recombination has been effective in integrating dsDNA into chromosomes, the efficiency reduces with the increase in the size of DNA insert. For example, when the homology arm targeting the lacZ gene in *E. coli* was increased from 1 kb to 2 kb, the recombination efficiency was as low as 25%, and the efficiency reduced with the size of the inserts greater than 2.5 kb and no recombinant colonies were observed (Rhiel et al., 2023).

CRISPR-Cas9 has increased the rates of editing, especially for shorter pieces of DNA. However, efficiency decreases with an increase in length of the inserted DNA to 60% for 3 kb, 40% for 5 kb, and 10% for 8 kb. This poses a challenge to genome engineering, especially when large DNA sequences are inserted. On the other hand, DNA transposons have been found to be efficient in delivering large DNA fragments of size ranging from a few kilo base pairs to tens of kilo base pairs. The integration patterns of transposons vary whereas Tn7 integrates large sequences by recognizing some unique sequences which are not present in all genomes, other transposons such as Tn5 and piggyBac are random in their integration. The integration process should be very accurate in order not to interfere with the host genes or other regulatory sequences which may have a negative impact on the host organism (Song & Bae, 2024).

To address these challenges, structured genomic-safe harbors (GSHs) have been employed. These specific regions in the genome enable the incorporation of foreign DNA without interfering with normal cellular functions, thus making them suitable for gene therapy and other related uses. The development of a technology that can efficiently and accurately ligate large DNA fragments into specific GSH regions would greatly expand the scope of genome engineering. The emergence of CRISPR-associated transposons (CAST) is considered a breakthrough in this field. The CAST systems include the I-F and V-K subtypes derived from *Vibrio cholerae* and *Scytonema hofmanni*, respectively; the crRNA sequence from CRISPR helps in precise targeting of DNA, thereby enabling accurate transposition. These systems work similarly to Tn7 transposons, which are well known for their site-specific integration but have higher accuracy due to the CRISPR system (Tou et al., 2023).

While retrons have been employed to synthesize ssDNA for recombineering, they operate through HDR, which is not very effective for large-scale genomic engineering. However, in the CAST system, there is no requirement for DSBs and homologous recombination, and the system is safer and more efficient and is useful in conditions where accurate and stable gene integration is important. By integrating the targeting capability of CRISPR with the transpositional capacity of MGEs, CAST provides a new approach to large genomic rearrangements, paving the way for new applications in therapeutics and synthetic biology (Tenjo-Castaño et al., 2022).

12.4.1 CAST in Prokaryotes

Strecker and colleagues were the first to characterize the type V-K CAST system from *Scytonema hofmannii* (ShCAST) and were able to design guide RNAs (gRNAs) for the purpose of the targeted DNA insertion into *E. coli*. This process works unidirectionally and requires a protospacer and a PAM sequence in the target location. They constructed a helper plasmid, pHelper, with the four ShCAST proteins and a donor plasmid, pDonor, containing the genetic material of interest flanked by LE and RE. When both plasmids were co-electroporated into *E. coli*, the efficiency of integration was reduced. The system uses Cas12k that works in conjunction with gRNA to bring the transposase complex and the DNA cargo (up to 10 kb) into the target site 60–66 base pairs downstream of the PAM sequence without the formation of DSBs.

Despite this optimization, ShCAST can efficiently integrate the LE/RE-flanked cargo, however, it is not free of problems, because the PAM and target sequences remain unchanged after the integration and thus cannot be integrated again at the same site. This is called the "target immunity" which prevents further insertions at the same site. ShCAST can also lead to off-target integrations due to the overexpression of the ShCAST through mechanisms that are independent to Cas12k. In addition, ShCAST transposes along with the pDonor plasmid, which is also observed in different *E. coli* strains (Chang et al., 2024).

To increase the efficiency of ShCAST, researchers enhanced the pHelper plasmid by using more strong promoters; they proved that increasing the activity of promoters increases the integration efficiency. This allowed the system to deliver multiplexed gene constructs to bacterial genomes containing multiple copies of the target element, as well as to integrate large DNA payloads at multiple genomic loci. To further enhance the on-target integration efficiency, the expression of Cas12k was elevated, leading to the development of the SHOT system. SHOT improves ShCAST through increasing Cas12k levels and optimizing the editing conditions, which enables the introduction of genes at loci that cannot be targeted using the original ShCAST (Zeng et al., 2023).

These enhancements show that ShCAST can be optimized for multiplexed, DSB-free large DNA fragment integration into prokaryotic genomes. However, integration efficiency is still dependent on the genomic target, and it is found that sequence adjacent to LE/RE ends are essential for integration. Furthermore, the efficiency of editing depends on such aspects as protein stoichiometry. It is possible to obtain increased editing efficiency and decreased off-target effects by modulating the expression of individual proteins; however, the problem of co-integration of the plasmid backbone persists. This co-integration problem was recently addressed in a study. Bacterial transposons are known for transposition through a cut-and-paste model or a copy-and-paste model. In the Tn7 transposon, TnsA is responsible for making a nick at the 5′ end of the donor DNA, while TnsB creates a nick at the 3′ end of donor DNA for cut-and-paste transposition. Tou et al. (2023) studied that lack of TnsA in type V-K CAST results in TnsB-mediated 3′ nicking only, leading to copy-and-paste integration and unwanted co-integration. To address this, they created the HELIX system through the integration of a nicking endonuclease (nAnil) to TnsB at the N-terminus which regained 5′ nicking functionality. HELIX was efficient in large DNA insertions into *E. coli* genomes in a cut-and-paste manner which greatly reduced co-integration while at the same time having a very high integration efficiency. Interestingly, the nAnil was fused at the C-terminal end to TnsB and no transposition occurred, demonstrating that the orientation of the fusion is crucial. HELIX derived from ShCAST brings a more efficient way to do type V-K transposition that can do programmable, unidirectional, cut-and-paste, and recombination-independent DNA insertions. However, the engineering of component proteins is remarkably delicate to changes, and thus, further optimization is essential to enhance its application in other type V-K CAST systems (Hwang et al., 2024; Tou et al., 2023).

12.4.2 INTEGRATE

Klompe et al. (2022) have shown that the type I-F3 CAST from *Vibrio cholerae* (VchCAST) could be used for programmable transposition in *E. coli*, which has spurred the development of the INTEGRATE system. This system encodes the required proteins and donor DNA across three separate plasmids: pQCascade, pTnsABC, and pDonor. INTEGRATE is capable of integration of large DNA fragments into the *E. coli* genome in forward and reverse orientation using the target site duplication (TSD) of 5 bp and precise insertion is 46–55 bp downstream of the target site. Similar systems based on type I-B CASTs have also been reported. Based on INTEGRATE, MUCICAT systems were developed using distinct type I-F CASTs for multicopy and multiplexed DNA insertions, which is possible due to the alteration of the crRNA design. However, multicopy integration was less efficient in specific *E. coli* strains and another type I-F CAST was not successful for cargo insertion (Klompe et al., 2022).

As evidenced in the work conducted by Vo and other researchers, the INTEGRATE system works best when all of its components are placed in a single expression vector (Vo et al., 2021). This all-in-one vector may improve integration, probably because it is easier to transform a single plasmid than co-delivery of three plasmids. It also minimizes off-target effects and makes it easier to achieve directional DNA insertion. Furthermore, this all-in-one system, when combined with an orthogonal type V-K ShCAST, allows for sequential insertions and solves the immunity problems related to

ShCAST. It can be multiplexed and can be implemented in the *E. coli* genome and can also be used for other bacterial species. Rubin and colleagues designed VcDART, another all-in-one system derived from *Vibrio cholerae*, which contains barcoded genetic payload for ET-seq to facilitate the monitoring of edited cells. These developments demonstrate the general applicability of all-in-one type I CAST systems in bacterial engineering. Thus, type I-F CASTs offer better target integration specificity and lower co-integration than type V-K CASTs due to the nature of their structures and actions. Both the CAST systems can be integrated into a single plasmid for increased efficiency and precision. While type V-K CASTs demonstrate higher rates of off-target integrations and co-integration issues, these problems can be solved using the HELIX design, addition of components such as pir, and better control of the protein ratio. Due to the small size of the type V-K CASTs, they need only four proteins compared to seven proteins in the case of the type I-F CASTs and, therefore, are easier to build and to deliver. This compactness indicates that type V-K CASTs may be used in the future for genome editing of prokaryotes (Hwang et al., 2024).

12.5 APPLICATION OF CAST IN BACTERIAL STRAINS

12.5.1 Metabolic Engineering

E. coli BL21(DE3) is a popular choice for protein production, but it tends to have problems with the accumulation of acid byproducts. To address this issue, Chang et al. (2024) developed a 10.3- kb CRISPR interference (CRISPRi) module that enables the knockdown of four acid-producing genes. When this CRISPRi module is introduced into *E. coli* BL21(DE3) using SHOT, these genes are knocked down and the acid production is minimized that, in turn, increases the yield of recombinant protein and thus this strain can be a good candidate for recombinant protein production (Gelsinger et al., 2023). Furthermore, the C12KGET system is used for the integration of genes into bacterial strain related to the vitamin B12 synthesis pathway, increasing vitamin B12 synthesis (Gelsinger et al., 2023).

Shewanella oneidensis has potential applications in bioenergy production but it has been found that its metabolic activity is dependent on three-carbon substrates. To overcome this limitation, Cheng and colleagues improved the ShCAST system by integrating two crucial genes so that the strain can metabolize six-carbon sugars including glucose and fructose (Cheng et al., 2022). Furthermore, MUCICAT system based on type I-F CAST was applied to enhance the gene copy number of glucose dehydrogenase (GDH) in *E. coli* BL21(DE3), thus improving the GDH production dramatically (Zhang et al., 2020). MUCICAT has also been used to improve the production of N-acetylglucosamine (GlcNAc) in *E. coli*, creating a collection of strains with 1–11 cassettes of GlcNAc. Notably, the strain with the five copies of GlcNAc cassette produced the highest titer of GlcNAc (Zhang et al., 2021).

12.5.2 Gene-Editing Approaches in Diverse Microbial Populations

The functions of many bacteria and archaea are still not very well known, mostly because these organisms are difficult to culture in a laboratory. It would be ideal to edit genes in a specific organism at a specific locus, especially within microbial populations, although this is not easy since most microorganisms are part of complex populations. Vo et al. (2021) solved this issue by transforming the all-in-one INTEGRATE plasmid from *E. coli* into a mixed bacterial population. This approach allowed for targeted and highly effective RNA directed transposition in many diverse sources of microbiome. Similarly, with the help of the VcDART system and the ET-seq method, selection-free tagging and tracking of genetic mutant fitness in synthetic soil conditions are possible. This technique allows the generation of loss and gain of function mutations that impose a very strong selective pressure to select two specific bacterial strains out of the community. Most importantly, VcDART enables strain-specific gene editing in an infant gut microbiome and the subsequent recovery of edited members from the rest of the population. The application of CRISPR-associated transposon systems (CASTs) paves the way for functional genomics, fitness, and gene function in microbial consortia, especially those that have non-culturable bacteria. Also, CASTs might be targeted to suppress methane-producing bacteria in the intestines of ruminant animals, which might lower methane release – a potent greenhouse gas and an agent of climate change (Chang et al., 2024).

One of the possible future developments of CASTs may be related to microbiome manipulation for human well-being. CASTs may apply precisely targeting and regulate the composition of the gut microbiota, which may provide novel approaches to managing gastrointestinal diseases, inclusive of inflammatory bowel disease, obesity, or even neurodevelopmental disorders associated with the gut-brain axis. CASTs could potentially enable the optimization of the human gut microbiome for the purpose of delivering personalized medicine solutions of benefit to each patient's unique microbial ecosystem.

A future application of CRISPR-associated transposon systems (CASTs) in plants may include improving microbiome engineering to improve plant growth and stress tolerance. By using CASTs, one could potentially introduce beneficial changes into the "desirable" microorganisms in the soil, e.g., rhizobacteria which are involved in nitrogen fixation, nutrient uptake, or disease resistance in plants, and it improves the symbiotic association.

12.5.3 CAST for Eukaryotic Genome Editing

Tou et al. (2023) further proved that the type V-K CAST system from Nostoc sp. N_7CAST (PCC7101) can facilitate DNA insertion in HEK293T cell lysate. They created a refined version, N_7HELIX, that contains pN_7HELIX to produce transposition proteins, in addition to pTarget, pDonor, psgRNA, and pS15 (which expresses the S15 ribosomal protein required for type V-K CAST activity). By delivering this system into HEK293T cells, it was possible to introduce a 2.6- kb cargo into the target plasmid, located 57–62 bp downstream of the PAM, with only 7.9% co-integrated products, which in fact, reflect a 5.3-fold improvement as compared to original N_7CAST system. However, the integration efficiency was very low (Tou et al., 2023). In another study, Lampe and coworkers assembled pDonor, pTarget, and other plasmids expressing the proteins and RNA components of a type I-E Cascade system isolated from *Pseudomonas sp.* When these plasmids were delivered into HEK293T cells, the transfer of cargo from pDonor to pTarget was as low as 0.1%. For this, researchers screened 18 more CAST systems and found a transposon from *Pseudoalteromonas sp.* (PseCAST). The modified PseCAST, through systematic engineering, was able to achieve a transposition rate of 3–5% for the cargo from pDonor to pTarget; however, this efficiency reduces with the increase in the size of cargo from 0.8 kb to 15 kb. Further optimization made it possible to introduce donor sequences into several genomic sites, the integration rate ranged from 0.005% to 0.01%. Moreover, PTC (Post Transposition Complex) unfolding, and disassembly were identified as major factors that determine the efficiency of transposition (Lampe et al., 2023). To address this issue, bacterial ClpX, an ATPase from the MuA transposon that unfolds protein substrates, was coexpressed, and the integration efficiency was increased to about 1%, which implies that PTC disassembly restricts integration into genomic sites in HEK293T cells. Even though this work gives a starting point for using CAST systems in eukaryotic genome engineering and suggests the possibility of editing efficiency by systematic improvement of the CAST systems and the identification of helper proteins, the efficiency of this technique is still relatively low compared to CRISPR-Cas9 and Cas12 systems. Moreover, not all the existing CAST systems allow the user to edit eukaryotic genomes (Gelsinger et al., 2023).

CAST systems look promising for genome engineering in the future if there can be enhanced integration efficiency, eukaryotic compatibility and precision. Further optimization of the transposition rates through systematic engineering and identification of new helper proteins may make the CAST systems more comparable to CRISPR technologies. Extending their application to eukaryotic genomes, especially multiplex genome editing would create new areas in synthetic biology, agriculture, and gene therapy. CAST systems may prove to be useful in gene therapy with further optimization of the system to deliver large pieces of DNA with high efficiency. Combining with other genome-editing methods may also generate more flexible platforms; Therefore, CAST could be viewed as an ideal direction for further study and innovation.

12.6 ENHANCING GENE INSERTION IN EUKARYOTES VIA THE PASTE APPROACH

The researchers from MIT have come up with a new genome-editing tool known as PASTE (Programmable Addition via Site-specific Targeting Elements) to address some limitations of previous CRISPR based approaches. PASTE uses CRISPR-Cas9 and viral integrases to insert large DNA fragments without causing DNA double- strand breaks and thus minimizes the risk of chromosomal rearrangements (Yarnall et al., 2023). Another advantage over other CRISPR based approaches is that this system can also selectively target non-dividing cells. PASTE works by using CRISPR-Cas9 to establish a "landing pad," which is a site on the genome where larger pieces of DNA can be inserted. The viral integrase enzyme then integrates the desired DNA into this pad allowing for the insertion of sequences of up to 36,000 base pairs. In experiments, PASTE was able to show gene insertion across various human cell lines, including hepatic cells, T cells, and lymphoblasts, with insertion efficiency of 5% to 60% with low risk of off-targets or unintended mutations. Further, in vivo testing in humanized mouse livers was performed, where gene integration into the target cells was observed in approximately 2.5% of human hepatocytes, thus underlining the therapeutic applicability of PASTE. The potential to deliver large DNA fragments or construct without causing DSBs makes PASTE a valuable tool for gene therapy of genetic diseases including cystic fibrosis, hemophilia, and Huntington's disease. In addition, the authors are also confident that PASTE can be scaled up to tackle even larger DNA sequences in future work. This versatility makes way for more advanced gene therapies and makes PASTE a much safer and more flexible option in comparison to previous CRISPR-based therapies. It has potential for therapeutic use, but it could also serve as a valuable research and biotechnological instrument for introducing precise and controlled DNA fragments into large numbers of target cells. It will probably be the subject of further research in the future to ascertain its effectiveness in a range of clinical and laboratory applications (Yarnall et al., 2023).

12.7 CRASPASE: A NOVEL RNA-GUIDED ENDONUCLEASE WITH DUAL ACTIVITY

In the last ten years CRISPR has evolved as a revolutionary molecular tool for crop improvement, functional genomics, and diagnostic tools as well as therapeutic purposes. Currently CRISPR tools are mostly aimed at modifying DNA or RNA sequences which, sometimes, could cause off-target effects. Hu et al. (2022) described a novel form of a CRISPR/Cas system called CRISPR/Cas-III-E or Craspase, in a recent study. After the discovery of CAST which uses CRISPR-associated transposons for targeted DNA insertions, the discovery of Craspase is another interesting element in the RNA guided system. Instead of being involved in transposon-based genome editing as CAST, Craspase uniquely possesses the enzymes of both RNA-guided nuclease to cut nucleic acids and protease to cut protein targets. This multifunctionality greatly minimizes cytotoxicity and side effects, putting it in the ranking of potential solutions for numerous applications in genomic manipulations. Different from commonly used CRISPR systems, it can achieve RNA and protein levels of operation that showcase a promising and multifunctional system with significant potential in future gene research and gene-editing therapies (Hu et al., 2022).

12.7.1 Mechanism of Action: Exploring its Dual Functionality

The Type III-E Cas7-11 system consists of a peculiar protein referred to as gRAMP, which is frequently associated with a tetratricopeptide repeat (TPR) structure within the Craspase HetF Associated with TPRs (CHAT) domain – a caspase-like protein containing N-terminal TPR motifs. Caspases are categorized as cysteine proteases and because of this relationship the gRAMP (Cas7-11)-TPR-CHAT complex is described as the CRISPR-guided caspase or Craspase (van Beljouw et al., 2021). Craspase is an RNA-guided endonuclease and protease that is regulated through RNA-mediated allosteric regulation (Hu et al., 2022). By employing cryo-EM and biochemical assays, Hu et al. identified that a highly conserved gating loop between Cas7.2 and Cas11 has a critical function in controlling Cas7-11's RNase activity as well as the protease activity of the TPR-CHAT domain. Cas7-11 cleaves its target RNA at two precise sites: one of them positioned at the third nucleotide (site 1) and the other one at the ninth nucleotide (site 2), which led to the scission of the RNA into three parts with a six nucleotides space between the two cut sites (van Beljouw et al., 2024). In its apo form, the N-terminal region of the gating loop prevents gRAMP RNA cleavage by occupying the site 1 position. When a target RNA anneals to the gRAMP, it contacts the crRNA spacer, which induces a gating loop movement that activates the RNase catalytic site 1. However, in both the resting state and the state where the CHAT protease is self-RNA-bound, the protein is inactive owing to the presence of auto-inhibition on the protease domain (Hu et al., 2022). However, if non-self-RNA with a non-matching protospacer flanking sequence (PFS) associates with the complex, structural changes occur in the CHAT domain. This leads to the widening of the cleft between the gRAMP and TPR-CHAT domains, allowing the positioning of the peptide substrate to the binding surface and hence, protein cleavage (Hu et al., 2022). Thus, using this RNA-guided structural regulation, Craspase properly organizes crRNA-guided RNase and non-self-RNA-dependent protease activities that will allow it to cut RNA or protein targets under certain conditions (Hu et al., 2022). Furthermore, newer studies conducted revealed that Cas7-11 can be designed to be a compact form that fits within a single AAV vector, which mediates high-level transcript knockdown in human cells (Kato, Okazaki, Schmitt-Ulms, et al., 2022). This suggests Cas7-11/Craspase could be used for RNA and protein editing in vivo with many applications including the use in crop modification and gene therapy. The mechanism of Craspase is shown in Figure 12.2.

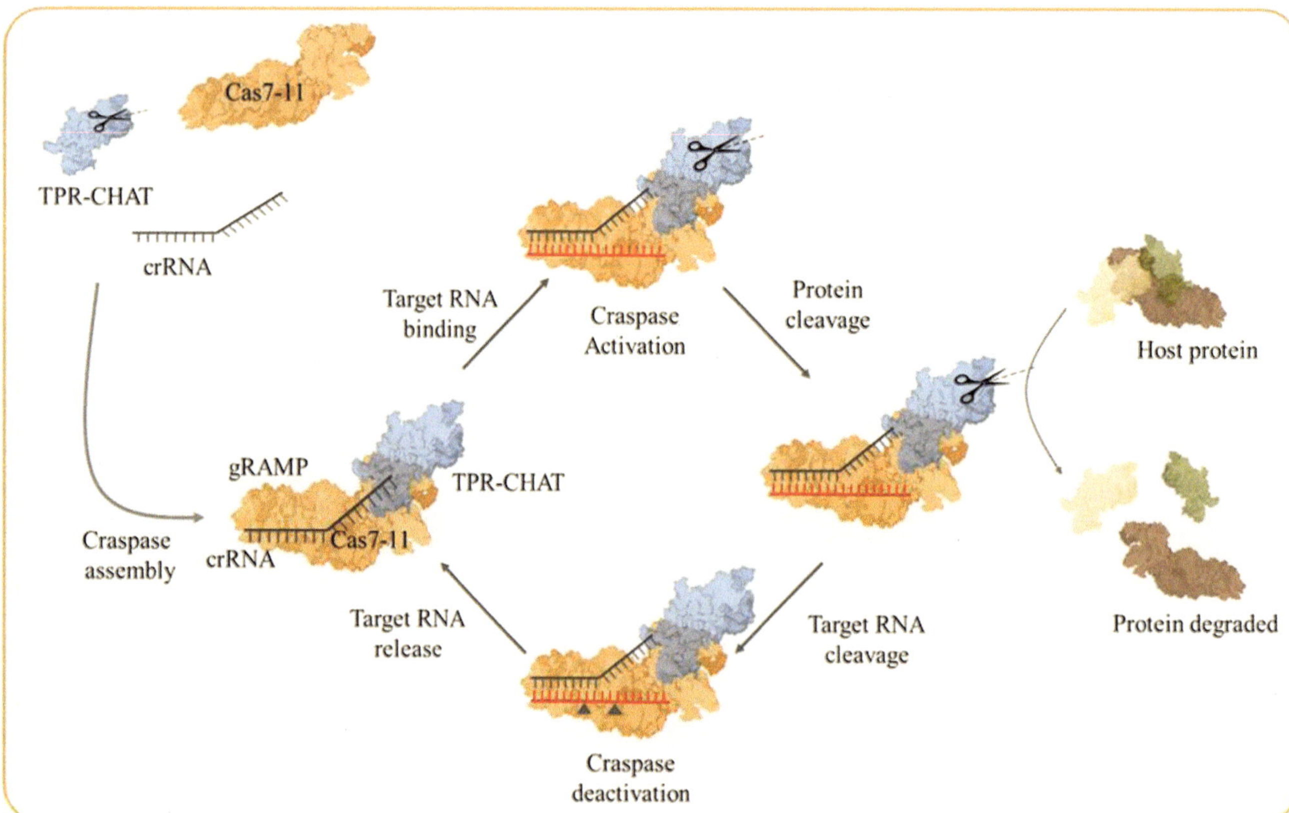

FIGURE 12.2 Mechanism of Craspase: Craspase works as a dual editor, switching between RNA and protein cleavage through complementarity and structural change. It can precisely target mRNAs and proteins linked to diseases. dCraspase retains binding ability without causing cleavage, allowing disease diagnostics and protein tracking.

12.7.2 Craspase as a Promising Tool for Advances in Gene Therapy and Biomedicine

Since the advent of CRISPR/Cas as a transformative genome-editing tool, research into human disease diagnostics and therapy has flourished and advanced impressively. A variety of CRISPR/Cas-based treatments for genetic conditions and cancers have made it into clinical trials. Precise and safe application concerns still exist and restrict its expanded use in clinical settings. A key shortcoming is the substantial rate of off-target changes noted with CRISPR/Cas9 as the principal CRISPR technique in therapeutic and basic research. Findings reveal that CRISPR/Cas9 edits might produce unwarranted mutations like significant genome deletions or structural alterations at multiple sites both on- and off-target. The recently discovered Craspase system presents a favorable option because it focuses on RNA and protein sequences rather than DNA. The research published by Hu et al. (2022) shows that Craspase uses a distinctive mechanism to control targeting of RNA and protein sequences while greatly lowering off-target impacts. Unlike competing systems, Craspase does not cause collateral RNA damage and is non-toxic to eukaryotic cells. Due to cytotoxicity and immunotoxicity being major issues for contemporary CRISPR systems such as CRISPR/Cas9 and CRISPR/Cas13 in gene therapy settings, Craspase presents valuable advantages. In addition to its use in gene therapy, Craspase demonstrates potential for diagnosing and treating infectious diseases, such as HIV and COVID-19, similarly to Cas9 and Cas12. To serve as a molecular imaging tool Craspase must be modified to stop its RNase and protease activities while still binding to its targets. It can closely monitor RNA and protein targets within human cells, depending on disease conditions like cancer, and provide new understanding of the mechanisms behind different human diseases. As a more versatile and secure solution, Craspase further extends the range of CRISPR technology to new areas beyond standard gene-editing applications (Yang and Zhang, 2023).

12.7.3 Potential Applications of Craspase in Future

Thus, Craspase with its dual targeting mechanism of RNA and protein sequences opens new horizons of further applications. Its dual activity opens new areas of application in disease diagnostics, enabling the design of precise diagnostic systems to identify RNA viruses such as SARS-CoV-2 or track cancer markers in real time. In the area of drug design for the treatment of metastatic malignancies, Craspase maintains precise control of the desired oncogenes and tumor suppressor genes at the RNA and protein level, excluding the potential risk of off-target effects and genomic instability. Researchers could also follow Craspase for functional genetics and proteomics analysis, since its bidirectional inactivation may reveal new aspects of RNA and protein interactions and regulate novel pathways contributing to various cell functions and diseases. Moreover, molecular imaging and tracking application form another interesting prospect; if the cleavage activity of Craspase is inactivated, it could be used to track the location of RNA and proteins in living cells – a significant contribution to understanding certain biological events in multicellular organisms. Craspase has further potential in the management of viral diseases since it can be developed to interfere with RNA virus replication by selectively targeting viral RNA and proteins to give new approaches to fight HIV, hepatitis C, and influenza viruses. Furthermore, in the context of the synthetic biology of proteins and enzymes, Craspase can be used to direct the synthesis and modifications of new biomolecules and metabolic pathways. Craspase potential also lies in the sphere of pharmacology, with the specificity of the reaction being remarkably high, and the off-target effects being negligible. Craspase seems to be a perfect substance for the development of highly individualized approaches to complex disease treatments, such as cancer, and autoimmune diseases. Altogether these potentially applicable fields shed light on the multifunctional potential of Craspase as well as its revolutionary role in future investigations and therapeutic approaches in different scientific disciplines.

12.8 EXPLORING THE OMEGAS: A NEW ERA OF RNA-GUIDED ENZYMES IN GENOME EDITING

Over the last ten years, scientists have converted the CRISPR systems from prokaryotes to gene-editing tools that can target DNA sequence with precision. In a recent study, researchers from MIT and Harvard have found a novel class of programmable DNA modifying systems that they refer to as OMEGAs or Obligate Mobile Element Guided Activity, that might have a natural function in the shuffling of small regions of DNA within bacterial chromosomes. OMEGA enzymes are derived from bacteria but have been modified to function in human cells, which suggests their prospect in gene therapy. The comparatively small size of these proteins, about 30% that of Cas9, makes them easier to deliver into cells, making them desirable substitutes for large previously reported related proteins. The findings from this study also reveal that natural RNA-guided enzymes are among the most abundant proteins on our planet and mark the beginning of a new field of research that has the potential for a revolutionary advancement in the use of new genome-editing tools (Tae, 2024). Information about the first RNA-directed OMEGA proteins was initially acquired through the study of IscB proteins, which are not related to CRISPR immunity and do not interact with RNA. The small RNA which is associated with each IscB protein directs the enzyme to target specific DNA sequences, which the team named ωRNAs. Subsequent studies showed that two other classes of small proteins, IsrBs and TnpBs ,that are among the most frequent genes in bacteria, also use ωRNAs as a guide to cleave DNA. These proteins, IscB, IsrB, and TnpB, are found in mobile genetic elements called transposons. This study also indicated that every time these transposons mobilize, they synthesize a new guide RNA, which makes it possible for the encoded enzyme to cleave DNA at specific locations. The TnpBs and IscBs are believed to be ancestral to the Cas9 and Cas12 CRISPR systems. The research team believes that there might have evolved other RNA-guided enzymes that they wish to discover (Badon et al., 2024).

12.9 COMPARISON BETWEEN OMEGA AND CRISPR/CAS SYSTEMS

These differences in structural organization and function of Cas9 and Cas12 suggest that Cas nucleases evolved from different ancestral proteins. Cas9 is believed to have evolved from the IscB protein, a member of the IS200/IS605 transposon family. In 2021, in a detailed analysis of the transposon superfamily, researchers came across another subset that encodes programmable transposon encoded RNA-guided nucleases, referred to as the TnpB family. These nucleases are collectively called obligate mobile element guided activity or OMEGA, and their distribution indicates that RNA-guided mechanisms are more common in prokaryotes. These analyses also demonstrate that these nuclease families had evolved independently and TnpB has been identified as a potential ancestor to both Cas12 and Fanzor, a eukaryotic RNA-guided DNA nuclease. Similarities in the structural characteristics of the OMEGA system and the CRISPR-Cas system have been discovered, which signifies that the OMEGA system could possibly have the same potential for research use. The OMEGA nucleases also consist of effector nuclease and ωRNA which is similar to the Cas effector nuclease and gRNA in the case of CRISPR. The effector proteins present in OMEGA system contain a RuvC nuclease domain which, in response to TAM sequence in the target genome, cleaves the target DNA. ωRNA is a ncRNA molecule that contains hairpin-like structures and variable sequences that help the OMEGA system to recognize its target sequence, as seen in the crRNA of CRISPR-Cas. TnpB belongs to the IS200/IS605 group, it is present in most archaea and bacteria, and is more abundant than Cas12 (Hirano et al., 2022). Because of their homology, TnpB has higher structural resemblance to Cas12 and has REC and NUC lobes divided into two halves (Badon et al., 2024). However, TnpB only possesses the catalytic domains and is even more compact, as it does not contain the nonessential domains of Cas12. Additionally, the TnpB gene, which is involved with ωRNA transposition, is positioned near upstream of ωRNA, similar to Cas12 gene which is also positioned upstream of the CRISPR array. This is opposite to the IscB and Cas9 systems where the ωRNA and CRISPR array is positioned upstream before the reading frame of their respective genes. The mechanism of action of TnpB and Cas12 are also similar: TnpB binds the target-adjacent motif (TAM) on the target DNA, as Cas12 binds with PAMs on the target DNA. The binding of ωRNA to the target DNA sequence stimulates R-loop formation, causing DNA cleavage. Nevertheless, TnpB is not capable of contacting the distal TAM at all, and binding is less stable as compared to Cas12, which forms a contact with the distal PAM for stable binding (Badon et al., 2024). Comparison between OMEGA, Cas12 and Fanzor is shown in Figure 12.3.

12.9.1 Cleavage Mechanism of CRISPR-Cas12 and OMEGA Effectors

The first step in the mechanism of different CRISPR systems is the interactions between crRNA and Cas effectors and between the crRNA and tracrRNA, which collectively define the target DNA recognition. Such complexes hybridize with the sequence that is complementary to the guide RNA, leading to DNA cleavage. Recent studies have addressed the characterization of the RNA-guided endonuclease function of the CRISPR-Cas12a system at the single-molecular level. First, the CRISPR-Cas12a recognizes the target sequence in the DNA and then binds to the PAM sequence, the crRNA causes conformational change in the CRISPR-Cas12a effector, leading to DNA recognition and subsequent cleavage.

Unlike the CRISPR-Cas9 system that possesses two domains of HNH and RuvC cleavage, the CRISPR-Cas12a model has just one RuvC domain to induce double-stranded DNA cleavage through the successive cleavage of the NTS (Non-Target Strand) and TS (Target Strand). The NUC domain, connected to the RuvC domain, helps to load the RuvC domain of the TS after the NTS is cleaved by RuvC domain. This sequential cleavage mechanism is further supported by the loss of cleavage functionality of both NTS and TS when the RuvC domain in the Cas12b effector is mutated. On the other hand, there are other Cas12c, Cas12m, Cas12k effectors reported which have non canonical cleavage pockets generally associated with a single divalent metal ion which mainly mediate transcriptional gene suppression by forming a complex with target DNA without causing double-stranded breaks. In particular, the Cas12m identified recently exhibited high transcriptional suppression by interacting with the REC and RuvC domains and binding to the NTS with high affinity (Badon et al., 2024).

Moreover, the OMEGA series of RNA-guided endonucleases, ISDra2, TnpB and SpuFanzor, has only one RuvC domain containing the specific D/E/D residues responsible for efficient DNA cleavage. Indeed, in the case of the TnpB effector, the binding to target DNA results in structural changes that lead to the release of the RuvC domain. How this activated effector cleaves the target DNA has also been elucidated and, as for the Cas12 family effectors, the structure of TnpB-ωRNA-target DNA indicates that it involves a more compact triple complex double-helix cutting system. Most importantly, OMEGA RNA binds to the RuvC domain to enhance the stability of the Fanzor effector for DNA cleavage. It seems to indicate that ωRNAs have a profound evolutionary role to play in cleavage in addition to the RuvC domain.

Significantly, both the CRISPR-Cas12 series and the ISDra2 TnpB effector present a non-specific trans-cleavage effect: non-specific ssDNA cleavage occurs after target recognition and cleavage. Following the guide RNA-based target recognition, each of the endonuclease effectors undergoes conformational changes that expose the RuvC domain to the solution for ssDNA cleavage. This character has been used in several different DNA-based detection systems (Badon et al., 2024).

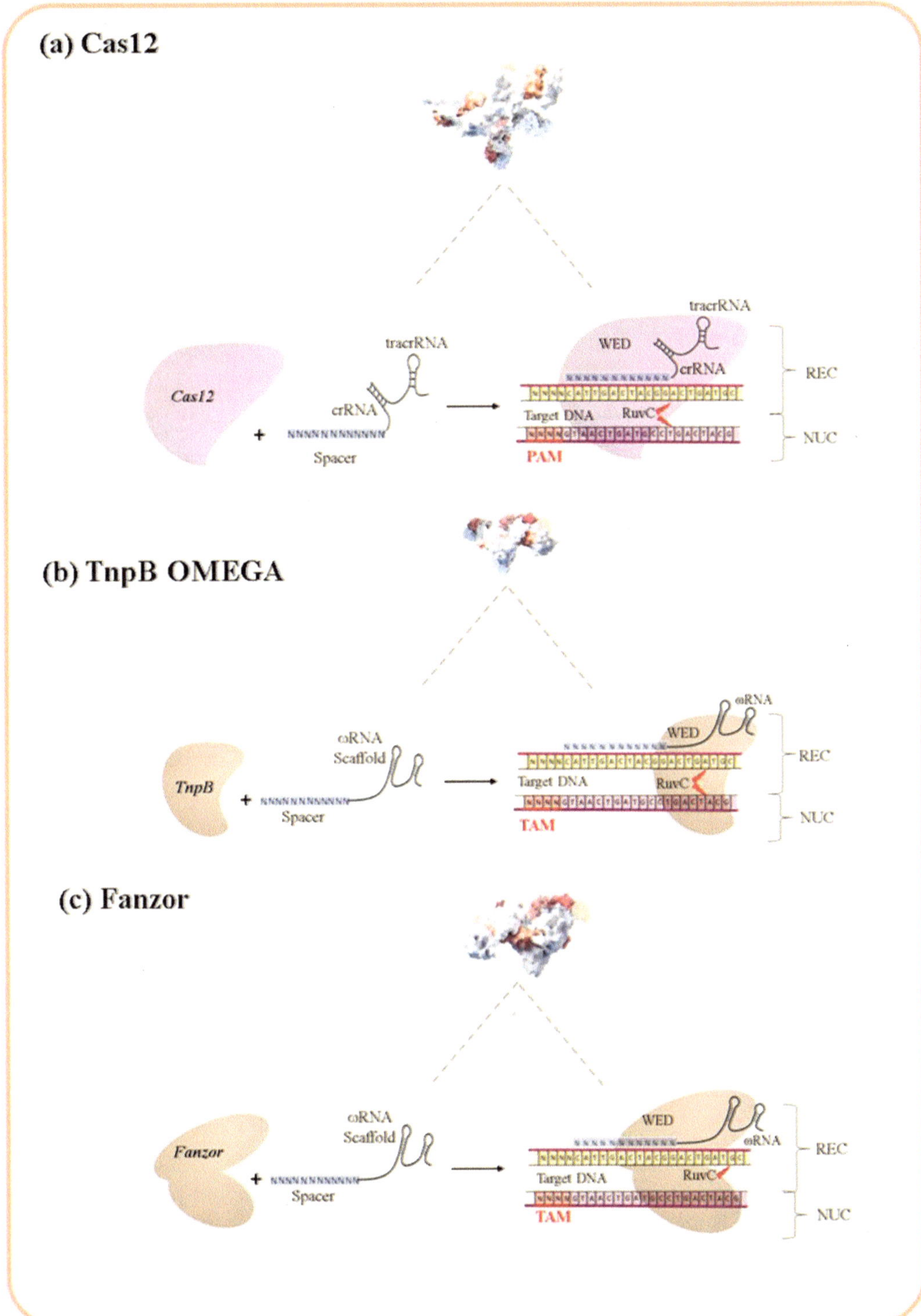

FIGURE 12.3 Comparison of Cas12, OMEGA and Fanzor (a) Cas12, originated from TnpB, combines with gRNA (tracrRNA and crRNA) and uses a similar WED-RuvC mechanism to make staggered DSBs at the G-rich PAM sites. TnpB, a compact precursor, forms a ribonuclease complex with ωRNA, and like Cas12, cleaves DNA via its RuvC domain. (b) OMEGA uses a similar mechanism, but is different in its recognition and cleavage property (c) Fanzor, evolutionarily originated from TnpB through horizontal gene transfer from prokaryotes to eukaryotes, makes a complex with ωRNA, the WED domain recognizes TAM sequences, the RuvC domain results in staggered double-strand breaks (DSBs). Cas12, originated from TnpB, combines with gRNA (tracrRNA and crRNA) and uses a similar WED-RuvC mechanism to make staggered DSBs at the G-rich PAM sites. TnpB, a compact precursor, makes a ribonuclease complex with the ωRNA, and similar to Cas12, cleaves DNA through its RuvC domain.

OMEGA elements do not employ gRNA like Cas9 or Cas12. To target specific genomic regions, they can be engineered or modified in several ways. Essentially, the DNA-binding domains of OMEGA can be modified or can be combined with proteins that specifically bind to a specific DNA sequence such as TALENs or zinc fingers, or linked to inactive Cas9 (dCas9) for target recognition. Homology-dependent targeting can also be applied by providing the necessary homologous sequences to direct the OMEGA element to the targeted position. These strategies are useful in improving the specificity of OMEGA in genome editing.

12.9.2 Off-Target Features of Cas12 and OMEGAs

It is imperative and essential to achieve high accuracy when implanting gene modifications or gene insertions into living organisms. Among the numerous methods of gene editing, the use of Cas12 has attracted considerable interest surpassing even the Cas9 system due to the increased ability to detect mismatches in the DNA/RNA heteroduplex formed by the guide RNA. This sensitivity leads to decreased probability of off-targets due to formation of mismatches in DNA/RNA heteroduplex by causing an unfavorable conformation or geometry for cleavage of the target DNA. Specifically, many Cas12 systems characterized here show sensitivity to the mismatches at the positions that were considered as the seed region proximal to the PAM sequence. To evaluate the target recognition capabilities of such RNA-guided endonucleases, some techniques can be employed, such as CIRCLE-seq, SITE-seq, CROss-seq, and Digenome-seq. which allow accurate identification of possible off-target position or candidates, thus reducing off-target effects and unintended mutations, even with Cas12 and OMEGA systems. Recently reported Cas12f based gene-editing experiments have shown that it is highly efficient for off-target detection and identifies small insertions and deletions (indels) and large deletions and translocations due to the double-strand breaks (DSBs). Furthermore, TnpB effectors possess comparably similar levels of sensitivity to mismatches in the DNA-RNA heteroduplex like CRISPR-Cas12. Especially in the proximal seed area of the TAM sequence, TnpB effectors show a high specificity to target DNA. This degree of selectivity strongly lowers the chances of off-target mutations, thereby increasing the efficiency of gene-editing tools. Collectively, ongoing investigations of these systems provide increased insights into the precision of gene editing for various applications in biotechnology and precision medicine (Kato, Okazaki, Kannan, et al., 2022).

12.9.3 Applications of OMEGA

In 2021, Altae-Tran and colleagues revealed that the prokaryotic IscB protein (≈400–500 amino acids) interacts with ωRNA to promote DNA cleavage in eukaryotic cells. This ωRNA, which is at least 250 nucleotides long, is contained within the left end (LE) of the transposon. The IscB-ωRNA complex cleaves dsDNA at 3′ target-adjacent motifs (TAMs). Out of six IscB proteins, compact OgeuIscB (496 aa) was found to incorporate indels into the genome of HEK293FT cells with frequency of 4.4% (Chang et al., 2024).

At the same time, Karvelis and Siksnys (2023) identified that the TnpB protein of *Deinococcus radiodurans* ISDra2 (408 amino acids) interacts with ωRNA, which is approximately 150 nucleotides in length and is located within the transposon right end (RE). This identified TnpB-ωRNA complex, therefore recognizes TAMs and possessed plasmid cleavage in vitro. When the plasmid containing TnpB and ωRNA was transduced into HEK293T cells, indels were created with a frequency of between 10–20%, which is similar to that observed for CRISPR-Cas9 and Cas12. TnpB also produces a staggered DNA cut and can be viewed as the structural and functional minimal core of Cas12 and its probable ancestor. Notably, TnpB has ωRNA biogenesis ability from its own mRNA indicating that all Cas12 guide RNAs share this feature (Chang et al., 2024).

Very recently, Xiang and collaborators reported new TnpB proteins, ISAam1 (369 AA) and ISYmu1 (382 AA), which can promote indels at different target sites in HEK293T cells (Xiang et al., 2024). These proteins demonstrated even higher efficiency than three Cas12f editors and similar efficiency comparable to SaCas9 (1053 amino acid residues), though this protein is considerably smaller. In general, mutations were mainly deletions, whereas insertions occurred much less – similar to the modes of action of CRISPR-Cas systems. Importantly, the compact ISAam1 editor was encapsulated into AAV8 and delivered into mice with the editing efficiency above 5%, which was superior to Un1Cas12f1 and SaCas9. This landmark experiment proved that the hypercompact OMEGA system can be delivered in vivo using AAVs and paved the way for TnpB as a human genome editor (Xu et al., 2023).

12.10 FANZOR: A NEW FRONTIER IN GENOME EDITING

Fanzor, which shares similarities to the OMEGA system TnpB, was first described as a eukaryotic TnpB-IS200/IS605 ortholog that modulates TnpB transposable element activity by methylation. However, recent molecular investigation has found that Fanzor produces a eukaryotic transposon-encoded protein which plays a role as a chimeric RNA-guided DNA endonuclease, making it a potential candidate for gene editing. Two primary types of Fanzor proteins have been identified: Fz1 and Fz2. Fz1 is present in fungi, arthropods, protist, plants, and eukaryotic viruses life forms. While Fz2 proteins are found more frequently in molluscs, choanoflagellates, viruses, and fungi. The phylogenetic trees revealed evolutionary relationship between Fanzor and TnpB may indicate that Fanzor is the eukaryotic OMEGA system. On that basis it has been suggested that the two types of Fanzor proteins were generated from distinct TnpB subtypes. The source of the OMEGA system is assumed to be horizontal gene transfer from prokaryotic organisms to eukaryotic organisms and eukaryotic viruses function as a means for the spread of this CRISPR-Cas-like system as shown in Figure 12.4. Perhaps,

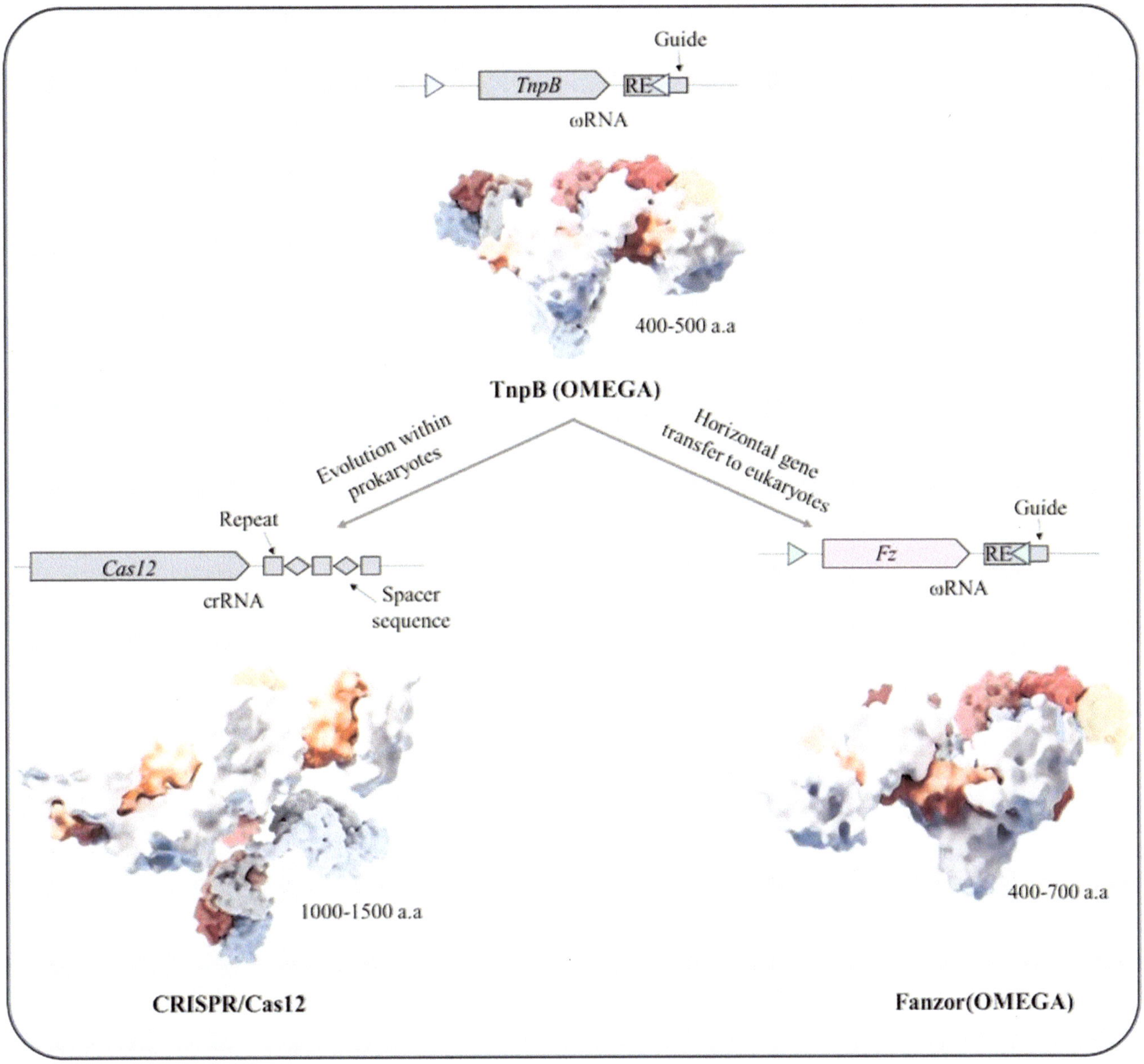

FIGURE 12.4 Evolutionary relationship between OMEGA, Cas12, and Fanzor, CRISPR/Cas systems are the ancestors of OMEGA. The ω protein (TnpB) later became associated with the CRISPR array, and evolved into CRISPR/Cas12 in prokaryotes and Fanzor in eukaryotes.

the TnpB proteins might have developed into Fanzor in eukaryotes over the years (Saito et al., 2023).

12.10.1 Structural and Functional Characterization of Fanzor: Comparisons with TnpB and Cas12

To determine the structural and functional relationships between Fanzor and TnpB and Cas12, Saito et al. (2023) studied several Fanzor orthologs isolated from soil fungi, Percolozoa, and marine molluscs. As it is the case with TnpB and Cas12, Fanzor has a bilobed structure, which contains both REC and NUC domains. At a molecular level, Fanzor also repeats TnpB by exhibiting a conserved endonuclease RuvC region and a ωRNA-recognition WED region. One conserved alternate glutamate residue, required for catalytic activity, is positioned in the RuvC domain. Conversely, some elements specific to each Fanzor ortholog were identified in the core domain, indicating that the Fanzor airships have evolved. Nevertheless, Fanzor has structural characteristics that seem to set it apart from TnpB and Cas12. For instance, it is missing several domains which are evident in the Cas12 family, but it has been suggested that ωRNA scaffolds may act as a substitute for the lost regions. Furthermore, the presence of a pseudoknot structure, which has been observed profoundly in TnpB and Cas12 ωRNAs, is absent in Fanzor. Another difference in structure relates to the WED domain of Fanzor where three α-helices are found to be essential for target DNA binding. Also, Fanzor is considerably more compact than TnpB and Cas12, mainly due to its main core domain and the complete absence of specific features (Awan et al., 2023).

12.10.2 Mechanism

Based on biochemical characterizations, it has been found that Fanzor functions as an RNA-guided DNA endonuclease similar to the TnpB and Cas12. Fanzor employs a small non-coding RNA, called ωRNA that navigates its endonuclease activity toward the target DNA. ωRNA is first transcribed, and then interacts with Fanzor's RuvC domain, which orients the effector complex in such a way that it can perform with the utmost efficiency. This stabilized complex allows Fanzor to recognize its target DNA sequence that binds to the ωRNA. The Fanzor WED domain binds to a specific target adjacent motif (TAM) on a DNA strand on which the sequence recognition involves binding to DNA. When the TAM is recognized and the ωRNA-DNA heteroduplex is generated, Fanzor cuts the target DNA, leading to staggered double-strand break (DSB). Interestingly, Fanzor does not seem to cleave other unrelated sequences as does Cas12, making it more specific and perhaps ideal for genome-editing applications. Several Fanzor orthologs were subjected to the sequencing of the cleaved products, analysis of which exposed differences in the cleavage patterns. The data implied that Fanzor proteins from different species recognize unique TAM sequences. While TnpB and Cas12 tend to be more T-biased for their target motifs in comparison to other reported CRISPR systems, there is a diverse range of motifs exhibited by Fanzor orthologs. Any alterations or mutations to the TAM region affect Fanzor's DNA cleavage activity, demonstrating the necessity of specific recognition of the TAM site. While the exact biological function of Fanzor is still ambiguous, its association with transposases present in eukaryotic genomes indicates that, similar to TnpB in prokaryotic genomes, Fanzor might be involved in transposon function and genome rearrangements in eukaryotes. More investigations are still required to explain its physiological role and evolutionary completion (Awan et al., 2023; Tae, 2024).

12.10.3 Potential of Fanzor as a Novel Genome-Editing Tool

While it is uncertain what further aspects of biology Fanzor governs, its discovery has been the source of enthusiasm in the academic sphere. The identification of Fanzor as an RNA-guided endonuclease in eukaryotic systems indicates that RNA-guided DNA targeting system may be widely used in all the kingdoms of the classification. This naturally poses interesting questions about the evolutionary history of such systems and extends the versatility of applications. Furthermore, the discovery of a functional RNA-guided endonuclease system in eukaryotes is highly significant for applications in genome editing because the system may overcome some of the current drawbacks of CRISPR-Cas, for example, in exciting immune reactions when transferred into human cells. Some of the first experiments revealed that Fanzor may be used as a genome-editing tool in human cells. With four types of Fz1 orthologs and eight guide RNA sequences used, the eight tested loci in human genome were targeted effectively. However, three of these orthologs displayed editing efficiency at least as high as AsCas12f1, a compact variant of the Cas12 family. However, the recombinant Fz1 produced from *Spizellomyces punctatus* (SpuFz1) demonstrated encouraging outcomes, which stimulated the optimization effort. Experiments including the addition of more ωRNA units or by altering specific amino acids significantly increased the nuclease activity of the engineered SpuFz1 as well as the interaction between the protein and target DNA. Based on these findings, Fanzor may be designed to achieve high accuracy and versatility for different genome-editing applications, especially where native CRISPR systems may prove suboptimal. Not only does Fanzor have these functional applications, but it also has structural attributes that are desirable for gene editing. Due to its relatively smaller size compared to Cas12 and TnpB, it is easier to transfer into target cells, because it may be better adapted to incorporate into viral vectors or nanoparticles. In addition, Fanzor creates DSBs with sticky ends that promote accurate DNA insertion by HDR because of the efficiency of sticky ends compared to blunt ends. The versatility of target-adjacent motifs (TAMs) across different Fanzor orthologs also broadens its targeting options. Further, Fanzor versatility at working temperature range 4–70°C increases the possibility of its application in diverse experimental and therapeutic conditions to become a game changer in genome editing in basic and clinical research areas (Karvelis & Siksnys, 2023).

12.10.4 Potential Applications of Fanzor in Genome Editing and Therapeutic Interventions

Different studies indicate that Fanzor is safer for human genome editing compared to other similar techniques with a high level of efficiency with the minimum activation of immune response. This characteristic makes Fanzor an interesting candidate for applications such as personalized cancer treatment. In such treatments, a patient's own T-cells, an important part of the body's immune system that identifies pathogens and immune system invaders, could be genetically altered to only recognize cancer cells. In addition, Fanzor might emerge as a viable candidate to transform the fields of regenerative medicine and beyond by focusing on stem cell applications for the treatment of numerous genetic diseases such as type 1 diabetes and neurodegenerative diseases. Furthermore, Fanzor may be applied to germline, to modify the DNA of the eggs and sperm for the eradication of various types of inherited diseases as well as for solutions to the problems of infertility. Although this approach could lead to disease prevention before conception, it has inherent risks and poses important ethical questions, making it necessary to encourage the formulation of clear standards. Besides, it can be used in targeted genome modification in plants making Fanzor a very potentially useful tool. These positive factors also include its small size and high specificity which could be used to increase nutrient densities in crops, improve crop resistance to diseases and increase yield. As climate change challenges progress, Fanzor can be used to create new crops that may be resistant to stress factors including water shortage, salty soil, and high or low temperatures. This could go a long way in enhance the sustainability of agriculture and food security in the world (Jiang et al., 2023).

12.11 CONCLUSION

In conclusion, we are in the age on the brink of a revolution in genome editing in which new technologies are not only revolutionizing our scientific future but our grasp on the capability of human life. The progression from retron-based gene editing and the Retron Library Recombineering right up to the efficiency of the CRISPR-associated transposons, or CAST, shows these are not simply tools: they are doors to a world full of possibility. The development of CRASPases marks a new era of targeted regulation of gene expression with higher precision than ever before and offers an interface between genomic and proteomic disciplines. Using OMEGA systems and Fanzor proteins we are provided with compact, efficient, and highly selective tools that enable us to confront biological problems directly. A world where not only genetic disorders are cured but genetic disorders do not exist, a world where plants that feed humanity adapt to climate change through biotechnology, and a world where ownership of life is taken to an entirely new level by using biotechnology to modify genes. It is almost already too much to imagine, and envisioning the possibilities of the effects it could have on humanity's health, food, and the environment are substantial. As researchers, we are now at the threshold of a new creation, possessing molecular weapons which gives us the potential to rewrite human genetic code. This is our time to break the mold, overcome expectations, and venture into areas not explored with reference to biotechnology. It will be my pleasure to have you on this exciting ride that is designed to bring together professionals from different disciplines all working and learning as one team in one environment.

12.12 FUTURE PERSPECTIVES

The future of genome editing looks quite promising. It will be possible to expect major advances in pharmacogenomics: targeted medical treatments with respect to specific tendencies of human DNA related to disease, which will increase the efficiency and decrease side-effects. It will be easier to predict the outcome when artificial intelligence is incorporated with genome editing and this will facilitate designing of better interventions. Guided RNA design can be optimized with AI, editing outcomes can be predicted by AI, and delivery methods improved by AI. The synergy that results will allow researchers to better predict outcomes and design more complex interventions that were once unimaginable. Genome editing is not just bright – but a canvas opens for collective imagination and ingenuity in the future of genome editing. Moreover, should we begin extending the enhancements beyond disease prevention – like changing intelligence, physical abilities and so on, we must be very careful. One issue is that it is sometimes difficult to distinguish between therapeutic uses and mere "enhancements" on which we are presented with ethical questions, which we can call "playing God." The strong ethical foundations needs to be set to prevent the ethics of the research and applications from losing its substantial moral base.

REFERENCES

Ahmad, A., Munir, A., Zafar, H., Zahoor, M. K., Hassan, S., & Khan, S. H. 2024. Tracking footprints of CRISPR-based genome editing. In K.A. Abd-Elsalam & A. Ahmad (Eds.), *Global regulatory outlook for CRISPRized* plants. Elsevier.

Altae-Tran, H., Kannan, S., Demircioglu, F. E., Oshiro, R., Nety, S. P., McKay, L. J., ... & Zhang, F. 2021. The widespread IS200/IS605 transposon family encodes diverse programmable RNA-guided endonucleases. *Science,* 374(6563), 57–65.

Arvas, Y. E., Marakli, S., Kaya, Y. & Kalendar, R. 2023. The power of retrotransposons in high-throughput genotyping and sequencing. *Frontiers in Plant Science,* 14, 1174339.

Awan, M. J. A., Awan, M. R. A., Amin, I. & Mansoor, S. 2023. Fanzor: A compact programmable RNA-guided endonuclease from eukaryotes. *Trends in Biotechnology,* 41(11), 1332–1334.

Badon, I. W., Oh, Y., Kim, H.-J. & Lee, S. H. 2024. Recent application of CRISPR-Cas12 and OMEGA system for genome editing. *Molecular Therapy,* 32, 32–43.

Brouha, B., Schustak, J., Badge, R. M., Lutz-Prigge, S., Farley, A. H., Moran, J. V. & Kazazian Jr, H. H. 2003. Hot L1s account for the bulk of retrotransposition in the human population. *Proceedings of the National Academy of Sciences,* 100, 5280–5285.

Buffington, J. D., Kuo, H.-C., Hu, K., Chang, Y.-C., Javanmardi, K., Voigt, B., Li, Y.-R., Little, M. E., Devanathan, S. K. & Xhemalce, B. 2024. Discovery and engineering of retrons for precise genome editing. *bioRxiv,* 2024.07. 21.604473.

Carattoli, A., Seiffert, S. N., Schwendener, S., Perreten, V. & Endimiani, A. 2015. Differentiation of IncL and IncM plasmids associated with the spread of clinically relevant antimicrobial resistance. *PloS One,* 10, e0123063.

Chang, C.-W., Truong, V. A., Pham, N. N. & Hu, Y.-C. 2024. RNA-guided genome engineering: Paradigm shift towards transposons. *Trends in Biotechnology,* 42, 970–985.

Cheng, Z.-H., Wu, J., Liu, J.-Q., Min, D., Liu, D.-F., Li, W.-W. & Yu, H.-Q. 2022. Repurposing CRISPR RNA-guided integrases system for one-step, efficient genomic integration of ultra-long DNA sequences. *Nucleic Acids Research,* 50, 7739–7750.

Crook, N., Abatemarco, J., Sun, J., Wagner, J. M., Schmitz, A. & Alper, H. S. 2016. In vivo continuous evolution of genes and pathways in yeast. *Nature Communications,* 7, 13051.

Ellington, A. J. & Reisch, C. R. 2022. Efficient and iterative retron-mediated in vivo recombineering in Escherichia coli. *Synthetic Biology,* 7, ysac007.

Frost, L. S., Leplae, R., Summers, A. O. & Toussaint, A. 2005. Mobile genetic elements: The agents of open source evolution. *Nature Reviews Microbiology,* 3, 722–732.

Gelsinger, D. R., Vo, P. L. H., Klompe, S. E., Ronda, C., Wang, H. & Sternberg, S. H. 2023. *Bacterial genome engineering using CRISPR RNA-guided transposases*. bioRxiv.

Harrison, E. & Brockhurst, M. A. 2012. Plasmid-mediated horizontal gene transfer is a coevolutionary process. *Trends in Microbiology,* 20, 262–267.

Hirano, S., Kappel, K., Altae-Tran, H., Faure, G., Wilkinson, M. E., Kannan, S., Demircioglu, F. E., Yan, R., Shiozaki, M. & Yu, Z. 2022. Structure of the OMEGA nickase IsrB in complex with ωRNA and target DNA. *Nature,* 610, 575–581.

Hou, S., Brenes-Álvarez, M., Reimann, V., Alkhnbashi, O. S., Backofen, R., Muro-Pastor, A. M. & Hess, W. R. 2019. CRISPR-Cas systems in multicellular cyanobacteria. *RNA Biology,* 16, 518–529.

Hu, C., Van Beljouw, S. P., Nam, K. H., Schuler, G., Ding, F., Cui, Y., Rodríguez-Molina, A., Haagsma, A. C., Valk, M. & Pabst, M. 2022. Craspase is a CRISPR RNA-guided, RNA-activated protease. *Science,* 377, 1278–1285.

Hwang, J., Ye, D.-Y., Jung, G. Y. & Jang, S. 2024. Mobile genetic element-based gene editing and genome engineering: Recent advances and applications. *Biotechnology Advances, May–June,* 72, 108343.

Jiang, K., Lim, J., Sgrizzi, S., Trinh, M., Kayabolen, A., Yutin, N., Bao, W., Kato, K., Koonin, E. V. & Gootenberg, J. S. 2023. Programmable RNA-guided DNA endonucleases are widespread in eukaryotes and their viruses. *Science Advances,* 9, eadk0171.

Karvelis, T. & Siksnys, V. 2023. Fanzors: Mysterious TnpB-like bacterial transposon-related RNA-guided DNA nucleases of Eukaryotes. *The CRISPR Journal,* 6, 310–312.

Kato, K., Okazaki, S., Kannan, S., Altae-Tran, H., Esra Demircioglu, F., Isayama, Y., Ishikawa, J., Fukuda, M., Macrae, R. K. & Nishizawa, T. 2022. Structure of the IscB–ωRNA ribonucleoprotein complex, the likely ancestor of CRISPR-Cas9. *Nature Communications,* 13, 6719.

Kato, K., Okazaki, S., Schmitt-Ulms, C., Jiang, K., Zhou, W., Ishikawa, J., Isayama, Y., Adachi, S., Nishizawa, T. & Makarova, K. S. 2022. RNA-triggered protein cleavage and cell growth arrest by the type III-E CRISPR nuclease-protease. *Science,* 378, 882–889.

Kaur, N. & Pati, P. K. 2024. Retron library recombineering: Next powerful tool for genome editing after CRISPR/Cas. *ACS Synthetic Biology,* 13, 1019–1025.

Kidwell, M. G. 2005. Transposable elements. In T. Ryan Gregory (Ed.), *The evolution of the genome*. Elsevier.

Klompe, S. E., Jaber, N., Beh, L. Y., Mohabir, J. T., Bernheim, A. & Sternberg, S. H. 2022. Evolutionary and mechanistic diversity of Type IF CRISPR-associated transposons. *Molecular Cell,* 82, 616–628. e5.

Lampe, G. D., King, R. T., Halpin-Healy, T. S., Klompe, S. E., Hogan, M. I., Vo, P. L. H., Tang, S., Chavez, A. & Sternberg, S. H. 2023. *Targeted DNA integration in human cells without double-strand breaks using CRISPR RNA-guided transposases*. bioRxiv.

Lim, K.-I. 2015. Recent advances in developing molecular tools for targeted genome engineering of mammalian cells. *BMB Reports,* 48, 6.

Liu, P., Panda, K., Edwards, S., Yi, H., Pandesha, P., Hung, Y.-H., Swanson, R., Klaas, G., Ye, X. & Veena, V. 2023. CRISPR-targeted transposable element insertion for efficient plant genome engineering. *Nature,* 26 June.

Liu, X., Zhang, L., Wang, H., Xiu, Y., Huang, L., Gao, Z., Li, N., Li, F., Xiong, W. & Gao, T. 2022. Target RNA activates the protease activity of Craspase to confer antiviral defense. *Molecular Cell,* 82, 4503–4518. e8.

Lopez, S. C., Crawford, K. D., Lear, S. K., Bhattarai-Kline, S. & Shipman, S. L. 2022. Precise genome editing across kingdoms of life using retron-derived DNA. *Nature Chemical Biology,* 18, 199–206.

Luqman-Fatah, A. & Miyoshi, T. 2023. Human LINE-1 retrotransposons: Impacts on the genome and regulation by host factors. *Genes & Genetic Systems,* 98, 121–154.

Mestre, M. R., González-Delgado, A., Gutiérrez-Rus, L. I., Martínez-Abarca, F. & Toro, N. 2020. Systematic prediction of genes functionally associated with bacterial retrons and classification of the encoded tripartite systems. *Nucleic Acids Research,* 48, 12632–12647.

Miller, S. M., Wang, T. & Liu, D. R. 2020. Phage-assisted continuous and non-continuous evolution. *Nature Protocols,* 15, 4101–4127.

Munir, A., Ali, M., Qari, S. H., Munawar, N., Saleem, M. S. & Ahmad, A. 2024. CRISPR workflow solutions: Cargos and versatile delivery platforms in genome editing. In K.A. Abd-Elsalam, A. Ahmad, & B. Zhang (Eds.), *CRISPRized horticulture crops*. Elsevier.

Ramakrishnan, M., Papolu, P. K., Mullasseri, S., Zhou, M., Sharma, A., Ahmad, Z., Satheesh, V., Kalendar, R. & Wei, Q. 2023. The role of LTR retrotransposons in plant genetic engineering: How to control their transposition in the genome. *Plant Cell Reports,* 42, 3–15.

Rhiel, M., Geiger, K., Andrieux, G., Rositzka, J., Boerries, M., Cathomen, T. & Cornu, T. I. 2023. T-CAST: An optimized CAST-Seq pipeline for TALEN confirms superior safety and efficacy of obligate-heterodimeric scaffolds. *Frontiers in Genome Editing,* 5, 1130736.

Rubin, B. E., Diamond, S., Cress, B. F., Crits-Christoph, A., Lou, Y. C., Borges, A. L., ... & Doudna, J. A. 2022. Species- and site-specific genome editing in complex bacterial communities. *Nature Microbiology,* 7(1), 34–47.

Saito, M., Xu, P., Faure, G., Maguire, S., Kannan, S., Altae-Tran, H., Vo, S., Desimone, A., Macrae, R. K. & Zhang, F. 2023. Fanzor is a eukaryotic programmable RNA-guided endonuclease. *Nature,* 620, 660–668.

Simon, A. J., Ellington, A. D. & Finkelstein, I. J. 2019. Retrons and their applications in genome engineering. *Nucleic Acids Research,* 47, 11007–11019.

Simon, A. J., Morrow, B. R. & Ellington, A. D. 2018. Retroelement-based genome editing and evolution. *ACS Synthetic Biology,* 7, 2600–2611.

Song, B. & Bae, S. 2024. Genome editing using CRISPR, CAST, and Fanzor systems. *Molecules and Cells,* 47(7), 100086.

Strecker, J., Ladha, A., Gardner, Z., Schmid-Burgk, J. L., Makarova, K. S., Koonin, E. V., & Zhang, F. 2019. RNA-guided DNA insertion with CRISPR-associated transposases. *Science,* 365(6448), 48–53.

Tae, H. 2024. *CRISPR, OMEGA, and Fanzor: Nixed blessings of genome editing technology*. digitalcommons.liberty.edu/honors/1385

Tenjo-Castaño, F., Montoya, G. & Carabias, A. 2022. Transposons and CRISPR: Rewiring gene editing. *Biochemistry,* 62, 3521–3532.

Tou, C. J., Orr, B. & Kleinstiver, B. P. 2023. Precise cut-and-paste DNA insertion using engineered type VK CRISPR-associated transposases. *Nature Biotechnology,* 41, 968–979.

Van Beljouw, S. P., Haagsma, A. C., Kalogeropoulos, K., Pabst, M. & Brouns, S. J. 2024. Craspase orthologs cleave a nonconserved site in target protein Csx30. *ACS Chemical Biology,* 19, 1051–1055.

Vo, P. L. H., Ronda, C., Klompe, S. E., Chen, E. E., Acree, C., Wang, H. H. & Sternberg, S. H. 2021. CRISPR RNA-guided integrases for high-efficiency, multiplexed bacterial genome engineering. *Nature Biotechnology,* 39, 480–489.

Wannier, T. M., Ciaccia, P. N., Ellington, A. D., Filsinger, G. T., Isaacs, F. J., Javanmardi, K., Jones, M. A., Kunjapur, A. M., Nyerges, A. & Pal, C. 2021. Recombineering and MAGE. *Nature Reviews Methods Primers,* 1, 7.

Widner, D. L. 2021. *Novel functions for large noncoding nucleic acids in bacteria.* Yale University Press.

Xiang, G., Li, Y., Sun, J., Huo, Y., Cao, S., Cao, Y., Guo, Y., Yang, L., Cai, Y. & Zhang, Y. E. 2024. Evolutionary mining and functional characterization of TnpB nucleases identify efficient miniature genome editors. *Nature Biotechnology,* 42, 745–757.

Xu, Y., Liu, T., Wang, J., Xiong, B., Liu, L. & Peng, N. 2023. Reprogramming an RNA-guided archaeal TnpB endonuclease for genome editing. *Cell Discovery,* 9, 112.

Yang, X. & Zhang, B. 2023. A review on CRISPR/Cas: A versatile tool for cancer screening, diagnosis, and clinic treatment. *Functional & Integrative Genomics,* 23, 182.

Yarnall, M. T., Ioannidi, E. I., Schmitt-Ulms, C., Krajeski, R. N., Lim, J., Villiger, L., Zhou, W., Jiang, K., Garushyants, S. K. & Roberts, N. 2023. Drag-and-drop genome insertion of large sequences without double-strand DNA cleavage using CRISPR-directed integrases. *Nature Biotechnology,* 41, 500–512.

Zeng, T., Yin, J., Liu, Z., Li, Z., Zhang, Y., Lv, Y., Lu, M.-L., Luo, M., Chen, M. & Xiao, Y. 2023. Mechanistic insights into transposon cleavage and integration by TnsB of ShCAST system. *Cell Reports,* 42.

Zhang, Y., Sun, X., Wang, Q., Xu, J., Dong, F., Yang, S., Yang, J., Zhang, Z., Qian, Y. & Chen, J. 2020. Multicopy chromosomal integration using CRISPR-associated transposases. *ACS Synthetic Biology,* 9, 1998–2008.

Zhang, Y., Yang, J., Yang, S., Zhang, J., Chen, J., Tao, R., Jiang, Y., Yang, J. & Yang, S. 2021. Programming cells by multicopy chromosomal integration using CRISPR-associated transposases. *The CRISPR Journal,* 4, 350–359.

13 Plant Genome Editing

Qamar U. Zaman, Darya Khan, Hua Feng Wang, Qiong Hu, and Rafaqat Ali Gill

13.1 INTRODUCTION

Conventionally, the generation of variations in plant genomes has relied on laborious and time-consuming breeding techniques. Plant breeding has been carried out by crossing and selecting plants with superior traits. However, conventional plant breeding techniques may not be sufficient to feed the world's population, which is projected to increase up to 10 billion by 2050 (Hickey et al., 2019). The demand for agricultural products has increased exponentially, while crop yields have remained relatively constant for the past three decades (Ahmad, 2023). Crop production is primarily affected by biotic and abiotic stresses. Plants continuously endure these stresses and respond by regulating complex molecular and physiological pathways and making changes in cellular adaptation. High-throughput sequencing (HTS) techniques helped the researchers to identify the regulatory changes underlying various complex mechanisms. With the availability of various plant genomes including rice (Kawahara et al., 2013), wheat (Aury et al., 2022), tomato (Wang et al., 2020b), potato (Reyes-Herrera et al., 2024), pitaya (Zaman et al., 2024), sweet cherry (Wang et al., 2020a), and apple (Zhang et al., 2019), it is easy for the researchers to explore the associated traits and the stress responsive genetic regulatory mechanisms for crop improvement through forward and reverse genetic engineering techniques. Transcription factors (TFs) are central regulators of gene expression levels and effectively respond to environmental factors. These TFs regulate gene expression by binding to *cis*-acting elements of a given gene under biotic and abiotic stresses. Change in the expression of certain genes are key events that help plants establish effective responses to adopt certain defensive position by regulating key hormones. There is convincing evidence that many genes play a role under multifactorial stresses and enhance plant tolerance (Rejeb et al., 2014). These stresses alter plant behavior by modulating the levels of microRNAs, heat shock proteins, and epigenetic variations (Dangi et al., 2018). Epigenetic variations may initiate adverse consequences on plant growth and development. Epigenetic changes can also result in modifications of epigenetic regulators (Zhang et al., 2022). A persistent stress stimulus is required to maintain changes in epigenetic patterns. Stress-induced epigenetic memory in plants seems to be transient and partially transferred to the next generations if the progeny is continuously stressed. These stresses imposed by climate change can leave severe consequences on the ecosystem. The primary causes of climate change are deforestation and anthropogenic activities, which add greenhouse gases to the environment. Rising temperatures and shifting ecological zones are drastically modifying the landscape of many countries, resulting in hunger and migration. Simultaneously, the global food supply chain is declining due to the effects of climate change on yields. In response to these challenges, conventional breeding approach can only tackle them to some extent. However, the use of genome-editing technology can assist plant breeders to generate crops that are highly tolerant of environmental challenges. Stress-resilient plants can be generated by overexpression of positive regulators or knocking out the negative regulators or the activity of key regulators involved in stress responses. Genome editing has developed into a practical approach to directly target undesirable traits in crops. Different sequence-specific nucleases-mediated genome-editing techniques, including zinc-finger (ZFN), transcription-like effector nucleases (TALENs), and clustered regularly interspaced short palindromic repeats (CRISPR), are used to target specific genes. Among these techniques, CRISPR opens new opportunities to precisely target the specific genes with minimal off-targets. In the last decade, CRISPR has expanded the role of genome editing and emerged as an effective tool for functional studies and gene discovery to improve plant traits and produce new genotypes (Cardi et al., 2023). It is an effective tool to generate DNA or RNA modification using specific Cas proteins at precise genomic positions. The effects of climate change are beginning to emerge, and the environmental activists have warned about changing climatic scenarios. Conventional breeding presents specific challenges, including seasonal dormancy, ploidy level, and long duration to induce desirable traits. However, recent developments in CRISPR-mediated genome editing demonstrated it as a leading technology to edit diploid or polyploid genomes with or without creating the double-stranded breaks (DSBs). These site-specific nucleases have significant potential for plant breeding, as they provide multiple mechanisms to manipulate the host genome structure and function, knockout, knock-in, single base conversion, and

 DOI: 10.1201/9781003500933-15

save time, compared to conventional plant breeding methods (Zaidi et al., 2020). By using advanced CRISPR/Cas variants, including Cas9 (Doudna and Charpentier, 2014), dCas9 (Moradpour and Abdulah, 2020), Mobile CRISPR (Zaman, Raza, Gill et al., 2023), base editing (Li, Wang et al., 2023), prime editing (Li, Wang, et al., 2023), and RNA editing (Kavuri et al., 2022), provides a robust and powerful tool for the precise modification of plant genomes (Molla et al., 2021).

To address the challenges associated with field crops due to climate change and low genetic diversity, plant breeders have been consistently working to develop new crop varieties to enhance tolerance against biotic and abiotic diseases and increase yield. In changing climate conditions, CRISPR provides a ray of hope for continuously developing and modifying the existing germplasm resources to create new varieties. CRISPR/Cas and its new variants have emerged as novel and effective technology for functional studies, inducing stress tolerance in crops, and de novo domestication of plants. The remarkable diversity of CRISPR-associated proteins with various PAM regions is providing an opportunity to identify novel protein for RNA guided gene regulation and generate transgene-free edits without off-targets effetcs (Zaman, Raza, Lozano-Juste et al., 2023).

13.2 MECHANISM OF CRISPR EDITING

CRISPR-Cas system consists of three basic steps: adaptation or spacer sequence acquisition (Jackson et al., 2017; Koonin et al., 2017; Makarova et al., 2015), expression or CRISPR RNA biogenesis (Brouns et al., 2008; Deltcheva et al., 2011; Hale et al., 2009), and RNA-guided interference (Bates, 2016 ; Semenova et al., 2011; Garneau et al., 2010; Sternberg, et al., 2014 ; Wright et al., 2016). A molecular memory is created during the spacer acquisition process by introducting the plasmid into the CRISPR locus as a new spacer (Fineran and Charpentier, 2012; Yosef et al., 2012). During the spacer integration (Yoganand et al., 2017), conserved Cas1 and Cas2 proteins form an integrase complex with two distal Cas1 dimers bridged by a Cas2 dimer (Nuñez et al., 2014). The protein complex creates a curved binding surface, splays the dual-forked DNA protospacer, and allows each terminal 3'-OH overhang to enter the channel of the Cas1 active site (Nuñez et al., 2015). In crRNA biogenesis or expression stage, the CRISPR locus is transcribed as a single pre-crRNA, which is subsequently processed into mature crRNA. Each mature crRNA consists of a single new spacer (Garneau et al., 2010; Jinek et al., 2012; Wright et al., 2016). In the last interference stage, mature crRNAs form an effector complex with Cas proteins, which recognize and cleave the invading nucleic acids bearing sequence complementary to spacer sequence of the crRNAs (Garneau et al., 2010; Jinek et al., 2012; Sashital et al., 2011; Sashital et al., 2012; Wiedenheft et al., 2011). These steps are facilitated by an array of short direct repeats and interspaced spacer sequences flanked by various Cas genes (Figure 13.1).

13.3 DIVERSITY OF CRISPR-ASSOCIATED PROTEINS FOR GENOME EDITING IN PLANTS

CRISPR has revolutionized genetic engineering by enabling precise targeting of desired sites within the plant genome using associated proteins such as Cas9, dCas9, Cas12, Cas13a, base editing and prime editing. These CRISPR-associated proteins have demonstrated unprecedented efficiency in inducing large fragment deletion, insertion, multiplex editing, and base conversion. They have enabled the generation of modified plants with desirable changes without off-target or deleterious mutations.

Cas variants have been identified and modified to facilitate knock-in or knockouts of gene expression. Among these, Cas9 is extensively used across a wide range of organism compared to Cas12a and Cas13a. Cas9 comprises two distinct lobes: the alpha-helical recognition lobe (REC) and the nuclease lobe (NUC), which includes the HNH conserved domain, RuvC nuclease domain, and C-terminal domain (CTD). The REC lobe exhibits a specific fold unique to Cas9 and contains sites for interacting with the protospacer adjacent motif (PAM). Cas9 nickase (H840A) forms a complex with an 83-nt sgRNA and target DNA containing NGG on 5′ strand. CRISPR/Cas9-mediated genome engineering holds immense promise for generating desirable modifications in the plant genome to produce mutants with no off-targets and\ transgene-free. Another variant, dCas9, was developed by Qi et al. (2013) through mutation of the Cas9 nickase (H840A) and the aD10A site in the RuvC domain. This dead version of Cas9 can no longer generate double-stranded breaks but retains its ability to target and bind DNA with precision, guided by sgRNA. This dCas9 variant enables single or multiple genes through transcriptional repression domains without introducing irreversible mutagenesis at the target site (Moradpour and Abdulah, 2020).

The Cas12a (Cpf1) system, belonging to Class 2 Type-V of the CRISPR system, is characterized by its size, which typically ranges between 1200–1500 amino acids (Shmakov et al., 2015; Zetsche et al., 2015). It consists of a recognition lobe (REC1/2) and a nuclease (NUC) lobe responsible for cleaving the DNA (Jianwei et al., 2023). Unlike the Cas9 system, CRISPR/Cpf1 does not require tracrRNA and RNase III for processing mature crRNA. Instead, it relies on TTTV PAM and facilitates a staggered cut to generate DSBs, using a short crRNA for potential genome targeting, typically located 4-nt upstream of the hairpin structures formed by the CRISPR repeats (Port et al., 2020). The stability of this crRNA is maintained through interactions between the RuvC and REC domains (Paul and Montoya, 2020).

Cas13a, also known as C2c2, belongs to the Type VI system, which is comprised of a higher eukaryotes and prokaryotes nucleotide binding (HEPN) domain along with two REC and NUC lobes (Zhao et al., 2022). The mature crRNA contains a 3′ spacer and a 5′ handle (Knott et al., 2017). RNA-guided and RNA-activated RNAs cleave the desired target RNAs in

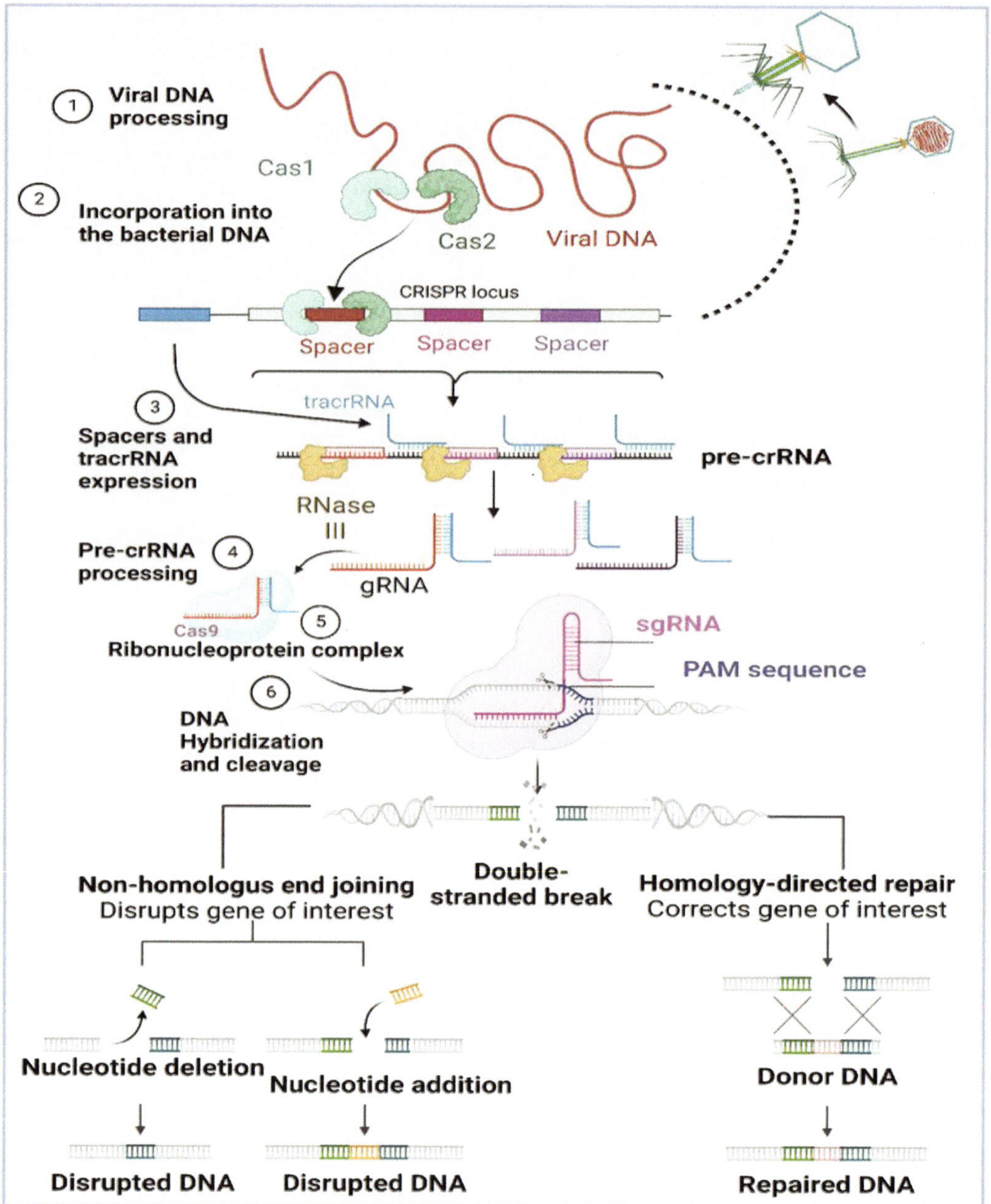

FIGURE 13.1 Mechanism of the CRISPR system. The CRISPR system is an adaptive immune system that uses RNA-guided nucleases to induce the heritable genetic modifications during the repair process via non-homologous end joining (NHEJ) or the homology-directed repair method (HDR).

a sequence-specific manner, similar to other endonucleases that target the DNAs (Kick et al., 2022). The crRNA form a target RNA duplex, which binds to the nuclease lobe and the Cas13a protein, inducing significant conformational changes in both the crRNA and Cas13 protein during RNA targeting (Liu et al., 2017). This allows Cas13a enzymes to generate functional crRNAs and target the RNAs effectively.

Base and prime editors primarily alter the identity of nucleobases at the desired target site without generating DSBs (Tan et al., 2022). Base editors are engineered ribonucleoproteins that facilitate single base substitution. DNA base editors comprise two classes: deoxycytidine deamination-derived editors, which convert C:G to T:A, and deoxyadenosine deamination-derived base editors, which modify A:T to G:C mutagenesis. CBEs carry a cytidine deaminase fused to Cas9, leading to C-to-T transitions by deaminating deoxycytidine to deoxyuridine, a nucleoside with similar base pairing properties as deoxythymidine. In ABEs, an adenosine deaminase is tethered to the Cas9 protein, catalyzing the oxidative deamination of deoxyadenosine to deoxyinosine (Tan et al., 2022). In contrast, prime editing involves a reverse transcriptase (RT) and its RT template (Lu et al., 2022). In this system, Cas9 nickase cleaves the strand containing the PAM, exposing the 3′ hydroxyl group, which then binds to the gRNA. Subsequently, reverse transcription occurs according to the RT template.

13.4 DELIVERY OF THE CRISPR SYSTEM FOR GENE TARGETING IN PLANTS

Currently, the delivery of the CRISPR system into the host cell is considered a major bottleneck in plant genome engineering. Several plant species have complex genome structures, especially polyploid crops (Zaman, Chu, et al., 2019; Zaman et al., 2021), compared to diploid genome. CRISPR reagents are delivered as plasmid DNA constructs using various transformation methods, including protoplast transformation, viral vectors, and mobile RNAs (Yang et al., 2023; Zaman, Raza, Gill et al., 2023), agrobacterium-mediated transformation (Zaman, Chu, et al., 2019), and the particle bombardment (Laforest and Nadakuduti, 2022). The agrobacterium-mediated transformation method remains the principal strategy for delivering CRISPR reagents into explant tissues. Explants are inoculated with agrobacterium expressing the constructed cassette integrated with its T-DNA. Mutant plants are regenerated with tissue culture method, except for *Arabidopsis*, where mutants can be generated using the floral dip transformation method (Ursache et al., 2021). Additionally, viral vectors and mobile RNAs can easily be introduced, efficiently delivering CRISPR reagents (Ali et al., 2015) into plants via agrobacterium (Zaman, Raza, Gill, et al., 2023), enabling tissue culture-free genome editing (Figure 13.2). The biolistic or particle bombardment method is an alternative strategy for transforming CRISPR reagents into host explants. In this method, gold-plated or tungsten microprojectiles coated with biomolecules physically breach the cell wall and membrane of the host cell at very high velocity. Plant genome editing mainly relies on tissue culture-based genetic transformation, which is time-consuming and a major bottleneck in fully realizing the potential of the CRISPR system for crop improvement. Recently, Yang et al. grafted the *Brassica rapa* wild-type shoot onto a transgenic root of Arabidopsis containing the Cas9/gRNA-TLS system (Yang et al., 2023; Zaman, Razam Gill, et al., 2023). TLS (tRNA-like sequence motif) assists the CRISPR expression cassette in moving it to the distal parts of the plant, including developing flowers. Cas9-TLS translates it into the recipient host tissues and targets desired genes. These mobile CRISPR-based strategies have several advantages and may bypass the tissue culture method, expanding the application of genome editing to low-regenerative plant species, clonally propagated species, and near-extinct plant species in the near future (Zaman, Raza, Gill, et al., 2023).

13.5 GENOME EDITING FOR CROP IMPROVEMENT

Ensuring food security for the exponentially growing global population in a changing climate poses adverse effects on agriculture. Conventional breeding can tackle these challenges to some extent, but these time-consuming and laborious methods take 7–10 years due to the limitations in the cropping system. Recently developed speed breeding methods can assist in growing crops 2–3 times per year under

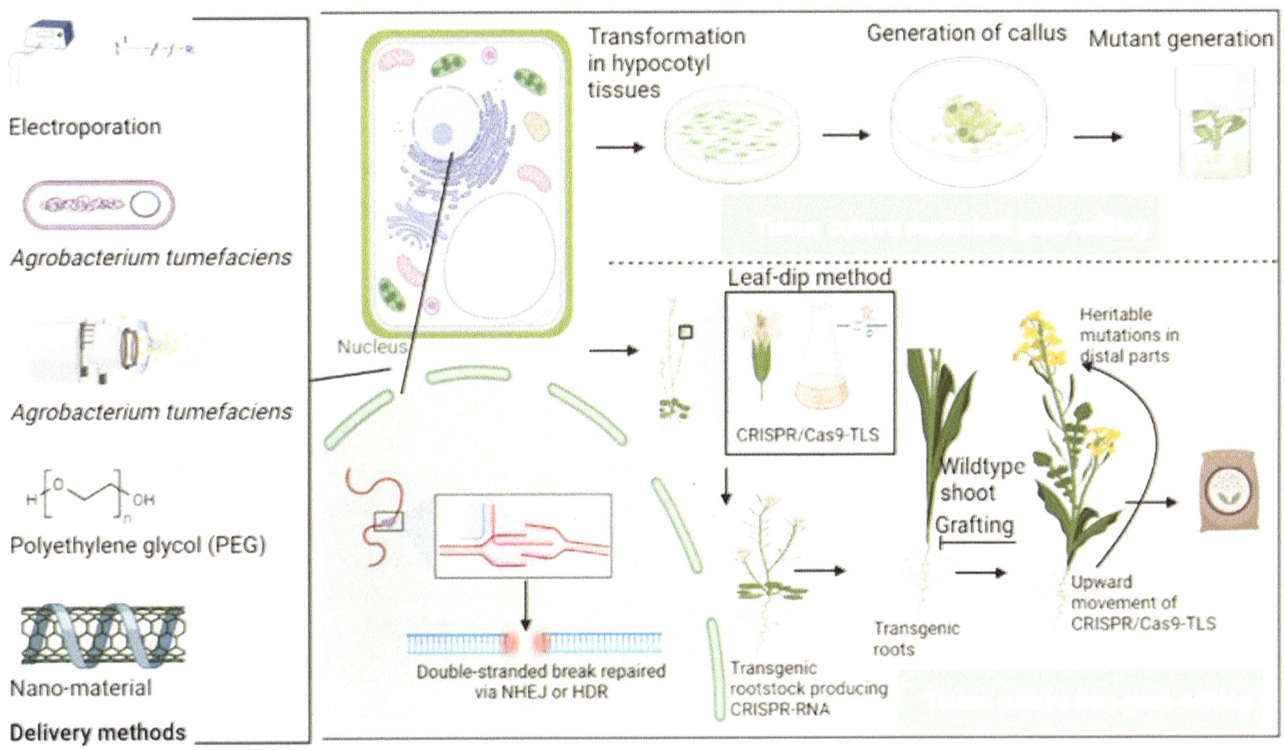

FIGURE 13.2 Delivery of CRISPR reagents into the host cell through tissue culture-dependent and tissue-culture-free methods.

controlled environments (Hickey et al., 2019). However, deploying speed breeding facilities on a large scale may not be feasible due to high cost of infrastructure (Wanga et al., 2021). Nevertheless, new biotechnological tools have opened up many opportunities to accelerate plant breeding and bridge the gap between new molecular tools and conventional breeding methods (Cardi et al., 2023). To improve quantitative and qualitative traits, breeders exploit genetic variations mainly induced by spontaneous mutations or during the evolution and domestication processes. Initially, transgenic experiments were developed to knockdown the expression of genes responsible for undesirable traits. However, some genome engineering approaches did not lead to complete gene knockout or produced transgenic mutants and have not been adapted in plant breeding so far. Several genome engineering techniques have been developed to induce genetic variation with high efficiency and precision in a wide range of plant species to accelerate plant breeding. The introduction of DSBs by MNs marked a paradigm shift in the mid-1990s (Jung et al., 2018). However, this method of genome editing was very laborious due to difficulties in redesigning the recognition motifs for specific genes. Among all the programmable endonucleases (MNs, ZFNs, TALENs, CRISPR) (Zaman, Li, et al., 2019), the CRISPR system is much easier to construct, low cost, highly efficient, and allows editing multiple genes at once (Zaman, Chu, et al., 2019; Zaman et al., 2021). CRISPR/Cas edited plants are generated using site-directed nucleases that can induce changes at desired loci. Incredible genetic variations have been achieved to knock out the expression of negative regulators of the trait in a genetic mechanism in diploid (Jung et al., 2018) and polyploid crops (Zaman, Li, et al., 2019) to enhance tolerance to biotic and abiotic stresses (Ahmad et al., 2023; Tariq et al., 2023; Zaidi et al., 2020), increase grain yield (Liu et al., 2021), nutritional enrichment (Kumar et al., 2022; Wang et al., 2019), and herbicide tolerance (Oz et al., 2021). Applying the retrieved knowledge from model crops enables the induction of genetic variations and improves existing germplasm resources. Mutations induced by the CRISPR system are identified in plants regenerated through tissue culture or the floral dipping method, which often carry chimeric mutagenesis. Mutation in germline cells is carried to the next generations and fixed if it contains homozygous mutations. In the next generations, mutants segregate in a Mendelian fashion. Mutants can be identified by amplifying the target region by polymerase chain reaction (PCR) and mismatches can be recognized. In some cases, engineered nucleases create off-target effects, which can lead to deleterious mutations. However, in the case of the CRISPR system, it strongly depends on the efficiency and specificity of gRNA (Li, Wang, et al., 2023). The efficiency of the sgRNA demonstrates and guides the CRISPR-associate proteins to edit the desired target. The specificity of the gRNA predicts whether it is unique to the target region or can cause off-target effects. Many computational tools are available that can predict specific and efficient gRNAs for genome editing in plants (Li, Chu, et al., 2023).

Genome editing refers to DNA modifications (insertion, deletion, substitution) that are generated in the plant genome more specifically than conventional plant breeding, which produces genetic variations randomly. CRISPR is a revolutionary tool that is changing the course of history. It has greatly influenced plant breeding, laboratory experiments, food production, and society as a whole. Genome editing mediated by the CRISPR system has the potential to transform agriculture entirely by generating biotic- and abiotic-tolerant plants under changing climate, with improved nutritional value and yield. Deregulation of genome-edited plants has already been initiated by many countries, and the European Union and China are also finalizing a decision on regulation. There is a need to modify existing germplasm resources to generate high-yield crops with better quality through genome editing, which can help mitigate the effects of climate change and contribute to achieving food security.

ACKNOWLEDGMENTS

The authors are grateful to the editors for editing and reviewing the book chapter. Figures were generated using BioRender (BioRender.com).

CONFLICT OF INTEREST

The authors declare no competing interests.

REFERENCES

Ahmad, M., 2023. Plant breeding advancements with "CRISPR-Cas" genome editing technologies will assist future food security. *Frontiers in Plant Science,* 14, 1133036. doi:10.3389/fpls.2023.1133036

Ahmad, N., Fatima, S., Mehmood, M.A., Zaman, Q.U., Atif, R.M., Zhou, W., Rahman, M.U. and Gill, R.A., 2023. Targeted genome editing in polyploids: lessons from Brassica. *Frontiers in Plant Science,* 14, 1152468. doi:10.3389/fpls.2023.1152468

Ali, Z., Abul-faraj, A., Li, L., Ghosh, N., Piatek, M., Mahjoub, A. and Mahfouz, M.M. 2015. Efficient virus-mediated genome editing in plants using the CRISPR/Cas9 System. *Molecular Plant,* 8(8), 1288–1291. doi:10.1016/j.molp.2015.02.011

Aury, J.M., Engelen, S., Istace, B., Monat, C., Lasserre-Zuber, P., Belser, C., Cruaud, C., Rimbert, H., Leroy, P., Arribat, S. and Dufau, I. 2022. Long-read and chromosome-scale assembly of the hexaploid wheat genome achieves high resolution for research and breeding. *Gigascience,* 11, giac034.

Bates, M., 2016. The CRISPR conundrum: as millions of dollars flow into developing the gene-editing tool, scientists continue to discover the technology? Potential and pitfalls. *IEEE Pulse,* 7(2), 17–21.

Ben Rejeb, I., Pastor, V. and Mauch-Mani, B., 2014. Plant responses to simultaneous biotic and abiotic stress: Molecular mechanisms. *Plants,* 3(4), 458–475.

Brouns, S.J.J., Jore, M.M., Lundgren, M., Westra, E.R., Slijkhuis, R.J.H., Snijders, A.P.L., Oost, J.V.D., 2008. Small CRISPR RNAs guide antiviral defense in prokaryotes. *Science,* 321(5891), 960–964.

Cardi, T., Murovec, J., Bakhsh, A., Boniecka, J., Bruegmann, T., Bull, S.E., Eeckhaut, T., Fladung, M., Galovic, V., Linkiewicz, A. and Lukan, T., 2023. CRISPR/Cas-mediated plant genome editing: Outstanding challenges a decade after implementation. *Trends in Plant Science*, 28(10), 1144–1165.

Dangi, A.K., Sharma, B., Khangwal, I., and Shukla, P., 2018. Combinatorial interactions of biotic and abiotic stresses in plants and their molecular mechanisms: Systems biology approach. *Molecular Biotechnology*, 60(8), 636–650. doi:10.1007/s12033-018-0100-9

Deltcheva, E., Chylinski, K., Sharma, C.M., Gonzales, K., Chao, Y., Pirzada, Z.A., Eckert, M.R., Vogel, J. and Charpentier, E., 2011. CRISPR RNA maturation by trans-encoded small RNA and host factor RNase III. *Nature, 471*(7340), pp. 602–607.

Doudna, J.A. and Charpentier, E., 2014. The new frontier of genome engineering with CRISPR-Cas9. Science, 346(6213), 1258096.

Fineran, P.C. and Charpentier, E., 2012. Memory of viral infections by CRISPR-Cas adaptive immune systems: acquisition of new information. *Virology*, 434(2), 202–209.

Garneau, J.E., Dupuis, M.È., Villion, M., Romero, D.A., Barrangou, R., Boyaval, P., Fremaux, C., Horvath, P., Magadán, A.H. and Moineau, S., 2010. The CRISPR/Cas bacterial immune system cleaves bacteriophage and plasmid DNA. *Nature*, 468(7320), 67–71.

Hale, C.R., Zhao, P., Olson, S., Duff, M.O., Graveley, B.R., Wells, L., Terns, R.M. and Terns, M.P., 2009. RNA-guided RNA cleavage by a CRISPR RNA-Cas protein complex. *Cell*, 139(5), 945–956.

Hickey, L.T., N. Hafeez, A., Robinson, H., Jackson, S.A., Leal-Bertioli, S.C., Tester, M., Gao, C., Godwin, I.D., Hayes, B.J. and Wulff, B.B., 2019. Breeding crops to feed 10 billion. *Nature Biotechnology*, 37(7), 744–754.

Jackson, S.A., McKenzie, R.E., Fagerlund, R.D., Kieper, S.N., Fineran, P.C. and Brouns, S.J., 2017. CRISPR-Cas: adapting to change. *Science*, 356(6333), p.eaal5056.

Jianwei, L., Jobichen, C., Machida, S., Meng, S., Read, R.J., Hongying, C., Jian, S., Yuan, Y.A. and Sivaraman, J., 2023. Structures of apo Cas12a and its complex with crRNA and DNA reveal the dynamics of ternary complex formation and target DNA cleavage. *PLoS Biology*, 21(3), e3002023. doi:10.1371/journal.pbio.3002023

Jung, C., Capistrano-Gossmann, G., Braatz, J., Sashidhar, N. and Melzer, S., 2018. Recent developments in genome editing and applications in plant breeding. *Plant Breeding*, 137(1), 1–9. doi:https://doi.org/10.1111/pbr.12526

Kavuri, N.R., Ramasamy, M., Qi, Y. and Mandadi, K., 2022. Applications of CRISPR/Cas13-based RNA editing in plants. *Cells,* 11(17), 2665. doi:10.3390/cells11172665

Kawahara, Y., de la Bastide, M., Hamilton, J.P., Kanamori, H., McCombie, W.R., Ouyang, S., Schwartz, D.C., Tanaka, T., Wu, J., Zhou, S. and Childs, K.L., 2013. Improvement of the Oryza sativa Nipponbare reference genome using next generation sequence and optical map data. *Rice,* 6, 1–10.

Kick, L.M., Von Wrisberg, M.K., Runtsch, L.S. and Schneider, S., 2022. Structure and mechanism of the RNA dependent RNase Cas13a from Rhodobacter capsulatus. *Communications. Biology,* 5(1), 71. doi:10.1038/s42003-022-03025-4

Knott, G.J., East-Seletsky, A., Cofsky, J.C., Holton, J.M., Charles, E., O'Connell, M.R. and Doudna, J.A., 2017. Guide-bound structures of an RNA-targeting A-cleaving CRISPR–Cas13a enzyme. *Nature Structural & Molecular Biology*, 24(10), 825–833. doi:10.1038/nsmb.3466

Koonin, E.V., Makarova, K.S. and Zhang, F., 2017. Diversity, classification and evolution of CRISPR-Cas systems. *Current Opinion in Microbiology,* 37, 67–78.

Kumar, D., Yadav, A., Ahmad, R., Dwivedi, U.N. and Yadav, K., 2022. CRISPR-based genome editing for nutrient enrichment in crops: a promising approach toward global food security. *Frontiers in Genetics,* 13, 932859. doi:10.3389/fgene.2022.932859

Laforest, L.C. and Nadakuduti, S.S., 2022. Advances in delivery mechanisms of CRISPR gene-editing reagents in plants. *Frontiers in Genetics,* 4, 830178. doi:10.3389/fgeed.2022.830178

Li, C., Chu, W., Gill, R.A., Sang, S., Shi, Y., Hu, X., Yang, Y., Zaman, Q.U. and Zhang, B., 2023. Computational tools and resources for CRISPR/Cas genome editing. *Genomics, Proteomics, & Bioinformatics*, 21(1), 108–126.

Li, C., Wang, L., Cseke, L.J., Vasconcelos, F., Huguet-Tapia, J.C., Gassmann, W., Pauwels, L., White, F.F., Dong, H. and Yang, B., 2023. Efficient CRISPR-Cas9 based cytosine base editors for phytopathogenic bacteria. *Communications Biology*, 6(1), 56. doi:10.1038/s42003-023-04451-8

Liu, L., Gallagher, J., Arevalo, E.D., Chen, R., Skopelitis, T., Wu, Q., Bartlett, M. and Jackson, D., 2021. Enhancing grain-yield-related traits by CRISPR–Cas9 promoter editing of maize CLE genes. *Nature Plants*, 7(3), 287–294. doi:10.1038/s41477-021-00858-5

Liu, L., Li, X., Ma, J., Li, Z., You, L., Wang, J. and Wang, Y., 2017. The molecular architecture for RNA-guided RNA cleavage by Cas13a. *Cell*, 170(4), 714–726.e710. doi:10.1016/j.cell.2017.06.050

Lu, C., Kuang, J., Shao, T., Xie, S., Li, M., Zhu, L. and Zhu, L., 2022. Prime editing: An all-rounder for genome editing. International Journal of Molecular Science, 23(17). doi:10.3390/ijms23179862

Makarova, K.S., Wolf, Y.I., Alkhnbashi, O.S., Costa, F., Shah, S.A., Saunders, S.J., Barrangou, R., Brouns, S.J., Charpentier, E., Haft, D.H. and Horvath, P., 2015. An updated evolutionary classification of CRISPR–Cas systems. *Nature Reviews Microbiology*, 13(11), 722–736.

Jinek, M., Chylinski, K., Fonfara, I., Hauer, M., Doudna, J.A., and Charpentier, E. 2012. A programmable dual-RNA–guided DNA endonuclease in adaptive bacterial immunity. *Science*, 337(6096), 816–821.

Molla, K.A., Sretenovic, S., Bansal, K.C. and Qi, Y., 2021. Precise plant genome editing using base editors and prime editors. *Nature Plants*, 7(9), 1166–1187. doi:10.1038/s41477-021-00991-1

Moradpour, M. and Abdulah, S. N. A., 2020. CRISPR/dCas9 platforms in plants: strategies and applications beyond genome editing. *Plant Biotechnology Journal*, 18(1), 32–44. doi:10.1111/pbi.13232

Nuñez, J.K., Kranzusch, P.J., Noeske, J., Wright, A.V., Davies, C.W. and Doudna, J.A., 2014. Cas1–Cas2 complex formation mediates spacer acquisition during CRISPR–Cas adaptive immunity. *Nature Structural & Molecular Biology*, 21(6), 528–534.

Nuñez, J.K., Lee, A.S.Y., Engelman, A. and Doudna, J.A., 2015. Integrase-mediated spacer acquisition during CRISPR-Cas adaptive immunity. *Nature,* 519(7542), 193–198.

Oz, M.T., Altpeter, A., Karan, R., Merotto, A. and Altpeter, F., 2021. CRISPR/Cas9-mediated multi-allelic gene targeting in sugarcane confers herbicide tolerance. *Frontiers in Genome Editing,* 3, 673566. doi:10.3389/fgeed.2021.673566

Paul, B. and Montoya, G., 2020. CRISPR-Cas12a: Functional overview and applications. *Biomedical Journal*, 43(1), 8–17. doi:https://doi.org/10.1016/j.bj.2019.10.005

Port, F., Starostecka, M. and Boutros, M., 2020. Multiplexed conditional genome editing with Cas12a in Drosophila. *Proceedings of the National Academy of Sciences*, 117(37), 22890–22899. doi:10.1073/pnas.2004655117

Qi, L.S., Larson, M.H., Gilbert, L.A., Doudna, J.A., Weissman, J.S., Arkin, A.P. and Lim, W.A., 2013. Repurposing CRISPR as an RNA-guided platform for sequence-specific control of gene expression. *Cell*, 152(5), 1173–1183. doi:10.1016/j.cell.2013.02.022

Reyes-Herrera, P.H., Delgadillo-Duran, D.A., Flores-Gonzalez, M., Mueller, L.A., Cristancho, M.A. and Barrero, L.S., 2024. Chromosome-scale genome assembly and annotation of the tetraploid potato cultivar Diacol Capiro adapted to the Andean region. *G3: Genes Genomes Genetics*, 14(9), jkae139.

Sashital, D., Jinek, M. and Doudna, J.A., 2011. An RNA-induced conformational change required for CRISPR RNA cleavage by the endoribonuclease Cse3. *Nature Structural & Molecular Biology*, 18(6), 680–687.

Sashital, D., Wiedenheft, B. and Doudna, J. (2012). Mechanism of foreign DNA selection in a bacterial adaptive immune system. *Molecular Cell*, 46(5), 606–615.

Semenova, E., Jore, M.M., Datsenko, K.A., Semenova, A., Westra, E.R., Wanner, B., Van Der Oost, J., Brouns, S.J. and Severinov, K., 2011. Interference by clustered regularly interspaced short palindromic repeat (CRISPR) RNA is governed by a seed sequence. *Proceedings of the National Academy of Science,* 108(25), 10098–10103.

Shmakov, S., Abudayyeh, O.O., Makarova, K.S., Wolf, Y.I., Gootenberg, J.S., Semenova, E. and Koonin, E.V., 2015. Discovery and functional characterization of diverse class 2 CRISPR-Cas systems. *Molecular Cell*, 60(3), 385–397. doi:10.1016/j.molcel.2015.10.008

Sternberg, S.H., Redding, S., Jinek, M., Greene, E.C., and Doudna, J.A. 2014. DNA interrogation by the CRISPR RNA-guided endonuclease Cas9. *Biophysical Journal*, 106(2), 695a.

Tan, J., Forner, J., Karcher, D. and Bock, R., 2022. DNA base editing in nuclear and organellar genomes. *Trends in Genetics*, 38(11), 1147–1169. doi:10.1016/j.tig.2022.06.015

Tariq, A., Mushtaq, M., Yaqoob, H., Bhat, B.A., Zargar, S.M., Raza, A., Ali, S., Charagh, S., Mubarik, M.S., Zaman, Q.U. and Prasad, P.V., 2023. Putting CRISPR-Cas system in action: A golden window for efficient and precise genome editing for crop improvement. *GM Crops & Food*, 14(1), 1–27. doi:10.1080/21645698.2023.2219111

Ursache, R., Fujita, S., Dénervaud Tendon, V. and Geldner, N., 2021. Combined fluorescent seed selection and multiplex CRISPR/Cas9 assembly for fast generation of multiple Arabidopsis mutants. *Plant Methods*, 17(1), 111. doi:10.1186/s13007-021-00811-9

Wang, J., Liu, W., Zhu, D., Hong, P., Zhang, S., Xiao, S., Tan, Y., Chen, X., Xu, L., Zong, X. and Zhang, L., 2020. Chromosome-scale genome assembly of sweet cherry (Prunus avium L.) cv. Tieton obtained using long-read and Hi-C sequencing. *Horticulture Research,* 7, article 122.

Wang, T., Zhang, H. and Zhu, H., 2019. CRISPR technology is revolutionizing the improvement of tomato and other fruit crops. *Horticulture Research*, 6(1), 77. doi:10.1038/s41438-019-0159-x

Wang, X., Gao, L., Jiao, C., Stravoravdis, S., Hosmani, P.S., Saha, S., Zhang, J., Mainiero, S., Strickler, S.R., Catala, C. and Martin, G.B., 2020. Genome of Solanum pimpinellifolium provides insights into structural variants during tomato breeding. *Nature Communications,* 11(1), 5817.

Wanga, M.A., Shimelis, H., Mashilo, J. and Laing, M.D., 2021. Opportunities and challenges of speed breeding: A review. *Plant Breeding*, 140(2), 185–194.

Wiedenheft, B., van Duijn, E., Bultema, J.B., Waghmare, S.P., Zhou, K., Barendregt, A., Westphal, W., Heck, A.J., Boekema, E.J., Dickman, M.J. and Doudna, J.A., 2011. RNA-guided complex from a bacterial immune system enhances target recognition through seed sequence interactions. *Proceedings of the National Academy of Sciences,* 108(25), 10092–10097.

Wright, A., Nuñez, J. and Doudna, J., 2016. Biology and applications of CRISPR systems: harnessing nature's toolbox for genome engineering. *Cell*, 164(1–2), 29–44.

Yang, L., Machin, F., Wang, S., Saplaoura, E. and Kragler, F., 2023. Heritable transgene-free genome editing in plants by grafting of wild-type shoots to transgenic donor rootstocks. *Nature Biotechnology*, 41(7), 958–967. doi:10.1038/s41587-022-01585-8

Yoganand, K. N. R., Sivathanu, R., Nimkar, S. and Anand, B., 2017. Asymmetric positioning of Cas1–2 complex and Integration host factor induced DNA bending guide the unidirectional homing of protospacer in CRISPR-Cas type I-E system. *Nucleic Acids Research*, 45(1), 367–381.

Yosef, I., Goren, M. G., and Qimron, U. (2012). Proteins and DNA elements essential for the CRISPR adaptation process in Escherichia coli. *Nucleic Acids Research*, 40(12), 5569–5576.

Zaidi, S. S.-e.-A., Mahas, A., Vanderschuren, H. and Mahfouz, M. M., 2020. Engineering crops of the future: CRISPR approaches to develop climate-resilient and disease-resistant plants. *Genome Biology*, 21(1), 289. doi:10.1186/s13059-020-02204-y

Zaman, Q.U., Chu, W., Hao, M., Shi, Y., Sun, M., Sang, S.F. and Hu, Q., 2019. CRISPR/Cas9-mediated multiplex genome editing of JAGGED gene in Brassica napus L. Biomolecules, 9(11). doi:10.3390/biom9110725

Zaman, Q.U., Hui, L., Nazir, M.F., Wang, G., Garg, V., Ikram, M., Raza, A., Lv, W., Khan, D., Khokhar, A.A. and You, Z., 2024. Chromosome-level genome assembly of autotetraploid Selenicereus megalanthus and gaining genomic insights into the evolution of trait patterning in diploid and polyploid pitaya species. bioRxiv, 2024–06. doi: https://doi.org/10.1101/2024.06.23.600268

Zaman, Q. U., Li, C., Cheng, H. and Hu, Q., 2019. Genome editing opens a new era of genetic improvement in polyploid crops. *Crop Journal*, 7(2), 141–150. doi:https://doi.org/10.1016/j.cj.2018.07.004

Zaman, Q.U., Raza, A., Gill, R.A., Hussain, M.A., Wang, H.F. and Varshney, R.K., 2023. New possibilities for trait improvement via mobile CRISPR-RNA. *Trends in Biotechnology*, 41(11), 1335–1338. doi:https://doi.org/10.1016/j.tibtech.2023.05.001

Zaman, Q.U., Raza, A., Lozano-Juste, J., Chao, L., Jones, M.G.K., Wang, H.F. and Varshney, R.K., 2023. Engineering plants using diverse CRISPR-associated proteins and deregulation of genome-edited crops. Trends in Biotechnology, doi:10.1016/j.tibtech.2023.10.007

Zaman, Q.U., Wen, C., Yuqin, S., Mengyu, H., Desheng, M., Jacqueline, B. and Qiong, H., 2021. Characterization of SHATTERPROOF homoeologs and CRISPR-Cas9-mediated genome editing enhances pod-shattering resistance in Brassica napus L. *CRISPR Journal*, 4(3), 360–370. doi:10.1089/crispr.2020.0129

Zetsche, B., Gootenberg, J.S., Abudayyeh, O.O., Slaymaker, I.M., Makarova, K.S., Essletzbichler, P., Volz, S.E., Joung, J., Van Der Oost, J., Regev, A. and Koonin, E.V., 2015. Cpf1 is a single RNA-guided endonuclease of a class 2 CRISPR-Cas system. *Cell,* 163(3), 759–771. doi:10.1016/j.cell.2015.09.038

Zhang, H., Zhu, J., Gong, Z. and Zhu, J.-K., 2022. Abiotic stress responses in plants. *Nat. Rev. Genet.,* 23(2), 104–119. doi:10.1038/s41576-021-00413-0

Zhang, L., Hu, J., Han, X., Li, J., Gao, Y., Richards, C.M., Zhang, C., Tian, Y., Liu, G., Gul, H. and Wang, D., 2019. A high-quality apple genome assembly reveals the association of a retrotransposon and red fruit colour. *Nature Communications*, 10(1), 1494.

Zhao, L., Qiu, M., Li, X., Yang, J. and Li, J., 2022. CRISPR-Cas13a system: A novel tool for molecular diagnostics. *Frontiers in Microbiology,* 13, 1060947. doi:10.3389/fmicb.2022.1060947

Part III

Applications of CRISPR in Plants

Genetic Improvement to Achieve Food Security

14 Genome Editing for Increasing Yield and Improving Nutritional Quality in Staple Crops

Utku Deniz, Selime Çelik, Emre Aksoy, and Ceyhun Kayıhan

14.1 INTRODUCTION

Climate change poses a formidable threat to global agriculture, significantly impacting staple crop production and yields. Marked by higher temperatures, changing rainfall patterns, and more frequent extreme weather events, changing climate presents unparalleled obstacles for maintaining stable crop growth. Staple crops, crucial for sustaining the world's growing population, are particularly vulnerable to these climatic shifts. Anomalies in temperature and precipitation patterns lead to a significant decline in staple crop yields, which may increase global food insecurity. Among staple crops, rice, wheat, potato, soybean, and maize are essential food items that form the foundation of a typical diet globally, but they are highly vulnerable to high temperatures and water deficits (Olabanji, Ndarana and Davis, 2020). For instance, wheat is highly sensitive to temperature changes. Studies project an approximately 6% decrease in wheat yields for every 1°C increase in global mean temperature. Elevated temperatures during critical growth stages of soybean can lead to reduced yields. Indeed, the estimates suggest that a 1°C increase in temperature may result in a 5–10% decline in soybean yields. Increased temperatures can accelerate maize crop development, potentially affecting yields. However, it cannot withstand temperatures higher than 35°C. Additionally, fluctuations in precipitation patterns, characterized by either more intense rainfall events or prolonged droughts, can significantly affect the cultivation of maize. Studies indicate that higher temperatures can reduce rice yields, and altered precipitation patterns may lead to an increased risk of flooding or drought, depending on the region. Therefore, understanding the intricate relationship between climate change and staple crop production is paramount for devising resilient strategies with the capacity to address these challenges and safeguard food security amidst an unpredictable climate future.

Hidden hunger poses a significant threat to populations in Southeast Asia, Africa, and Latin America, affecting millions with nutrient deficiencies that often go unnoticed. In Southeast Asia alone, an estimated 266 million people suffer from hidden hunger, while the numbers rise to around 416 million in Africa and 37 million in Latin America (Lowe, 2021). Hidden hunger, primarily caused by insufficient uptake of essential micronutrients such as zinc, iron, vitamin A, and iodine, can lead to severe health consequences. Insufficient iron, for example, contributes to anemia, affecting cognitive development, physical growth, and overall productivity. The immune system is weakened by zinc deficiency, rendering individuals more vulnerable to infections. Vitamin A deficiency is associated with visual impairments and an increased risk of infectious diseases. Iodine deficiency can lead to iodine-deficiency disorders, including goiter and impaired mental development. Addressing hidden hunger is vital not only for individual health but also for the holistic welfare and economic vitality of these regions. Comprehensive strategies, including biofortification and genome-editing approaches, are imperative to combat hidden hunger and enhance the nutritional resilience of vulnerable populations.

Traditionally, breeders have primarily emphasized yield improvements in staple crops, inadvertently leading to unintended consequences for seed mineral content. Several studies have shown that while successful in enhancing yields, these breeding efforts have unintentionally decreased the nutrient content of seeds (Maqbool et al., 2020). For instance, in pursuit of higher yields, some maize varieties have shown decreased levels of essential minerals like zinc and iron. Recognizing the critical importance of addressing both yield and nutrient content, there is now a growing demand from consumers and farmers for crop varieties that offer not only increased yields but also improved nutritional profiles. This shift in focus necessitates novel breeding strategies to pyramid multiple traits, ensuring that staple crop cultivars not only meet the demands for productivity but also contribute to combating nutrient deficiencies and promoting overall human health. The integration of advanced technologies, including genome editing and precision breeding, offers promising avenues for developing cultivars that strike a balance between enhanced yields and improved nutrient content, aligning with the evolving preferences and needs of both farmers and consumers.

Traditional plant breeding strategies have proven effective in creating staple crop varieties with increased yields, but they are associated with limitations such as a relatively slow pace of progress, dependence on natural genetic variation, and challenges in addressing complex traits. Unlike transgenics, conventional breeding may lack precision in targeting specific genes, leading to the unintentional introduction of undesirable traits. Additionally, it may involve time-consuming processes of crossing and selecting desired traits over multiple generations. Transgenic technologies offer a more precise approach by introducing specific genes into crops to confer

DOI: 10.1201/9781003500933-17

desired traits. However, public acceptance of transgenics varies globally due to concerns about environmental impact, potential health risks, and ethical considerations. In some countries, such as several European nations, public resistance to genetically modified organisms (GMOs) has led to restrictive regulations and limited commercial cultivation of transgenic crops. Examples include the rejection of genetically modified maize in several European countries despite its potential benefits in terms of pest resistance and improved yields. Public perception, regulatory frameworks, and ethical considerations play significant roles in shaping the acceptance of transgenics in agriculture (Kawall, Cotter, and Then, 2020).

The landscape of plant breeding has been revolutionized by the emergence of gene editing as a transformative tool, offering a precise alternative to traditional transgenic methods. Particularly notable is the acceptance of transgene-free genome-edited plants in regions, including the European Union, where public resistance to GMOs prevails. Unlike traditional transgenics, gene editing allows for the precise modification of specific target genes without introducing foreign DNA. This precision addresses concerns associated with transgenics and aligns more closely with conventional breeding practices, assuaging public apprehensions. Genome editing has potential advantages over both transgenics and conventional breeding since it allows for simultaneous targeting of multiple genes, speeds up the breeding process, and minimizes unintended effects. Additionally, genome editing enables the development of crops with enhanced traits, including enhanced nutritional content, climate resilience, and disease resistance, fostering a more widely accepted and sustainable approach to crop improvement.

Genome editing is a powerful and versatile tool employed in various staple crops to enhance both yield and nutritional quality. By precisely modifying specific genes, scientists have been able to develop superior new crop varieties. This transformative approach aligns seamlessly with the points drawn in the United Nations Sustainable Development Goals (SDGs). Genome editing in plants plays a crucial role in addressing multiple SDGs, including climate change adaptation, alleviation of hidden hunger by developing biofortified crops, hunger eradication through enhanced crop productivity, and ensuring food security for the growing global population. The precision of genome editing facilitates the development of resilient crops, supporting eco-friendly farming methods and presenting solutions to the complex challenges outlined in the UN SDGs.

This chapter aims to provide a solid exploration of CRISPR-Cas9-based genome editing in staple crops, focusing on key seven agricultural commodities, including wheat, soybean, maize, rice, canola, potato, and tomato. The overarching goal is to delve into the transformative capability of genome-editing technology in these essential crops, elucidating its applications in increasing yields, enhancing nutrient content, and decreasing anti-nutrient levels. By offering an in-depth examination of the latest advances in gene-editing techniques, this chapter seeks to contribute valuable insights into the sustainable improvement of staple crops, addressing challenges related to global food security, nutritional deficiencies, and environmental sustainability. Through a comprehensive exploration of CRISPR-Cas9 applications across diverse staple crops, this chapter endeavors to shed light on the promising prospects and implications of precision genome editing in agriculture. Studies involving genome editing for yield improvement are provided in Table 14.1, while those exploring nutrient content modifications are

TABLE 14.1
Studies focused on crop yield by using the CRISPR-Cas9 gene editing system

Crop	Trait	Target gene	Results	Reference
Wheat	Grain yield	*TaARE1*	Delayed senescence, higher yield, and enhanced nitrogen use efficiency	Guo et al., 2021
	Grain yield	*TaARE1*	Delayed senescence, increased yield, and enhanced nitrogen use efficiency	Zhang et al., 2021
	Grain yield	*TaARF12*	Increased grain number per spike and yield	Kong et al., 2023
	Grain shape and weight	*TaIAA21*	Increased grain area and weight	Jia et al., 2021
	Grain weight	*TaGASR7*	Increased grain weight	Zhang et al., 2016
	Grain yield	*TaCKX2-D1*	Increased grain number per spikelet and yield	Zhang et al., 2019
	Grain yield	*TaSPL13*	Decreased flowering time, tiller number, and plant height but increased grain size and number	Gupta et al., 2023
	Grain shape and weight	*TaPpd*	Increased grain area, weight and spikelet per spike	Errum et al., 2023
	Grain shape and weight	*TaGW7*	Reduced grain length Increased grain width and weight,	Wang et al., 2019

TABLE 14.1 (Continued)
Studies focused on crop yield by using the CRISPR-Cas9 gene editing system

Crop	Trait	Target gene	Results	Reference
	Grain shape and weight	*TaGW2*	Increased grain area and weight	Wang et al., 2018
	Spike number per plant	*TaDUO-B1*	Increased spike number per plant and grain yield	Wang et al., 2022
Rice	Grain Weight	*GW2,* *GW5,* *TGW6*	Increased grain area and weight	Xu et al., 2016
	Grain shape and weight	*OsSNB*	Increased grain area and weight	Ma et al., 2019
	Grain number and yield	*OsLOGL5*	Increased grain number and yield	Wang et al., 2020
	Abscisic acid content	*PYL1, PYL4, PYL6*	Improved growth and yield	Miao et al., 2018
	Tiller number	*LAZY1*	Increased tiller number	Miao et al., 2013
	Tiller number and yield	*OsAAP3*	Increased tiller number and grain yield	Lu et al., 2018
	Panicle number and grain yield	*OsIPA1*	Increased panicle number, size and grain yield	Song et al., 2022
	Grain yield	*OsKRN2*	increased grain yield	Chen et al., 2022
	Grain number per panicle; Plant height, erect panicle, grain size; Seed size; Plant height and tiller number	*Gn1a;* *DEP1;* *GS3;* *IPA1;*	Enlarged panicles, increased yield, grain size, dwarf plant with more tillers	Li et al., 2016
Maize	Grain yield	*ZmKRN2*	Increased grain yield	Chen et al., 2022
	Grain yield	*GRGs*	Increased plant growth and yield	Lorenzo et al., 2023
	Grain yield	*ARGOS8*	Higher grain yield under drought conditions	Shi et al., 2017
	Leaf architecture and grain yield	*ZmRAVL*	Narrower leaf architecture and increased grain yield	Tian et al., 2019
Soybean	Plant architecture	*GmSPL9* genes	Increased node number on main stem, branch number, total node number per plant and dry weight	Bao et al., 2019
	Seed number per pod	*GmJAGGED1*	Increased number of seeds per pod	Cai et al., 2021
	Plant height	*GmLHY1a,GmLHY1b,GmLHY2a,* and *GmLHY2b*	Reduced plant height and internode length	Cheng et al., 2019
	Flowering time	*E1*	Early flowering phenotype	Han et al., 2019
Tomato	Fruit size, number of vase	*CLV3, WUS*	Increased fruit size, increased number of vase, improved yield	Rodríguez-Leal et al., 2017
	Flowering time	*SP5G*	Accelerated flowering, improved yield	Soyk et al., 2017

detailed in Table 14.2, presenting a comprehensive overview of the current research landscape in the field.

14.2 WHEAT

Different research groups caused mutations in the genes related to plant development and yield increment in wheat by using CRISPR-Cas9-based genome editing. These precisely controlled mutations led to a marked increase in plant yield and yield components although some of them could alter other plant parameters. Using the CRISPR-Cas9 genome-editing system, mutations were introduced into three copies of the bread wheat (*Triticum aestivum*) *abnormal cytokinin response1 repressor1* (*TaARE1*) gene by Guo et al. (2021). TaARE1, located in the chloroplast, serves as a negative regulator of nitrogen use efficiency (NUE). The resulting

TABLE 14.2
Studies focused on nutritional qualities by using the CRISPR-Cas9 gene-editing system

Crop	Trait	Target gene	Results	Reference
Wheat	Grain hardness; amylose biosynthesis	*Pinb; Waxy*	Harder grain	Zhang et al., 2021
	Gliadin biosynthesis	*α-gliadin* genes	Reduced gliadin content	Sánchez-León et al., 2018
	Gliadin biosynthesis	*α/β-gliadin* genes	Reduced gliadin content	Yu et al., 2023
	Asparagine biosynthesis	*TaASN2*	Reduced free asparagine concentration	Raffan et al., 2021
	Phytic acid content	*TaIPK1*	Increased iron and zinc concentration	Ibrahim et al., 2022
Rice	Nutritional content and grain shape	*OsGW2*	Increased protein, iron, zinc, potassium, phosphorus, calcium content Increased grain area	Achary & Reddy, 2021
	Amylose content	*Waxy; OsISA1*	Reduced amylose and amylopectin content	Shufen et al., 2019
	Amylose and resistant starch content	*OsSBEIIb*	Increased amylose and resistant starch content	Sun et al., 2017
	Amylose content	*Waxy*	Reduced amylose content	Zhang et al., 2018
Maize	Amylose content	*Waxy*	Reduced amylose content	Gao et al., 2020
Soybean	Seed storage proteins	*Glyma.20g148400, Glyma.03g163500,* and *Glyma.19g164900*	–	Li et al., 2019
	Fatty acid composition	*GmFAD2–1A GmFAD2–1B*	Increased oleic acid content, decreased linoleic and α-linolenic acid content in seeds	Do et al., 2019
	Fatty acid composition	*GmFAD2–1A GmFAD2–2A*	Increased oleic acid content, and decreased linoleic acid content in seeds	Wu et al., 2020
	Isoflavone content	*GmF3H1, GmF3H2* and *GmFNSII-1*	Increased leaf isoflavone content	Zhang et al., 2020
	Beany flavor in soybean seeds	*GmLox1, GmLox2,* and *GmLox3*	Lipoxygenase-free soybean seeds	Wang et al., 2020
Canola	Fatty acid composition	*BnaFAE1* genes	Lower erucic acid content, and higher oleic acid content in seeds	Liu et al., 2022
	Fatty acid composition	*BnFAD2*	Increased oleic acid content in seeds	Liu et al., 2022
Tomato	Sugar content	*SlINVINH1*	Increased glucose and fructose content in fruit	Kawaguchi et al., 2021
	Sugar content	*SlINVINH1* and *SlVPE5*	Increased glucose, fructose and TSS content in fruit	Wang et al., 2021
	Lycopene content	*SGR1, LCY-E, Blc, LCY-B1,* and *LCY-B2*	Increased lycopene content in fruit	Li et al., 2018
	GABA content	*SlGAD2* and *SlGAD3*	Increased GABA content in fruit	Nonaka et al., 2017
	GABA content	*GABA-TP1,GABA-TP2, GABA-TP3, CAT9* and *SSADH*	Increased GABA content in fruit	Li et al., 2018
	Vitamin C content	*GGP1*	Increased vitamin C content	Li et al., 2018
	Accumulation of provitamin D3	*Sl7-DR2*	Increased provitamin D3 (7-DHC)	Li et al., 2022
Potato	Starch composition	*StGBSSI*	Abolished or strongly impaired amylose accumulation	Veillet et al., 2019
	Starch composition	*StGBSS*	Absence of amylose accumulation	Andersson et al., 2018
	Starch composition	*SBE1 and SBE2*	Diverse starch phenotypes	Tuncel et al., 2019
	Starch composition	*SBE1 and SBE2*	Diverse starch phenotypes	Zhao et al., 2021
	Anti-nutrient content	*St16DOX*	Complete elimination of SGA accumulation in potato hairy roots	Nakayasu et al., 2018
	Anti-nutrient content	*StSSR2*	Reduced SGA accumulation in peels, tuber fleshes, and leaves	Zheng et al., 2021
	Polyphenol oxidase activity	*StPPO2*	Reduced enzymatic browning	González et al., 2020

are1 mutants exhibited enhanced NUE, delayed senescence, and increased grain yield under nitrogen-limiting conditions, showing a yield increase of approximately 3–8% and 6–10% under low and high nitrogen conditions, respectively. In a parallel study, Zhang et al. (2021) also focused on the *TaARE1* gene, generating transgene-free mutants with delayed senescence, increased yield in normal growth conditions, and enhanced nitrogen deficiency tolerance under hydroponic conditions due to increased NUE. Boosting NUE not only empowers wheat plants to extract and utilize nitrogen from the soil but also diminishes the need for excessive nitrogen fertilizer. Such a benefit not only improves resource use, thereby cutting production expenses for farmers but also mitigates the environmental repercussions of nitrogen runoff. Optimizing nitrogen uptake and use directly contributes to better crop yields. Wheat varieties with increased NUE can produce higher yields with the same or even lower nitrogen inputs, leading to improved food security and economic benefits for farmers.

Turning attention to another gene with a similar focus, Kong et al. (2023) targeted the *TaARF12* gene which is a wheat auxin response factor and responsible for plant architecture (Li et al., 2022). The resulting mutant exhibited an increase in grain number per spike up to 14.7%, grain weight up to 3.9%, and grain yield up to 11.1%. Similarly, Jia et al. (2021) targeted the *TaIAA21* gene, a repressor in the auxin signaling pathway and important for controlling grain size and weight, using the CRISPR-Cas9 system. The resulting mutant displayed significant increases in grain length, grain width, and grain weight. Targeting another grain-regulatory gene, Zhang et al. (2016) aimed to disrupt the *TaGASR7* gene, which controls grain length and weight as a negative regulator. The resulting mutants showed a significant increase in grain weight. In a similar study, Zhang et al. (2019) disrupted another grain-regulatory gene, *TaCKX2-D1*, encoding cytokinin oxidase/dehydrogenase which plays a role in spike formation. The mutants showed significantly increased grain number/spikelet and spikelet number/spike, eventually leading to a significant increase in wheat yield.

Transitioning to another grain-regulatory gene, Gupta et al. (2023) mutated the *TaSPL13* gene, which encodes an important *SQUAMOSA promoter-binding protein-like (SPL)* protein playing a vital role in plant growth and development. Mutants exhibited decreased flowering time, tiller number, and plant height, but increased grain size and number. Errum et al. (2023) shifted attention to the *TaPpd* genes, which have a role in modifying flowering time. The double mutant of *Ta-Ppd-B1* and *TaPpd-D1* genes showed an increase in grain weight, area, and spikelets per spike by 2%, 8%, and 5%, respectively. Even though these numbers may seem low, these mutant lines can be used as parents in further breeding programs to enhance the yield.

Exploring genes affecting grain shape and weight, Wang et al. (2019) aimed to knock out the *TaGW7* gene, which encodes a TONNEAU1-recruiting motif (TRM) protein that controls grain shape and weight. Editing homologs of the gene in the B and D genomes resulted in increased grain weight and width, but decreased length. In a related study, Wang et al. (2018) aimed to knock out all three copies of the *GRAIN WIDTH and WEIGHT2 (GW2)* gene, responsible for grain weight in cereal crops. The loss-of-function mutations in all copies of the *TaGW2* gene exhibited an increase in grain weight and area. In another study, Wang et al. (2022) targeted the *DUO-B1* gene, encodes an APETALA2/ethylene response factor (AP2/ERF) transcription factor that regulates spike architecture. The loss-of-function mutants showed increased grain number per spikelet, spikelet number per spike, and spike number per plant, finally resulting in increased grain yield.

Wheat grains contain two types of starch granules: hard, vitreous ones, and soft, floury ones. The proportion of these granules contributes to the overall hardness of the grain. The Grain Hardness Index (GHI) is a measure used to characterize the hardness or softness, meaning the texture, of wheat grains. High GHI values indicate hard wheat, which is preferred for making products like bread, while low GHI values indicate soft wheat, suitable for products like cakes and pastries. The index is a valuable tool for wheat breeders and millers to select wheat varieties that meet the requirements of various food applications. For modifications related to grain hardness and amylose biosynthesis, Zhang et al. (2021) mutated the *Pinb*, which regulates grain hardness, and *Waxy* (*Wx*), which regulates amylose biosynthesis, in wheat. The obtained mutants displayed a three-fold increase in grain hardness index and significantly reduced grain amylose content. High amylose content in wheat grains can affect starch properties and, consequently, impact factors such as gelatinization and retrogradation during cooking and cooling processes. These properties may indirectly influence the overall quality of wheat-based products, including those made from hard or soft wheat. Therefore, double mutants of *Pinb* and *Wx* were great examples of wheat lines that can be used in baking better bread products.

Sánchez-León et al. (2018) targeted *α-gliadin* genes, involved in gliadin biosynthesis in wheat. Gliadin in wheat provides elasticity to dough, influencing bread texture. Essential for gluten formation, it enhances baking quality. The mutants of *α-gliadin* genes showed an 82% reduction in overall gliadin content in bread wheat and a 69% reduction in durum wheat. Similarly, Yu et al. (2023) used CRISPR-Cas9 to knock out *α/β-gliadin* genes, resulting in a 62% reduction in ω-gliadin and a 52% reduction in γ-gliadin content in the mutants. Reducing gliadin levels in wheat is crucial for addressing gluten intolerance or sensitivity in individuals. Gliadin is a component of gluten, and its consumption can trigger adverse reactions in some people, leading to conditions like celiac disease. Breeding wheat with low gliadin levels helps create gluten-free or gluten-reduced products, offering safe alternatives for those with gluten-related health concerns. This approach aligns with the growing demand for diverse and inclusive food options to cater to varying dietary needs and health considerations.

Other gene-editing studies in wheat focus on altering the amino acid levels or decreasing the anti-nutrient levels in the grain. For example, Raffan et al. (2021) employed

CRISPR-Cas9 to knock out the asparagine synthetase gene, *TaASN2*, resulting in reduced free asparagine concentration in the grains of the mutant lines. Low levels of free asparagine in wheat grain are highly desirable for numerous reasons, especially when considering food safety and human well-being. The presence of asparagine in abundance within wheat can result in the creation of acrylamide when cooking techniques involving high temperatures are used, like baking and frying. Unfortunately, acrylamide is a chemical compound that forms in various foods when subjected to heat and has been labeled a possible carcinogen for humans. Therefore, wheat lines with low levels of free asparagine are preferred by breeders since lowering free asparagine levels in wheat aligns with efforts to protect consumer health.

Phytic acid, or phytate, is another notorious anti-nutrient in plants. It is the storage form of phosphorus in plants, and it can form complexes with essential minerals, reducing their bioavailability. In a study, disrupting the phytic acid biosynthesis gene called *inositol pentakisphosphate 2-kinase 1 (IPK1)*, a 1.5-2.1-fold increase in iron and a 1.6-1.9-fold increase in zinc concentration were achieved in wheat grains (Ibrahim et al., 2022). Although some studies in different crops showed a decrease in the germination percentage and yield parameters of the *ipk* mutants, this study did not determine these parameters from the created mutant wheat lines. Therefore, it is essential to conduct comprehensive assessments of germination percentage and yield parameters in genome-edited mutant wheat lines before considering their introduction to the market.

14.3 RICE

Rice genome-editing studies intended to enhance yield have gained substantial attention in agricultural research. Targeted modifications often involve genes responsible for plant architecture, grain size, and stress responses. Xu et al. (2016) selected the genes negatively regulating grain weight in rice, including *Grain Width 2 (GW2), Grain Width 5* (*GW5*), and *Thousand-Grain Width 6* (*TGW6*). Multiple gene-editing strategies using the CRISPR-Cas9 system resulted in single, double, and triple mutants. Among the generated mutants, the double mutant *gw5tgw6* displayed 11.69%, 8.47%, and 12.68% increases in grain length, width, and weight, respectively. The triple mutant *gw2gw5tgw6* exhibited even more significant increases, ranging from 24.21% to 25.32% in grain length, 19.79% to 20.46% in width, and 27.13% to 29.84% in weight. This study highlights the importance of targeting multi-copy genes to achieve the desired phenotypic improvements in grain characteristics.

In another study, Ma et al. (2019) targeted the *OsSNB* gene, a novel grain width gene, in rice. The grain width, length, and overall weight were considerably higher in knockout mutant. Additionally, the mutants generated by targeting the *OsLOGL5* gene, encoding a cytokinin-activating enzyme, showed a significant increase in grain number per spike and total grain yield (Wang et al., 2020). Similarly, in another study, *PYL* (*PYL1*, *PYL4*, and *PYL6*) genes, encoding ABA receptors with roles in inhibiting rice growth, were mutated with the CRISPR-Cas9 system (Miao et al., 2018). The mutants exhibited enhanced growth, leading to an increased grain yield in rice. These studies collectively emphasize the significance of using gene-editing techniques on negative regulators to enhance crop yields.

Focusing on the tillering number, Miao et al. (2013) targeted the *LAZY1* gene, a negative regulator of polar auxin transport affecting the tiller angle in rice. The resulting mutants exhibited a tiller-spreading phenotype with an increased tiller number. Likewise, Lu et al. (2018) aimed to manipulate the *OsAAP3* gene, encoding an amino acid transporter known to be negatively correlated with the tiller number. The resulting mutant exhibited inhibited *OsAAP3* expression, leading to an increase in tiller number and enhanced grain yield per plant. In a study conducted by Song et al. (2022), the *Ideal Plant Architecture 1* (*IPA1*) gene, encoding a SQUAMOSA promoter binding protein-like (SPL) transcription factor OsSPL14 was disrupted by CRISPR-Cas9. OsSPL14 is a regulator of rice plant architecture; therefore, the resulting *ipa1* mutants showed increased panicle number and size, which led to increased grain yield. In another study, Chen et al. (2022) manipulated the rice *Kernel Row Number 2* (*OsKRN2*) gene, which negatively regulates grain number via control of secondary panicle branches, in rice. The CRISPR knockout mutant of the *OsKRN2* gene increased grain yield by 8%. CRISPR-Cas9 system was utilized by Li et al. (2016) to disrupt four yield-related genes simultaneoulsy in rice, namely *GRAIN NUMBER 1a* (*Gn1a*), *DENSE AND ERECT PANICLE* (*DEP1*), *Grain Size* (*GS3*), and *IDEAL PLANT ARCHITECTURE1* (*IPA1*), which act as regulators of grain number, panicle architecture, grain size, and plant architecture, respectively. The resulting mutants displayed distinct phenotypes, with the *gn1a* mutant showing large panicles and increased numbers of flowers, the *dep1* mutant increasing rice yield by 15–20%, *gs3* mutants displaying larger grain size, and ipa*1* mutants exhibiting a dwarf phenotype with more tillers.

In some recent studies, researchers started to focus on increasing the nutrient quality while decreasing the anti-nutrients in the rice grains. A particular study by Achary and Reddy (2021) aimed to knock out the *GW2* gene, which determines the grain weight in cereal crops. The resulting *OsGW2* mutation not only displayed enhanced grain protein content, with a 13–16% increase, but also showed accumulations of essential elements such as Fe (9–11% increase), Zn (13–15% increase), K (6–7% increase), P (9–11% increase), and Ca (8–10% increase) in the rice endosperm. Additionally, an enlargement in grain size was observed, with a 42–44% increase in width and a 27–32% increase in length. Moreover, Shufen et al. (2019) applied the CRISPR-Cas9 system to edit the *Waxy* (*Wx*) gene, responsible for amylose content reduction and *ISA1* gene, involved in amylopectin synthesis in rice. The mutants exhibited reduced amylose content from 14.6% to 2.6%, a decrease of 82%, a shrunken endosperm, and increased grain length, but decreased grain weight. Similarly, Sun et al. (2017) targeted the *SBEIIb* gene, encoding starch-branching enzyme in rice, to increase amylose and resistant starch content using CRISPR-Cas9. The resulting mutants showed a significant increase in amylose and resistant starch content by 25% and 9.8%, respectively. Zhang et al. (2018) reported the development of new rice lines with lower amylose content by mutating the *Waxy* gene using the CRISPR-Cas9 system. Taken together, these recent studies employing advanced gene-editing techniques emphasize the potential for enhancing the nutrient quality of rice grains by targeting specific genes. Strategies such as knocking out specific genes involved in anti-nutrients may lead to pleitropic phenotype enhancement.

14.4 MAIZE

The application of genome-editing techniques stands as a transformative strategy for elevating both yield and nutrient quality in maize. By intensely targeting various genes pivotal in growth and development processes, these advances aim to unlock the full potential of maize crops, fostering improvements in productivity and nutritional content. For instance, Chen et al. (2022) knocked out the *Kernel Row Number 2* (*ZmKRN2*) gene in maize and they observed a 10% grain yield increase in the mutant lines. In a second study, Lorenzo et al. (2023) used a CRISPR-Cas9 gene-editing system to manipulate growth-related genes (GRGs) that act as negative regulators of plant growth in maize. The resulting mutants displayed improved plant growth with increased yield. In a third study, Tian et al. (2019) disrupted the *RAVL* gene, a B3-domain transcription factor that indirectly affects the *Upright Plant Architecture2* (*UPA2*) gene which is responsible for upright plant architecture. The mutants exhibited reduced leaf angle and increased grain yield under high planting densities, in other words, greater yield but in less space due to narrower leaf architecture. This property is particularly essential in regions where limited farming land is available, as it offers the potential to maximize crop production efficiency and address the challenges of population growth and land scarcity.

In another study, the *ARGOS8* gene, which is a negative regulator of ethylene responses, was knocked out (Shi et al., 2017). As a novel finding, the resulting mutant showed enhanced grain yield under drought stress, suggesting the potential significance of manipulating ethylene responses in bolstering climate change resilience and improving drought tolerance in maize crops. This research holds promise for developing more robust and adaptive varieties critical for safeguarding food security in response to changing environmental conditions.

Some other studies focused on the manipulation of genes related to nutritional content or nutritional availability in maize. For example, Gao et al. (2020) manipulated the *Waxy* (*Wx*) gene, responsible for elongating the glucose polymers in amylose starches, in maize. The knockout *wx* mutant seeds had ~100% amylopectin whereas, unmutated wild-type seeds were made of ~25% amylose and ~75% amylopectin. Although altering the anti-nutrient contents is essential in maize grains, nobody worked on decreasing the phytate levels by using CRISPR-Cas9 in maize. Some mutagenesis studies developed maize lines with low phytate levels. For example, a study involved creating a *lpa1-7* mutant in maize using ethylmethane sulfonate (EMS), resulting in an 80% decrease in phytate concentration compared to the control plants (Cerino Badone et al., 2012). The mutant phenotype was determined to arise from a mutation in the *ZmMRP4* gene. However, pleiotropic effects, such as slow growth and changes in leaves, were observed in the *lpa1-7* mutant. Shi et al. (2007) demonstrated that the *LPA1* gene responsible for the *lpa1* mutant in maize consists of 11 exons and encodes MRP4, an MRP ABCC transporter. Additionally, they transferred the *LPA1* gene into maize plants under the maize embryo-specific promoter *Glb*. Their experiments revealed that transgenic plants germinated normally, showed no differences in vegetative growth from the control plants in the field, and did not exhibit a statistically significant decrease in seed weight. When the *Ole* promoter was used, an approximately 85% decrease in phytate was detected in the transgenic lines. However, a decrease in seed dry weight and germination difficulties were also noted. These studies prove that gene editing can be used to decrease the phytic acid in maize grains while the pleiotropic effects on plant growth, yield, and quality should be assessed in the field.

14.5 SOYBEAN

Soybean, a crucial economic contributor, serves as a staple in food, vegetable oil, and animal feed production. However, as a paleopolyploid species, studies on gene function face hurdles attributed to genetic redundancy and low genetic transformation frequencies. Despite these obstacles, CRISPR-Cas9 gene-editing system has been successfully applied to produce targeted mutations in hairy roots, somatic embryos, and stable transgenic plants of soybean (Zheng et al., 2020).

Crucial for both culinary and industrial uses, soybean seeds stand as a significant source of vegetable proteins. The major storage proteins in soybean seeds, namely conglycinins and glycinins, make up roughly 70% of the total protein content. These proteins play a central role in shaping the bioactive composition of soybean seeds and, thus, impact the quality of soy-based products. The CRISPR-Cas9 technology effectively induced mutations in three out of nine soybean seed storage protein genes (Li et al., 2019). The editing efficiency was measured at 5.8% for *Glyma.20g148400*, 3.8% for *Glyma.03g163500*, and remarkably high at 43.7% for *Glyma.19g164900*. These developed lines await further analyses of storage protein, total protein, and amino acid profiles as well as yield and yield parameters.

Soybean oil, a significant food commodity, exhibits a suboptimal fatty acid composition with approximately 24% monounsaturated fatty acids. In comparison, oils like canola (61%) and olive (75%) surpass this monounsaturated content. Oils high in monounsaturated fats are favored for their health benefits, prolonged shelf life, and enhanced oxidative stability. Improving soybean oil through partial hydrogenation leads to the production of trans-fatty acids associated with LDL cholesterol and heart disease. The conversion of monounsaturated oleic acid (C18:1) to polyunsaturated linoleic acid (C18:2) is catalyzed by the

FAD2 family, and there are at least five members of the FAD2 family in soybean. Of these five members, FAD2-1A and FAD2-1B primarily influence oleic acid levels in soybean oil. Previously, RNA interference (RNAi) silencing of *GmFAD2* in soybean led to a reduction in its activity, causing an accumulation of oleic acid and a decrease in the synthesis of linoleic acid. Do et al. (2019) employed CRISPR-Cas9 technology to achieve similar outcomes. The genotyping of T_0 lines revealed that all tested transformation events harbored mutations in either or both *GmFAD2* genes, with 40% of T_0 events exhibiting null mutations in both *GmFAD2–1A* and *GmFAD2–1B* genes. The analysis of T_1 seeds from CRISPR-edited plants, which are homozygous for both *GmFAD2* genes, revealed a significant elevation in oleic acid content exceeding 80%, coupled with a decrease in linoleic acid to 1.3–1.7%. Wu et al. (2020) also focused on the FAD2 family and created both single and double mutant knockout lines by utilizing CRISPR-Cas9 technology. In the study, the editing efficiencies were 95% and 55.56% for *GmFAD2-1A* and *GmFAD2-2A* genes, respectively. The mutants exhibited increases in oleic acid content from 17.10% to 73.50%, in linoleic acid content from 62.91% to 12.23%, and in protein content from 37.69% to 41.16% in the T_2 generation, compared to the JN38 control variety. Additionally, the oleic acid content increased from 19.15% to 72.02%, whereas linoleic acid content decreased from 56.58% to 17.27%, and protein content increased from 37.52% to 40.58%, in the T_3 generation. These studies are important for developing high oleic acid soybean lines, with reduced levels of polyunsaturated fatty acids, which are desired for their enhanced stability, extended shelf life, and improved nutritional profile, making them suitable for various food and industrial applications.

Soybean not only yields high-quality vegetable oils and proteins but also contains various physiologically active substances beneficial to human health. Isoflavonoids, a crucial group of secondary metabolites, are particularly abundant in leguminous plants, with soybeans exhibiting an isoflavonoid content approximately 100 times higher than other legumes. Among these, isoflavone synthase (IFS) is a key enzyme in isoflavonoid synthesis, sharing a common substrate with flavanone-3-hydroxylase (F3H) and flavone synthase II (FNS II). In a study, CRISPR-Cas9-mediated multiplex gene-editing technology was utilized to target simultaneously *GmF3H1, GmF3H2,* and *GmFNSII-1* in soybean hairy roots and plants (Zhang et al., 2020). The triple gene mutation efficiency was found to be 44.44% in the T_0 transgenic plants, which was significantly higher than the efficiency of triple gene mutation in hairy roots, which range from 6.25% to 12.50%. Metabolomic analysis of T_0 triple-mutant leaves showed a substantial increase in isoflavone content. The targeted mutagenesis was also stably inherited in the progeny. In comparison to the wild type, the leaf isoflavone content was almost doubled in the homozygous triple mutants in the T_3 generation, and the isoflavone content was found to be significantly elevated in the seeds of triple mutants in T_2 generation.

In mature soybean seeds, lipoxygenase proteins (LOX1, LOX2, and LOX3) catalyze the oxidation of unsaturated fatty acids, producing conjugated unsaturated fatty acid hydroperoxides that contribute to an unpleasant beany flavor, which hampers human consumption. Treatments such as heat, microwave processing, and organic solvent extraction have traditionally been employed in the food industry to mitigate this flavor, adding to the cost of soybean production and processing. A promising solution to this challenge is the generation of lipoxygenase-free soybean varieties, eliminating the beany flavor without incurring additional costs. In a study, a CRISPR-Cas9 strategy targeting three *GmLox* genes (*GmLox1, GmLox2,* and *GmLox3*) was implemented, resulting in the generation of positive T_0 transgenic plants with various combinations of sgRNAs and mutations (Wang et al., 2020). Colorimetric assay revealed that plants carrying different combinations of mutations lost the corresponding lipoxygenase activities. The lipoxygenase-free mutants could be used for soybean beany flavor reduction without restriction by regulatory frameworks governing transgenic organisms.

The plant architecture, which is determined by stem growth habit, node number, plant height, internode length, branch number, leaf size, and shape, significantly impacts grain yield in crops, including soybean. In plants, the *SQUAMOSA PROMOTER BINDING PROTEIN-LIKE* (*SPL*) transcription factor family, mostly regulated by *miR156*, influences the transition from juvenile to adult phases. *GmmiR156b* directly cleaves *SPL* gene products, notably impacting soybean plant architecture. Bao et al. (2019) employed CRISPR-Cas9-based multiple gene editing to target four *SPL9* genes simultaneously, resulting in T_4 generation soybean mutants with altered plant architecture traits. The mutants, including a quadruple mutant, reveal that the *GmSPL9* genes may have redundant or independent roles depending on the traits examined. Interestingly, the same study suggested that *GmSPL9b* regulates the expression of all four *GmSPL9* genes, including itself, in soybean. These findings enhance the understanding of CRISPR-Cas9 system applications and provide insights into the regulation of soybean plant architecture.

The number of seeds per pod (NSPP) is a crucial determinant of soybean yield. The *Ln* gene, encoding *GmJAGGED1* (*GmJAG1*), is a key regulator of NSPP and leaflet shape in soybean. Loss of function mutations in *GmJAG1* and *Ln* significantly increases NSPP and has been extensively utilized in soybean breeding. In a study, CRISPR-Cas9-based genome-editing technology was applied to modify *GmJAG1* in the low-latitude spring soybean cultivar Huachun 6. In this study, a CAS9-free homozygous mutant, named *gmjag*, was identified in the T_1 segregation population, exhibiting loss of start codons and small deletions in both *GmJAG1* and *GmJAG2* (Cai et al., 2021). Trait analysis revealed that *gmjag* displayed narrow leaflets. Field trials of this mutant soybean line showed no significant changes in growth period, plant height, number of branches, nodes, or total pods compared with nontransgenic Huachun 6. However, *gmjag* line exhibited the development of 4-seeded pods, a trait absent in Huachun 6. Moreover, the percentage of three-seeded pods increased significantly, while two-seeded pods decreased. The average NSPP and total seeds per plant of *gmjag* were significantly higher than Huachun 6, leading to an increased yield of 8.81% and 8.67% in the spring and summer trials, respectively. These findings demonstrate the effectiveness of CRISPR-Cas9 knockout of *GmJAG1* in introducing beneficial traits into varieties in low latitudes.

Plant height, a crucial trait impacting soybean yield, was targeted in a study using the CRISPR-Cas9 system to modify four *LATE ELONGATED HYPOCOTYL* (*LHY*) genes in soybean (Cheng et al., 2019). Analysis of the transgene-free quadruple mutant in the T_2 generation revealed significantly shorter height and internodes. Additionally, the quadruple mutant exhibited lower relative expression levels of genes in the gibberellic acid metabolic pathway. These results indicate that *GmLHY*, through direct or indirect regulation, influences plant height by modulating key components of the gibberellic acid pathway. Manipulating these genes presents an opportunity for enhancing plant height and internodes in soybean, thus contributing to an increased yield due to the relatively shorter stem length.

Soybean, a vital short-day crop, faces limitations in transgenic material creation and breeding efficiency due to its sensitivity to photoperiod and narrow geographical adaptability. *E1*, among the 11 identified genes influencing soybean growth period, plays an important role in breeding by inhibiting flowering and enabling adaptation to diverse environments. CRISPR-Cas9 was employed by Han et al. (2019) to generate targeted mutants of the *E1* gene. Two novel mutations resulted in conspicuous early flowering under long-day conditions, attributed to decreased *E1* gene expression. The flowering dates revealed significantly earlier flowering in mutants. Under short-day conditions, when mutants flowered, wild-type plants did not, and no significant difference was observed. Homozygous transgene-free mutants, lacking T-DNA elements, also exhibited early flowering. The created photo-insensitive soybean transformation receptor serves as a foundation for breeding transgenic receptors suitable for high latitudes.

Similar to maize, altering the anti-nutrient contents is essential in soybean seeds. Although some mutagenesis studies developed soybean lines with low phytate levels, there are no genome-editing studies focusing on decreasing the phytate levels. Therefore, research efforts should prioritize the development of soybean varieties with decreased phytic acid through precise genome-editing techniques. This can be achieved by targeting essential genes involved in phytic acid biosynthesis, such as *GmIPK1* (Inositol 1,3,4,5,6-pentakisphosphate 2-kinase) and *GmITPK1* (Inositol tetrakisphosphate 1-kinase), which play crucial roles in the phytic acid synthesis pathway. Focusing on these genes holds promise for achieving more efficient and targeted reductions in phytic acid content in soybean seeds.

14.6 CANOLA

Oilseed rape, being a high-value commodity crop, has undergone modification and traditional genetic engineering, despite the regulatory challenges associated with such approaches (Boniecka, 2024). The interest in precision breeding through genome-editing technologies for oilseed rape is evident, potentially offering more efficient and targeted genetic improvements.

Rapeseed, a globally significant crop, is an important supplier of high-quality edible oil and vital nutrients. The fatty acid profile, particularly erucic acid (EA), influences the quality of vegetable oils, making low erucic acid (LEA) a sought-after breeding trait in rapeseed. Currently, breeders possess multiple *B. napus* varieties that exhibit high seed oil content (SOC), but their effective use in breeding is hindered because many of these high SOC varieties also contain high erucic acid (EAC). Fatty acid elongase 1 (FAE1) is a crucial enzyme in EA synthesis, with two functional homologous copies present in *B. napus*. In a study, CRISPR-Cas9 was utilized for the targeted mutations on *BnaFAE1* in three *B. napus* varieties with high EA and high SOC (Liu et al., 2022). The mutations resulted in more than 10% decrease of EA in the one mutant and nearly zero EA in the double mutant. Oleic acid content was elevated to different degrees, while knockout of *BnaA08.FAE1* or/and *BnaC03.FAE1* caused a mild decrease in SOC without significant effects on other agronomic traits.

In rapeseed breeding, enhancing seed oleic acid content is a crucial quality trait. In another study, CRISPR-Cas9 strategy was employed to target the double loci of *BnFAD2*, exhibiting high regeneration and transformation efficiency in rapeseed with 64.35% total editing efficiency (Liu et al., 2022). The relative proportion of seed oleic acid in three single-locus-edited lines was below 80%, but it increased up to 15%, compared to the wild type. The oleic acid proportion in seed oil significantly increased, exceeding 85% in the T_1 and T_4 generations for double locus-edited mutants. Despite the wild-type plants having more siliques per plant, the mutant showed comparable seed yield and oil content, emphasizing the potential of this method to enhance oleic acid content in rapeseed.

14.7 TOMATO

As tomato production faces the need for enhanced quality, productivity, and added-value traits within limited space, along with an extended shelf life, genome editing has surfaced as a promising tool. Various genes associated with tomato fruit yield and quality have been identified, and successful production of high-quality individuals has been achieved through genome-editing techniques.

High sugar content is an important trait in tomatoes, often achieved through water stress, albeit at the cost of reduced fruit size and yield. Cell wall invertase plays a crucial role in sucrose unloading in tomato fruits, with invertase inhibitor (*SlINVINH1*) regulating this process. In a study, CRISPR-Cas9 was used for the knockout of *SlINVINH1* in tomatoes, and the resulting tomato lines exhibited increased fructose and glucose contents in mature fruits, with comparable fruit weight to the original cultivar (Kawaguchi et al., 2021).

Soluble sugars are key determinants of fruit sweetness and a significant component of total soluble solids (TSS) in tomatoes. The TSS content is crucial for the cost-effective production of processed tomatoes and tomato sauce since elevating the soluble sugar content in tomatoes enhances their natural sweetness and improves overall efficiency in tomato sauce production. Wang et al. (2021) employed CRISPR-Cas9 to obtain knockout lines of two genes, *SlINVINH1* and *SlVPE5*, that inhibit soluble sugar accumulation in tomato fruits. Compared to the wild-type, significant increases in glucose, fructose, and total soluble solid (TSS) contents were demonstrated in the knockout lines. Simultaneous knockout of both genes further enhanced glucose, fructose, and TSS levels, suggesting a synergistic effect. This study provides a practical foundation for improving tomato fruit sweetness and processing quality by targeting *SlINVINH1* and *SlVPE5*.

Lycopene is a bioactive compound known for its potential health benefits, including reducing the risk of chronic diseases. Synthesized during fruit ripening, lycopene is a key component of carotenoid metabolism, contributing to the color and quality of ripe tomatoes. Understanding and enhancing lycopene accumulation in fruit are crucial for improving both the visual and functional properties of tomatoes. Li et al. (2018) uniquely employed CRISPR-Cas9 to regulate five genes associated with the carotenoid metabolic pathway, *SGR1*, *LCY-E*, *Blc*, *LCY-B1*, and *LCYB2*, simultaneously offering a novel approach to enhance lycopene content in fruits. This study utilized a bidirectional approach, enhancing lycopene biosynthesis and inhibiting its conversion to β- and α-carotene. As a result, a remarkable 5.1-fold increase in lycopene content in tomato fruit was obtained. The homozygous mutations demonstrated stable transmission to subsequent generations, highlighting CRISPR-Cas9 as a highly efficient, low-off-target mutation tool for significantly improving lycopene content with stable heredity in tomatoes.

Tomatoes have higher levels of the non-proteinogenic amino acid gamma-aminobutyric acid (GABA) compared to other major crops. GABA has garnered significant attention as a health-promoting functional compound in recent years. Elevating GABA levels in tomatoes could enhance their blood pressure-lowering function. GABA is accumulated intensively in green-stage fruits; however, its levels decline in red-stage fruits. Glutamate decarboxylase (GAD) plays a crucial role in biosynthesis of GABA, and deleting its autoinhibitory domain enhances GAD activity. Among the five GAD genes in the tomato genome, *SlGAD2* and *SlGAD3* genes are expressed during fruit development. Nonaka et al. (2017) used CRISPR/Cas9 technology to delete the autoinhibitory domain of two *GAD* genes, *SlGAD2* and *SlGAD3*, resulting in a 7- to 15-fold increase in GABA accumulation in tomato fruits. It was found that the plant and fruit size, thus yield, are varied, yet this study marks the first use of CRISPR-Cas9 to enhance GABA content in tomato, offering insights for genetic modifications in other crops.

Similarly, Li et al. (2018) also focused on GABA accumulation in tomato, by using a multiplex *pYLCRISPR-Cas9* system to manipulate the GABA shunt by targeting five key genes. Single plant transformation yielded 53 genome-edited plants with representations of single to quadruple mutants. The GABA accumulation significantly increased in leaves and fruits of genomically edited lines, with quadruple mutants exhibiting a 19-fold higher GABA content in leaves compared to wild-type plants, and single and double-mutants exhibiting 1.34-fold to 3.50-fold increase in GABA content in red fruit. These findings showcase the precision of the multiplex CRISPR-Cas9 system for editing tomato genomic sequences, enabling the effective creation of multisite knockout mutations and offering insights into plant metabolic engineering regulations.

Enhancing the nutritional value of wild plants is a sought-after goal. Traditional breeding to domesticate wild plants boosts productivity but may decrease fitness and genetic diversity, limiting growth in challenging conditions. To expedite the domestication of *S. pimpinellifolium*, Li et al. (2018) used a multiplex CRISPR-Cas9 strategy to target *SlGGP1*, a gene involved in vitamin C biosynthesis, in the LA1589 accession. Mutations in *SlGGP1* resulted in significantly increased foliar ascorbic acid content, showing the effectiveness of CRISPR-Cas9 in boosting the amount of foliar ascorbic acid in wild tomatoes. This innovative approach seeks to accelerate the domestication process, addressing challenges associated with conventional breeding, including reduced fitness and genetic diversity.

Insufficient vitamin D is a global health concern, with limited dietary sources, particularly in plant-based diets. Tomatoes, during cholesterol and steroidal glycoalkaloid synthesis, naturally produce 7-DHC, a precursor to vitamin D3. Recent studies have revealed that there is a duplicate pathway in Solanaceous species where specific isoforms of enzymes involved in phytosterol and brassinosteroid biosynthesis also contribute to cholesterol production for the formation of specialized stress chemicals like SGAs. This partial separation of pathways allows metabolic flexibility. The loss-of-function of the gene that encodes Sl7-DR2, a key enzyme in cholesterol biosynthesis, facilitates the accumulation of 7-DHC without affecting other pathways. Addressing this, Li et al. (2022) employed genome editing to enhance provitamin D3 accumulation in tomatoes, offering a biofortified food source. For this purpose, the activity of Sl7-DR2, a key enzyme converting 7-DHC to cholesterol in tomatoes, was disrupted by CRISPR-Cas9-based gene editing. This led to increased 7-DHC content in both leaves and green fruit, potentially providing a biofortified source of provitamin D3. Despite the loss of Sl7-DR2 activity, there were no adverse effects on plant growth, development, or yield. Additionally, the 7-DHC levels in ripe fruit, while lower, remained sufficient for potential vitamin D3 synthesis when exposed to UV-B light. This innovation provides a novel plant-based solution to combat vitamin D insufficiency, especially pertinent for individuals following plant-centric, vegetarian, or vegan diets.

Rodríguez-Leal et al. (2017) leveraged CRISPR-Cas9 technology to engineer quantitative trait variation by mutagenizing cis-regulatory regions, unveiling a broad spectrum of quantitative variation. Using CRISPR-Cas9 to modify the cis-elements of *CLAVATA3* (*CLV3*) and *WUSCHEL* (*WUS*) genes associated with fruit size resulted in the expected enlargement of the meristem in cultivar M82, subsequently leading to an increase in fruit size. Since these genes play crucial roles in plant development, particularly in the regulation of meristem activity, which is essential for proper growth and fruit development, their mutations can lead to pleiotropic effects. Therefore, plant growth and yield should be investigated in these genome-edited tomato lines.

The evolution of plants led to the development of flowering triggered by seasonal changes in day length. Yet, day-length sensitivity in crops restricts their geographical cultivation range. Therefore, modifying the photoperiod response has been crucial for crop domestication. In tomato, the loss of day-length-sensitive flowering is attributed to the florigen paralog and flowering repressor *SELF-PRUNING 5G* (*SP5G*). Induced to high levels during long days in wild species, *SP5G* expression is downregulated in cultivated tomato due to cis-regulatory variation. CRISPR-Cas9-engineered mutations in *SP5G* lead to rapid flowering, enhancing the compact determinate growth habit in field tomatoes, resulting in an early-yielding plant in a single generation (Soyk et al., 2017). This demonstrates the potential of gene editing to quickly improve yield traits in crop breeding.

14.8 POTATO

The challenges in studying and breeding cultivated potato stem from its tetraploid nature and extensive heterozygosity. The tetrasomic inheritance adds complexity to traditional crossbreeding methods, posing obstacles in potato research and breeding programs. However, specific applications of genome editing offer the possibility to add one or a few traits to a commercial variety, thereby avoiding the need for traditional cross-breeding (Andersson et al., 2018). CRISPR-Cas9-based genome-editing technology can be used to study and develop commercially important traits in potato. These traits, which may be challenging to incorporate through traditional breeding methods, are now within reach through precise genome-editing techniques.

In higher plants, photosynthetic cells produce transitory starch in chloroplasts and export sucrose to heterotrophic organs, where it is converted to starch for long-term storage in amyloplasts. Starch, composed of two glucose polymers, amylose, and amylopectin, has physicochemical properties determined by their ratio, influencing parameters like pasting temperature and viscosity. Altering potato starch composition by reducing amylose (or amylopectin) content could have industrial applications, such as reducing the need for post-harvest treatments. In a study, the *StGBSI* gene, responsible in amylose production, were knocked out in two elite tetraploid potato cultivars by expressing CRISPR-Cas9 components (Veillet et al., 2019). Additionally, the concept for CRISPR-Cas9 base editing in the tetraploid potato was proved since the induction of DNA substitutions in the KTGGL-encoding locus using a cytidine base editor (CBE) was both precise and efficient. This resulted in generation of a variation in the amino acid sequence and thus a loss-of-function allele, showcasing the potential of base editing for genome engineering in this species.

In another study, CRISPR-Cas9 technology was further advanced as a breeding method in potato by employing ribonucleoprotein (RNP) delivery to protoplasts, to reduce unintended inserts originating from plasmid DNA (Andersson et al., 2018). The successful development of the amylopectin starch trait, a well-studied marker, demonstrated the functionality of the RNP technology. The potential of CRISPR-Cas9 RNP technology for potato breeding was highlighted by generating potatoes with amylopectin starch by knocking out the gene of amylose synthesize enzyme.

Starch properties for industrial purposes were targeted before but recently the focus has shifted toward modifying starch properties to alleviate potential negative effects on individuals such as Type II diabetes and obesity. Prior studies reducing starch-branching enzyme (SBE) activity in potatoes have resulted in extreme starch phenotypes. Despite their resistance to digestion, the resulting transgenic plants exhibited enhanced processability with altered tuber composition and dry matter content. Tuncel et al. (2019) aimed to employ CRISPR-Cas9-mediated targeted mutagenesis on SBEs in tetraploid potatoes to generate tuber starches with distinctive properties. The near elimination of both SBEs has severe impacts on starch, whereas the near elimination of either SBE1 or SBE2 has different impacts on starch composition, leading to the development of diverse starch phenotypes.

In another study, mutations were introduced into one or two branching enzyme genes (*SBE1* and *SBE2*) in tetraploid potato, aiming to enhance the amylose ratio and elongate amylopectin chains in starch (Zhao et al., 2021). Achieving a mutation frequency of up to 72% with RNP transfection of potato protoplasts, diverse mutants with various starch granule abnormalities were obtained. Selected mutants exhibited reductions in tuber SBE1 and/or SBE2 proteins consistent with genotype analysis expectations. Severe reduction in both SBE isoforms led to an extreme starch phenotype, similar to previous reports for low-SBE potato tubers. Additionally, a severe reduction in SBE2 alone displayed normal amylopectin structure and minor branching reductions but showed a significant increase in starch granule initiation. These findings suggest that CRISPR-Cas9 based gene editing of *SBE* genes can be resulted in new starch properties without introducing foreign DNA.

The anti-nutrients in potato tubers such as steroidal glycoalkaloids (SGAs) can negatively impact the quality. SGAs, present in various potato tissues, contribute a bitter taste and can be toxic, so managing its level in tubers is a crucial for breeding. Despite efforts to reduce SGAs through the silencing of SGA biosynthetic genes, the SGA-free potato has not been generated so far. Nakayasu et al. (2018) successfully knocked out *St16DOX* gene, encoding a steroid

16α-hydroxylase in SGA biosynthesis, using CRISPR-Cas9, leading to the complete reduction of SGA accumulation in roots. In another study, the gene *StSSR2*, responsible for the biosynthesis of SGAs in potatoes, was edited using the CRISPR-Cas9 system, resulting in a mutation efficiency of 46% (Zheng et al., 2021). Mutated potatoes exhibited a significant reduction in SGA content compared to the wild-type, with the lowest SGA concentrations recorded at 54% of the wild-type level in peel, 66% in tuber flesh, and 44% in leaves.

Polyphenol Oxidases (PPOs) play a role in enzymatic browning by converting phenolic substrates to quinones, leading to the formation of dark-colored precipitates in fruits and vegetables. This process can result in undesirable changes in organoleptic properties and the loss of nutritional quality in plant-derived products. In potato, PPOs are encoded by a multi-gene family with varied expression patterns. In a study, González et al. (2020) explored the use of the CRISPR-Cas9 system to induce mutations in the *StPPO2* gene of the tetraploid cultivar Desiree. Up to 68% of regenerated plants exhibited mutations in at least one allele of the target gene, with 24% of edited lines carrying mutations in all four alleles. Mutations induced in the four alleles of the *StPPO2* gene led to lines with a reduction up to 69% in tuber PPO activity and a 73% in enzymatic browning compared. These findings highlight the potential of the CRISPR-Cas9 system for developing potato varieties with reduced enzymatic browning in tubers by specifically editing a single member of the *StPPO* gene family.

14.9 CONCLUSION

Increasing agricultural yield is a cornerstone of sustainable farming practices and a pivotal strategy in the global quest to eliminate hunger. With the world population steadily increasing and climate change imposing unpredictable challenges, enhancing crop yield becomes imperative for ensuring food security. Climate change, characterized by extreme weather events, altered precipitation patterns, and rising temperatures, poses a threat to traditional agricultural practices. By focusing on improving crop yield, we can develop resilient plant varieties that withstand adverse environmental conditions, contributing to the stability of global food production. Genome editing, a cutting-edge technology distinct from conventional transgenic methods, allows precise modification of target genes without off-target mutations. This precision ensures the creation of genetically modified plants that are free of foreign DNA, addressing concerns associated with transgenic technologies. Integrating genome editing into plant breeding efforts offers a promising avenue to develop climate-resilient crops with increased yields, paving the way for sustainable agriculture and playing a pivotal role in eradicating hunger worldwide.

Furthermore, the application of genome editing holds immense potential in addressing hidden hunger, particularly prevalent in Southeast Asia, Africa, and Latin America. Biofortification through genome editing offers a targeted solution by enhancing the nutritional content of staple crops. For instance, editing genes associated with micronutrient uptake and storage can pave the way to the development of biofortified crop varieties with elevated levels of essential nutrients and vitamins, including zinc, iron and vitamin A. These nutrient-enriched crops exhibit the potential for significantly improving the nutritional well-being of communities relying heavily on staple foods. By employing genome-editing technologies, scientists can precisely engineer crops to combat specific nutrient deficiencies, contributing to a more sustainable and health-conscious approach in the global fight against hidden hunger.

Decreasing the levels of anti-nutrients in crop grains and seeds is crucial for enhancing mineral bioavailability, benefiting both human and animal nutrition. Anti-nutrients, such as phytic acid, can form complexes with essential minerals in the digestive tract, limiting their absorption by the human body and animals. By using genome-editing techniques to reduce anti-nutrient content in crops, such as targeting genes responsible for phytate biosynthesis and accumulation, we can enhance the availability of key minerals like iron and zinc. This approach complements biofortification efforts, where crops are engineered to accumulate higher levels of essential nutrients. Combining genome editing to decrease anti-nutrients with biofortification for increased nutrient content creates a synergistic strategy that can significantly improve the overall nutritional quality of crops. This approach holds promise for addressing nutrient deficiencies in both human and animal diets, contributing to better health outcomes and sustainable agricultural practices.

In conclusion, the evolving landscape of plant breeding is marked by the transformative potential of genome editing, offering an accurate and targeted approach to tackle the intricate obstacles of global food security and nutritional deficiencies. Harnessing the power of genome editing leads to the simultaneous alteration of multiple genes, presenting a promising avenue to increase yields without compromising nutrient levels. By fine-tuning the genetic makeup of crops, we can develop varieties that not only withstand the difficulties presented by climate change but also offer enhanced nutritional content. This approach signifies a crucial shift in the paradigm of plant breeding, highlighting the necessity for a nuanced and sustainable strategy to address the needs of an expanding global population. As we delve into this era of precision agriculture, genome editing emerges as a key tool in shaping the future of resilient, nutrient-rich crops, paving the way for a more food-secure and healthier world.

REFERENCES

Achary, V.M.M. and Reddy, M.K. (2021) CRISPR-Cas9 mediated mutation in GRAIN WIDTH and WEIGHT2 (GW2) locus improves aleurone layer and grain nutritional quality in rice. *Scientific Reports*, 11(1), 21941. https://doi.org/10.1038/s41598-021-00828-z.

Andersson, M. et al. (2018) Genome editing in potato via CRISPR-Cas9 ribonucleoprotein delivery. *Physiologia Plantarum*, 164(4), 378–384. https://doi.org/10.1111/ppl.12731.

Bao, A. et al. (2019) CRISPR/Cas9-mediated targeted mutagenesis of GmSPL9 genes alters plant architecture in soybean. *BMC Plant Biology*, 19(1), 131. https://doi.org/10.1186/s12870-019-1746-6.

Boniecka, J. (2024) CRISPR/Cas-based precision breeding of oilseed rape (Brassica napus L.) – recent improvements. In A. Ricroch et al. (Eds.), A roadmap for plant genome editing. Springer Nature Switzerland, pp. 291–307. https://doi.org/10.1007/978-3-031-46150-7_18.

Cai, Z. et al. (2021) CRISPR/Cas9-mediated gene editing of GmJAGGED1 increased yield in the low-latitude soybean variety Huachun 6. *Plant Biotechnology Journal*, 19(10), 1898–1900. https://doi.org/10.1111/pbi.13673.

Cerino Badone, F. et al. (2012) Study of low phytic acid1-7 (lpa1-7), a new zmmrp4 mutation in maize. *Journal of Heredity*, 103(4), 598–605. https://doi.org/10.1093/jhered/ess014.

Chen, W. et al. (2022) Convergent selection of a WD40 protein that enhances grain yield in maize and rice. Science, 375(6587). https://doi.org/10.1126/science.abg7985.

Cheng, Q. et al. (2019) CRISPR/Cas9-mediated targeted mutagenesis of GmLHY genes alters plant height and internode length in soybean. *BMC Plant Biology*, 19(1), 562. https://doi.org/10.1186/s12870-019-2145-8.

Do, P.T. et al. (2019) Demonstration of highly efficient dual gRNA CRISPR/Cas9 editing of the homeologous GmFAD2–1A and GmFAD2–1B genes to yield a high oleic, low linoleic and α-linolenic acid phenotype in soybean. *BMC Plant Biology*, 19(1), 311. https://doi.org/10.1186/s12870-019-1906-8.

Errum, A. et al. (2023) CRISPR/Cas9 editing of wheat Ppd-1 gene homoeologs alters spike architecture and grain morphometric traits. *Functional & Integrative Genomics*, 23(1), 66. https://doi.org/10.1007/s10142-023-00989-2.

Gao, H. et al. (2020) Superior field performance of waxy corn engineered using CRISPR-Cas9. *Nature Biotechnology*, 38(5), 579–581. https://doi.org/10.1038/s41587-020-0444-0.

González, M.N. et al. (2020) Reduced enzymatic browning in potato tubers by specific editing of a polyphenol oxidase gene via ribonucleoprotein complexes delivery of the CRISPR/Cas9 System. Frontiers in Plant Science, 10. https://doi.org/10.3389/fpls.2019.01649.

Guo, M. et al. (2021) Genetic manipulations of TaARE1 boost nitrogen utilization and grain yield in wheat. *Journal of Genetics and Genomics*, 48(10), 950–953. https://doi.org/10.1016/j.jgg.2021.07.003.

Gupta, A. et al. (2023) CRISPR-induced miRNA156-recognition element mutations in TaSPL13 improve multiple agronomic traits in wheat. *Plant Biotechnology Journal*, 21(3), 536–548. https://doi.org/10.1111/pbi.13969.

Han, J. et al. (2019) Creation of early flowering germplasm of soybean by CRISPR/Cas9 technology. *Frontiers in Plant Science*, 10. https://doi.org/10.3389/fpls.2019.01446.

Ibrahim, S. et al. (2022) CRISPR/Cas9 mediated disruption of Inositol Pentakisphosphate 2-Kinase 1 (TaIPK1) reduces phytic acid and improves iron and zinc accumulation in wheat grains. *Journal of Advanced Research*, 37, 33–41. https://doi.org/10.1016/j.jare.2021.07.006.

Jia, M. et al. (2021) TaIAA21 represses TaARF25-mediated expression of TaERFs required for grain size and weight development in wheat. *The Plant Journal*, 108(6), 1754–1767. https://doi.org/10.1111/tpj.15541.

Kawaguchi, K. et al. (2021) Functional disruption of cell wall invertase inhibitor by genome editing increases sugar content of tomato fruit without decrease fruit weight. *Scientific Reports*, 11(1), 21534. https://doi.org/10.1038/s41598-021-00966-4.

Kawall, K., Cotter, J. and Then, C. (2020) Broadening the GMO risk assessment in the EU for genome editing technologies in agriculture. *Environmental Sciences Europe*, 32(1), 106.

Kong, X. et al. (2023) Grain yield improvement by genome editing of <scp> TaARF12 </scp> that decoupled peduncle and rachis development trajectories via differential regulation of gibberellin signalling in wheat. *Plant Biotechnology Journal*, 21(10), 1990–2001. https://doi.org/10.1111/pbi.14107.

Li, A. et al. (2022) Wheat breeding history reveals synergistic selection of pleiotropic genomic sites for plant architecture and grain yield. *Molecular Plant*, 15(3), 504–519. https://doi.org/10.1016/j.molp.2022.01.004.

Li, C. et al. (2019) Mutagenesis of seed storage protein genes in Soybean using CRISPR/Cas9. *BMC Research Notes*, 12(1), 176. https://doi.org/10.1186/s13104-019-4207-2.

Li, J. et al. (2022) Biofortified tomatoes provide a new route to vitamin D sufficiency. *Nature Plants*, 8(6), 611–616. https://doi.org/10.1038/s41477-022-01154-6.

Li, M. et al. (2016) Reassessment of the four yield-related genes Gn1a, DEP1, GS3, and IPA1 in rice using a CRISPR/Cas9 system. Frontiers in Plant Science, 7. https://doi.org/10.3389/fpls.2016.00377.

Li, R. et al. (2018) Multiplexed CRISPR/Cas9-mediated metabolic engineering of γ-aminobutyric acid levels in Solanum lycopersicum. *Plant Biotechnology Journal*, 16(2), 415–427. https://doi.org/10.1111/pbi.12781.

Li, T. et al. (2018) Domestication of wild tomato is accelerated by genome editing. *Nature Biotechnology*, 36(12), 1160–1163. https://doi.org/10.1038/nbt.4273.

Li, X. et al. (2018) Lycopene is enriched in tomato fruit by CRISPR/Cas9-mediated multiplex genome editing. Frontiers in Plant Science, 9. https://doi.org/10.3389/fpls.2018.00559.

Liu, H. et al. (2022) CRISPR/Cas9-mediated editing of double loci of BnFAD2 increased the seed oleic acid content of rapeseed (Brassica napus L.). Frontiers in Plant Science, 13. https://doi.org/10.3389/fpls.2022.1034215.

Liu, Y. et al. (2022a) CRISPR/Cas9-targeted mutagenesis of BnaFAE1 genes confers low-erucic acid in Brassica napus. Frontiers in Plant Science, 13. https://doi.org/10.3389/fpls.2022.848723.

Lorenzo, C.D. et al. (2023) BREEDIT: A multiplex genome editing strategy to improve complex quantitative traits in maize. *The Plant Cell*, 35(1), 218–238. https://doi.org/10.1093/plcell/koac243.

Lowe, N.M. (2021) The global challenge of hidden hunger: Perspectives from the field. *Proceedings of the Nutrition Society*, 80(3), 283–289.

Lu, K. et al. (2018) Blocking amino acid transporter Os AAP 3 improves grain yield by promoting outgrowth buds and increasing tiller number in rice. *Plant Biotechnology Journal*, 16(10), 1710–1722. https://doi.org/10.1111/pbi.12907.

Ma, X. et al. (2019) A novel rice grain size gene OsSNB was identified by genome-wide association study in natural population. *PLOS Genetics*, 15(5), e1008191. https://doi.org/10.1371/journal.pgen.1008191.

Maqbool, A. et al. (2020) Biofortification under climate change: the fight between quality and quantity. In S. Fahad et al. (Eds.), *Environment, climate, plant and vegetation* growth, pp. 173–227. Springer.

Miao, C. et al. (2018) Mutations in a subfamily of abscisic acid receptor genes promote rice growth and productivity. *Proceedings of the National Academy of Sciences*, 115(23), 6058–6063. https://doi.org/10.1073/pnas.1804774115.

Miao, J. et al. (2013) Targeted mutagenesis in rice using CRISPR-Cas system. *Cell Research*, 23(10), 1233–1236. https://doi.org/10.1038/cr.2013.123.

Nakayasu, M. et al. (2018) Generation of α-solanine-free hairy roots of potato by CRISPR/Cas9 mediated genome editing of the St16DOX gene. *Plant Physiology and Biochemistry*, 131, 70–77. https://doi.org/10.1016/j.plaphy.2018.04.026.

Nonaka, S. et al. (2017) Efficient increase of γ-aminobutyric acid (GABA) content in tomato fruits by targeted mutagenesis. *Scientific Reports*, 7(1), 7057. https://doi.org/10.1038/s41598-017-06400-y.

Olabanji, M.F., Ndarana, T. and Davis, N. (2020) Impact of climate change on crop production and potential adaptive measures in the olifants catchment, South Africa. *Climate*, 9(1), 6.

Raffan, S. et al. (2021) Wheat with greatly reduced accumulation of free asparagine in the grain, produced by CRISPR/Cas9 editing of asparagine synthetase gene TaASN2. *Plant Biotechnology Journal*, 19(8), 1602–1613. https://doi.org/10.1111/pbi.13573.

Rodríguez-Leal, D. et al. (2017) Engineering quantitative trait variation for crop improvement by genome editing. *Cell*, 171(2), 470–480.e8. https://doi.org/10.1016/j.cell.2017.08.030.

Sánchez-León, S. et al. (2018) Low-gluten, nontransgenic wheat engineered with CRISPR/Cas9. *Plant Biotechnology Journal*, 16(4), 902–910. https://doi.org/10.1111/pbi.12837.

Shi, J. et al. (2007) Embryo-specific silencing of a transporter reduces phytic acid content of maize and soybean seeds. *Nature Biotechnology*, 25(8), 930–937.

Shi, J. et al. (2017) ARGOS 8 variants generated by CRISPR-Cas9 improve maize grain yield under field drought stress conditions. *Plant Biotechnology Journal*, 15(2), 207–216. https://doi.org/10.1111/pbi.12603.

Shufen, C. et al. (2019) Editing of rice Isoamylase Gene ISA1 provides insights into its function in starch formation. *Rice Science*, 26(2), 77–87. https://doi.org/10.1016/j.rsci.2018.07.001.

Song, X. et al. (2022) Targeting a gene regulatory element enhances rice grain yield by decoupling panicle number and size. *Nature Biotechnology*, 40(9), 1403–1411. https://doi.org/10.1038/s41587-022-01281-7.

Soyk, S. et al. (2017) Variation in the flowering gene SELF PRUNING 5G promotes day-neutrality and early yield in tomato. *Nature Genetics*, 49(1), 162–168. https://doi.org/10.1038/ng.3733.

Sun, Y. et al. (2017) Generation of high-amylose rice through CRISPR/Cas9-mediated targeted mutagenesis of starch branching enzymes. Frontiers in Plant Science, 8. https://doi.org/10.3389/fpls.2017.00298.

Tian, J. et al. (2019) Teosinte ligule allele narrows plant architecture and enhances high-density maize yields. *Science*, 365(6454), 658–664. https://doi.org/10.1126/science.aax5482.

Tuncel, A. et al. (2019) Cas9-mediated mutagenesis of potato starch-branching enzymes generates a range of tuber starch phenotypes. *Plant Biotechnology Journal*, 17(12), 2259–2271. https://doi.org/10.1111/pbi.13137.

Veillet, F. et al. (2019) The Solanum tuberosum GBSSI gene: A target for assessing gene and base editing in tetraploid potato. *Plant Cell Reports*, 38(9), 1065–1080. https://doi.org/10.1007/s00299-019-02426-w.

Wang, B. et al. (2021) Enhanced soluble sugar content in tomato fruit using CRISPR/Cas9-mediated SlINVINH1 and SlVPE5 gene editing. *PeerJ*, 9, e12478. https://doi.org/10.7717/peerj.12478.

Wang, C. et al. (2020) A cytokinin-activation enzyme-like gene improves grain yield under various field conditions in rice. *Plant Molecular Biology*, 102(4–5), 373–388. https://doi.org/10.1007/s11103-019-00952-5.

Wang, J. et al. (2020) Generation of seed lipoxygenase-free soybean using CRISPR-Cas9. *The Crop Journal*, 8(3), 432–439. https://doi.org/10.1016/j.cj.2019.08.008.

Wang, W. et al. (2018) Transgenerational CRISPR-Cas9 activity facilitates multiplex gene editing in Allopolyploid Wheat. *The CRISPR Journal*, 1(1), 65–74. https://doi.org/10.1089/crispr.2017.0010.

Wang, W. et al. (2019) Gene editing of the wheat homologs of TONNEAU 1-recruiting motif encoding gene affects grain shape and weight in wheat. *The Plant Journal*, 100(2), 251–264. https://doi.org/10.1111/tpj.14440.

Wang, Y. et al. (2022) Improving bread wheat yield through modulating an unselected AP2/ERF gene. *Nature Plants*, 8(8), 930–939. https://doi.org/10.1038/s41477-022-01197-9.

Wu, N. et al. (2020) Construction and analysis of GmFAD2-1A and GmFAD2-2A soybean fatty acid desaturase mutants based on CRISPR/Cas9 technology. *International Journal of Molecular Sciences*, 21(3), 1104. https://doi.org/10.3390/ijms21031104.

Xu, R. et al. (2016) Rapid improvement of grain weight via highly efficient CRISPR/Cas9-mediated multiplex genome editing in rice. *Journal of Genetics and Genomics*, 43(8), 529–532. https://doi.org/10.1016/j.jgg.2016.07.003.

Yu, Z. et al. (2023) CRISPR-based editing of the ω- and γ-gliadin gene clusters reduces wheat immunoreactivity without affecting grain protein quality. Plant Biotechnology Journal [Preprint]. https://doi.org/10.1111/pbi.14231.

Zhang, J. et al. (2018) Generation of new glutinous rice by CRISPR/Cas9-targeted mutagenesis of the Waxy gene in elite rice varieties. *Journal of Integrative Plant Biology*, 60(5), 369–375. https://doi.org/10.1111/jipb.12620.

Zhang, J. et al. (2021) Increasing yield potential through manipulating of an ARE1 ortholog related to nitrogen use efficiency in wheat by CRISPR/Cas9. *Journal of Integrative Plant Biology*, 63(9), 1649–1663. https://doi.org/10.1111/jipb.13151.

Zhang, P. et al. (2020a) Multiplex CRISPR/Cas9-mediated metabolic engineering increases soya bean isoflavone content and resistance to soya bean mosaic virus. *Plant Biotechnology Journal*, 18(6), 1384–1395. https://doi.org/10.1111/pbi.13302.

Zhang, S. et al. (2021) CRISPR/Cas9-mediated genome editing for wheat grain quality improvement. *Plant Biotechnology Journal*, 19(9), 1684–1686. https://doi.org/10.1111/pbi.13647.

Zhang, Y. et al. (2016) Efficient and transgene-free genome editing in wheat through transient expression of CRISPR/Cas9 DNA or RNA. *Nature Communications*, 7(1), 12617. https://doi.org/10.1038/ncomms12617.

Zhang, Z. et al. (2019) Development of an Agrobacterium-delivered CRISPR/Cas9 system for wheat genome editing. *Plant*

Biotechnology Journal, 17(8), 1623–1635. https://doi.org/10.1111/pbi.13088.

Zhao, X. et al. (2021) Amylose starch with no detectable branching developed through DNA-free CRISPR-Cas9 mediated mutagenesis of two starch branching enzymes in potato. *Scientific Reports*, 11(1), 4311. https://doi.org/10.1038/s41598-021-83462-z.

Zheng, N. et al. (2020) CRISPR/Cas9-based gene editing using egg cell-specific promoters in arabidopsis and soybean. Frontiers in Plant Science, 11. https://doi.org/10.3389/fpls.2020.00800.

Zheng, Z. et al. (2021) Editing sterol side chain reductase 2 gene (StSSR2) via CRISPR/Cas9 reduces the total steroidal glycoalkaloids in potato. *All Life*, 14(1), 401–413. https://doi.org/10.1080/26895293.2021.1925358.

15 Multiplex Gene Editing and Its Applications

Expanding the Horizons of CRISPR-Cas

Aftab Ahmad, Annena Ijaz, Adeen Ijaz, Sameer H. Qari, Abdul Ghaffar, Ahmed Al-Harrasi, and Baohong Zhang

15.1 INTRODUCTION

Since the dawn of the twenty-first century, scientists have been using genome-editing tools such as ZFNs and TALENs for precise editing of genes for basic as well as applied applications. Recently CRISPR-Cas has revolutionized the field of gene editing by altering genes in a simple, precise, and efficient way. Multiplex gene editing with CRISPR-Cas has enabled researchers to simultaneously target multiple genes, which was not possible with ZFN and TALENs. Plant scientists have been using CRISPR-Cas to introduce site-specific mutations in genes for improved crop production, enhancing disease resistance and climate resilience in crops while also accelerating functional genomics (Mishra et al., 2018; Wu et al., 2018; Munir et al., 2024). CRISPR-Cas holds immense potential to fulfill the early promises of genetic engineering such as food security and sustainable agriculture (Ahmad et al., 2024).

CRISPR-Cas technology has emerged as a fascinating and powerful tool for genome editing in model and crop plants, contributing to food security and achieving zero hunger. CRISPR allows the precise editing of targeted DNA sequences in the genome to introduce desired traits in the crops (Razzaq et al., 2019). A 20 bp gRNA complementary to the target site is required to guide the Cas9 to the target site and subsequently introduce a double-stranded break (DSB). The targeted breaks are repaired by a DNA repair mechanism either by non-homologous end joining (without homologous template) or homologous end joining (requires a sequence of donor DNA template). Homologous end joining is the best way to perform targeted modifications in the eukaryotic genome (Pandey et al., 2019; Aslam et al., 2022).

CRISPR-based genomic manipulations have been successfully demonstrated in different plants, with proof-of-concept studies including yield improvement, disease resistance, nutritional enhancement, and metabolic engineering. For example, studies in model plants such as *Arabidopsis* and *Nicotiana* and several crops like wheat, maize, rice, soya bean, chili pepper, and tomatoes have been reported using CRISPR Cas system for increased production and enhanced nutritional quality (Osakabe & Osakabe, 2017). In *Arabidopsis*, several genes such as AtPDS3, AtFLS2, AtADH, AtCHL1, AtTT4, and AtAP1 have been precisely targeted with CRISPR-Cas (Li et al., 2013; Feng et al., 2014). Similarly, in *Nicotiana benthamiana*, genes such as NbPDS, NbPCNA, and NbPDR6 have been efficiently edited with CRISPR-Cas (Song et al., 2016). In tomato, CRISPR has been successfully used to achieve morphological changes in the compound leaves by editing ARGONAUTE 7 gene (Brooks et al., 2014). In addition, other genes such as SHORT-ROOT (SHR), Ripening inhibitor (RIN), and Anthocyanin 1 (ANT1) have been targeted to study root development, fruit ripening, and anthocyanin biosynthesis, respectively, in tomatoes (Čermák et al., 2015). CRISPR-Cas technology has been used to develop partially gluten-free wheat by precise editing of more than 40 genes, despite the complex genome of wheat (Sánchez León et al., 2018). In maize, phytic acid, an anti-nutritional compound was reduced by targeting two genes necessary for its biosynthesis using CRISPR-Cas (Liang et al., 2014). CRISPR has been used to improve the grain size of maize under drought stress by enhancing the promotor sequence of ARGOS8. In addition, CRISPR has been used to edit the ZmGA20ox3 gene of maize, leading to semi-dwarf maize plants that are useful for the creation of new germplasm with improved lodging resistance and suitable for high-density planting (Zhang et al., 2019). In soybean, CRISPR-Cas has been used to edit GmPDS11 and GmPDS18 genes for improved carotenoid biosynthesis (Du et al., 2016). These studies demonstrate the enormous potential of CRISPR-Cas to develop new crop varieties with improved traits by successfully targeting the key genes controlling specific agronomic traits (Lee et al., 2019).

Polygenic traits are influenced by the contribution of multiple genes. One significant advantage of the CRISPR/Cas system is its ability to multiplex genome editing which enables simultaneous editing of multiple genes in the genome. Multiplex gene editing through CRISPR-Cas is a valuable tool for the functional analysis of large gene families, polyploid genomes, functionally redundant genes, and interconnected genes involved in biochemical pathways all require. Multiplex gene editing can be used to enhance breeding efficiency by introducing mutations at different loci in a single generation (Wu et al., 2024). Multiplex gene editing is of key importance in constructing genetic circuits, epigenome regulations and manipulating polygenic traits. Multiplex gene editing with CRISPR-Cas has been successfully demonstrated in different crops for improving desired traits such as disease resistance,

DOI: 10.1201/9781003500933-18

herbicide tolerance, metabolic engineering, hormone biosynthesis, plant development, production of seedless plant species, and molecular farming. Multiple genomic editing allows the editing of coding as well as non-coding regions (promoters) of the genes. It has enabled researchers to study interaction among different members of a gene family with different functions (Abdelrahman et al., 2021).

15.2 CRISPR-CAS SYSTEMS FOR MULTIPLEX GENE EDITING

15.2.1 CRISPR-Cas9

CRISPR/Cas9 is a powerful system for multiplex gene editing enabling simultaneous editing of multiple genes. It enables researchers to decipher the functions of gene families that regulate biological networks (Ma et al., 2015). Success in the multiplex gene editing with CRISPR-Cas9 depends upon the selection of gRNA and target sites. Although the CRISPR-Cas9 system shows high mutation efficiency for multiplex gene editing, it may not completely suppress gene functions particularly those involving long non-coding RNAs. Multiplex gene editing also enables the deletion of chromosomal fragments between the target sites with improved mutation efficiency, however, it requires a careful selection of gRNAs for an efficient deletion.

Multiplex genome editing with CRISPR-Cas9 has been successfully demonstrated in different crops to enhance disease resistance and improve agronomic traits, including plant height, flowering time, and salinity tolerance. For example, CRISPR-Cas9 was used to produce a double mutant (SlPelo/SlMlo1) in tomatoes, resulting in transgene-free plants with improved resistance against powdery mildew disease and tomato leaf curl virus (Pramanik et al., 2021). Similarly, multiplex gene editing has been reported in wheat to develop resistance against powdery mildew disease by targeting the three homologs of the EDR1 gene (Zhang et al., 2017). In addition, CRISPR-Cas9 has been used to edit four LNK2 genes of soybean, that are involved in flowering. The study resulted in transgene-free soybean plants with accelerated flowering under long photoperiod. These findings provided a deeper understanding of the mechanism underlying the regulatory networks of flowering time in soybean (Li et al., 2021). Another study has demonstrated the potential of a multiplex CRISPR-Cas9 approach in *Solanum lycopersicum* by targeting the five genes (Li et al., 2018). All these studies highlight the potential of multiplex CRISPR-Cas9 gene editing in genetic improvement of crops.

15.2.2 The dCas9 System

CRISPR/dCas9 technology was first developed by Qi and colleagues in 2013. The dCas9-gRNA complex uses a dead-Cas9 (dCas9) enzyme that is unable to cleave the DNA but can specifically bind to the target site (Qi et al., 2013). The nuclease activity of Cas9 can be abolished by mutating specific amino acids in it, thus converting it into dead or deactivated Cas9 (dCas9). dCas9 is particularly useful in the regulation of genes at the transcriptional level such as CRISPRi and CRISPRa, without introducing any permanent mutations in the DNA. The dCas9-gRNA complex in CRISPRi and CRISPRa recruits transcriptional repressors and activators, respectively, to the target sites in the DNA to regulate the transcription of a gene. In addition, dCas9 is also useful for epigenetic regulation and base editing applications of CRISPR-Cas. dCas9 can be fused with epigenetic regulators such as methyl transferases or demethylases to install precise epigenetic changes at the target site in the genome. Similarly, dCas9 can be fused with adenosine deaminase or cytosine deaminase to introduce base editing the target site in the genome. In the multiplex CRISPR-dCas9 approach, multiple gRNAs are used to recruit dCas9 fused with functional domains to control the expression of multiple genes. During base editing, dCas9 (fused with deaminases), can be used to edit a single nucleotide in the entire genome to achieve a desirable trait in the plant (Minkenberg et al., 2017). For example, multiplex CRISPR-dCas9 has been used to induce multiple point mutations leading to herbicide tolerance in plants (Shimatani et al., 2017).

Xu et al. (2019) developed a toolkit for rice that combines high-fidelity dCas9 variants coupled to a tandem-arrayed tRNA-modified sgRNA thus increasing the efficiency of this method by 2–5-fold. Another study demonstrated the repression of *Arabidopsis* genes using the CRISPR-dCas9 system (Abdelrahman et al., 2021). Similarly, Roca-Paixão et al. (2019) reported the use of dCas9 in *Arabidopsis* to enhance drought resistance by fusing dCas9 with the histone acetyltransferase 1 (AtHAT1), thereby activating abscisic acid (ABA)-responsive element binding protein 1/ABRE binding factor (AREB1/ABF2). The edited plants showed a higher survival rate as compared with control plants under drought stress.

15.2.3 CRISPR/Cas12

CRISPR-Cas12a is the second most widely used CRISPR system, characterized as a Type V endonuclease that requires a T-rich PAM sequence. It is a simple and versatile system, particularly effective for precise gene modifications using multiplex gene editing. In plants, Cas12a is more effective with higher targeting efficiency than Cas9 (Zhang et al., 2021). In contrast to Cas9, Cas12a does not require transactivating crRNA as it only requires a short crRNA to bind and cleave at the target sequence in the genome. In addition, Cas12a also possesses RNase activity for processing its crRNA array, which has been exploited for multiplex genome-editing applications (Tang et al., 2018). Unlike Cas9, which produces blunt end breaks, Cas12a generates staggered end breaks at the cleavage site of the target sequence.

The most widely used Cas12a orthologs in plants include LbCas12a, AsCas12a, and FnCas12a. Among these, LbCas12a has shown higher editing activity, with 87% editing efficiency in cotton (Li et al., 2019), maize (Lee et al., 2019), *Arabidopsis* (Bernabé et al., 2019), and rice (Hu et al., 2016).

Similarly, CRISPR/LbCas12a has been successfully used to produce gossypol-free cotton by editing the coding sequence of GhPGF and GhCLA1 genes in cotton, providing a valuable germplasm resource for molecular breeding of cotton. In addition, LbCas12a has been demonstrated to edit maize genome with 100% mutation efficiency. In rice, CRISPR-Cas12 was used to induce heritable mutations with editing efficiency of 77.8% and 92.8% at two target sites in T_0 plants (Malzahn et al., 2019). Cas12a has been effectively used for gene editing in woody plants, advancing research of forest genetics and fulfilling the deficiency of woody plant mutants (Yang et al., 2023). These findings highlight the potential of CRISPR-Cas12a in multiplex gene editing in plants.

15.2.4 CRISPR/Cas13

CRISPR-Cas13 is a Class 2, Type VI, single effector-based RNA targeting system. It is a novel and versatile system that can be used for multiplex gene editing at the post-transcriptional level. Unlike DNA editing systems such as CRISPR-Cas9 and CRISPR-Cas12a, Cas13 enables transient and reversible gene regulation without introducing permanent mutations in the genome. It makes CRISPR-Cas13 particularly useful for functional genomics and therapeutic applications where permanent gene editing is undesirable. Cas13 effectors can be effectively used to target multiple RNAs simultaneously by introducing multiple gRNAs, facilitating diverse applications in plants such as disease resistance (especially against RNA viruses such as potyvirus), functional genomics, abiotic stress tolerance, transcriptome engineering, and synthetic biology (Hillary et al., 2023). In plants, Cas13 has been used for RNA editing with high specificity. CRISPR-Cas13 identifies the target mRNA with high specificity and efficiency, enabling cleavage of target RNA and blocking protein production. To evaluate the potential of the CRISPR/Cas13 system, a comparative assessment between RNAi and Cas13-mediated RNA degradation was performed in rice protoplast. The study demonstrated that Cas13 has higher efficiency in degrading specific splice variant transcripts (Abudayyeh et al., 2017).

15.2.5 CRISPR/Cas14

CRISPR/Cas14, is a new member of Class 2 CRISPR-Cas systems with potential advantages over other members of this class as it does not require protospacer adjacent motifs (PAM) near their target site. Cas14 protein is small, almost half the size of Cas9 protein, having a molecular weight of 40-70 KD and consisting of only 400-700 amino acids. Cas14 has been characterized as a single-strand DNA cleavage protein without any requirement of PAM. Due to its compact size as compared with Cas9, it has become an ideal system to target ssDNA in plants. Some 24 different Cas 14 proteins have been identified. However, Cas14a has been demonstrated to show resistance against plant ssDNA viruses that belong to Geminiviradae (Manghwar et al., 2019). CRISR-Cas14 also holds enormous potential for multiplex gene-editing applications in plants.

15.3 STRATEGIES FOR ACHIEVING MULTIPLEX GENOME EDITING

Scientists have developed various strategies to efficiently express multiple gRNAs by taking advantage of multiplex genome editing. Multiplex gene editing enables researchers to develop CRISPR-edited organisms, carrying precise mutations at multiple target sites in the genome in one generation (Najera et al., 2019; Sultan et al., 2022). Although editing efficiency in multiplex gene-editing studies is variable, scientists are focused on improving the editing efficiency, particularly through the development of a rapid multiplex gene-editing toolkit that involves the cloning of multiple gRNAs in a single binary vector. Although multiple rounds of cloning can be used during multiplex gene editing, to clone individual gRNAs in a vector, however, it is a laborious process with a limited number of gRNAs that can be assembled (Li et al., 2013; Zhou et al., 2014). Researchers have developed alternate strategies such as the Golden Gate method, Gibson Assembly and Infusion cloning, to efficiency clone multiple gRNAs in a single cloning step.

Golden Gate cloning is an advanced cloning method that relies on Type IIS restriction enzymes such as *BsaI*, to generate non palindromic sticky ends in a binary vector to clone multiple fragments in a single step. Golden Gate cloning facilitates the cloning of multiple gRNAs in a single binary vector tailored for multiplex CRISPR/Cas9 gene-editing toolkits. To simplify cloning and processing of gRNAs in multiplex gene editing, several strategies have been developed to encode, transcribe and process multiple numbers of gRNAs in a single step. These strategies include assembling multiple gRNA with their own promotor (RNA polymerase III promoters, such as U6 or U3) and terminator (Lowder et al., 2015; Zhang et al., 2016) for individual gRNA, expressing multiple sgRNAs as a single transcript using a single promotor and terminator, using tRNA processing enzymes (Xie et al., 2015), or employing self-processing ribozymes. Each strategy has its own advantages and disadvantages, depending on the editing efficiency and number of gRNAs that can be cloned and expressed. One significant advantage of the strategies that process transcript to produce mature gRNA, is the ability to use a Pol II promoter instead of a Pol III promoter, relieving sequence limitations and enabling tissue-specific expression. In the following section, we have summarized different strategies used for cloning multiple gRNAs in multiplex gene editing through CRISPR-Cas. Figure 15.1 shows various strategies for the expression of gRNA cassettes in plants.

15.3.1 Individual Expression Cassettes for Each gRNA

The most widely used method for multiplex genome editing in CRISPR-Cas involves expressing multiple gRNAs under the control of individual RNA polymerase promoters (RNA polymerase III, such as U6 and U3) and terminators. When gRNAs are driven by RNA polymerase III, the expression cassette is constructed in three steps: first, constructing an

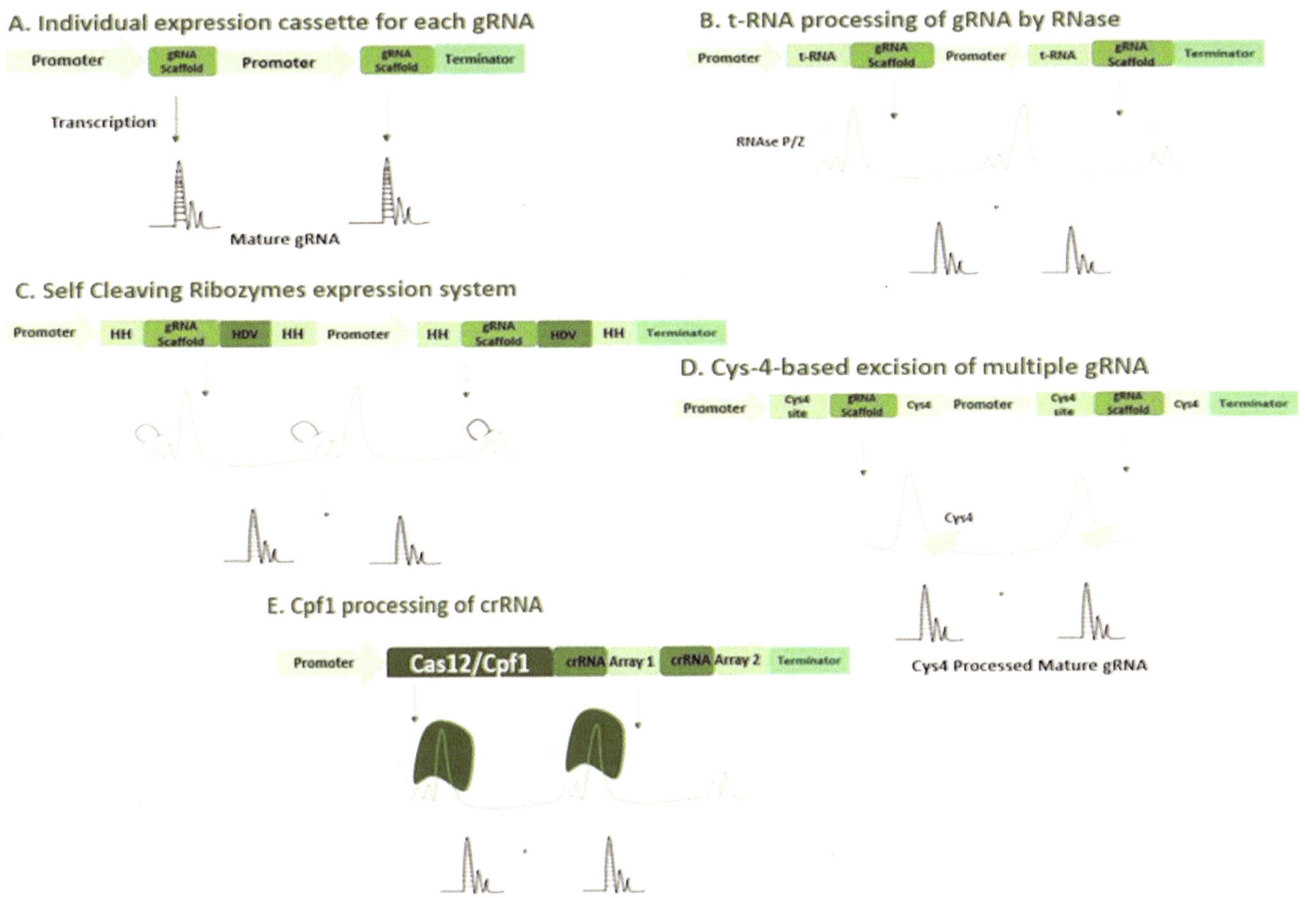

FIGURE 15.1 Various strategies for the expression of gRNA cassettes in plants: (a) Expression of multiple gRNA under the RNA polymerase promoter to generate individual gRNA from individual promoter. (b) An engineering approach to process t-RNA-gRNA transcript into functional gRNAs by endogenous RNase P and RNase Z, cleaving the 5′ and 3′ ends of the pre-tRNAs. Arrays of this form have been expressed through Pol II or Pol III promoters. (c) Processing of gRNA arrays involves flanking each gRNA with self-cleaving enzymes, such as Hammerhead or HDV ribozymes, that can be transcribed through Pol II or Pol III promoters. (d) gRNA arrays can be processed by the exogenous expression of Csy4, allowing each gRNA to be flanked by its own recognition site (Cys site). This enzyme recognize 28-nt stem–loop sequence in RNA and cuts after the 20th nucleotide. (e) Unlike Cas9, Cas12a has gRNA self-cleaving feature and it process crRNA arrays and remove direct repeats without requiring elements for posttranscriptional processing.

intermediate vector containing individual sgRNAs with their promoters and terminators, followed by digestion with a restriction enzyme (Type IIS) and subsequent recovery of sgRNA cassettes for Golden Gate cloning into a binary vector (Lowder et al., 2015). A significant limitation of this method is the repetitive promotor sequences which make the final expression cassette large and cause plasmid instability in *Agrobacterium* and *E. coli*. In addition, balancing between insert size and cloning efficiency is also a potential concern in this strategy. This method has been reported to express six to eight gRNAs in a single cassette for multiplex genome editing (Ding et al., 2020). Multiple gRNAs can be assembled using RNA Polymerases (II and III) and Type II restriction enzymes such as *BsaI*. Using the *BsaI* restriction enzyme, scientists have demonstrated the assembly of multiple gRNAs (up to 24 gRNAs) in plants, including *Arabidopsis* (Stuttmann et al., 2021). Although both RNA Polymerase II and III have been reported to express gRNAs in CRISPR-Cas studies, RNA Poly II promotors have high processivity, are often inducible, and are easily tuned. Many scientists have developed CRISPR toolboxes using this approach for the expression of multiple gRNA in plants. The simplicity and efficiency of expressing individual gRNA cassettes underscore the importance of this method in multiplex genome editing. The efficiency of multiplexing can be further enhanced by screening better promotors (Arimbasseri et al., 2013).

15.3.2 t-RNA Processing System and Processing of gRNA by RNase

A tRNA-based assembly of multiple gRNAs is a sophisticated method to introduce precise mutations at multiple target sites in the host genome. This system relies on the natural properties of the tRNA to deliver and process multiple gRNAs during multiplex gene editing through CRISPR-Cas. Using this system, tRNA sequences are used to separate multiple gRNAs, and these tRNA-gRNA units are expressed as a single polycistronic transcript. Endogenous RNases (RNase Z and RNase P) recognize the tRNA sequences and cleave the primary transcript at specific sites to release the mature gRNAs from the transcript. Mature gRNAs drive the Cas9 to the different target sequences in host genome to introduce precise editing. Xie et al. (2015) demonstrated the t-RNA based multiplex CRISPR-Cas9 by targeting eight genes in rice. A single Pol III promotor and terminator was used to express multiplex tRNA-gRNA units in a single cassette, allowing RNase P and RNase Z to recognize and cleave the tRNA-gRNA at the 5′ and 3′ ends respectively. Cleavage of the primary transcript at 5' and 3' ends of the tRNA results in mature and functional gRNA that can guide Cas9 to the target site in the genome. Although this strategy was mainly employed in rice with high editing efficiency, the authors observed the reduction in efficiency of this system when more than six gRNA expressed simultaneously. Despite this, the system remains effective for multiplex genes in plants. For example, Qi et al. (2016) demonstrated an endogenous tRNA-gRNA based CRISPR-Cas system by targeting a gene with two gRNAs in maize with 100% efficiency. Similarly, Wang et al. (2016) demonstrated this system for transient expression of multiple gRNAs in wheat protoplast, targeting multiple genes. A single polycistronic cassette carrying multiple tRNA-gRNA units was expressed under wheat U3 snoRNA promoter for multiplex gene editing. These findings indicate that the tRNA-based multiplex CRISPR-Cas strategy is a robust approach for targeting multiple genes with broad-range applications in metabolic engineering, knockout of multiple genes, synthetic biology, and biotic and abiotic stress tolerance in plants.

15.3.3 Self-Cleaving Ribozymes

Self-cleavage ribozymes are an advanced method to facilitate production of multiple gRNAs leading to multiplex gene editing with CRISPR-Cas. It relies on the self-cleavage properties of ribozymes to produce mature individual gRNAs from a single transcript with recognizable ribozyme cleavage sites. This approach was initially developed in yeast to produce multiple individual gRNAs from single transcript using ribozyme's self-catalytic properties to cleave RNA molecule at specific sites (Lee et al., 2016). The ribozyme-flanked sgRNAs cassette will undergo self-catalyzed cleavage with the ribozyme's intrinsic nuclease activity, releasing mature sgRNA to interact with Cas9. The two commonly used ribozymes in this approach include Hammerhead ribozyme (HHR) and Hepatitis Delta Virus (HRD) ribozyme. Scientists have developed an artificial gene known as Ribozyme-gRNA-Ribozyme (RGR) unit. Ribozymes require conserved sequences or secondary structures to catalyze the self-cleavage. Using this approach for multiplex gene editing, each gRNA is flanked by a small hammerhead-type (HH) ribozyme (having less than 50 bps) on its 5′ end and a hepatitis delta virus (HDV) ribozyme on its 3′ end. This enzyme does not require another specific sequence at cleavage site (Gao & Zhao, 2014).

Initially, this strategy was used for expressing single gRNA in CRISPR-Cas mediated gene editing. However, studies have demonstrated that this method was effective for multiplex gene editing by joining several RGR cassettes and expressing them under single promotor and terminator (He et al., 2017). Both Pol II and Pol III promoters were effective for transcription during this multiplex approach. This RGR system has been successfully demonstrated in multiple organisms including an efficient expression of multiple gRNAs in plants (Gao et al., 2015). For example, targeting six target sites in three rice genes resulted in mutation frequencies ranged from 53.8% to 100%, with most of the mutations being biallelic in T0 plants (Tang et al., 2016).

15.3.4 Csy4-Based Excision of gRNAs

Csy4 based excision for multiplex gene editing is a method to process a single precursor RNA transcript to produce individual mature gRNAs, targeting multiple genomic loci in an organism. It uses a Csy4 RNA-endoribonuclease from CRISPR-Cas system of *Pseudomonas aeruginosa* to recognize and cleave at specific sites known as Csy4 recognition sites (CRS), in the RNA (Čermák et al., 2017). It allows the simultaneous production of multiple gRNAs from a single transcript by cleaving at CRS sites during multiplex gene editing. Multiple gRNAs can be assembled and expressed as single precursor RNA using a Pol III promoter, with each gRNA flanked by the CRS of Csy4, that processes pre-crRNAs in some native CRISPR systems. Csy4 recognizes a 28-nt stem–loop structure in the RNA transcript and cleaves after the 20th nucleotide. The presence of CRS at 3′ end of each gRNA in a single transcript allows it to be cleaved by Csy4, to release mature individual gRNAs (Haurwitz et al., 2010). Examples include McCarty et al. (2020) expressed 12 sgRNA in *S. cerevisiae* driven by a single Pol II promoter with co-expression of Csy4. Cys4-mediated multiplexing has also been successfully demonstrated in tomato, tobacco, wheat, and barley. However, the disadvantage of this strategy is that Csy4 is cytotoxic at high concentration, so its use is undesirable. Moreover, this system requires expression of Csy4 endoribonuclease in addition to Cas 9 and gRNA. The primary focus of researcher in multiplex gene editing is to develop a system that does not require any additional nuclease because it may compromise editing efficiency (Prajapati & Tyagi, 2024).

15.3.5 CRISPR/Cpf1 System for Expressing Multiple crRNA

Although CRISPR-Cas9 has been successfully demonstrated for multiplex gene editing in different organisms, however, there are significant challenges in its practical applications, such as the size of Cas9 protein, the requirement of GC-rich PAM, and additional sequences such as tracrRNA. To address these limitations, CRISPR-Cas12 represents an alternate and versatile system for multiplex gene editing. Compared with Cas9, which requires a dual RNA (crRNA and tracrRNA) to recruit Cas9 to the target site, the Cas12 system requires a shorter and single RNA (crRNA) to drive Cas12 to the target sequence. Similarly, Cas12 requires a T-rich PAM as compared with Cas9, which relies on G-rich PAM to introduce DSB at the target site. In addition, Cas12 is also relatively small in size, making it easier to deliver in animal and plant systems (Fonfara et al., 2016). Cas12 also differs in the structure and nature of the DSB it makes at the target site, typically producing staggered ends rather than blunt ends by Cas9. CRISPR-Cas12 was first developed for gene editing in plants in 2016. Cas12 also possesses RNase activity to process precursor crRNA into mature crRNA, making CRSIPR-Cas12 an efficient tool for multiple gene editing. Using this intrinsic ability to process precursor crRNA into mature crRNA, multiple gRNAs can be produced from a single RNA transcript, both constitutively and inducibly. CRISPR-Cas12 system has been considered a powerful technique for genome editing in plants (Alok et al., 2020). Targeted mutagenesis has been successfully demonstrated in tobacco and rice using CRISPR/Cas12a (Tang et al., 2018). CRISPR-Cas12a has also been demonstrated for multiplex gene editing in plants for a broad range of applications, including disease resistance, abiotic stress tolerance, and functional genomics.

15.4 APPLICATIONS OF MULTIPLEX GENOME EDITING

Developments in CRISPR-Cas technology, especially multiplex gene editing, have emerged as a powerful tool for functional genomics in plants to reveal the molecular basis of complex cellular processes and traits controlled by multiple genes. In functional genomic studies, a single gene mutation sometimes manifests a weaker phenotype, while generating double or triple mutants is quite challenging as it is labor-intensive and time-consuming. Multiplex gene editing through CRISPR-Cas offers an alternate and efficient method to characterize multiple genes involved in complex traits and cellular pathways (Table 15.1). Multiplex gene editing with CRISPR-Cas allows scientists to delete genomic fragments such as introns, exons, or even chromosomal fragments by using multiple gRNAs. Furthermore, multiplex gene editing using the CRISPRi and CRSIPRa approach enables precise regulation of multiple genes without introducing permanent mutations in the genome (Minkenberg et al., 2017).

15.4.1 Knockout of Multiple Genes

Gene knockout represents the most significant application of the CRISPR-Cas9 system. A single gRNA can drive the Cas9 to the targeted location in the gene to introduce a DSB subsequently repaired with NHEJ. Repair of DSB through NHEJ leads to indels at the DSB site. When indels are introduced in the coding region, it may lead to premature stop codon in the gene consequently resulting in non-functional protein. The CRISPR/Cas system has been adapted extensively for simultaneous editing of multiple genes in the host organisms. It enables editing of multiple genes with high efficiency in a single transformation event (Ma et al., 2015). Several studies have successfully demonstrated the potential of multiplex gene editing in plants. For example, Li et al. (2021) has demonstrated the potential of CRISPR-Cas9 mediated multiplex gene editing in soyabean by targeting four genes (LNK2) that are involved in flowering. Editing these genes in soyabean accelerated flowering time thus providing insights into the soybean flowering mechanism. Similarly, Ashraf et al. (2023) demonstrated the potential of multiplex knockout to develop broad-spectrum resistance against cotton leaf curl virus in tobacco plants using CRISPR-Cas12a system. The author targets the three regions of virus genome. Additionally, Wolabu et al. (2020) investigated multiplex gene editing in Alfafa (*Medicago sativa*) by targeting Medicago sativa stay-green (MsSGR) gene with multiple gRNAs. The study achieved 75% mutagenesis efficiency, significantly higher than sgRNA. Knockout of MsSGR gene leads to enhanced greening in seeds, stem, pods, and leaves.

Moreover, multiplex gene editing in tomatoes also improved horticultural trait such as fruit color. Interestingly, multiplex CRISPR-Cas9 was used to target genes involved in carotenoid production such as Phytoene Synthase 1 (PSY1), MYB transcription factor (MYB12) and STAY-GREEN 1 (SGR1). Simultaneous editing of these genes in tomatoes leads to altered fruit color and enhanced commercial value (Yang et al., 2023).

15.4.2 Multiplex Gene Editing for Disease Resistance

Crop production has decreased significantly due to biotic stress such as viruses, bacteria, and fungi, posing a serious challenge to food security. Multiplex gene editing emerged as a powerful tool to disrupt disease-causing genes, such as S genes, to develop disease-resistant plants. For example, rice bacterial blight (BB) causes 10–20% yield loss, and under favorable conditions for it, it may lead to the complete loss of crop. The bacteria exploit the SWEET gene of rice, encoding the sucrose transporters in rice, diverting the transfer of sugar from rice to sustain their growth. Targeted knockout for three S genes, SWEET11, SWEET13 and SWEET14, using multiplex gene editing improved resistance against bacterial blight disease in rice caused by *Xanthomonas oryzae*. The edited rice plants showed enhanced disease resistance against bacterial blight (Oliva et al., 2019).

Researchers have demonstrated multiplex gene editing to improve disease resistance against fire blight disease caused by *Erwinia amylovora* in apples. Editing of disease-specific DPIM genes such as DIPM-1, DIPM-2, and DIPM-4, resulted in improved resistance against fire blight disease in apples (Malnoy et al., 2016). *Pseudomonas syringae* is a causal agent of bacterial speck disease in tomatoes, causing heavy losses in crop production worldwide. Ortigosa et al. (2019) introduced a mutation in the COR stomatal co-receptor AtJAZ2 gene in tomato (SlJAZ2) that prevented the reopening of stomata, thereby blocking the host entry through stomata. Similarly, Late blight disease is one of the most common diseases of potatoes caused by the *Phytophthora*

TABLE 15.1
Significant studies demonstrating the CRISPR-Cas Multiplex Genome editing in plants for enhanced crop production

	Species	Number of grna	Number of targeted genes	Notes on mutation	Application	Mutation efficiency	Transformation strategy	Mechanism	Reference
1	Rice (*Oryza sativa*)	Four gRNAs	Three genes: SWEET11, SWEET13 and SWEET14	Bacterial blight disease resistance	Knock out	100% mutation efficiency	Agrobacterium-mediated transformation	NHEJ	(Oliva et al., 2019)
2	*Nicotiana benthamiana*	Three gRNAs	Four different ORFs of CLCV (C1, V1 and overlapping region of C2 and C3).	Cotton leaf curl multan virus resistance	Knock out	21.7%), crRNA2 (24.9%) and crRNA3 (55.6%	Agrobacterium-mediated transformation	NHEJ	(Ashraf et al., 2023)
3	Alfalfa (*Medicago sativa*)	Four gRNAs	Exon 1,2 & 3 of Medicago sativa stay-green (MsSGR) gene	Chlorophyll retention with a consistently deep green appearance of leaves	Knock out	75% mutagenesis efficiency	Agrobacterium-mediated transformation	NHEJ	(Wolabu et al., 2020)
4	Tomato (*Solanum lycopersicum*)	Six gRNAs	Three genes: PSY1, MYB12, and SGR1	Biosysnthesis of fruit pigments	Knock out	7.69–92.30% in To plants	Agrobacterium-mediated transformation	NHEJ	(Yang et al., 2023)
5	Potato	Two gRNAs	Three genes: StDND1, StCHL1, StDMR6–1	Resistance aginst late blight disease	Knock out	Reduced lesion size in mutant lines	Agrobacterium-mediated transformation	NHEJ	(Kieu et al., 2021).
6	Apples	Three gRNAs	Three genes: DIPM-1, DIPM-2, and DIPM-4	Resistance aginst fire blight disease	Knockout	0.5 to 6.9% indel mutagenesis efficiency)	Agrobacterium-mediated transformation	NHEJ	(Malnoy et al., 2016)
7	Soyabean	Three gRNAs	Four genes: LNK2a, LNK2b, LNK2c and LNK2d	Accelerated flowering time under long photoperiod	Knock out	High mutation efficiency	Agrobacterium-mediated transformation	NHEJ	(Li et al., 2021)
8	Rice (*Oryza sativa*)	49 gRNAs	49 cytokinin metabolism-associated genes	Manipulation of cytokinin genes for improved crop production	Metabolic engineering	69.4% co-editing rate	Agrobacterium-mediated transformation	NHEJ	(Wu et al., 2024)
9	Tomato (*Solanum lycopersicum*)	Six sgRNA	Five genes: GABA-TP1, GABA-TP2, GABA-TP3, CAT9, and SSADH	Manipulation of GABA metabolic pathway and production of improved traits	Metabolic engineering	19-fold higher GABA in mutant lines & 56.82% editing efficiency in T2 plants	Agrobacterium-mediated transformation	NHEJ	(Li et al., 2018)
10	Tomato (*Solanum lycopersicum*)	Six sgRNAs	Five genes: SGR1,LCY-E, Blc, LCY-B1, LCY-B2	Increase the accumulation of lycopene	Metabolic engineering	95.83% mutation efficiency	Agrobacterium-mediated transformation	NHEJ	(Li et al., 2018)
11	Soyabean	Three gRNAs	Three genes: GmF3H1, GmF3H2 and GmFNSII-1	Increase the isoflavone content and resistance against soyabean mosaic virus	Metabolic engineering	44.44% mutation efficiency in To plants	Agrobacterium-mediated transformation	NHEJ	(Zhang et al., 2020)

12	Rice (Oryza sativa)	Four sgRNA	Four target genes: OsALS1, OsGS1, OsTubA2, OsACC	Herbicide resistance	Base alteration	56.25% to 89.58% editing efficiency	Agrobacterium-mediated transformation	NHEJ	(Yan et al., 2021)
13	Rice (Oryza sativa)	Four gRNAs	Four genes: OsAOS1, OsJAR1, OsJAR2 and OsCOI2	Genetic Variations in rice, indels mutations in rice	Base alteration	8.3 % to 78.6% editing efficiency	Agrobacterium-mediated transformation	NHEJ	(Ren et al., 2018)
14	Cotton (*Gossypium hirsutum*)	Three gRNAs	Two genes: GhCLA and GhPEBP	Improved developmental processes	Base alteration	26.67% to 57.78% editing efficiency	Agrobacterium-mediated transformation	NHEJ	(Qin et al., 2020)
15	Rice (Oryza sativa)	21 pegRNAs	Six genes: OsCDC48, OsALS, OsDEP, OsEPSPS, OsLDMAR, OsGAPDH	Primarily indels	Prime genome editing optimzation	2.6%-21.8% editing efficiency	Agrobacterium-mediated transformation	NHEJ	(Lin et al., 2020)
16	Arabidopsis	Two gRNA for each gene, three gRNA for FWA	Three genes: Floral development gene APETALA3 (AP3), stem cell regulator CLAVATA3 (CLV3), FWA promoter	Regulation of methylated loci for creating meiotically heritable epialleles	Epigenetic modification	300–500 fold upregulation in plants	Agrobacterium-mediated transformation	NHEJ	(Papikian et al., 2019)
17	Arabidopsis	Two gRNA	Two genes: AREB1 and RD29A	Enhanced drought stress tolerance, High chlorophyll content	Histone modification	2–3 fold increase expression in plants	Agrobacterium-mediated transformation	NHEJ	(Bournaud et al., 2019)
18	*N. benthamiana*	Three gRNAs	One gene: PDS (phytoene desaturase gene)	Transcriptional gene activation and repression	Transcriptional regulation	14 fold transcriptional activation	Agrobacterium-mediated transformation	NHEJ	(Piatek et al., 2015)
19	Arabidopsis	Three gRNAs	Two genes: Anthocyanin Pigment1 (AtPAP1) & miR319	Transcription gene activation, Improved Phenotype	Transcriptional regulation	2–7.5 fold increase in transcriptional activation	Agrobacterium-mediated transformation	NHEJ	(Lowder et al., 2015)
20	Arabidopsis	Three gRNAs	Three genes: Arabidopsis Cleavage Stimulating Factor64 (AtCSTF64) gene, microRNA genes, miR159A and miR159B	Transcription gene repression	Transcriptional regulation	60% reduced transcription expression	Agrobacterium-mediated transformation	NHEJ	(Lowder et al., 2015)
21	Rice (*Oryza sativa*)	Three gRNAs	Three target sites in two genes (OsYSA) and (OsROC5)	Targeted deletions at specific sites	Transcriptional regulation	33.3% to 53.3% Mutation frequecny and 100% transformation efficiency	Agrobacterium-mediated transformation	NHEJ	(Lowder et al., 2015)

(continued)

TABLE 15.1 (Continued)
Significant studies demonstrating the CRISPR-Cas Multiplex Genome editing in plants for enhanced crop production

	Species	Number of grna	Number of targeted genes	Notes on mutation	Application	Mutation efficiency	Transformation strategy	Mechanism	Reference
22	Tobacco	Three gRNAs	Two genes: NbFLS2 and NbBAK1	Targeted deletions at specific sites, strong activation of GUS expression	Transcriptional regulation	100% efficiency	Agrobacterium-mediated transformation	NHEJ	(Lowder et al., 2015)
23	Arabidopsis	Three gRNAs	Three genes: WRKY30, RLP23 and CDG1	Transcriptional gene activation	Transcriptional regulation	80-fold for WRKY30, 37-fold for RLP23 and 192-fold for CDG1 increased gene expression	Polyethylene glycol (PEG)-mediated transfection	NHEJ	(Li et al., 2017)
24	Tomato (*Solanum lycopersicum*)	Eight gRNAs	LC QTL region, SlCLV3, S and SP promtor	Generating beneficial regulatory variants, Manipulation of yield components)	Transcriptional regulation	70% mutation efficiency	Agrobacterium-mediated transformation	NHEJ	(Rodríguez-Leal et al., 2017)
25	Potato (*Solanum tuberosum*)	Four gRNAs	Four targets of StGBSSI gene: (sgGBSS1, sgGBSS2, sgGBSS3 and sgGBSS4)	Impaired amylose synthesis	Base alteration	90% mutation efficiency	Agrobacterium-mediated transformation	NHEJ	(Veillet et al., 2019)
26	Rice (Oryza sativa)	Three gRNAs	Three genes: TMS5, Pi21, and Xa13	Enhanced bacterial blight resistance	Knock out	47%-79.9% efficiency	Agrobacterium-mediated transformation	NHEJ	(Li et al., 2019)
27	Rice (Oryza sativa)	Eight gRNAs	Four genes: MPK1, MPK5, MPK2, MPK6	Functional discovery of inaccessible genes	Knock out	45–86% biallelic mutations	Agrobacterium-mediated transformation	NHEJ	(Minkenberg et al., 2017)
28	Rice (Oryza sativa)	Five gRNA	Three genes: Hd2 Hd4 Hd5; Five target sites Hd2-1, Hd2-2, Hd4-1, Hd4-2 and Hd5	Development of early-maturing rice varieties	Single base insertions and deletions	77.8% editing efficiency	Agrobacterium-mediated transformation	NHEJ	(Li et al., 2017)

infestans. It affects various parts of the plant, such as the tuber, leaves, and stem of the potato plant. CRISPR-edited potato plants with mutated S genes (StDND1, StCHL1, and StDMR6-1) showed enhanced late blight resistance with reduced lesion size (Kieu et al., 2021). These examples highlight the potential of multiplex gene editing by targeting S genes in plants to develop disease resistance in crops, thereby improving global food security.

15.4.3 Engineered Plants' Metabolic Pathways

Multiplex gene editing is a versatile tool for metabolic engineering by editing two or more genes in plants. Recently Wu et al. (2024) explored the efficiency of multiplex genome editing by targeting the 40 genes involved in the cytokinin (CK) metabolic pathway of rice. Edited rice lines exhibited improved traits such as increased nutrient absorption, stem enlargement, and improved stress response. The authors also identified a novel mechanism by manipulating several genes to enhance the crop yield. The findings expressed the power of ultra-multiplex genome editing as a novel tool for plant metabolic engineering enabling new insights on large gene families, biochemical pathways, and functionally redundant genes.

Multiplex CRISPR-Cas9 has also been applied to engineer metabolic traits in tomatoes, specifically targeting the gamma-aminobutyric acid (GABA) shunt metabolic pathway. In this study, mutations were introduced in five genes of the GABA pathway, including GABA-TP1, GABA-TP2, GABA-TP3, CAT9, and SSADH; however, successful editing was achieved in only four genes. Editing of GABA-TP2 was challenging due to the presence of high GC content at the targeted site. Edited plants displayed three different types of knockouts, i.e., single, double, and quadruple knockout. Out of 53 knockout lines, one line was quadruple, showing a 19-fold increase in GABA contents compared to the wild-type plants. Recovered knockout lines also exhibited single and double knockout of GABA-TP1 and/or GABA-TP3 characterized by large DNA deletions at the targeted sites resulting in mixed heterozygous and homozygous mutants (Li et al., 2018). Li et al. (2018) also used a multiplex strategy to target the five genes of tomatoes involved in lycopene production. The edited lines showed an increased lycopene production by inhibiting its conversion to β- and α-carotene. Likewise, Chen et al. (2019) demonstrated the metabolic engineering in rice by targeting 13 different genes involved in the gibberellin metabolic pathway using the CRISPR-Cas9 system. Gibberellin plays an important role in plant growth and development including flower development and internode elongation. Among 13 genes targeted, knockout mutants were obtained for 11 genes, with two genes including OsKO2 and OsGA20ox1 were resistant to knockout. Edited plants showed GA-deficient dwarf phenotype with potential applications in agriculture such as lodging resistance and increase yield.

15.4.4 Base Alteration

Base editing involves the targeted modification of a single nucleotide at multiple sites in the genome to improve desired traits with agronomic importance. Unlike the traditional method of CRISPR-Cas9, base editing doesn't require double-stranded breaks at the target site. The technique relies on the fusion of nCas9 with deaminase domains, thus facilitating conversion from A > G or C > T without introducing DSB at the target site. The two most widely used base editors are cytosine base editors (CBE) and adenine base editors (ABE). SaBE4 GAM is a fourth-generation cytosine base editor developed from *Staphylococcus aureus*. Cas9-derived BE4-GAM, is an efficient base editing system with reduced off-target effects and minimized nonspecific interactions with the phosphate group (Li et al., 2020). UGI further enhances the editing efficiency by limiting the formation of unwanted by-products.

Advances in base editing systems with their applications in plants have been comprehensively reviewed by Azameti and Dauda (2021). Multiplex base editing represents a powerful system to install precise mutations at the target sites in the genome. Yan et al. (2021) demonstrated the highest editing efficiency in rice with TadA9, a variant of ABE. Targeted base editing at the four sites conferred herbicide resistance in the mutated lines of rice. Additionally, Ren et al. (2018) successfully installed precise base conversions in Pi-d2 and OsFLS2 genes of rice, leading to improved blast tolerance in the edited plants. Similarly, Qin et al. (2020) demonstrated the use of the GhBE3 system for base editing at multiple sites in the GhCLA and GhPEBP genes of cotton, achieving editing efficiencies ranging from 26.67% to 57.78% in edited plants. Although base editing holds tremendous potential for the improvement of agricultural crops, several challenges remain that limit the applications of BE systems for the genetic improvement of plants.

Prime editing is another revolutionary approach that enables precise base editing with all possible base conversions without introducing DSBs in the genome (Anzalone et al., 2019). This system uses nCas protein fused with reverse transcriptase guided by pegRNA, allowing synthesis and integration of desired DNA sequence at the target site (Anzalone et al., 2022). It allows researchers to precisely install desired modifications at the target site including deletion, replacement, integration, or inversion. PE offers several advantages over base editing such as reduced off-targets, enhanced efficiency, and greater precision. PE holds great potential for genetic improvement of desired traits in plants (Lin et al., 2020).

15.4.5 Epigenome Editing

Epigenetic engineering is a promising approach for gene regulation that involves covalent modification of histones or DNA to regulate gene expression without introducing mutations in DNA. Commonly employed epigenetic modifications include DNA methylation and histone modification, which play a significant role in plant development, increasing yield, and stress tolerance in plants. For example, DNA methylation involves adding methyl group at the C5 position of cytosine residues allowing gene silencing and contributing to the regulatory processes such as genome stability, gene silencing, transposon mobility, and genome imprinting (Lee et al., 2020). Various tools have been employed for precise epigenetic modifications in the genome, such as TALEN, ZFN, and CRISPR/Cas. CRISPR-dCas9 has emerged as an effective tool for targeted epigenetic modifications. dCas9 binds specifically to the target site, having no catalytical activity, while its fusion with DNA methyltransferase transforms dCas9 into an important epigenetic regulator. Several epigenomic modifications have been reported in plants contributing to developmental processes such as flowering time, fruit ripening, and starch metabolism (Shin et al., 2022). For example in *Arabidopsis*, fusion of dCas9 with tobacco DRM methyltransferase catalytic domain induced epigenetic modifications at FWA (Flowering Wageningen) locus (Papikian et al., 2019). Similarly, precise hypermethylation enhanced resistance to pathogens like *Alternaria brassicicola* in *Arabidopsis* (López Sánchez et al., 2016).

Histone modifications, which involve the post-translation modification at the N terminal region of histones, regulate gene expression and chromatin dynamics. These modifications include precise methylation, acetylation, and phosphorylation, each contributing to gene transcription or gene silencing (Gardiner et al., 2022). For example, acetylation of the lysine residue of histone weakens the interaction between DNA and histone by neutralizing the positive charge of histone, thereby making DNA accessible to the transcriptional factors (Pandey et al., 2002). CRISPR-dCas9-mediated precise histone modifications in *Arabidopsis* enhanced drought tolerance in edited plants. The mutant lines exhibited a high chlorophyll content and higher survival rate than control plants as this histone modification triggers the abscisic acid (ABA) responsive element binding protein1/ABRE binding factor (AREB1/ABF2) and RD29A gene (Bournaud et al., 2019). Another study by de Melo et al. (2020) demonstrated that CRISPRa-based activation of AREB-1 in *Arabidopsis* resulted in drought tolerance in edited plants. The edited plants showed low levels of malonaldehyde, indicating reduced oxidative stress.

15.4.6 Transcriptional Regulation

Manipulation of gene expression without introducing permanent mutations in the DNA has proven to be a versatile tool for the programmed regulation of genes. CRISPRi and CRISPRa can be used individually or in combination for functional characterization of genes. Transcriptional activator or repressor domains fused with dCas9 can be precisely recruited to the target site by gRNA (Zhao et al., 2016). Regulation of multiple genes with CRISPRi and CRISPRa is an emerging tool for metabolic engineering in plants (Minkenberg et al., 2017). Applications of CRISPRi and CRISPRa have been successfully demonstrated in bacteria, plants and cell lines (McCarty et al., 2020). For example, targeting the promoter region of the phytoene desaturase gene in tobacco resulted in enhanced expression by 14-fold. The authors demonstrated the potential of dCas9/EDLL (plant specific activator) and dCas9/SRDX (plant repressor) to modulate the expression of the target gene (Piatek et al., 2015).

Lowder et al. (2015) successfully demonstrated a multiplex CRISPRa system in *Arabidopsis* by designing the gRNAs and fusing the dCas9 with the transcriptional activator domain (VP64), resulting in the activation of AtFIS2. Similarly, another study demonstrated the activation of three genes, including WRKY30, RLP23, and CDG1, in *Arabidopsis* using the CRISPRa system (Li et al., 2017). Another milestone study demonstrated the applications of the CRISPR-Cas9 system to generate cis-regulatory elements for gene regulation in tomatoes. The authors demonstrated targeted mutations in the promotor region of tomato genes to create beneficial quantitative variations beneficial for improving traits such as fruit size, branching and inflorescence in tomato (Rodríguez-Leal et al., 2017).

15.5 CHALLENGES IN MULTIPLEX GENE EDITING

15.5.1 Potential Off-Target Effects

Site-directed nucleases (SDNs) such as CRISPR/Cas systems have the potential to introduce precise modifications in the models as well as crop plants (Feng et al., 2014). Multiplex gene editing with CRISPR-Cas is a versatile bioengineering tool to introduce precise genetic modifications in the genomes; however, off-targets remain a major concern, especially in therapeutic and translational applications. Off-targets occur when gRNA binds nonspecifically at the genomic site other than the target site due to the sequence similarity. For example, the CRISPR/Cas9 system can tolerate 3–5 mismatches in 20 bp targeted DNA sequence, leading to the binding as well as cleavage by the Cas9 at the unintended sites in the genome (Manghwar et al., 2020). The outcome of off-target effects may lead to loss of functional gene activity and chromosomal rearrangement, thus limiting the use of genome-editing applications; however, several studies have demonstrated that CRISPR is very precise in plants. In addition, several strategies have been developed to eliminate the off-targets, such as transient transformation, RNPs for clean editing, and using double gRNA with a nCas9. Double nickase strategy has been successfully demonstrated to minimize off-targets (Shen et al., 2014). Similarly, the designing of highly specific gRNA reduces the off-targets, leading to improved efficiency and specificity of CRISPR-Cas. Compared with Cas9, CRISPR-Cas12 has been found to have higher specificity with fewer off-target effects. In addition, fusing dCas9 with a dimerization-dependent FokI results in cleavage at the target site only when both FokI nucleases are in proximity, thus reducing off-targets. The use of purified RNPs with a unique gRNA proved to be an effective strategy for reducing off-targets (Tsai et al., 2014).

15.5.2 The Choice of CRISPR Reagents

CRISPR cargos can be delivered as DNA (Cas9 and gRNA as a plasmid), RNA (in-vitro transcribed mRNA of Cas9 and gRNA), or RNPs (purified Cas9 protein with mRNA of gRNA). Several factors influence the commercialization of CRISPR crops, including the specificity of the gRNA, the format of CRISPR cargos, and the method used to repair the DSB (Gordon et al., 2021). Various methods have been used to deliver CRISPR reagents in plants, such as agrobacterium-mediated transformation, the particle bombardment method, and PEG-mediated protoplast transformation. Although Agrobacterium-mediated transformation remains one of the most widely used methods for delivering CRISPR reagents in plants due to its simplicity and high editing efficiency, however, it always results in permanent integration of Cas cassette in the plant genome, leading to increased off-targets (Lino et al., 2018). Therefore, the transient expression of Cas9 and the use of RNPs have become attractive methods for producing transgene-free CRISPR-edited crops with better public acceptance (Chuang et al., 2021). Biolistic is a valuable tool for the transformation of plant species that are not easy to transform with agrobacterium. It also allows transient transformation of CRISPR reagents, thus minimizing the off-targets. However, its applications in plants are limited by low efficiency and random integration of transgene in the plant genome. In conclusion, the choice of CRISPR reagents and their delivery format significantly influence the public acceptance and regulatory status of CRISPR-edited crops (Sandhya et al., 2020; Miller et al., 2021).

15.5.3 Regulatory Challenges

CRISPR/Cas has revolutionized precise genetic improvement of plants with broad-spectrum applications in functional genomics, disease resistance, and abiotic stress tolerance, leading to sustainable agriculture and food security (Ahmad et al., 2021). It has enabled researchers to overcome the challenges and concerns associated with GMOs and fulfill the early promises of genetic engineering. CRISPR crops are appearing in the world markets, and some countries, such as the USA, Argentina, and Brazil, have revised their strict regulatory oversights of GMOs, especially for transgene-free CRISPR crops (Georges & Ray, 2017). Transgene-free CRISPR crops are indistinguishable from conventionally bred crops because they contain only precise changes at the target site without integrating any foreign DNA into the genome. So different countries such as the USA, Japan, Argentina, and Russia consider SDN1 and SDN2 edited crops as non-GMOs, thus deregulating these crops for commercialization. However, the EU and New Zealand still consider all CRISPR-edited crops as GMOs, regulating all CRISPR crops under strict GMO regulatory measures (Turnbull et al., 2021). This mosaic patttern in global regulatory measures of CRISPR-edited crops arises because many countries are not signatories of the Cartagena Protocol of Biosafety. Similarly, this difference arises from different definitions of genetic modifications in different countries.

Multiplex gene editing in crops involves precise genetic modifications simultaneously at multiple target sites in the genome that may present unique challenges for regulatory agencies. The challenges may arise from various aspects, such as potential off targets, complexity of the gene editing, traceability of editing, and regulatory challenges of the CRISPR-edited crops (Zannoni, 2019). Here we summarize the key challenges in regulatory oversights of multiplex gene-edited crops:

1. Multiplex gene editing introduces multiple genetic modifications simultaneously, making it challenging and complex for regulatory agencies to characterize these modifications compared with single gene editing.
2. Different countries have different definitions of genetic modifications, so this mosaic pattern in regulatory oversights of CRISPR-edited crops may lead to an uncertain future for these crops (Wolt et al., 2016).
3. Multiplex gene-editing in crops may result in increased off-targets in the genome. Therefore, researchers and regulatory agencies would need a comprehensive analysis of off-targets and their impact on human health and the environment.
4. A comprehensive analysis of the footprints of Cas in multiplex gene-edited crops will ensure consumer trust, public acceptance, and food safety. Similarly, appropriate labeling may further enhance public acceptance of these crops (Ahmad et al., 2023).

To address these challenges, the world community needs a universal and harmonized definition of genetic modifications and CRISPR-edited crops. In addition, a universal and scalable regulatory framework for CRISPR-edited crops is also required to facilitate developments in CRISPR technology and streamline the commercialization and trade of CRISPR-edited crops. Moreover, rapid and advanced tracking methods of multiplex gene editing are also crucial for the compliance of regulatory and safety assessments (Zhang et al., 2020). Furthermore, a comprehensive and science-based risk and safety assessment of CRISPR edited crops will streamline the public acceptance of these crops. Finally, these challenges require a coordinated effort from all stakeholders, including scientists, regulatory agencies, policy makers, politicians, governance, the public and AI tools to ensure safe and responsible development of CRISPR edited crops (Entine et al., 2021).

15.6 AI and Multiplex Gene Editing with CRISPR

In the last 10 years, CRISPR has emerged as a simple, precise, and innovative technology that has revolutionized the biological sciences with groundbreaking inventions in agriculture, diagnostics, precision medicine, and human health. Multiplex gene editing allows simultaneous editing of multiple genes, thus reducing the time and resources required compared with editing each gene individually. The rise of AI in the last decade is another revolution of the twenty-first century with tremendous potential to address the challenges associated with CRISPR-Cas, such as off-targets, integrative modeling, precise designing of gRNA, data analysis, improved efficiency, and enhanced specificity (Maserat, 2022). The integration of AI with CRISPR represents the powerful convergence of technologies with tremendous potential to transform biotechnology. AI can significantly contribute to multiplex gene editing through various applications such as precise identification of target sites, efficient design of gRNA, enhancing specificity, predicting off-targeting, and predicting editing outcomes (Hatem et al., 2022; Vilhekar & Rawekar, 2024). Here are the key areas where AI can contribute to the multiplex gene editing through CRISPR-Cas.

1- *Precise identification of the target site*: Precise identification of target sites specifically in multiplex gene editing is crucial for achieving desired mutations in the genome. Machine learning and AI algorithms can predict potential target sites with greater efficiency and minimum off targets. AI tools can perform rapid sequence alignment to identify best target sites for gene editing (Lee, 2023).
2- *gRNA design*: gRNA design plays an important role in the gene-editing applications of CRISPR-Cas. Several gRNA tools, such as CRISPR-direct, CHOPCHOP, GuideMaker, CRISPR seek, and E-CRISP, are available for the efficient design of gRNA (Li et al., 2023). AI is playing a transformative role in the efficient design of gRNA based on sequence features and training AI models to identify patterns that correlate with high editing efficiency (Wang et al., 2020). Similarly, AI tools can simultaneously evaluate gRNA design parameters such as GC contents, and secondary structure its efficiency. Optimization of these factors contributes to the efficiency as well as stability of the gRNA (Baisya et al., 2022)
3- *Predicting off-targets*: Off-targets are the unintended mutations induced by the CRISPR-Cas on sites other than the target sites. Off-targets are major concerns in CRISPR applications especially in human health and therapeutics. AI has a vital role in predicting off-targets in CRISPR applications. For example, machine learning algorithms can be trained on large data sets of on-target and off-target sites to predict potential off-targets with better accuracy. Deep learning such as convolutional neural networks (CNNs) and recurrent neural networks (RNNs) can analyze interaction between gRNA and genomic DNA to predict off-targets (Chen et al., 2019; Konstantakos et al., 2022). AI models can also analyze sequence features such as sequence composition, nucleotide composition and secondary structure of the gRNA to predict off-targets (Niu et al., 2021). In addition, AI models can also predict off-targets based on genomic sequence contexts such as DNA methylation and chromatin accessibility. Several AI-powered tools have been developed to predict off-targets, such as DEEPCRISPR, GUIDES, CRISPR-OFF, and CCTOP. These AI powered tools can simulate off-targets in CRISPR applications before they can be analyzed practically (Baisya et al., 2022; Chen et al., 2023)
4- *Optimizing multiplex gene editing*: Optimizing multiplex gene editing involves simultaneous editing at multiple target sites in the genome. AI plays a vital role in the optimization and enhancing precision and efficiency of these complex studies. AI contributes in designing multiple gRNAs that work together efficiently without interfering with each other. In addition, AI helps in optimizing efficient delivery systems, predicting off-targets, and simulating outcomes. Several AI-based tools, such as CRISPR-ai and DeepCRISPR, have been developed to optimize gRNA for multiplex gene editing and predict Cas activity and efficiency (Ma et al., 2022).
5- *Editing outcomes*: Predicting editing outcomes in CRISPR involves using AI tools to anticipate the results of the genetic modifications at the target site. Indels are the common outcomes of CRISPR-based gene editing. AI models can be trained on large datasets to precisely predict the likelihood and characteristics of the indels for a new genomic site (Chen et al., 2024). AI tools can use specific sequence features to predict the editing outcomes at the new target sites in the genome. In addition, AI tools can also predict functional consequences of the editing outcomes, such as how genetic modifications may influence gene expression, biological networks, metabolic pathways, and protein functions. Examples include Guide-seq, Deep-CRISPR, and CRISPETa, which can predict indel profiles, genome-wide off-targets, and editing efficiency in gene-editing studies (Chaudhari et al., 2020).
6- *Integrating multi-omics data with CRISPR-Cas*: Integrating multi-omics data with CRISPR-Cas gene editing holds immense potential to characterize gene functions, gene interactions, and biological pathways in living systems (Razzaq et al., 2021). AI can integrate data from various sources, such as genomics, transcriptomics, proteomics, epigenomics, and metabolomics, with CRISPR cleavage patterns to predict how gene editing will influence the cellular states, metabolism, and phenotype, providing greater insights into the long-term effects of the genetic modifications (Zhang et al., 2016; Shulse et al., 2019). For example, an AI-driven tool, CRISPRAI, combines various omics data with CRISPR-Cas to enhance the precision of CRISPR. AI-driven integration of multi-omics data with CRISPR-Cas will not only enhance

our understanding of the gene-editing process, gene functions, and regulatory networks but will also lead to more precise and efficient gene editing (Dixit et al., 2024). In addition, it will also accelerate developments in research, diagnostics, therapeutics and human health (Das et al., 2023).

15.7 CONCLUSION

Multiplex CRISPR-mediated gene editing serves as a versatile tool for the expression of multiple genes simultaneously. Various strategies have been employed for the assembly, expression, and processing of multiple guide RNA arrays in plants facilitating the multiplex editing system without compromising the efficiency of mutagenesis. It has the potential to revolutionize plant genome editing with the development of several potential applications in plant development by knocking out multiple genes to disrupt gene function, metabolic engineering for understanding the dynamics of synthetic metabolic pathways and transcriptional regulation to facilitate the gain of functions of multiple gene by base editing and prime editing. Several systems such as physical, viral, or nanoparticles allow the robust delivery of CRISPR reagents into plants. Occurrence of off-targets effects may be associated with multiplex genome editing as the number of targets increased. However, several AI modeling algorithms for off-target detection have been developed and updated for eliminating the percentage of off-target occurrence thereby designing gRNA at the accurate targeted site. Multiplex genome editing accelerates the genetic improvement of agriculture crops and the future of multiplex genome editing and will have a profound impact on new plant biology discoveries.

REFERENCES

Abdelrahman, M., Wei, Z., Rohila, J. S., & Zhao, K. (2021). Multiplex genome-editing technologies for revolutionizing plant biology and crop improvement. *Frontiers in Plant Science,* 12, 721203.

Abudayyeh, O. O., Gootenberg, J. S., Essletzbichler, P., Han, S., Joung, J., Belanto, J. J., & Zhang, F. (2017). RNA targeting with CRISPR-Cas13. *Nature*, 550(7675), 280–284.

Ahmad, A., Abd-Elsalam, K. A., Ali, M., Majeed, H. N., Arif, A., Munir, A., Khan, Z., & Munawar, N. (2024). Regulatory triggers of CRISPR-edited crops. *In Global Regulatory Outlook for CRISPRized Plants* Academic Press, 91-112. https://doi.org/10.1016/b978-0-443-18444-4.00016-8

Ahmad, A., Ghouri, M. Z., Munawar, N., Ismail, M., Ashraf, S., & Aftab, S. O. (2021). Regulatory, ethical, and social aspects of CRISPR crops. In A. Ahmad, Z. H. Khan, & Z. Khan (Eds.), *CRISPR crops: The future of food security*, Springer, pp. 261–287.

Ahmad, A., Munawar, N., Khan, Z., Qusmani, A. T., Khan, S. H., Jamil, A., Ashraf, S., Ghouri, M. Z., Aslam, S., Mubarik, M. S., Munir, A., Sultan, Q., Abd-Elsalam, K. A., & Qari, S. H. (2021). An outlook on global regulatory landscape for genome-edited crops. *International Journal of Molecular Sciences,* 22(21), 11753. https://doi.org/10.3390/ijms222111753

Ahmad, A., Jamil, A., & Munawar, N. (2023). GMOs or Non-GMOs? The CRISPR Conundrum. *Frontiers in Plant Science,* 14, 1232938.

Alok, A., Sandhya, D., Jogam, P., Rodrigues, V., Bhati, K. K., Sharma, H., & Kumar, J. (2020). The rise of the CRISPR/Cpf1 system for efficient genome editing in plants. *Frontiers in Plant Science,* 11, 494295.

Anzalone, A. V., Gao, X. D., Podracky, C. J., Nelson, A. T., Koblan, L. W., Raguram, A., & Liu, D. R. (2022). Programmable deletion, replacement, integration and inversion of large DNA sequences with twin prime editing. *Nature Biotechnology*, 40(5), 731–740.

Anzalone, A. V., Randolph, P. B., Davis, J. R., Sousa, A. A., Koblan, L. W., Levy, J. M., & Liu, D. R. (2019). Search-and-replace genome editing without double-strand breaks or donor DNA. *Nature*, 576(7785), 149–157.

Arimbasseri, A. G., Rijal, K., & Maraia, R. J. (2013). Transcription termination by the eukaryotic RNA polymerase III. *Biochimica et Biophysica Acta (BBA)-Gene Regulatory Mechanisms*, 1829(3–4), 318–330.

Ashraf, S., Ahmad, A., Khan, S. H., Jamil, A., Sadia, B., & Brown, J. K. (2023). LbCas12a mediated suppression of Cotton leaf curl Multan virus. *Frontiers in Plant Science,* 14, 1233295.

Aslam, S., Munir, A., Aslam, H. M. U., Khan, S. H., & Ahmad, A. (2022). Genome editing advances in soybean improvement against biotic and abiotic stresses. In S. H. Wani et al. (Eds.), *Soybean improvement: Physiological, molecular and genetic perspectives*, pp. 241–274. Springer International Publishing.

Azameti, M. K., & Dauda, W. P. (2021). Base editing in plants: applications, challenges, and future prospects. *Frontiers in Plant Science,* 12, 664997.

Baisya, D., Ramesh, A., Schwartz, C., Lonardi, S., & Wheeldon, I. (2022). Genome-wide functional screens enable the prediction of high activity CRISPR-Cas9 and-Cas12a guides in Yarrowia lipolytica. *Nature Communications*, 13(1), 922.

Bernabé-Orts, J. M., Casas-Rodrigo, I., Minguet, E. G., Landolfi, V., Garcia-Carpintero, V., Gianoglio, S., & Orzaez, D. (2019). Assessment of Cas12a-mediated gene editing efficiency in plants. *Plant Biotechnology Journal*, 17(10), 1971–1984.

Bournaud, C., Lourenço-Tessutti, I. T., Noriega, D. D., paes de Melo, B., de Almeida-Engler, J., & Grossi-de-Sa, M. F. (2019). Improved drought stress tolerance in Arabidopsis by CRIspR/dCas9 fusion with a Histone AcetylTransferase. *Science Reports,* 9(1) 1–9.

Brooks, C., Nekrasov, V., Lippman, Z. B., & Van Eck, J. (2014). Efficient gene editing in tomato in the first generation using the clustered regularly interspaced short palindromic repeats/CRISPR-associated9 system. *Plant Physiology*, 166(3), 1292–1297.

Čermák, T., Baltes, N. J., Čegan, R., Zhang, Y., & Voytas, D. F. (2015). High-frequency, precise modification of the tomato genome. *Genome Biology,* 16, 1–15.

Čermák, T., Curtin, S. J., Gil-Humanes, J., Čegan, R., Kono, T. J., Konečná, E., & Voytas, D. F. (2017). A multipurpose toolkit to enable advanced genome engineering in plants. *The Plant Cell*, 29(6), 1196–1217.

Chaudhari, H. G., Penterman, J., Whitton, H. J., Spencer, S. J., Flanagan, N., Lei Zhang, M. C., & Kernytsky, A. (2020).

Evaluation of homology-independent CRISPR-Cas9 off-target assessment methods. *The CRISPR Journal*, 3(6), 440–453.

Chen, L., Liu, G., & Zhang, T. (2024). Integrating machine learning and genome editing for crop improvement. *In aBIOTECH*, 5, 262–267.

Chen, Q., Chuai, G., Zhang, H., Tang, J., Duan, L., Guan, H., & Liu, Q. (2023). Genome-wide CRISPR off-target prediction and optimization using RNA-DNA interaction fingerprints. *Nature Communications*, 14(1), 7521.

Chen, X., Tian, X., Xue, L., Zhang, X., Yang, S., Traw, M. B., & Huang, J. (2019). CRISPR-based assessment of gene specialization in the gibberellin metabolic pathway in rice. *Plant Physiology*, 180(4), 2091–2105.

Chuang, Y. F., Phipps, A. J., Lin, F. L., Hecht, V., Hewitt, A. W., Wang, P. Y., & Liu, G. S. (2021). Approach for in vivo delivery of CRISPR/Cas system: a recent update and future prospect. *Cellular and Molecular Life Sciences*, 78, 2683–2708.

Das, J., Kumar, S., Mishra, D. C., Chaturvedi, K. K., Paul, R. K., & Kairi, A. (2023). Machine learning in the estimation of CRISPR-Cas9 cleavage sites for plant system. *Frontiers in Genetics*, 13, 1085332.

de Melo, B. P., Lourenço-Tessutti, I. T., Paixão, J. F. R., Noriega, D. D., Silva, M. C. M., de Almeida-Engler, J., & Grossi-de-Sa, M. F. (2020). Transcriptional modulation of AREB-1 by CRISPRa improves plant physiological performance under severe water deficit. *Scientific Reports*, 10(1), 16231.

Ding, T., Huang, C., Liang, Z., Ma, X., Wang, N., & Huo, Y. X. (2020). Reversed paired-gRNA plasmid cloning strategy for efficient genome editing in Escherichia coli. *Microbial Cell Factories*, 19, 1–13.

Dixit, S., Kumar, A., Srinivasan, K., Vincent, P. D. R., & Ramu Krishnan, N. (2024). Advancing genome editing with artificial intelligence: opportunities, challenges, and future directions. *Frontiers in Bioengineering and Biotechnology*, 11, 1335901.

Du, H., Zeng, X., Zhao, M., Cui, X., Wang, Q., Yang, H., & Yu, D. (2016). Efficient targeted mutagenesis in soybean by TALENs and CRISPR/Cas9. *Journal of Biotechnology*, 217, 90–97.

Entine, J., Felipe, M. S. S., Groenewald, J. H., Kershen, D. L., Lema, M., McHughen, A., & Wray-Cahen, D. (2021). Regulatory approaches for genome edited agricultural plants in select countries and jurisdictions around the world. *Transgenic Research*, 30(4), 551–584.

Feng, Z., Mao, Y., Xu, N., Zhang, B., Wei, P., Yang, D. L., & Zhu, J. K. (2014). Multigeneration analysis reveals the inheritance, specificity, and patterns of CRISPR/Cas-induced gene modifications in Arabidopsis. *Proceedings of the National Academy of Sciences*, 111(12), 4632–4637.

Fonfara, I., Richter, H., Bratovič, M., Le Rhun, A., & Charpentier, E. (2016). The CRISPR-associated DNA-cleaving enzyme Cpf1 also processes precursor CRISPR RNA. *Nature*, 532(7600), 517–521.

Gao, Y., Zhang, Y., Zhang, D., Dai, X., Estelle, M., & Zhao, Y. (2015). Auxin binding protein 1 (ABP1) is not required for either auxin signaling or Arabidopsis development. *Proceedings of the National Academy of Sciences*, 112(7), 2275–2280.

Gao, Y., & Zhao, Y. (2014). Self-processing of ribozyme-flanked RNAs into guide RNAs in vitro and in vivo for CRISPR-mediated genome editing. *Journal of Integrative Plant Biology*, 56(4), 343–349.

Gardiner, J., Ghoshal, B., Wang, M., & Jacobsen, S. E. (2022). CRISPR–Cas-mediated transcriptional control and epi-mutagenesis. *Plant Physiology*, 188(4), 1811–1824.

Georges, F., & Ray, H. (2017). Genome editing of crops: a renewed opportunity for food security. *GM Crops & Food*, 8(1), 1–12.

Gordon-Kamm, W., Barone, P., Svitashev, S., D Sander, J., Kumar, S., & Jones, T. (2021). Strategies for CRISPR/Cas9-mediated genome editing: From delivery to production of modified plants. In M. R. Willmann (Ed.), *Genome editing for precision crop breeding*. Burleigh Dodds Science Publishing.

Hatem, Y., Hammad, G., & Safwat, G. (2022). Artificial intelligence for plant genomics and crop improvement. *Egyptian Journal of Botany*, 62(2), 291–303.

Haurwitz, R. E., Jinek, M., Wiedenheft, B., Zhou, K., & Doudna, J. A. (2010). Sequence-and structure-specific RNA processing by a CRISPR endonuclease. *Science*, 329(5997), 1355–1358.

He, Y., Wang, R., Dai, X., & Zhao, Y. (2017). On improving CRISPR for editing plant genes: ribozyme-mediated guide RNA production and fluorescence-based technology for isolating transgene-free mutants generated by CRISPR. *Progress in Molecular Biology and Translational Science*, 149, 151–166.

Hillary, V. E., & Ceasar, S. A. (2023). A review on the mechanism and applications of CRISPR/Cas9/Cas12/Cas13/Cas14 proteins utilized for genome engineering. *Molecular Biotechnology*, 65(3), 311–325.

Hu, X., Wang, C., Liu, Q., Fu, Y., & Wang, K. (2016). Targeted mutagenesis in rice using CRISPR-Cpf1 system. *Journal of Genetics and Genomics= Yi Chuan Xue Bao*, 44(1), 71–73.

Kieu, N. P., Lenman, M., Wang, E. S., Petersen, B. L., & Andreasson, E. (2021). Mutations introduced in susceptibility genes through CRISPR/Cas9 genome editing confer increased late blight resistance in potatoes. *Scientific Reports*, 11(1), 4487.

Konstantakos, V., Nentidis, A., Krithara, A., & Paliouras, G. (2022). CRISPR–Cas9 gRNA efficiency prediction: an overview of predictive tools and the role of deep learning. *Nucleic Acids Research*, 50(7), 3616–3637.

Lee, J. H., Mazarei, M., Pfotenhauer, A. C., Dorrough, A. B., Poindexter, M. R., Hewezi, T., & Stewart Jr, C. N. (2020). Epigenetic footprints of CRISPR/Cas9-mediated genome editing in plants. *Frontiers in Plant Science*, 10, 1720.

Lee, K., Zhang, Y., Kleinstiver, B. P., Guo, J. A., Aryee, M. J., Miller, J., & Wang, K. (2019). Activities and specificities of CRISPR/Cas9 and Cas12a nucleases for targeted mutagenesis in maize. *Plant Biotechnology Journal*, 17(2), 362–372.

Lee, K., Zhu, H., Yang, B., & Wang, K. (2019). An Agrobacterium-mediated CRISPR/Cas9 platform for genome editing in maize. In Y. Qi (Ed.), *Plant genome editing with CRISPR systems: Methods and protocols*, Springer, pp. 121–143.

Lee, M. (2023). Deep learning in CRISPR-Cas systems: a review of recent studies. *Frontiers in Bioengineering and Biotechnology*, 11, 1–14, 1226182.

Lee, R. T. H., Ng, A. S. M., & Ingham, P. W. (2016). Ribozyme mediated gRNA generation for in vitro and in vivo CRISPR/Cas9 mutagenesis. *PLoS One*, 11(11), e0166020.s

Li, B., Rui, H., Li, Y., Wang, Q., Alariqi, M., Qin, L., & Jin, S. (2019). Robust CRISPR/Cpf1 (Cas12a)-mediated genome editing in allotetraploid cotton (Gossypium hirsutum). *Plant Biotechnology Journal*, 17(10), 1862.

Li, C., Chu, W., Gill, R.A., Sang, S., Shi, Y., Hu, X., Yang, Y., Zaman, Q.U. and Zhang, B. (2023). Computational tools and resources for CRISPR/Cas genome editing. *Genomics, Proteomics & Bioinformatics*, 21(1), 108–126.

Li, J. F., Norville, J. E., Aach, J., McCormack, M., Zhang, D., Bush, J., & Sheen, J. (2013). Multiplex and homologous recombination–mediated genome editing in Arabidopsis and

Nicotiana benthamiana using guide RNA and Cas9. *Nature Biotechnology*, 31(8), 688–691.

Li, R., Li, R., Li, X., Fu, D., Zhu, B., Tian, H., & Zhu, H. (2018). Multiplexed CRISPR/Cas9-mediated metabolic engineering of γ-aminobutyric acid levels in Solanum lycopersicum. *Plant Biotechnology Journal*, 16(2), 415–427.

Li, S., Shen, L., Hu, P., Liu, Q., Zhu, X., Qian, Q., et al. (2019). Developing disease-resistant thermosensitive male sterile rice by multiplex gene editing. *Journal of Integrative Plant Biology,* 61, 1201–1205. doi: 10.1111/jipb.12774

Li, X., Wang, Y., Chen, S., Tian, H., Fu, D., Zhu, B., et al. (2018). Lycopene is enriched in tomato fruit by CRISPR/Cas9-mediated multiplex genome editing. *Frontiers in Plant Science,* 9, 559. doi: 10.3389/fpls.2018.00559

Li, Y., Zhu, J., Wu, H., Liu, C., Huang, C., Lan, J., & Xie, C. (2020). Precise base editing of non-allelic acetolactate synthase genes confers sulfonylurea herbicide resistance in maize. *The Crop Journal*, 8(3), 449–456.

Li, Z., Cheng, Q., Gan, Z., Hou, Z., Zhang, Y., Li, Y., & Dong, L. (2021). Multiplex CRISPR/Cas9-mediated knockout of soybean LNK2 advances flowering time. *The Crop Journal*, 9(4), 767–776.

Li, Z., Zhang, D., Xiong, X., Yan, B., Xie, W., Sheen, J., & Li, J. F. (2017). A potent Cas9-derived gene activator for plant and mammalian cells. *Nature Plants*, 3(12), 930–936.

Liang, Z., Zhang, K., Chen, K., & Gao, C. (2014). Targeted mutagenesis in Zea mays using TALENs and the CRISPR/Cas system. *Journal of Genetics and Genomics*, 41(2), 63–68.

Lin, Q., Zong, Y., Xue, C., Wang, S., Jin, S., Zhu, Z., & Gao, C. (2020). Prime genome editing in rice and wheat. *Nature Biotechnology*, 38(5), 582–585.

Lino, C. A., Harper, J. C., Carney, J. P., & Timlin, J. A. (2018). Delivering CRISPR: a review of the challenges and approaches. *Drug Delivery*, 25(1), 1234–1257.

López Sánchez, A., Stassen, J. H., Furci, L., Smith, L. M., & Ton, J. (2016). The role of DNA (de) methylation in immune responsiveness of Arabidopsis. *The Plant Journal*, 88(3), 361–374.

Lowder, L. G., Zhang, D., Baltes, N. J., Paul III, J. W., Tang, X., Zheng, X., & Qi, Y. (2015). A CRISPR/Cas9 toolbox for multiplexed plant genome editing and transcriptional regulation. *Plant Physiology*, 169(2), 971–985.

Ma, E., Chen, K., Shi, H., Stahl, E. C., Adler, B., Trinidad, M., & Doudna, J. A. (2022). Improved genome editing by an engineered CRISPR-Cas12a. *Nucleic Acids Research*, 50(22), 12689–12701.

Ma, X., Zhang, Q., Zhu, Q., Liu, W., Chen, Y., Qiu, R., & Liu, Y. G. (2015). A robust CRISPR/Cas9 system for convenient, high-efficiency multiplex genome editing in monocot and dicot plants. *Molecular Plant*, 8(8), 1274–1284.

Manghwar, H., Li, B., Ding, X., Hussain, A., Lindsey, K., Zhang, X., & Jin, S. (2020). CRISPR/Cas systems in genome editing: methodologies and tools for sgRNA design, off-target evaluation, and strategies to mitigate off-target effects. *Advanced Science*, 7(6), 1902312.

Manghwar, H., Lindsey, K., Zhang, X., & Jin, S. (2019). CRISPR/Cas system: recent advances and future prospects for genome editing. *Trends in Plant Science*, 24(12), 1102–1125.

Malnoy, M., Viola, R., Jung, M. H., Koo, O. J., Kim, S., Kim, J. S., & Nagamangala Kanchiswamy, C. (2016). DNA-free genetically edited grapevine and apple protoplast using CRISPR/Cas9 ribonucleoproteins. *Frontiers in Plant Science,* 7, 1904.

Malzahn, A. A., Tang, X., Lee, K., Ren, Q., Sretenovic, S., Zhang, Y., & Qi, Y. (2019). Application of CRISPR-Cas12a temperature sensitivity for improved genome editing in rice, maize, and Arabidopsis. *BMC Biology,* 17, 1–14.

Maserat, E. (2022). Integration of artificial intelligence and CRISPR/Cas9 system for vaccine design. *Cancer Informatics,* 21, 11769351221140102.

McCarty, N. S., Graham, A. E., Studená, L., & Ledesma-Amaro, R. (2020). Multiplexed CRISPR technologies for gene editing and transcriptional regulation. *Nature Communication,* 11, 1281.

Miller, K., Eggenberger, A. L., Lee, K., Liu, F., Kang, M., Drent, M., & Jiang, S. (2021). An improved biolistic delivery and analysis method for evaluation of DNA and CRISPR-Cas delivery efficacy in plant tissue. *Scientific Reports*, 11(1), 7695.

Minkenberg, B., Wheatley, M., & Yang, Y. (2017). CRISPR/Cas9-enabled multiplex genome editing and its application. *Progress in Molecular Biology and Translational Science,* 149, 111–132.

Minkenberg, B., Xie, K., & Yang, Y. (2017). Discovery of rice essential genes by characterizing a CRISPR-edited mutation of closely related rice MAP kinase genes. *The Plant Journal*, 89(3), 636–648.

Mishra, R., Joshi, R. K., & Zhao, K. (2018). Genome editing in rice: recent advances, challenges, and future implications. *Frontiers in Plant Science,* 9, 1361.

Munir, A., Amin, I., Zahoor, M. K., Majeed, H. N., Almoammar, H., Ghaffar, A., & Ahmad, A. (2024). Multiplex genome editing in plants through CRISPR-Cas. In K. A. Abd-Elsalam et al. (Eds.), *CRISPRized horticulture crops*, pp. 127–142. Academic Press. https://doi.org/10.1016/b978-0-443-13229-2.00012-0

Najera, V. A., Twyman, R. M., Christou, P., & Zhu, C. (2019). Applications of multiplex genome editing in higher plants. *Current Opinion in Biotechnology,* 59, 93–102.

Niu, M., Lin, Y., & Zou, Q. (2021). sgRNACNN: identifying sgRNA on-target activity in four crops using ensembles of convolutional neural networks. *Plant Molecular Biology,* 105, 483–495.

Oliva, R., Ji, C., Atienza-Grande, G., Huguet-Tapia, J. C., Perez-Quintero, A., Li, T., & Yang, B. (2019). Broad-spectrum resistance to bacterial blight in rice using genome editing. *Nature Biotechnology*, 37(11), 1344–1350.

Ortigosa, A., Gimenez-Ibanez, S., Leonhardt, N., & Solano, R. (2019). Design of a bacterial speck resistant tomato by CRISPR/Cas9-mediated editing of Sl JAZ 2. *Plant Biotechnology Journal*, 17(3), 665–673.

Osakabe, Y., & Osakabe, K. (2017). Genome editing to improve abiotic stress responses in plants. *Progress in Molecular Biology and Translational Science,* 149, 99–109.

Pandey, P. K., Quilichini, T. D., Vaid, N., Gao, P., Xiang, D., & Datla, R. (2019). *Versatile and multifaceted CRISPR/Cas gene editing tool for plant research.* In Seminars in Cell & Developmental Biology (vol. 96, pp. 107–114). Academic Press.

Pandey, R., MuÈller, A., Napoli, C. A., Selinger, D. A., Pikaard, C. S., Richards, E. J., & Jorgensen, R. A. (2002). Analysis of histone acetyltransferase and histone deacetylase families of Arabidopsis thaliana suggests functional diversification of chromatin modification among multicellular eukaryotes. *Nucleic Acids Research*, 30(23), 5036–5055.

Papikian, A., Liu, W., Gallego-Bartolomé, J., & Jacobsen, S. E. (2019). Site-specific manipulation of Arabidopsis loci using CRISPR-Cas9 SunTag systems. *Nature Communications*, 10(1), 729.

Piatek, A., Ali, Z., Baazim, H., Li, L., Abulfaraj, A., Al-Shareef, S., & Mahfouz, M. M. (2015). RNA-guided transcriptional regulation in planta via synthetic dC as9-based transcription factors. *Plant Biotechnology Journal*, 13(4), 578–589.

Prajapati, R., & Tyagi, K. (2024). CRISPR/Cas-Mediated Multiplex Genome Editing in Plants and Applications. In S. K. Upadhyay (Ed.), *Applications of Genome Engineering in Plants*, Wiley online, pp. 20–39.

Pramanik, D., Shelake, R. M., Park, J., Kim, M. J., Hwang, I., Park, Y., & Kim, J. Y. (2021). CRISPR/Cas9-mediated generation of pathogen-resistant tomato against tomato yellow leaf curl virus and powdery mildew. *International Journal of Molecular Sciences*, 22(4), 1878.

Qi, L. S., Larson, M. H., Gilbert, L. A., Doudna, J. A., Weissman, J. S., Arkin, A. P., & Lim, W. A. (2013). Repurposing CRISPR as an RNA-guided platform for sequence-specific control of gene expression. *Cell*, 152(5), 1173–1183.

Qi, W., Zhu, T., Tian, Z., Li, C., Zhang, W., & Song, R. (2016). High-efficiency CRISPR/Cas9 multiplex gene editing using the glycine tRNA-processing system-based strategy in maize. *BMC Biotechnology*, 16, 1–8.

Qin, L., Li, J., Wang, Q., Xu, Z., Sun, L., Alariqi, M., & Jin, S. (2020). High-efficient and precise base editing of C• G to T• A in the allotetraploid cotton (Gossypium hirsutum) genome using a modified CRISPR/Cas9 system. *Plant Biotechnology Journal*, 18(1), 45–56.

Rodríguez-Leal, D., Lemmon, Z. H., Man, J., Bartlett, M. E., & Lippman, Z. B. (2017). Engineering quantitative trait variation for crop improvement by genome editing. *Cell*, 171(2), 470–480.

Razzaq, A., Saleem, F., Kanwal, M., Mustafa, G., Yousaf, S., Imran Arshad, H. M., & Joyia, F. A. (2019). Modern trends in plant genome editing: an inclusive review of the CRISPR/Cas9 toolbox. *International Journal of Molecular Sciences*, 20(16), 4045.

Razzaq, M. K., Aleem, M., Mansoor, S., Khan, M. A., Rauf, S., Iqbal, S., & Siddique, K. H. (2021). Omics and CRISPR-Cas9 approaches for molecular insight, functional gene analysis, and stress tolerance development in crops. *International Journal of Molecular Sciences*, 22(3), 1292.

Ren, B., Yan, F., Kuang, Y., Li, N., Zhang, D., Zhou, X., & Zhou, H. (2018). Improved base editor for efficiently inducing genetic variations in rice with CRISPR/Cas9-guided hyperactive hAID mutant. *Molecular Plant*, 11(4), 623–626.

Roca-Paixão, J. F., Gillet, F. X., Ribeiro, T. P., Bournaud, C., Lourenço-Tessutti, I. T., Noriega, D. D., ... & Grossi-de-Sa, M. F. (2019). Improved drought stress tolerance in Arabidopsis by CRISPR/dCas9 fusion with a Histone AcetylTransferase. *Scientific Reports, 9*(1), 8080.

Sánchez-León, S., Gil-Humanes, J., Ozuna, C. V., Giménez, M. J., Sousa, C., Voytas, D. F., & Barro, F. (2018). Low-gluten, nontransgenic wheat engineered with CRISPR/Cas9. *Plant Biotechnology Journal*, 16(4), 902–910.

Sandhya, D., Jogam, P., Allini, V. R., Abbagani, S., & Alok, A. (2020). The present and potential future methods for delivering CRISPR/Cas9 components in plants. *Journal of Genetic Engineering and Biotechnology*, 18(1), 25.

Shen, B., Zhang, W., Zhang, J., Zhou, J., Wang, J., Chen, L., & Skarnes, W. C. (2014). Efficient genome modification by CRISPR-Cas9 nickase with minimal off-target effects. *Nature Methods*, 11(4), 399–402.

Shimatani, Z., Kashojiya, S., Takayama, M., Terada, R., Arazoe, T., Ishii, H., & Kondo, A. (2017). Targeted base editing in rice and tomato using a CRISPR-Cas9 cytidine deaminase fusion. *Nature Biotechnology*, 35(5), 441–443.

Shin, H., Choi, W. L., Lim, J. Y., & Huh, J. H. (2022). Epigenome editing: Targeted manipulation of epigenetic modifications in plants. *Genes & Genomics,* 44, 1–9.

Shulse, C. N., Cole, B. J., Ciobanu, D., Lin, J., Yoshinaga, Y., Gouran, M., ... Dickel, D. E. (2019). High-throughput single-cell transcriptome profiling of plant cell types. *Cell Reports,* 27(7), 2241–2247.

Song, G., Jia, M., Chen, K., Kong, X., Khattak, B., Xie, C., & Mao, L. (2016). CRISPR/Cas9: a powerful tool for crop genome editing. *Crop Journal,* 4, 75–82.

Stuttmann, J., Barthel, K., Martin, P., Ordon, J., Erickson, J. L., Herr, R., & Bonas, U. (2021). Highly efficient multiplex editing: one-shot generation of 8× Nicotiana benthamiana and 12× Arabidopsis mutants. *The Plant Journal*, 106(1), 8–22.

Sultan, Q., Ashraf, S., Munir, A., Khan, S. H., Munawar, N., Abd-Elsalam, K. A., & Ahmad, A. (2022). Beyond genome editing: CRISPR approaches. In A. Ahmad, S. H. Khan, & Z. Khan (Eds.), The CRISPR/Cas tool kit for genome editing, Springer, pp. 187–218. https://doi.org/10.1007/978-981-16-6305-5_6

Tang, X., Liu, G., Zhou, J., Ren, Q., You, Q., Tian, L., & Zhang, Y. (2018). A large-scale whole-genome sequencing analysis reveals highly specific genome editing by both Cas9 and Cpf1 (Cas12a) nucleases in rice. *Genome Biology,* 19, 1–13.

Tang, X., Zheng, X., Qi, Y., Zhang, D., Cheng, Y., Tang, A., & Zhang, Y. (2016). A single transcript CRISPR-Cas9 system for efficient genome editing in plants. *Molecular Plant*, 9(7), 1088–1091.

Turnbull, C., Lillemo, M., & Hvoslef-Eide, T. A. (2021). Global regulation of genetically modified crops amid the gene edited crop boom–a review. *Frontiers in Plant Science,* 12, 630396.

Veillet, F., Chauvin, L., Kermarrec, M. P., Sevestre, F., Merrer, M., Terret, Z., & Chauvin, J. E. (2019). The Solanum tuberosum GBSSI gene: a target for assessing gene and base editing in tetraploid potato. *Plant Cell Reports*, 38(9), 1065–1080.

Vilhekar, R. S., & Rawekar, A. (2024). Artificial intelligence in genetics. Cureus, 16(1). www.cureus.com/articles/177320

Wang, J., Zhang, X., Cheng, L., & Luo, Y. (2020). An overview and metanalysis of machine and deep learning-based CRISPR gRNA design tools. *RNA Biotechnology*, 17(1), 13–22.

Wang, W., Akhunova, A., Chao, S., & Akhunov, E. (2016). Optimizing multiplex CRISPR/Cas9-based genome editing for wheat. *BioRxiv,* 051342.

Wu, S., Kyaw, H., Tong, Z., Yang, Y., Wang, Z., Zhang, L., & Qin, G. (2024). A simple and efficient CRISPR/Cas9 system permits ultra-multiplex genome editing in plants. *The Crop Journal*, 12(2), 569–582.

Wu, W. Y., Lebbink, J. H., Kanaar, R., Geijsen, N., & Van Der Oost, J. (2018). Genome editing by natural and engineered CRISPR-associated nucleases. *Nature Chemical Biology,* 14(7), 642–651.

Wolabu, T. W., Cong, L., Park, J. J., Bao, Q., Chen, M., Sun, J., & Wang, Z. Y. (2020). Development of a highly efficient multiplex genome editing system in outcrossing tetraploid alfalfa (Medicago sativa). *Frontiers in Plant Science,* 11, 1063.

Wolt, J. D., Wang, K., & Yang, B. (2016). The regulatory status of genome-edited crops. *Plant Biotechnology Journal*, 14(2), 510–518.

Xie, K., Minkenberg, B., & Yang, Y. (2015). Boosting CRISPR/Cas9 multiplex editing capability with the endogenous tRNA-processing system. *Proceedings of the National Academy of Sciences*, 112(11), 3570–3575.

Xu, W., Song, W., Yang, Y., Wu, Y., Lv, X., Yuan, S., & Yang, J. (2019). Multiplex nucleotide editing by high-fidelity Cas9 variants with improved efficiency in rice. *BMC Plant Biology*, 19, 1–10.

Yan, D., Ren, B., Liu, L., Yan, F., Li, S., Wang, G., & Zhou, H. (2021). High-efficiency and multiplex adenine base editing in plants using new TadA variants. *Molecular Plant*, 14(5), 722–731.

Yang, T., Ali, M., Lin, L., Li, P., He, H., Zhu, Q., & Deng, L. (2023). Recoloring tomato fruit by CRISPR/Cas9-mediated multiplex gene editing. *Horticulture Research*, 10(1), uhac214.

Yang, Y., Wang, D., Lü, P., Ma, S., & Chen, K. (2023). Research progress on nucleic acid detection and genome editing of CRISPR/Cas12 system. *Molecular Biology Reports*, 50(4), 3723–3738.

Zannoni, L. (2019). Evolving regulatory landscape for genome-edited plants. *The CRISPR Journal*, 2(1), 3–8.

Zhang, D., Hussain, A., Manghwar, H., Xie, K., Xie, S., Zhao, S., & Ding, F. (2020). Genome editing with the CRISPR-Cas system: an art, ethics and global regulatory perspective. *Plant Biotechnology Journal*, 18(8), 1651–1669.

Zhang, P., Du, H., Wang, J., Pu, Y., Yang, C., Yan, R., & Yu, D. (2020). Multiplex CRISPR/Cas9-mediated metabolic engineering increases soya bean isoflavone content and resistance to soya bean mosaic virus. *Plant Biotechnology Journal*, 18(6), 1384–1395.

Zhang, Q., Zhang, Y., Lu, M. H., Chai, Y. P., Jiang, Y. Y., Zhou, Y., & Chen, Q. J. (2019). A novel ternary vector system united with morphogenic genes enhances CRISPR/Cas delivery in maize. *Plant Physiology*, 181(4), 1441–1448.

Zhang, Y., Bai, Y., Wu, G., Zou, S., Chen, Y., Gao, C., & Tang, D. (2017). Simultaneous modification of three homoeologs of Ta EDR 1 by genome editing enhances powdery mildew resistance in wheat. *The Plant Journal*, 91(4), 714–724.

Zhang, Y., Ren, Q., Tang, X., Liu, S., Malzahn, A. A., Zhou, J., & Qi, Y. (2021). Expanding the scope of plant genome engineering with Cas12a orthologs and highly multiplexable editing systems. *Nature Communications*, 12(1), 1944.

Zhang, Z., Mao, Y., Ha, S., Liu, W., Botella, J. R., & Zhu, J. K. (2016). A multiplex CRISPR/Cas9 platform for fast and efficient editing of multiple genes in Arabidopsis. *Plant Cell Reports*, 35, 1519–1533.

Zhao, Y., Zhang, C., Liu, W., Gao, W., Liu, C., Song, G., & Xie, C. (2016). An alternative strategy for targeted gene replacement in plants using a dual-gRNA/Cas9 design. *Scientific Reports*, 6(1), 23890.

Zhou, H., Liu, B., Weeks, D. P., Spalding, M. H., & Yang, B. (2014). Large chromosomal deletions and heritable small genetic changes induced by CRISPR/Cas9 in rice. *Nucleic Acids Research*, 42(17), 10903–10914.

16 Genome-Editing Tools for Biotic Stress Tolerance in Plants

Florencia Alvarez and Jorge Alberto Zavala

16.1 INTRODUCTION

It is expected that the world population will reach 10 billion in 2050, consequently increasing cultivated land area and declining forested areas (FAO, 2018). However, new climatic conditions encourage the development of new insect pests, thus currently traditional and new plagues are threatening to put crop production at risk. These changes, in combination with climate change will probably reduce the productivity of crops. Furthermore, fighting these emerging pests will require intensification of the use of plant protection agrochemicals, with the consequent increase of toxic by-products. However, new breeding technologies have been developed in the last decade, and will help the world to face urgent challenges related to food production and crop protection.

Plant breeding has been used for many years as one of the most efficient strategies to address these challenges in crop production. Providing crops' genetic pathways related to the production of chemical defenses against insects will allow them to defend themselves without the need for external agents. Generating crops resistant to herbivores involves modifying the genetic information of plant genomes to express new traits against insect attack. Traditionally, crops have been modified by selecting for random mutations that increase yield and quality of crops. Although previously the possibility of performing gene editing was constrained by the emergence of spontaneous mutations, in the last few years marker-assisted breeding has accelerated the incorporation of mutations in different crop species. However, new biotechnology tools appeared in the 1980s that allowed manipulation of plant genomes by transferring foreign DNA to modify plant traits, such as resistance against herbivory (Barton et al. 1983; Herrera-Estrella et al. 1983). The transfer of exogenous DNA to higher plants can be performed by different methods. The typical and widely used methodologies for the development of transgenic crops were genetic transformation using *Agrobacterium tumefaciens* and particle bombardment (biolistics). These two methods incorporate the foreign DNA randomly into the genome of plants to be transformed, thus the transgenic plants generated need to be selected for the appropriate phenotype. By using these methods, traditional transgenic plants received one or a few new genes resulting from DNA recombination, which is not much when compared with the nearly 50,000 genes of the soybean genome (Schmutz et al. 2010). The most successful ones that are still used commercially are transgenic crops that express the *Bacillus thuringiensis* (BT) Cry protein, which is toxic to some herbivorous insects. Although BT crops, such as cotton, corn or soybean, have been successfully used to avoid lepidopteran damage for a long time, in the past few years insect pests have become resistant to Cry proteins. Since approximately 75% of the soybean crops planted today contain these kinds of genetic modifications, plant transgenesis has had a very important impact on the agriculture of the world.

In the context of this first generation of biotechnological tools, our ability to edit genomes has increased dramatically in the last decade thanks to a set of new technologies. Among these, two deserve special mention: DNA synthesis and CRISPR (Clustered Regularly-Interspaced Short Palindromic Repeats) genetic editing. DNA synthesis allows us to write genetic instructions in the laboratory. In recent years, our ability to synthesize DNA has grown almost exponentially so that now we can write increasingly longer DNA fragments at a lower cost (Wang et al. 2018). Although the length of DNA chains that can be synthesized at once using purely chemical methods is not unlimited, so-called modular assembly techniques have recently allowed the creation of ever longer constructs by joining DNA fragments with each other as if they were Lego parts (Engler et al. 2008; Vazquez-Vilar et al. 2018). As a result, the ability to transfer complex instructions to plants has grown greatly, and the price and effort required to do so have decreased.

The second major technological breakthrough was CRISPR gene-editing technology. Being able to write genetic instructions is of little use if we cannot insert them correctly into the corresponding genome. CRISPR/Cas9 proteins allow us to go through the genome and introduce small modifications or incorporate new information into it with precision. The novelty of CRISPR technology is the possibility to incorporate new information at the specific location where an instruction could be functional. Thus, combining the growing capability of DNA synthesis with the power and precision of CRISPR for genomic editing makes a new starting point for a technological revolution.

The development of crops capable of defending themselves against pest attacks is considered one of the challenges facing future plant biotechnology. The success of this development will significantly depend on how new plant breeding technologies are developed and integrated into agricultural practices. In the present chapter, a description is provided regarding how CRISPR/Cas technology works and the strategies for minimizing off-target effects for the creation of genetically modified plants with defensive traits against biotic stress. Additionally, various uses of CRISPR to modify crops to resist insect pests, by increasing resistance against herbivores and diseases, are presented. Finally, perspectives

DOI: 10.1201/9781003500933-19

and opportunities for the development of new crops using CRISPR/Cas technology are proposed.

16.2 CRISPR TECHNOLOGY

16.2.1 How CRISPR/Cas9 Works

CRISPR sequences cooperate with associated nucleases (Cas) to set up the bases of an adaptive defense system in prokaryotes against virus and plasmids (Mojica et al. 2005, 2009). Two small RNAs (crRNA and tracrRNA) guide the Cas protein toward the DNA target sequence, and a chimeric single guide RNA (gRNA) is programmed to promote a genome-editing application (Jinek et al. 2012). Cas9 derived from *Streptococcus pyogenes* (SpCas9) was the first one characterized, and is the most used endonuclease in genome editing. According to the CRISPR RNA (crRNA) processing mechanism, Cas9 belongs to type II CRISPR-Cas system, in which a trans-activating crRNA (tracrRNA) complementary to the repeat sequences in pre-crRNA triggers processing by the double-stranded (ds) RNA-specific ribonuclease RNase III in the presence of the Cas9 protein. Thus, Cas9 is responsible for crRNA-guided silencing of foreign DNA. The other two systems (I and III) specialized Cas endonucleases process the pre-crRNAs, and, once mature, each crRNA assembles into a large multi-Cas protein complex capable of recognizing and cleaving nucleic acids complementary to the crRNA.

The CRISPR-Cas9 system has been developed as a versatile genome-editing tool to perform on multiple target sites, with efficacy, simplicity, and low cost compared to other site-specific nucleases such as zinc-finger nucleases (ZFNs) and transcription activator-like effector nuclease (TALEN). Unlike these, CRISPR is relying on simple Watson-Crick base pairing between the target DNA sequence and its gRNA that edits the genome without any sophisticated protein engineering steps (Munaweera et al. 2022). After recognition, the site-specific nucleases, in this case Cas9, introduce specific double-stranded breaks (DSBs) at or near the target site. The two nuclease domains of Cas9 (HNH and RuvC) cleave the DNA strands that are complementary and non-complementary to the 20-nucleotide (nt) guide sequence in crRNAs, respectively. In addition, the PAM-interacting (PI) domain is critical for the Cas9 function as it is required for the recognition of a protospacer adjacent motif (PAM) (5′-NGG-3′), located just downstream of the genomic sequence target, which is a strict requirement for Cas9 - mediated DNA cleavage (Nishimasu et al. 2014). However, novel Cas variants have been engineered to recognize alternative PAM sequences and widen the potential applications of precision genome editing (Kleinstiver et al. 2015, 2019; Hu et al. 2018).

DSBs are further resolved by a cellular DNA repair mechanism, either via non-homologous end-joining (NHEJ) or via homology-directed repair (HDR), which often results in mutations and thus genome editing, while the inability to prepare the DSB will result in cell death. The NHEJ repair is an error-prone pathway which creates random insertions and deletions (indels), and results in frameshift mutations and targeted gene knockouts. Therefore, this technique has become a popular way to disrupt genes by creating small indels at specific points in the target genes (Mishra et al. 2018).

Homologous templates are another important component to consider for the process. A homologous template is a key component of HDR-mediated genome editing when producing targeted gene knock-in or other specific mutations. It usually contains intended mutations or insertions flanked by homologous regions. Templates can be plasmids (up to Kb modification) or single-stranded oligodeoxynucleotides (ssODN) (50–100 nt modification). It is suggested that sequencing around the region of interest should be carried out because cell-specific mutations and SNPs can influence gRNA targeting up to sixfold merely one mismatch in 100 bases. The HDR-mediated insertion in plants occurs at very reduced efficiency compared with the NHEJ-mediated insertion. In recent years, several strategies have been deployed in crop plants to improve HDR editing. Some strategies to stabilize the donor templates include include use of germinivirus constructs, *in planta* gene targeting systems, manipulating DNA repair pathways, employing site-specific nucleases, and chemical modification to stabilize the donor templates. Many studies have attempted to increase HDR efficiency by transiently inhibiting proteins central to NHEJ using short interfering (si) RNAs or small molecules that achieved different percentage efficiency HDR-mediated DSB repair. Recently, to improve the efficiency of precise genome editing by HDR, the research group of Nobel Prize-winner Svante Pääbo introduced mutations in genes necessary for NHEJ as well as another end-joining pathway as MMEJ (microhomology-mediated end joining) that resulted in DSB repair almost exclusively by HDR, while indels, large deletion/rearrangements and off-target editing events are largely abolished (Riesenberg et al. 2023).

16.2.2 Strategies for Minimizing Off-Target Effect

Although CRISPR/Cas systems exhibit tremendous potential, off-target effects remain a major challenge. The off-target effects occur when Cas9 acts on untargeted genomic sites and creates cleavages that may lead to adverse outcomes. The off-target sites are often gRNA-dependent, since Cas9 is known to tolerate up to three mismatches between gRNA and genomic DNA. Methods to assess off-target effects of CRISPR/Cas9 have quickly evolved in the last decade. However, limitations still remain in balancing the accuracy versus sensitivity of these new techniques (efficiency and specificity are the two critical parameters that determine the success of CRISPR/Cas9-based editing of the genome). In this scenario, *in silico* tools are useful to search for potential off-target sites in the whole genome and calculate the likelihood of an off-target editing. Nevertheless, accumulated studies have proved that gRNA-independent off-target effects also exist, urging unbiased experimental detection and validation. To date, many methods have been developed to nominate or detect the off-target sites of CRISPR/Cas9, which laid the basis for the successful upgrades of CRISPR-Cas9 derivatives with enhanced precision. CRISPR/Cas9 off-target effects can be predicted by *in silico* tools, such as CRISPR-P, CasOT, Cas-OFFinder, FlashFry, DeepCRISPR, etc. In addition, these off-target effects can be identified by cell culture-based methods, which avoid the prohibited costs of WGS (Guo et al. 2023). Some of the strategies to reduce off-target effects are Cas9 improvement, gRNA improvement, delivery method improvement, and DSB-independent gene editing. It has been suggested that the Cas9-mediated DSB generation is the main source of CRISPR/CAS9 off-target effects. Interestingly, there are new versions of gene editors that do not create DSBs and usually exhibit greater specificity of genome editing. For example, base editors are Cas9 nickases (mutated to cleave only one DNA strand) fused to adenine or cytosine deaminases that convert A·T to G·C or C·G to T·A base pairs, respectively. Optimized base editors can correct single-point mutations. In addition, prime editors are Cas9 nickases fused to reverse transcriptases. Prime editors use a prime-editing guide RNA (pegRNA) that has both a spacer for recognition of the DNA target and a 3'-extension that encodes a template for reverse transcriptase to synthesize the edited DNA product. Because base and prime editors do not create dsDNA breaks, they cause fewer off-target editing than Cas9-mediated HDR (Chen et al. 2019; Biswas et al. 2021).

In addition to gRNA length adjustment, it is also now appreciated that chemically modified nucleic acids can substitute for unmodified nucleotides in gRNAs, providing protection from exonuclease degradation and improving gene-editing efficiency (Munaweera et al. 2022). For example, one study incorporated 2'-O-methyl-3'-phosphonoacetate (MP) into the ribose-phosphate backbone of gRNA to enhance on-target specificity while dramatically reducing off-target activity. Incorporation of bridged nucleic acids (2',4'-BNA^{NC}[N-Me] or locked nucleic acids (LNA) into the gRNAs also reduces the kinetics of Cas9 nuclease reactivity, thereby improving the specificity of genome editing. The replacement of ribonucleotides by deoxyribonucleotides also reduced off-target effects (Rozners, 2022).

The off-target activity with Cpf1 cleavage is also comparatively lower than Cas9. In a CRISPR-Cpf1 system, a T-rich protospacer adjacent motif (PAM) creates DSBs at the distal region of the recognition site producing cohesive ends; it may boost the chance to repair a DSB through HDR (Alok et al. 2020). In addition, the targeted insertion frequency was increased by Cpf1 nuclease in rice (Begemann et al. 2017).

16.3 FUNCTIONAL USE OF CRISPR/CAS TECHNOLOGY

16.3.1 CRISPR/Cas as a New Breeding Technique (NBT)

The CRISPR technology is considered a precision genome-modification tool within the new breeding techniques (NBTs) which are aimed at shortening the breeding cycles and accelerating crop research. CRISPR/Cas has significant potential for plant breeding programs in association with molecular marker-assisted selection (MAS) based on SNP (single nucleotide polymorphism) deep genotyping. Genome-editing NBTs differ from conventional genetic modification (GM) techniques in various ways. For example, CRISPR/Cas can be used for precise genetic manipulation without introducing exogenous DNA such as antibiotic-resistant genes, thus eliminating the fear that foreign DNA may be present in the final product. Whereas classical GM crop production requires the insertion of foreign DNA (transfer DNA, or T-DNA from *Agrobacterium* species), some genome-editing protocols do not require T-DNA insertion, such as CRISPR via a ribonucleoprotein (RNP) complex or via virus-based DNA replicons to induce precisely targeted edits in the crop plant DNA. Further elaborations of this methodology, including protoplast delivery of preassembled CRISPR-Cas RNPs transient expression of programmable nucleases through agroinfiltration, and site-specific integration of a CRISPR array in other chromosomal locations, followed by removal via segregation, have paved the way toward developing transgene-free CRISPR plants. The improvements in HDR and the use of transgene-free approaches allow the breeders to have non-GM improved crop varieties through genome engineering. A transgene-free or non-GM version of CRISPR-based editing such as the use of viral vectors for the delivery of CRISPR-Cas machinery, ribonucleoproteins (RNPs), or transient expression of CRISPR-Cas9 system through dsDNA cassettes or RNA transcripts could be potentially used for the introduction of identified SNPs in crop plants (Chen et al. 2019).

Engineering crop resistance against disease by manipulating susceptibility (S) genes is a new challenge. Pathogens exploit the effector proteins to alter the S genes in the host plant, rendering the host defense inactivated. Thus, altering the S genes by employing precise genome-editing tools provides a great opportunity to develop broad-spectrum and resistant crop species. Over the past few years, with the dawn of next-generation sequencing and innovative CRISPR technology, the identification and targeting of S genes have become a method of choice for enhancing disease resistance in plants. Recently, Wang et al. (2022) provided one of the successful examples, targeting the wheat S gene Pst-Induced Protein Kinase 1 (TaPsIPK1) that can further be used in wheat to confer stripe rust disease resistance by genetic modifications (Wang et al. 2022). Upon infection, the obligate biotrophic fungal pathogen *Puccinia striiformis* f. sp. *tritici* (Pst) induces a receptor-like cytoplasmic kinase, TaPslPK1, that specifically interacts with the fungal effector PsSpg1, which is vital for Pst virulence. The association of TaPsIPK1/PsSpg1 boosted the auto-phosphorylation and enhanced the receptor's nuclear translocation where the phosphorylation of the transcription factor (TaCBF1d) occurs and steers the gene expression. CRISPR-Cas9 inactivation of *TaPslPK1* in wheat confers broad-spectrum resistance against Pst stimulating the effector-triggered immunity (ETI), along with HR response and H_2O_2 production, which suggests that the wheat receptor negatively affects both pathogen-induced immunity (PTI) and ETI (Wang et al. 2022).

The above-mentioned approaches would momentously profit from prime editing and homologous recombination procedures. Nonetheless, these methods still demand further improvement to enable editing larger chunks within promoters and/or genes. Recent innovations in genome-editing tools have improved the way to disrupt S genes. New studies suggested that coupling of CRISPR-Cas9 with base editors to introduce single-base alterations/SNPs in the S genes can avoid the pleiotropic effects. In addition, allele engineering might also be used to produce hypomorphic alleles that decrease the susceptibility with uncertain results about the pleiotropic effects. Also, these changes can be accomplished via base editors to make single- nucleotide modifications and lead to the substitutions of amino acids. Similar results can be obtained through homologous recombination of amino acids or through homologous recombination to replace larger sequences. Moreover, by employing multiple guide RNAs (gRNAs), the CRISPR technology can perform precise deletion of a certain cis-element in the promoter region of a particular S gene. Additionally, simultaneous mutation in effector-binding elements via multiplex CRISPR confers broad-spectrum resistance.

Breeding of crops over millennia for yield and productivity has led to reduced genetic diversity. As a result, beneficial traits of wild species, such as disease resistance and stress tolerance, have been lost. Thus, CRISPR-Cas9 genome engineering strategies can be used in molecular breeding programs to combine agronomically desirable traits with useful traits present in wild lines (Khan et al. 2019). For example, at least six loci important for key domestication traits have been identified in tomato (*Solanum lycopersicum*). Taking this into account, researchers used a multiplex CRISPR-Cas9 approach to generate loss-of-function alleles that enabled *de novo* domestication of wild *Solanum pimpinellifolium* (Zsögön et al. 2018). In a complementary way, editing of genes in cultivated crops based on the respective variation in crop wild relatives would be a tangible approach to introduce variability (Tyagi et al. 2020).

16.3.2 Biotechnological Strategies For Plant Disease Control

Since the first successful completion of genome-based editing in the model plant *Arabidopsis* using CRISPR/Cas9 in 2013, a large number of agronomic trait-related genes have been efficiently edited in rice (*Oryza sativa* L.) (Jiang et al. 2013). In addition, by 2020, more than 120 edited events have been reported just for crop quality improvement (Liu et al. 2021).

Bacterial leaf blight (BB) is one of the most important rice diseases responsible for devastating crop losses among farmers in Asia and Africa. Key aspects of the mechanisms underlying BB disease have been elucidated and provided an efficient roadmap for resistance breeding. Different approaches using CRISPR-Cas technology have been reported as potential biotechnological solutions for disease control (Figure 16.1).

It is known that *Xanthomonas oryzae* pv. *oryzae* (Xoo) strains secrete a suite of Transcription Activator-Like effectors (TALEs) into rice xylem parenchyma cells via type III secretion systems to induce the expression of one or several host SWEET sucrose uniporter genes (*OsSWEET*11, *OsSWEET*13 and *OsSWEET*14), which function to confer disease susceptibility (S). Activation of target *OsSWEET* genes elevates sugar transport to where the pathogen needs it, enhancing its virulence and leading to disease susceptibility. As TALEs bind to effector-binding elements (EBEs) in the promoters of *OsSWEET* genes (a 21-gene family in rice), CRISPR/Cas9-mediated gene editing was used to disrupt the sequences of the EBEs in rice varieties resulting in the desired resistance to bacterial blight (Figure 16.1A) (Oliva et al. 2019; Xu et al. 2019; Schepler-Luu et al. 2023). The finding that only half of the six class III SWEET susceptibility factors are actually used by the known Xoo strains indicates a limited ability of Xoo to evolve new TALEs that target these promoters. This observation is in line with the strict continental isolation, where American Xoo strains lack TALEs and hence are poorly virulent (Schepler-Luu et al. 2023). More than 40 genes for bacterial blight have been identified and several have been associated with modular TAL effectors. Complete genome sequence of Xoo enables identification of TAL effectors, information that can then be used together with the TAL effector recognition code to identify the respective TAL EBEs throughout the rice genome.

The information generated about the TAL effector was used by scientists from the University of Missouri. They chose to target specific EBE polymorphisms based on their sequence-level characterization of TALEs in 63 Xoo strains from diverse areas, and the mutation they discovered in the *OsSWEET* promoter regions was present in three rice varieties. Furthermore, they applied a multiplexed CRISPR-Cas9 genome-editing approach to systematically interfere with *OsSWEET* gene induction at all known major TALE EBEs, and the engineered Kitaake rice lines showed resistance to all currently known Xoo strains (Oliva et al. 2019). In addition, a diagnostic kit (SWEET^R kit 1.0) that can identify the promoter variants was developed, which is best suited to defeat the local pathogen population and monitor variant strain emergence. These results allow breeders and pathologists to work together to provide selected lines to farmers, with an overall strategy of deploying multiple lines in a region to disfavor the evolution of novel strains (Eom et al. 2019; Varshney et al. 2019).

The rulings by the United States Department of Agriculture (USDA) and the Colombian Agricultural Institute (ICA) clear the way for field tests of bacterial blight-resistant rice variety developed through gene editing (geneticliteracyproject.org). Although this disease has not yet been detected in crops in Colombia, rice blight is included in the list of main pests and preventive detection tests are carried out. The USA, Canada, and many South and Central American countries have implemented regulations that considered transgene-free, genome-edited lines to be equivalent to conventionally breeding lines. At this time, an increased number of governments decided to follow the decision of many South

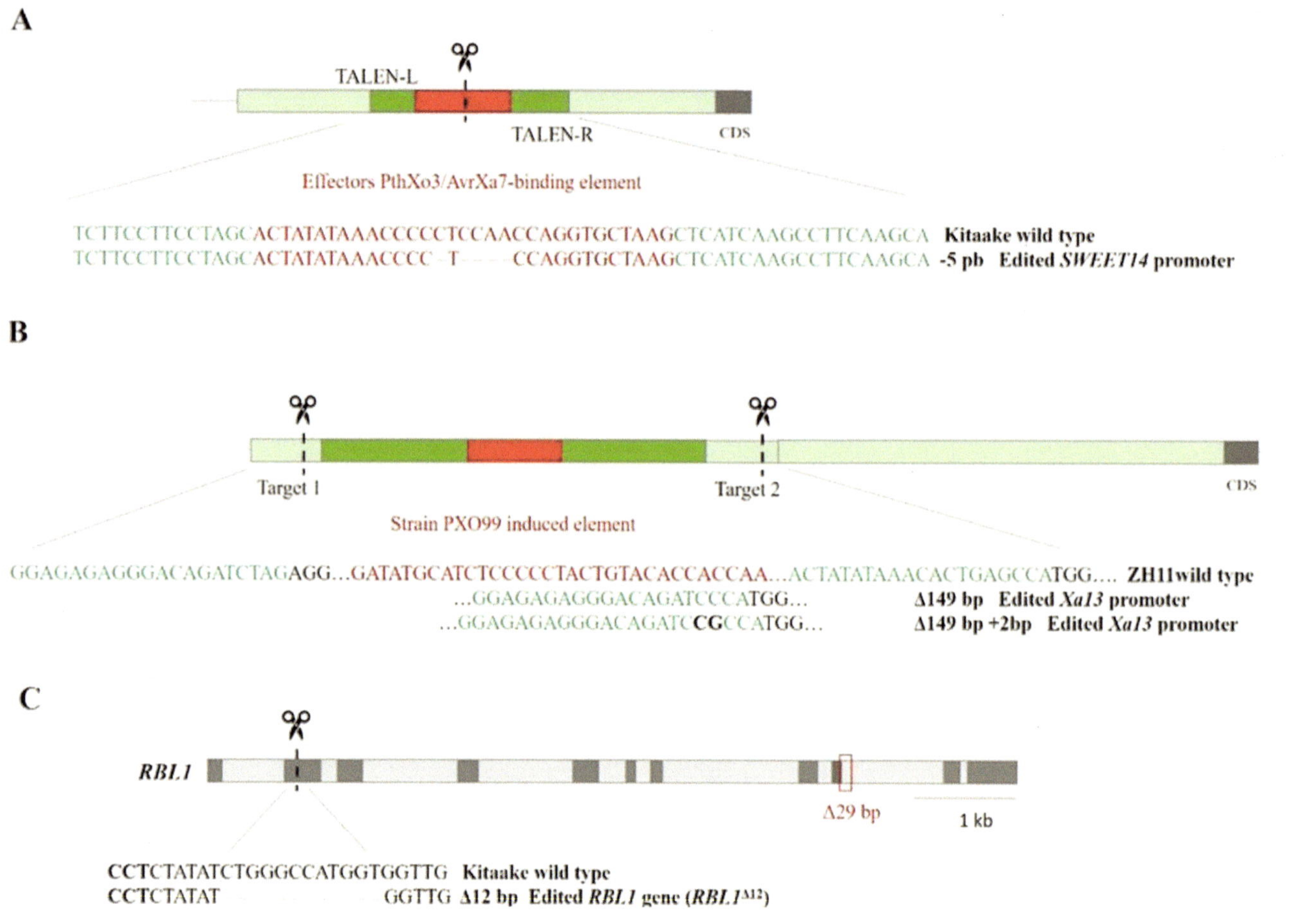

FIGURE 16.1 Biotechnological strategies to control rice blight using CRISPR/Cas. (A, B) Editing of effector-binding elements (EBEs) (shown in red) in the promoters (shown in green) of *SWEET14* and *Xa13* genes, respectively (Oliva et al. 2019; Li et al. 2020). (C) Editing of *RBL1* gene (shown in gray) to obtain the allele $RBL1^{\Delta 12}$. The deletion causes an in-frame deletion of the ninth exon (shown in dark gray) resulting in a 19-amino-acid truncation at the conserved C terminus of RBL1. (Sha et al. 2023.)

American countries to implement a case-by-case review of genome-edited crops. However, the European Court of Justice (ECJ) classified all genome-edited lines as GMO (Buchholzer and Frommer 2023).

A second strategy is the editing of the promoter region of the *Xa13* gene, a pluripotent recessive gene that regulates rice bacterial blight resistance and participates in anther development. Direct editing of the CDS regions could improve the resistance to rice but cause changes in important agronomic traits and a sterile phenotype. However, a deletion of its promoter sequence could improve rice disease resistance without affecting rice fertility (Figure 16.1B) (Li et al. 2020).

A third approach to generate bacterial blight-resistant rice varieties was focused on editing a LMM (lesion mimic mutant) gene called *RBL1* from a mutagenized rice population. A 29-base-pair deletion in *RBL1* caused a broad-spectrum resistance to rice blight and rice blast but caused an approximately 20-fold reduction in yield. *RBL1* encodes a cytidine diphosphate diacylglycerol synthase that is required for phospholipid biosynthesis. To optimize the *RBL1* gene for disease resistance, the research team employed genome-editing techniques to target multiple sites within its genomic region. Thus, they found that a 12-bp deletion in the second exon of *RBL1* not only restored plant growth but also maintained broad-spectrum disease resistance against multiple Xoo strains as well as *Magnaporthe oryzae* and *Ustilaginoidea virens* but did not decrease yield in a model rice variety (Figure 16.1C) (Sha et al. 2023). LMM genes often encode negative regulators of immunity, which are proteins with diverse functions including the regulation of gene transcription, protein translation and modification, metabolism, as well as vesicular trafficking. LMMs spontaneously develop necrotic lesions resembling hypersensitive responses triggered by resistance genes without being challenged by pathogens and they usually confer more durable and broader-spectrum resistance than typical R genes (Sha et al. 2023).

16.3.3 CRISPR/Cas9 System Applied to Molecular Interactions Between Plants and Insect Herbivores

Another important use of CRISPR/Cas9 technology is in increasing resistance of crop plants against insect attack. Gene editing using CRISPR/Cas9 system has made it possible to

TABLE 16.1
Overview of CRISPR/Cas9-mediated targeted mutagenesis of gene candidates involved in plant-insect interaction

Disrupted gene/s	Description	Host plant	Insect target	Molecular mechanism	Reference
LecRK-I.1	Lectin receptor kinase	*Arabidopsis thaliana*	*Pieris brassicae* (large white or cabbage butterfly)	HR-like response and SA signaling in response to oviposition	Groux et al. 2021
OsCOI2	COI jasmonate receptor	rice	none	COI2 is essential for JA-induced accumulation of phytoalexins	Inagaki et al. 2023
MpAOS	Allene oxidase synthase	*Marchantia polymorpha*	*Tetranychus urticae* (spider mite)	OPDA signaling pathways in liverwort regulates biosynthesis of sesquiterpenes and other secondary metabolites	Koeduka et al. 2022
CYP71A1	Tryptamine 5-hydroxylase (cytochrome P450)	rice	*Nilaparvata lugens* (brown planthopper-BPH)	Disruption of *OsCYP71A1* blocks serotonin biosynthesis, increases SA levels, conferring resistance to BPH	Lu et al. 2018
IncA07-IncD09	Long non-coding RNAs (IncRNAs)	cotton	*Bemisia tabaci* (whitefly)	Decrease JA content may increase susceptibility to insect infestation	J. Zhang et al. 2022
CaM2	Calmodulin 2	tomato	*Helicoverpa armigera* (cotton bollworm)	Ca^{+2} signals induce *CaM* genes and trigger interaction with ERF16, leading to JA burst	Hu et al. 2022
MYC1-MYC2	bHLH transcription factors (MYC subfamily)	tomato	Broad range of herbivores and pathogens	Suppression of constitutive SGA byosinthesis	Swinnen et al. 2022
NtMYC2a	Same as above	tobacco		MYC2 acts as a "master switch" for JA-induced nicotine accumulation	Sui et al. 2021
GmUGT	UDP-glycosyltransfrerase	soybean	*Helicoverpa armigera,*	Alters flavonoid biosynthesis pathway	Y. Zhang et al. 2022
1636/1639	β-1,3-glucanases	barley	*Rhopalosiphum padi* L (cereal pest)	Double mutants showed reduced glucanase activity but had no significant effect on callose deposition	Kim et al. 2020

provide information related to molecular networks triggered by herbivory and the plant response to insect attacks (Table 16.1). In most of the cases, the CRISPR-Cas gene-editing reagents were introduced via transformation with disarmed *Agrobacterium tumefaciens* harboring plasmid vector encoded gRNA cassette(s), Cas9 endonuclease and hygromycin phosphotransferase (HPT) or neomycin as selectable markers, and kanamycin as a bacterial resistance gene. Conventional breeding was used to generate null-segregant progeny that contained the intended mutations without the introduced exogenous DNA.

Plants perceive herbivory through pattern recognition receptors including the yet unknown herbivore/egg-associated molecular patterns (HAMPs/EAMPs) or through damage-assiciated molecular patterns (DAMPs), which are plant cytosolic or apoplastic molecules freely released upon tissue injury (Erb and Reymond 2019; Gouhier-Darimont et al. 2019). Specific defense reprogramming proceeds via signaling networks that include phytohormones, transcription factors, and secondary metabolites.

16.3.3.1 Phytohormone Pathways

In *Arabidopsis*, salicylic acid (SA) and reactive oxygen species (ROS) accumulate in response to oviposition by the specialist insect *Pieris brassicae* (Large White butterfly), followed by induction of defense genes and localized necrosis, which activates a similar response to the recognition of pathogen-associated molecular patterns (PAMPs) associated with PAMP-triggered immunity (PTI) (Reymond 2013). PAMPs are generally perceived by receptor-like kinases at the plasma membrane. The L-type lectin receptor kinases LecRK-I.8 is crucial for the establishment of egg extract-induced systemic acquired resistance (SAR), which indicates that the response of *A. thaliana* to *P. brassicae* eggs shares similarities with the plant's response to phytopathogens (Gouhier-Darimont et al. 2013, 2019). The observation that *LecRK-I.8*- mutant does not display a complete lack of pathogenesis-related gene (*PR-1*) induction suggests that other receptor-like kinases are required for the egg's plant perception (Gouhier-Darimont et al. 2019). Generating large deletion of *LecRK-I* clusters by CRISPR/Cas9 technology may be a useful strategy to test the role of these receptors in responses to insect oviposition. The same research group demonstrated that the disruption of *LecRK-I.1*, a close homolog of *LecRK-I.8*, reduced the hypersensitive-like response (HR), *PR1* expression, and SA accumulation after egg-extract treatment. Moreover, double knockout of both *LecRK-I.1* and *LecRK-I.8* did not result in a further reduction of egg extract-elicited symptoms, suggesting that both genes function in the same transduction pathway or are part of the same complex (Groux et al. 2021). More studies of this region in other Brassicaceae using gene editing and germplasm screening may demonstrate a role for clade I LecRKs in plant perception of insect oviposition.

One of the most studied defensive hormones against insect attack is jasmonic acid (JA). Jasmonate and its derivatives are a class of oxygenated lipid-based hormones. They are ubiquitous in the plant kingdom and play critical roles in plant development and responses to abiotic stresses and pathogens. The biosynthesis of JA is initiated with the release of α-linolenic acid (18:3) from the lipid membrane by phospholipases in chloroplasts. After a series of reactions catalyzed by 13-lipoxygenases (LOX), allene oxidase synthase (AOS), and allene oxidase cyclase (AOC), 12-oxophytodienoic acid (OPDA) is produced in chloroplasts. OPDA is transported into peroxisomes and reduced by OPDA reductase 3 and then oxidized by acyl-CoA oxidase 1 to produce JA. JA is finally transported into the cytoplasm and conjugated to Ile by JA - resistant 1 (JAR1) (Wang et al. 2021).

Jasmonyl-isoleucine (JA-Ile), the bioactive form of JA, regulates many physiological and development processes as well as responses to environmental stresses. The receptor for JA-Ile is coronatine insensitive 1 (COI1), an F-box protein of the E3 ubiquitin ligase complex components. COI1 recognizes JA-Ile by forming a receptor complex with the jasmonate zinc-finger inflorescence meristem-domain (JAZ) protein, which serves as a repressor of JA responses. Upon JA-Ile accumulation, JAZ proteins are recruited by COI1, ubiquitinated and rapidly degraded in a proteasome-dependent manner, thus freeing transcriptional factors, such as MYC2, MYC3, MYC4, MYB21, and MYB24, and inducing the expression of JA-responsive genes. In eudicot species such as *Arabidopsis* and tomato, JA-Ile is perceived by a single receptor encoded by *COI1* that interacts with members of the JAZ family. In contrast, monocots such as rice, maize, and wheat have two groups of *COI* genes due to gene duplication. In rice, there are three COI genes: *OsCOI1a* and *OsCOI1b* that are functionally equivalent to the dicotyledons, and *OsCOI2* whose function remains unclear. Previous studies indicated that *OsCOI1a* plays a role in defense responses against rice leaffolder caterpillars (Ye et al. 2017) and resistance to the rice stripe virus, which is transmitted by small brown planthopper (*Laodelphax striatellus*) (Yang et al. 2020).

Recently, genome editing of rice *COI* genes revealed specialized functions for *OsCOI2* in defense response, fertility, and senescence, having a primary role in JA-signaling in rice leaves (Inagaki et al. 2023) and crown root growth inhibition (Nguyen et al. 2023). *OsCOI2* is essential for the JA-induced accumulation of phytoalexins, such as sakuranetin, phytocassanes, and momilactones (Inagaki et al. 2023).

Silencing expression of *OsCOI1* and *OsAOS* attenuated ratoons resistance to rice leaf folder (LF) *Cnaphalocrocis medinalis* but showed no significant effect on the resistance to brown planthopper (BPH) *Nilaparvata lugens*. In addition, these two specialist herbivores induced different suites of defense responses in parent rice plants: LF induced higher OsCOI1 and *OsAOS* transcript levels and JA accumulation, while BPH induced higher *OsPAL1* transcript level and SA accumulation (Deng et al. 2022). The differential activation of signaling pathways elicited by LF and BPH is in line with the prior reports that the JA-mediated defense signaling pathway has a central role in chewing herbivore attack, whereas the SA signaling pathway is more specific to piercing - sucking herbivores (Nguyen et al. 2016). Following this observation, a CRISPR-Cas9 mutation in *CYP71A1*, a rice gene that encodes a cytochrome P450 monooxygenase that converts tryptamine to serotonin, resulted in the suppression of serotonin synthesis and an accumulation of SA levels conferring resistance to BPH (Figure 16.2), while this mutant had similar levels of JA and H_2O_2 with respect to the wild type line (Lu et al. 2018). The broad-spectrum resistance to this insect pest of rice produced by the *CYP71A1* knockout may be exploited in future breeding programs as an alternative or complementary approach to BT transgenic crops.

In spite of 450 million years of independent evolution, the COI1 receptor is functionally conserved between the bryophyte *Marchantia polymorpha* and the eudicot *Arabidopsis thaliana*, but COI1 responds to different ligands in each species. Liverwort (*M. polymorpha*) neither synthesizes nor perceives JA-Ile, and it has been found that COI1 harbors a single amino acid substitution, allowing it to bind to two isomeric forms of the JA-Ile precursor dinor-OPDA but not to JA-Ile (Monte et al. 2018). CRISPR/Cas9-mediated disruption of the allene oxidase synthase (*MpAOS*) gene results in defective OPDA accumulation and reduced defense against the generalist spider mite (*Tetranychus urticae*) (Koeduka et al. 2022). See Figure 16.2.

16.3.4 Early Signaling Events, Transcriptional Factors, and Secondary Metabolites

Early signaling steps following insect perception include depolarization of the plasma transmembrane potential (Vm), rise in cytosolic Ca^{2+}, production of reactive oxygen species (ROS), and mitogen-activated protein kinase (MAPK) activity (Erb and Reymond, 2019). The wounding of a local leaf in *Arabidopsis*, either by insect feeding or by mechanical damage, induces an elevation in cytosolic calcium concentrations ($[Ca^{2+}]_{cyt}$) within seconds at the site of wounding. This change in $[Ca^{2+}]_{cyt}$ propagates signaling to distal leaves through the vasculature, in a process that requires the function of glutamate receptor-like (GLR) genes, especially GLR3.3 and GLR3.6. Although Ca^{2+} signal-regulated JA biosynthesis has been reported, there have been many uncertainties regarding how Ca^{2+} signals initiate JA synthesis during herbivore defense. To address this aspect, a knockout calmodulin gene *CaM2* was generated by CRISPR/Cas9-induced mutation. In addition, the gene that encodes ethylene response factor 16 (*ERF16*), a transcription factor that is rapidly induced by herbivory and functions as a critical regulator of JA biosynthetic genes in tomato, was silenced by virus-induced gene silencing (VIGS) in wild type and glr3.3/3.5 mutants, which are analogs of

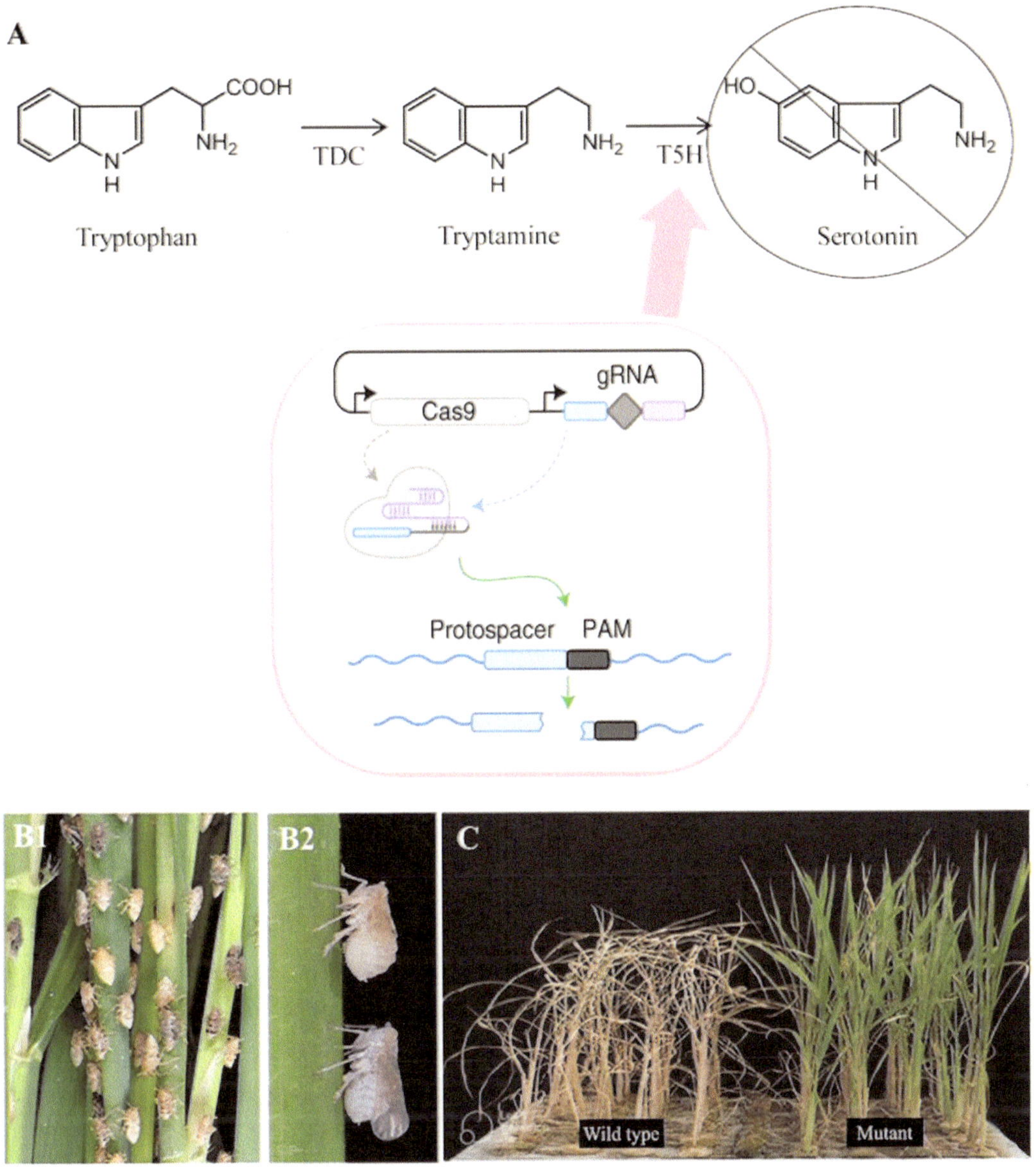

FIGURE 16.2 Mutation in *CYP71A1* confers resistance to brown planthopper (BPH). (A) CRISPR/Cas system for *CYP71A1* knockout mutant, resulting in the absence of serotonin caused by loss of tryptamine 5-hydroxylase (T5H) activity). (B1) Population of BPH (*Nilaparvata lugens*). (B2) Close-up of nymphs (https://apps.lucidcentral.org/pppw). (C) Performance of wild type (Jiazhe line) and mutant rice plants in response to BPH infestation. (A: adapted from Walton et al. (2021); C: Lu et al. (2018).)

glr3.3/3.6 in *Arabidopsis*. The results showed that Ca^{+2} signals induce the expression of *CaM* genes and trigger interaction between ERF16 and *CaM2*, which further enhances the transcriptional activity of ERF16, resulting in rapid ERF16 activation and a JA burst (Hu et al. 2022). These findings reveal a crucial role of interactions between Ca^{+2} signals and ERF16 in regulating JA biosynthesis in plant defense responses.

Myelocytomatosis (MYC) is a subfamily of the basic helix-loop-helix (bHLH) transcription factor family in plants and are essential core regulators in the JA signaling pathway. Members of the Solanaceae family, such as tomato (*Solanum lycopersicum*), potato (*Solanum tuberosum*), native tobacco (*Nicotiana attenuata*), and tobacco (*Nicotiana tabacum*), synthesize structurally diverse specialized metabolites, including phenylpropanoid-polyamine conjugates (PPCs) and steroidal glycoalkaloid (SGA). In tomato and potato, CRISPR/Cas9-mediated mutagenesis has been successfully used to validate genes involved in SGA accumulation. Double *myc1* and *myc2* loss-of-function tomato hairy roots displayed suppressed constitutive expression of SGA biosynthesis (Swinnen et al. 2022). The biosynthetic precursor for SGAs is cholesterol. CRISPR/Cas9 (VQR) genome editing of an endogenous G-box targeted by MYC1/2 in the promoter of the cholesterol biosynthesis gene STEROL C-5(6) desaturase 2 (C5-SD2) leads to decreased constitutive expression of this gene but did not affect its JA inducibility. Moreover, nicotine biosynthesis and transport genes expression, as well as leaf nicotine content, were significantly reduced in *NtMYC2* knockout tobacco plants edited by CRISPR-Cas9, demonstrating that MYC2 acts as a "master switch" to coordinate JA-induced nicotine accumulation against herbivory (Sui et al. 2021).

Both cell wall lignification and flavonoids contribute to resistance against leaf-chewing insects as a physical barrier or by influencing insect behavior and development, respectively. Using the CRISPR/Cas9 system to create UDP-glycosyltransferase (UGT) gene knockout mutants (a 33-bp deletion and a single-bp insertion at the target site, resulting in a frameshift mutation) demonstrated an improvement in soybean resistance against leaf-chewing insects by altering the flavonoid biosynthesis pathway in soybean (*Glycine max*). UGT protein can transfer glycosyl moieties from UDP-sugars to flavonoids during flavonoid biosynthesis. Loss-of-function of *GmUGT*, which catalyzes the final step of flavonoid biosynthesis to form their glycoside derivatives, affects the flavonoid content in soybean attacked by *Helicoverpa armigera* (Y. Zhang et al. 2022).

Callose deposition is induced in plants when they are attacked by herbivores and pathogens. Callose plugging of aphid damaged phloem sieve tubes is expected to reduce aphid access to the phloem sap, while aphid-induced upregulation of callose-degrading β-1,3-glucanase genes in the host plant might counteract this negative effect on aphid performance. In barley, two β-1,3-glucanase genes (1636 and 1639) were previously found to be upregulated by the cereal pest *Rhodalosiphum padi* in susceptible barley genotypes. Mutant lines generated by CRISPR/Cas9 showed significantly reduced glucanase activity, and bacterial flagellin inducer resulted in significantly more callose formation in the leaves of double mutants, compared to control and single mutant lines. However, no effect of these modified plant traits was found on barley resistance to *R. padi* (Kim et al. 2020).

16.4 FUTURE PERSPECTIVES

Plant synthetic biology is an emerging field that combines plant biology with engineering principles to design and produce new devices that exhibit predictable behaviors. Although the first transgenic plant was generated more than 30 years ago, currently, BT crops are the only commercially used genetically modified plants for insect resistance. The application of new technologies in crop protection against pests, either insect pests or diseases, is crucial for the agricultural system of food production. New advances in biotechnology offer opportunities to design plants with novel functions.

The CRISPR/Cas system holds great potential for enhancing plant design and synthetic biology. For instance, artificial DNA sequences, such as promoters, genes, transcriptional regulatory elements, and genome assemblies, can be inserted into plant genomes to modify cell or plant behavior and generate novel functions. An important goal of synthetic biology is to construct regulatory circuits that manipulate plant behavior, producing novel traits that improve crop productivity. In this regard, dCas9-mediated gene regulation via multiplex gene activation, repression, and epigenome editing provides unprecedented possibilities for designing synthetic transcription factors, which could be used to build increasingly complex, programmable, efficient gene circuits.

Since cross-breeding with different varieties is common during commercial product development and seed multiplication, it allows for the exclusion of CRISPR-Cas reagents and potential off-target effects within significantly reduced timelines compared to standard cross-breeding approaches. Several strategies have been devised to enhance the specificity of Cas9-base editors and deliver base editors via ribonucleoprotein (RNP).

Furthermore, there is a recent suggestion that the CRISPR/Cas system could facilitate the accelerated domestication of wild plants, potentially bypassing the thousands of years typically required for crop domestication. This alternative approach would enable the production of new types of food products that were previously unknown to humanity.

REFERENCES

Alok, A., Sandhya, D., Jogam, P., et al. 2020. The rise of the CRISPR/Cpf1 system for efficient genome editing in plants. *Frontiers in Plant Science,* 11, 1–9.

Barton, K.A., Binns, A.N., Matzke, A.J., Chilton, M.D. 1983. Regeneration of intact tobacco plants containing full-length copies of genetically engineered T-DNA, and transmission of T-DNA to R1 progeny. *Cell,* 32, 1033–1043.

Begemann, M.B., Gray, B.N., January, E., et al. 2017. Precise insertion and guided editing of higher plant genomes using Cpf1 CRISPR nucleases. *Science Reports,* 7, 1–6.

Biswas, S., Zhang, D., Shi, J. 2021. CRISPR/Cas systems: Opportunities and challenges for crop breeding. *Plant Cell Reports,* 40, 979–998.

Buchholzer, M., Frommer, W.B. 2023. An increasing number of countries regulate genome editing in crops. *New Phytologist,* 237, 12–15.

Chen, K., Wang, Y., Zhang, R., et al. 2019. CRISPR/Cas genome editing and precision plant breeding in agriculture. *Annual Review of Plant Biology,* 70, 667–697.

Deng, Q.Q., Ye, M., Wu, X.B., et al. 2022. Damage of brown planthopper (BPH) Nilaparvata lugens and rice leaf folder (LF) Cnaphalocrocis medinalis in parent plants leads to distinct resistance in ratoon rice. *Plant Signaling & Behavior,* 17, 1–10.

Engler, C., Kandzia, R., Marillonnet, S. 2008. A one pot, one step, precision cloning method with high throughput capability. *PLoS One*, 3(11), e3647.

Eom, J.S., Luo, D., Atienza-Grande, G., et al. 2019. Diagnostic kit for rice blight resistance. *Nature Biotechnology,* 37, 1372–1379.

Erb, M., Reymond, P. 2019. Molecular interactions between plants and insect herbivores. *Annual Review of Plant Biology,* 70, 527–557.

FAO. 2018. The future of food and agriculture: Alternative pathways to 2050. Food and Agriculture Organization of the United Nations. Available at: www.fao.org/3/I8429EN/i8429en.pdf

Gouhier-Darimont C., Schmiesing, A., Bonnet, C., et al. 2013. Signalling of Arabidopsis thaliana response to Pieris brassicae eggs shares similarities with PAMP-triggered immunity. *Journal of Experimental Botany,* 64, 665–674.

Gouhier-Darimont, C., Stahl, E., Glauser, G., Reymond, P. 2019. The Arabidopsis lectin receptor kinase LecRK-I.8 is involved in insect egg perception. *Frontiers in Plant Science,* 10, 1–10.

Groux, R., Stahl, E., Gouhier-Darimont, C., et al. 2021. Arabidopsis natural variation in insect egg-induced cell death reveals a role for LECTIN RECEPTOR KINASE-I.1. *Plant Physiology,* 185, 240–255.

Guo, C., Ma, X., Gao, F., Guo, Y. 2023. Off-target effects in CRISPR/Cas9 gene editing. *Frontiers in Bioengineering and Biotechnology,* 11, 1–11.

Herrera-Estrella, L., Depicker, A., Van Montagu, M., Schell, J. 1983. Expression of chimeric genes transferred into plant cells using a Ti-plasmid-derived vector. *Nature,* 303, 209–213.

Hu, C., Wu, S., Li, J., et al. 2022. Herbivore-induced Ca2+ signals trigger a jasmonate burst by activating ERF16-mediated expression in tomato. *New Phytologist,* 236, 1796–1808.

Hu, X., Meng, X., Liu, Q., et al. 2018. Increasing the efficiency of CRISPR-Cas9-VQR precise genome editing in rice. *Plant Biotechnology Journal,* 16, 292–297.

Inagaki, H., Hayashi, K., Takaoka, Y., et al. 2023. Genome editing reveals both the crucial role of OsCOI2 in jasmonate signaling and the functional diversity of COI1 homologs in rice. *Plant Cell Physiology,* 64, 405–421.

Jiang, W., Zhou, H., Bi, H., et al. 2013. Demonstration of CRISPR/Cas9/sgRNA-mediated targeted gene modification in Arabidopsis, tobacco, sorghum and rice. *Nucleic Acids Research,* 41, 1–12.

Jinek, M., Chylinski, K., Fonfara, I., et al. 2012. A Programmable Dual-RNA – guided DNA endonuclease in adaptive bacterial immunity. *Science,* 337, 816–822.

Khan, M.Z., Zaidi, S.S., Amin, I., Mansoor, S. 2019. A CRISPR way for fast-forward crop domestication. *Trends Plant Science,* 24, 293–296.

Kim, S.Y., Bengtsson, T., Olsson, N., et al. 2020. Mutations in two aphid-regulated β-1,3-glucanase genes by CRISPR/Cas9 do not increase barley resistance to Rhopalosiphum padi L. *Frontiers in Plant Science,* 11, 1–14.

Kleinstiver, B.P., Prew, M.Sv, Tsai, S.Q., et al. 2015. Engineered CRISPR-Cas9 nucleases with altered PAM specificities. *Nature,* 523, 481–485.

Kleinstiver, B.P., Sousa, A.A., Walton, R.T., et al. 2019. Engineered CRISPR–Cas12a variants with increased activities and improved targeting ranges for gene, epigenetic and base editing. *Nature Biotechnology,* 37, 276–282.

Koeduka, T., Takaishi, M., Suzuki, M., et al. 2022. CRISPR/Cas9-mediated disruption of ALLENE OXIDE SYNTHASE results in defective 12-oxo-phytodienoic acid accumulation and reduced defense against spider mite (Tetranychus urticae) in liverwort (Marchantia polymorpha). *Plant Biotechnology,* 39, 191–194.

Li, C., Li ,W., Zhou, Z., et al. 2020. A new rice breeding method: CRISPR/Cas9 system editing of the Xa13 promoter to cultivate transgene-free bacterial blight-resistant rice. *Plant Biotechnology Journal,* 18, 313–315.

Liu, Q., Yang, F., Zhang, J., et al. 2021. Application of CRISPR/Cas9 in crop quality improvement. *International Journal of Molecular Science,* 22, 1–16.

Lu, H.P., Luo, T.V., Fu, H.W., et al. 2018. Resistance of rice to insect pests mediated by suppression of serotonin biosynthesis. *Nature Plants,* 4, 338–344.

Mishra, R.V., Joshi, R.K., Zhao, K. 2018. Genome editing in rice: Recent advances, challenges, and future implications. *Frontiers in Plant Science,* 9, 1361.

Mojica, F.J.M., Díez-Villaseñor, C., García-Martínez, J., Almendros, C. 2009. Short motif sequences determine the targets of the prokaryotic CRISPR defence system. *Microbiology,* 155, 733–740.

Mojica, F.J.M., Díez-Villaseñor, C., García-Martínez, J., Soria, E. 2005. Intervening sequences of regularly spaced prokaryotic repeats derive from foreign genetic elements. *Journal of Molecular Evolution,* 60, 174–182.

Monte, I., Ishida, S., Zamarreño, A.M., et al. 2018. Ligand-receptor co-evolution shaped the jasmonate pathway in land plants. *Nature Chemical Biology,* 14, 480–488.

Munaweera, T.I.K., Jayawardana, N.U., Rajaratnam, R.V., Dissanayake, N. 2022. Modern plant biotechnology as a strategy in addressing climate change and attaining food security. *Agriculture & Food Security,* 11, 1–28.

Nguyen, D., Rieu, I., Mariani, C., van Dam, N.M. 2016. How plants handle multiple stresses: Hormonal interactions underlying responses to abiotic stress and insect herbivory. *Plant Molecular Biology,* 91, 727–740.

Nguyen, H.T., Cheaib, M., Fournel, M., et al. 2023. Genetic analysis of the rice jasmonate receptors reveals specialized functions for OsCOI2. *PLoS One,* 18, 1–11.

Nishimasu, H., Ran, F.A., Hsu, P.D., et al. 2014. Crystal structure of Cas9 in complex with guide RNA and target DNA. *Cell,* 156, 935–949.

Oliva, R., Ji, C., Atienza-Grande, G., et al. 2019. Broad-spectrum resistance to bacterial blight in rice using genome editing. *Nature Biotechnology,* 37, 1344–1350.

Reymond, P. 2013. Perception, signaling and molecular basis of oviposition-mediated plant responses. *Planta,* 238, 247–258.

Riesenberg, S., Kanis, P., Macak, D., et al. 2023. Efficient high-precision homology-directed repair-dependent genome editing by HDRobust. *Nature Methods,* 20, 1388–1399.

Rozners, E. 2022. Chemical modifications of CRISPR RNAs to improve gene-editing activity and specificity. *Journal of the American Chemical Society,* 144, 12584–12594.

Schepler-Luu, V., Sciallano, C., Stiebner, M., et al. 2023. Genome editing of an African elite rice variety confers resistance against endemic and emerging Xanthomonas oryzae pv. *oryzae strains. eLife,* 12, 1–26.

Schmutz, J., Cannon, S.B., Schlueter, J., et al. 2010. Genome sequence of the paleopolyploid soybean. *Nature,* 463, 178–183.

Sha, G., Sun, P., Kong, X., et al. 2023. Genome editing of a rice CDP-DAG synthase confers multipathogen resistance. *Nature,* 618, 1017–1023.

Sui, X., He, X., Song, Z., et al. 2021. The gene NtMYC2a acts as a 'master switch' in the regulation of JA-induced nicotine accumulation in tobacco. *Plant Biology,* 23, 317–326.

Swinnen, G., De Meyer, M., Pollier, J., et al. 2022. The basic helix–loop–helix transcription factors MYC1 and MYC2 have a dual role in the regulation of constitutive and stress-inducible specialized metabolism in tomato. *New Phytologist,* 236, 911–928.

Tyagi, S., Kesiraju, K., Saakre, M., et al. 2020. Genome editing for resistance to insect pests: An emerging tool for crop improvement. *ACS Omega,* 5, 20674–20683.

Varshney, R.K., Godwin, I.D., Mohapatra, T., et al. 2019. A SWEET solution to rice blight. *Nature Biotechnology,* 37, 1280–1282.

Vazquez-Vilar, M., Orzaez, D., Patron, N. 2018. DNA assembly standards: Setting the low-level programming code for plant biotechnology. *Plant Science,* 273, 33–41.

Walton, R.T., Hsu, J.Y., Joung, J.K., Kleinstiver, B.P. 2021. Scalable characterization of the PAM requirements of CRISPR-Cas enzymes using HT-PAMDA. *Nature Protocols,* 16, 1511–1547.

Wang, L., Jiang, S., Chen, C., et al. 2018. Synthetic genomics: From DNA synthesis to genome design. *Angewandte Chemie International Edition,* 57, 1748–1756.

Wang, N., Tang, C., Fan, X., et al. 2022. Inactivation of a wheat protein kinase gene confers broad-spectrum resistance to rust fungi. *Cell,* 185, 2961–2974.e19.

Wang, Y., Mostafa, S., Zeng, W., Jin, B. 2021. Function and mechanism of jasmonic acid in plant responses to abiotic and biotic stresses. *International Journal of Molecular Sciences,* 22(16), 8568.

Xu, Z., Xu, X., Gong, Q., et al. 2019. Engineering broad-spectrum bacterial blight resistance by simultaneously disrupting variable TALE-binding elements of multiple susceptibility genes in rice. *Molecular Plant,* 12, 1434–1446.

Yang, Z., Huang, Y., Yang, J., et al. 2020. Jasmonate signaling enhances RNA silencing and antiviral defense in rice. *Cell Host Microbe,* 28, 89–103.e8.

Ye, M., Song, Y.Y., Baerson, S.R., et al. 2017. Ratoon rice generated from primed parent plants exhibit enhanced herbivore resistance. *Plant, Cell & Environment,* 40, 779–787.

Zhang, J., Li, J., Saeed, S., et al. 2022. Identification and functional analysis of lncRNA by CRISPR/Cas9 during the cotton response to sap-sucking insect infestation. *Frontiers in Plant Science,* 13, 1–15.

Zhang, Y., Guo, W., Chen, L., et al. 2022. CRISPR/Cas9-mediated targeted mutagenesis of GmUGT enhanced soybean resistance against leaf-chewing insects through flavonoid biosynthesis. *Frontiers in Plant Science,* 13, 1–14.

Zsögön, A., Čermák, T., Naves, E.R., et al. 2018. De novo domestication of wild tomato using genome editing. *Nature Biotechnology*, 36, 1211–1216.

17 Genome-Editing Tools for Abiotic Stress Resistance in Plants

Şükran Yıldız, Selin Doğru, Ilayda Reis, Dilek Tekdal, and Faheem Shahzad Baloch

17.1 INTRODUCTION

The use of genome-editing techniques against abiotic stress in agricultural products has become the focus of great interest in recent years. Genome-editing techniques are based on the principle of changing genes to ensure that plants are resistant to environmental stresses. Abiotic stress factors are various external factors that negatively affect the growth and development of plants. These factors can often disrupt the physiological functions of plants and reduce crop yields. Abiotic stress factors include various stress factors, such as drought, extreme cold, high radiation, salinity, extreme temperature, and chemical compounds in the soil. To combat abiotic stresses, attempts have been made to increase plant productivity by applying some traditional strategies in the past. These include studies such as developing abiotic stress-resistant plant varieties with traditional breeding methods, improving irrigation management, adopting appropriate agricultural practices to increase soil fertility, and developing fertilization techniques. However, some of these studies take time, and some reduce soil fertility or do not seriously impact crop productivity.

In addition to traditional agricultural studies, many genetic studies are carried out to increase the productivity of agricultural products with the developing technology. Genome-editing techniques in these studies enable the DNA of plants to be altered. In this way, unwanted genetic features can be corrected, or desired features can be added. Genome-editing techniques enable effective identification and manipulation of genetic traits, making crops more resistant to abiotic stresses. To balance the ongoing increase in the human population and the parallel increase in food demand, the use of genome-editing techniques in agriculture and the increase in research in this field are critical for the productivity of agricultural products.

17.2 GENOME-EDITING TECHNIQUES USED IN AGRICULTURE

Changing the genetic sequence in a clear, convenient, impeccable and precise way is called genome editing (Sah, 2017). DNA sequences in the genome of all organisms contain a series of messages and instructions. What is desired using genome-editing techniques is indirectly exchanging messages by changing the genetic sequences. This change is possible by activating the cell's natural DNA repair mechanisms after creating cuts or breaks in the DNA, enabling the desired changes to be made. Genome-editing techniques such as zinc finger nuclease (ZFN), transcription activator-like effector nucleases (TALENs), and clustered regularly interspaced short palindromic repeats (CRISPR) are now available. However, CRISPR, one of the fastest, cheapest and most accessible methods, looks suitable for this job (Vidyasagar, 2018; Tastan et al., 2020; Lopes et al., 2021). Gene-editing technology techniques are useful to obtain new varieties of plants to withstand various environmental stresses by simply changing a few nucleotides in the plant's DNA without introducing foreign genes into the plant genome (Malzahn et al., 2017).

17.3 ZINC FINGER NUCLEASE (ZFN)

In recent years, molecular biology has made remarkable advances, and many organisms' genome sequences have been elucidated. In addition, databases containing information on many genes and their expressions have been compiled (Wingender et al., 2001). In addition to studies that will lead to a better understanding of genome structure and function, modern biotechnology has provided various tools that allow controlled change of DNA sequences in plant genomes. One such genome-editing tool is zinc finger nuclease (ZFN). ZFNs, consisting of DNA binding and nuclease domains, recognize specific DNA sequences and thus provide targeted cleavage (Urnov et al., 2010).

Recent advances in zinc finger (ZF) technology make it possible to design and/or select ZF proteins that recognize an 18-bp target sequence within the plant and mammalian genomes (Moore et al., 2001; Beerli and Barbas III, 2002; Hurt et al., 2003). Zinc finger, combining the nonspecific cleavage domain of the seal restriction enzyme and ZF protein nucleases (ZFNs), causes site-specific double-strand breaks (DSBs) within the organism's genome (Li et al., 1992; Kim et al., 1996). Cys2His2 ZF proteins bind DNA by inserting it into the central region of the double helix. Each zinc finger primarily binds to a triple nucleotide sequence within DNA substrates. Amino acid positions in this helix cause sequence-specific interactions with ZF motifs (Pavletich and Pabo, 1991; Kim and Berg, 1996). More ZFs can be added to these

DOI: 10.1201/9781003500933-20

ZF motifs. This allows greater specificity to be achieved. This way, ZF and DNA binding motifs provide an excellent framework for designing ZFNs with site-specific sequence specifications due to their modular nature and structure (Chandrasegaran and Smith, 1999; Kandavelou et al., 2004).

Numerous studies have been conducted using the ZFN genome-editing tool in plants, animals, and microorganisms, and double-strand breaks in DNA have been successfully achieved. For example, using the ZFN genome-editing tool, a mutation frequency of 2% was obtained in the model organism *Arabidopsis*, resulting in deletions ranging from 1 to 80 bp (De Pater et al., 2009). At the same time, a pre-integrated reporter gene containing the ZFN cleavage site was used to examine targeted mutagenesis. This construct was integrated into *Arabidopsis* and subsequent co-culture with an *Agrobacterium* strain harboring the corresponding ZFN expression cassette, resulting in GUS staining as targeted (Tovkach et al., 2009).

17.4 TRANSCRIPTION ACTIVATOR-LIKE EFFECTOR NUCLEASES (TALENS)

Transcription Activator-Like Effector Nucleases (TALENs) technology offers a new, faster, and perfect way to improve crops. TALENs have emerged as alternatives to ZFNs by introducing targeted DSBs for genome editing. TALENs have a structure similar to ZFNs, containing a customizable DNA binding domain and a nonspecific *FokI* nuclease domain fused to it. This DNA binding domain is expressed in *Xanthomonas* spp. It consists of highly conserved repeats derived from transcription activator-like effector proteins (TALEs). They are highly effective tools to alter gene transcription in host plant cells (Boch and Bonas, 2010).

TALENs create double-strand breaks in the targeted gene region in the designated organism and genome, subsequently allowing specific genetic modifications via non-homologous end joining (NHEJ) (Moehle et al., 2007) or homology-directed repair (HDR) (Lieber, 2010). This genome-editing technology has significant benefits for improving agricultural products. As a result of the studies conducted with TALEN, there is also the ability to eliminate unwanted marker genes in transgene integration. Other applications of TALEN include deleting large regions of highly repetitive DNA and producing marker-free agricultural products. Developing gene function and regulation tools is imperative so plants can cope with environmental stresses.

Many studies have been conducted on plants using TALENs, which are used very effectively for genome editing. There are also studies on increasing crop yield and developing plants that are more resistant to environmental stresses. For example, Wendt et al. (2013) used the TALEN system to use a single vector system targeting an endogenous barley gene promoter and applied this vector to immature embryos. Research has primarily resulted in NHEJ leading to small deletions and has achieved a conversion rate of 20% (Wendt et al., 2013). Gurushidze et al. (2014) observed differences in the genome using transgenic barley cells containing *green fluorescent protein* (GFP) lines using two different TALEN arms. The study focused on knocking out GFP in transgenic haploid barley cells in which both TALEN arms were introduced into individual cells. As a result of the study, it was determined that a mutation induction frequency of up to 20% was reached in transgenic barley lines. Interestingly, researchers have found that even in plants where only one of the two TALEN arms is stably integrated, nucleotide losses in the GFP sequence indicate a transient activity of the other TALEN of interest in plants (Gurushidze et al., 2014).

17.5 CLUSTERED REGULARLY INTERSPACED SHORT PALINDROMIC REPEATS (CRISPR)

The CRISPR-Cas system was discovered as the immune system in bacteria and archaea to defend against invading plasmid and/or phage DNA (Sovová et al., 2017; Albitar et al., 2018; Ishino et al., 2018). The CRISPR-Cas system generally consists of two main parts: (1) Cas proteins (involved in protecting invading nucleotides) and (2) CRISPR array consisting of conserved repetitive domains known as direct repeats and sequences known as variable spacers. The CRISPR sequence helps bacteria or archaea to remember the invader during a repeat invasion (Wang et al., 2020). The CRISPR-Cas system works in a specific order by recognizing foreign nucleotides and degrading them. CRISPR-Cas as a defense mechanism has three steps: (1) adaptation: nucleotide fragments of the invading phage are captured by the host organism and inserted into the CRISPR array; (2) expression/crRNA biogenesis: CRISPR array yields a large precursor crRNA (pre-crRNA), cleaved by RNases to produce a mature guide crRN; and (3) interference: finally, mature crRNA directs effector complexes that capture and degrade invading nucleotides and engage the immune system by degrading them (Hille and Charpentier, 2016; Wang et al., 2020).

Scientists working on the CRISPR system discovered in prokaryotes have managed to redesign it under laboratory conditions and used it as a genome-editing tool for different organisms. Thus, a system that allows additions, deletions or changes in the targeted region in the genome of the organism to be used has emerged. In prokaryotes, two different structures, crRNA and tracrRNA, would form a complex with the Cas protein and complement the target region. To make these structures more usable and more superficial and able to be used in genome editing, scientists have obtained a single guide RNA (sgRNA). Cas9 DNA endonuclease and sgRNA create double-strand breaks in the targeted gene region (Mali et al., 2013). The 20-base-long sgRNA determines the specificity of the CRISPR-Cas9 mechanism and acts as a protospacer sequence that allows precise targeting of the desired portion of DNA. The Cas9 molecule acts as molecular scissors and can split target DNA at specific locations (Shalem et al., 2014).

Using CRISPR technology, the yield of agricultural products can be increased, and therefore, more crops can be obtained in less area. Many factors affect efficiency. However, manipulating cytokinin homeostasis is a practical way to increase grain yield. Considering this literature, a group of researchers observed that editing the C terminus of *Oryza sativa* L. LOGL5, which encodes the cytokinin activation enzyme in rice, increases grain yield under various environmental conditions (Wang et al., 2019). Similarly, it has been determined that high-yield phenotypes occur in wheat, a plant with high economic value, by inactivating the gene encoding cytokinin oxidase/dehydrogenase (CKX), an enzyme that catalyzes the degradation of cytokinin (Zhang et al., 2019).

The CRISPR/Cas system has revolutionized the improvement of crops against abiotic stresses by playing important roles. Many genes responsible for abiotic stresses function simultaneously. The CRISPR system plays a vital role in understanding the function of genes by activating or repressing genes that work against abiotic stress (Osakabe et al., 2016). For example, using CRISPR/Cas9 to produce OsDST varieties in the indica mega rice variety MTU1010 is crucial for improving drought and salt tolerance (Santosh Kumar et al., 2020). The Slagamous-LIKE6 (*SlAGL6*) gene was targeted for heat tolerance using the CRISPR/Cas9 genome-editing tool in a study conducted on tomatoes. As a result, it led to increased fruit set under heat stress (Klap et al., 2017).

17.6 GENOME-EDITING STUDIES TO DEVELOP ABIOTIC STRESS VARIETIES AND RESISTANT PLANTS

17.6.1 Salinity Stress in Plants

17.6.1.1 Salt Stress and Its Effects on Plants

Salt stress is an abiotic factor significantly affecting plant growth and development (Doğru and Canavar, 2020). Plants cannot absorb water easily in saline soils due to high osmotic pressure and are also exposed to high concentrations of ions, such as sodium, carbonate and chloride (Korkmaz and Durmaz, 2017). Accumulation of high doses of sodium ions also causes the formation of rectifier oxygen species in plants, and as a result, lipid peroxidation, membrane damage, disorders in photosynthetic activities, and, finally, plant death may occur. At the same time, chlorophyll will also be negatively affected as food intake is restricted due to increased salt stress (Gürsoy, 2023). Salt stress in agricultural ecosystems, with a wide variety of primary and more complex physiological imbalances caused by salinity stress, impairs food production, yield and quality of crops. In addition, permanent and irrecoverable soil degradation may occur, depending on the geo-hydro-morphological and climatological conditions of salinity and periods of extreme salinity. Therefore, every disorder caused by salinity should be considered, as each one will have significant consequences (Ondrasek et al., 2022).

17.6.1.2 Salinity Stress Tolerance Mechanism

The ability of plants to survive in an environment containing high concentrations of various soluble salts is called salt tolerance. Salt tolerance is a mechanism the plant develops against salt stress and varies depending on the plant species, habitat, and environmental conditions. According to their reactions to salt, plants are divided into two groups: halophytes can adapt to high salt concentrations and survive, and glycophytes are plants sensitive to salt and cannot survive in salty environments (Yilmaz, 2011). There are mechanisms that plants have developed against salt tolerance. One of these is the function of the SOS (Salt-Overly-Sensitive) signaling pathway in maintaining ion homeostasis and controlling salt tolerance in plants. The SOS signaling pathway and SOS proteins play a role in preventing the plant from damage that may occur as a result of excess ion accumulation. Another mechanism is the biosynthesis of osmolytes and osmotic adjustment. In this mechanism, plants make the necessary osmotic adjustment by synthesizing protective compounds such as proline, glycine, and soluble sugars in response to osmotic stress resulting from exposure to high salt. In the latter mechanism, high salt levels cause changes in the expression of many genes in plants, usually seen as increased expression. Therefore, to develop tolerance to salinity in plants, changing the expression of regulatory genes such as protein kinases, protein phosphatases and transcription factors is a fundamental mechanism due to both the activation of stress signaling and the regulation of many downstream genes (Sevgi and Leblebici, 2023).

17.6.1.3 Some Genetic Studies Based on Salinity Stress in Plants

Genetic studies under salt stress in plants have generally been conducted on gene expression. In a study, the potential effects of MYB4 on salt tolerance were investigated in potatoes transformed with the *Osmyb4* gene. The study performed gene expression analyses via total RNA isolation and qRT-PCR. As a result of this study, it was observed that the expression amounts of some stress-related genes increased, and it was concluded that the MYB4 transcription factor may be effective in regulating the stress tolerance mechanism in potatoes (Aydın, 2020). Again, in a study on potatoes, the expression level of the *Asg1* gene, which is related to stomatal resistance, was examined in potatoes under salinity stress. Expression-level analysis of the gene was performed by total RNA isolation and real-time PCR application. As a result, it was determined that the *Asg1* gene plays a role in the plant protection mechanism against salt stress (Uranbey et al., 2017). Genome analysis of the Phospholipase D (*PLD*) gene family was carried out in the bean (*Phaseolus vulgaris* L.) plant under salt and drought stress. As a result of *in silico* analyses, it was determined that there was a significant increase in gene expression of 4 *PLD* genes in the *P. vulgaris* genome under drought and salt stress compared to the control group. (Isıyel et al., 2022). In 2019, a gene-editing study was conducted in the rice (*Oryza sativa* L.) plant to increase crop yield under salinity stress. In this study, abscisic acid signaling network genes were edited to increase grain yield using the CRISPR/Cas9 technique, and thanks to the study, rice resistant to growth in salty environments was obtained (Farhat et al., 2019). In another study conducted in 2023, the roles of CRISPR against drought and salinity stress in plants were investigated. As a result of the study, it was concluded that CRISPR application is a groundbreaking technique for making crops resistant to abiotic stresses, developing mutant plants, and editing genomes (Angon et al., 2023).

17.6.2 Heavy Metal Stress in Plants

17.6.2.1 Heavy Metal Stress and Its Effects on Plants

Heavy metals have concentrations higher than 5 g/cm^3 (Sevgi end Leblebici, 2022). Heavy metals can be divided into two classes: (1) essential, and (2) unnecessary elements for the growth and development of plants. Variable concentrations of elements in both classes affect plant growth directly or indirectly. Heavy metals that affect plant development have properties that can change the physical and chemical structure of the soil, accumulating on the surfaces of the above-ground organs of the plant, entering from the soil into the roots and from the roots into other tissues and cells of the plant, causing morphological, molecular, physiological and biochemical changes (Acar et al., 2022). At the same time, the effect of the damage of heavy metals to the plant depends on the oxidation stage. Heavy metals entering the plant cause toxic effects, causing the death of the plant in the long term. Toxic heavy metals are As, Cd, Cr, Cu, Hg, Ni, Pb, and Zn (Özyiğit, 2021).

17.6.2.2 Heavy Metal Stress Tolerance Mechanism

When plants are exposed to heavy metals for a long time, they try to remove them from their bodies through various methods. For example, defoliation and shedding of bark in trees are ways of removal such as changing the structure of some metals and metalloids by the plant, the plant accumulating heavy metals in the cuticle layer and producing peroxidase (POD) against stress (Yalçın, 2014). Depending on the genotype, plants have developed tolerance mechanisms against stress to survive outside these pathways. One of these mechanisms is the antioxidant defense system. The antioxidant defense system is divided into two classes: (1) enzymatic, and (2) non-enzymatic. The enzymatic defense system has the task of clearing reactive oxygen species (ROS) formed in cells. The non-enzymatic defense system reduces molecules such as glutathione, ascorbate, flavonoids, lignins, and organic acids. Another mechanism is the binding system of metals to ligands. In this system, ligands enable the transport and storage of metals in plant tissues and contribute to tolerance to heavy metal stress by forming a complex structure called chelate by binding to two or more ligands (Acar et al., 2022).

17.6.2.3 Some Genetic Studies Based on Heavy Metal Stress in Plants

Genetic studies conducted on heavy metal stress in plants include research such as epigenetic controls, identification of gene families and gene expression analyses. The genetic and epigenetic effects of corn (*Zea mays* L.) seeds under heavy metal stress were examined in a study. Zinc (Zn) was used as a heavy metal, and *Z. mays* was examined using genotoxic and epigenetic analyses, mitotic index analyses, and the *coupled restriction enzyme digestion and random amplification* (CRED-RA) technique. As a result of these analyses, it was observed that zinc caused abnormalities in the mitotic phase, changes in DNA methylation through its epigenetic effect, and changes in phytohormone content (Erturk et al., 2015). Another study identified the gene family involved in heavy metal stress in potato (*Solanum tuberosum* L.) plants. As a result of the study, the *NRAMP* gene family was identified and found to be involved in heavy metal stress. It has also been determined that there are similar conserved regions among *NRAMP* gene family members (Tian et al., 2021). Studies have been conducted in which cadmium-resistant plants were developed with the CRISPR/Cas9 technique, and gene regulations were conducted for heavy metal stress tolerance. Transporter genes such as *OsLCT1* and *OsNramp5* are involved in cadmium absorption in rice. Studies have been conducted to reduce the amount of elements such as cadmium and lead in rice by manipulating and gene editing with CRISPR/Cas9 (Yadav et al., 2023).

17.6.3 Deficient Oxygen Stress in Plants

17.6.3.1 Deficient Oxygen Stress and Its Effects on Plants

Heavy rains caused by climate changes cause plants to remain under water; in this case, plants cannot receive the required amount of oxygen for their tissues and cells (Limami et al., 2014). In the case of oxygen deficiency, the plant tries to survive by anaerobic respiration. The decrease in the amount of oxygen causes a decrease in the synthesis of ATP, which is synthesized by the plant and provides energy (Pucciariello and Perata, 2012). However, the energy is too low to support respiration and causes cell and tissue death in the plant (Limami et al., 2014). Fluctuations in the amount of oxygen in plants lead to changes in the redox balance and the formation of reactive oxygen/nitrogen species. Although these reagents help plants respond to stress, their excessive accumulation also causes plant damage and death (Sasidharan et al., 2021).

17.6.3.2 Tolerance Mechanisms Against Oxygen Stress in Plants

Plants have developed some morphological adaptations in response to insufficient oxygen. These are thick root hypoderm, aerenchyma, lentils to provide oxygen exchange, stem elongation and small roots emerging from the stem. In addition to morphological adaptations, plants also have physiological adaptations to oxygen stress. These physiological changes are seen as an increase in the stomatal resistance of the plant, a decrease in photosynthesis and root conductance, aging of the leaves, and an improvement in phytohormones (Nasrullah, 2022). Plants have also created adaptations in their metabolism against oxygen deficiency. These are generally anaerobic carbon metabolism. It has also been determined that nitrogen metabolism is an adaptation to low oxygen stress in plants (Limami et al., 2014). When it detects insufficient oxygen in the plant, it produces reactive oxygen derivatives (ROS) in response (Pucciariello and Perata, 2012).

17.6.3.3 Some Genetic Studies Based on Oxygen Stress in Plants

In a study conducted by Christianson et al. (2009), they examined the response of the *Arabidopsis thaliana* (L.) Heynh seeds to low oxygen stress during the germination period. As a result of this study, it was observed that the NAC domain transcription factor *ANAC102* gene, which is induced by low oxygen during the germination period of the seed, showed overexpression, and this gene proved to be an essential regulator against oxygen stress during the germination period of the *A. thaliana* plant (Christianson et al., 2009). In another study conducted on the *Arabidopsis* plant, the expression levels of *RAP2.12*, *RAP2.2*, *RAP2.3*, and ethylene response factor VII genes were examined under oxygen stress. High expression levels of these genes have been observed to increase the survival probability of *Arabidopsis* plants in low oxygen conditions and even provide tolerance to osmotic and oxidative stresses (Papdi et al., 2015). Another study using highly tolerant, moderately tolerant, and intolerant plants was conducted to look at comparative gene expression levels using the gene set enrichment analysis (GSEA) method. As a result of this study, conducted under low oxygen stress, they observed that each species rearranged energy metabolic pathways, such as glycolysis, fermentation and nitrogen metabolism, and starch metabolism, and that plant species created new metabolic pathways to cope with stress (Hwang et al., 2020).

17.6.4 Soil pH Effects

17.6.4.1 Soil pH and Its Effects on Plants

Soil pH is the main component of soil chemistry as it determines the bioavailability and mobility of elements such as plant nutrients and phytotoxic compounds (Penn and Camberato, 2019). Soil pH range varies depending on plant type. Lower or upper pH limit values may slow plant development or reduce product yield. Soil pH is essential for plants because the elements they need become soluble according to the pH value in the soil. For example, while creating a toxic effect by increasing elements such as Al, Fe, and Mn in acid soil and causing a decrease in elements such as K, Mg, and P, Alkaline soils do not benefit the plant because they cause phosphorus, an essential plant nutrient, to become insoluble with calcium (Bilen, 1993).

17.6.4.2 pH Stress-Related Tolerance System and Genetic Studies in Plants

The pH range must be between 6 and 8 for plants to produce crops. The pH needs to remain optimal for the sustainability of product yield and the continuation of vital events in the plant. Studies have observed that plants increase their endogenous melatonin (MT) levels due to the stress they experience due to fluctuations in the pH level to which they are tolerant. They also determined that MT has a healing effect on pH stress in the plant and a protective role in crop yield (Farag et al., 2020). In a study conducted on tomato (*Solanum lycopersicum* L.) plant, it was determined that silicon (Si) and salicylic acid (SA) increased in plant seedlings in their tolerance mechanism against pH stress and prevented reactive oxygen accumulation and lipid peroxidation (Khan et al., 2019). The low pH and aluminum tolerance study on rice beans (*Vigna umbellata* Thunb.) in 2016 examined the Formate Dehydrogenase (FDH) expression level. The study found that overexpression of FDH improved plant performance under low pH stress (Lou et al., 2016). In a study conducted in 2022, regulatory gene identification research was conducted using the CRISPR/Cas9 technique in rice varieties in lands adapted to acidic soils (Osakabe et al., 2016). As a result of the study, *ART1*, *DRO1*, *SUB1A*, *PSTOL1*, and *SPDT* genes were found and revealed that they contain mechanisms related to the physiological response to soil pH and change the root growth angle in response to soil pH (Barrero et al., 2022).

17.6.5 Effect of UV on Plants

Investigation of the ozone layer every day causes the amount of UV rays reaching the Earth to increase. The number of UV-B rays reaching the Earth is increasing, and, in parallel, the rate of global warming is also increasing. This negatively affects plant growth and development. The ultraviolet (UV) light spectrum and UV-B stress include wavelengths of the light spectrum between 200–400 nm. UV is examined in three groups according to wavelength ranges (Tevini, 2004). The most dangerous UV-C for living things is between 200–280 nm and does not reach the Earth under normal conditions. UV-B between 280–320 nm, which causes harmful effects on plants. UV-A between 320–400 nm has the most minor effect. Plants make several changes in their leaf structure and cells to protect against UV-B radiation. These changes are due to plants producing compounds that absorb UV-B. Flavonoids, in particular, absorb UV-B and accumulate in the leaf epidermis. These compounds act as an effective filter to prevent UV-B radiation from damaging the DNA and the photosynthetic apparatus. Flavonoids are also known for their ability to scavenge free radicals. Plant tolerance to UV-B depends on developing these protection mechanisms and involves leaf morphology and physiological and biochemical processes in leaf cells (Láposi et al., 2009). Sensitivity or resistance to UV-B radiation varies among families and has been reported to be quite different, even among species within the same family (Yao and Liu, 2007; Ren et al., 2010).

Increasing UV-B radiation reduces the amount of ATPase in the cell membrane; however, it has been determined that membrane permeability increases and unsaturated fatty acids are affected (An et al., 2000). High UV radiation affects light reactions and the loss of Calvin cycle enzymes and stomatal closure, disrupting chloroplast integrity (Keiller et al., 2003).

17.6.5.1 UV Stress on Plant and Genome Editing

UV Resistance Locus 8 (UVR8) is a UV-B photoreceptor crucial in regulating various aspects of plant photomorphogenesis.

Although the physiological functions of UVR8 have been extensively studied in *Arabidopsis*, the functions of Tomato UVR8 (SlUVR8) remain largely unknown. In order to gain insight into the physiological functions of SlUVR8, the CRISPR-Cas9 gene-editing approach is used to generate sluvr8 knockout mutant lines. The investigation revealed that SlUVR8 regulates hypocotyl elongation and anthocyanin accumulation in seedlings exposed to UV-B radiation. These findings suggest that SlUVR8 plays a role in the early stages of tomato seedling development (Yin et al., 2017; Deng et al., 2018).

Furthermore, it was observed that SlUVR8 is essential for acclimatizing Tomato plants to low levels of UV-B radiation and enhancing their tolerance to elevated UV-B stress. The crucial roles of SlUVR8 in regulating Tomato seedling development and UV-B stress tolerance indicate that manipulating this photoreceptor could potentially offer a powerful tool to enhance the performance of tomato plants in natural environments where high doses of UV-B radiation are present (Yin et al., 2017; Deng et al., 2018).

In conclusion, the study sheds light on the previously unknown physiological functions of the Tomato UVR8 (SlUVR8) photoreceptor. The findings highlight the importance of SlUVR8 in regulating seedling development and stress responses in Tomato plants, particularly concerning UV-B radiation. This knowledge opens up new avenues for future research and presents opportunities for using SlUVR8 as a target for improving Tomato plant performance in environments with high UV-B exposure (Yin et al., 2017; Deng et al., 2018).

17.6.6 Effects of Water Scarcity on Plants and Drought-Resistance Strategies

Water scarcity causes reductions in plants' growth process, leading to shrinkage of leaves and reduced photosynthetic production. Lack of water also negatively affects the transport of substances in the xylem and phloem, causing fruits to remain small, seeds in grains not being able to become plump, and product quality to decrease. Water stress in plants has a significant effect on enzyme activity and amount. In addition, while Abscisic acid (ABA), a plant hormone, increases 40-fold in the leaves, this increase is less in other organs, including the root. Abscisic acid prevents water loss by closing the stomata and allows water to be used in the root system by reducing the development of the top organs of the plant. Plant development and growth depend on the continuity of cell division, growth, and differentiation. Water stress affects metabolic events and cell development in plants. While the growth and division of cells decrease with water stress, it causes the plant's above-ground parts to shrink and decrease (Hsiao, 1973). When water potential is low, chlorophyll formation is prevented, and the photosynthetic efficiency of the plant decreases (Neumann et al., 1988). Likewise, water stress reduces the activity of many enzymes, causing plants' metabolic events to slow down. For example, nitrogen fixation in legumes is reduced by the decrease in the enzyme nitrate reductase (Begg and Turner, 1976). Due to decreased water potential, turgor pressure in leaf cells decreases and abscisic acid level increases. In this case, the stomata close, which reduces the plant's photosynthetic activity (Davies et al., 1994).

17.6.7 Effects of High Temperature Stress on Plants and Genome Editing

The growth and productivity of plants occur within a specific temperature range. Any increase or decrease outside this temperature range can adversely affect the development and productivity of plants. The most suitable temperature range for plants varies depending on the plant species, and this feature is called the preferred temperature of plants. In addition, one of the abiotic factors plants face is heating stress. Heating stress is a critical situation that negatively affects the growth and development of plants. Heating stress tremendously impacts agricultural production and can significantly reduce crop yields. For example, every 1°C increase in atmospheric temperature reduces wheat yield by 6%, rice yield by 10–20%, and corn yield by 21–31%. This shows the negative impact of climate change and increasing air temperatures on agricultural production. Therefore, the impact of climate change and increasing temperatures should be considered in plant cultivation and agricultural production. To cope with these effects, farmers and agronomists should select heat stress-resistant varieties of plants and implement appropriate agricultural practices. It is also essential to study sustainable agricultural methods and innovative solutions to combat climate change (Asseng et al., 2015; Yang et al., 2017; Wang et al., 2019).

Some molecular processes that respond to heat stress are governed by heat shock transcription factors (HSFs) and heat shock proteins (HSPs). These factors and proteins play essential roles in helping plants respond to heat stress. It can also increase the heat tolerance of plants by providing resistance to ROS (reactive oxygen species) compounds and improving the antioxidant properties in the plant. Metabolite production and temperature-dependent gene expression are crucial for increasing plant temperature tolerance. Thanks to these processes, plants can better adapt to heat stress and increase their chances of survival. One of the newest and most effective technologies to understand in more detail the molecular processes associated with heat stress and improve the heat tolerance of plants is the CRISPR-Cas9 method. This method is the most advanced of all the genome-editing techniques and allows changes to be made directly to the plant genome. CRISPR-Cas9 is an essential tool for regulating genes that help plants become more resistant to heat stress and for increasing the plant's heat tolerance by interfering with various molecular targets (Awasthi et al., 2015; Duan et al., 2016; Parmar et al., 2017).

It has been determined that CRISPR-edited mutant plants have improved agronomic properties and are resistant to abiotic stresses. A study examined the effect of the *BRZ1* gene as a component of heat tolerance in tomato plants. As a result of the research, it was determined that *BRZ1* positively regulates the formation of ROS in the apoplastic region of tomato plants. As a result, these mutant plants appeared to have reduced temperature tolerance and respiratory burst oxidase homolog 1 (*RBOH1*) levels, along with reduced hydrogen peroxide production in the apoplast. These findings were confirmed with CRISPR-Cas9-based bzr1 mutants (Yin et al., 2018).

17.6.8 Effects of Cold Stress on Plant Developmental Stages

Cold sensitivity in plants affected by low temperatures depends on the developmental phase of the plant. The seeds of these plants are pretty tolerant to long-term low temperatures due to their low moisture content and dormancy. However, when water meets the seed, the seed may become sensitive to cold. Cold may delay water absorption or disrupt membrane integrity during initial absorption, increasing electrolyte (ion) leakage and preventing germination. All these effects show that cold can negatively affect the vital processes of plants. Every stage of the reproductive phase, namely the formation of reproductive organs, flowering, and seed formation, is sensitive to cold stress. Research has shown that pollen germination and tube formation are suppressed at low temperatures. Studies by Clarke and Siddique (2004) revealed that cold stress negatively affects the functions of reproductive organs.

Additionally, any defects in the male or female parts of the flower may also result from cold stress, resulting in flower abscission. According to the results of a study conducted by Nayyar et al. (2005), cold stress is one of the most critical factors of flower abscission. Therefore, it is vital for the reproductive stage to be protected against cold stress and to grow under suitable conditions (Nayyar et al., 2005).

PRPs are proline-rich proteins that, in addition to helping combat low temperatures, can reduce nutrient loss, increase antioxidant activity, and aid chlorophyll production. These proteins help increase the rice's ability to withstand cold. A protein called OsPRP1 has been used to increase cold tolerance in the rice plant. As a result of removing OsPRP1 from the genetic structure using CRISPR/Cas9, the cold resistance of rice was increased. This study provides a new approach to cold tolerance for rice plants (Shakiba et al., 2017).

17.7 CONCLUSION

Considering that the world population is increasing, there is a need to increase agricultural products in parallel. However, climate crises and the accompanying environmental stresses cause plant productivity losses. In order to meet the increasing food demand, the benefits of the Green Revolution to humanity are gradually decreasing. It has become necessary to obtain plants resistant to adverse environmental conditions. Producing new plant varieties that cannot be obtained using traditional methods is laborious and takes time. Scientists have produced revolutionary genome-editing tools in order to obtain plant varieties that are more innovative, faster, and more cost-effective, with higher yields and more resistance to environmental stresses. Genome-editing techniques such as ZFN, TALEN and the most rational of the moment, CRISPR, are unique tools designed to produce the plants of our age and future. In order to meet food demand using these technologies, more studies are needed to develop plants that are resistant to environmental stresses and can produce high yields.

REFERENCES

Acar, O., ÖztÜrk, E., Dinler, B.S. et al. 2022. *Tarım bilimleri alanında multidisipliner güncel çalışmalar*, vol. 6, pp. 165–192. İksad Publishing House. (In Turkish)

Albitar, A., Rohani, B., Will, B. et al. 2018. The application of CRISPR/Cas technology to efficiently model complex cancer genomes in stem cells. *Journal of Cellular Biochemistry,* 119, 134–140.

An, L., Feng, H., Tang, X., et al. 2000. Changes of microsomal membrane properties in spring wheat leaves (Triticum aestivum L.) exposed to enhanced ultraviolet-B radiation. *Journal of Photochemistry and Photobiology Biology,* 57, 60–65.

Angon, P., Mondal, S., Akter, S. et al. 2023. Roles of CRISPR to mitigate drought and salinity stresses on plants. *Plant Stress,* 8, 100169.

Asseng, S, . Ewert, F., Martre, P. et al. 2015. Rising temperatures reduce global wheat production. *Nature Climate Change,* 5, 143–147.

Awasthi, R., Bhandari, K., and Nayyar, H. 2015. Temperature stress and redox homeostasis in agricultural crops. *Frontiers in Environmental Science,* 3, 11.

Aydın, G. 2020. Oryza sativa Osmyb4 geni aktarılmış transgenik patateste Osmyb4 gen ifadesinin tuzluluk toleransına etkisi. *Anadolu Journal of Agricultural Sciences,* 35, 115–123. (In Turkish)

Barrero, L.S., Willmann, M.R., Craftv E.J. et al. 2022. Identifying genes associated with abiotic stress tolerance suitable for CRISPR/Cas9 editing in upland rice cultivars adapted to acid soils. *Plant Direct,* 6, 12.

Beerli, R.R. and Barbas III, C.F. 2002. Engineering polydactyl zinc-finger transcription factors. *Nature Biotechnology*, 20(2), 135–141.

Begg, J.E. and Turner, N.C. 1976. Crop water deficits. *Advances in Agronomy,* 28, 161–217.

Bilen, S., and Sezen, Y. 1993. Toprak reaksiyonunun bitki besin elementleri elverişliliği üzerine etkileri. *Atatürk Üniversitesi Ziraat Fakültesi Dergisi*, 24(2), 156–166. (In Turkish)

Boch, J. and Bonas, U. 2010. Xanthomonas AvrBs3 family-Type III effectors, *Discovery and Function,* 48, 419–436.

Chandrasegaran, S. and Smithv, J. 1999. Chimeric restriction enzymes, What is next? *Biological Chemistry,* 380, 841–848.

Christianson, J.A., Wilson, I.W., Llewellyn, D.J. et al. 2009. The low-oxygen-induced NAC domain transcription factor ANAC102 affects viability of Arabidopsis seeds following low-oxygen treatment1[W][OA]. *Plant Physiology,* 149, 1724–1738.

Clarke, H.J. and Siddique, K.H.M. 2004. Response of chickpea genotypes to low temperature stress during reproductive development. *Field Crops Research,* 90, 323–334.

Davies, W.J., Tardieu, F., and Trejo, C.L. 1994. How do chemical signals work in plants that grow in drying soil? *Plant Physiology,* 104, 309–314.

Deng, L., Wang, H., Sun, C. et al. 2018. Efficient generation of pink-fruited tomatoes using CRISPR/Cas9 system. *Journal Genetics and Genomics,* 20, 51–54.

De Pater, Sv, Neuteboom, L.W., Pinas, J.E. et al. 2009. ZFN-induced mutagenesis and gene-targeting in Arabidopsis through Agrobacterium-mediated floral dip transformation. *Plant Biotechnology Journal,* 7, 821–835.

Doğru, A. and Canavar, S. 2020. Bitkilerde tuz toleransının fizyolojik ve biyokimyasal bileşenleri. *Academic Platform Journal of Engineering and Science*, 8(1), 155–174. (In Turkish)

Duan, Y.B., Li J., Qin, R.Y. et al. 2016. Identification of a regulatory element responsible for salt induction of rice OsRAV2 through ex situ and in situ promoter analysis. *Plant Molecular Biology*, 90(1–2), 49–62.

Erturk, F.A., Agar, G., Arslan, E. et al. 2015. Analysis of genetic and epigenetic effects of maize seeds in response to heavy metal (Zn) stress. *Environmental Science and Pollution Research,* 22, 10291–10297.

Farag, M., Elkelish, A., Dafea, M., Khan, M. et al. 2020. Role of melatonin in plant tolerance to soil stressors: Salinity, pH and heavy metals. *Molecules*, 25(22): 5359.

Farhat, S., Jain, N., Singh, N. et al. 2019. CRISPR-Cas9 directed genome engineering for enhancing salt stress tolerance in rice. *Seminars in Cell and Developmental Biology,* 96, 91–99.

Gürsoy, M. 2023. *Tuz stresinin bitki gelişimine etkilerine genel bakış*. Paper presented at 9th International Zeugma Conference on Scientific Research, pp. 508–513.

Gurushidze, M., Hensel, G., Hiekel, S. et al. 2014. True-breeding targeted gene knock-out in barley using designer TALE-nuclease in haploid cells. *PLoS ONE,* 9, 3.

Hille, Fv and Charpentier, E. 2016. CRISPR-Cas: Biology, mechanisms and relevance. *Philosophical Transactions of the Royal Society of London. Series B, Biological Sciences*, 371, 1707.

Hsiao, T.C. 1973. Plant responses to water stress. *Annual Review of Plant Physiology,* 24, 519–570.

Hurt, J.A., Thibodeauc, A.S., Hirsh, C.O. et al. 2003. Highly specific zinc finger proteins obtained by directed domain shuffling and cell-based selection. *Proceedings of the National Academy of Sciences,* 100, 12271–12276.

Hwang, J.H., Yu, S., Lee, B. et al. 2020. Modulation of energy metabolism is important for low-oxygen stress adaptation in Brassicaceae species. *International Journal of Molecular Sciences,* 21, 1787.

Ishino, Y., Krupovic, M. and Forterre, P. 2018. History of CRISPR-Cas from encounter with a mysterious repeated sequence to genome editing technology. *Journal of Bacteriology,* 200, 7.

Isıyel, M., Öner, B.M., Yaprak, E. et al. 2022. Fasulye bitkisinde phospholipase d gen ailesinin tuz ve kuraklık stresi altında genom çaplı karakterizasyonu. *European Journal of Science and Technology Special Issue,* 34, 585–593. (In Turkish)

Kandavelou, K., Mani, M., Durai, S. et al. 2004. Engineering and applications of chimeric nucleases. *Restriction Endonucleases,* 413–434.

Keiller, D.R., Mackerness, S.A.H. and Holmes, M.G. 2003. The action of a range of supplementary ultraviolet (UV) wavelengths on photosynthesis in Brassica napus L. in the natural environment: effects on PS II, CO(2) assimilation and level of chloroplast proteins. *Photosynthesis Research,* 75, 139–150.

Khan, A., Kamran, M., Imran, M. et al. 2019. Silicon and salicylic acid confer high-pH stress tolerance in tomato seedlings. *Scientific Reports,* 9, 1.

Kim, C. and Berg, J.M. 1996. A 2.2 Å resolution crystal structure of a designed zinc finger protein bound to DNA. *Nature Structural Biology*, 3(11), 940–945.

Kim, Y.G., Cha, J. and Chandrasegaran, S. 1996. Hybrid restriction enzymes: zinc finger fusions to Fok I cleavage domain. *Proceedings of the National Academy of Sciences of the USA,* 93, 1156.

Klap, C., Yeshayahou, E., Bolger, A.M. et al. 2017. Tomato facultative parthenocarpy results from SlAGAMOUS-LIKE 6 loss of function. *Plant Biotechnology Journal,* 15, 634–647.

Korkmaz, H. and Durmaz, A. 2017. Bitkilerin abiyotik stres faktörlerine verdiği cevaplar. *GÜFBED,* 7, 192–207. (In Turkish)

Láposi, R., Veres, S., Lakatos, G., Olah, V. et al. 2009. Responses of leaf traits of European beech (Fagus sylvatica L.) saplings to supplemental UV-B radiation and UV-B exclusion. *Agricultural and Forest Meteorology,* 149, 745–755.

Li, L., Wv, L.P. and Chandrasegaran, S. 1992. Functional domains in Fok I restriction endonuclease. *Proceedings of the National Academy of Sciences*, 89(10), 4275–4279.

Lieber, M.R. 2010. NHEJ and its backup pathways in chromosomal translocations. *Nature Structural and Molecular Biology,* 17, 393–395.

Limami, A.M., Diab, H. and Lothier, J. 2014. Nitrogen metabolism in plants under low oxygen stress. *Planta,* 239, 531–541.

Lopes, J.H., da Silva, V.C.H., dos Santos, J.C. et al. 2021. *Introduction to genome editing in plants.* In CRISPR technology in genomic plant editing. pp. 11–46-.Embrapa Brasilia,.

Lou, H.Q., Gong, Y.L., Fan, W. et al. 2016. A formate dehydrogenase confers tolerance to aluminum and low pH. *Plant Physiology.* 171, 294–305.

Mali, P., Esvelt, K.M. and Church, G.M. 2013. Cas9 as a versatile tool for engineering biology. *Nature Methods,* 10, 957–963.

Malzahn, A., Lowder, L., and Qi, Y. 2017. Plant genome editing with TALEN and CRISPR. *Cell and Bioscience*, 7(1), 1–18.

Moehle, E.A., Rock, J.M., Lee, Y.L. et al. 2007. Targeted gene addition into a specified location in the human genome using designed zinc finger nucleases. *Proceedings of the National Academy of Sciences of the USA,* 104, 3055–3060.

Moore, M., Klug, A. and Choo, Y. 2001. Improved DNA binding specificity from poly zinc finger peptides by using strings of two-finger units. *Proceedings of the National Academy of Sciences,* 98, 1437–1441.

Nasrullah, S., Umar, M., Sun, L. et al. 2022. Flooding tolerance in plants: From physiological and molecular perspectives. *Brazilian Journal of Botany*, 45(4), 1161–1176.

Nayyar, H., Bains, T. and Kumar, S. 2005. Low temperature induced floral abortion in chickpea: Relationship to abscisic acid and cryoprotectants in reproductive organs. *Environmental and Experimental Botany,* 53, 39–47.

Neumann, P.Mv, Van Volkenburgh, E. and Cleland, R.E. 1988. Salinity stress inhibits bean leaf expansion by reducing turgor, not wall extensibility. Plant Physiology, 88, 233–237.

Ondrasek, G., Rathod, S., Manohara, K.K. et al. 2022. Salt stress in plants and mitigation approaches. *Plants,* 11, 717.

Osakabe, Y., Watanabe, T., Sugano, S.S. et al. 2016. Optimization of CRISPR/Cas9 genome editing to modify abiotic stress responses in plants. *Scientific Reports*, 6(1), 26685.

Özyiğit, İİ. 2021. Tarım topraklarında ağır metaller; kökenleri, yayılışları ve etkileri. *Ereğli Journal of Agricultural Science*. 1(1), 46–71.

Papdi, C., Pérez-Salamó, I., Joseph, M.P. et al. 2015. The low oxygen, oxidative and osmotic stress responses synergistically act through the ethylene response factor VII genes RAP2.12, RAP2.2 and RAP2.3. *The Plant Journal,* 82, 772–784.

Parmar, N., Singh, K.H., Sharma, D. et al. 2017. Genetic engineering strategies for biotic and abiotic stress tolerance and quality enhancement in horticultural crops: A comprehensive review. *3 Biotech,* 7, 1–35.

Pavletich, N.P. and Pabo, C.O. 1991. Zinc Finger-DNA recognition: Crystal structure of a Zif268-DNA complex at 2.1 *Å*. *Science,* 252, 809–817.

Penn, C.J. and Camberato, J.J. 2019. A critical review on soil chemical processes that control how Soil pH affects phosphorus availability to plants. *Agriculture,* 9, 120.

Pucciariello, C. and Perata, P. 2012. How plants sense low oxygen. *Plant Signaling and Behavior*, 7(7), 813–816.

Ren, J., Yao, Y., Yang, Y. et al. 2010. Differences in growth and physiological traits of two poplars originating from different altitudes as affected by UV-B radiation and nutrient availability. *Physiologia Plantarum,* 138, 278–288.

Sah, N.K. 2017. Scope of plant biotechnology in the developing countries. *Plant Biotechnology,* 1, 45–66.

Santosh Kumar, V.V., Verma, R.K., Yadav, S.K. et al. 2020. CRISPR-Cas9 mediated genome editing of drought and salt tolerance (OsDST) gene in indica mega rice cultivar MTU1010. *Physiology and Molecular Biology of Plants,* 26, 1099–1110.

Sasidharan, R., Schippers, J.H.M. and Schmidt, R.R. 2021. Redox and low-oxygen stress, signal integration and interplay. *Plant Physiology,* 186, 66–78.

Sevgi, B. and Leblebici, S. 2023. Tuz stresinin bitkiler üzerindeki etkileri ve geliştirilen tolerans mekanizmaları. *Düzce University Journal of Science and Technology,* 11, 1498–1516. (In Turkish)

Sevgi, K. and Leblebici, S. 2022. Bitkilerde ağır metal stresine verilen fizyolojik ve moleküler yanıtlar. *Journal of Anatolian Environmental and Animal Sciences,* 7, 528–536. (In Turkish)

Shakiba, E., Edwards, J.D., Jodari, F. et al. 2017. Genetic architecture of cold tolerance in rice (Oryza sativa) determined through high resolution genome-wide analysis. *PLOS ONE,* 12, 0172133.

Shalem, O., Sanjana, N.E., Hartenian, E. et al. 2014. Genome-scale CRISPR-Cas9 knockout screening in human cells. *Science,* 343, 84–87.

Sovová, T., Kerins, G., Demnerová, K. et al. 2017. Genome editing with engineered nucleases in economically important animals and plants: State of the art in the research pipeline. *Current Issues in Molecular Biology,* 21, 41–62.

Tastan, C., Yasar, S., Tanyolac, M.B. et al. 2020. CRISPR-Of-Things: Applications and challenges of the most popular gene editing tool in the fields of health, agriculture and environment. *International Journal of Innovative Approaches in Science Research,* 4, 153–190.

Tevini, M. 2004. Plant responses to ultraviolet radiation stress. *In Chlorophyll a fluorescence*. pp.605–621. Springer, Netherlands.

Tian, W., He, G., Qin, L. et al. 2021. Genome-wide analysis of the NRAMP gene family in potato (Solanum tuberosum): Identification, expression analysis and response to five heavy metals stress. *Ecotoxicology and Environmental Safety,* 208, 111661.

Tovkach, A., Zeevi, V. and Tzfira, T. 2009. A toolbox and procedural notes for characterizing novel zinc finger nucleases for genome editing in plant cells. *The Plant Journal,* 57, 747–757.

Uranbey, S., Köm, D., Akdoğan, G. et al. 2017. Tuz stresinin patateste stoma ıdayanıklılığı ile ilgili Asg1 geni ifade düzeyine etkileri. *Mediterranean Agricultural Sciences,* 30, 235–238. (In Turkish)

Urnov, F.D., Rebar, E.J., Holmes, M.C. et al. 2010. Genome editing with engineered zinc finger nucleases. *Nature Reviews Genetics*, 11(9), 636–646.

Vidyasagar, A. 2018. *What is CRISPR? Live Science. chrome-extension://efaidnbmnnnibpcajpcglclefindmkaj/*https://www.ccsoh.us/cms/lib/OH01913306/Centricity/Domain/2695/Science%20Summer%20Assignment%20%202019%20Part%202.pdf.

Wang, F., Wang, L., Zou, X. et al. 2019. Advances in CRISPR-Cas systems for RNA targeting, tracking and editing. *Biotechnology Advances,* 37, 708–729.

Wang, H., Cimen, E., Singh, N. et al. 2020. Deep learning for plant genomics and crop improvement. *Current Opinion in Plant Biology,* 54, 34–41.

Wang, Y., Wang, L., Zhou, J. et al. 2019. Research progress on heat stress of rice at flowering stage. *Rice Science,* 26, 1–10.

Wend, T., Holmv P.B., Starker, CvG. et al. 2013. TAL effector nucleases induce mutations at a pre-selected location in the genome of primary barley transformants. *Plant Molecular Biology,* 83, 279–285.

Wingender, E., Chen, X., Fricke, E. et al. 2001. The TRANSFAC system on gene expression regulation. *Nucleic Acids Research*, 29(1), 281–283.

Yadav, R., Tripathi, M.K., Tiwariv S. et al. 2023. Genome editing and improvement of abiotic stress tolerance in crop plants. *Life,* 13, 1456.

Yalçın, B.E. 2014. *Investigating phenolic content, antioxidant activity and bioavailability of raw/steam cooked buckwheat, black chickpea and brown lentil.* (M.Sc. Thesis). Istanbul Technical University, Istanbul, Turkey.

Yang ,H., Huang, T., Ding, M. et al. 2017. High temperature during grain filling impacts on leaf senescence in Waxy Maize. *Agronomy Journal,* 109, 906–916.

Yao, X. and Liu, Q. 2007. Changes in photosynthesis and antioxidant defenses of Picea asperata seedlings to enhanced ultraviolet-B and nitrogen supply. *Physiologia Plantarum,* 129, 364–374.

Yilmaz, E., Tuna, A.L. and Bürün, B. 2011. Bitkilerin tuz stresi etkilerine karşı geliştirdikleri tolerans stratejileri. *Journal of Science,* 7, 47–66. (In Turkish)

Yin, R., Ulm, R. and Ch, R.U. 2017. How plants cope with UV-B: From perception to response. *Current Opinion in Plant Biology,* 37, 42–48.

Yin, Y., Qin, K., Song, X. et al. 2018. BZR1 transcription factor regulates heat stress tolerance through FERONIA receptor-like kinase-mediated reactive oxygen species signaling in tomato. *Plant and Cell Physiology*, 59(11), 2239–2254.

Zhang, Z., Lei, H., Ajay, H. et al. 2019. Development of an Agrobacterium-delivered CRISPR/Cas9 system for wheat genome editing. *Plant Biotechnology Journal,* 17, 1623–1635.

18 Abiotic Stress Tolerance Through CRISPR/Cas9-Mediated Genome Editing in Tomato

Burcu Arıkan, Elif Pulat, Neslihan Turgut Kara, and Özgür Çakır

18.1 INTRODUCTION

Solanaceae is a very large family in the plant kingdom with several agronomically important plants. In the Solanum genus, just after the potato, the second most important vegetable with the production of approximately 100 million tons of fresh fruit is the cultivated tomato (*Solanum lycopersicum*) (FAO, 2021). Tomato is an economically important crop because of its valuable nutritional and rich content of antioxidants, β-carotene (vitamin A), ascorbic acid (vitamin C), potassium, and lycopene. In addition to these features, tomato is being used in scientific studies because it possesses the special characteristics of a model organism. It could be said that the fruit development and ripening data are obtained from tomatoes (Seymour et al., 2013; Pesaresi et al., 2014; Xia et al., 2021).

We know that the world's population is continuously growing while the demand of people for food will not cease to increase. Agricultural productivity is negatively influenced by abiotic stressors. Since plants are immobile organisms in nature, they are susceptible to stress and should adapt to environmental changes to keep their cellular activities on track. Abiotic stresses influence and decrease crop productivity. Drought, salinity, high/low temperature, heavy metals, etc., can be regarded as abiotic stresses and they cause major difficulties in agriculture. If the molecular mechanisms underlying the abiotic tolerance could be understood, agriculturally abiotic stress tolerant crops might be developed and cultivated. With this aspect in mind, traditional plant breeding methods have been applied by humans to develop plants with desired traits, but modern techniques enable us to generate all kinds of new traits, including tolerance to abiotic stressors (Moose and Mumm, 2008; Ahmar et al., 2020).

Tomato has an evolution that could be explained by natural selection, variations caused by climate change and artificial breeding and selection for tomatoes have been carried out for more than 10,000 years (Chandrasekaran et al., 2021). Over the past decade, genome-editing technologies that could modify plant genomes accurately at specific sites to engineer new DNA sequences/genes of desired features have served as a valuable tool in crop improvement (Li et al., 2022). Three main genome-editing methods have been described up to now, zinc finger nucleases (ZFNs) (Kim et al., 1996), transcription activator-like effector nucleases (TALENs) and clustered regularly interspaced short palindromic repeats (CRISPR)/CRISPR associated protein 9 (Cas9) (Jinek et al., 2012; Cong et al., 2013). When compared to ZFNs and TALENs, the CRISPR/Cas9 system has some advantages and these advantages led to the quick abandonment of ZFNs and TALENs. The outstanding features of CRISPR/Cas9 could be listed as being simple, stable, effective, and non-rigid. The CRISPR/Cas9 system is giving researchers the ability to discover and modify new traits robustly and that is being used for genome editing in many organisms, including crop plants. Being simple in design seems to be the main reason for choosing the CRISPR/Cas9 system over the other genome-editing tools (Baloglu et al., 2022).

CRISPR/Cas9 is a natural system of bacteria and archaea that defends them against bacteriophages. These organisms keep the sequences from previous bacteriophage infections to create a memory for similar future infections, giving immunity to bacteria. After infection, the DNA sequences of phages integrate into the CRISPR repeat sequences in the bacterial genome to create a memory. Briefly, Cas9 is a DNA endonuclease that performs a cleavage on the targeted site of DNA by using a CRISPR RNA (crRNA) and trans-activating crRNA (tracrRNA). crRNA and tracrRNA together form chimeric RNA which is around 20 bp long and called single guide RNA (sgRNA). These are the main parts of the CRISPR/Cas9 system (Li et al., 2022). Editing of a genome occurs through the Cas protein performing double-strand breaks (DSB) on a DNA which is recognized by sgRNA. A short protospacer adjacent motif (PAM) is a region recognized by the Cas protein. PAM is located adjacent to the target DNA sequence at the 3' end. The PAM sequences are 2–6 bp long. The 5'NGG3' sequence is canonical for the Cas9 enzyme from *Streptococcus pyogenes* (Anders et al., 2014). sgRNA gives rise to the cleavage of target sequence. These DSBs are then repaired by two mechanisms: non-homologous end joining (NHEJ) and homolog recombination (HR). NHEJ and HR are used to create gene knockouts and gene modification/insertion, respectively (Huang and Puchta, 2019; Baloglu et al., 2022). The NHEJ repair system either inserts or deletes a sequence in a coding region, thus creating a frameshift mutation but the HR repair system inserts a desired sequence which is taken from a donor (Baloglu et al., 2022). CRISPR systems have two classes, and both classes

DOI: 10.1201/9781003500933-21

also have three different subclasses, Class I (subclass I, III, and IV) and Class II (subclass II, V, VI) (Chandrasekaran et al., 2021). While Class I CRISPR/Cas systems use complex multiple Cas proteins, Class II CRISPR/Cas systems use a single large Cas protein. There are challenges in using CRISPR/Cas9 systems as well as benefits, which are called off-target mutations. It has been determined that the possible off-target mutation frequency is much less in plants than in mammals (Fu et al., 2013; Kuscu et al., 2014; Zhang and Showalter, 2020). To reduce the risk of possible off-target mutation, a sgRNA sequence can be chosen to hybridize with only the target sequence but nowhere else in the target genome (Baltes and Voytas, 2015; Zhang and Showalter, 2020) using a paired Cas9 enzyme (Ran et al., 2013) or finally using a catalytically inactive fusion Cas9 to the *Fok*I nuclease to develop a specific genome modification (Guilinger et al., 2014; Tsai et al., 2014).

Genome-editing systems have given rise to new prospects for crop development with new traits. The CRISPR/Cas systems were used to generate stress-tolerant plants with enhanced traits, can regulate genes, and resist pathogens (Mushtaq and Molla, 2021; Ghosh and Dey, 2022). Researchers have edited the genomes of several crop plants, including rice (*Oryza sativa*) (Sun et al., 2017; J. Zhang et al., 2018; Kim et al., 2019; Zhang et al., 2019; Zhu et al., 2019; Zafar et al., 2020), maize (*Zea mays*) (Liang et al., 2014; Qi et al., 2016; Liu et al., 2021; Jiang et al., 2022), soybean (*Glycine max*) (Jacobs et al., 2015; Liu et al., 2019; Summra, 2022), wheat (*Triticum aestivum*) (Su et al., 2011; Liang et al., 2018; Sánchez-León et al., 2018; Y. Zhang et al., 2018; Jouanin et al., 2020; J. Li et al., 2021; Smedley et al., 2021; Habtamu and Sewnet, 2022), potato (*Solanum tuberosum*) (Andersson et al., 2017; Kusano et al., 2018; Nakayasu et al., 2018; González et al., 2019; Tiwari et al., 2022; Toinga-Villafuerte et al., 2022), and tomato (*Solanum lycopersicum*) (Čermák et al., 2015; Ito et al., 2015; Deng et al., 2018; Jung et al., 2018; R. Li et al., 2018; X. Li et al., 2018; Swinnen et al., 2020). Tomato, as a model organism, is a central plant with outstanding characteristics and an ideal nominee to study genome editing. Here in this chapter, we review recent advances in the application of CRISPR/Cas9 genome-editing systems to improve tomato against abiotic stresses as well as the challenges and potential advantages to develop tomato with elevated agronomic properties.

18.2 ABIOTIC STRESSES

A decrease in food supply due to global climate change is expected to occur in the future. The most common abiotic stresses are drought, temperature, and salinity, which we feel more severely due to climate change. Most of the world's arable land is prone to one or more of the abiotic stresses, resulting in yield losses. These types of stress are one of the main problems for crop growth, development, yield, and sustainability of agriculture all over the world. Plants have evolved various strategies to cope with environmental stresses through rapid and adaptive changes at transcriptional and post-transcriptional levels, including epigenetic regulation of gene expression (Miryeganeh, 2021; Halder et al., 2022).

Data obtained from genome-editing studies have also been used to elucidate the mechanisms of plant stress response. To date, CRISPR/Cas has been efficiently applied in genome editing in many plant species. Although CRISPR/Cas has only recently become the tool of choice for genetic manipulation in plants, rapid advances have been made (Li et al., 2022; Nagamine et al., 2022). In this context, recent studies that will contribute to understanding abiotic stress response mechanisms using CRISPR/Cas9 technology in tomatoes are summarized in Table 18.1.

18.2.1 Drought Stress

Several studies were conducted to understand the drought tolerance in tomatoes. In a recent study, the function of SIMAPK6, one of the 16 mitogen-activated protein kinases in tomato, was studied by using CRISPR-Cas. MAPKs are considered one of the enzymes that help plants tolerate drought. By examining the *SIMAPK6* CRISPR/Cas knockouts in dwarf tomato plants, it was found that the key regulatory element of axillary bud growth and tomato architecture is SIMAPK6. The researchers showed that gibberellin synthesis genes (*GA20ox3* and *GA3ox1*) and strigolactone synthesis genes (*SICCD7* and *SICCD8*) were positively regulated, and axillary bud development-related genes *Ls*, *BL*, and *BRC1b/TCP8* and *GAI* were negatively regulated by SIMAPK6 (Y. Li et al., 2021). In another study, Wang et al. (2022) characterized the *SlERF.B1* gene, which is an ethylene-responsive factor, as overexpressed in tomato and *Arabidopsis*. To understand SlERF.B1's role in abiotic stress, transgenic tomatoes overexpressing *SlERF. B1* were obtained by CRISPR/Cas9-mediated mutagenesis. The overexpression of this transcription factor in tomatoes led to a lower mannitol and drought stress tolerance, while also negatively regulating tomato salt tolerance (Wang et al., 2022). The relationship between abiotic stresses and *SlARF4* (*AUXIN RESPONSE FACTOR 4*) gene, which is a key element for several physiological mechanisms, was studied by creating *SlARF4* mutated lines of tomatoes by CRISPR/Cas9 to compare them with wild type and ARF4 antisense silenced line (ARF4-as). It is found that the downregulation of *SlARF4* caused tolerance to drought and salinity (Bouzroud et al., 2020). Also, knockout mutants of *SlARF4* when compared to control plants, were able to retain much less water (Chen et al., 2021). Another example of CRISPR/Cas application in tomato about drought tolerance is the knock-out line of the *SlLBD40* gene. This gene is a member of LATERAL ORGAN BOUNDARIES DOMAIN (LBD) and *SlLBD40* knockout tomatoes created by CRISPR/Cas9 showed enhanced drought tolerance (Liu et al., 2020).

Recently, drought resistance provided by mutations in the SlHB8 transcription factor encoding gene was demonstrated in a study conducted using loss of function and overexpression mutants. Knockout of *SlHB8* by the CRISPR/Cas9 system, presented resistance to drought

TABLE 18.1
Studies using CRISPR/Cas9 technology for tolerance against abiotic stress

Tomato	Stress	CRISPR system	Target gene	Function/phenotype	Result	Reference
2012 BC1F2,a test cross population, 2012 BC2F2 and F2 population derived from 2012x Marmande cross	Heat stress	CRISPR/Cas9	SlAGL6	Knockout mutant is able to produce fruit under high heat conditions that set back fertilization-dependent fruit set	Mutation caused facultative parthenocarpy allowing fruit production under heat stress, pointing out a new function of SlAGL6	(Klap et al., 2017)
Solanum lycopersicum cv. "Ailsa Craig"	Drought stress	CRISPR/Cas9	SlMAPK3	Severe wilting symptom, higher hydrogen peroxide content, lower antioxidant enzyme activities suffered more membrane damage	SlMAPK3 is involved in drought response by protecting cell membranes from oxidative damage and expression of stress-related genes	(Wang et al., 2017)
*Solanum lycopersicum*cv. "Ailsa Craig"	UV-B	CRISPR/Cas9	UVR8	*sluvr8* knockout mutants are generated	SlUVR8 has functions in tomato seedling development, acclimation and tolerance to strong UV-B stress	(H. Li et al., 2018)
Solanum lycopersicum "Ailsa Craig"	Chilling stress	CRISPR/Cas9	SlCBF1	Knockout of SlCBF1	Significantly increased IAA contents but decreased MeJA, ABA and zeatin riboside contents	(R. Li et al., 2018)
Solanum lycopersicum L. cv. Condine Red	Heat Stress	CRISPR/Cas9 Virus-induced gene silencing	BZR1	Reduced growth not responsive to 24-epibrassinolide (EBR) with regard to the promotion of plant growth	BZR1 (BRASSINAZOLE RESISTANT 1) regulates heat stress responses in tomato	(Yin et al., 2018)
Solanum lycopersicum cv. "Ailsa Craig"	Heat stress	CRISPR/Cas9	MAPK3	Less severe wilting and less membrane damage, lower ROS contents, higher activities and transcript levels of antioxidant enzymes, increased expressions of genes	Mutants are more tolerant to heat stress compared to wild-type	(Yu et al., 2019)
Solanum lycopersicum, L.	Salinity and osmotic stress	CRISPR/Cas9 Two single-guide (sg) RNAs *Agrobacterium tumefaciens* mediated genetic transformation	SlARF4	Promoted root development and density, increased soluble sugars content maintained chlorophyll content at high levels, reduced stomatal conductance, increased leaf relative water content, increased Abscisic acid (ABA) content	SlARF4 downregulation increases tolerance to high salt and drought	(Bouzroud et al., 2020)
Solanum lycopersicum cv. M82 (sp^{-}/sp^{-})	Drought	CRISPR/Cas9	GID1 (Gibberellin-insensitive Dwarf1)	High leaf water content	Drought resistance increased	(Illouz-Eliaz et al., 2020)

Solanum lycopersicum cv. "Micro Tom"	Drought stress	CRISPR/Cas9	SlLBD40	Wilted slightly, SlLBD40 overexpressing plants suffered severe wilting. knockout of SlLBD40 increased tomato water-retention.	SlLBD40, involved in JA signaling, negative regulator of drought tolerance, knockout of SlLBD40 enhanced drought tolerance	(L. Liu et al., 2020)
Solanum lycopersicum cv. "Micro Tom"	Drought stress	pYAO-based CRISPR/Cas9 system	WHIRLY2 (SlWHY2) gene	Increased wilting, lower fresh weight, chlorophyll content, quantum yield of photosystem I, maximal photochemical efficiency of PSII, higher ROS accumulation, lower AOX (alternative oxidase) activity.	SlWHY2, essential role in maintaining mitochondrial function and enhancing drought tolerance in tomato	(Meng et al., 2020)
Solanum lycopersicum L. "Micro-Tom"	Chilling stress	CRISPR/Cas9	SBPase	SBPase knockout mutant accumulated higher levels of reactive oxygen species (ROS) than wild type.	GSH biosynthesis and AsA-GSH recycling are inhibited by induction of oxidative stress, SBPase functions in responding to lower temperatures in tomato, optimally	(M. Wang, et al., 2020)
Solanum lycopersicum-*TS21Solanum lycopersicum*-Salt-sensitive line TS-670	Salt	CRISPR/Cas9	HAK20	When the Na+/K+ ratio of roots and shoots was evaluated, *SlHAK20* knockout mutants presented a differential ratio compared to wild-type (TS-21) under salt stress. These mutants were more susceptible to salt stress indicating the role of SlHAK20 in salt tolerance response. Two loss of function mutations were created in HAK20 genes of salt-sensitive (TS-670) tomato, where both mutations resulted in more sensitive plants to salinity. Both strong and weak alleles of HAK20 were shown to have an important role in the salt tolerance of tomatoes.		(Z. Wang et al., 2020)
Solanum lycopersicum cv. Condine Red	Heat stress	CRISPR/Cas9	CPK28	Induction of ROS accumulation and protein oxidation levels, lower ascorbate peroxidase (APX) and other antioxidant enzyme activity.	Enhanced thermotolerance	(Hu et al., 2021)
Solanum lycopersicum	heat, chilling, drought, heavy metal toxicity, nutrient deficiency, and short photoperiod	Multiplex CRISPR/Cas9 system	class II GRX	Knockout all four members of the class II glutaredoxin (GRX) gene family	Higher sensitivity to several abiotic stresses signifies the role of GPX in plant stress tolerance	(Kakeshpour et al., 2021)

(*continued*)

TABLE 18.1 (Continued)
Studies using CRISPR/Cas9 technology for tolerance against abiotic stress

Tomato	Stress	CRISPR system	Target gene	Function/phenotype	Result	Reference
Dwarf tomato plants (*Solanum lycopersicum*)	Drought	CRISPR/Cas9	SlMAPK6	*GA20ox3* and *GA3ox1, SICCD7 and SICCD8 positively regulatedLs*, *BL* and *BRC1b/TCP8* and *GAI* negatively regulated by MAPK6-knockout.	Regulatory element of axillary buds growth and tomato architecture	(Y, Li et al., 2021)
Hongkwang and 15T01 local tomato varieties	Salt	Multiplex CRISPR/Cas9 system	SlHyPRP1	Non-functional negative-response domain(s) of SlHyPRP1 resulted in salinity tolerance		(Tran et al., 2021)
Solanum lycopersicum L.	Drought	CRISPR/Cas9	ALD1, FMO1	*Slald1* mutants of a pip biosynthetic gene and *Slfmo1* mutants of a hydroxylated modification gene were generated	*Slald1* mutants showed higher drought tolerance related to many physiological changes, whereas *Slfmo1* mutants showed drought susceptibility	(P. Wang et al., 2021)
Solanum lycopersicum L. "Micro-Tom"	Heat Stress	CRISPR/Cas9	NRX1	Mutant was extremely sensitive to heat stress with higher electrolyte leakage, malondialdehyde contents, and H_2O_2 concentration	SlNRX1 alleviates heat stress-induced oxidative damage positively regulates the transcription of antioxidants and HSPs	(Cha et al., 2022)
Solanum lycopersicum cv. "Micro Tom"	Salt	CRISPR/Cas9	SlABIG1	Increased chlorophyll content, photosynthetic capacity, root dry weight, and proline; decreased amounts of ROS, MDA, and Na+.	Knockout of SlABIG1 increased tomato salt tolerance	(Ding et al., 2022)
Solanum lycopersicum	Cadmium toxicity	CRISPR/Cas9	E3 ubiquitin ligase gene SI1	Knockout and overexpression mutants are produced. Cadmium tolerance is higher in overexpressing plants, and lower in knockout mutants.	Sl1 provides Cd tolerance by relieving oxidative stress and lowering heavy metal accumulation in tomatoes	(Liu et al., 2022)
Solanum lycopersicum plants in M82 background (sp/sp)	Salt	CRISPR/Cas9	ABA-IMPORTING TRANSPORTER 1.1	ait1.1 mutants had faster radicle emergence, increased germination, and partial resistance to ABA		(Shohat et al., 2022)
Solanum lycopersicum, L.	Salt and drought stress	CRISPR/Cas9 *Agrobacterium tumefaciens*	SlERF.B1 (AP2/ERF transcription factor)	SlERF.B1 overexpression resulted in higher plant sensitivity to osmotic stress, increased ROS accumulation, and lowered expression of stress-responsive genes		(Wang et al., 2022)

Solanum lycopersicum cv. "Ailsa Craig"	Salt and drought	CRISPR/Cas9	MYC family	*SlMYC* loss of functions mutants are generated to evaluate responses to NaCl and mannitol, mutant plants presented lower tolerance to both stress conditions, with higher ROS content. Indicates the possibility for tomato MYC1 to have a regulatory impact during salt and drought responses.		(Feng et al., 2023)
Solanum lycopersicum, "Micro-Tom" background	Aluminum	CRISPR/Cas9	Formate dehydrogenase (SlFDH)	FDH knockout lines are generated which show lower tolerance to Al stress, therefore stating a relationship between FDH expression regulation and Al tolerance		(He et al., 2023)
Solanum lycopersicum cv. "Ailsa Craig" (TS-21)	Salt	CRISPR/Cas9	SOS2	Knockout mutations created in SOS2 caused reduced salt tolerance, which is a strong indication of the role of SOS2 in tomato salt stress tolerance		(Hong et al., 2023)
Solanum lycopersicum cv. "Ailsa Craig"	Drought	CRISPR/Cas9	SlWRKY52	Knockout mutants of WRKY52 were more susceptible to drought conditions because of their positive regulatory effect on drought stress-related genes		(Jia et al., 2023)
Solanum lycopersicum "Ailsa Craig"	Chilling stress	CRISPR/Cas9	SlNPR1	MDA, H_2O_2, and O^{2-} accumulation, low soluble protein and proline content, lower antioxidant enzyme activity	SlNPR1 makes tomato plants more tolerant to low-temperature conditions	(Shu et al., 2023)
Solanum lycopersicum L.	Heat stress	CRISPR/Cas9	MYB41	CRISPR/Cas-mediated mutations of MYB41 caused higher ion leakage, MDA content, and ROS levels, along with lower photosynthetic rate and downregulation of HSP gene expressions. MYB41 positively directly regulated SlHSP90.3 expression, resulting in higher thermotolerance.		(J. Wang et al., 2023)

(continued)

TABLE 18.1 (Continued)
Studies using CRISPR/Cas9 technology for tolerance against abiotic stress

Tomato	Stress	CRISPR system	Target gene	Function/phenotype	Result	Reference
Solanum lycopersicum cv. "Ailsa Craig"	Salt	CRISPR/Cas9	SlGATA17	Sl-GATA17 overexpression lines were generated by CRISPR/Cas9 gene editing to validate GATA17 function in tomatoes under abiotic stress. *SlGATA17-7* and *SlGATA17-9* mutants had insertions and deletions causing frameshift and early termination of transcription. These knockout mutants showed higher tolerance to salt, whereas overexpression lines showed sensitivity.		(Y. Wang et al., 2023)
Solanum lycopersicum cv.. "Micro Tom"	Drought	CRISPR/Cas9	SlHB8	Knockout mutants of HB8, have higher drought tolerance due to increased antioxidant enzyme activity		(Y. Yang et al., 2023)
Solanum lycopersicum cv. "Ailsa Craig"	Salt	CRISPR/Cas9	Put2	Put2 (Polyamine uptake transporter) overexpression lines and mutant lines containing a frameshift mutation were created by CRISPR/Cas9. After 7 days of salt stress, overexpression mutants showed heightened tolerance to salt, and *put2* mutants with frameshift mutation were hypersensitive to salt treatment. Results indicate the positive regulatory role of Put2 in salt stress tolerance.		(Zhong et al., 2023)
Solanum lycopersicum cv. "Ailsa Craig"	Cold	CRISPR/Cas9	BBX31	*SlBBX31* mutants with insertion and deletions were created as allele plants. Both alleles were more susceptible to cold temperatures than wild types. SlBBX31 is shown to be a possible positive regulator of cold stress response.		(Zhu et al., 2023)

conditions due to the closed stomata, lower oxidative damage, and higher antioxidant enzyme activity (Y. Yang et al., 2023). Another transcription factor (TF) WRKY52, was also recently found to positively regulate drought stress resistance. Overexpression lines of this TF showed enhanced tolerance to drought resulting from its involvement in ABA synthesis-related stomatal closure and reduced water loss. Whereas CRISPR/Cas9-mutated *SlWRKY52* lines were more susceptible to drought stress, supporting the idea that WRKY52 positively regulates drought tolerance-related genes in tomato (Jia et al., 2023).

18.2.2 Salt Stress

In a study conducted with *Solanum lycopersicum* cv. (Micro-Tom), expression analysis of the HD-ZIP II transcription factor subfamily has been investigated with RNA-seq analysis and it has been found that knocking out of a salt-responsive gene *SlABIG1* enhances the salinity tolerance of tomato plants. Therefore, they have concluded that *SlABIG1* works as a negative regulator of salt stress (Ding et al., 2022). Similar to this research, (Tran et al., 2021) made an excision in the negative response domains of SlHyPRP1 protein with the help of multiplex CRISPR/Cas9 systems resulting in the high salinity tolerance of the Hongkwang variety and 15T01 local tomato varieties at different stages of plant growth. It has been suggested that domain editing of multidomain proteins may bring the possibility of engineering crops with high tolerance to stress conditions. Additionally, it has been shown that plant hormones such as abscisic acid (ABA) play crucial roles in the stress responses of plants. ABA is mostly known for its role in dormancy. In addition, it limits the germination of plants when they encounter abiotic stress conditions. In a study conducted with *Solanum lycopersicum* in M82 background (sp/sp), the role of an ABA importer during salinity has been shown. In the study, they induced mutagenesis in *ABA-IMPORTING TRANSPORTER 1.1* (AIT1.1) using CRISPR/Cas 9 technology with sgRNAs. As a result, while germination of wild-type tomatoes was restricted, *ait1.1* allele showed normal seed longevity, indicating a potential candidate for future studies with the aim of seed germination improvement under salinity (Shohat et al., 2022).

18.2.3 Heat and Chilling Stress

Researchers make great efforts to elucidate the mechanisms of heat tolerance. In the study conducted by Yin et al. (2018), the relationship between BRASSINAZOLE RESISTANT 1 (BZR1), which participates in BR-mediated development processes, and stress tolerance was investigated. In this study conducted with the CRISPR-bzr1 mutant, it was observed that the plants did not respond to 24-epibrassinolide (EBR) and growth did not occur. Its mutation in BZR1 impaired the induction of RBOH1, H_2O_2 production in the apoplast, and heat tolerance. Exogenous H_2O_2 administration, on the other hand, restored heat tolerance in these tomato *bzr1* mutants. Conversely, overexpression of BZR1 increased apoplastic H_2O_2 production and heat stress responses. Also, when RBOH1 was silenced, BZR1-mediated heat tolerance disappeared. Their results showed that BZR1 binds to the FERONIA2 (FER2) and FER3 promoters and induces its expression. Silencing FER2/3 suppressed BZR1-dependent BR signaling for induction of RBOH1 transcripts, apoplastic H_2O_2 accumulation, and heat tolerance. These results indicate that BZR1 regulates heat stress responses in tomato through RBOH1-dependent reactive oxygen species (ROS) signaling, at least in part mediated by FER2 and FER3. Since the tomato is a chill-sensitive garden plant, the main restriction in production is cold stress. Highly conserved C-rebinding factors (CBFs) are parts of the cold response system found in most species. Shu et al. (2023) determined that the expression of the non-expressor of pathogenesis-related gene 1 (*NPR1*) gene, which is a master transcription factor in terms of biotic stress and plays a role in resistance to cold stress, increases significantly at low temperatures. At the same time, the effects of cold stress intensified in the CRISPR-npr1 mutant they created. To understand the role of calcium-dependent protein kinases (CPKs) in plant thermotolerance, Hu et al. (2021) produced tomato *cpk28* mutants using the CRISPR/Cas9 gene-editing approach. Due to the findings obtained in this study, CPK28 is demonstrated as an attractive target for the development of heat-stress tolerant crop varieties.

ROS-based tolerance mechanisms are an alternative strategy used by plants to overcome abiotic stresses. Multiplex rearrangement of four GRX (Glutaredoxins) genes (*GRXS14, GRXS15, GRXS16,* and *GRXS17*) in tomato resulted in the generation of binary, triple, and quadruple mutants that exhibit hypersensitive phenotypes under abiotic stresses, such as drought, heat, heavy metal toxicity, nutrient deficiency, and cooling (Kakeshpour et al., 2021). Another group designed a heat stress-sensitive CRISPR/Cas9-mediated *slnrx1* mutant. They found out then that the *slnrx1* mutant was extremely sensitive to heat stress with higher electrolyte leakage, malondialdehyde (MDA) contents, and H_2O_2 concentration compared to wild-type tomato plants, suggesting that SlNRX1 negatively regulates heat stress-induced oxidative damage (Cha et al., 2022). CRISPR/Cas9 and RNAi-mediated mutations of SlMYB41, which are known to be associated with tomato thermotolerance, resulted in higher ion leakage, more MDA content, and higher ROS levels. Compared to wild-type, mutants had lower photosynthetic rates and downregulation of gene expressions related to heat stress, such as HSPs (heat shock proteins). Normally, MYB41 is known to be a heat-induced TF, here, it was also shown to positively regulate *SlHSP90.3* expression directly. Therefore, resulting in higher thermotolerance (Wang et al., 2023).

18.2.4 Other Abiotic Stresses

There has also been research conducted to use CRISPR/Cas9 gene-editing technologies to fight against various other stressors for tomato plants. In a study with *Solanum lycopersicum* cv. "Ailsa Craig" cultivar, CRISPR/Cas9 systems are used to generate knock-out mutants of the gene *UVR8*. The *SlUVR8* gene was previously shown to have roles in plant seedling development and tolerance against UV-B stress. In this study, *UVR8* is successfully manipulated, using a CRISPR/Cas9 system, and the obtained data from the *sluvr8* knockout mutants pointed out the importance of *UVR8* in UV-B tolerance of the plant, thus, opening the possibility of generating higher acclimation of plants to UV-B using CRISPR/Cas systems for future applications (X. Liu et al., 2020). Another study used CRISPR/Cas9 in *S. lycopersicum* cv. "Micro-Tom" plants to create a deletion mutant of the gene *PHO1*, which is involved in phosphate translocation. During phosphate deficiency, *PHO1* is upregulated to hasten the transport of P_i from the roots to the shoots of the plant. The deletion mutants verified the role of PHO1 in alleviating the effects of phosphate deficiency in tomato and presented future potential for the use of CRISPR/Cas9 systems against nutritional deficiencies (Zhao et al., 2019).

Heavy metals are also critical stress agents for crop plants. Targeting four different CGFS-type glutaredoxin (GRX) genes in tomato, a multiplex CRISPR/Cas9 system was used to create various mutants of all these GRX genes. When the mutants of three GRX genes were treated with cadmium (Cd), it was observed by phenotypic analysis that the mutants showed greater sensitivity to Cd stress, compared to the wild-type plants. In the same study, the effects of nutrient deficiencies and shorter photoperiod were also analyzed for these three mutants, resulting in the detection of impaired growth mostly in the double and triple mutants and delayed flowering time in both single and multiple mutants (Kakeshpour et al., 2021). Knockout and overexpression of E3 ubiquitin ligase gene *Sl1* by CRISPR/Cas9 in another study have proved the role of *Sl1* in cadmium tolerance of tomato. Cadmium tolerance of the *Sl1* overexpressing plants was higher than wild-type plants, whereas *Sl1* knockout mutants presented lower tolerance to the adverse effects of the heavy metal (Liu et al., 2022).

A recent study also pointed out a potential use of CRISPR/Cas systems in enhancing plant stress tolerance by controlling unfolded protein response (UPR). This approach represents an important opportunity to combat endoplasmic reticulum (ER) stress caused by various and multiple environmental stressors. UPR induces the expression of key genes such as ER chaperones to regulate protein folding and decreases protein synthesis to combat ER stress. Using CRISPR/Cas-mediated mutations of UPR-associated genes, ER stress-related pathways and roles of UPR components in plants under stress have been better understood over the years. Aside from studying ER stress-related pathway elements, this system can also be used to edit cis-regulatory elements and promoters for targeted control of gene expression, aiming to enhance desirable traits or grant stress tolerance (Vu et al., 2023).

18.3 TRANSGENE-FREE EDITING OF TOMATO

Genomes could be altered by the methods which are described as DNA-free genome-editing systems. Up to now, scientists have studied in order to generate transgene-free tomato plants by introducing new features via CRISPR/Cas9 technologies (Nekrasov et al., 2017; Danilo et al., 2019; Veillet et al., 2019; T. Yang et al., 2023). To achieve this, they used certain CRISPR/Cas9 techniques to produce transgene-free plants. The main aim of using these techniques is to create desirable new traits that harbor no foreign DNA sequences. The main parts of this system—which are gRNA and Cas9—are introduced into a plasmid with an appropriate selection marker to constitute a functional CRISPR gene-editing construct (Bhattacharjee et al., 2023). The selection marker could be an antibiotic or herbicide resistance gene. *Agrobacterium*-mediated or biolistic transformation is then applied to introduce the construct to the plant cells. The follow-up work to be done on stable transformation is genetic segregation to eliminate the transgenic plants. It is a simple and powerful but time-consuming technique. Another way of generating a transgene-free edited plant is to use *Agrobacterium*-mediated or biolistic transformation for transient expression. PCR is applied afterwards to select the transgene-free plants (Gu et al., 2021). Zhang et al (2016) created a transient expression of the CRISPR/Cas9 system called TECCDNA which uses RNA vectors in immature wheat embryos. Created plants are then sequenced directly. Chen et al. (2018) used the *Agrobacterium tumefaciens* transformation method to get a transient expression of the CRISPR/Cas9 construct in tobacco. Another technology is haploid induction editing in maize, developed by Kelliher et al. (2019). Ribonucleoproteins can be used to deliver Cas9 and sgRNAs as a preassembled construct (Woo et al., 2015; Svitashev et al., 2016; Liang et al., 2017; Park et al., 2019). Polyethylene glycol-calcium is used for successful transfection in this system. Hence this system is DNA-free, and scientists were able to create DNA-free genome-edited plants. Transgene-killer technology is another approach to produce transgene-free genome-edited plants which were first developed by He et al. (2018, 2019). In this system *barnase* gene and *CMS2* are added to the CRISPR/Cas9 construct. Plant cells are killed by *barnase* activity, it is a nuclease expressed under the rice early embryo *REG2* gene promotor. *CMS2* is regulated by a 35S promotor and stimulates male sterility to ensure that after transformation, the target genes are edited and when the plants begin to reproduce, pollens and embryos are killed by this system (He et al. 2018, 2019).

18.4 NOVEL CRISPR TECHNIQUES FOR GENE EDITING IN TOMATO

To overcome the challenges of CRISPR/Cas9 disadvantages, novel techniques are constantly being developed. In CRISPR/Cas9 research in tomato, these new technologies are used such as base editing (Veillet et al., 2019; Hunziker et al., 2020; Kawaguchi et al., 2021) and prime editing (Lu et al., 2021). To summarize, the challenge that the scientists face in the CRISPR/Cas9 method is having off-target impacts due to the double-strand breaks Cas9 creates. To subdue this, a base editing method was developed, allowing the transition changes of bases (C/G to T/A by cytosine base editor or A/T to G/C by adenine base editor) (Lu et al., 2021; Sretenovic et al., 2021). It seems that an adenine base editor is preferred instead of a cytosine base editor due to fewer off-target events. Base editing offers precise genome editing in the target DNA sequence.

Nowadays, a new technique called prime editing is being used as a more precise method. This method consists of a new CRISPR/Cas9 complex and it includes prime editing guide RNA and a special Cas9 fused to reverse transcriptase. Prime editing involves two nicks in DNA sequences. After having the first nick, prime editing guide RNA is reverse transcribed. Modified DNA is added to the nicked sequence and heteroduplex DNA is formed. Then another nick is created in the other strand to make sure that modified DNA is going to be used in mismatch repair (Huang et al., 2021; Tiwari et al., 2023). Another approach could be multiplex editing which is the practice of performing multiple gene, protein, and metabolite examinations at the same time on a specific level of accuracy. It is a tool that lets scientists modify two or more DNA sequences precisely (Kumar et al., 2023).

18.5 FUTURE PROSPECTS OF CRISPR-MEDIATED EDITING OF TOMATO FOR ABIOTIC STRESS TOLERANCE

CRISPR/Cas9 is a promising technology to overcome the challenges of climate change and related abiotic stresses with the accelerating research in tomato (Kong et al., 2012; Li et al., 2016; Wang et al., 2017; R. Li et al., 2018; Yin et al., 2018; Li et al., 2019; Bouzroud et al., 2020; L. Liu et al., 2020; X. Liu et al., 2020; Y. Li et al., 2022; Z. Wang et al., 2021; Tran et al., 2021). Due to the accumulating knowledge on the molecular mechanisms underlying these stresses, there will be more opportunities that result in a crop yield of sufficient quality for tomatoes globally. CRISPR/Cas technology is excellent for genome-edited plant breeding because it can deal with multiple genes at once with maximum accuracy and cost-effectiveness compared to traditional methods. Abiotic stress-resistant plants can be produced more easily thanks to future developments in CRISPR/Cas and related technologies. These developments will also make it easier to identify the critical genes governing biotic and abiotic stress pathways (Sardar et al., 2023). The CRISPR-Cas9 led to new editing methods such as base editing (BE), and prime editing (PE) technologies which provide new opportunities for a better understanding of genes and their functions. Base and prime, transgene-free, and multiplex genome-editing technologies seem to be the future of CRISPR/Cas9-mediated crop plants resistant to abiotic stresses. CRISPR/Cas-mediated mutant libraries are also useful for tomato improvement (Kumar et al., 2023). To create a set of mutant lines with the fewest transformation trials and in the shortest amount of time, pooled CRISPR libraries were transformed to alter tomatoes (Jacobs et al., 2017).

There are some limitations to CRISPR/Cas technology alongside its numerous advantages. The off-target effect, cytotoxicity of nucleases such as Cas9 and derivatives, NHEJ repair system-created gene knockouts, handicaps of gene delivery systems and genome instability could be regarded as some challenges that should be evaluated case by case (Das et al., 2023). For this technology to be applied efficiently, it must be practiced in detail on the plant where the genome editing will be made. In this way, it will be possible to increase sensitivity, ensure that there are no off-target effects, and apply the system to different cell types (Das et al., 2023).

18.6 CONCLUSION

The abiotic stress threat is increasing worldwide because of the climate change. Due to the growing populations of the world, humankind must keep on track for crop improvement and productivity. Conventional plant breeding systems which have been used for over 10,000 years, are time-consuming and uncertain. Therefore, there was a need for a rapid and reliable method for plant breeding purposes. With the progress in molecular biology, genome-editing tools have arisen and differed from others because of the benefits compared to traditional methods, but especially among them CRISPR/Cas9 technology is a tool that has been widely used to edit genomes to create plants carrying the desired trait. The distinct features of CRISPR/Cas9 (fast, easy to handle, powerful, and rigid), made it preferable to other genome-editing tools. Tomato is an appropriate plant for CRISPR/Cas9 studies because of its biological and agronomical properties. Following the examples of CRISPR/Cas9 systems being successfully used to modify genes that have critical roles in regulating tomato plants' tolerance to abiotic stress conditions, it is possible to assume that CRISPR technologies will provide a useful and promising tool for generating stress tolerance to tomatoes in future applications.

REFERENCES

Ahmar, S., Saeed, S., Khan, M.H.U., Ullah Khan, S., Mora-Poblete, F., Kamran, M., et al. 2020. A revolution toward gene-editing technology and its application to crop improvement. *International Journal of Molecular Science*, 21(16), 5665.

Anders, C., Niewoehner, O., Duerst, A., and Jinek, M. 2014. Structural basis of PAM-dependent target DNA recognition by the Cas9 endonuclease. *Nature*, 513(7519), 569–573. doi: 10.1038/nature13579.

Andersson, M., Turesson, H., Nicolia, A., Fält, A.S., Samuelsson, M., and Hofvander, P. 2017. Efficient targeted multiallelic mutagenesis in tetraploid potato (Solanum tuberosum) by transient CRISPR-Cas9 expression in protoplasts. *Plant Cell Reports*, 36(1), 117–128. doi: 10.1007/s00299-016-2062-3.

Baloglu, M.C., Celik Altunoglu, Y., Baloglu, P., Yildiz, A.B., Türkölmez, N., and Özden Çiftçi, Y. 2022. Gene-editing technologies and applications in legumes: Progress, evolution, and future prospects. *Frontiers in Genetics,* 13, 859437. doi: 10.3389/fgene.2022.859437.

Baltes, N.J., and Voytas, D.F. 2015. Enabling plant synthetic biology through genome engineering. *Trends in Biotechnology*, 33(2), 120–131. doi: 10.1016/j.tibtech.2014.11.008.

Bhattacharjee, S., Bhowmick, R., Kant, L. et al. 2023. Strategic transgene-free approaches of CRISPR-based genome editing in plants. *Molecular Genetics and Genomics,* 298, 507–520. doi:10.1007/s00438-023-01998-3

Bouzroud, S., Gasparini, K., Hu, G., Barbosa, M.A.M., Rosa, B.L., Fahr, M., et al. 2020. Down regulation and loss of auxin response factor 4 function using CRISPR/Cas9 alters plant growth, stomatal function and improves tomato tolerance to salinity and osmotic stress. *Genes*, 11(3), 272.

Čermák, T., Baltes, N.J., Čegan, R., Zhang, Y., and Voytas, D.F. 2015. High-frequency, precise modification of the tomato genome. *Genome Biology,* 16, 232. doi:10.1186/s13059-015-0796-9.

Cha, J.-Y., Ahn, G., Jeong, S.Y., Shin, G.-I., Ali, I., Ji, M.G., et al. 2022. Nucleoredoxin 1 positively regulates heat stress tolerance by enhancing the transcription of antioxidants and heat-shock proteins in tomato. *Biochemical and Biophysical Research Communications,* 635, 12–18. doi: 10.1016/j.bbrc.2022.10.033.

Chandrasekaran, M., Boopathi, T., and Paramasivan, M. 2021. A status-quo review on CRISPR-Cas9 gene editing applications in tomato. *International Journal of Biological Macromolecules,* 190, 120–129. doi: 10.1016/j.ijbiomac.2021.08.169.

Chen, L., Li, W., Katin-Grazzini, L., Ding, J., Gu, X., Li, Y., Gu, T., Wang, R., Lin, X., Deng, Z., McAvoy, R.J., Gmitter, F.G. Jr, Deng, Z., Zhao, Y., and Li, Y. 2018. A method for the production and expedient screening of CRISPR/Cas9-mediated non-transgenic mutant plants. *Horticulture Research,* 2, 13. doi: 10.1038/s41438-018-0023-4.

Chen, M., Zhu, X., Liu, X., Wu, C., Yu, C., Hu, G., et al. 2021. Knockout of Auxin response factor SlARF4 improves tomato resistance to water deficit. International Journal of Molecular Sciences, 22(7). doi: 10.3390/ijms22073347.

Cong, L., Ran, F.A., Cox, D., Lin, S., Barretto, R., Habib, N., et al. 2013. Multiplex genome engineering using CRISPR/Cas systems. *Science*, 339(6121), 819–823. doi:10.1126/science.1231143.

Danilo, B., Perrot, L., Mara, K. et al. 2019. Efficient and transgene-free gene targeting using Agrobacterium-mediated delivery of the CRISPR/Cas9 system in tomato. *Plant Cell Reports*, 38, 459–462. doi: 10.1007/s00299-019-02373-6

Das, T., Anand, U., Pal, T., Mandal, S., Kumar, M., Radha, Gopalakrishnan, A.V., Lastra, J.M.P. de la, and Dey, A. 2023. Exploring the potential of CRISPR/Cas genome editing for vegetable crop improvement: An overview of challenges and approaches. Biotechnology and Bioengineering, 120(5), 1215–1228. doi:10.1002/bit.28344

Deng, L., Wang, H., Sun, C., Li, Q., Jiang, H., Du, M., et al. 2018. Efficient generation of pink-fruited tomatoes using CRISPR/Cas9 system. *Journal of Genetics and Genomics*, 45(1), 51–54. doi: 10.1016/j.jgg.2017.10.002.

Ding, F., Qiang, X., Jia, Z., Li, L., Hu, J., Yin, M., et al. 2022. Knockout of a novel salt responsive gene SlABIG1 enhance salinity tolerance in tomato. *Environmental and Experimental Botany,* 200, 104903. doi: 10.1016/j.envexpbot.2022.104903.

FAO (2021). Crop description and climate [Online]. Available at: www.fao.org/land-water/databases-and-software/crop-information/tomato/en/ (accessed November 21, 2022).

Feng, Y., Zeng, S., Yan, J., Li, K., and Xu, H. 2023. Genome-wide analysis and expression of MYC family genes in tomato and the functional identification of slmyc1 in response to salt and drought stress. *Agronomy*, 13(3), 757. doi: 10.3390/agronomy13030757.

Fu, Y., Foden, J.A., Khayter, C., Maeder, M.L., Reyon, D., Joung, J.K., et al. 2013. High-frequency off-target mutagenesis induced by CRISPR-Cas nucleases in human cells. *Nature Biotechnology*, 31(9), 822–826. doi: 10.1038/nbt.2623.

Ghosh, S., and Dey, G. 2022. Biotic and abiotic stress tolerance through CRISPR-Cas mediated genome editing. *Journal of Plant Biochemistry and Biotechnology*, 31(2), 227–238. doi: 10.1007/s13562-021-00746-1.

González, M.N., Massa, G.A., Andersson, M., Turesson, H., Olsson, N., Fält, A.S., et al. 2019. Reduced enzymatic browning in potato tubers by specific editing of a polyphenol oxidase gene via ribonucleoprotein complexes delivery of the CRISPR/Cas9 system. *Frontiers in Plant Science,* 10, 1649. doi: 10.3389/fpls.2019.01649.

Gu, X., Liu, L., and Zhang, H. 2021. Transgene-free genome editing in plants. *Frontiers in Genome Editing,* 3, 805317. doi:10.3389/fgeed.2021.805317.

Guilinger, J.P., Thompson, D.B., and Liu, D.R. 2014. Fusion of catalytically inactive Cas9 to FokI nuclease improves the specificity of genome modification. *Nature Biotechnology*, 32(6), 577–582. doi: 10.1038/nbt.2909.

Habtamu, K., and Sewnet, G. 2022. Gene editing improves the agronomic important traits of wheat – CRISPR-Cas9 and Cas12/Cpf1. In A. Mahmood-ur-Rahman (Ed.), *Wheat. IntechOpen*.

Halder, K., Chaudhuri, A., Abdin, M.Z., Majee, M., and Datta, A. 2022. Chromatin-based transcriptional reprogramming in plants under abiotic stresses. *Plants*, 11(11), 1449.

He, Q., Jin, J., Li, P., Zhu, H., Wang, Z., Fan, W., and Yang, J.L. 2023. Involvement of SlSTOP1 regulated SlFDH expression in aluminum tolerance by reducing NAD+ to NADH in the tomato root apex. *Plant Journal*, 113(2), 387–401. doi: 10.1111/tpj.16054.

He, Y., Zhu, M., Wang, L., Wu, J., Wang, Q., Wang, R., and Zhao, Y. 2018. Programmed self-elimination of the CRISPR/Cas9 construct greatly accelerates the isolation of edited and transgene-free rice plants. *Molecular Plant,* 11, 1210–1213. doi: 10.1016/j.molp.2018.05.005.

He, Y., Zhu, M., Wang, L., Wu, J., Wang, Q., Wang, R., and Zhao, Y. 2019. Improvements of TKC technology accelerate isolation of transgene-free CRISPR/Cas9-edited rice plants. *Rice Science,* 26, 109–117. doi: 10.1016/j.rsci.2018.11.001.

Hong, Y. et al. 2023. Natural variation in SlSOS2 promoter hinders salt resistance during tomato domestication. *Horticulture Research*, 10(1). doi: 10.1093/hr/uhac244.

Hu, Z., Li, J., Ding, S., Cheng, F., Li, X., Jiang, Y., et al. 2021. The protein kinase CPK28 phosphorylates ascorbate peroxidase and enhances thermotolerance in tomato. *Plant Physiology*, 186(2), 1302–1317. doi: 10.1093/plphys/kiab120.

Huang, T.K., and Puchta, H. 2019. CRISPR/Cas-mediated gene targeting in plants: Finally a turn for the better for homologous recombination. *Plant Cell Reports*, 38(4), 443–453. doi: 10.1007/s00299-019-02379-0.

Huang, T.K., and Puchta, H. 2021. Novel CRISPR/Cas applications in plants: From prime editing to chromosome engineering. *Transgenic Research,* 30, 529–549. doi:10.1007/s11248-021-00238-x

Hunziker, J., Nishida, K., Kondo, A. et al. 2020. Multiple gene substitution by Target-AID base-editing technology in tomato. *Science Reports Transgenic Research,* 10, 20471. doi:10.1038/s41598-020-77379-2

Illouz-Eliaz, N., Nissan, I., Nir, I., Ramon, U., Shohat, H., and Weiss, D. 2020. Mutations in the tomato gibberellin receptors suppress xylem proliferation and reduce water loss under water-deficit conditions. *Journal of Experimental Botany*, 71(12), 3603–3612. doi: 10.1093/jxb/eraa137.

Ito, Y., Nishizawa-Yokoi, A., Endo, M., Mikami, M., and Toki, S. 2015. CRISPR/Cas9-mediated mutagenesis of the RIN locus that regulates tomato fruit ripening. *Biochemical and Biophysicaly Research Communications*, 467(1), 76–82. doi: 10.1016/j.bbrc.2015.09.117.

Jacobs, T.B., LaFayette, P.R., Schmitz, R.J., and Parrott, W.A. 2015. Targeted genome modifications in soybean with CRISPR/Cas9. *BMC Biotechnology*, 15(1), 16. doi: 10.1186/s12896-015-0131-2.

Jacobs, T.B., Zhang, N., Patel, D., and Martin, G.B. 2017. Generation of a collection of mutant tomato lines using pooled CRISPR libraries. *Plant Physiology*, 174(4), 2023–2037. doi: 10.1104/pp.17.00489.

Jia, C., Wang, J., Guo, B., Li, X., Yang, T., et al. 2023. A group III WRKY transcription factor, SlWRKY52, positively regulates drought tolerance in tomato. *Environmental and Experimental Botany,* 215, 105513. https://doi.org/10.1016/j.envexpbot.2023.105513

Jiang, Y., Sun, K., and An, X. 2022. CRISPR/Cas system: Applications and prospects for maize improvement. *ACS Agricultural Science and Technology,* 2(2), 174–183. doi: 10.1021/acsagscitech.1c00253.

Jinek, M., Chylinski, K., Fonfara, I., Hauer, M., Doudna, J.A., and Charpentier, E. 2012. A programmable dual-RNA-guided DNA endonuclease in adaptive bacterial immunity. *Science*, 337(6096), 816–821. doi:10.1126/science.1225829.

Jouanin, A., Gilissen, L.J.W.J., Schaart, J.G., Leigh, F.J., Cockram, J., Wallington, E.J., et al. 2020. CRISPR/Cas9 gene editing of gluten in wheat to reduce gluten content and exposure—reviewing methods to screen for Coeliac safety. Frontiers in Nutrition, 7. doi: 10.3389/fnut.2020.00051.

Jung, Y.J., Lee, G.-J., Bae, S., and Kang, K.K. 2018. Reduced ethylene production in tomato fruits upon CRISPR/Cas9-mediated LeMADS-RIN mutagenesis. *Horticultural Science & Technology*, 36(3), 396–405. doi: 10.12972/kjhst.20180039.

Kakeshpour, T., Tamang, T.M., Motolai, G., Fleming, Z.W., Park, J.-E., Wu, Q., et al. 2021. CGFS-type glutaredoxin mutations reduce tolerance to multiple abiotic stresses in tomato. *Physiologia Plantarum*, 173(3), 1263–1279. doi: 10.1111/ppl.13522.

Kawaguchi, K., Takei-Hoshi, R., Yoshikawa, I. et al. 2021. Functional disruption of cell wall invertase inhibitor by genome editing increases sugar content of tomato fruit without decrease fruit weight. *Science Reports,* 11, 21534. doi:10.1038/s41598-021-00966-4

Kelliher, T., Starr, D., Su, X., Tang, G., Chen, Z., Carter, J., Wittich, P., Dong, S., Green, J., Burch, E., McCuiston, J., Gu, W., Sun, Y., Strebe, T., Roberts, J., Bate, N., and Que, Q. 2019. One-step genome editing of elite crop germplasm during haploid induction. *Nature Biotechnology*, 37, 287–292. doi: 10.1038/s41587-019-0038-x

Kim, Y.-A., Moon, H., and Park, C.-J. 2019. CRISPR/Cas9-targeted mutagenesis of Os8N3 in rice to confer resistance to Xanthomonas oryzae pv. oryzae. *Rice*, 12(1), 67. doi: 10.1186/s12284-019-0325-7.

Kim, Y.G., Cha, J., and Chandrasegaran, S. 1996. Hybrid restriction enzymes: Zinc finger fusions to Fok I cleavage domain. *Proceedings of the National Academy of Sciences*, 93(3), 1156–1160. doi:10.1073/pnas.93.3.1156.

Klap, C., Yeshayahou, E., Bolger, A.M., Arazi, T., Gupta, S.K., Shabtai, S., et al. 2017. Tomato facultative parthenocarpy results from SlAGAMOUS-LIKE 6 loss of function. *Plant Biotechnology Journal*, 15(5), 634–647. doi: 10.1111/pbi.12662.

Kong, F., Wang, J., Cheng, L., Liu, S., Wu, J., Peng, Z., and Lu, G. 2012. Genome-wide analysis of the mitogen-activated protein kinase gene family in Solanum lycopersicum. *Gene*, 499(1), 108–120. doi:10.1016/j.gene.2012.01.048

Kumar, M., Prusty, M.R., Pandey, M.K., Singh, P.K., Bohra, A., Guo, B., and Varshney, R.K. 2023. Application of CRISPR/Cas9-mediated gene editing for abiotic stress management in crop plants. *Frontiers in Plant Science,* 14, 1157678. doi:10.3389/FPLS.2023.1157678/BIBTEX

Kusano, H., Ohnuma, M., Mutsuro-Aoki, H., Asahi, T., Ichinosawa, D., Onodera, H., et al. 2018. Establishment of a modified CRISPR/Cas9 system with increased mutagenesis frequency using the translational enhancer dMac3 and multiple guide RNAs in potato. *Science Reports*, 8(1), 13753. doi: 10.1038/s41598-018-32049-2.

Kuscu, C., Arslan, S., Singh, R., Thorpe, J., and Adli, M. 2014. Genome-wide analysis reveals characteristics of off-target sites bound by the Cas9 endonuclease. *Nature Biotechnology*, 32(7), 677–683. doi: 10.1038/nbt.2916.

Li, H., Li, Y., Deng, H., Sun, X., Wang, A., Tang, X., et al. 2018. Tomato UV-B receptor SlUVR8 mediates plant acclimation to UV-B radiation and enhances fruit chloroplast development via regulating SlGLK2. *Science Reports*, 8(1), 6097. doi: 10.1038/s41598-018-24309-y.

Li, J., Li, Y., and Ma, L. 2021. Recent advances in CRISPR/Cas9 and applications for wheat functional genomics and breeding. *aBIOTECH*, 2(4), 375–385. doi: 10.1007/s42994-021-00042-5.

Li, J., Ouyang, B., Wang, T., Luo, Z., Yang, C., Li, H., Sima, W., Zhang, J., and Ye, Z. 2016. HyPRP1 gene suppressed by multiple stresses plays a negative role in abiotic stress tolerance in tomato. *Frontiers in Plant Science,* 7, 202218. doi:10.3389/FPLS.2016.00967.

Li, R., Fu, D., Zhu, B., Luo, Y., and Zhu, H. 2018. CRISPR/Cas9-mediated mutagenesis of lncRNA1459 alters tomato fruit ripening. *Plant Journal*, 94(3), 513–524. doi: 10.1111/tpj.13872.

Li, R., Liu, C., Zhao, R., Wang, L., Chen, L., Yu, W., et al. 2019. CRISPR/Cas9-Mediated SlNPR1 mutagenesis reduces tomato plant drought tolerance. *BMC Plant Biology*, 19(1), 38. doi: 10.1186/s12870-018-1627-4.

Li, R., Zhang, L., Wang, L., Chen, L., Zhao, R., Sheng, J., et al. 2018. Reduction of tomato-plant chilling tolerance by CRISPR–Cas9-mediated SlCBF1 mutagenesis. *Journal of Agricultural and Food Chemistry*, 66(34), 9042–9051. doi: 10.1021/acs.jafc.8b02177.

Li, X., Wang, Y., Chen, S., Tian, H., Fu, D., Zhu, B., et al. 2018. Lycopene is enriched in tomato fruit by CRISPR/Cas9-mediated multiplex genome editing. Frontiers in Plant Science, 9. doi: 10.3389/fpls.2018.00559.

Li, Y., Wu, X., Zhang, Y., and Zhang, Q. 2022. CRISPR/Cas genome editing improves abiotic and biotic stress tolerance of crops. Frontiers in Genome Editing, 4. doi: 10.3389/fgeed.2022.987817.

Li, Y., Yue, N., Basit, A., Li, Y., Zhang, D., Qin, L., et al. 2021. Targeted editing of SlMAPK6 using CRISPR/Cas9 technology to promote the development of axillary buds in tomato plants. Journal of Agricultural Science, 13(2). doi: 10.5539/jas.v13n2p11.

Liang, Z., Chen, K., Li, T., Zhang, Y., Wang, Y., Zhao, Q., Liu, J., Zhang, H., Liu, C., Ran, Y., and Gao, C. 2017. Efficient DNA-free genome editing of bread wheat using CRISPR/Cas9 ribonucleoprotein complexes. *Nature Communications*, 18(8), 14261. doi: 10.1038/ncomms14261.

Liang, Z., Chen, K., Zhang, Y., Liu, J., Yin, K., Qiu, J.-L., et al. 2018. Genome editing of bread wheat using biolistic delivery of CRISPR/Cas9 in vitro transcripts or ribonucleoproteins. *Nature Protocols*, 13(3), 413–430. doi: 10.1038/nprot.2017.145.

Liang, Z., Zhang, K., Chen, K., and Gao, C. 2014. Targeted mutagenesis in Zea mays using TALENs and the CRISPR/Cas system. *Journal of Genetics and Genomics*, 41(2), 63–68. doi: 10.1016/j.jgg.2013.12.001.

Liu, C.-X., Yang, T., Zhou, H., Ahammed, G.J., Qi, Z.-Y., and Zhou, J. 2022. The E3 Ubiquitin Ligase gene Sl1 is critical for cadmium tolerance in Solanum lycopersicum L. *Antioxidants*, 11(3). doi: 10.3390/antiox11030456

Liu, J., Gunapati, S., Mihelich, N.T., Stec, A.O., Michno, J.M., and Stupar, R.M. 2019. Genome editing in soybean with CRISPR/Cas9. *Methods in Molecular Biology*, 1917, 217–234. doi: 10.1007/978-1-4939-8991-1_16.

Liu, L., Gallagher, J., Arevalo, E.D., Chen, R., Skopelitis, T., Wu, Q., et al. 2021. Enhancing grain-yield-related traits by CRISPR-Cas9 promoter editing of maize CLE genes. *Nature Plants*, 7(3), 287–294. doi: 10.1038/s41477-021-00858-5.

Liu, L., Zhang, J., Xu, J., Li, Y., Guo, L., Wang, Z., et al. 2020. CRISPR/Cas9 targeted mutagenesis of SlLBD40, a lateral organ boundaries domain transcription factor, enhances drought tolerance in tomato. *Plant Science*, 301, 110683. doi: 10.1016/j.plantsci.2020.110683.

Liu, X., Zhang, Q., Yang, G., Zhang, C., Dong, H., Liu, Y., et al. 2020. Pivotal roles of tomato photoreceptor SlUVR8 in seedling development and UV-B stress tolerance. *Biochemical and Biophysical Research Communications*, 522(1), 177–183. doi:10.1016/j.bbrc.2019.11.073.

Lu, Y., Tian, Y., Shen, R., Yao, Q., Zhong, D., Zhang, X., and Zhu, J. K. 2021. Precise genome modification in tomato using an improved prime editing system. *Plant Biotechnology Journal*, 19(3), 415–417. doi:10.1111/PBI.13497

Meng, C., Yang, M., Wang, Y., Chen, C., Sui, N., Meng, Q., et al. 2020. SlWHY2 interacts with SlRECA2 to maintain mitochondrial function under drought stress in tomato. *Plant Science*, 301, 110674. doi: 10.1016/j.plantsci.2020.110674.

Miryeganeh, M. 2021. *Plants' epigenetic mechanisms and abiotic stress*. Genes (Basel), 12(8). doi: 10.3390/genes12081106.

Moose, S.P., and Mumm, R.H. 2008. Molecular plant breeding as the foundation for 21st century crop improvement. *Plant Physiology*, 147(3), 969–977. doi: 10.1104/pp.108.118232.

Mushtaq, M., and Molla, K.A. 2021. CRISPR technologies for plant biotechnology innovation, In A. Ricroch, S. Chopra, and M. Kuntz (Eds.), Plant biotechnology: Experience and future prospects (pp. 51–67). Springer International Publishing.

Nagamine, A., Takayama, M., and Ezura, H. 2022. Genetic improvement of tomato using gene editing technologies. *Journal of Horticultural Science and Biotechnology*, 98(1). doi: 10.1080/14620316.2022.2091489.

Nakayasu, M., Akiyama, R., Lee, H.J., Osakabe, K., Osakabe, Y., Watanabe, B., et al. 2018. Generation of α-solanine-free hairy roots of potato by CRISPR/Cas9 mediated genome editing of the St16DOX gene. *Plant Physiology and Biochemistry*, 131, 70–77. doi: 10.1016/j.plaphy.2018.04.026.

Nekrasov, V., Wang, C., Win, J. et al. 2017. Rapid generation of a transgene-free powdery mildew resistant tomato by genome deletion. *Science Reports*, 7, 482. doi: 10.1038/s41598-017-00578-x

Park, J., Choi, S., Park, S., Yoon, J., Park, A.Y., and Choe, S. 2019. DNA-Free genome editing via Ribonucleoprotein (RNP) delivery of CRISPR/Cas in lettuce. *Methods in Molecular Biology*, 1917, 337–354. doi: 10.1007/978-1-4939-8991-1_25.

Pesaresi, P., Mizzotti, C., Colombo, M., and Masiero, S. 2014. Genetic regulation and structural changes during tomato fruit development and ripening. Frontiers in Plant Science, 5. doi: 10.3389/fpls.2014.00124.

Qi, W., Zhu, T., Tian, Z., Li, C., Zhang, W., and Song, R. 2016. High-efficiency CRISPR/Cas9 multiplex gene editing using the glycine tRNA-processing system-based strategy in maize. *BMC Biotechnology*, 16(1), 58. doi: 10.1186/s12896-016-0289-2.

Ran, F.A., Hsu, P.D., Wright, J., Agarwala, V., Scott, D.A., and Zhang, F. 2013. Genome engineering using the CRISPR-Cas9 system. *Nature Protocols*, 8(11), 2281–2308. doi: 10.1038/nprot.2013.143.

Sánchez-León, S., Gil-Humanes, J., Ozuna, C.V., Giménez, M.J., Sousa, C., Voytas, D.F., et al. 2018. Low-gluten, nontransgenic wheat engineered with CRISPR/Cas9. *Plant Biotechnology*, 16(4), 902–910. doi: 10.1111/pbi.12837.

Sardar, A. 2023. Genetic amelioration of fruit and vegetable crops to increase biotic and abiotic stress resistance through CRISPR genome editing. *Frontiers in Plant Science,* 14, 1260102.

Seymour, G.B., Østergaard, L., Chapman, N.H., Knapp, S., and Martin, C. 2013. Fruit development and ripening. *Annual Review of Plant Biology*, 64, 219–241. doi: 10.1146/annurev-arplant-050312-120057.

Shohat, H., Cheriker, H., Cohen, A., and Weiss, D. 2022. *ABA-IMPORTING TRANSPORTER 1.1 activity in the endosperm of tomato seeds restrains germination under salinity stress*. bioRxiv. 2022.2002.2028.482276. doi: 10.1101/2022.02.28.482276.

Shu, P., Li, Y., Xiang, L., Sheng, J., and Shen, L. 2023. SlNPR1 modulates chilling stress resistance in tomato plant by alleviating oxidative damage and affecting the synthesis of ferulic acid. *Scientia Horticultura,* 307, 111486. doi: 10.1016/j.scienta.2022.111486.

Smedley, M.A., Hayta, S., Clarke, M., and Harwood, W.A. 2021. CRISPR-Cas9 based genome editing in wheat. *Current Protocols*, 1(3), e65. doi: 10.1002/cpz1.65.

Sretenovic, S., Liu, S., Li, G., Cheng, Y., Fan, T., Xu, Y., et al. 2021. Exploring C-To-G base editing in rice, tomato, and poplar. *Frontiers in Genome Editing,* 3, 756766. doi: 10.3389/fgeed.2021.756766

Su, Z., Hao, C., Wang, L., Dong, Y., and Zhang, X. 2011. Identification and development of a functional marker of TaGW2 associated with grain weight in bread wheat (Triticum aestivum L.). *Theoretical and Applied Genetics*, 122(1), 211–223. doi: 10.1007/s00122-010-1437-z.

Summra, S. 2022. Role of CRISPR/Cas9 in soybean (Glycine max. L.) quality improvement. In O. Takuji, T. Yoshihiko, O. Norikuni, S. Takashi and T. Sayuri (Eds.), Soybean. IntechOpen.

Sun, Y., Jiao, G., Liu, Z., Zhang, X., Li, J., Guo, X., et al. 2017. Generation of high-amylose rice through CRISPR/Cas9-mediated targeted mutagenesis of starch branching enzymes. *Frontiers in Plant Science,* 8, 298. doi: 10.3389/fpls.2017.00298.

Svitashev, S., Schwartz, C., Lenderts, B. et al. 2016. Genome editing in maize directed by CRISPR-Cas9 ribonucleoprotein complexes. *Nature Communications,* 7, 13274. doi: 10.1038/ncomms13274

Swinnen, G., Jacobs, T., Pauwels, L., and Goossens, A. 2020. CRISPR-Cas-mediated gene knockout in tomato. *Methods in Molecular Biology,* 2083, 321–341. doi: 10.1007/978-1-4939-9952-1_25.

Tiwari, J.K., Buckseth, T., Challam, C., Zinta, R., Bhatia, N., Dalamu, D., et al. 2022. CRISPR/Cas genome editing in potato: Current status and future perspectives. Frontiers in Genetics, 13. doi: 10.3389/fgene.2022.827808.

Tiwari, J.K., Singh, A.K., and Behera, T.K. 2023. CRISPR/Cas genome editing in tomato improvement: Advances and applications. *Frontiers in Plant Science,* 14, 1121209. doi:10.3389/fpls.2023.1121209.

Toinga-Villafuerte, S., Vales, M.I., Awika, J.M., and Rathore, K.S. 2022. CRISPR/Cas9-mediated mutagenesis of the granule-bound starch synthase gene in the potato variety Yukon Gold to obtain amylose-free starch in tubers. *International Journal of Molecular Science*, 23(9), 4640.

Tran, M.T., Doan, D.T.H., Kim, J., Song, Y.J., Sung, Y.W., Das, S., et al. 2021. CRISPR/Cas9-based precise excision of SlHyPRP1 domain(s) to obtain salt stress-tolerant tomato. *Plant Cell Reports*, 40(6), 999–1011. doi: 10.1007/s00299-020-02622-z.

Tsai, S.Q., Wyvekens, N., Khayter, C., Foden, J.A., Thapar, V., Reyon, D., et al. 2014. Dimeric CRISPR RNA-guided FokI nucleases for highly specific genome editing. *Nature Biotechnology*, 32(6), 569–576. doi: 10.1038/nbt.2908.

Veillet, F., Perrot, L., Chauvin, L., Kermarrec, M.-P., Guyon-Debast, A., Chauvin, J.-E., Nogué, F., and Mazier, M. 2019. Transgene-free genome editing in tomato and potato plants using Agrobacterium-mediated delivery of a CRISPR/Cas9 cytidine base editor. *International Journal of Molecular Science,* 20, 402. doi:10.3390/ijms20020402

Vu, B.N., Vu, T.V., Yoo, J.Y., Nguyen, N.T., Ko, K.S., Kim, J.-Y., and Lee, K.O. 2023. CRISPR-Cas mediated unfolded protein response control for enhancing plant stress resistance. *Frontiers in Plant Science,* 14, 1271368. doi: 10.3389/fpls.2023.1271368

Wang, J., Chen, C., Wu, C., Meng, Q., Zhuang, K., and Ma, N. 2023. SlMYB41 positively regulates tomato thermotolerance by activating the expression of SlHSP90. 3. *Plant Physiology and Biochemistry,* 204, 108106.

Wang, L., Chen, L., Li, R., Zhao, R., Yang, M., Sheng, J., et al. 2017. Reduced drought tolerance by CRISPR/Cas9-mediated SlMAPK3 mutagenesis in tomato plants. *Journal of Agricultural and Food Chemistry*, 65(39), 8674–8682. doi: 10.1021/acs.jafc.7b02745.

Wang, M., Ding, F., and Zhang, S. 2020. Mutation of SlSBPASE aggravates chilling-induced oxidative stress by impairing glutathione biosynthesis and suppressing ascorbate-glutathione recycling in tomato plants. Frontiers in Plant Science, 11. doi: 10.3389/fpls.2020.565701.

Wang, P., Luo, Q., Yang, W., Ahammed, G.J., Ding, S., Chen, X., et al. 2021. A novel role of pipecolic acid biosynthetic pathway in drought tolerance through the antioxidant system in tomato. *Antioxidants,* 10(12), 1923.

Wang, Y., Xia, D., Li, W., Cao, X., Ma, F., Wang, Q., et al. 2022. Overexpression of a tomato AP2/ERF transcription factor SlERF.B1 increases sensitivity to salt and drought stresses. *Scientia Horticultura,* 304, 111332. doi: 10.1016/j.scienta.2022.111332.

Wang, Y. et al. 2023. SlGATA17, A tomato GATA protein, interacts with SlHY5 to modulate salinity tolerance and germination. *Environmental and Experimental Botany,* 206, 105191. doi: 10.1016/j.envexpbot.2022.105191.

Wang, Z. et al. 2020. Loss of salt tolerance during tomato domestication conferred by variation in a Na+/K+ transporter. *EMBO Journal*, 39(10), e103256. doi: 10.15252/EMBJ.2019103256.

Wang, Z., Hong, Y., Li, Y., Shi, H., Yao, J., Liu, X., Wang, F., Huang, S., Zhu, G., and Zhu, J.K. 2021. Natural variations in SlSOS1 contribute to the loss of salt tolerance during tomato domestication. *Plant Biotechnology Journal*, 19(1), 20–22. doi: 10.1111/pbi.13443.

Woo, J.W., Kim, J., Kwon, S.I., Corvalán, C., Cho, S.W., Kim, H., Kim, S.G., Kim, S.T., Choe, S., and Kim, J.S. 2015. DNA-free genome editing in plants with preassembled CRISPR-Cas9 ribonucleoproteins. *Nature Biotechnology*, 33(11), 1162–1164. doi: 10.1038/nbt.3389.

Xia, X., Cheng, X., Li, R., Yao, J., Li, Z., and Cheng, Y. 2021. Advances in application of genome editing in tomato and recent development of genome editing technology. *Theoretical and Applied Genetics*, 134(9), 2727–2747. doi: 10.1007/s00122-021-03874-3.

Yang, T., Ali, M., Lin, L., Li, P., He, H., Zhu, Q., Sun, C., Wu, N., Zhang, X., Huang, T., Li, C.B., Li, C., and Deng, L. 2023b. Recoloring tomato fruit by CRISPR/Cas9-mediated multiplex gene editing. *Horticulture Research*, 10(1). doi:10.1093/HR/UHAC214

Yang, Y., Zhang, X., Zhong, Q., Liu, X., Guan, H., Chen, R., Hao, Y., and Yang, X. 2023. Photosynthesis response and transcriptional analysis: Dissecting the role of SlHB8 in regulating drought resistance in tomato plants. *International Journal of Molecular Science,* 24, 15498. doi: 10.3390/ijms242015498

Yin, Y., Qin, K., Song, X., Zhang, Q., Zhou, Y., Xia, X., et al. 2018. BZR1 transcription factor regulates heat stress tolerance through FERONIA receptor-Like Kinase-mediated reactive oxygen species signaling in tomato. *Plant Cell Physiology*, 59(11), 2239–2254. doi: 10.1093/pcp/pcy146.

Yu, W., Wang, L., Zhao, R., Sheng, J., Zhang, S., Li, R., et al. 2019. Knockout of SlMAPK3 enhances tolerance to heat stress involving ROS homeostasis in tomato plants. *BMC Plant Biology*, 19(1), 354. doi: 10.1186/s12870-019-1939-z.

Zafar, K., Sedeek, K.E.M., Rao, G.S., Khan, M.Z., Amin, I., Kamel, R., et al. 2020. Genome editing technologies for rice improvement: Progress, prospects, and safety concerns. Frontiers in Genome Editing, 2. doi: 10.3389/fgeed.2020.00005.

Zhang, A., Liu, Y., Wang, F., Li, T., Chen, Z., Kong, D., et al. 2019. Enhanced rice salinity tolerance via CRISPR/Cas9-targeted mutagenesis of the OsRR22 gene. *Molecular Breeding*, 39. doi: 10.1007/s11032-019-0954-y.

Zhang, J., Zhang, H., Botella, J.R., and Zhu, J.K. 2018. Generation of new glutinous rice by CRISPR/Cas9-targeted mutagenesis of the waxy gene in elite rice varieties. *Journal of Integrative Plant Biology*, 60(5), 369–375. doi: 10.1111/jipb.12620.

Zhang, Y., Li, D., Zhang, D., Zhao, X., Cao, X., Dong, L., et al. 2018. Analysis of the functions of TaGW2 homoeologs in wheat grain weight and protein content traits. *Plant Journal*, 94(5), 857–866. doi: 10.1111/tpj.13903.

Zhang, Y., Liang, Z., Zong, Y. et al. 2016. Efficient and transgene-free genome editing in wheat through transient expression of CRISPR/Cas9 DNA or RNA. *Nature Communications,* 7, 12617. doi:10.1038/ncomms12617

Zhang, Y., and Showalter, A.M. 2020. CRISPR/Cas9 genome editing technology: A valuable tool for understanding plant cell wall biosynthesis and function. Frontiers in Plant Science, 11. doi: 10.3389/fpls.2020.589517.

Zhao, P., You, Q., and Lei, M. 2019. A CRISPR/Cas9 deletion into the phosphate transporter SlPHO1;1 reveals its role in phosphate nutrition of tomato seedlings. *Physiologia Plantarum*, 167(4), 556–563. doi: 10.1111/ppl.12897.

Zhong, M. et al. 2023. Genome-wide identification and characterization of the polyamine uptake transporter (PUT) gene family in tomatoes and the role of Put2 in response to salt stress. *Antioxidants*, 12(2), 228. doi: 10.3390/ANTIOX12020228/S1.

Zhu, Y., Lin, Y., Chen, S., Liu, H., Chen, Z., Fan, M., et al. 2019. CRISPR/Cas9-mediated functional recovery of the recessive rc allele to develop red rice. *Plant Biotechnology Journal*, 17(11), 2096–2105. doi: 10.1111/pbi.13125.

Zhu, Y., Zhu, G., Xu, R., et al. 2023. A natural promoter variation of SlBBX31 confers enhanced cold tolerance during tomato domestication. *Plant Biotechnology Journal [Preprint].* doi: 10.1111/pbi.14016.

19 Advancing Metabolic Engineering in Plants Using CRISPR/Cas

Ahmad Munir, Noor Asma, Sidra Ashraf, Paulo Zaini, Nayla Munawar, Baohong Zhang, and Aftab Ahmad

19.1 INTRODUCTION

Plants produce an enormous number of secondary metabolites that have diverse chemistries. Some of these components act as a form of communication between the plant and the surrounding environment to survive and adapt to it (Twaij and Hasan, 2022). Due to having the advantageous characteristics as well as increased efficiency, the secondary metabolites from the plants have significantly improved the quality of human lives. They are utilized in food production, serve as a source of medication, and are employed in the manufacturing of fertilizers and pesticides (Wawrosch and Zotchev, 2021). As the knowledge of the biosynthetic pathways expands, the possibility of engineering secondary metabolism continues to expand. Some of the highlights include the shikimate pathway of aromatic compounds and the isoprenoid pathway of terpenoids, as some of the specific pathways that can be engineered. Moreover, it is possible to enhance specific plant secondary metabolites of interest, such as monoterpenoids, carotenoids, sesqui-, di-, and triterpenes by targeting various functional components within specific metabolic pathways (Yeshi et al., 2022).

The control of various pathways involving biosynthesis in primary metabolism is mainly at the genetic level. However, in secondary metabolism, environmental stimuli appear to be the main cause of activators, some of which are the reactive oxygen species ROS, which triggers biosynthesis of certain metabolites, for instance, apocarotenoids. In contrast to the regulation found in primary metabolism, regulation of secondary metabolism occurs at the enzymatic level and through the compartmentalization of metabolic pathways. For example, in the isoquinoline alkaloids biosynthesis, particularly berberine biosynthesis, the regulation occurs at the enzymatic/transporter level. To increase the yield of secondary metabolites, attempts at in-vitro production or through genetic manipulation of the native genes through genome editing have been employed (Marchev et al., 2020). Still, while the gene modification strategy is promising and can be effective for specific applications, the overall probability of success is still not very high. Moreover, some off-target effects may arise, which can be toxic to the organism, further reinforcing the need to fine-tune the methods used in gene-editing to maximize the impact only on the targeted gene (Miladinovic et al., 2021).

Genome sequencing technologies are revolutionizing the plant genetic engineering, especially in breeding programs. The advances in molecular biology techniques, particularly the use of molecular markers, have enhanced the development of genetic and linkage maps. These maps are useful in mapping quantitative trait loci (QTLs), which are regions of the genome that are associated with traits of agricultural importance. Once a specific gene is found to be involved in a specific cellular process, biologists have a range of highly specific genome-editing tools to manipulate the genes with maximum efficiency. Some of the most used genome editing tools include zinc finger nucleases (ZFNs) and transcription activator-like effector nucleases (TALENs), while the CRISPR-Cas system has gained much praise for its efficiency and versatility in plant genome editing. Through these tools, the productivity of plants can be optimized, particularly in the production of secondary metabolites—compounds produced by plants that serve functions beyond growth, development, or reproduction. These metabolites have significant utility for humans, including applications in medicine and the production of various chemicals. Discussions like the scarcity of resources, production, and the fact that these valuable compounds can come in diversified forms could all greatly benefit from enhanced capabilities of gene editing (Salwan and Sharma, 2020). In this chapter, we discuss the current state of these innovative technologies in enhancing the synthesis of secondary metabolites with special focus on the CRISPR-Cas system limitations and opportunities. Traditional perspectives of farmers and their practices are also discussed as well as broader themes of resource utilization and sustainability of agriculture.

Medicinal plants as an important source of natural products with potential curative and therapeutic effects have contributed immensely to human health. These bioactive compounds have broad applications across multiple industries, including healthcare (e.g., nutraceuticals, functional foods), agriculture (e.g., insecticides), and manufacturing (e.g., fragrances, colors, and antioxidants). In response to adaptational pressures in their environment, plants have evolved intricate metabolic pathways which lead to synthesis of vast array of secondary metabolites

DOI: 10.1201/9781003500933-22

(Courdavault et al., 2021). These metabolites can be broadly categorized into two main groups: while primary metabolites are involved in primary processes that are vital for life such as growth and development, secondary or specialized metabolites have a variety of adaptive benefits and are deeply tied to plant defense systems (Tian et al., 2024).

Plants are undoubtedly the most diversified source of specialized metabolites with estimated genetic potential of more than 100,000 distinct secondary metabolites recorded as occurring in roughly estimated 50,000 plant species. These specialized metabolites encompass a diverse array of chemical classes and are primarily classified into three main groups based on their biosynthetic pathways: this includes terpenes, nitrogenous compounds, and phenolics, and their concentrations differ significantly from those found in younger plants. These metabolites are divided into more specific categories including volatiles, sterols, carotenoids, glycosides, alkaloids, glucosinolates, and tannins which are composed of biochemical entities with specialized physiological functions (Nabavi et al., 2020).

Unlike primary metabolites, specialized metabolites are not directly involved in fundamental biochemical processes in plants; instead, they are involved in assisting plants to overcome specific environmental adversities and physiological restraints. Apart from their functions in plant physiology, these compounds are pharmacologically goldmines. They are diverse in pharmacological activities such as antitumor, antimicrobial, anticancer, antibiotics, and other related actions. The examples include taxol, artemisinin, withaferin A, asiaticoside, withanone, and madecassoside among the other therapeutic fractions, as well as the essential oils. These compounds originate from medicinal plants, including taxus, withania, centella, artemisia, and cymbopogon. All these metabolites not only prove that plants are quite strong and versatile to face most environmental challenges but also open windows of opportunities for developing new medical products and therapeutic approaches (Sharma et al., 2022).

The use of these medicinal herbs has been recorded for centuries, which clearly demonstrates that herbal practices are rooted deep in the history of mankind. These plants are still considered to have rather powerful curative effects not only providing people with natural medicines but also efficient treatments today. Thus, in the current era of biotechnology, a continuous process of research on the bioactive compounds derived from the medicinal plants assumes a crucial role in enhancing the knowledge base and advances in diverse areas of human endeavors right from therapeutic agents, food processing, agro-chemicals, and cosmo-care. These include Salvia miltiorrhiza, Cannabis sativa, Dendrobium officinale, and Opium poppy plant species with high efficiency for transformation and good reference genomes. Scientists are increasingly focusing on novel synthesis approaches, including the identification of key biosynthetic genes for genetic manipulation and optimization of metabolic pathways in medicinal plants. These efforts aim to develop superior germplasm lines with enhanced quality and yield (Guo et al., 2024).

The emergence of new advanced genome engineering systems has made a big change in the future of the field, where scientists can get genomic data from the number of plant species and genomic data can also be manipulated in a precise manner. These pioneering tools offer the ability to edit genes with pinpoint precision at specific loci in the genome and redirect the functions of pre-specified zones. Highly specific approaches, such as generating DNA double-strand breaks (DSBs) by sequence-specific nucleases, can be used to introduce desired DNA modifications at targeted genomic loci, which are then repaired using either the donor-dependent homology-directed repair (HDR) pathway or the error-prone non-homologous end joining (NHEJ) pathway. Breakthrough in the plant gene-editing domain has been witnessed by the first-generation sequence-specific nucleases like meganuclease (MN), zinc-finger nucleases (ZFNs), and transcriptional activator-like effector nucleases (TALENs), these involve complexities in protein engineering (Sabzehzari et al., 2020). The third generation of CRISPR/Cas systems that are based on DNA-RNA recognition is clearly superior to the earlier approaches initially developed through the analysis of protein-DNA interactions. Techniques such as CRISPR/Cas9 and other systems involve the generation of DSBs at the target site, with ease and low cost making it possible to perform precise manipulation of the genome, including gene deletion, gene insertion, or gene substitution (Wang and Doudna, 2023).

This chapter provides a review of how these advanced technologies are being used for the improvement of secondary metabolites, specifically the CRISPR-Cas system and the challenges that are associated with it. Not only does this investigation present the epitome of what modern agriculturally applied biotechnology is or can be, but it also sets out the roadmap as to where this discipline is set to travel in the field of plant-based medicines.

19.2 OVERVIEW OF CRISPR

CRISPR/Cas is an adaptive system for phage immunity in bacteria and archaea that originates from prokaryotes. CRISPR/Cas system is a programmable RNA-directed enzyme for targeted DNA editing since its function is based on the interaction of DNA with RNA and the sequence-specific endonuclease activity of Cas proteins. Compared with these earlier gene-editing systems, ZFN and TALEN, this third-generation genome-editing technology has the following advantages: Its fabrication needs no protein engineering, costs much less, and is readily flexible in constructing the multiplex mutation system (Zhang et al., 2022).

sgRNA is a combination of tracrRNA and crRNA which is a part of the CRISPR/Cas system that belongs to type II SpCas9, and it is the most used Cas9 protein from Streptococcus pyogenes and it has a very wide application in different species such as plants. The sgRNA and Cas together form a complex that targets the DNA sequence, cuts the DNA, and recognizes a protospacer adjacent motif (PAM) site at the position next to the sgRNA binding region. Another widely used Cas enzyme is the Cas12a, which belongs to class 2, type V, RNA-guided DNA endonuclease that contains a single nuclease domain, uses a single small gRNA molecule, and has RNAase activity to process crRNA, has DNA ends that are offset, and targets T-rich PAM sites (Guo et al., 2022).

The principle of the CRISPR/Cas system is to create DSB at the specific chromosomal locus and repair the cleaved fragments in DSBs by incorporating DNA alterations through the donor-dependent HDR pathway or the error-prone NHEJ pathway. Thus, when the two ends of the DSB are joined by NHEJ, small nucleotide insertions and deletions (indels) are incorporated into the targeted sequences (e.g. genes and promoters) resulting in frameshift mutations that disrupt the original structure and functionality. However, unlike the NHEJ, the repairs that are enabled by HDR involve the accurate integration or substitution of a predetermined DNA sequence with the help of exogenous homology donor templates. The advances in the technology of CRISPR/Cas have been rapidly developing, making it expand beyond the conventional targeted mutagenesis induced by DSBs. These developments include base editors, high precision achieved by HDR and prime editors and transcriptional regulation (Kim et al., 2021; Ramawat and Arora, 2021).

In addition, the CRISPR/Cas system has the capability to modulate genes at the epigenetic level via the modulation of transcription control elements such as promoters, enhancers, terminal, and other related sequences. It is important to note that base editing and prime editing strategies are not based on the traditional CRISPR/Cas methods of repairing a DSB in the gene sequence. The base editors are further subdivided into cytosine base editors and adenine base editors based on the target bases and delivered bases. These base editing tools consist of either a nicking Cas9 (nCas9) or a dead Cas9 (dCas9) fused to a specific deaminase. They temporarily change the DNA by replacing C-G with T-A or A-T with G-C, depending on the specific deaminases. Moreover, the prime editing tools composed of two main components: a hybrid molecule composed of Cas9 nickase (H840A) reverse transcriptase fused together, along with a second molecule known as prime editing guide RNA (pegRNA). The Cas9 nickase variant comes from H840, which harbors an additional RuvC functioning domain and cuts at the non-target DNA strand. Reverse transcriptase then collectively with pegRNA template to alter the required DNA (Kantor et al., 2020).

The primer editor tool has developed into a powerful instrument to assist in precise base exchange, deletion of DNA, and incorporation of DNA. The CRISPR/Cas system also involves gene regulation via two main methods: Control of gene expression was further advanced in the form of activating transcription or CRISPRa and inhibiting it, known as CRISPR interference or CRISPRi (Zhang et al., 2021). However, some of the current developments have led to the advanced CRISPR-based techniques with the aim of combining epigenetic modifiers to dCas9. These strategies have been made to epigenetically manipulate the target genes to achieve specific changes in DNA methylation pattern and histone modifications (J. Wang et al., 2021).

After the first application of CRISPR/Cas system for GE in 2012, further improvements have been made to refine new CRISPR-based systems permitting the genetic modification of different organisms, including plants (Montecillo et al., 2020). The basic mechanism of CRISPR-mediated GE involves the recognition of an adjacent protospacer motif (PAM), binding, and the induction of double-stranded breaks (DSBs) at the targeted locus, facilitated by two essential components: the Cas enzyme and a single guide RNA (sgRNA), which is chimeric in nature. The DNA repair pathway that is engaged depends on the desired result, and the type of CRISPR tool also affects the result: the NHEJ pathway results in the formation of indels, while the HR pathway enables the insertion or replacement of the desired sequences (A. Tariq, et al., 2023). To date, the primary utility of CRISPR/Cas in plant GE is in the creation of germinal, heritable, and transgene-free KO phenotypes of the intended gene. The process of plant GE can be divided into seven steps: Targeted intervention steps include:

1. identification of cells or tissues that are expected to provide the required phenotypic change upon intervention;
2. selection of thes CRISPR tool based on the desired outcome;
3. design of the sgRNA that complement the selected sequence;
4. synthesis of CRISPR vectors or components;
5. delivery of CRISPR components into the plant cells;
6. regeneration of edited plants;
7. analysis of edited plants.

However, some basic components for all organisms are needed, but there are certain parameters related to plant GE. For example, the techniques that are employed in delivery and genetic transformation differ based on the explant type, species or cultivar, such as Agrobacterium transfer, biolistic particles, electric impulses, chemical uptake (polyethylene glycol or PEG), ribonucleoprotein (RNP) complex, magnetofection, sound waves, and vectors, or DNA and RNA viruses. Furthermore, the regeneration and screening processes need to be optimized for each of the plant varieties and this may prove to be complex for some of the species, such as recalcitrant or difficult species as far as transformation is concerned (Huang et al., 2022).

19.3 USING CRISPR/CAS FOR METABOLIC ENGINEERING IN PLANTS

A bacterial and archaeal antiviral defense system, CRISPR/Cas, is derived from prokaryotes. The CRISPR/Cas system is RNA-programmable for genetic regulation, as it depends on the DNA-RNA interaction and sequence-specific nucleic acid cleavage binding activities. This third-generation genome-editing technology has more advantages than the earlier gene-editing systems, such as ZFN, as their construction involves cumbersome protein engineering, high cost, and lack of flexibility to develop multiple mutations systems (Doudna et al., 2021). A component of the CRISPR-Cas9 system is the single-guide RNA (sgRNA), which is a fusion of a transactivating RNA (tracrRNA) and a CRISPR RNA (crRNA).

The CRISPR/Cas system used here is based on Type II SpCas9 (the Cas9 protein from *Streptococcus pyogenes*) which is the most used Cas9 protein and has a broad use in different species including plants. The sgRNA and Cas complex targets the specific DNA sequence, cuts the DNA and also detects a protospacer adjacent motif (PAM) site next to the sgRNA target site. Another Cas enzyme that is widely used is Cas12a, which is a Class 2, type V, RNA-guided DNA nuclease with a single nuclease domain that uses one small gRNA molecule, RNAase activity for the processing of crRNA, produces DNA ends, and recognizes T-rich PAM sites. Thus, the basis of the CRISPR/Cas tool is to create DSB at a specific chromosomal site and to repair the broken segments of DSBs by the introduction of DNA changes with the help of the donor-dependent HDR pathway or the erroneous NHEJ pathway (Shelake et al., 2023). Thus, NHEJ replaces the DNA sequences at the ends of the DSB with inaccurate short nucleotide insertions and deletions (indels) in the targeted regions like genes and promoters, resulting in frame shift mutations that disrupt the proper functioning of the genes and other regulatory elements. On the other hand, the repairs that involve HDR include the accurate incorporation or exchange of a specific nucleotide sequence with the help of external homologous donor templates. Owing to the fast progress of the CRISPR/Cas technology, it has evolved to a new level beyond the targeted mutagenesis associated with DSBs in plants. These developments include base editors, high precision editing via HDR and prime editors, and transcriptional regulation (Mipeshwaree Devi et al., 2023). Also, the CRISPR/Cas system has the potential to regulate gene expression through transcriptional regulation using promoters, enhancers, terminators, and other related parts. The strategies of base editing and prime editing do not use the typical CRISPR/Cas systems for seeking out and fixing a DSB in the gene sequence. The base editing tools are categorized into cytosine base editors (CBEs) and adenine base editors (ABEs). These base editing tools include either a nicking Cas9 (nCas9) or a catalytically dead Cas9 (dCas9) fused with a particular deaminase. The deaminases cause temporary changes in the DNA base by converting C•G to T•A or A•T to G•C depending on the specific function of the deaminases. Moreover, the prime editing tools comprised two main components: a Cas9 nickase that has been mutated at position 840 and a reverse transcriptase, and a second component named the prime editing guide RNA (pegRNA). The Cas9 nickase variant (H840), having a RuvC functioning domain makes a nick to the non-target DNA strand. Reverse transcriptase then uses pegRNA template to edit the required DNA. The primer editor tool has emerged from a simple tool that enables the accurate base substitution and formation of DNA deletion to complex DNA insertion. The CRISPR/Cas system is also involved in gene regulation via two main methods: CRISPR-mediated transcriptional regulation which encompasses transcriptional activation or CRISPRa and transcriptional repression or CRISPRi. Furthermore, due to recent developments, there are novel strategies using CRISPR for methodologies that seek to target epigenetic modifiers with dCas9. Such strategies have been employed to change the epigenome to achieve desired changes in DNA methylations and histones (Dey, 2021). The pipeline to using CRISPR/Cas-based metabolic engineering is shown in Figure 19.1.

The use of plants which yield plant secondary metabolites is one of the significant practices in agriculture. The use of CRISPR technology in metabolic engineering in food crops is the key in the production of nutraceuticals containing plant secondary metabolites (PSMs) that can be incorporated into value-added foods. Thus, by developing new superfoods with better nutritional values, there is a huge possibility that the health of the entire world can be improved by increasing the accessibility of missing nutrients in widely used crops. CRISPR technologies have been found to be quite useful in increasing the level of PSMs in different classes of food crops, including carotenoids and vitamin precursors, oils, alkaloids and flavonoids, phenolics, terpenoids, non-protein amino acids, aromatics, and also decreasing levels of toxic metabolites (Yadav et al., 2021).

However, it is noteworthy that the use of CRISPR technology for metabolic engineering in food crops has certain limitations. Some plants have intricate metabolic processes that might be disrupted by CRISPR making off-target effects a possibility. Therefore, the challenge of regulatory acceptance and public perception of genetically modified crops must be solved so that the application of CRISPR food crops in the market can be effective.

19.3.1 Carotenoids Biosynthesis

Carotenoids are bioactive compounds belonging to the isoprenoid biosynthesis pathway that plays a role in different plant processes as photosynthesis, phytohormones biosynthesis, photoprotection, and pigments. These compounds are found in different plant organs and in addition to the physiological activity in plants, they also have significant functions in the human body. Due to the antioxidant capacity and vitamin A synthesis potential, carotenoids are linked to various positive impacts on human health and quality of life, such as improved vision and lowered incidence of chronic diseases (Li et al., 2020).

Among all the carotenoids, the following are regarded as the most important with appreciable properties: beta carotene, zeaxanthin, lutein, lycopene, and phytoene. These compounds provide a shield against such diseases as age-related diseases and assist in maintaining a healthy status of the body. Even more beneficial, there is β-carotene in foods that helps in the synthesis of vitamin A in the human body, hence making it more beneficial.

The application of CRISPR/Cas for genome editing has been helpful in changing the level of carotenoids within various plants. For example, the β-carotene levels in the rice endosperm have been increased through editing the OsOr gene and β-carotene levels in banana fruit through the targeting of

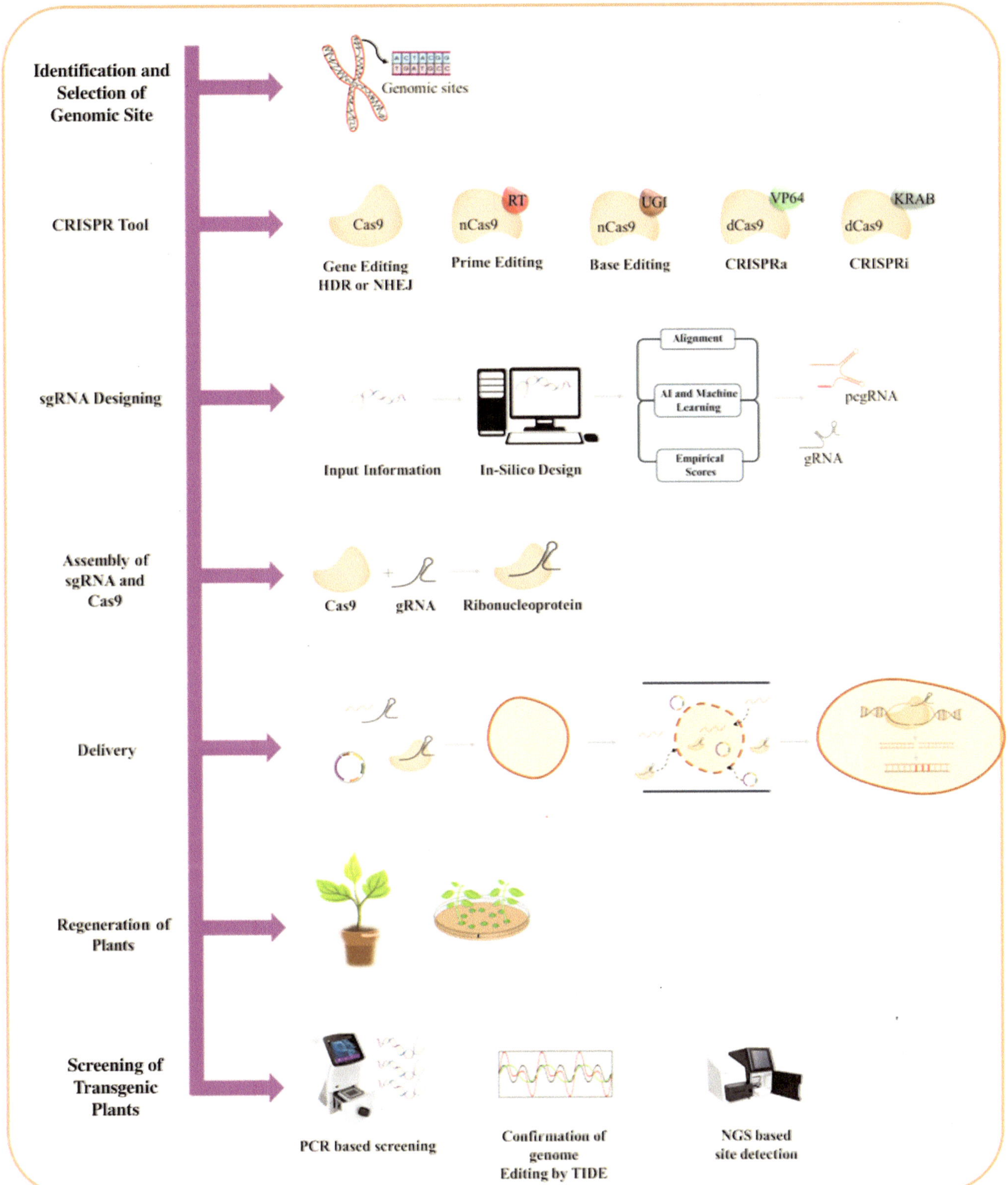

FIGURE 19.1 The pipeline of CRISPR/Cas-based metabolic engineering in plants.

the fifth exon of the lycopene epsilon-cyclase (LCYϵ) gene. Other important genes have also been edited using CRISPR/Cas9 to change the colors and develop higher accumulation of carotenoids in plants. Downregulation of CCD7 and CCD8 has been achieved through CRISPR/Cas to increase the synthesis of strigolactone and carotenoids (Stra et al., 2023).

For example, in horticultural plants like white-flowered Ipomoea nil, the CRISPR/Cas system for knocking out InCCD4 has been efficient. This genetic change led to pale yellow petal color that was also accompanied by increased total carotenoid concentration. In addition, pCRISPR system has also been used for the enhancement of carotenoids and flavonoids accumulation (Mipeshwaree Devi et al., 2023). Indeed, it has been possible to increase such plants through multiplex genome editing of genes that play a role in carotenoid biosynthesis and other metabolic pathways.

Tomato has proven to be very sensitive to the CRISPR/Cas9 method where several genes that determine carotenoid synthesis have been targeted. Specifically, this particular intervention resulted in a very significant reduction of the incidences of the disease by 5%. This work proved that through the introduction of CRISPR/Cas, there was 1-fold increase in lycopene accumulation, thus establishing the role of this technology in enhancing the favorable attributes. Moreover, from the study in which the SlSGR1 gene was silenced in *Solanum lycopersicum*, there were significant improvements in chlorophyll and carotenoids levels. The enhancement of lycopene content also fits the theory that SlSGR1 knockout influences several genes involved in carotenoid production (Li et al., 2018).

Therefore, precise CRISPR/Cas mediated genome editing has been used to enhance the carotenoid accumulation in numerous plants, altering the carotenoid biosynthetic pathways and enhancing the synthesis of strigolactones. These have brought about changes in plant architecture, changes in color pigments, changes in nutritional quality, which shows that CRISPR/Cas technology has the potential for plant genetic improvement and crop breeding.

19.3.2 Oils and Fatty Acids Biosynthesis

In oil crop research, CRISPR technology has been applied to modify some of the genes related to fatty acid (FA) synthesis. The primary objective is to enhance the production and nutritional value of edible oils. The factors that influence the quality of oil include fatty acid, N-S compounds, phenolics and sterols, flavonoids, and amongt all these, the fatty acids are essential.

The levels of monounsaturated fatty acids including oleic acid are usually elevated because this enhances the oxidative stability of the oils and the nutraceutical content. On the other hand, polyunsaturated fatty acids affect the product negatively by making it spoil faster and thus leads to shorter shelf life. In all the oilseed crops, such as allotetraploid peanuts, rapeseed, camelina, and soybean, specific genes like fatty acid desaturase 2 (FAD2) have been targeted by the CRISPR/Cas system to enhance the levels of oleic acid and oil stability (Ullah et al., 2021).

For example, in camelina, researchers have used CRISPR for the deletion of three alleles of the fatty acid elongase 1 (FAE1) gene. This modification resulted in the reduction of the amount of some VLCFAs (Very Long-Chain Fatty Acids) and the enhancement of oleate to linoleate ratio in the seed oil. It is not only used in regulating the fatty acid profile of the oil in respect to the consumers' choice and requirement of nutrition but also enhances the market value of the oil on the shelves; it also enhances the shelf life of the oil and its usability in frying (Bansal and Durrett, 2016).

This specific genetic engineering shows that the technology of CRISPR can be used to enhance the quality of the oil in oilseed crops. The objective of the CRISPR-KO for the genes BnSFAR4 and BnSFAR5 in rapeseed was to enhance seed oil yield without the loss of the vital seed features such as germination and seed health (Raipuria et al., 2021). Oilseeds are valued for their nutritionally rich content of bioactives, such as flavonoids, PUFA, organosulphur, phenolic compounds, and phytosterols. These compounds are used in the production of drugs, pesticides, and cosmetics among others, hence the value of oilseeds across the various industries.

For instance, Camelina sativa has very high PUFA, especially linolenic acid, which makes up 30–40% of the oil, and hence it is challenging to manipulate fatty acid profiles in oilseed crops. While linolenic acid is good for human consumption because it falls under the category of PUFA, the factor of rancidity comes into play because, with high PUFA content, it tends to go rancid quickly, making it hard to store or use. On the other hand, the oils containing monounsaturated fatty acids (MUFA), such as oleic acid are in high demand due to the improved oxidative stability and well-known health benefits (Song et al., 2023).

In response, scientists have considered the CRISPR/Cas9 system as a mode of genetic engineering that is more precise. As a result of the following genetic engineering of the fatty acid desaturase 2 (FAD2) gene, it has been able to decrease PUFA (Polyunsaturated Fatty Acids) and increase oleic acid in Camelina sativa. Besides, this change enhances the shelf life of the oil, and makes it more suitable for use in cooking and other industrial applications. This concept demonstrates that certain genetic manipulations may enhance the nutritional and functional quality of oilseed crops to fit the consumers' and industries' requirements. In agricultural biotechnology applications, the CRISPR/Cas methodologies hold a significant role in the enhancement of oil-bearing crop plants for desired nutritional use. In allotetraploid crops, including peanut (Arachis hypogaea L.), the scientists have used CRISPR/Cas to genetically edit the AhFAD2 genes that are associated with the fatty acid profile. The specific changes enhance the content of oleic acid in the developed peanut varieties that in turn assists the breeding programs to come up with oils that have improved stability and health benefits, as observed by Yuan et al. (2019). Similarly, in soybean, one of the most important legume crops around the globe, CRISPR/Cas9 has been applied to introduce genetic changes on the microsomal omega-6 desaturase gene known as FAD2-2. The kind of genetic modification aimed at has been proven to improve oleic acid to the highest levels of 65% and decrease the linoleic acid content up to 16% (Rauf et al., 2023). Such alterations are not only enhancing the nutritional quality of the oil but also the oxidative stability for many other applications such as storage for a longer period and many other industrial uses.

Moreover, in oil palm (Elaeis guineensis), the CRISPR/Cas9 multiplex genome-editing technique has been used to edit the FAD2 and PAT genes which are involved in the synthesis of fatty acids. Bahariah et al., in their study, have noted that due to the development of single and double gene knockouts, the high oleic acid oil palm varieties have been able to accumulate high levels of oleic acid. This advance also offers a good market for the oil palm and at the same time helps in the sustainability of the palm oil through the reduction in the process of hydrogenation which is unhealthy for the consumers (Bahariah et al., 2023). These examples emphasize the high potential of the use of CRISPR/Cas systems to enhance the nutritional properties of oils and beneficial characteristics of the oilseed plants for people and other branches of industry. Other similar innovations are still emerging and continue to shape the crop genetic engineering's future, creating new opportunities to develop improved and healthier agricultural products.

19.3.3 Enhancing Ascorbic Acid (Vitamin C) Production

In plants' physiology, researchers are discovering various methods of enhancing the levels of nutritional elements in plants, especially vitamin C in human beings. This nutrient is essential in human diet and is an antioxidant and a critical immunity strengthener. Regarding this study, it is focused on small parts of plant genes known as open reading frames (ORFs) in the ascorbate pathway. Among the developments are that where these uORF sections in particular genes are altered slightly in specific genes, like that of GDP-l-galactose phosphorylase in plants such as lettuce and wild relatives to tomatoes, then synthesis of vitamin C is boosted. This enzyme is engaged in the formation of ascorbic acid (Li et al., 2018). However, one cannot ignore the fact that the application of the ultra-modern methods, including CRISPR/Cas9 gene-edited technology, has expandeded the opportunity further. Studies conducted recently have centered on a gene that is known as PLP, which codes for photoreceptor that is sensitive to blue light. It appears that this protein is meant to inhibit the synthesis of vitamin C and because of the disruption of the PLP gene, scientists have been able to enhance the level of vitamin C in tomatoes by three-fold. Such groundbreaking findings are not only helpful in enhancing the understanding of how plants regulate the concentration of vitamin C within cells and other plant cells, but also for enhancing biofortification of various crops so that our fruits and vegetables are even more nutritious for us to consume (Mishra et al., 2023). Such researches may have significant implications for the world's nutrition, especially in the areas where proper availability of nutrients remains a concern. Apart from its role in human nutrition, ascorbate has many roles in plant physiology. It is involved in the detoxification of free radicals and protects the plants from oxidative damage and harm that may be occasioned by ROS. It also plays a role in several other cellular activities such as cell division, cell growth, development of the cell membrane and synthesis, and the function of hormones, among others. Consequently, it is believed that enhancement of ascorbate levels in crops by genetic engineering is expected to enhance yield, stress tolerance, and the nutritional quality of the crops meant for human consumption (Sabzehzari et al., 2020).

19.3.4 Increasing Terpene Production

Saponins are a vast class of triterpenes or steroid glycosides, and they are found in plants of the Medicago family. These compounds include haemolytic sapogenins which exhibit high biological and pharmacological activity but soyasapogenol saponins do not exhibit such properties. However, it seems that when the CRISPR/Cas was applied in Medicago truncatula, significant advances have been made in regulating saponin synthesis. Specifically, they have succeeded in knocking in a mutation in the CYP93E2 gene using the CRISPR/Cas gene-editing system. This mutant interrupts the biosynthesis of soyasapogenols but boosts the synthesis of such haemolytic sapogenins that are valued. This breakthrough illustrates that the CRISPR/Cas technology can be applied to fine-tune the saponin biosynthesis pathway for several applications (Sabzehzari et al., 2020). Currently chicory is one of the most developed agricultural crops due to inulin production apart from being used for culinary and medicinal purposes in ancient times. It is a natural sweetener and prebiotic that is derived from a polymer of fructose and is preferred for some of its health benefits as it enhances gut health and enhances calcium uptake. Among them there is one issue: bitter compounds, sesquiterpene lactones (STLs) which can be present in chicory and could potentially have a negative effect on the taste of products containing inulin (Mipeshwaree Devi et al., 2023). To counter this problem, the scientific community has turned to the complex molecular scissors: CRISPR/Cas9. They have focused on the genes concerned in the first step of the STL synthesis pathway in chicory, that is, Germacrene A synthase (CiGAS). This is important in the synthesis of STLs. In the latest advances made by the scientists, they have successfully achieved the goal of knocking out the four genes that synthesize CiGAS enzyme and in the process eradicating the bitter taste (Cankar et al., 2023).

Thus, chicory plants can produce inulin and do not have a bitter taste; as a result, the market value of the chicory plant and the use of inulin as an organic sweetening agent rise. This advance goes hand in hand with the chance of improving the quality and taste of foods containing inulin, besides opening up more opportunities for the functional application of the prebiotic inulin, which now poses a threat to the food industries and consumers' health, in a way that is beneficial (Mipeshwaree Devi et al., 2023).

Salvia miltiorrhiza, or Chinese medicinal herb, includes diterpene compounds, that are tanshinones: cryptotanshinone, tanshinone IIA, and tanshinone I and phenolic acids: rosmarinic acid, salvianolic acid, and lithospermic acid. The study used CRISPR/Cas9 to generate SmCPS1 knockout plants since this gene is essential in the synthesis of tanshinone. Homozygous mutants did not produce tanshinones, but the phenolic metabolites were not affected as in the wild type plants. However, in the chimeric mutants, there was an increase in tanshinone. Thus, CRISPR/Cas9 emerges as a powerful tool for dissecting and manipulating pathways that govern the production of SSYs, which will significantly expand our understanding of complex biochemical pathways (Shi et al., 2021).

19.3.5 Increasing the Aroma Profile and the Nutraceutical Potential

It is a well-known fact that plants produce numerous volatile substances, these are synthesized in specific tissues, glands, or ducts that are localized in different regions of the plant body. These compounds have a high utility value because of their application in industries such as fragrances, cosmetics, drugs and chemicals, bakery, and many others. Rice, an important grain crop, is classified into aromatic or non-aromatic based on the occurrence of some particular scent. The aroma in rice is mainly attributed to 2-acetyl-1-pyrroline (2-AP) and that is controlled by the BADH2 gene on chromosome 8. The practical application of the CRISPR/Cas9 system showed that it is possible to change the expression level of BADH2 gene, and, thus, the content of 2-AP in rice. This modification was done by directing the RNA polymerase II to the intron 2 splice donor site (GT) of the OsBADH2 gene, thus changing the gene's splicing site and an increase of 2-AP in the mutants (Abbas et al., 2023). New alleles of OsBADH2 have been produced using CRISPR/Cas9, which has made the formerly non-aromatic rice varieties have a moderate fragrance. The same approach has also been applied to develop aromatic maize and fragrant sorghum by targeting the BADH2 orthologs to increase the generation of 2-AP. CRISPR technology is now helping to transform how foods are made more nutritional through changes in the inheritable characteristics such as sweetness, bitterness, and taste. These changes not only make healthy foods appetizing but might also benefit consumers in adopting healthier diets. Besides the flavor modulation, CRISPR is applied for the reduction of toxic compounds, for instance, steroidal glycoalkaloids in tomatoes and potatoes, and for the development of low phytate soybeans and low gluten wheat. Furthermore, it also allows the enhancement of production of desirable products such as the γ-aminobutyric acid (GABA) in tomatoes (Shelake et al., 2023). The launch of a high-GABA tomato on the market in 2021 is a perfect example, it has been proved to enhance the health of the human heart and the mental health of a person. These advances illustrate the idea of applying traditional genetic engineering to metabolic engineering to create plant-based nutraceuticals to support health and wellness and the preventive healthcare approach (Singh, 2023).

19.3.6 Amplifying Flavonoid Biosynthesis

Flavonoids, secondary metabolites in plants, serve dual roles: they have functions and roles in defense and also in pigmentation of fruits and flowers and origin of their color and smell. Flavonoids are well-known antioxidants, anti-inflammatories, and anticancer agents with subclasses such as flavones, flavonols, flavanones, flavanols, and anthocyanins. Pigmentation in some varieties of rice has been interfered with using CRISPR/Cas technology to target anthocyanins. Scientists were also able to knock out genes like OsF3'H, OsDFR, OsLDOX, and OsTTG1 and manipulated the pathway of anthocyanin production (H. Tariq et al., 2023; Jung et al., 2019).

In tomatoes, the fruit color is the net result of the action of many compounds, such as carotenoids, flavonoids, and chlorophyll-derived products. By employing the CRISPR/Cas9 gene-editing technique, the authors were able to edit multiple genes that include PSY1, MYB12, and SRG1; thereby creating tomato varieties that have red fruits, with the ability to produce varieties of color shades ranging from yellow, brown, pink, light yellow, pink-brown, yellow-green, and light green. In cultivated apples, the dihydrochalcone (DHC) which is phloridzin and its precursor, phloretin, has useful therapeutic properties for the treatment of degenerative diseases such as Alzheimer's diseases. Similarly, using CRISPR/Cas9 to knock down the MdPGT1 gene, the foliar phloridzin levels were reduced while the growth was normal, where knock down lines depicted a great decrease in the growth retardation (Mikhaylova, 2024). For Tartary buckwheat, the FTMYB45 transcription factor, which is a negative regulator of flavonoids, was subjected to precise editing to enhance the flavonoid content. Soybean plants were genetically modified using the CRISPR/Cas9 system to minimize the function of the GmUGT gene, thus improving resistance to the chewing insect and increasing the flavonoid production. In a similar manner, down-regulation of the F3'H gene in red poinsettias led to the synthesis of pelargonidin and in grapevines, removal of the negative controller VvBZIP36 boosted the formation of anthocyanin. Also, the use of multiplex CRISPR/Cas9 system was aimed at certain enzymes involved in isoflavonoid biosynthesis in soybeans, and it has resulted in the elevated level of isoflavones in seeds and improved resistance to SMV. This information has raised the possibilities of using CRISPR/Cas9 in understanding the complexity of flavonoid metabolism with regard to crops and human beings (Mipeshwaree Devi et al., 2023).

19.3.7 Alkaloids and Their Manipulation Through CRISPR/Cas9

The Opium poppy (Papaver somniferum L.) primarily produces BIAs (benzyl isoquinoline alkaloids) which are secondary metabolites with major pharmacological importance and research interest. The elimination of 4'OMT and 4'OMT2 genes by employing the CRISPR/Cas9 technology has led to a reduction in morphine and thebaine, which proves that the CRISPR/Cas9 can be used to manipulate the BIAs metabolism and biosynthesis in Papaver species. In the case of tobacco, the generation of alkaloids includes enzymes belonging to the BBL (Berberine Bridge Like) group of enzymes. Using CRISPR/Cas editing technique, the BBL gene family was knocked out to get a non-transgenic Nicotiana tabacum which is devoid of nicotine (Alagoz et al., 2016).

Caffeine, a purine alkaloid which is present in tea, contributes to over 90% of the total alkaloid content. The CsHB1 is among the transcription factor genes from the HD-Zip class of proteins that regulates the genes of caffeine biosynthesis, including the yhNMT1 gene. When CsHB1 was downregulated by CRISPR/Cas, the content of caffeine reduced in tea callus. Tomatoes have been reported to contain steroidal glycoalkaloids (SGAs) such as tomatine and dehydrotomatine; they have a defensive function against several insects and pathogens. Knockdown of SlS5aR2 gene using CRISPR/Cas9 that encodes a protein involved in the synthesis of C5a and contributes to the biosynthesis of a-tomatine resulted in low levels of a-tomatine and high amounts of dehydrotomatine (Bhushan et al., 2023).

In the Solanum tuberosum containing high levels of SGAs, it has been revealed that silencing of StSSR2 gene using CRISPR/Cas9 technique has led to decreased SGA levels in both leaves and tubers (Liu et al., 2017).

Symphytum officinale commonly used as a pain killer, anti-inflammatory and for tissue healing has poisonous compounds referred to as pyrrolizidine alkaloids (PAs). In response, researchers used the CRISPR/Cas9 approach to silence the HSS gene that is responsible for the production of PA through homospermidine synthase enzyme. By this, they established plants with markedly low levels of PA; the reduction of PA content was due to single gene modification while double modification led to the complete elimination of these toxic alkaloids (Zakaria et al., 2021).

Thus, using CRISPR/Cas9 technology, it became possible to increase or decrease the production of alkaloids, which can open up vast potential for the application of alkaloids in practice in fields such as pharmacology, agriculture, and the food industry. Further research can be focused on the elaboration of these methods and evaluation of their applicability for other plant species and alkaloids.

19.4 ENHANCING SECONDARY METABOLITE PRODUCTION IN MEDICINAL PLANTS THROUGH THE CRISPR/CAS TECHNOLOGY

Medicinal plants produce secondary metabolites but generally in small amounts. This is because there is a cost-benefit trade-off and the plants need to go through normal growth and reproduction and modulate the toxicity of the PSMs through synthesis or storage. The isolation of pharmacologically active plant secondary metabolites (PSMs) from medicinal plants is difficult and costly since the compounds are complex and time-consuming to synthesize chemically and the processes of their synthesis are not well understood. Since extraction from plants is still possible, the issue of over-collection seems to arise because of the scarcity of the raw materials. To address these challenges, GE tools, especially those that employ CRISPR technology, have been used by researchers. CRISPR-based GE has been applied to medicinal plants to generate protocols for plant regeneration, to analyze the enzymatic activities during the metabolic process, to modify several agronomical traits and to enhance the production of phytochemicals (Arun et al., 2023; Naik et al., 2023).

Thus, new possibilities of increasing the yield of plant secondary metabolites in medicinal plants are opened up with the help of the CRISPR-based genetic engineering tools. Because this approach also positively contributes to preservation of the natural plant resource base, the relative availability of these important chemicals for medicinal use is enhanced. Future work could be directed to enhance these genetic engineering techniques and to explore the applicability of these techniques in medicinal plants (Das et al., 2024).

19.4.1 CRISPR Advances for Enhanced Therapeutic Properties of Salvia Miltiorrhiza

Salvia miltiorrhiza is a diploid herb with 16 chromosomes, belonging to the Labiatae family, found to have great medicinal value. Its use has been attributed to the fact that it has a short generation time, has genome fully sequenced and has a well-established tissue culture-genetic transformation system. This plant contains lipid-soluble tanshinones and water-soluble phenolic acids which are believed to have therapeutic impacts on cerebral and inflammation diseases (Guo et al., 2022).

In 2017, a significant advance was achieved by employing the CRISPR/Cas9 system to precisely modify the CPS1 gene responsible for diterpene synthase in the tanshinone biosynthetic pathway (Camerlengo et al., 2022; Diez-Fernandez and Häberle, 2017). This achievement paved the way to use of CRISPR tools to modify metabolic pathways of S. miltiorrhiza. Another study employed CRISPR techniques to knock out genes which are found in S. miltiorrhiza for various purposes. For example, attempts have been made to change the genes associated with phenolic acid biosynthesis such as rosmarinic acid synthase, multiple laccase family genes, and the basic leucine zipper 2 transcription factor. Moreover, knockout of two cytochrome P450 family enzymes CYP76AK2, and CYP76AK3 validated their role as activators of tanshinone synthesis. From the previous research on the mutants of MYB98 and bZIP1, it has been identified that they regulate the tanshinone and phenolic acid biosynthesis at the transcriptional level. Such discoveries show how CRISPR technology helps in improving the knowledge and regulation of metabolic processes in Salvia miltiorrhiza (Mipeshwaree Devi et al., 2023; Sharma et al., 2019).

19.4.2 CRISPR Advances in Artemisia Annua

Artemisinin, a sesquiterpene endoperoxide originally isolated from *Artemisia annua* sweet wormwood in 1972 remains the key treatment for malaria. Although there have been efforts to engineer microbial production and develop heterologous plant expression systems, A. annua remains the primary source, despite being limited. Recently, the use of artemisinin has been extended to other diseases and therefore its consumption has risen.

Artemisinin is biosynthesized in the glandular secreting trichomes or the GSTs of A. annua. A variety of CRISPR tools are employed in the identification of the roles of transcriptional factors (TFs) in GST development. For instance, targeted mutagenesis of enzymes such as squalene synthase (SQS) which is crucial for artemisinin synthesis is severely affected. Furthermore, the CRISPR/Cas9 knockout or repression of a gene called TAR2 in the R2R3 MYB family is beneficial to the development of GSTs by increasing their numbers to produce more artemisinin (El-Desoky et al., 2022).

Based on the above findings, the roles of TFs in regulation of GST formation and artemisinin biosynthesis can be further explored using the CRISPR technology for the purpose of increasing the yield of artemisinin. Because this research could have implications for malaria and the treatment of other diseases, further studies on CRISPR in A. annua are important to discover better methods to produce artemisinin (Alok et al., 2020).

19.4.3 Studies Involving CRISPR Technology in the Panax Genus for Enhanced Therapeutic Efficiency

Ginseng is referred to as the king of oriental medicine as it is widely used to help people fight numerous stressors and treat diseases including cancer, inflammation, diabetes, and nervous disorders. These benefits are derived from the roots and consist of ginsenosides as the major effective ingredients. The more recent studies in ginseng have included the whole-genome sequencing of different Panax species, including P. ginseng, P. japonicus, P. notoginseng, P. quinquefolius, and P. stipuleanatus, in which they have identified genes that are actively involved in the synthesis of ginsenosides. Choi and colleagues' (2022) study aimed at generating knockout lines of two CYP genes (CYP716A47 and CYP716A53v2) in P. ginseng through CRISPR technology (Dan et al., 2024; Fan et al., 2020). This genetic shift changed the biosynthetic pathway of saponin, thus formulating only PPD saponin (Protopanaxadiol-type Saponins) group saponin and eliminating PPT saponin (Polyphyllin/Paris Polyphylla Saponin). In another study, CRISPR was demonstrated as a tool to improve certain ginsenosides by knocking down the PPT group biosynthesis and at the same time, upregulating the PPD group saponin Rg3 (Choi et al., 2022). This entailed the overexpression of enzymes such as PAL and Pq3-O-UGT2, and CRISPR knockout of SE, which led to an extraordinary enhancement of the Rg3 level by 21-fold. These works shed light on the importance of the targeted genetic changes in the further investigation and improvement of ginseng's medicinal value as well as the potential for discovering new applications for the plant in the field of medicine.

19.4.4 Using Tobacco as an Ideal Plant Model for Metabolic Engineering

In metabolic engineering, various plants such as *N. benthamiana* and *N. tabacum* are being developed as an important source of producing secondary metabolites or any other valuable products of industrial importance. Some of the benefits include a non-food nature, a high rate of growth, and well-known procedures such as transient agroinfiltration, and genetic transformation that are suitable for molecular farming (Alagoz et al., 2016). In the past, tobacco was applied in the treatment of diseases such as Syphilis and the Black Death. Today, it is used in engineering for plant secondary metabolites (PSMs), such as artemisinic acid and taxadiene, by the introduction of synthetic genetic elements or by altering its own metabolism (Molina-Hidalgo et al., 2021). For instance, Herman et al. (2021) have transformed tobacco and incorporated α-1,3-fucosyltransferase and β-1,2-xylosyltransferase that has given rise to knockout mutants that boost glycoprotein expression important for Hepatitis B vaccine production. CRISPR/Cas9 technology is more dominant in tobacco, which is used to modify flavonoid biosynthetic pathways to enhance the accumulation of certain flavonoids, such as eriodictyol and quercetin. Also, the CRISPR/Cas9-based knockout of berberine bridge enzyme-like family genes has been useful in lowering the nicotine content and rationalizing downstream processing. These advances highlight the possibility of using tobacco as an expressive platform for metabolic engineering by encompassing the plant growth habits alongside the flexibility of the CRISPR system (Zhang et al., 2023).

19.5 MULTIPLEX GENOME MODIFICATION IN MEDICINAL PLANTS

19.5.1 Exploring Laccase Genes in Salvia Miltiorrhiza for Enhanced Therapeutic Potential

CRISPR/Cas editing platform-based methods are great for studying genes that perform similar tasks. Therefore, if the target gene is one of the multiple genes with related functions or if there are multiple variants of the gene, then it becomes crucial to deliver multiple sgRNAs and Cas proteins into the same system. A study used CRISPR/Cas9 to knock out multiple, especially the laccase gene family in Salvia miltiorrhiza, which is a medicinal plant in traditional Chinese medicine (Munir et al., 2024; Zhou et al., 2024).

Previous studies show that *Salvia miltiorrhiza* has bioactives that are beneficial for cardiovascular and cerebral disorders: tanshinones which are phenolic compounds consist of rosmarinic acid (RA) and salvianolic acid B (SAB). Certain enzymes such as laccases are used in this plant in the chemical reactions that are employed in producing SAB. Twenty-nine laccases (SmLACs) were identified in S. miltiorrhiza; of these, SmLAC7 and SmLAC20 were chosen to investigate their roles in SAB biosynthesis (Zou et al., 2021). There might be number of laccase genes and that is why it is challenging to define how these genes affect the characteristics of plants (Yang et al., 2022).

19.5.2 Immunogenic Glycoproteins in Nicotiana Tabacum for Biopharmaceutical Applications

A study used a multiplex CRISPR/Cas9 system to simultaneously knock out four fucosyltransferase (FucT) and two xylosyltransferase (XylT) genes from Nicotiana tabacum. These enzymes are involved in the synthesis of plant-derived glycoproteins that can activate the immune processes. That is why, through knocking out FucT and XylT, the study aimed at creating a more secure background for generating recombinant proteins in plant-based systems. The study proved that these immunogenic residues have been knocked out effectively, thus paving the way for a safe approach in the manufacture of therapeutic proteins (Munir et al., 2024).

19.6 ESSENTIAL TRANSCRIPTION FACTOR FAMILIES REGULATING METABOLIC PATHWAYS IN MEDICINAL PLANTS

Stress response in plants is followed by the synthesis of a few secondary metabolites through signal transduction pathways, which includes flavonoids, alkaloids, terpenoids, polyphenols, and other related compounds. These pathways are controlled by Transcription factors (TFs) that bind to specific sequences in the gene promoters and therefore control the activity of genes. This general information of how TFs control metabolic processes may be helpful in either finding out or enhancing secondary metabolites in the plants or synthesizing some of them if they are yet to be discovered. Generating knockout lines of TF genes is aimed at understanding the genes and their role in such pathways. CRISPR/Cas9 can be employed to regulate pathways responsible for the synthesis of secondary metabolites by precisely targeting TFs, thereby enhancing secondary metabolite synthesis in plants (M. Wang et al., 2021). This shows how they are involved in controlling the synthesis of secondary metabolites in medicinal plant species and offers methods on how to harness genome editing for increasing the production of certain metabolites (Wang et al., 2017).

19.7 CHALLENGES OF CRISPR/CAS IN PLANT METABOLIC ENGINEERING

19.7.1 Off-Targets and Unintended Consequences

The CRISPR/Cas system, a revolutionary gene-editing tool, provides a precise method of modifying DNA sequences. However, it has some disadvantages, such as the problem of off-target effects where other similar genetic sequences are also altered. In a study reported by Modrzejewski et al. (2020), the following factors led to the above-mentioned unintended change. A logistic regression analysis has pinpointed five critical factors affecting off-target precision: Some aspects in CRISPR/Cas gene editing are: (1) the total number of mismatches, and the structural incorporation of the sgRNA; (2) the location of these mismatches in the genetic sequence; (3) the GC content of the sequence; (4) the selection of the nuclease variant; and (5) the method used to deliver the CRISPR/Cas components into the cells.

Optimizing the sgRNA design by increasing mismatch tolerance and refining their structure has been proven to reduce off-target events. For instance, introducing even a single mismatch can decrease off-target activity by up to 59%, with greater reductions observed when four or more mismatches are present. Notably, mismatches occurring within the seed region—specifically the first eight nucleotides adjacent to the protospacer adjacent motif (PAM)—are particularly efficient at minimizing these effects. However, the influence of factors such as GC content, the use of different nuclease types like Cas9 nickases, and various delivery methods on off-target effects is not yet fully understood, necessitating further investigation. The ongoing challenge is to enhance the sgRNA design to achieve more precise targeting and minimize unintended genetic alterations, thus optimizing the effectiveness and safety of CRISPR/Cas applications (Gohil et al., 2021).

19.7.2 Delivery Efficiency of the CRISPR/Cas9 Tool

The major limitation in employing CRISPR/Cas9 system in plant biotechnology is the possibility of delivering Cas9 protein and the sgRNA into the plant cells. Challenges come in the form of low efficiency of the transformation procedures and the selection of the right DNA plasmid vectors Cas9 and gRNAs. There might be new ways of addressing such limitations and enhancing the efficiency of delivery. One of them is the transduction of the Cas9 protein prior to its expression into the cells (Luther et al., 2018). Such a transfection method ensures that the Cas9 system is inside the cells prior to the transfection that will enable it to work in conjunction with the sgRNA upon its delivery (Munir et al., 2024). This may possibly eliminate the need to provide both at once and therefore may help in simplifying the transformation process. Another approach employs delivery of Cas9 and sgRNA in a single complex outside the cell which is called the ribonucleoprotein (RNP) complex. This approach entails direct uptake of the preformed complex into the cell, which is less time-consuming and may be more accurate since the components of CRISPR remain active as soon as they get into the cell. Similarly, the idea also holds for the use of nanoparticles for the targeted delivery of CRISPR/Cas components into cells. It also enhances the chances of delivery of the CRISPR system to the intended genomic loci within the cellular system. All these strategies outlined here provide potential approaches to addressing the current approaches to delivering CRISPR/Cas9 in plant systems and designing how to improve the process and outcomes of genetic microsurgery (Lino et al., 2018).

19.7.3 Driving Machinery behind the CRISPR/Cas System

Application of the CRISPR/Cas system in native plant systems needs significant advances in the working parameters of the system, such as promoters, terminators, selectable markers, and the methods of transformation. Medicinal plants, however, have the following disadvantages because their endogenous gene expression control mechanisms for the synthesis of valuable bioactive products are closely related, which hampers the plant's ability to produce these compounds and therefore influences the efficiency of CRISPR/Cas-based gene editing. These are the promoters and terminators that define how the CRISPR/Cas components are to be transcribed in plants, therefore, enhancing their function and compatibility with plant-specific regulatory networks could offer better gene editing (Shivram et al., 2021). The selectable markers that are used to identify cells that have been successfully transformed with CRISPR/Cas components also need improvement so they can work in the different biochemical environments of various plants. Transformation protocols should also be optimized with reference to the plant cell walls and the potential for tissue regeneration because they vary within and between species and influence the CRISPR/Cas-based gene-editing efficiency. Improvement of these interactions by sophisticated engineering of the CRISPR/Cas system and its components could lead to the increased precision and efficiency of their function, and thus open up the potential of several revolutionary advances in the production of medicinal and other valuable plant-derived products (Ahmad et al., 2024).

19.7.4 Complexities in the Genotype to Phenotype Linkage

CRISPR/Cas9 is a widely known gene-editing tool that is considered efficient and relatively cost-effective compared to other gene-editing tools. However, one of the major limitations that it encounters is the difficulty in correlating the changes made in the DNA sequence (genotype) to the outcomes that are phenotypically visible in an organism. This difficulty occurs mainly because phenotypes are not only defined by genotypes but also influenced by environmental factors. In response, researchers are seeking better tools that could help in narrowing down the gap between genotype and phenotype. Some of these breakthroughs include the CRISPR (MAGIC) systems as well as High Throughput Biosensors. MAGIC, or Multiplexed Accurate Genome Editing with CRISPR, is a system intended to increase the accuracy and feasibility of CRISPR editing, which allows for the application of multiple edits (Mipeshwaree Devi et al., 2023).

It permits more complex manipulation of the genetic code to be linked with the resultant phenotypes. High-throughput biosensors can generate biological responses in gene-edited organisms and provide data about the impact of changes in DNA on the phenotype. These sensors prove very useful when screening and analyzing large number of gene-edited cells or organisms to quickly identify the changes that lead to the desired phenotypes. Incorporation of these tools in the CRISPR/Cas9 process helps in understanding the impact of specific inherited changes on the organism's phenotype. According to the study done by Lian et al. (2019), these technologies enhance the efficiency of gene editing and, at the same time, the understanding of phenotypic effects of these modifications, considering its possible applications in research and therapy.

19.7.5 Effective Transformation and Regeneration Strategies

The crucial need for efficient plant transformation and regeneration systems is vital to fully harness the potential of gene-editing tools such as CRISPR/Cas9 in plant biotechnology. It is clear that many plant species have trouble in achieving high transformation efficiency due to poor tissue culture or transformation protocols. Overcoming these limitations is paramount to the further development of the use of gene editing in plant biotechnology. In vitro tissue culture and protoplast regeneration are the important methods which have been employed, which include the process of growing plant cells or tissues under controlled conditions. One of the advantages of using in vitro culture systems is that the researcher can directly control and observe the implications of the CRISPR/Cas9 system on the plant cells and be sure that the cells contain the intended genetic modifications before the cells are prompted to form whole new plants (Alok et al., 2018). This approach not only increases the efficiency of the transformation process but also increases the successful regeneration rates of the modified plants. The controlled conditions of the in vitro systems lower the biological variation that one comes across in direct plant transformations and minimizes some of the natural barriers that hinder cell uptake of the foreign DNA. These developments play an important role in addressing some of the limitations present in species that usually do not respond well to conventional transformation methods, hence paving the way for the expansion of the use of CRISPR technology to as many plants as possible (Mipeshwaree Devi et al., 2023).

19.7.6 Cell Repair Mechanism

Most of the gene-editing tools, especially those involving CRISPR/Cas9 systems, use the process of introducing double-stranded breaks (DSBs) in the DNA, which is crucial to bring about genetic changes. Essentially, the processes that repair these breaks are vital to the fate of any genetic manipulations done. There are two primary pathways through which cells repair DSBs: such as homologous recombination (HR) and non-homologous end joining (NHEJ). Homologous recombination is a high-fidelity repair mechanism that needs a homologous DNA sequence to direct the repair event and is typically used for gene targeting when the DNA sequence to be targeted or inserted is available. This method is suitable for cases where the modifications must be made on a particular gene sequence and the alteration or replacement of such sequences leads to the regulation of metabolic pathways (Sultan et al., 2022; Nishida and Kondo, 2021). Non-homologous end joining is another kind of repair technique that is more likely to lead to errors and does not involve the use of a DNA template. NHEJ can lead to the formation of indels at the site of the break (Aslam et al., 2022). These indels can cause frameshift mutations, which can result in the knockout or re-activation of genes or affect metabolic pathways, which can be a problem in metabolic engineering where specific genetic modifications are desirable. It is important to note that all these gene-editing outcomes, including gene arrangement, knockout, and replacement play a critical role in increasing the production of secondary metabolites within plants. Thus, gene editing provides highly effective tools for altering genomes to increase the yield of metabolites, but the possibility of NHEJ indicates that it is necessary to be very accurate in the use of these technologies. The precision serves to confirm that improvements made to metabolic pathways are positive changes and do not compromise the plant's genotype or phenotype (Sabzehzari et al., 2020).

19.8 PROSPECTS AND PERSPECTIVES ON THE USE OF THE CRISPR/CAS SYSTEM TO ENHANCE SPECIALIZED METABOLITES IN MEDICINAL PLANTS

Medicinal plants therefore are important in many aspects of economic activities, in the fields of pharmacology and biomedicine, in nutrition, as well as in industry. Modifying their metabolic routes appears to be an effective strategy for increasing the synthesis of specialized metabolites to meet increasing needs. CRISPR/Cas9 gene editing has provided the breakthrough in this area to make the desired changes in plant genome and epigenome, transcription, and translation. Newer techniques in CRISPR/Cas include HDR and single nucleotide base editing, which have broadened the reach of the technology. This enables gene editing through knockout, knockdown using CRISPR interference, activation using CRISPR activation, and the establishment of desired DNA sequences originating from improved plant species. Such approaches help to improve the genome-editing methodologies, such as using synthetic or endogenous promoters, and terminators for the large-scale production of desirable metabolites.

Multiplex genome editing involves editing several genes at the same time, addressing problems that relate to the regulation of the genes. Furthermore, understanding the significance of transcription factors and the changes in the expression of genes can be considered as the prospects for the higher production of metabolites. CRISPR/Cas technology is an effective tool for functional studies, which reveals new directions in medicinal plants, for example, the cannabinoid and terpenoid biosynthesis in Cannabis sativa. Some of the important components that should be in place for efficient plant gene editing include regeneration and transformation systems. Nevertheless, constant difficulties in using CRISPR-based methods for altering endogenous genes have been acknowledged, but the system has received widespread acceptance. Post-transcriptional regulation through cis-regulatory elements and untranslated regions provides efficient ways of fine-tuning gene expression. However, the CRISPR/Cas system turns out to be more economical and efficient for gene editing and has the prospect for further developments in the field of medicinal plants.

19.9 CONCLUSION

CRISPR/Cas9 gene editing in medicinal plants has some limitations due to the lack of massive genomic and transcriptomic data. Comprehensive sequence data and functional studies on metabolic pathways are essential for the performance of the coordination of enzymatic activities and the control of complex metabolic networks. This knowledge is crucial to achieving an appropriate use of CRISPR/Cas9 techniques in medicinal plants. Genomics and transcriptomics studies open up the opportunity to describe the intricate process of the biosynthesis of certain metabolites. Based on this knowledge, researchers can identify some of the genes that are associated with the production of metabolites with bioactivity, with nutritional and pharmacological significance. The CRISPR/Cas9 has the possibility to modify these essential genes to improve the production of these vital metabolites in medicinal plants. The overall enhancement of knowledge in genome, transcriptome, and proteome processes related to the biosynthesis of specialized metabolites, along with advances in gene editing that is still ongoing, makes it possible to synthesize these metabolites at a reasonable cost in pharmaceutical quality medicinal plants. Nevertheless, it is obligatory to conduct the R&D under the legal and ethical standards, which will help the use of gene-editing technologies in the sphere of medicinal plants safely.

REFERENCES

Abbas, F., Zhou, Y., O'Neill Rothenberg, D., Alam, I., Ke, Y. and Wang, H.-C. 2023. Aroma components in horticultural crops: chemical diversity and usage of metabolic engineering for industrial applications. *Plants,* 12, 1748.

Ahmad, A., Munir, A., Zafar, H., Zahoor, M. K., Hassan, S. and Khan, S. H. 2024. Tracking footprints of CRISPR-based genome editing. In K. A. Abd-Elsalam and A. Ahmad (Eds.), Global regulatory outlook for CRISPRized plants. Elsevier.

Alagoz, Y., Gurkok, T., Zhang, B., and Unver, T. 2016. Manipulating the biosynthesis of bioactive compound alkaloids for next-generation metabolic engineering in opium poppy using CRISPR-Cas 9 genome editing technology. *Scientific Reports,* 6, 30910.

Alok, A., Jain, P., Kumar, J., Yajnik, K., and Bhalothia, P. 2020. Genome engineering in medicinally important plants using CRISPR/Cas9 tool. In V. Singh and P. K. Dhar (Eds.), Genome engineering via CRISPR-Cas9 system. Elsevier.

Alok, A., Kumar, J., Jogam, P., and Sandhya, D. 2018. CRISPR/Cas9-mediated gene editing tool and fathomless genetic and metabolic engineering applications in plants. In S. K. Yadav, V. Yadav, and S. K. Singh (Eds.), Recent trends and techniques in plant metabolic engineering (pp. 167–179). Springer.

Arun, M., Halka, J., and Kowsalya, K. 2023. CRISPR/Cas9: A novel genetic tool to manipulate plant secondary metabolite pathways. In R. Singh and N. Kumar (Eds.), Genetic manipulation of secondary metabolites in medicinal plant. Springer.

Aslam, S., Munir, A., Aslam, H. M. U., Khan, S. H. ,and Ahmad, A. 2022. Genome editing advances in soybean improvement against biotic and abiotic stresses. In S. H. Wani, N. u R. Sofi, M. A.

Bhat, and F. Lin (Eds.), Soybean improvement: Physiological, molecular and genetic perspectives. Springer.

Bahariah, B., Masani, M. Y. A., Fizree, M. P. M. A. A., Abd Rasid, O., and Parveez, G. K. A. 2023. Multiplex CRISPR/Cas9 gene-editing platform in oil palm targeting mutations in EgFAD2 and EgPAT genes. *Journal of Genetic Engineering and Biotechnology,* 21, 3.

Bansal, S. and Durrett, T. P. 2016. *Camelina sativa: An ideal platform for the metabolic engineering and field production of industrial lipids.* Biochimie,120, 9–16.

Bhushan, S., Sharma, D., Rakshant, Kaul, S., Dhar, M. K., and Sharma, M. 2023. Recent strategies to engineer alkaloid biosynthesis in medicinal plants. In A. Husen and M. Iqbal (Eds.), Medicinal plants: Their response to abiotic stress. Springer.

Camerlengo, F., Frittelli, A., and Pagliarello, R. 2022. CRISPR towards a sustainable agriculture. *Encyclopedia,* 2, 538–558.

Cankar, K., Van Laere, K., and Bosch, D. 2023. Genome editing for reduction of bitterness and for production of medicinal terpenes in Cichorium species. In A. Ricroch et al. (Eds.), *A roadmap for plant genome editing.* Springer..

Choi, H. S., Koo, H. B., Jeon, S. W., Han, J. Y., Kim, J. S., Jun, K. M., and Choi, Y. E. 2022. Modification of ginsenoside saponin composition via the CRISPR/Cas9-mediated knockout of protopanaxadiol 6-hydroxylase gene in Panax ginseng. *Journal of Ginseng Research,* 46, 505–514.

Courdavault, V., O'connor, S. E., Jensen, M. K., and Papon, N. 2021. Metabolic engineering for plant natural products biosynthesis: New procedures, concrete achievements and remaining limits. *Natural Product Reports,* 38, 2145–2153.

Dan, L., Gao, T., Zhang, Y., Gong, C., Liu, C., Li, Y., Song, X., and Zhang, D. 2024. Panacis majoris Rhizoma: A comprehensive review of phytochemistry, pharmacology and clinical applications. *Records of Natural Products,* 18. Online.

Das, S., Kwon, M., and Kim, J.-Y. 2024. Enhancement of specialized metabolites using CRISPR/Cas gene editing technology in medicinal plants. *Frontiers in Plant Science,* 15, 1279738.

Dey, A. 2021. CRISPR/Cas genome editing to optimize pharmacologically active plant natural products. *Pharmacological Research,* 164, 105359.

Diez-Fernandez, C. and Häberle, J. 2017. Targeting CPS1 in the treatment of Carbamoyl phosphate synthetase 1 (CPS1) deficiency, a urea cycle disorder. *Expert Opinion on Therapeutic Targets,* 21, 391–399.

Doudna, J., Cofsky, J., Soczek, K., Knott, G., and Nogales, E. 2021. CRISPR-Cas9 bends and twists DNA to read its sequence. *Nature Structural and Molecular Biology,* 29(4), 395–402.

El-Desoky, A. H., Atia, M. A., and Omer, E. A. 2022. Metabolic engineering: New approaches in pharmaceutical production. In T. Aftab and K. R. Hakeem (Eds.), Metabolic engineering in plants. Springer.

Fan, G., Liu, X., Sun, S., Shi, C., Du, X., Han, K., Yang, B., Fu, Y., Liu, M., and Seim, I. 2020. The chromosome level genome and genome-wide association study for the agronomic traits of Panax notoginseng. *Iscience,* 23.

Gohil, N., Bhattacharjee, G., Lam, N. L., Perli, S. D., and Singh, V. 2021. CRISPR-Cas systems: challenges and future prospects. *Progress in Molecular Biology and Translational Science,* 180, 141–151.

Guo, M., Chen, H., Dong, S., Zhang, Z., and Luo, H. 2022. CRISPR-Cas gene editing technology and its application prospect in medicinal plants. *Chinese Medicine,* 17, 33.

Guo, M., Lv, H., Chen, H., Dong, S., Zhang, J., Liu, W., He, L., Ma, Y., Yu, H., and Chen, S. 2024. Strategies on biosynthesis and production of bioactive compounds in medicinal plants. *Chinese Herbal Medicines,* 16, 13–26.

Herman, X., Far, J., Courtoy, A., Bouhon, L., Quinton, L., De Pauw, E., Chaumont, F., and Navarre, C. 2021. Inactivation of N-acetylglucosaminyltransferase I and α1, 3-fucosyltransferase genes in nicotiana tabacum BY-2 cells results in glycoproteins with highly homogeneous, high-mannose N-glycans. *Frontiers in Plant Science,* 12, 634023.

Huang, J., Zhou, Y., Li, J., Lu, A., and Liang, C. 2022. CRISPR/Cas systems: delivery and application in gene therapy. *Frontiers in Bioengineering and Biotechnology,* 10, 942325.

Jung, Y. J., Lee, H. J., Kim, J. H., Kim, D. H., Kim, H. K., Cho, Y.-G., Bae, S., and Kang, K. K. 2019. CRISPR/Cas9-targeted mutagenesis of F3′ H, DFR and LDOX, genes related to anthocyanin biosynthesis in black rice (Oryza sativa L.). *Plant Biotechnology Reports,* 13, 521–531.

Kantor, A., Mcclements, M. E., and Maclaren, R. E. 2020. CRISPR-Cas9 DNA base-editing and prime-editing. *International Journal of Molecular Sciences,* 21, 6240.

Kim, S., Ji, S., and Koh, H. R. 2021. CRISPR as a diagnostic tool. *Biomolecules,* 11, 1162.

Li, C., Swofford, C. A., and Sinskey, A. J. 2020. Modular engineering for microbial production of carotenoids. *Metabolic Engineering Communications,* 10, e00118.

Li, R., Li, R., Li, X., Fu, D., Zhu, B., Tian, H., Luo, Y., and Zhu, H. 2018. Multiplexed CRISPR/Cas9-mediated metabolic engineering of γ-aminobutyric acid levels in Solanum lycopersicum. *Plant Biotechnology Journal,* 16, 415–427.

Lian, J., Schultz, C., Cao, M., Hamedirad, M., and Zhao, H. 2019. Multi-functional genome-wide CRISPR system for high throughput genotype–phenotype mapping. *Nature Communications,* 10, 5794.

Lino, C. A., Harper, J. C., Carney, J. P., and Timlin, J. A. 2018. Delivering CRISPR: a review of the challenges and approaches. *Drug Delivery,* 25, 1234–1257.

Liu, Q., Guo, Q., Akbar, S., Zhi, Y., El Tahchy, A., Mitchell, M., Li, Z., Shrestha, P., Vanhercke, T., and Ral, J. P. 2017. Genetic enhancement of oil content in potato tuber (Solanum tuberosum L.) through an integrated metabolic engineering strategy. *Plant Biotechnology Journal,* 15, 56–67.

Luther, D., Lee, Y., Nagaraj, H., Scaletti, F., and Rotello, V. 2018. Delivery approaches for CRISPR/Cas9 therapeutics in vivo: Advances and challenges. *Expert Opinion on Drug Delivery,* 15, 905–913.

Marchev, A. S., Yordanova, Z. P., and Georgiev, M. I. 2020. Green (cell) factories for advanced production of plant secondary metabolites. *Critical Reviews in Biotechnology,* 40, 443–458.

Mikhaylova, E. V. 2024. Insight into the flavonoids enrichment in plants by genome engineering. In S. K. Upadhyay (Ed.), *Applications of genome engineering in plants* (pp. 242–259). Wiley.

Miladinovic, D., Antunes, D., Yildirim, K., Bakhsh, A., Cvejić, S., Kondić-Špika, A., Marjanovic Jeromela, A., Opsahl-Sorteberg, H.-G., Zambounis, A., and Hilioti, Z. 2021. Targeted plant improvement through genome editing: from laboratory to field. *Plant Cell Reports,* 40, 935–951.

Mipeshwaree Devi, A., Khedashwori Devi, K., Premi Devi, P., Lakshmipriyari Devi, M., and Das, S. 2023. Metabolic engineering of plant secondary metabolites: prospects and its technological challenges. *Frontiers in Plant Science,* 14, 1171154.

Mishra, S., Sharma, A., and Srivastava, A. K. 2023. Ascorbic acid: A metabolite switch for designing stress-smart crops. *Critical Reviews in Biotechnology*, 44(7), 1350–1366..

Modrzejewski, D., Hartung, F., Lehnert, H., Sprink, T., Kohl, C., Keilwagen, J., and Wilhelm, R. 2020. Which factors affect the occurrence of off-target effects caused by the use of CRISPR/Cas: A systematic review in plants. *Frontiers in Plant Science,* 11, 574959.

Molina-Hidalgo, F. J., Vazquez-Vilar, M., D'andrea, L., Demurtas, O. C., Fraser, P., Giuliano, G., Bock, R., Orzáez, D., and Goossens, A. 2021. Engineering metabolism in Nicotiana species: a promising future. *Trends in Biotechnology,* 39, 901–913.

Montecillo, J. A. V., Chu, L. L., and Bae, H. 2020. CRISPR-Cas9 system for plant genome editing: current approaches and emerging developments. *Agronomy,* 10, 1033.

Munir, A., Ali, M., Qari, S. H., Munawar, N., Saleem, M. S., and Ahmad, A. 2024. CRISPR workflow solutions: cargos and versatile delivery platforms in genome editing. In K. A. Abd El-Salam, A. Ahmad, and B. Zhang (Eds.), CRISPRized horticulture crops. Elsevier.

Munir, A., Amin, I., Zahoor, M. K., Majeed, H. N., Almoammar, H., Ghaffar, A., and Ahmad, A. 2024. Multiplex genome editing in plants through CRISPR-Cas. In K. A. Abd El-Salam, A. Ahmad, and B. Zhang (Eds.), CRISPRized horticulture crops. Elsevier.

Nabavi, S. M., Šamec, D., Tomczyk, M., Milella, L., Russo, D., Habtemariam, S., Suntar, I., Rastrelli, L., Daglia, M., and Xiao, J. 2020. Flavonoid biosynthetic pathways in plants: versatile targets for metabolic engineering. *Biotechnology Advances,* 38, 107316.

Naik, B. J., Vasamsetti, B. M. K., Kim, S.-C., Gunti, M., Swamy, M. K., and Mekapogu, M. 2023. Genome editing in medicinal plants using CRISPR/Cas9 tool. In M. K. Swamy and A. Kumar (Eds.), Phytochemical genomics: Plant metabolomics and medicinal plant genomics. Springer.

Nishida, K. and Kondo, A. 2021. CRISPR-derived genome editing technologies for metabolic engineering. *Metabolic Engineering,* 63, 141–147.

Raipuria, R. K., Watts, A., Meena, N. L., and Watts, A. 2021. Genome editing in oilseed crops. In O. M. Gupta (Ed.), *Genome editing in plants.* CRC Press.

Ramawat, K. G. and Arora, J. 2021. Medicinal plants domestication, cultivation, improvement, and alternative technologies for the production of high value therapeutics: An overview. In H. M. Ekiert et al. (Eds.), *Medicinal plants: Domestication, biotechnology and regional importance* (pp. 1–29). Springer.

Rauf, S., Fatima, S., and Ortiz, R. 2023. Modification of fatty acid profile and oil contents using gene editing in oilseed crops for a changing climate. *GM Crops and Food,* 14, 1–12.

Sabzehzari, M., Zeinali, M., and Naghavi, M. R. 2020. CRISPR-based metabolic editing: next-generation metabolic engineering in plants. *Gene,* 759, 144993.

Salwan, R. and Sharma, V. 2020. Molecular and biotechnological aspects of secondary metabolites in actinobacteria. *Microbiological Research,* 231, 126374.

Sharma, A., Mistry, V., Kumar, V., and Tiwari, P. 2022. Production of effective Phyto-antimicrobials via metabolic engineering strategies. *Current Topics in Medicinal Chemistry,* 22, 1068–1092.

Sharma, M., Koul, A., Sharma, D., Kaul, S., Swamy, M. K., and Dhar, M. K. 2019. Metabolic engineering strategies for enhancing the production of bio-active compounds from medicinal plants. In M. S. Akhtar and M. K. Swamy (Eds.), Natural bio-active compounds, vol. 3: Biotechnology, bioengineering, and molecular approaches (pp. 287–316). Springer.

Shelake, R. M., Jadhav, A. M., Bhosale, P. B., and Kim, J.-Y. 2023. Unlocking secrets of nature's chemists: potential of CRISPR/Cas-based tools in plant metabolic engineering for customized nutraceutical and medicinal profiles. *Plant Physiology and Biochemistry,* 203, 108070.

Shi, M., Du, Z., Hua, Q., and Kai, G. 2021. CRISPR/Cas9-mediated targeted mutagenesis of bZIP2 in Salvia miltiorrhiza leads to promoted phenolic acid biosynthesis. *Industrial Crops and Products,* 167, 113560.

Shivram, H., Cress, B. F., Knott, G. J., and Doudna, J. A. 2021. Controlling and enhancing CRISPR systems. *Nature Chemical Biology,* 17, 10–19.

Singh, H. 2023. Enhancement of plant secondary metabolites by genetic manipulation. In R. Singh and N. Kumar (Eds.), Genetic manipulation of secondary metabolites in medicinal plant. Springer.

Song, H., Taylor, D. C., and Zhang, M. 2023. Bioengineering of soybean oil and its impact on agronomic traits. *International Journal of Molecular Sciences,* 24, 2256.

Stra, A., Almarwaey, L. O., Alagoz, Y., Moreno, J. C., and Al-Babili, S. 2023. Carotenoid metabolism: new insights and synthetic approaches. *Frontiers in Plant Science,* 13, 1072061.

Sultan, Q., Ashraf, S., Munir, A., Khan, S. H., Munawar, N., Abd-Elsalam, K. A,. and Ahmad, A. 2022. Beyond genome editing: CRISPR approaches. In A. Ahmad, S. H. Khan and Z. Khan (Eds.), The CRISPR/Cas tool kit for genome editing (pp. 187–218). Springer.

Tariq, A., Mushtaq, M., Yaqoob, H., Bhat, B. A., Zargar, S. M., Raza, A., Ali, S., Charagh, S., Mubarik, M. S., and Zaman, Q. U. 2023. Putting CRISPR-Cas system in action: a golden window for efficient and precise genome editing for crop improvement. *GM Crops and Food,* 14, 1–27.

Tariq, H., Asif, S., Andleeb, A., Hano, C., and Abbasi, B. H. 2023. Flavonoid production: current trends in plant metabolic engineering and de novo microbial production. *Metabolites,* 13, 124.

Tian, Y., Kong, L., Li, Q., Wang, Y., Wang, Y., An, Z., Ma, Y., Tian, L., Duan, B., and Sun, W. 2024. Structural diversity, evolutionary origin, and metabolic engineering of plant specialized benzylisoquinoline alkaloids. *Natural Product Reports,* 41, 1787–1810.

Twaij, B. M. and Hasan, M. N. 2022. Bioactive secondary metabolites from plant sources: types, synthesis, and their therapeutic uses. *International Journal of Plant Biology,* 13, 4–14.

Ullah, N., Shahzad, K., and Wang, M. 2021. The role of metabolic engineering technologies for the production of fatty acids in yeast. *Biology,* 10, 632.

Wang, J., Yang, J., Li, D., and Li, J. 2021. Technologies for targeting DNA methylation modifications: basic mechanism and potential application in cancer. *Biochimica et Biophysica Acta (BBA)-Reviews on Cancer,* 1875, 188454.

Wang, J. Y. and Doudna, J. A. 2023. CRISPR technology: a decade of genome editing is only the beginning. *Science,* 379, eadd8643.

Wang, M., Qiu, X., Pan, X., and Li, C. 2021. Transcriptional factor-mediated regulation of active component biosynthesis in medicinal plants. *Current Pharmaceutical Biotechnology,* 22, 848–866.

Wang, R., Xu, S., Wang, N., Xia, B., Jiang, Y., and Wang, R. 2017. Transcriptome analysis of secondary metabolism pathway, transcription factors, and transporters in response to methyl jasmonate in Lycoris aurea. *Frontiers in Plant Science,* 7, 1971.

Wawrosch, C. and Zotchev, S. B. 2021. Production of bioactive plant secondary metabolites through in vitro technologies—status and outlook. *Applied Microbiology and Biotechnology*, 105, 6649–6668.

Yadav, B., Jogawat, A., Rahman, M. S., and Narayan, O. P. 2021. Secondary metabolites in the drought stress tolerance of crop plants: A review. *Gene Reports,* 23, 101040.

Yang, Q.-Q., Hua, W.-P., Zou, H.-L., Yang, J.-X., Wang, X.-Z., Zhang, T., Wang, D.-H., Zhu, X.-J., and Cao, X.-Y. 2022. Overexpression of SmLAC25 promotes lignin accumulation and decreases salvianolic acid content in Salvia miltiorrhiza. *Plant Science,* 325, 111462.

Yeshi, K., Crayn, D., Ritmejerytė, E., and Wangchuk, P. 2022. Plant secondary metabolites produced in response to abiotic stresses has potential application in pharmaceutical product development. *Molecules,* 27, 313.

Yuan, M., Zhu, J., Gong, L., He, L., Lee, C., Han, S., Chen, C., and He, G. 2019. Mutagenesis of FAD2 genes in peanut with CRISPR/Cas9 based gene editing. *BMC Biotechnology,* 19, 1–7.

Zakaria, M. M., Schemmerling, B., and Ober, D. 2021. CRISPR/Cas9-mediated genome editing in comfrey (Symphytum officinale) hairy roots results in the complete eradication of pyrrolizidine alkaloids. *Molecules,* 26, 1498.

Zhang, J., Xing, J., Mi, Q., Yang, W., Xiang, H., Xu, L., Zeng, W., Wang, J., Deng, L., and Jiang, J. 2023. Highly efficient transgene-free genome editing in tobacco using an optimized CRISPR/Cas9 system, pOREU3TR. *Plant Science,* 326, 111523.

Zhang, Y., Iaffaldano, B., and Qi, Y. 2021. CRISPR ribonucleoprotein-mediated genetic engineering in plants. *Plant Communications*, 2.

Zhang, Y., Ma, X., and Xie, X. 2022. *CRISPR*. Springer.

Zhou, Y.-M., Dong, X.-R., Xu, D., Tang, J., and Cui, Y.-L. 2024. Therapeutic potential of traditional Chinese medicine for interstitial lung disease. *Journal of Ethnopharmacology,* 318, 116952.

Zou, H., Guo, X., Yang, R., Wang, S., Li, L., Niu, J., Wang, D., and Cao, X. 2021. MiR408-SmLAC3 module participates in salvianolic acid B synthesis in Salvia miltiorrhiza. *International Journal of Molecular Sciences,* 22, 7541.

20 Global Regulatory Framework of CRISPR-Edited Plants and Transgene-Free CRISPR Crops

Muhammad Faheem, Anam Niamat, Aftab Ahmad, Nayla Munawar, Kamel Abd Elsalam, Ahmed Al-Harrasi, and Paulo Zaini

20.1 INTRODUCTION

The world's population is continuously increasing and it is expected to reach 10 billion by the year 2050 (Friedrichs et al., 2019). The growing population demands an increased amount of food production, hindered by climate change. To meet this challenge of food security, many advances in plant breeding have been made, such as use of biotechnological tools and genetic engineering techniques. However, despite these advances in plant breeding to improve crop productivity, researchers have been facing severe challenges to achieve the sustainable production and zero-hunger goals. Genetically engineered crops have contributed significantly to meet the food security challenge but their social acceptance and regulatory oversights are still a matter of debate in some parts of the world such as the EU and New Zealand (Qaim and Kouser, 2013). New technologies and innovations are needed to address this challenge of food security and food safety for everyone. Recently new techniques, especially CRISPR-Cas have emerged as powerful tools for precise genetic improvement of plants. Gene editing through CRISPR-Cas is a revolutionary technology that has changed the genomic landscape of agricultural crops. CRISPR applications are continuously growing in agricultural crops for precise modification in their genomes to improve the desired traits, such as better yield and productivity, disease resistance, drought and salt tolerance and improved photosynthesis. CRISPR-Cas has been successfully used as a proof-of-concept in studies to improve grains, vegetables and fruit crops as well (Friedrichs et al., 2019; Sultan et al., 2022).

CRISPR-Cas has emerged as a remarkable gene-editing technology which has revolutionized the field of agriculture, medicine, human health, and synthetic biology. In nature, CRISPR-Cas exists as a part of the bacterial immune system, which provides immunity to bacteria against invading viruses. The main strength of the CRISPR-Cas defense system is its precise nature while cleaving DNA of invading bacteriophages. In 2012, an engineered CRISPR-Cas system was used for the first time as a gene-editing tool to modify an eukaryotic genome (Nidhi et al., 2021). CRISPR-Cas is a simple, precise, and powerful tool which allows scientists and breeders to precisely modify genes responsible for desirable traits in crops (Bacha and Iqbal, 2023). Several CRISPR-Cas systems, such as CRISPR-Cas9, CRISPR-Cas12, CRISPR-Cas13 and CRISPR-Cas3, have been identified with diverse applications from agriculture to medicine and human health. Different CRISPR-Cas systems have been successfully used to precisely manipulate agricultural crops for desirable traits, such as blight resistance in potatoes (Tussipkan and Manabayeva, 2021), improved yield and grain size in wheat (Khan et al., 2021; Nazir et al., 2022), and rice (Zhou et al., 2022), disease resistance in cotton (Liu et al., 2021), shelf life in tomatoes (Chaudhuri, Halder and Datta, 2022), and herbicide tolerance in plants (Bigini et al., 2023).

Regulation and the legal status of GMOs have become an integral part of their approval for human consumption. Although GM crops are cultivated globally on a large scale and the area under cultivation of GM crops is continuously increasing, they face limited commercialization in the EU and New Zealand due to strict regulatory measures. In addition, public perceptions of GM crops is also facing challenges due to, for example, the presence of antibiotic resistance genes. In contrast to GM crops, which always contain foreign genes to introduce desired traits in plants, CRISPR-Cas systems can be used to install desired modifications without introducing any foreign gene into the host plant. Moreover, researchers can introduce as small as a single nucleotide modification into the genome for genetic improvement of a plant. Therefore, some countries such as the USA, Argentina, Japan, and Australia have introduced a relaxed regulatory framework for transgene-free CRISPR-edited crops. However, the EU and New Zealand are still following strict regulatory measures for all gene-edited crops and these crops are considered as GM crops which require strict regulatory and risk assessment before commercialization. Transgene-free gene editing in crops without introducing any foreign gene in the host plant is a tremendous achievement in the genetic improvement of agricultural crops, with promising benefits, especially for food security in developing countries. However, transgene-free gene editing in crops is not free of limitations. One of the most challenging aspects is screening of edited plants without any marker gene for the selection of edited plants (Gao, 2021). In addition, while removing the transgene through crossing is feasible for many crops, it can be time-consuming as well. The goal of transgene-free

DOI: 10.1201/9781003500933-23

gene editing with CRISPR is to develop crops that do not contain any foreign DNA, consequently addressing concerns associated with public perception, legal status, and regulations of conventional GMOs (Bhattacharjee et al., 2023).

The global regulatory horizon of CRISPR-edited crops is a mosaic with different countries using different regulatory triggers and frameworks to oversee the regulation of CRISPR-edited crops (Bacha and Iqbal, 2023). A combination of scientific, ethical, public, and commercial concerns influences these regulatory frameworks of CRISPR-edited crops. International organizations such as the Codex Alimentarius Commission are striving to establish international regulations for gene-edited crops. This is the main international body that adopts rules regarding food safety. However, due to diverse genome-editing approaches, types of genetic modifications, and priorities in different countries, finding a consensus on these standards is challenging (Zhang et al., 2020). In addition, public perception and acceptability of CRISPR-edited crops are also important in making regulatory decisions in some countries, as they obtain public opinion and collect feedback from stakeholders during the regulatory oversight of gene-edited crops. To ensure the safe and appropriate use of CRISPR in genetic improvement of agricultural crops, all stakeholders, such as regulatory agencies, researchers, social scientists, and political and governing authorities, need to work collaboratively (Zhang et al., 2020).

The future of CRISPR-edited crops is linked with their legal and regulatory status because their classification as conventional GM crops will certainly affect their commercialization in different countries. Regulatory approval is frequently required before commercialization of CRISPR-edited crops, however, countries like Argentina and the USA have introduced new guidelines for companies and researchers to determine in advance if their final product will be regulated or deregulated (Bhattacharjee et al., 2023). Regulatory oversights have various impacts on CRISPR-edited crops. For example, the process of securing regulatory approval can be time-consuming and resource-intensive, which influences how quickly these crops can reach the market. It has also a direct impact on market access, research agendas, consumer acceptance, international trade, intellectual property rights, environmental safety, and the larger societal and ethical context in which these crops are used. As CRISPR technology is rapidly advancing, regulatory frameworks will continue to evolve to handle emerging issues and possibilities, ultimately influencing the commercialization of gene-edited crops at the global level. The objective of this chapter is to highlight the mechanism of double-strand break (DSB) repair and regulatory triggers of CRISPR-edited crops. In addition, we also discussed the regulatory classification and global regulatory framework of CRISPR-edited crops.

20.2 CRISPR-CAS AND ITS DEVELOPMENTS

CRISPR-Cas is a revolutionary gene-editing technology for precise modification of genomes with tremendous applications. In this section, we briefly discuss some important CRISPR-Cas systems, comprising CRISPR-Cas9, CRISPR-Cas12, CRISPR-Cas13 and their potential applications. CRISPR-Cas9 is often recognized as the first and most widely used CRISPR-Cas system to introduce precise genetic modifications in the genome. This revolutionary tool has not only enabled scientists to precisely modify the DNA, but also facilitated applications beyond gene editing such as CRISPRi, CRISPRa, epigenome editing, and genome imaging. CRISPR-Cas9 has two components: a guide RNA (gRNA) and a Cas9 endonuclease protein. The gRNA contains nucleotide sequence complementary to the target site and recruits the Cas9 to the target site. Once the gRNA/Cas9 complex binds the target site with an adjacent PAM, Cas9 creates the double-stranded break (DSB) at the target site. The CRISPR-Cas9 system is the most widely used system for gene-editing applications such as gene knockout, gene insertion, base editing, and prime editing. CRISPR-Cas12 is also characterized as a single effector CRISPR system to create DSB in the genome. However, compared with Cas9, it possesses several unique characteristics, such as recognizing 5'-TTTN-3' PAM to target T-rich region, smaller Cas nuclease, and it requires only crRNA to generate DSB. Like CRISPR-Cas9, CRISPR-Cas12 has also been successfully used in gene editing and diagnostic applications (Gupta et al., 2022). The CRISPR-Cas13 has been characterized as an RNA-targeting system with a broad range of applications in the RNA world. For example, CRISPR-Cas13 has potential applications in RNA imaging, RNA editing, RNA-based therapeutics, and RNA-based detection of viruses (Makarova et al., 2020). Some other newly characterized CRISPR-Cas systems include CRISPR-Cas14, CRISPR-Cas7-11, CRISPR-Cas3, CRISPR-CasX, and Fanzor (Aman et al., 2020). All these systems are developing with tremendous potential applications in different fields of the life sciences.

CRISPR-Cas-based techniques such as prime editing (PE) or base editing (BE) have enabled researchers to precisely edit, insert, or replace desired nucleotides in the genome. PE is a CRISPR-Cas9-based approach which was first introduced in 2019 (Tabassum et al., 2021). It consists of three components: nCas, pegRNA, and reverse transcriptase. nCas is the modified version of Cas9 which produces a nick instead of DSB in the target DNA sequence, thus it is known as Cas9 nickase. The pegRNA is also known as prime editing guide RNA, and is a modified form of gRNA, containing 20 nucleotides for recruiting nCas9 to the target site and an additional sequence with the desired nucleotides to be copied in the target site. The reverse transcriptase domain produces cDNA with homology arms to the nick, using the sequence provided in the pegRNA, to copy or replace the desired sequences into the target site in the host genome. Once the nick is produced by nCas9 and reverse transcriptase produces the desired strand to be copied at the target site, the cell repair system uses the modified strand to repair the nick, thus copying the modified strand into the host genome. Modified PE systems have been engineered to improve the efficiency and guaranteed insertion of the desired template at the target site. PE has the potential to reduce off-target effects as it does not produce DSB and has broadened the scope of genome editing in crop genetic improvement (Tabassum et al., 2021). BE also uses nCas9 and a modified version of guide RNA to replace the desired bases at the target site in the genome. However, the PE system is better than BE to introduce site-specific base editing in the genome. BE systems (ABE and CBE) use deaminases, an enzyme to catalyze the reaction that involves the conversion of one nucleotide to another nucleotide present in the target site. Both ABE and CBE have been successfully used to correct the desired point mutations that are particularly linked with the single nucleotide change in the field of functional genomics (Naeem et al., 2020). PE and BE systems have been successfully used to edit plants with desired traits, which highlights the potential of both systems in the genetic improvement of plants.

CRISPRi and CRISPRa are the engineered CRISPR/Cas9 systems, which use dead-Cas9 (dCas9) to control the gene expression rather than introducing permanent mutations in the genome. dCas9 is an engineered version of Cas9 which contains an inactive nuclease domain, but it can bind to the target site due to its active DNA binding site. Therefore, dCas9 can be used to recruit transcription activators, repressors, or other effector proteins such as epigenetic effectors to the target site in the genome. CRISPR interference (CRISPRi) and CRISPR activation (CRISPRa) are two related, but distinct techniques derived from the CRISPR/Cas9 that silence and activate the gene expression respectively. CRISPRi is involved in silencing or downregulation of the expression of specific genes by recruiting repressor domains, such as the KRAB domain, without changing the natural genome sequence. The CRISPRi technique used gRNA and dCas9, which does not produce DSB in the target DNA sequence. The dCas9/gRNA complex works by binding to the specific regulatory sequence of the gene (promoter) and blocks the transcription of the associated gene from this specific promoter, thus silencing the expression of the gene by interference. However, in CRISPRa, the expression of the gene is enhanced or increased with the help of dCas9 which is additionally fused with the transcriptional activation domain of the VP64 regulatory protein (Kazi and Biswas, 2021). In this case, the dCas9-VP64/gRNA complex binds with the promoter sequence and recruits other transcription activators to enhance the expression of the specific gene associated with that promoter. Both of these techniques have applications in gene function research and future therapeutic processes (Kazi and Biswas, 2021).

20.3 OUTCOMES OF CRISPR-BASED MODIFICATIONS

Generally, gene-editing systems rely on creating a DSB in the genome and its subsequent repair through the cell's repair systems. Once the DSB is created by Cas9 at the target site, the cell uses either the Non-Homologous End Joining (NHEJ) or the Homology Directed Repair (HDR) pathway to repair the DSB. NHEJ is a DNA repair strategy that cells use to repair DSB without involving any repair or donor template. It typically operates during the G1 and early S phases of the cell cycle. NHEJ involves joining the broken DNA ends together, which frequently results in the insertions or deletions (indels) at the repair site. NHEJ is used strategically in genome editing to induce gene knockouts or gene disruptions by introducing indels at the target site of a gene. This repair method is prone to errors, leading to non-functional proteins (Yu et al., 2020). HDR, on the other hand, is a precise DNA repair mechanism that cells adopt to repair DSBs, using a homologous DNA repair template with homology arms to the DSB. HDR is most active during the S and G2 phases of the cell cycle (Yang et al., 2020). This procedure produces precise alterations such as gene substitution, gene insertion, and base editing at the target site in the genome. HDR is used in genome editing to apply targeted and controlled modifications to the target gene. This can include insertion or replacement of specific sequences, correction of mutations, or introduction of point mutations.

Various ways exist in the domain of genome editing and CRISPR technology to introduce precise changes into the genome of an organism. Based on the DSB repair process and their repair outcomes, site-directed Nucleases (SDNs) are classified into SDN1, SDN2, and SDN3 systems (Organisms et al., 2020; Van Vu et al., 2022). SDN1, SDN2 and SDN3 are also considered a regulatory classification of CRISPR- based modifications. SDN1 produces single-nucleotide changes, small insertions, or deletions at the targeted genomic location using NHEJ as a repair system of DSB. SDN1 is useful in the creation of deletions or precise sequence modifications in the genome, notably in the removal of specific DNA sequences between two target locations (Organisms et al., 2020). NHEJ indels can cause frame-shift mutations, gene inactivation, or non-functional protein synthesis. SDN2 involves larger insertions and deletions than SDN1. This category represents more significant modifications to the DNA sequence at the targeted location. SDN2 makes use of a DNA repair template that is complementary to the target DNA sequence and uses micro homology-mediated end joining repair pathway (MMEJ). This repair template directs the cell's repair machinery to make specific genomic modifications at the target site. SDN2 is frequently used to introduce particular genetic changes, such as gene insertions, substitutions, or exact sequence adjustments, making it excellent for applications that require great accuracy (Singh et al., 2023). From a regulatory point of view, SDN1 and SDN2 modified crops are generally considered deregulated, especially in the USA and Argentina. SDN3, on the other hand, uses the HDR pathway to introduce larger fragments of DNA or gene(s) at the target site in the host genome. The CRISPR-Cas components cause double-strand breaks (DSBs) in the target DNA during SDN3 (Sharma et al., 2023) and the cell's repair machinery uses HDR system to insert the donor template with long homology arms at the target site. SDN3 is often used to insert gene(s) at the target site and all SDN3 modified crops are always considered as GM crops. Figure 20.1 shows the outcomes after DSB through SDN.

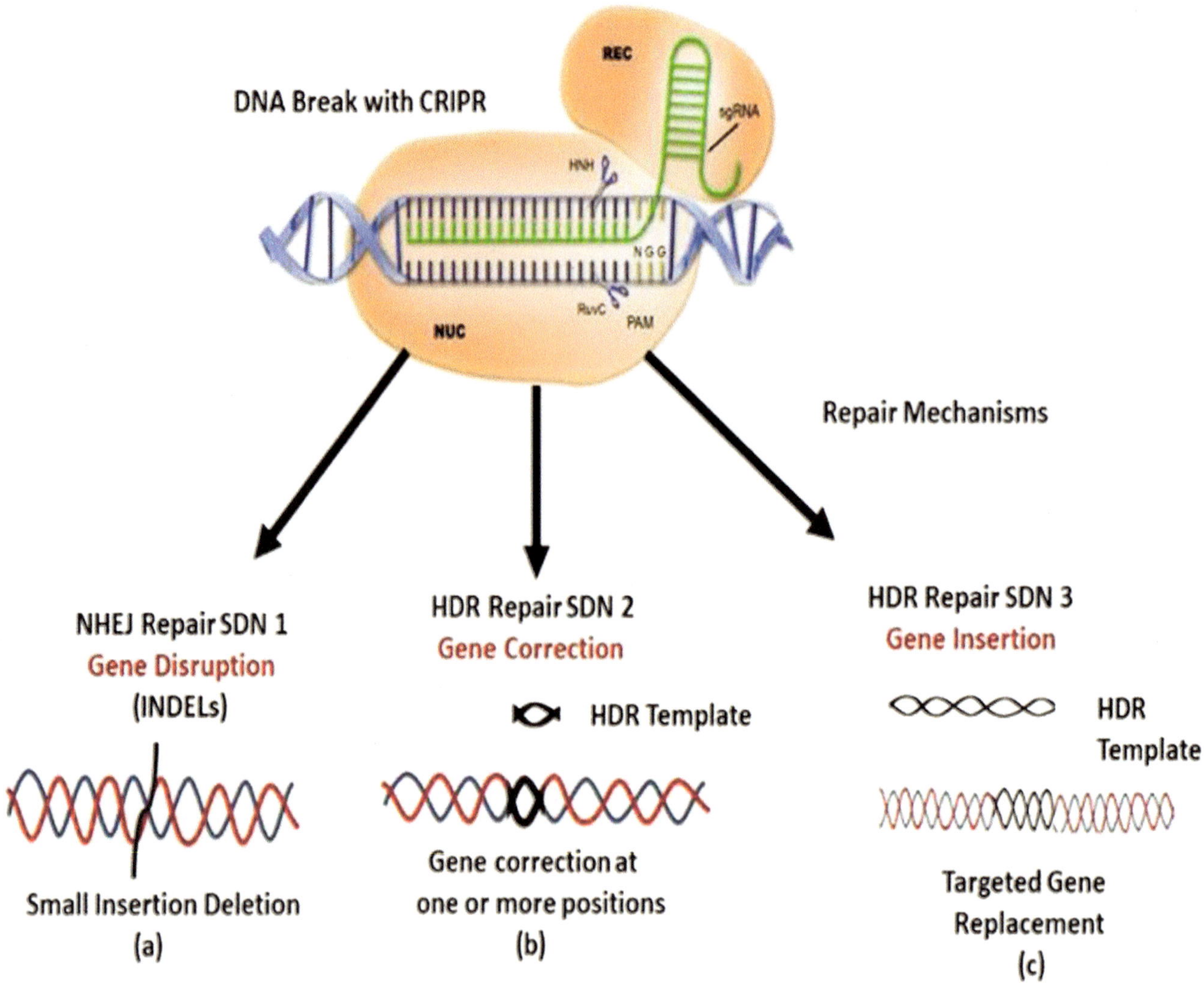

FIGURE 20.1 Overview of outcomes after double-stranded break through the SDN mechanism.

20.4 EMERGING APPLICATIONS OF CRISPR-CAS-MEDIATED GENE EDITING IN PLANTS

CRISPR-mediated gene editing has revolutionized the field of plant biotechnology, offering precise and efficient tools for targeted genetic modifications with a broad range of applications. The process begins with the designing of a gRNA that matches the targeted gene sequence, guiding the Cas protein to the desired/targeted genomic location. Once at the target site, the Cas protein induces a double-stranded break in the DNA, natural repair machinery of the host plant comes into play, either introducing small indels or incorporating a donor DNA template to facilitate targeted gene insertion or replacement. This powerful and versatile technology has been employed in models as well as crop plants for a range of applications, including the enhanced tolerance in plants against abiotic and biotic stress, functional genomics, improved nutritional profile, better yield, and improving complex traits like photosynthesis and N-fixation. In the following section, we briefly discuss a few emerging applications of CRISPR in the genetic improvement of plants. Table 20.1 shows the implications of CRISPR Cas9 in agriculture.

20.4.1 Enhancing Biotic Stress Tolerance in Plants Through CRISPR

Biotic stress is the adverse impact produced by living organisms, i.e. pests and pathogens (including viruses, bacteria, and fungi) on plant health. These organisms directly feed on plants, compete for food, and cause disease in plants. CRISPR technology has emerged as a promising tool for enhancing biotic stress tolerance in plants, providing a precise and efficient means for researchers to fortify crops against various pathogens and pests (Mohd Azman et al., 2022). By selectively targeting genes associated with susceptibility to biotic stress, researchers can modify plant genomes to increase the natural defense mechanisms of plants against these pathogens (Ghosh and Dey, 2022). CRISPR-Cas9 has been successfully used to enhance disease resistance in rice, particularly against the devastating rice blast fungus. The specific susceptibility genes, such as Bsr-d1, Pi21, and

TABLE 20.1
Implications of CRISPR/Cas9 in agriculture

Aspect	Description	References
Genome editing precision	CRISPR/Cas9 allows for precision crop DNA alteration, targeting specific genes for trait improvement while reducing unwanted modifications	(Bacha and Iqbal, 2023)
Improved crop traits	CRISPR/Cas9 is used by scientists to improve desired features including productivity, calorie content, insect and disease resistance, as well as stress tolerance	(Bacha and Iqbal, 2023)
Reduced chemical usage	CRISPR-edited crops with greater disease resistance require fewer chemical insecticides and fungicides, which benefits the environment	(Khumsupan et al., 2019).
Accelerated crop breeding	CRISPR speeds up the production of superior crop varieties compared to traditional breeding approaches	(Maharajan et al., 2022)
Sustainable agriculture	CRISPR aids in the development of resource-efficient crops that use less water and fertilizer, so helping to sustainability	(Mattick et al., 2023)
Addressing nutritional deficiencies	Genome editing has the potential to strengthen crops with critical vitamins and minerals, thereby treating malnutrition in specific areas	(Wikandari et al., 2021).
Reduced food waste	Post-harvest trait editing increases shelf life and decreases spoiling, resulting in less food waste	(Kumari et al., 2022)
Public and ethical concerns	CRISPR in agriculture raises issues of intellectual property, equitable availability of technology, and ethical concerns	(Blix, 2023)
Regulatory challenges	The deployment and commercialization of CRISPR-edited crops are hampered by complex laws and public views	(Almeida and Ranisch, 2022)

ERF922, have been successfully modified, developing rice varieties with increased resistance to this pathogen (Wang et al., 2016; Zhou et al., 2022). Similarly, CRISPR has also been employed to fortify plants against insect pests. By targeting genes involved in the interaction between plants and pests, researchers have created genetically modified crops with improved resistance. For example, cotton has been made resistant to insects with the use of CRISPR via the successful expression of the Bt gene from bacteria in cotton which express an endotoxin protein that kills the pest, in this way, it increased the quality of cotton by protecting it from pests (Liu et al., 2021). This technology has also been used to improve crop quality, yield performance, and stress tolerance in various crops (Ricroch, Clairand, and Harwood, 2017). Moreover, CRISPR technology has also enhanced crop resilience and productivity to face global food security challenges. Rasheed et al. (2021) highlighted the potential of CRISPR-Cas9 to improve crop yield and quality, particularly in the context of biotic and abiotic stress tolerance. This technology offers a faster, cheaper, and more precise alternative to conventional breeding methods, making it a promising avenue for crop improvement (Rasheed et al., 2021). Mutezo et al. (2021) further emphasized the potential of CRISPR-Cas9 to develop non-transgenic, genome-edited crop plants that can withstand the harmful effects of microorganisms.

20.4.2 Improving Abiotic Stress Tolerance with CRISPR

Abiotic stress is the condition in which the growth of crops is inhibited by non-living factors. It mainly results from adverse physical or chemical environmental conditions, i.e., extreme conditions of temperature, pH, drought, as well as soil salinity. CRISPR has not only been explored to improve biotic stress tolerance in crops but also used to characterize new genes involved in abiotic stress mechanism in plants. Genes involved in abiotic stress such as salinity and drought have been successfully edited by the CRISPR technology, resulting in more resilient crops such as salt-resistant grain crops (Altaf et al., 2022). Salinity is a critical abiotic stress that affects agricultural productivity, particularly in areas with high soil salinity. CRISPR technology has shown promise in enhancing salt tolerance in crops like rice and wheat, which are particularly susceptible to salinity stress (Khan et al., 2021; Nazir et al., 2022). By modifying genes related to ion transport and salt tolerance, CRISPR-driven modifications helped these crops thrive in saline environments, thereby expanding arable land and ensuring food security in challenging soil conditions. This technology offers a potential solution to the global issue of salinity stress in agriculture (Raza et al., 2022). Additionally, drought also impacts crop yields in areas of extreme temperatures. CRISPR-Cas9 has been used to engineer plant resilience to drought conditions. The gene responsible for drought stress responses, i.e. Dehydration-Responsive Element-Binding protein 1 (DREB1), has been genetically modified through the CRISPR-Cas system to cope with dehydration due to higher temperatures (Erdoğan et al., 2023). Furthermore, CRISPR-Cas9 has been used to enhance crop tolerance to a range of heavy metals (Kumar et al., 2023). CRISPR technology has also been used to enhance tolerance to extreme temperatures, particularly in crops like tomatoes. This is crucial in the context of climate change, where rising temperatures pose a significant threat to global crop production. It has also been applied to understand gene function in response to abiotic stresses and improve tolerance to temperature, drought, and salinity stresses (Liu., 2021).

20.4.3 CRISPR and the Functional Genomics of Plants

CRISPR technology plays a huge role in the field of plant functional genomics, enabling precise gene manipulation for their functional characterization. This has been particularly valuable in understanding the role of genes in plant natural product biosynthesis and in the application of metabolomic engineering in plants. One illustrative example is the study of *Arabidopsis thaliana*, a model plant widely used in functional genomics research. This small plant, with its 120-megabase genome and 20,000 genes, has been instrumental in advancing our understanding of plant growth and development. CRISPR has been employed to knock out or modify specific genes in this model plant, revealing their roles in processes such as flowering time regulation, response to environmental stimuli, and nutrient uptake. These findings contribute to a deeper understanding of the fundamental biological mechanisms governing plant growth and development (Guo et al., 2021; Ahmad et al., 2024). Furthermore, the plant has been used to enhance genome annotation and study proteome dynamics and cell functions. The evolutionary, systematic, and spatio-temporal context of *A. thaliana* has also been explored using CRISPR, providing insights into its natural variation and adaptive evolution (Koch, 2018).

Additionally, CRISPR has significant implications in functional genomics studies in crop plants, particularly in rice, by enabling the elucidation of gene functions associated with disease resistance, grain quality, and stress responses. This has paved the way for the development of rice varieties with enhanced agronomic traits and resilience to environmental challenges (Romero and Gatica-Arias, 2019). The technology's potential for improving crop yield under biotic and abiotic stresses has been highlighted, with CRISPR-Cas9 offering high efficiency, accuracy, and low risk of off-target effects (Liu et al., 2021). The precision and efficiency of CRISPR technology have accelerated functional genomics research, allowing scientists to investigate the roles of individual genes in diverse plant species more rapidly and accurately than previous methods such as T-DNA insertions, activation tagging, insertional mutagenesis, and transposon-mediated insertions. As CRISPR applications in plant genomics continue to expand, they hold the promise of unlocking new knowledge that will harness crop improvement and address global food security challenges.

20.4.4 Use of Crispr in Improving Complex Traits in Plants

CRISPR technology has proven to be a groundbreaking tool in the pursuit of improving complex traits, such as photosynthesis, N-fixation, and architecture in plants. Photosynthesis is a multifaceted process crucial for plant growth and biomass production, enhancing its efficiency, and has significant implications for global food security. Researchers are using CRISPR to precisely target and modify genes associated with various aspects of the photosynthetic pathway, intending to optimize plant performance under diverse environmental conditions (Khumsupan, Donovan, and McCormick, 2019; Ahmad et al., 2021). CRISPR-Cas9 is implemented to improve agronomic traits in plants, including photosynthesis, through gene knockout and single-base editing (Rao and Wang, 2021). One notable example is the manipulation of genes involved in the Calvin cycle, a key component of photosynthesis responsible for carbon fixation in rice and other plants. By using CRISPR to modify specific genes in crops like rice or wheat, scientists aim to enhance the carbon assimilation and overall photosynthetic efficiency, potentially leading to increased crop yields (Thapa and Bhusal, 2020; Aslam et al., 2022). This targeted approach allows for the fine-tuning of complex metabolic processes, offering a level of precision crucial for engineering plants with improved photosynthetic traits. Furthermore, CRISPR has been employed to optimize the regulation of photosynthetic genes, ensuring that plants can efficiently adapt to fluctuating environmental conditions such as light intensity, temperature, and water availability. This genetic fine-tuning holds the promise of developing crops that maintain optimal photosynthetic rates across diverse climates, contributing to sustainable agriculture in the face of climate change (Thapa and Bhusal, 2020). Similarly, researchers have been using CRISPR to improve N-fixation in plants and altering plant architecture to enhance productivity and efficiency (Tiwari et al., 2020) Figure 20.2 shows applications of CRISPR-based systems in plants.

20.5 HOW CRISPR CROPS ARE DIFFERENT FROM GM CROPS

CRISPR technology provides unique benefits in precise manipulation of plants genomes which have altered the genomic landscape of agricultural crops in a way that was not possible before (Munir et al., 2024). In this section, we summarize some of these benefits that CRISPR offers in agriculture. Although genetic engineering has been used for genetic improvement of plants such as Bt crops and herbicide resistance in plants, as well as in other GMO examples, CRISPR-Cas technology has been recognized for its unrivaled precision and specificity in performing targeted DNA sequences (Bigini et al., 2023). It has enabled scientists to create precise and controlled genetic modifications with as small as a single nucleotide change and without introducing any foreign gene in the host genome. This precision reduces the likelihood of unintended outcomes in the gene-edited crop by minimizing undesirable changes in the genome. The degree of control in CRISPR-based techniques is unprecedented, allowing for the fine-tuning of desired features in the host plants. Similarly, transgene-free gene editing with mRNA and ribonucleoprotein complexes (RNPs) as CRISPR reagents has enabled the gene editing of crops which are considered as non-GM crops because such crops can be generated through conventional breeding. For example, SDN1 and SDN2 crops are indistinguishable from their conventional counterparts because these crops can also be developed with conventional crossing. In addition, for SDN1 and SDN2 transgene-free crops, it is difficult to track which

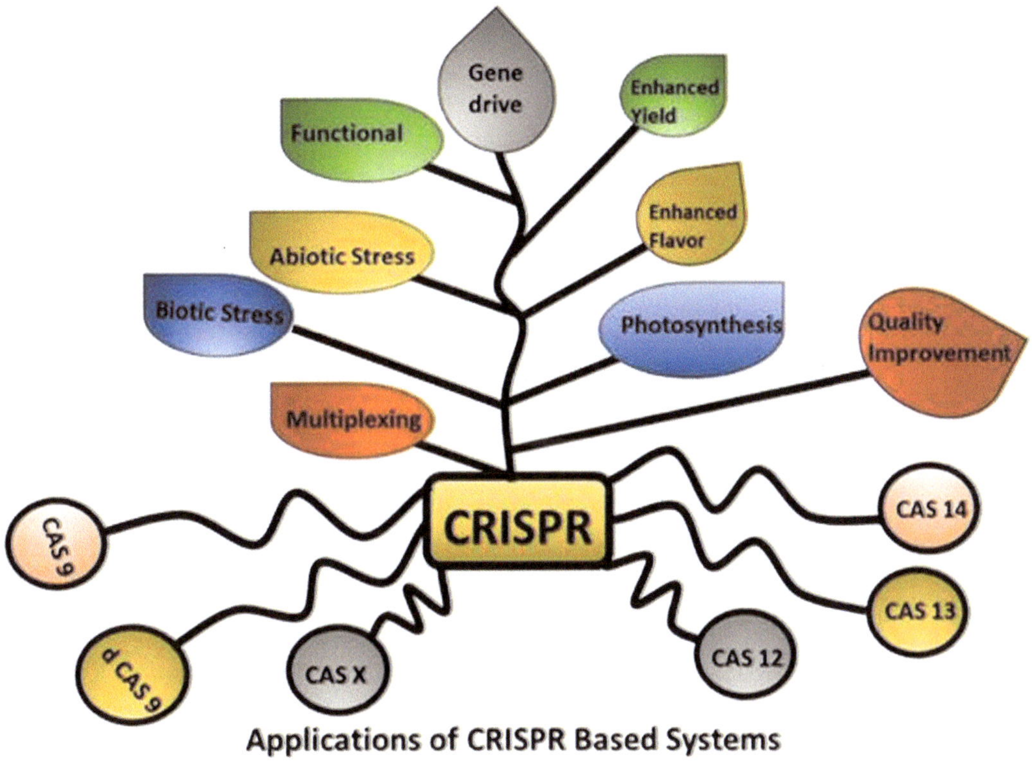

FIGURE 20.2 Applications of the CRISPR-based systems.

techniques have been used to develop these crops. Moreover, compared with GMOs, SDN1 and SDN2, transgene-free crops have higher public acceptance and relaxed regulatory measures because of the precise, controlled, and very small modifications in the genome. Traditional breeding procedures, such as selective breeding, might take years or even decades to introduce the desired features into the crops. CRISPR, on the other hand, produces comparable results in a short time compared to traditional methods. This quick turnaround is critical in tackling pressing agricultural issues, such as crop diseases, pests, and the changing environmental conditions, as well as guaranteeing food security and sustainability. Despite these promising applications and benefits, the challenges and limitations of CRISPR-Cas9, such as off-target effects and gene drive, need to be carefully considered.

20.6 POTENTIAL RISKS ASSOCIATED WITH CRISPR CROPS

CRISPR holds tremendous potential to fulfill the early promises of genetic engineering. CRISPR has been used extensively with proof-of-concept studies reported for model and crop plants with various applications including disease resistance, yield improvement, abiotic stress tolerance, and insect pest resistance. Although CRISPR use is intended to be precise based on gRNA design, off-targets and biosafety concerns have been raised about its potential for generating unintended mutations or its misuse for bioweapons. To address these concerns, researchers have been working on increasing the specificity and stability of CRISPR as well as safety procedures to restrict the use of DNA sequences considered dangerous (Ben Ouagrham-Gormley, and Fye-Marnien, 2019). Despite these efforts, the safety and security challenges associated with CRISPR-Cas technology remain a topic of ongoing discussion and research. In this section, we discuss off-targets, biosafety issues, and risks associated with potential release of CRISPR-based gene drives into the environment.

Off-target effects represent a concern in the application of CRISPR technology to create gene-edited crops. Off-target effects occur when the CRISPR system unintentionally introduces genetic changes at the genomic sites other than the intended target site, potentially leading to undesired modifications in the host genome. While significant efforts are made to minimize off-target effects, the risk remains, and it is crucial to thoroughly assess and mitigate these risks before deploying CRISPR-modified crops on a large scale (Patsali, Kleanthous, and Lederer, 2019). However, it has been reported by several researchers that compared with gene-editing experiments with animals, CRISPR is much more precise in plants with almost no off-targets. As a result, off-targets are not a major concern in CRISPR-edited plants; however, it remains a major challenge in therapeutic and health applications. Most of the off-target effects in CRISPR-edited crops are not expected to be critically harmful to humans or the environment, suggesting that existing regulations for genetically modified crops are sufficient for risk assessment (Liu et al., 2021).

Several attempts have been made to improve efficiency and specificity of CRISPR techniques to minimize off-target effects. For example, advanced CRISPR variants such as high-fidelity Cas9 and improved base editing and prime editing systems have been reported with minimal unintended modifications in the host plant. Rigorous testing and comprehensive analysis of the edited crops are essential steps in ensuring the safety and reliability of the final product before their widespread adoption. Monomeric Cas9/sgRNA systems and truncated gRNA sequences have been used as methods to destabilize off-target binding without sacrificing on-target cleavage efficiencies. These advances are crucial for the responsible and sustainable use of genome editing in agriculture (Chen, 2019).

CRISPR-based gene drive is a technique that aims to manipulate the inheritance of a specific gene within a population. Unlike traditional Mendelian inheritance, where a gene has a 50% chance of being passed on to offspring, CRISPR gene drive systems are designed to increase the likelihood of a particular gene being inherited to nearly 100%. This can result in the rapid spread of desired genetic traits through a population in just a few generations (Bier, 2021). Gene drive, a genetic engineering technique using CRISPR technology, has the potential to address public health and ecological challenges but also raises significant ethical and ecological concerns. It can be used for population management in conservation, but its spread dynamics and ecological risks need to be carefully considered (Kuzma, 2020).

One major issue associated with gene drive, particularly when using CRISPR, is the potential for unintended consequences and the irreversible genomic modifications in an entire population or ecosystem. The off-targets in CRISPR gene drives can also lead to unintended modifications in the target as well as non-target population, posing risks to the biodiversity and the ecological balance. The release of CRISPR gene drives into the environment raises questions about the predictability of their behavior and the ability to contain or reverse their effects, once released. As a result, the responsible and ethical use of CRISPR-based gene drives has become a focal point in discussions surrounding genetic engineering and environmental stewardship (Nestor and Wilson, 2020). To deal with these challenges especially off-target effects, CRISPR technology allows for the development of transgene-free edited crops, a distinct benefit over conventional GMOs (Ahmad et al., 2021). Transgene-free crops, containing no foreign DNA from unrelated species, alleviate worries about gene flow, environmental effects, and public perception. This feature increases the likelihood of broader consumer acceptability and easier regulatory processes.

20.7 TRANSGENE-FREE CRISPR CROPS

Transgene-free CRISPR crops are GM crops that are produced using CRISPR Cas technology and have no transgene or foreign DNA present in the host plant. The production of transgene-free crops is the main goal of genome-editing studies to address the concerns associated with conventional GMOs such as presence of antibiotic markers or transgenes. Several techniques have been used to attain this goal and each of these techniques has its versatility as well as benefits (Son and Park, 2022). These techniques include use of messenger RNA (mRNA) and RNPs as CRISPR reagents to develop gene-edited plants, and Transgene Killer CRISPR (TKC).

The use of messenger RNA (mRNA) as a CRISPR reagent is one of the promising strategies for developing transgene-free crops. In this approach, instead of introducing plasmid DNA into the host genome, synthesized mRNA or in-vitro transcribed mRNA encoding the Cas9 and gRNA are introduced into plant cells. The mRNA carrying the specific information of a Cas9 gene is translated within the plant cell, using the cellular machinery of the host, and converts the mRNA to protein. Thus allowing the desired genetic modification to occur without the integration of foreign DNA into the genome (Tsanova et al., 2021). Due to its short half-life and as it is easily degradable through a natural process, mRNA serves as the transient carrier of the genetic information in plants and has the advantage of temporary gene expression. For example, mRNA has been used as a CRISPR reagent to develop transgene-free potatoes that are resistant to late blight disease (Tussipkan and Manabayeva, 2021).

Another promising strategy for transgene-free genome editing is the use of ribonucleoprotein complexes (RNPs) as CRISPR cargos. RNPs are made up of the Cas protein (e.g. Cas9) and in-vitro synthesized guide RNA (sgRNA) molecule (Wada et al., 2020). These complexes are assembled in vitro before being transfected to plant cells. RNPs are extremely effective in inducing specific genomic changes without introducing foreign DNA. The advantage of using RNP is their transient nature, that is, when their job is done, they decay spontaneously, leaving no sign of exogenous genetic material in the genome of a plant. This transient nature of RNPs provides excellent accuracy and control in genome editing, contributing to the production of transgene-free crops with desirable traits.

Several examples of transgene-free crops that have been produced using RNPs. RNPs have been incorporated in the protoplast of tobacco as well as rice plants via polyethylene glycol calcium (PEG-Ca^{2+})-mediated transfection to produce transgene-free plants (Koto and Kongsawat, 2023). Successful incorporation of RNP into the embryonic maize and wheat cells has also been done through particle bombardment-mediated transfection, having subsequent regeneration of plant cells from these embryonic cells (Gu, Lijing, and Huawei, 2021).

Additionally, Transgene Killer CRISPR (TKC) is another promising strategy to generate transgene-free crops. This strategy used the precision and efficiency of the CRISPR-Cas9 system without the insertion of transgenes in gene-edited crops. It used synthetic gRNA and the Cas9 protein complex that are transient and easily degradable by a natural process of plant cells, alleviating the transgene from the gene-edited plants. Transgene-free rice has been produced using TKC by Yubing et al. (2019). In this research, incorporation of two gene cassettes in the construct of CRISPR-Cas9 was used.

These cassettes were suicide genes, i.e., Barnase and CMS2. Barnase is the bacterial gene that expresses a protein toxin having nuclease activity, that kills plant cells containing transgenes. On the other hand, CMS2 was involved in the disruption of mitochondrial function during the development of male gametophytes, resulting in the sterility of male gametophytes of rice. This strategy allowed CRISPR-Cas9 to do gene editing at the stages of callus formation, vegetative growth as well as transformation, resulting in the removal of embryos and pollen-containing transgenes and also ensuring that the harvested crop seeds are free from foreign genes (Yubing et al., 2019; Gu, Lijing and Huawei, 2021).

20.8 THE SIGNIFICANCE OF TRANSGENE-FREE CRISPR CROPS FOR FOOD SECURITY

CRISPR crops without transgenes hold great promise for tackling global food security challenges. The significance of these crops stems from their ability to supply sustainable and accessible food sources, especially in developing countries while alleviating regulatory concerns about GMOs. One of the key benefits of transgene-free CRISPR crops is that they have a lower environmental impact. These crops reduce worries about gene flow, unforeseen ecological consequences, and the possible disturbance of native plant populations, by eliminating the introduction of foreign DNA from gene-edited crops. Additionally, transgene-free crops encounter fewer regulatory challenges than regular GMOs, thereby simplifying regulatory processes. Streamlined approvals of transgene-free crops allow for faster development and implementation of new crop varieties, assisting in meeting the world's expanding food needs. Furthermore, consumer acceptance of transgene-free CRISPR crops is often higher than that of traditional GMOs because the absence of transgenes boosts consumer faith in the safety of crop consumption and food products developed from these crops (He and Zhao, 2019). This improved acceptance is critical for the broad adoption of gene-edited crops and their viability as a solution to food security concerns.

The ability of CRISPR technology to produce transgene-free crops also speeds up the production process of crops with desirable traits, enabling researchers and breeders to rapidly incorporate traits in crops such as disease resistance, greater nutritional content, as well as increased yield. This speed is especially essential since it allows for a timely reaction to emerging agricultural concerns like pests or climate-related problems. Transgene-free CRISPR crops are tailored to solve specific regional or local food security challenges (Sharma et al., 2023). Crops can be tailored to grow in a variety of environmental situations, ensuring that they are well suited to the distinct needs of different places. This adaptation helps to make the global food supply more secure and sustainable. These crops also help to reduce food waste due to the degradation of crops during storage and transportation by the improvement in post-harvest attributes such as shelf life (Sharma et al., 2023). Food waste reduction is especially important since it helps to ensure that more of the food produced reaches consumers, addressing food security on multiple levels (Munaweera et al., 2022). To understand the relaxed regulation of transgene-free CRISPR crops especially in the USA, it is necessary to understand the mechanisms associated with regulatory triggers of CRISPR-edited crops.

20.9 REGULATORY TRIGGERS FOR CRISPR CROPS

Regulatory triggers are critical in regulating the approval for commercialization and usage of CRISPR-edited crops for food and feed. To evaluate and identify gene-edited crops, three types of regulatory triggers are currently used, including process-based, product-based, and plant with novel traits (PNT)-based regulatory triggers. Process-based regulatory triggers are the criteria for CRISPR-edited plants including crops that bring about regulatory oversights during the process of development, commercialization, and testing stages of CRISPR-edited crops. The main goal of establishing these triggers is to maintain the safety, efficacy, as well as regulatory compliance of CRISPR-edited crops. A process-based regulatory trigger is concerned with the procedures and techniques employed in the production of gene-edited crops. Regulatory bodies use a process-based approach to evaluate specific genetic engineering procedures, regardless of the qualities of the ultimate product. A regulatory body, for example, might require that all CRISPR-edited crops are regulated, independent of the genetic modifications introduced during the process of development. This strategy seeks to ensure that safety assessments and regulatory scrutiny are done consistently throughout the genome-editing process (Zhang et al., 2020). The EU and New Zealand in particular are following the process-based trigger to regulate CRISPR crops and consider all CRISPR crops as GMOs, thereby regulating them under conventional GMO assessments.

Product-based triggers seek to identify and regulate gene-edited crops based on their distinct characteristics and possible risks, regardless of the process or technique used to develop the crop. Using this method, regulatory bodies evaluate the final product of genome editing to see if it exhibits specific traits that necessitate regulatory assessments. For example, if a gene-edited plant exhibits traits that cannot be obtained through normal breeding, such as novel disease resistance or increased nutritional content, it will be subject to regulatory assessments. Several countries such as the USA, Argentina, Russia, and Japan have been using product-based triggers to regulate gene-edited crops. PNT is a regulatory trigger used in Canada to regulate novel traits or phenotypes, regardless of the process being used, i.e., mutagenesis, genetic engineering, or gene editing, whether it is gene knockout, site-directed gene insertion, or replacement of allele. To establish whether an edited crop qualifies as a PNT, it must have at least a 30% increase or decrease of a novel trait after the genome editing (GE) as compared to conventional variety. If the GE crop exceeds the set value, it will be subjected to regulatory restrictions, whereas those that fall below it will not be considered as PNT. PNT-based triggers seek to establish a compromise between crop safety and regulatory load with

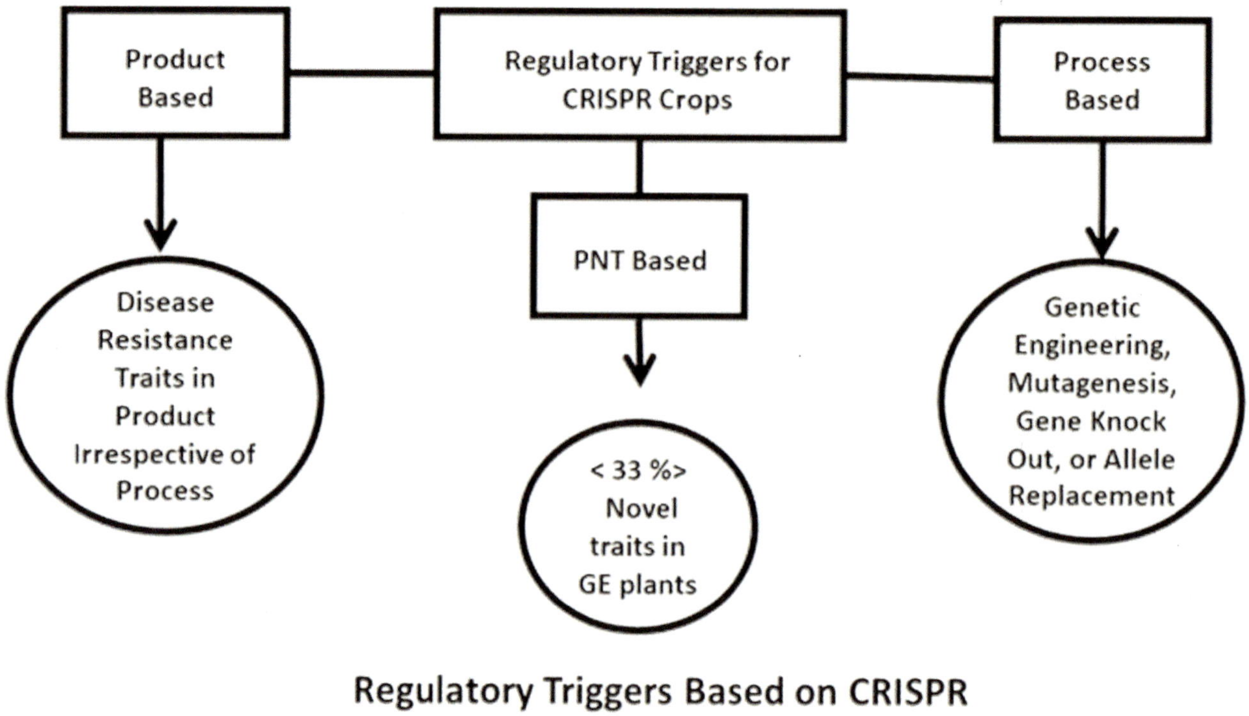

FIGURE 20.3 An overview of the process-based, product-based, and PNT-based regulatory triggers.

the minimum genetic modifications (Entine et al., 2021). Figure 20.3 shows an overview of the process-based, product-based, and PNT-based regulatory triggers (Entine et al., 2021).

20.10 GLOBAL REGULATORY OVERSIGHT OF CRISPR CROPS

21.10.1 The United States

The United States (USA) uses a product-based regulation approach for CRISPR-edited crops. Crops that do not contain transgenes and closely resemble traditionally cultivated varieties, such as SDN1 and SDN2, crops face less stringent regulatory assessments. Multiple regulatory bodies are involved in the oversight of CRISPR-edited crops, including the United States Department of Agriculture (USDA), the Environmental Protection Agency (EPA), and the Food and Drug Administration (FDA). The regulatory process is adaptable and varies depending on the qualities of the crop and its intended application. In contrast to the European Union (EU), the United States takes a more lenient approach, attempting to balance innovation with safety and sustainability. Environmental impact and public safety are considered in regulatory decisions. Ongoing conversations and research help to shape the growing regulatory framework, which addresses scientific and ethical concerns (Zhang et al., 2020; Ahmad et al., 2021). In 2023, the USDA provided a set of revised regulations for different CRISPR-edited crop varieties. In accordance with Executive Order 13874, the USDA made a rule which is known as the SECURE rule to make an update for overseeing genetically modified crops. According to the rule, if the crop is made by a normal breeding method, then there is no need for regulation. The USDA is currently focusing on the risk that the genetically modified crop can pose rather than the method by which it is made. This fast procedure is helping the USDA to keep an eye on the potential risks associated with genetically modified crops, making an easy regulatory framework for researchers and companies working at smaller scales (Hoffman, 2021).

20.10.2 China

China has been a leader in CRISPR research, developing a regulatory framework for CRISPR-edited crops. Notably, in China, several CRISPR-edited crops have been approved for field trials and commercialization. Ethical considerations, such as public participation and openness, are essential components of regulatory oversights in China. China's legislative actions and CRISPR technical breakthroughs have the potential to affect the global adoption and commercialization of CRISPR-edited crops. The country's commitment to being at the forefront of scientific and biotechnological breakthroughs is reflected in the country's regulatory evolution (Zhang et al., 2020). To assure food safety and trade of crops, China has built a regulatory framework for controlling and analyzing the risk of gene-edited crops, which is regulated based on individual cases. The Regulation provided by the State Council on Administration of Agricultural Genetically Modified Organisms Safety, describes GMOs as organisms with modified genetic structures developed with genome-editing technology for use in agriculture. As a result, GM crops fall under the

GMO regulatory framework, which serves as an interim policy (Zhang et al., 2020). In January 2022, the Ministry of Agriculture and Rural Affairs (MARA) of China issued the "Guidelines for safety evaluation of genetically modified plants for agricultural use." These recommendations describe a risk assessment procedure in which a low-risk finding in GMOs leads to a small-scale test. After having successful results in the test, GMOs become eligible for a safety certificate for commercial production. The Chinese government has formally endorsed these recommendations, which aim to encourage research, development, and commercial application while remaining compliant with existing GM rules. This presents a more efficient approach, enabling the issuance of a safety certificate following a small-scale test. Although China retains its fundamental legal foundation for GMOs, recent regulatory modifications suggest an aggressive attempt to advance rules on GM crops (Tachikawa and Matsuo, 2023).

20.10.3 India

India is the fourth-largest country harvesting genetically modified crops, with an estimated 11.6 million hectares (Gupta et al., 2021). The first transgenic crop approved for commercial use by the Government of India (GOI) and the Genetic Engineering Approval Committee (GEAC) was Bt cotton and it has made India a major exporter of cotton (Singh, 2021). India has a strict regulatory approach to CRISPR-edited crops, with a strong emphasis on biosafety concerns. The GEAC oversees the approval procedure, which involves a thorough examination of field trials and releases. Indian regulation of GM crops began in 1982 with the formation of the National Biotechnology Board, which evolved into the Department of Biotechnology (DBT) in 1986. In 2020, the DBT issued draft guidelines outlining a risk assessment framework for regulating GMOs. These criteria classify plants with one or more base-pair modifications as "Group 1," whereas inserting large foreign DNA into is a high-risk group, totally deregulating Site-Directed Nuclease (SDN) type 1 and 2 in line with the United States (Gupta et al., 2021). Recently, India has issued revised guidelines for CRISPR-edited crops, to ensure the sustainability as well as safety of the environment. These include assessment of risk, monitoring, and labeling of GMOs, showing a cautious approach to adopt emerging technology for genetic modification of crops (Caradus, 2023).

20.10.4 Canada

Canada has the fifth-largest growing area for genetically modified (GM) crops, totaling 11 million hectares. The regulatory procedure for GM crops in Canada adheres to the PNT framework, which employs a product-based risk assessment methodology with a case-by-case evaluation. PNT is a regulatory trigger used in Canada to check novel traits or phenotypes, regardless of the process being used, i.e., mutagenesis, gene knockout, site-directed gene insertion, or replacement of allele. To establish whether a gene-edited crop qualifies as a genetically modified organism (GMO), PNT-based triggers rely on specified criteria that the plant must have at least a 30% increase or decrease of novel trait, compared to the conventional variety. If the gene-edited crop exceeds the set PNT, it will be subject to GMO restrictions, whereas those that fall below set criteria will not be subject to any restriction. PNT-based triggers seek to establish a compromise between crop safety and regulatory load with minimum genetic modifications (Entine et al., 2021). Since the early 1990s, the Canadian Food Inspection Agency (CFIA), as an independent regulator, has authorized herbicide-tolerant canola cultivars created by site-directed mutagenesis. Canada has shown a strong interest in genome editing by conducting international workshops on the issue and exempting SDN-1 and SDN-2 from review and regulation due to their lack of novelty (Gupta et al., 2021).

20.10.5 Japan

As part of its integrated innovation policy, the Japanese cabinet tried to clarify the regulation of GMOs under the Cartagena Act and the Food Sanitation Act in March 2019. An expert group from the Ministry of Environment (MOE) recommended that organisms produced from SDN-1 not be regulated. The MOE issued a comprehensive guide to help interested parties determine the status of genome-editing technology under the Cartagena Act. Allowances to use GM crops will be granted if the absence of foreign nucleic acid sequence in the genome has been proved and only the genome-edits exist (Menz et al., 2020). As a result, organisms generated from SDN-1 are exempt from regulation. For SDN-2, the absence of HDR templates must be shown. Anyhow, no information is provided on how the absence should be proved. In March 2019, the MOE revised its stance, noting that organisms transformed via the insertion of extracellularly processed nucleic acids are generally classified as Living Transformed Organisms (LMOs) and must comply with the Cartagena Act. SDN-3 applications are typically classified as LMO, but it is unclear how allele swaps will be governed (Menz et al., 2020). Recently, Japan has approved genetically modified tomatoes using CRISPR technology that have good texture, are resistant to biotic stress, and contain increased nutrient quantity (Nagamine, Takayama, and Ezura, 2022).

20.10.6 New Zealand

New Zealand has adopted a process-based regulatory strategy for CRISPR-edited crops. The Hazardous Substances and New Organisms (HSNO) Act of 1996 established strict regulations for GMOs in New Zealand. The Act defines a GMO as any organism whose genes or genetic material have been modified using in vitro methods. The broad application of this definition encompasses several processes of development, including somaclonal variation, cell fusion, and chemical and physical mutagenesis, that were in use at the time the Act was passed (Fritsche et al., 2018). As genome editing in plants became more widely known, the Environmental Protection Authority (EPA) in New Zealand decided to classify some plants as non-GMOs. This decision was expressed in a 2013 EPA report. Under the Act of HSNO, plants that were modified with Site-Directed Nuclease Type 1 (SDN-1) were initially considered equivalent to those that had undergone chemical mutagenesis, meaning they were not subject to GMO control. This classification, however, was challenged, and in the Kershen case (2015), a High Court ruling overturned it. This prompted changes to the law, which made it clear that any mutagenesis method developed after 1996 would be regarded as a GMO (Menz et al., 2020).

20.10.7 Australia

Australia has set strong guidelines and follows a process-based regulatory approach to oversee genetically modified crops. The Office of Gene Technology Regulator (OGTR) is responsible for checking and evaluating GMOs (Jones et al., 2022). Developers submit the details of applications, potential risks, and molecular characteristics associated with genetically modified crops. The regulatory process in Australia consists of process evaluation, assessment of risk as well as consultancy with the public before the GM crop is approved for its release in the marketplace for human consumption (Redden, 2021). This regulatory framework reflects the commitment of Australia to balance innovation with public health concerns and environmental protection.

20.10.8 The European Union

The European Union (EU) uses process-based GMO legislation, which assesses risks based on the techniques used to create a GMO rather than the end products. The European Court of Justice (ECJ) defines a GMO as any organism that has modified genetic material, including CRISPR/Cas-directed mutagenesis, regardless of the size of DNA modification. ECJ case C-520/1650 states that organisms generated using SDN approaches (SDN-1, SDN-2, and SDN-3) are subject to the regulations of GMOs (Ahmad et al., 2021). SDN-1 includes single-nucleotide changes, SDN-2 includes bigger insertions/deletions with no foreign DNA insertion, and SDN-3 includes considerable gene mutations. Regulation (EC) No 1829/2003 governs how EU member states regulate genetically modified foods and feed products. Despite continuous debate, the EU currently prohibits the growing of GM crops, demonstrating its cautious approach to novel breeding techniques (Ahmad et al., 2021).

20.10.9 The United Kingdom

After Brexit, the United Kingdom has introduced its own framework of regulation for genetically modified crops using CRISPR technology. This framework is dependent upon the process- and product-based assessment of CRISPR-edited crops. The Department of Environment Food and Rural Affairs (DEFRA) started the consulting process in January 2021, nationwide (Firmansyah, 2021). During this process, the UK gathered information regarding the regulation of genome-editing technologies in agriculture. The aim of this information was to discuss the problems regarding the use of genome-editing technologies in the UK, that is, whether it should be regulated in the same way as GMOs or not (Firmansyah, 2021). However, the UK is now following a less stringent process to regulate CRISPR-edited crops, compared to the EU, aiming to foster innovation in agriculture as well as in CRISPR-edited crops. The government of the UK has applied two legislative amendments for CRISPR-edited crops. According to these amendments, GM crops will be exempted from the regulatory process if the product does not contain any foreign DNA or the modification of the genome is performed via natural process or traditional breeding (Marone, Mastrangelo, and Borrelli, 2023).

20.10.10 Africa

South Africa, Ethiopia, Kenya, Malawi, Nigeria, and Sudan are among the few African countries that currently farm GMCs. South Africa is the continent's leader in genetically modified crops production, because of its solid regulatory environment. Despite legislation allowing for prospective cultivation in Burkina Faso, South Africa is the only African country that has approved GM white maize for direct consumption. The barriers to GM crops' adoption in Africa are diverse, encompassing social, political, and legislative reasons. GM cotton has just received environmental release clearances in Kenya, Malawi, and Nigeria. Ghana and Uganda are moving closer to approving commercial production of GM crops, indicating good moves toward wider adoption (Qaim, 2020). These discoveries include both classic GM crops and those generated using innovative breeding techniques, such as those classified as SDN-1, SDN-2, and SDN-3. A cautious approach is required to avoid overregulation, which may hinder progress (Smyth, 2020).

20.11 CAN CRISPR CROPS AVOID THE GMO CURSE

CRISPR holds tremendous potential to reshape the genomic landscape of agri-food crops and contribute to the future of global food security. Many researchers have demonstrated CRISPR applications in plants with proof-of-concept studies to improve disease resistance, tolerance to biotic and abiotic stresses, and enhanced crop yield and productivity. Although CRISPR-edited crops have been approved for commercialization by some countries, such as the USA, Argentina, and Japan, their regulation is still being debated regarding their safety, risk assessment, and public perception but there is still no practical evidence of health risk associated with CRISPR-edited crops. So far, the regulatory framework of CRISPR crops varies in different jurisdictions, and the differences stem from different definitions of genetic modification. For example, the USA has been following product-based regulation for CRISPR-edited crops and introduced exemptions for transgene-free SDN1 and SDN2 CRISPR-edited crops if conventional breeding could also achieve those modifications (Zhang et al., 2020; Ahmad et al., 2021). In contrast, the EU uses process-based regulation and, at present, still considers all CRISPR-edited crops as GMOs (Ahmad et al., 2021). The dichotomy in the regulatory and legal status of CRISPR-edited crops may limit market access and hence limit applications to solve challenges of food security and sustainable agriculture.

It is important to note that CRISPR-edited crops are not the same as GM crops. Compared with GM crops, CRISPR crops are precise, have controlled modifications, and are indistinguishable from conventionally bred crops, especially SDN1 and SDN2 crops. GM crops are usually transgenic, containing genes from other species such as the Bt gene, antibiotic-resistance genes, and CaMV 35S promoter. So, their public acceptance is low in many parts of the world (Liu et al., 2021). Although GM crops are safe and there is no reported evidence that GM crops are not safe for human health or the environment, however, their public acceptance is low due to the transgenes present in them. This dichotomy in regulatory triggers and risk assessments of GM crops has resulted in low acceptance of GM crops, especially in the EU and New Zealand. In contrast, CRISPR crops are different from GM Crops in their genetic modifications, preciseness, and being free of transgenes. Hence, it is important to consider that CRISPR crops are not like GM crops in their genetic modifications and the world community should consider them differently. However, if CRISPR-edited crops are treated the same as GM crops for their legal status, regulations, and risk assessment, these crops will end up the same as GM crops. In addition, public acceptance of CRISPR-edited crops will be low as well in several parts of the world. Therefore, we need a universal, reproducible, evidence-based, and scalable regulatory system for CRISPR-edited crops for their better public acceptance and commercialization. In addition, all stakeholders such as scientists, private companies, regulatory authorities, governance bodies, farmers, policy makers, and politicians should collaborate to develop a universal regulatory system for CRISPR-edited crops. With better public acceptance and awareness of CRISPR technology, gene-edited crops may play a greater role in addressing the challenge of food security and sustainable development in developing countries.

20.12 CONCLUSION

CRISPR has the potential to bring in a new era of sustainable, robust, and nutritious crops, due to its breakthrough capabilities in precise genome editing and broad uses in plant agriculture. As it has various applications in plant agriculture, it has been successfully used to produce GMCs including GM rice. Furthermore, it has also been utilized to modify complex traits like photosynthesis and also has applications in functional genomics. The difficulty of transgene-free genome editing emphasizes the need to resolve regulatory concerns as well as societal acceptance. The importance of transgene-free CRISPR crops for food security cannot be emphasized enough, especially in underdeveloped nations where agricultural production is crucial to providing access to nutritious food. The effect of regulatory control on CRISPR crops cannot be overstated, influencing not just their research and commercialization, but also public perception and worldwide trade. The regulatory frameworks imposed by different countries regulate, accept, and make the commercial use of GMCs for the people. However, ongoing research, public involvement, and ethical vigilance are required to ensure that CRISPR technology continues to be a force for good in agriculture and beyond.

REFERENCES

Ahmad, A., Abd-Elsalam, K. A., Ali, M., Majeed, H. N., Arif, A., Munir, A., Khan, Z. and Munawar, N. (2024). Regulatory triggers of CRISPR-edited crops. In Elsevier EBooks (pp. 91–112). Elsevier. https://doi.org/10.1016/b978-0-443-18444-4.00016-8

Ahmad, A., Ghouri, M.Z., Munawar, N., Ismail, M., Ashraf, S. and Aftab, S. O. (2021a). Regulatory, ethical, and social aspects of CRISPR crops. In A. Ahmad, S.H. Khan, and Z. Khan (Eds.), CRISPR crops: The future of food security (pp. 261–287). Springer.

Ahmad, A., Munawar, N., Khan, Z., Qusmani, A.T., Khan, S.H., Jamil, A., Ashraf, S., Ghouri, M.Z., Aslam, S., Mubarik, M.S., Munir, A., Sultan, Q., Abd-Elsalam, K.A. and Qari, S.H. (2021). An outlook on global regulatory landscape for genome-edited crops. *International Journal of Molecular Sciences*, 22(21), 11753.

Almeida, M. and Ranisch, R. (2022). Beyond safety: Mapping the ethical debate on heritable genome editing interventions. *Humanities and Social Sciences Communications*, 9(1), 1–14.

Altaf, M.A., Mandal, S., Behera, B., Mangal, V., Naz, S., Kumar, R., Kumar, A., Ghorai, M., Singh, B., Dey, A., Tiwari, R.K., Lal, M.K. and Aftab, T. (2022). Salinity stress tolerance in Solanaceous crops: Current understanding and its prospects in genome editing. *Journal of Plant Growth Regulation*, 42(7), 4020–4036.

Aman, R., Mahas, A. and Mahfouz, M. (2020). Nucleic acid detection using CRISPR/Cas biosensing technologies. *ACS Synthetic Biology*, 9, 1226–1233.

Aslam, S., Munir, A., Hafiz, Khan, S. H. and Ahmad, A. (2022). Genome editing advances in soybean improvement against biotic and abiotic stresses. In Springer EBooks (pp. 241–274). Springer. https://doi.org/10.1007/978-3-031-12232-3_13

Bacha, S.A.S. and Iqbal, B. (2023). Advancing agro-ecological sustainability through emerging genetic approaches in crop improvement for plants. *Functional & Integrative Genomics*, 23, 145.

Ben Ouagrham-Gormley, S. and Fye-Marnien, S.R. (2019). Is CRISPR a security threat? In S.K. Singh and J.H. Kuhn (Eds.), Defense aganst biological attacks, Springer.

Bhattacharjee, S., Bhowmick, R., Kant, L. and Paul, K. (2023). Strategic transgene-free approaches of CRISPR-based genome editing in plants. *Molecular Genetics and Genomics*, 298, 507–520.

Bier, E. (2021). Gene drives gaining speed. *Nature Reviews Genetics*, 23, 5–22.

Bigini, F., Lee, S.H., Sun, Y.J., Sun, Y. and Mahajan, V.B. (2023). Unleashing the potential of CRISPR multiplexing: Harnessing

Cas12 and Cas13 for precise gene modulation in eye diseases. *Vision Research,* 213, 108317.

Blix, T.P.B. (2023). *Genome-edited salmon: A sustainable and socially acceptable solution to aquaculture?* [online] munin.uit.no.

Caradus, J.R. (2023). Processes for regulating genetically modified and gene edited plants. *GM Crops & Food,* 14, 1–41.

Chaudhuri, A., Halder, K. and Datta, A. (2022). Classification of CRISPR/Cas system and its application in tomato breeding. *Theoretical and Applied Genetics*, 135(2), 367–387.

Chen, S.-J. (2019). Minimizing off-target effects in CRISPR-Cas9 genome editing. *Cell Biology and Toxicology,* 35, 399–401.

Entine, J., Felipe, M.S.S., Groenewald, J.-H., Kershen, D.L., Lema, M., McHughen, A., Nepomuceno, A.L., Ohsawa, R., Ordonio, R.L., Parrott, W.A., Quemada, H., Ramage, C., Slamet-Loedin, I., Smyth, S.J. and Wray-Cahen, D. (2021). Regulatory approaches for Genome Edited agricultural plants in select countries and jurisdictions around the world. *Transgenic Research*, 30(4), 551–584.

Erdoğan, İ., Cevher-Keskin, B., Bilir, Ö., Hong, Y. and Tör, M. (2023). Recent developments in CRISPR/Cas9 genome-editing technology related to plant disease resistance and abiotic stress tolerance. *Biology*, 12(7), 1037.

Firmansyah, A. (2021). Review of the current and post-Brexit UK legal framework on the regulation of genetic technologies. *Bioethica*, 7(2), 51–68.

Friedrichs, S., Takasu, Y., Kearns, P., Dagallier, B., Oshima, R., Schofield, J. and Moreddu, C. (2019). *Meeting report of the OECD Conference on "genome editing": Applications in agriculture implications for health, environment and regulation*. Springer.

Fritsche, S., Poovaiah, C., MacRae, E. and Thorlby, G. (2018). A New Zealand perspective on the application and regulation of gene editing. *Frontiers in Plant Science,* 9, 1323.

Gao, C. (2021). Genome engineering for crop improvement and future agriculture. *Cell,* 184, 1621–1635.

Ghosh, S. and Dey, G. (2022). Biotic and abiotic stress tolerance through CRISPR-Cas mediated genome editing. *Journal of Plant Biochemistry and Biotechnology,* 31, 227–238.

Gu, X., Lijing, L. and Huawei, Z. (2021). Transgene-free genome editing in plants. *Frontiers in Genome Editing,* 3, 805317.

Guo, M., Liu, X., Jiang, Y., Yu, J. and Meng, T. (2021). Identification of Arabidopsis genes associated with cold tolerance based on integrated bioinformatics analysis. *Journal of Plant Interactions*, 16(1), 344–353.

Gupta, R., Ghosh, A., Chakravarti, R., Singh, R., Ravichandiran, V., Swarnakar, S. and Ghosh, D. (2022). Cas13d: A new molecular scissor for transcriptome engineering. *Frontiers in Cell and Developmental Biology,* 10, 866800.

Gupta, S., Kumar, A., Patel, R. and Kumar, V. (2021). Genetically modified crop regulations: Scope and opportunity using the CRISPR-Cas9 Genome Editing approach. *Molecular Biology Reports*, 5(48), 4851–4863.

He, Y. and Zhao, Y. (2019). Technological breakthroughs in generating transgene-free and genetically stable CRISPR-edited plants. *aBIOTECH*, 1(1), 88–96.

Hoffman, N.E. (2021). Revisions to USDA biotechnology regulations: The SECURE rule. *Proceedings of the National Academy of Sciences*, 118(22), e2004841118.

Jones, M.G.K., Fosu-Nyarko, J., Iqbal, S., Adeel, M., Romero-Aldemita, R., Arujanan, M., Kasai, M., Wei, X., Prasetya, B., Nugroho, S., Mewett, O., Mansoor, S., Awan, M.J.A., Ordonio, R.L., Rao, S.R., Poddar, A., Hundleby, P., Iamsupasit, N. and Khoo, K. (2022). Enabling trade in gene-edited produce in Asia and Australasia: The developing regulatory landscape and future perspectives. *Plants (Basel, Switzerland),* 11(19), 2538.

Kazemian, P., Yu, S.-Y., Thomson, S.B., Birkenshaw, A., Leavitt, B.R. and Ross, C.J.D. (2022). Lipid-nanoparticle-based delivery of CRISPR/Cas9 genome-editing components. *Molecular Pharmaceutics*, 19(6), 1669–1686.

Kazi, T.A. and Biswas, S.R. (2021). *CRISPR/dCas system as the modulator of gene expression*. ScienceDirect.

Khan, I., Khan, S., Zhang, Y., Zhou, J., Akhoundian, M. and Jan, S.A. (2021). CRISPR-Cas technology-based genome editing for modification of salinity stress tolerance responses in rice (Oryza sativa L.). *Molecular Biology Reports*, 48(4), 3605–3615.

Khumsupan, P., Donovan, S. and McCormick, A.J. (2019). CRISPR/Cas in Arabidopsis: Overcoming challenges to accelerate improvements in crop photosynthetic efficiencies. *Physiologia Plantarum*, 166(1), 428–437.

Koch, M.A. (2018). The plant model system Arabidopsis set into an evolutionary, systematic and spatio-temporal context. *Journal of Experimental Botany*, 70(1), 55–67.

Koto, R. and Kongsawat, C. (2023). Improving plant molecular farming via genome editing. In C. Kole et al. (Eds.), *Tools and techniques of plant molecular farming* (pp. 63–88). Springer.

Kumar, M., Prusty, M.R., Pandey, M.K., Singh, P.K., Bohra, A., Guo, B. and Varshney, R.K. (2023). Application of CRISPR/Cas9-mediated gene editing for abiotic stress management in crop plants. *Frontiers in Plant Science,* 14, 1157678.

Kumari, C., Sharma, M., Kumar, V., Sharma, R., Kumar, V., Sharma, P., Kumar, P. and Irfan, M. (2022). Genome editing technology for genetic amelioration of fruits and vegetables for alleviating post-harvest loss. *Bioengineering*, 9(4), 176.

Kuzma, J. (2020). Engineered gene drives: Ecological, environmental, and societal concerns. In A. Chaurasia et al. (Eds.), *GMOs* (pp. 371–399). Springer.

Lee, T.-M., Lin, J.-Y., Tsai, T.-H., Yang, R.-Y. and Ng, I.-S. (2022). Clustered regularly interspaced short palindromic repeats (CRISPR) technology and genetic engineering strategies for microalgae towards carbon neutrality: A critical review. *Bioresource Technology,* 368, 128350.

Liu, Q., Yang, F., Zhang, J., Liu, H., Rahman, S., Islam, S., Ma, W. and She, M. (2021). Application of CRISPR/Cas9 in crop quality improvement. *International Journal of Molecular Sciences*, 22(8), 4206.

Maharajan, T., Krishna, T.A., Rakkammal, K., Ceasar, S.A. and Ramesh, M. (2022). Application of CRISPR/Cas system in cereal improvement for biotic and abiotic stress tolerance. *Planta,* 256, 106.

Makarova, K.S., Wolf, Y.I., Iranzo, J., Shmakov, S.A., Alkhnbashi, O.S., Brouns, S.J., Charpentier, E., Cheng, D., Haft, D.H. and Horvath, P. (2020). Evolutionary classification of CRISPR–Cas systems: A burst of class 2 and derived variants. *Nature Reviews Microbiology,* 18, 67–83.

Marone, D., Mastrangelo, A.M. and Borrelli, G.M. (2023). From transgenesis to genome editing in crop improvement: Applications, marketing, and legal issues. *International Journal of Molecular Sciences*, 24(8), 7122.

Mattick, J.S., Amaral, P.P., Carninci, P., Carpenter, S., Chang, H.Y., Chen, L.-L., Chen, R., Dean, C., Dinger, M.E. and Fitzgerald, K.A. (2023). Long non-coding RNAs: Definitions, functions, challenges and recommendations. *Nature Reviews Molecular Cell Biology,* 24, 430–447.

Menz, J., Modrzejewski, D., Hartung, F., Wilhelm, R. and Sprink, T. (2020). Genome edited crops touch the market: A view on the global development and regulatory environment. *Frontiers in Plant Science,* 11, 586027.

Mohd Azman, A.A., Iskandar, M.F., Muhammad Zaidi Leong, M.N.H.L. and Mohd Fauzi, N.A. (2022). *The enhancement of plant disease resistance using CRISPR/Cas9 technology.* eprints.uthm.edu.my.

Munaweera, T.I.K., Jayawardana, N.U., Rajaratnam, R. and Dissanayake, N. (2022). Modern plant biotechnology as a strategy in addressing climate change and attaining food security. *Agriculture & Food Security*, 11(1), 1–28.

Munir, A., Amin, I., Zahoor, Z. K., Majeed, H. N., Almoammar, H., Ghaffar, A. and Ahmad, A. (2024). Multiplex genome editing in plants through CRISPR-Cas. In Elsevier EBooks (pp. 127–142). Elsevier. https://doi.org/10.1016/b978-0-443-13229-2.00012-0

Mutezo, W., Sedibe, M.M., Mofokeng, A., Shargie, N. and Soko, T. (2021). The Application of CRISPR/Cas9 technology in the management of genetic and nongenetic plant traits. *International Journal of Agronomy,* 2021, 1–9.

Naeem, M., Majeed, S., Hoque, M. Z. and Ahmad, I. (2020). Latest developed strategies to minimize the off-target effects in CRISPR-Cas-mediated genome editing. *Cells,* 9, 1608.

Nagamine, A., Takayama, M. and Ezura, H. (2022). Genetic improvement of tomato using gene editing technologies. *Journal of Horticultural Science and Biotechnology,* 98, 1–9.

Nazir, R., Mandal, S., Mitra, S., Ghorai, M., Das, N., Jha, N.K., Majumder, M., Pandey, D.K. and Dey, A. (2022). Clustered regularly interspaced short palindromic repeats (CRISPR)/ CRISPR-associated genome-editing toolkit to enhance salt stress tolerance in rice and wheat. *Physiologia Plantarum,* 174(2), e13642.

Nestor, M.W. and Wilson, R.L. (2020). Beyond Mendelian genetics: Anticipatory biomedical ethics and policy implications for the use of CRISPR together with gene drive in humans. *Journal of Bioethical Inquiry,* 17, 133–144.

Nidhi, S., Anand, U., Oleksak, P., Tripathi, P., Lal, J.A., Thomas, G., Kuca, K. and Tripathi, V. (2021). Novel CRISPR–Cas systems: An updated review of the current achievements, applications, and future research perspectives. *International Journal of Molecular Sciences*, 22(7), 3327.

Organisms, E.P.O.G.M., Naegeli, H., Bresson, J.L., Dalmay, T., Dewhurst, I.C., Epstein, M.M., Firbank, L.G., Guerche, P., Hejatko, J. and Moreno, F.J. (2020). Applicability of the EFSA Opinion on site-directed nucleases type 3 for the safety assessment of plants developed using site-directed nucleases type 1 and 2 and oligonucleotide-directed mutagenesis. *EFSA Journal,* 18, e06299.

Patsali, P., Kleanthous, M. and Lederer, C.W. (2019). Disruptive technology: CRISPR/Cas-based tools and approaches. *Molecular Diagnosis & Therapy*, 23(2), 187–200. https://doi.org/10.1007/s40291-019-00391-4.

Qaim, M. (2020). Role of new plant breeding technologies for food security and sustainable agricultural development. *Applied Economic Perspectives and Policy*, 42(2), 129–150.

Qaim, M. and Kouser, S. (2013). Genetically modified crops and food security. *PLoS ONE*, 8(6), e64879.

Rao, M.J. and Wang, L. (2021). CRISPR/Cas9 technology for improving agronomic traits and future prospective in agriculture. *Planta*, 254(4), 1–16.

Rasheed, A., Gill, R.A., Hassan, M.U., Mahmood, A., Qari, S., Zaman, Q.U., Ilyas, M., Aamer, M., Batool, M., Li, H. and Wu, Z. (2021). A critical review: Recent advancements in the use of CRISPR/Cas9 technology to enhance crops and alleviate global food crises. *Current Issues in Molecular Biology*, 43(3), 1950–1976.

Raza, A., Tabassum, J., Fakhar, A., Sharif, R., Chen, H., Zhang, C., Ju, L., Fotopoulos, V., Siddique, M., Singh, R., Zhuang, W. and Varshney, R.K. (2022). *Smart reprograming of plants against salinity stress using modern biotechnological tools.* Institute of Agriculture, University of Western Australia, pp.1–28.

Redden, R. (2021). Genetic modification for agriculture—proposed revision of GMO regulation in Australia. *Plants*, 10(4), 747.

Ricroch, A., Clairand, P. and Harwood, W. (2017). Use of CRISPR systems in plant genome editing: Toward new opportunities in agriculture. *Emerging Topics in Life Sciences*, 1(2), 169–182.

Romero, F.M. and Gatica-Arias, A. (2019). CRISPR/Cas9: Development and application in rice breeding. *Rice Science,* 26(5), 265–281.

Sharma, P., Pandey, A., Malviya, R., Dey, S., Karmakar, S. and Gayen, D. (2023). Genome editing for improving nutritional quality, post-harvest shelf life and stress tolerance of fruits, vegetables, and ornamentals. *Frontiers in Genome Editing,* 5, 1094965.

Singh, P. (2021). *Genetically modified crops in India: Politics, policies, and political economy*. ScienceDirect.

Singh, S., Chaudhary, R., Deshmukh, R. and Tiwari, S. (2023). Opportunities and challenges with CRISPR-Cas mediated homologous recombination based precise editing in plants and animals. *Plant Molecular Biology,* 111, 1–20.

Smyth, S.J. (2020). Regulatory barriers to improving global food security. *Global Food Security*, 26, 100440.

Son, S. and Park, S. R. (2022). Challenges facing CRISPR/Cas9-based genome editing in plants. *Frontiers in Plant Science,* 13, 902413.

Sultan, Q., Ashraf, S., Munir, A., Khan, S. H., Munawar, N., Abd-Elsalam, K. A. and Ahmad, A. (2022). Beyond genome editing: CRISPR approaches. In A. Ahmad, S.H. Khan, and Z. Khan (Eds.), The CRISPR/Cas tool kit for genome editing (pp. 187–218). Springer. https://doi.org/10.1007/978-981-16-6305-5_6

Tabassum, J., Ahmad, S., Hussain, B., Mawia, A.M., Zeb, A. and Ju, L. (2021). Applications and potential of genome-editing systems in rice improvement: Current and future perspectives. *Agronomy*, 11(7), 1359.

Tachikawa, M. and Matsuo, M. (2023). Divergence and convergence in international regulatory policies regarding genome-edited food: How to find a middle ground. *Frontiers in Plant Science,* 14, 1105426.

Thapa, R. and Bhusal, N. (2020). Designing rice for the 22nd century: Towards a rice with an enhanced productivity and efficient photosynthetic pathway. *Turkish Journal of Agriculture - Food Science and Technology*, 8(12), 2623–2634.

Tiwari, J.K., Buckseth, T., Singh, R.K., Kumar, M. and Kant, S. (2020). Prospects of improving nitrogen use efficiency in potato: Lessons from transgenics to genome editing strategies in plants. *Frontiers in Plant Science,* 11, 597481.

Tsanova, T., Stefanova, L., Topalova, L., Atanasov, A. and Pantchev, I. (2021). DNA-free gene editing in plants: A brief overview. *Biotechnology & Biotechnological Equipment,* 35, 131–138.

Tussipkan, D. and Manabayeva, S.A. (2021). Employing CRISPR/Cas technology for the improvement of potato and other tuber crops. *Frontiers in Plant Science,* 12, 747476.

Van Vu, T., Das, S., Hensel, G. and Kim, J.-Y. (2022). Genome editing and beyond: What does it mean for the future of plant breeding? *Planta,* 255, 130.

Wada, N., Ueta, R., Osakabe, Y. and Osakabe, K. (2020). Precision genome editing in plants: State-of-the-art in CRISPR/Cas9-based genome engineering. *BMC Plant Biology,* 20, 1–12.

Wang, F., Wang, C., Liu, P., Lei, C., Hao, W., Gao, Y., Liu, Y.-G. and Zhao, K. (2016). Enhanced rice blast resistance by CRISPR/Cas9-targeted mutagenesis of the ERF transcription factor gene OsERF922. *PloS One,11(4),* e0154027.

Wikandari, R., Manikharda, Baldermann, S., Ningrum, A. and Taherzadeh, M.J. (2021). Application of cell culture technology and genetic engineering for production of future foods and crop improvement to strengthen food security. *Bioengineered,* 12(2), 11305–11330.

Yang, H., Ren, S., Yu, S., Pan, H., Li, T., Ge, S., Zhang, J. and Xia, N. (2020). Methods favoring homology-directed repair choice in response to CRISPR/Cas9 induced-double strand breaks. *International Journal of Molecular Sciences,* 21, 6461.

Yu, W., Lescale, C., Babin, L., Bedora-Faure, M., Lenden-Hasse, H., Baron, L., Demangel, C., Yelamos, J., Brunet, E. and Deriano, L. (2020). Repair of G1 induced DNA double-strand breaks in S-G2/M by alternative NHEJ. *Nature Communications,* 11, 5239.

Yubing, H., Min, Z., Lihao, W., Junhua, W., Qiaoyan, W., Rongchen, W. and Yunde, Z. (2019). Improvements of TKC technology accelerate isolation of transgene-free CRISPR/Cas9-Edited rice plants. *Rice Science*, 26(2), 109–117.

Zhang, D., Hussain, A., Manghwar, H., Xie, K., Xie, S., Zhao, S., Larkin, R.M., Qing, P., Jin, S. and Ding, F. (2020). Genome editing with the CRISPR-Cas system: An art, ethics and global regulatory perspective. *Plant Biotechnology Journal*, 18(8), 1651–1669.

Zhang, Y., Pribil, M., Palmgren, M. and Gao, C. (2020). A CRISPR way for accelerating improvement of food crops. *Nature Food,* 1(4), 200–205.

Zhou, Y., Xu, S., Jiang, N., Zhao, X., Bai, Z., Liu, J., Yao, W., Tang, Q., Xiao, G., Lv, C., Wang, K., Hu, X., Tan, J. and Yang, Y. (2022). Engineering of rice varieties with enhanced resistances to both blast and bacterial blight diseases via CRISPR/Cas9. *Plant Biotechnology Journal*, 20(5), 876–885.

21 Innovation, Governance, and Crop Genome Editing

Krishna Ravi Srinivas

21.1 INTRODUCTION

Genome editing is a technology that has evoked a mixed response, of great hope and fear with concerns about responsibly using it. Crop genome editing is not as controversial as human genome editing and is projected to be a great opportunity for the humanity to address issues related to food security, climate change, and increase in food production in the context of population growth and other issues. Assessing the claims and counter-claims on the benefits and risks of crop genome editing, Rocks et. al. point out the need to temper big promises and consider the technology as an opportunity for reimagining and redesigning programs to facilitate rural development in Africa (Rock et al., 2023, p. 126). They caution against repeating the mistakes of the past and suggest that "co-development of technologies with farmers" (p. 126) is a necessary step. A key lesson from the Genetic Modification Revolution in agriculture is that it cannot be assumed that a technology will be widely adopted and accepted solely on the claims of its proponents and promoters and skepticism and resistance to technology have to be anticipated and factored in. The GM revolution in agriculture, despite all claims is confined mostly to soy, maize, cotton, and canola and two traits (herbicide tolerance, herbicide tolerance clubbed with insect resistance). The much touted Golden Rice is yet to be widely adopted and is entangled in a messy legal battle in the Philippines (Normile, 2024).

Over the years many studies have shown that regulation of agricultural biotechnology is not uniform across countries and although it can be claimed that science and technology are universal, there is much and significant divergence in regulation (Turnbull et al., 2021). The same seems to be true of crop genome editing because, while countries like the USA and Canada have approved the commercialization of crop genome editing, Europe is yet to shift from its regulation of GM crops with a new regime for genome-edited crops, and countries like India have not come out with a comprehensive policy on crop genome editing. It is likely that Europe will opt for a liberal regulatory regime for crop genome editing compared to the current one on GM crops. But to what extent this will be accepted and whether European consumers will accept crop genome-edited food products unresistingly are not clear.

A key lesson from the governance and regulation of agricultural biotechnology is while science-based regulation is necessary, that alone may not be sufficient as it can be questioned and challenged. The transatlantic divide over genetically modified organisms (GMO) regulation in food and agriculture is much more than a divide over process-based regulation and product-based regulation. It was also a divide over consumers' sovereignty, adopting precautionary principle and labeling of GM food products. Of course, in developing countries like India, there are other issues and factors like prices of the seeds, Intellectual Property Rights over seeds and regulatory capacity (Aga, 2022). Emerging technologies pose novel challenges for governance and crop genome editing is no exception (Selfa et al., 2023).

Ethical issues/concerns, environmental sustainability, acceptance by consumers, and other issues also matter and regulations based on technical risks or traditional cost-benefit ratios are necessary but not sufficient. That's why governance is needed, not just innovations and regulation. According to the World Economic Forum (WEF):

> Governance models describe a set of mechanisms and rules intended to influence the development, dissemination or use of biotechnology, as well as relevant economic models, with a view to maximizing the positive and minimizing the negative impacts. Such mechanisms and rules can include hard-governance tools (government policy and regulation), soft-governance tools (standards, guidelines, norms, codes of conduct) and other forms of formal or informal organization among stakeholders.
>
> (World Economic Forum, 2018, p. 7)

The WEF situates biotechnology in a larger context as part of bio-innovation and points out the need for systematic governance solutions (World Economic Forum, 2018, p. 17). This approach takes a holistic view of innovation and governance and eschews the traditional view on innovation and goes beyond a technocratic perspective.

This chapter argues that for crop genome editing to be harnessed well and responsibly and this calls for new approaches by adopting and adapting new concepts and practices such as responsible innovation, inclusive innovation,

DOI: 10.1201/9781003500933-24

and stakeholder engagement. Using these concepts will enable better use of the technology in a responsible manner so as to maximize gains for society. For this to happen, there is a need to understand innovation and governance in light of recent developments in technology and innovation governance, particularly in the context of emerging technologies.

21.2 INNOVATION AND GOVERNANCE IN AGRICULTURAL BIOTECHNOLOGY

Agricultural biotechnology has gone beyond traditional genetic engineering-based technologies. Today a wide variety of technologies, including what are called New Plant Breeding Technologies (NPBT), are available. Synthetic biology is another emerging technology that has huge potential. Crop genome editing is considered a key technology with well-defined advantages over the earlier generation of biotechnology. The other chapters in this volume amply make this clear. But there are also concerns about the impacts of crop genome editing on human health and the environment (FAO 2022, pp. 15–18). These certainly can be addressed through regulation and better practices in developing varieties besides adopting tailor-made technical solutions.

But to harness the technology effectively and to ensure that gains are maximized and risks and negative impacts are minimized, it is important to go beyond technology regulation to governance. Governance is a broader term and technology governance is defined as:

> Technology governance pertains to formal government activities, but also to the activities of firms, civil society organisations and communities of practice. In its broadest sense, it represents the sum of the many ways in which individuals and organisations shape technology and how, conversely, technology shapes social order.
>
> www.oecd.org/sti/science-technology-innovation-outlook/technology-governance/

Various approaches to technology governance in emerging technologies have been discussed in the literature. For example, the Organisation for Economic Co-operation and Development (OECD) has proposed, *inter alia*, anticipatory governance and responsible innovation (OECD 2024a, 2024b). The National Academy of Medicine's Committee on Emerging Science, Technology, and Innovation (CESTI) has identified JUSTICE, AUTONOMY, FAIRNESS, COLLECTIVE GOOD, and, INDIVIDUAL GOOD as five guiding ethical principles in governance for emerging technologies (National Academy of Medicine 2023). These ethical values are translated into policy goals resulting in policy tools. For example, Diversity and Inclusion are policy goals reflecting Justice and Fairness and the corresponding policy tool is dedicated to funding for equity (Mathews et al., 2022).

The CESTI's work is more relevant for health technologies but the insights from it can also be used in governance of agricultural biotechnology. For example, accessible and affordable food is a collective good, and policy tools and policy goals to realize that can be established. Similarly, sustainability is related to the collective good, while public engagement is related to justice. But translating these into policies and regulations is a challenging task as there are multiple stakeholders with varied interests. For example, incentivizing innovation through strong Intellectual Property Rights (IPRs) can adversely impact access and be a constraint to collective good and individual good as far as smallholder farmers are concerned. To overcome this, the state had to step in by fixing the prices of seeds (Aga, 2022).

Given the likely environmental impacts and other broader societal impacts, in 2021 some environmental non-governmental organizations (NGOs) formulated an approach based on five principles for governance of genome-edited crops. The six principles are:

> 1) risk avoidance and delivery of tangible societal benefits, 2) robust, inclusive societal Engagement, 3) effective, science-based government regulation, 4) voluntary best practices to supplement and complement regulatory oversight, 5) transparency on gene-edited products in the environment, and, 6) inclusive access to technology and resources.
>
> (Gordon et al., 2021)

This was proposed in the US context, and, based on the experience in regulating GMOs in agriculture. But the principles are applicable elsewhere although in terms of priority and capacity, there will be differences. But as there are many countries that have not commercialized GMOs in agriculture, for them, if they are to adopt genome editing in agriculture, the primary challenge will be their inability to use the technology, adoption and regulation. Regulatory regimes evolve over a period of time and are impacted not just by technological advances but by other factors as well. (See discussion in Section 21.3.) So translating these six principles into practice will be constrained by previous experience and regulation of GMOs in agriculture and regulatory capacity.

But we have to understand innovation governance of genome-edited crops in the larger context of transformations and sustainability in the global food system and whether they are appropriate solutions for challenges like climate change. According to the Food and Agriculture Organization (FAO), gene editing in agriculture can be part of a consolidated effort which includes other interventions also. An approach to problem solving, involving representatives from the public sector, the private sector, and communities which are to be impacted by technological intervention is necessary. In this way, manyof the problems that arose on account of the introduction of technologies, in the past, might be obviated (FAO, 2022, p. 46). Echoing a similar view, Lemarié and Marette point out that use of New Genomic Technologies in Agriculture can reinforce the current industrial food and intensive agriculture system or it could be used for enhancing the diversity of agriculture through adopting alternative practices and crops (Lemarié and Marette, 2022).

But this is a challenging task as there can be different policy options and their respective implications. Changes in policy do not occur in a vacuum. While change can be incremental or a departure from the past, there are implications when a novel technology is introduced for adoption, particularly in countries where the previous generation of the technology has not been adopted or well received.

A study done for the European Parliament identified four broad policy options for Europe:

1. No revision of the European GMO Directive.
2. An amendment to the European GMO Directive to exempt genome-edited crops and plants without any foreign DNA present in the product.
3. A level-based policy including broader considerations where the level of risk assessment is based on genetic changes.
4. A level-based policy including broader considerations where the level of approval system is based on societal values (EPRS, 2021).

The study examined the pros and cons of each of the options and also engaged with stakeholders on the options and their preferences.

From an innovation governance perspective, governance for genome-edited crops should eschew extreme options, a purely technocratic perspective that ignores societal concerns and stakeholders' perspectives and a perspective that takes a too precautionary view on risks from genome-edited crops, and burdens the governance regime with impossible demands and expectations. Striking a middle way is not easy but is possible. Some of the tools that can be used for developing such a perspective in governance include stakeholder engagement, technology assessment and foresight studies, responsible research and innovation, socio-economic impact assessment and anticipatory governance. By now there is a good amount of literature on each of these. For reasons of space, I am not elaborating this further.

21.3 REGULATION AND GOVERNANCE IN CROP GENOME EDITING

A global survey of regulation of genome-edited crops shows seven approaches in regulation:

1. Genome-edited crops as conventional crops (except in the case of SDN3).
2. Policy not in place but moving toward it.
3. Discussions are ongoing but without any official opinion made public.
4. Genome-edited crops are regulated like GMOs and under a strict policy.
5. Genome-edited crops are covered by Special Risk Assessment under GMO regulation.
6. Special Risk Assessment regime in place, and
7. Special Risk Assessment regime being planned (Sprink and Wilhelm, 2024).

The regulation of genome-edited crops will be more challenging and complex than regulating GM crops because in the latter it was just a question of a trait or stacking of traits and a relatively technically less sophisticated technology. In case of the genome-edited crops, the technology landscape is complex and is evolving. Recently it was announced that the Environmental Protection Agency (EPA), the U.S. Food and Drug Administration (USFDA), and the U.S. Department of Agriculture (USDA) are to clarify overlapping biotechnology regulatory frameworks and the Coordinated Framework for the Regulation of Biotechnology; https://natlawreview.com/article/epa-usda-and-fda-clarify-overlapping-biotechnology-regulatory-frameworks. According to the official announcement, "EPA, FDA, and USDA intend to clarify their regulatory approaches to genome-edited plants and ensure their regulatory approaches are risk-proportionate" https://usbiotechnologyregulation.mrp.usda.gov/eo14081-section8c-plan-reg-reform.pdf.

In the meantime, regulation in Europe is set to be revised with the European Commission planning to revise the regulations so as not to treat genome-edited crops as GMOs and regulate them differently. A proposal from the European Commission lists the following options:

1. "Authorisation with adapted risk assessment and detection method requirements.
2. Authorisation with incentives for products containing modified traits that have the potential to contribute to sustainability.
3. Authorisation with the requirement that products do not contain modified traits that can be detrimental to sustainability.
4. Notification of products that could also occur naturally or be produced by conventional breeding.
5. Combination of option 4 and one of the other options" (European Commission, 2023, p. 72).

Of these five options, the proposal suggests a combination of option 4 (for products which could be naturally occurring or can be produced by conventional breeding), and option 2 for all other products. It is argued that this combination ensures that new genome techniques (NGT) plants and food/feed products derived from them are equal to their conventional counterparts in safety. Furthermore it is contended that this option will be apt for introducing a wide variety of NGT plants and derived food or feed products without incurring an unwanted regulatory burden. It is expected that these plants will have traits that will contribute to sustainable agri food system (p. 72). What will be the final outcome of this move is not clear as of now, but there is clarity that genome-edited crops will not be treated as GMOs and sustainability can be a criterion.

The difference is while the USA is trying to take a coordinated approach to update the coordinated framework, Europe is embarking upon a major revision that can result in deregulation (with some restrictions) for genome-edited crops. But this need not necessarily mean that there will be harmonization between the two regulatory regimes or the underlying processes. The FAO has pointed out that after creating national regulatory regimes, other issues in governance such as trade impacts, access and distribution, and intellectual property rights can be dealt with. At the multilateral level, through different specialized mechanisms and institutions, various aspects in governance can be addressed (FAO, 2022, p. 3). But the FAO's view on international harmonization is too simplistic and optimistic.

A regulatory approach may facilitate or hinder international harmonization and this depends on many factors. For example, even if a country is in favor of harmonization, other issues such as tariffs, sanitary and phytosanitary/technical barriers to trade (SPS/TBT) issues and policies on self-reliance in food will make a difference. But regulation of genome-edited crops cannot be reduced to two simple categories: product-based and process-based, because regulation will cover new genomic technologies (NGTs) and these are not amenable to such simple classification and approaches. Moreover, as the regulatory regime for genome-edited crops is under development and evolving, and since there will be more technological options in the future, the regulations have to be agile, proportionate to risk, and have to consider sustainability impacts also.

According to a report from the Congressional Research Service (CRS), there are many new issues that may be considered by the US Congress, such as the co-existence of genome-edited crops with organic and conventional crops, the long-term impacts on ecosystems and biodiversity, labeling disclosures, and standards and efficacy of regulation (CRS, 2023). Some of the issues are common to all countries but approaches will vary. Issues related to global trade and lack of harmonized regulation among countries have been of concern even in the case of GM crops. In the case of genome-edited crops, while these are matters of concern, there are others such as difficulties in detectability and traceability when genome-edited crops are on the market, issues in implementation of co-existence policies, and proving patent infringements (EPRS, 2022, p. 11). The report acknowledges that there is divergence in views on regulating genome-edited crops in Europe, indicating the need for debate (EPRS, 2022, p. 24).

Acknowledging that there are issues such as patents on genome-edited crops with traits, labeling and definition of GMO, an editorial in *Nature Plants* stated:

> The EU should be commended for attempting to tackle the question of genetic modification and editing in a more scientific way. However, fears that persist in wide sections of the population are not necessarily restricted to this scientific basis. As long as the broad public does not agree on the definition of a GMO, markets based on a GMO-free label will try to defend their grounds against any technology. And farmers will fear the effects of patent restrictions as long as patent law is not incorporated in an adequate way.
>
> (*Nature Plants*, 2023, p. 1170)

In the context of the USA, a report from the National Academies of Sciences pointed out the gaps, redundancies, and complexities of the current regulatory framework and cautioned that this complexity can cause uncertainty and lack of predictability with potential loss of confidence by the public on the oversight of biotechnology products (NASEM, 2017, p. 6). Thus, in both Europe and the USA, there are debates on the regulation of genome-edited crops and on the issues that can be broadly classified as technical and non-technical. While technical issues can be addressed by regulators using consensus among scientists and by standardizing the terminology, non-technical issues are trickier and more challenging. While defining a GMO is primarily a technical issue, the regulatory approaches are different. According to Hilbeck, Meyer, Wynne et al., while the USA chose to adopt an existing regulatory framework instead of developing a new one specific to GMOs, the EU chose to develop a new one which mandated specific and prior risk assessment. The US approach was called a product-based approach as it was concerned not about the technologies and processes used to produce the GMO but about the novel features of GM end products (Hilbeck et al., 2020).

A survey of global agri-biotechnology regulation done in 2021 pointed out processes specified in respective biosafety laws determine whether a GM crop will be commercialized or not. For most of the countries that have earlier implemented authorization for gene-edited products, harmonization seems to be possible. But differences in defining what constitutes a GMO, such as a conservative threshold for defining organisms edited using the SDN-2 technique as the GMO, can make further harmonization difficult (Turnbull et al., 2021, p. 14). The authors also observe that deciding on exceptions in regulations and biosafety law honors "the societal values of risk assessment and risk management for the ultimate goal of preserving human, animal and environmental health" (p. 14).

Thus, the biosafety laws and regulatory frameworks that are to regulate genome-edited crops are likely to be more complex, technology- and process-based, and hence there is a need to think beyond product and process categorization in regulation and for regulation. However, a comparative analysis of governance of agri-biotechnology in Germany, the UK, and the USA points out that analyzing how opportunity structures and regulatory cultures interact can be useful in interpreting the evolving dynamism of regulating genome editing in agriculture (Ely et al., 2023). From another vantage point, Kuzma et al. argue that the narrow regulatory remits of agencies like USDA and EPA limits the review of gene-edited products and assessing them in terms of sustainability parameters. Calling for strengthening the oversight process for agricultural biotechnology, they suggest many options in policy and regulation (Kuzma et al., 2023, p. 11).

Thus, even a preliminary analysis of regulatory changes and challenges in the USA and the EU in genome-edited crops reveals that it is going to more complex with more demands on the regulatory system in terms of expectations and contributions. For example, sustainability, including environmental impacts, is a major concern. Stakeholders are also more active as the stakes are high and even as proponents highlight the huge potential of the technology to contribute to food security and address the challenges of climate change, skeptics and critics point out issues like affordable access to the technology, the role of intellectual property rights, making it more appropriate and relevant to the needs in developing countries, particularly that of small farmers, issues in harmonization and implications for global trade.

For example, according to Trump et al.:

> The divergence in biotechnology capabilities and influence could exacerbate global inequities, fostering a world where access to the benefits of biotechnology—whether in health, agriculture, or industry—is unevenly distributed. This situation could lead to growing disparities in health outcomes, economic prosperity, and overall quality of life between nations. Furthermore, the concentration of power in a few early adopter nations might stifle global collaboration, hinder knowledge sharing, and create a more fragmented and competitive global biotech landscape.
>
> (Trump et al., 2023)

This negative perception is unwarranted although we find that in the literature some of these issues have been highlighted. There is also a sense of technological determinism in the argument put forth by Trump et al. and perhaps they are exaggerating the potential negative impacts. A fragmented global biotech landscape can happen on account of many factors, and capability in biotechnology is not the determining factor in that. A bigger question is how responsive the technological options to newer challenges such as climate change, environmental sustainability, and affordability will be.

The real challenge in regulation is twofold: (1) how to address the technical issues in light of the potential of the technologies, biosafety norms and risk assessment; and (2) how to make regulation more responsive to other concerns, such as the impact on the environment, sustainability, accessibility, and winning the public trust in regulation as well as in genetically modified crops and products derived/based on them.

The solutions are: (1) responsible innovation; (2) inclusive innovation; and (3) socio-economic considerations, to address the second challenge which is more critical for adoption of genome-edited crops than it was for the adoption of GM crops in agriculture. These three solutions, taken together, can go a long way in ensuring that crop genome editing is widely accepted and helps in addressing some of the concerns, including public engagement, making technologies for genome-edited crops accessible and affordable. My contention is not that these three are the magic bullets that can address all issues related to the second challenge. Rather the point is these three can help in finding a way forward in ensuring that genome-edited crops are accepted by society as credible solutions. For reasons of space, I discuss inclusive innovation only in this chapter but in detail.

21.4 INCLUSIVE INNOVATION

Inclusive innovation is a concept and practice that focus on making innovations work for and accessible to all stakeholders, meeting different needs and in diverse contexts. A definition by Foster and Heeks, often cited, states:

> Inclusive innovation explicitly conceives development in terms of active inclusion of those who are excluded from the mainstream of development. Differing in its foundational view of development, inclusive innovation therefore refers to the inclusion within some aspect of innovation of groups who are currently marginalized.
>
> (Foster and Heeks, 2013, p. 4)

On the other hand, Cozzens and Sutz (2012) highlight that innovation has to be inclusive in terms of process and problems and solutions.

Inclusion can be conceptualized in many ways and which aspects in inclusive innovation need to be prioritized? There are no easy answers to that because the process of inclusion can be active or passive (Johnson and Andersen, 2012). Inclusive innovation is also linked with inclusive growth and inclusive development. According to UNESCAP, designing for inclusive outcomes is needed and so is a dose of pragmatism. Further promoting inclusion is also an issue of trade-offs; www.unescap.org/kp/2021/frontiers-inclusive-innovation-formulating-technology-and-innovation-policies-leave-no-one.

Inclusive innovation is not a homogeneous concept, nor backed by a single worldview or ideology. Pointing this out, the CGIAR's Independent Science for Development Council (ISDC) suggests that the CGIAR should advance inclusivity through partnerships, institutional transformations, evidence-based strategies and assessing gains in inclusive innovation. It acknowledges that there is no universal framework for inclusive innovation and hence suggests a learning-while-doing agenda (ISDC, 2022, p. 22). Interestingly, underscoring the need for inclusivity, the U.S Patent & Trademark Office (USPTO) stresses the importance of inclusivity and suggests that all sectors in the innovation ecosystem should take effective steps to diversify and expand to enable every American to get skills and experience that will allow fuller utilization of their inventive abilities (USPTO, 2023 , p. 6). Inclusivity is to be achieved by multiple means and strategies and inclusive innovation is feasible only if inclusivity is at the core of the steps taken to promote and facilitate it. On the other hand, inclusive innovation can be conceptualized so as to be linked/combined with equity, and access (Srinivas, 2020). By now, there is significant literature on inclusive innovation (Srinivas, 2024).

In the case of crop genome editing, it has been pointed out that it can help small farmers, women, and, youth to increase productivity, and adapt to climate change. For this to succeed, understanding the needs and preferences, including gendered needs, desirable traits, and extension services is essential and so is the development and deployment of appropriate varieties and other innovations (Falck-Zepeda et al., 2022). Another way to conceptualize inclusive innovation is by growing underutilized crops and the relevant genetic resources so that such an use of genomics becomes inclusive and integrative. For this, stakeholder engagement, participation in plant breeding, and capacity building are essential (Shorinola et al., 2024). Interestingly, according to the USDA Science and Research Strategy 2023–2026, there are calls for an Innovation Culture that is transformative by building trust involving diversity, equity, and inclusion (USDA, 2023). Thus, from the above, it is clear that inclusive innovation and inclusivity are gaining traction among *inter alia*, UN agencies, departments/agencies of the U.S government and are now part and parcel of many a policy and strategy. Hence the need for inclusive innovation in outcomes of crop genome editing makes sense. In fact, there are some initiatives that are related to inclusive innovation and crop genome editing in one way or another.

After studying 30 projects, Beumer and de Roij point out that there are two approaches: spacecraft and helicopter (Beumer and de Roij, 2023). The two approaches they identified as the spacecraft and helicopter approach but their orientation is different. This is related to conceptualizing inclusion in innovation. In one approach, i.e., spacecraft, users are involved at a later stage and in the other approach, i.e., helicopter, they are involved right from the beginning. They correspond to passive and active inclusion respectively.

These two approaches also indicate that inclusion can be conceptualized in many ways. Farmers can be considered only as consumers or end users or as participants in developing varieties based on their needs. Traditional plant breeding is a top-down approach and in this farmers are considered as users whose needs are to be met by innovator/plant breeders. In contrast, participatory plant breeding is farmer-centric and depends on farmers' choices and needs including crops/varieties. Although this is a broad differentiation there are many options to make plant breeding more inclusive and participatory. In other words there is more to participation than the spacecraft approach and helicopter approach. In the spacecraft approach farmers are engaged with after the crops have been developed. This is less inclusive but inclusive nevertheless. In contrast, the helicopter approach is more inclusive and smallholder farmers are involved in different stages, including identification of research priorities. They are inclusive in all steps of the ladder of inclusive innovation.

It is not clear from the interviews they have conducted or projects reviewed as to how many of the projects they reviewed and studied used participatory plant breeding. Perhaps none. Participatory plant breeding is certainly an approach that is more inclusive than the spacecraft approach. I have not come across any reference to adoption of participatory plant breeding in crop genome editing in the literature. There are many examples and case studies on participatory plant breeding in traditional plant breeding. By now it is an established approach that has been defined, tested, and applied (Ceccarelli and Grando, 2020). So extending that to inclusive innovation in crop genome editing is feasible.

On the other hand, inclusive innovation can also be conceptualized as part of democratizing crop genome editing, in theory and practice, as it opens up new avenues in applying this technology. Democratizing is obviously a broader term than inclusion. Here it is not an issue of democratizing a particular technology, it is about democratizing science, technology, and innovation. Taking such a broad view Montenegro de Wit argues that:

> It is not enough to promote public engagement in biotechnology issues. Epistemic justice suggests that the poor, the historically marginalized, and the socially excluded must participate in meaningful ways to make decisions about biotech research, investment, regulation, and their intersections with farm and food policy. Doing so recognizes that those who stand to benefit most from a technology should engage with those who may be adversely affected by it. It also reveals the global heterogeneity in who is at risk and how risk is defined.
>
> (Montenegro de Wit, 2020, p. 22)

She further points out that we should pay attention to pitfalls and contradictions in inclusion and consider asymmetries of power and status in participatory spaces. Although she has not mentioned it clearly, the scope for inclusion tilting toward one set of stakeholders and their needs cannot be ignored.

In such a situation we are confronted with issues on making inclusion more inclusive, participatory, and democratic. Of course, this also raises the question of overloading the innovation process, i.e., plant breeding, participatory or otherwise, with so many expectations and norms that it will not be able to fulfill them and end up with experiments that are mired in internal contradictions and conflicts. Between making inclusion a matter of ticking the correct boxes and picking the "right" stakeholder to stretching inclusivity to extremes, there are approaches that can strike a golden mean. One approach in this regard is highlighted to point out that inclusive innovation in crop genome editing can go beyond spacecraft and helicopter approaches.

But there are a few caveats such as there may be trade-offs in terms of participation and priority in needs, inclusion in terms of identities vs. inclusion in terms of poverty. Moreover, as the technology itself is unevenly adopted or being developed, where inclusion matters most, such as in less developed and developing countries, inclusive innovation in crop genome editing may be elusive on account of various factors such as lack of capacity. In fact, as Beumer and Sanne de Roij (2023) point out, the number of projects are limited and confined to only some countries and the large developing countries, like China, India, and Brazil, do not figure in the list they have provided. Inclusive innovation goes beyond traditional cost-benefit approaches but there are no methodologies or indicators to assess inclusive innovation. There is a need for developing indicators and methodologies to assess inclusive innovation, in terms of processes and outcomes. This is all the more important in the case of emerging technologies such as crop genome editing.

Another issue is, should we assume that a process with a high level of inclusion will always result in outcomes that meet the needs of most stakeholders? A process that facilitates a high level of participation from the beginning may result in a high level of participation by some of the stakeholders or a group of stakeholders rather than equal participation by all stakeholders. For example, an inclusive process may be inclusive in terms of including the poor but less inclusive in terms of gender. So it can be argued that if we have to advocate inclusive innovation in crop genome editing, we have to be sensitive to such issues rather than assuming that inclusive innovation will really address all issues in participation, sensitivity to users' needs, and the development of the most appropriate solutions. But the beauty of inclusive innovation is it is not bound by or confined to a single worldview or approach, nor is it a concept that has been tested in all possible contexts. For example, some of the ideas suggested by Tarjem (2022), such as "feminist crop breeding" and "feminist crops" are worth exploring and she has discussed crop genome editing also.

Living Labs or Living Laboratories are defined as "Living Labs (LLs) are open innovation ecosystems in **real-life environments** using **iterative feedback processes** throughout a **lifecycle approach** of an innovation to create **sustainable impact"** https://enoll.org/about-us/what-are-living-labs/.

They can be real or virtual and a key feature is that they use feedback processes in a collaborative manner. They are, in one sense, testing grounds or pilot studies to test, revise, and further better an innovation or solution(s). Living Labs are obviously based on participation and driven by collaboration for innovation. On the other hand, the concept is relatively new and means different things in different contexts. For example, a living lab irrespective of all tall claims may just be a testing ground for facilitating adoption of an innovation or it can be one that is anchored in principles of Responsible Research and Innovation (Szüdi et al., 2023).

Gardezi et al. suggest a methodology:

> (a) face-to-face interviews and surveys with farmers, (b) multidimensional field data collection and analysis, (c) a quasi-field experiment and serious games to test the effectiveness of sensor-driven performance-based payment for improving ecosystem services, (d) design workshops, and (e) extension and outreach of PA tools and knowledge to farmers and rural communities.
>
> (Gardezi et al., 2024)

They proposed it in the context of precision agriculture and hence it cannot be simply applied to crop genome editing. Instead the following changes can be made: (1) collect data based on, *inter alia*, farmers' needs and use participatory plant breeding; (2) address inclusion in terms of gender, vulnerability, and marginality; (3) take into account emerging challenges such as climate change, the food system and environmental sustainability; and (4) assess the accessibility of the innovation and affordability. The authors have discussed Living Labs from an inclusive and responsible innovation perspective. For reasons of space, this chapter deals with inclusive innovation only and hence although the responsible innovation aspect is relevant, it is not elaborated. Moreover, responsible research and innovation (RRI) is a more appropriate approach than mere responsible innovation. Social labs are well suited for facilitating participatory research, if certain aspects are taken into account and some prerequisites in implementation are met.

21.5 CONCLUSION

Crop genome editing can be considered a technological option in the lineage of the Green Revolution. The Green Revolution was the first agricultural modernization project based on technology, initiated in the 1960s amidst fears of over-population, scarcity of food, and endemic hunger. But within two decades, the unwanted side-effects of the Green Revolution became obvious and the need for sustainable agriculture was understood. This resulted in another look at traditional agricultural practices, including organic culture. But the technology-oriented approach continued through genetically modified crops/plants. Crop genome editing is the successor to genetic engineering approach to plant breeding and varietal development. But as an editorial in *Nature Biotechnology* points out, it can succeed where GM crops could not, if some issues are addressed and through collaboration and other means the technologies can be harnessed for the greater common good (*Nature Biotechnology*, 2024).

In the last six decades or so, global agriculture has been transformed on a massive scale and despite all the gains in yields and productivity there is still hunger and malnutrition, although not as widespread as before in the 1960s/1970s. The societal response to agricultural technology also has changed and today environmental sustainability has become as important as, if not more than, gains in productivity. Climate smart agriculture is again touted as the need of the hour and crop genome editing is expected to contribute to this in a significant way.

But as pointed out in the previous sections, today, innovation governance is needed for effective harnessing of crop genome editing. Innovation governance that is agile, sensitive to the needs of stakeholders and inclusive, is desirable and feasible. To make it work, we have to go beyond traditional approaches to regulation in technology, particularly agricultural biotechnologies.

Fortunately, today the techno-centric approach is not advocated by UN agencies like FAO or by many policy research think tanks like EPRS. The shift has been more toward recognizing the need for a balanced and inclusive approach. But that is easier said than done and actualizing it in an emerging technology is a real challenge. However, the challenge has been recognized and incorporated in at least some of the programs in crop genome editing, such as COST-PlantEd https://plantgenomeediting.eu/ and Program 2 of ARC Training Center for Future Crops Development https://futurecropscentre.edu.au/program-2/. Interestingly, the proposal from the European Commission on regulating genome-edited crops has indicated a preference for an option "Authorisation with incentives for products containing modified traits that have the potential to contribute to sustainability" (European Commission, 2023).

This is necessary but not sufficient as more such programs and proposals for similar incentives are needed in developing countries as well. It is gratifying that in these options stakeholder engagement and responsible innovation figure prominently in the two projects cited above. Yet there is a need for guidance on good practices in stakeholder engagement and how to embed values of responsible innovation in such programs. Institutions like the European Commission and CGIAR can work together in this. By now there is a lot of literature on the potential of crop genome editing and, as pointed out in the earlier sections, regulations that are conducive to research and development in crop genome editing are being adopted in more countries, including Europe. But the gap between innovation and governance cannot be met solely by a technocratic approach to regulation. Some of the concepts and practices mentioned in this chapter can help in reducing the gap. Adopting them is not a panacea for all regulatory and governance challenges in crop genome editing. They will be useful in overcoming some of the important challenges which could make or break the large-scale adoption of crop genome editing globally.

REFERENCES

Aga, A. 2022. *Genetically modified democracy: Transgenic crops in contemporary India*. Yale University Press.

Beumer, K., de Roij, S. 2023. Inclusive innovation in crop gene editing for smallholder farmers: Status and approaches. *Elementa: Science of the Anthropocent,* 11, 1. https://doi.org/10.1525/elementa.2022.00089

Ceccarelli, S., Grando, S. 2020. Participatory plant breeding: Who did it, who does it and where? *Experimental Agriculture*, 56(1), 1–11. doi:10.1017/S0014479719000127

Cozzens, S., Sutz, J. 2012. *Innovation in informal settings: A research agenda*. IDRC.

CRS (Congressional Research Service). 2023. Gene-edited plants: Regulation and issues for Congress September 7, 2023. CRS. Available at: https://crsreports.congress.gov/product/pdf/R/R47683/2

Ely, A., et al. 2023. Governing agricultural biotechnologies in the United States, the United Kingdom, and Germany: A trans-decadal study of regulatory cultures. *Science, Technology and Human Values*, 48(6), 1292–1328. https://doi.org/10.1177/01622439221122513

EPRS (European Parliamentary Research Service) (Scientific Foresight Unit (STOA)). 2021. *Regulating genome editing: Societal hopes and fears*. EPRS.

EPRS (European Parliamentary Research Service) (Scientific Foresight Unit (STOA)). 2022. *Genome-edited crops and 21st century food system challenges*. EPRS.

European Commission. 2023. *Proposal for a Regulation of the European Parliament and of the Council on plants obtained by certain new genomic techniques and their food and feed, and amending Regulation (EU) 2017/625 Brussels, 5.7.2023 COM(2023)* 411 final 2023/0226 (COD). European Commission.

Falck-Zepeda, J.B., et al. 2022. Bio-innovations: Genome-edited crops for climate-smart food systems. In J. B. Falck-Zepeda et al. (Eds.). 2022 global food policy report: Climate change and food systems (pp. 90–99). International Food Policy Research Institute (IFPRI). https://doi.org/10.2499/9780896294257_10

FAO. 2022. Gene editing and agrifood systems. FAO https://doi.org/10.4060/cc3579en

Foster, C., Heeks, R. 2013 Conceptualising inclusive innovation: Modifying systems of innovation frameworks to understand diffusion of new technology to low-income consumers. *European Journal of Development Research,* 25, 333–355 https://doi.org/10.1057/ejdr.2013.7

Gardezi, M., et al. 2024. The role of living labs in cultivating inclusive and responsible innovation in precision agriculture. *Agricultural Systems,* 216, 103908. https://doi.org/10.1016/j.agsy.2024.103908

Gordon, D.R., Jaffe, G., Doane, M. et al. 2021. Responsible governance of gene editing in agriculture and the environment. *Nature Biotechnology,* 39, 1055–1057. https://doi.org/10.1038/s41587-021-01023-1

Hilbeck, A., Meyer, H., Wynne, B. et al. 2020. GMO regulations and their interpretation: How EFSA's guidance on risk assessments of GMOs is bound to fail. *Environmental Sciences Europe,* 32, 54. https://doi.org/10.1186/s12302-020-00325-6

Independent Science for Development Council (ISDC). 2022. Transformation through inclusive innovation. CGIAR Independent Advisory and Evaluation Service. Available at: https://cgspace.cgiar.org/server/api/core/bitstreams/4a940d4b-816c-42b9-9d2e-0e9e62e5d30e/content

Johnson, B., Andersen, A.D. 2012. *Learning, innovation and inclusive development*. Aalborg University Press.

Kuzma, J., et al. 2023. Parameters, practices, and preferences for regulatory review of emerging biotechnology products in food and agriculture. *Frontiers in Bioengineering and Biotechnology,* 11, 1256388. doi: 10.3389/fbioe.2023.1256388

Lemarié, S., Marette, S. 2022. The socio-economic factors affecting the emergence and impacts of new genomic techniques in agriculture: A scoping review. *Trends in Food Science and Technology,* 129, 38–48. Doi:10.1016/j.tifs.2022.07.013

Mathews, D.J.H., et al. 2022. Imagining governance for emerging technologies. *Issues in Science and Technology*, 38(3), 40–46.

Montenegro de Wit, M. 2020. Democratizing CRISPR? Stories, practices, and politics of science and governance on the agricultural gene editing frontier. Elementa: Science of the Anthropocene, 8(9). https://doi.org/10.1525/elementa.405

NASEM (National Academies of Sciences, Engineering, and Medicine). 2017. Preparing for future products of biotechnology. The National Academies Press. doi:10.17226/24605

NASEM (National Academies of Sciences, Engineering, and Medicine and National Academy of Medicine). 2023. *Toward equitable innovation in health and medicine: A framework*. The National Academies Press. https://doi.org/10.17226/27184.

Nature Biotechnology. 2024. Cis-editing for all. *Nature Biotechnology,* 42, 821. https://doi.org/10.1038/s41587-024-02299-9

Nature Plants. 2023. EU rethinks genome editing. *Nature Plants,* 9, 1169–1170. https://doi.org/10.1038/s41477-023-01505-x

Normile, D. 2024. What a Philippine court ruling means for transgenic Golden Rice, once hailed as a dietary breakthrough. *Science*. .www.science.org/content/article/what-philippine-court-ruling-means-transgenic-golden-rice-once-hailed-dietary *doi: 10.1126/science.zl7058i*

OECD. 2024a. Framework for anticipatory governance of emerging technologies; OECD Science, Technology and Industry Policy Papers. OECD Publishing. https://doi.org/10.1787/0248ead5-en.

OECD. 2024b. *Neurotechnology toolkit: To support policymakers in implementing the OECD Recommendation on Responsible Innovation in Neurotechnology*. OECD Publishing. www.oecd.org/sti/emerging-tech/neurotech-toolkit.pdf

Rock, J.S., et al. 2023. Beyond the genome: Genetically modified crops in Africa and the implications for genome editing. *Development and Change,* 54, 117–142. https://doi.org/10.1111/dech.12750

Selfa, T., et al. 2023. Gene editing in agriculture: Social concerns, public engagement, and governance. *Science, Technology and Human Values*, 48(6), 1249–1259. https://doi.org/10.1177/01622439231194817

Shorinola, O., et al. 2024. Integrative and inclusive genomics to promote the use of underutilised crops. *Nature Communications,* 15, 320. https://doi.org/10.1038/s41467-023-44535-x

Sprink, T., Wilhelm, R. 2024. Genome editing in biotech regulations worldwide. In A. Ricroch, D. Eriksson, D. Miladinović, J. Sweet, K. Van Laere, E. Woźniak-Gientka (Eds.), A roadmap for plant genome editing. Springer. https://doi.org/10.1007/978-3-031-46150-7_25

Srinivas, K.R. 2020. *Access, Equity and Inclusion and Science, Technology and Innovation Policy. RIS Policy Brief 94. Research and Information System (RIS)* http://ris.org.in/sites/default/files/Policy%20brief-94%20Dr%20Ravi%20K%20Srinivas.pdf

Srinivas, K.R. 2024. Access, Equity and Inclusion (AEI) in science, technology and innovation: Contribution of RIS to theory and practice in four decades of RIS, vol. II, pp. 163–182. RIS. Available at: https://ris.org.in/sites/default/files/Publication/Conceptulalism-Framework-Vol-2.pdf

Szüdi, G., Lampert, D., Hajdinjak, M., Asenova, D., Alves, E., Bidstrup, M.V. 2023. Evaluation of RRI institutionalisation endeavours: Specificities, drivers, barriers, and good practices based on a multi-stakeholder consultation and Living Lab experiences. In E. González-Esteban, R.A. Feenstra, L.M. Camarinha-Matos (Eds.), Ethics and responsible research and innovation in practice: Lecture notes in computer science, vol. 13875. *Springer* https://doi.org/10.1007/978-3-031-33177-0_4

Tarjem, I. 2022. Feminist crops: A more-than-human concept for advancing feminist crop breeding for development. *Catalyst: Feminism, Theory, Technoscience*, 8(2), 1–26. https://catalystjournal.org/index.php/catalyst/article/view/37243/30135

Trump, B.D., Cummings, C.L., Loschin, N., Keisler, J.M., Wells, E.M. Linkov, I. 2023. The worsening divergence of biotechnology: The importance of risk culture. *Frontiers in Bioengineering and Biotechnology,* 11, 1250298. doi: 10.3389/fbioe.2023.1250298

Turnbull, C., Lillemo, M., Hvoslef-Eide, T.A.K. 2021. Global regulation of genetically modified crops Amid the gene edited crop boom – a review. *Frontiers in Plant Science,* 12, 630396. doi: 10.3389/fpls.2021.6303

USDA (United States Department of Agriculture). 2023. *USDA science and research strategy 2023—2026.USDA.*

USPTO (United States Patent and Trademark Office). 2023. National strategy for inclusive innovation. United States Patent and Trademark Office. Available at: www.uspto.gov/sites/default/files/documents/NationalStrategy.pdf

World Economic Forum. 2018. *Bio-innovation in the food system: Towards a new chapter in multistakeholder collaboration/ WEF.* www3.weforum.org/docs/WEF_Bio-Innovation_Dialogue_WP.pdf

22 Combining CRISPR and AI to Facilitate Food Security and Sustainable Development Goals

Anam Niamat, Aftab Ahmad, Muhammad Faheem, Nayla Munawar, and Paulo Zaini

22.1 INTRODUCTION

The demand for food to feed the continuously increasing world population is alarming. Expanding crop productivity with limited land available in face of climate change and increasing biotic and abiotic stresses is urging scientists to emphasize the importance of food security as well as sustainability in agriculture. To meet the challenge of increasing demand of food, there is a need for a diversified strategies in developing countries. These countries can increase food security by improved food processing which helps to ensure that food supply is secure and available. This is done by decreasing food loss by improving the process of food preservation, nutritional composition, and security. Sustainability of the food supply, on the other hand, is the long-term storage of food by increasing its shelf life so that food can be available for longer periods for the human population. Both security and sustainability are major factors to deal with food production as well as storage around the world, requiring long-term and effective solutions or strategies. This can be achieved by adopting new technologies and analyzing the major challenges in food security and sustainability (Mrabet, 2023).

Inefficient irrigation is one of the rising challenges in agriculture, affecting the yield due to insufficient availability of fresh water as well as salts in the soil. The introduction of effective strategies such as drip irrigation helps to improve the irrigation process and ultimately increase the sustainability of crops (Srivastav et al., 2021). Moreover, the excessive use of pesticides, fertilizers, and fungicides also increases water and environmental pollution which is directly or indirectly harming humans as well as the ecosystem. Here we discuss only a few of the many agricultural concerns that are associated with food security and sustainability. These dilemmas highlight the need for a global agricultural revolution to ensure the security of food to feed almost 10 billion people by the year 2050 (Aslam et al., 2022; Nations et al., 2022).

The discovery of CRISPR-Cas as a gene-editing tool in eukaryotes has been considered a breakthrough technology in human health and agriculture, due to its widespread applications (Munir et al., 2024). This game-changing technology is becoming an essential component in various fields with its ability to precisely edit the genome of plants and animals. Gene modification by CRISPR has profound implications for agriculture, for example, in improving tolerance to water scarcity. The stress regulatory genes in plants can be precisely modified such as Dehydration Response Element Binding protein (DREB) in wheat, which has been successfully edited using CRISPR technology, as an example of increased food security as well as sustainability (Kumar et al., 2022; Sultan et al., 2022).

Artificial intelligence (AI) is an emerging field with tremendous potential to impact food security. AI combined with CRISPR holds an immense potential to improve efficiency and specificity of CRISPR-Cas systems for precise genetic modifications of the food crops and improve the process of food production. AI is run by human intelligence and has multiple fields including language processing, machine learning, and computer vision. All these fields have a common goal of creating an online system that can gather data from the databases process it, have the ability to adopt new emerging information, and make predictions and effective decisions (Sarker, 2022). AI in agriculture is helping researchers to understand the complex nature of agricultural processes, providing valuable solutions to increase the food yield. It does so by a quick review of the literature present on different databases and providing evidence-based solutions to help protect and maintain the different crops around the world. It not only provides information but also helps to check and analyze the conditions of soils and plants in a very precise manner, making easy informed decisions for farmers and scientists. This ability of AI has made it a valuable asset in agriculture and its integration with CRISPR expands its impact on food security and sustainability (Javaid et al., 2022).

Leveraging CRISPR with AI has various potential applications in protecting crops from biotic and abiotic stresses in a very efficient and user-friendly way, ultimately improving food security and sustainability in a more efficient way, as compared to using either technology on its own. CRISPR provides precise gene editing for specific traits like photosynthesis, disease resistance, and nutritional quality improvement. On the other hand, AI synergistically affects the process by providing a more precise review of the literature regarding these traits and helping scientists deal with these problems more efficiently, as it makes the use of CRISPR more targeted and safer. AI can analyze a large amount of

DOI: 10.1201/9781003500933-25

genomic data in a very short time to predict the on-target effects of CRISPR on gene editing and help to prevent off-targeting. For instance, AI assisted analyzing and identifying the genetic markers that are associated with decreased crop yield (e.g., DREB in the case of wheat drought). In addition, the combination of these two technologies provides real-time information on genome of crops and precise targeting which results in improved crop quality, yield, and environmental conditions (Patil, 2023; Dixit et al., 2024). Furthermore, AI is also helping to check the off-target effects by providing information about bioinformatics tools, such as CRISPR Direct, which helps to design efficient guide-RNA to make CRISPR Associated System (Cas), an endonuclease, more efficient to produce double-stranded breaks at the targeted sequence. In this way, leveraging CRISPR with AI represents a more effective strategy to deal with the challenge of the increasing demand for food and the devastating effects of food scarcity around the world.

The increasing demand for food production, along with an expanding population, weather uncertainties, and resource restrictions, requires new methods of food production to ensure food security, especially in developing countries. The aim of this chapter is to highlight the transformative potential of combining two cutting-edge technologies, CRISPR and AI, in addressing the rising challenge of global food security and sustainability in agriculture. We will start with a brief introduction to CRISPR and AI separately and then explore the synergistic effects of these two mind-blowing techniques. We also discuss practical applications of CRISPR and AI in agriculture in the management of pests and diseases, enhancing nutritional contents, management of resources, and regulatory triggers regarding the safe use of these techniques. it finish by exploring how integration of CRISPR and AI offers a revolutionary approach, that not only can guarantee food security but also can pave the way for a sustainable and strong future for global food production in agriculture.

22.2 CRISPR TECHNOLOGY

CRISPR-Cas9 is an efficient genome-editing technique that has revolutionized molecular biology and has potential applications in a wide range of sectors. It is a natural RNA-mediated adaptive immune mechanism in bacteria and archaea that has been adapted to consist of a single guide-RNA (sgRNA) and an endonuclease such as Cas9. This technique relies on producing a double-stranded break (DSB) in the target DNA using the endonuclease (Cas9 in case of CRISPR-Cas9 system), recruited by the sgRNA which helps to stabilize and guide the endonuclease to the target site in the genome. Once DSB is created at the target site by the Cas9, it can be fixed by using non-homologous end joining (NHEJ) or homology-directed repair (HDR) pathways, enabling efficient modification of that genome site. It is a dynamic and accessible platform for modifying the genetic content in diverse biological systems (Khanzadi and Khan, 2019; Ahmad et al., 2024). The mechanism of CRISPR-cas9 system is shown in Figure 22.1.

CRISPR technology emerged as a game-changer in agriculture, providing unbeatable accuracy for precise genetic alterations and development of different high yielding crops (Chen et al., 2019). CRISPR allowed scientists to accurately modify the DNA of crops, permitting the introduction of advantageous features such as greater yield, better nutritional value, better resistance to diseases, insect pests, and environmental stressors. CRISPR, compared to traditional breeding strategies, enables precise modifications of the agriculturally important traits in an efficient manner, avoiding the complex, lengthy, and uncertain processes of conventional breeding. Furthermore, CRISPR technology holds great potential to address the critical challenges associated with food security, such as climate change by developing climate-resilient crop varieties that are capable of growing under harsh conditions. CRISPR also reduced agriculture's adverse environmental impact through minimizing the demand for chemical inputs and encouraging sustainable farming methods. In addition, rapid domestication of crops using CRISPR, directed evolution through CRISPR and CRISPR-based gene drives in crops can help scientists to overcome the major challenges of food security. For example, recently it has been predicted that future CRISPR crops could be colored for an efficient killing of weeds in the crops. Such an achievement complements the technology's potential to produce crops with increased nutritional contents to combat food shortage in underdeveloped countries. Applications of CRISPR technology in agriculture are expected to revolutionize agricultural crop production, contributing to global food security and sustainability (Zhu, et al., 2020; Sampath et al., 2023).

22.3 AI IN AGRICULTURE

AI plays an important role in agricultural data collection and analysis, allowing for efficient real-time analysis. It has been used efficiently in precision agriculture to model behavior in livestock and optimize irrigation and fertilization techniques. The use of AI in agriculture has transformed farming operations, ensuring profitable harvests and facilitating smart farming. Artificial intelligence technologies such as expert systems, the processing of natural languages, as well as machine vision have dramatically increased both the quantity and the standard of agricultural system. The use of the Internet of Things (IoT), drones, and sensors in smart agriculture has greatly enhanced farming techniques. Drones equipped with AI, sensors, and IoT have been used to monitor farm animals, spray crops, and analyze crop health (Ghazali, 2022; Elbasi et al., 2023).

Artificial intelligence-powered decision support systems are progressively being employed in agriculture to enhance the administration of farms and productivity. For example, AgroDSS, a cloud-based system, allows farmers to input their data and apply various data analyzing tools to make decisions

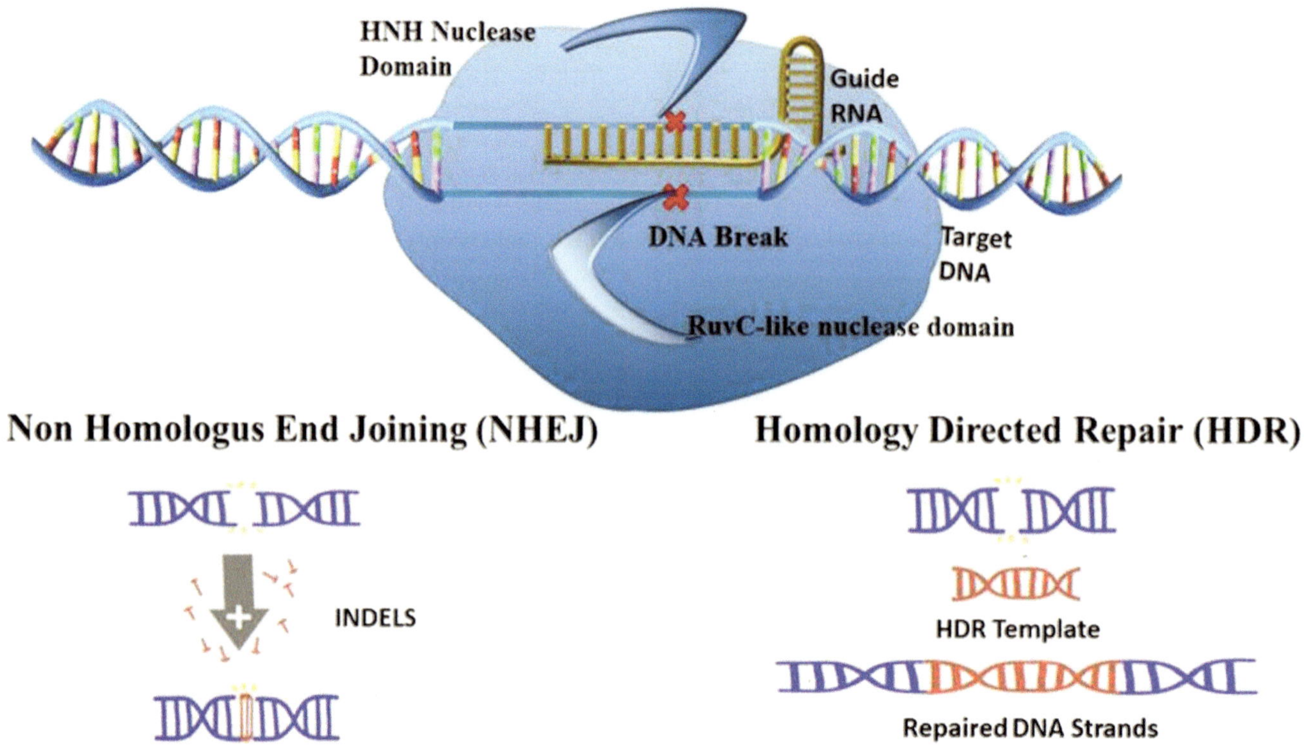

FIGURE 22.1 The mechanism of CRISPR-Cas9 system. where (a) in fig In blue, Cas9 endonuclease, an enzyme having HNH and RuvC domains which produce cuts on the complementary and non-complementary strands respectively. Guide RNA (gRNA), shown in yellow, guides the Cas9 enzyme to the target sequence. Crosses on DNA indicate the site at which Cas9 produces a break or cut. where (b) in fig The repair mechanism for broken DNA strands using the cell's natural processes NHEJ repair system. where (a) in fig The repair mechanism for broken DNA strands using thr cell's natural processes HDR-based repair system.

(Balogun et al., 2022). Similarly, intelligent agents, such as e-learning, monitoring, and weather information agents, are integrated into decision support systems to propose future scenarios to aid farmers to make efficient decisions. These systems can also predict and manage crop irrigation processes, fertilizers and their effective use, and pesticide requirements based on environmental factors. Furthermore, machine learning has been used to model animal behaviors, optimize irrigation, and recommend fertilizer levels to improve agricultural production (Sudduth, 2020). These examples demonstrate the potential of AI-driven decision support systems in agriculture and future food security. Following is the leveraging potential of AI and CRISPR in agriculture. Applications of AI in agricultural improvements are shown in Figure 22.2.

22.4 INTEGRATING CRISPR AND AI FOR PRECISION AGRICULTURE

CRISPR technology with its unique and precise nature has huge potential in agriculture to edit the genome of different crops, resulting in improvements in traits which ultimately lead to increased yield of food (Maximiano et al., 2021). The potential of CRISPR-Cas has been demonstrated in model systems and crop plants, with published examples of edited plants challenged with biotic and abiotic stresses. The major applications of CRISPR-Cas in plants include increased yield, biotic stress tolerance, bio-fortification, and salt stress tolerance. The successful integration of CRISPR with AI has more potential and applications to increase the production and sustainability of food, by optimizing the process of gene selection, the selection of target sites and optimal gRNA design, besides making predictions about editing outcomes done with CRISPR-Cas. For this purpose, various AI tools have been developed recently that help researchers to select the specific gene of interest for genetic improvement of crops with CRISPR-Cas. For instance, CRISPOR is an example of an AI tool that is being used to align guide-RNAs by providing information about the presence of similar sequences in the target gene, minimizing the off-target effects that could occur using CRISPR technology (Concordet and Haeussler, 2018). The design of gRNA is crucial in CRISPR-based genome editing, as it contains complementary sequences to the target site and guides the endonuclease enzyme to bind and produce cuts at the target sequence. CRISPOR applies an algorithm to find the most suitable and optimized sequence having minimal off-target effects, through analysis of sequence homology in the genome of an organism. In this way, it increases the on-target activity of CRISPR-based systems and increases the efficiency of the gene-editing process as a whole (Li et al., 2022). DeepCRISPR, a deep learning-based technology, improves the precision of CRISPR-Cas9 genome editing by

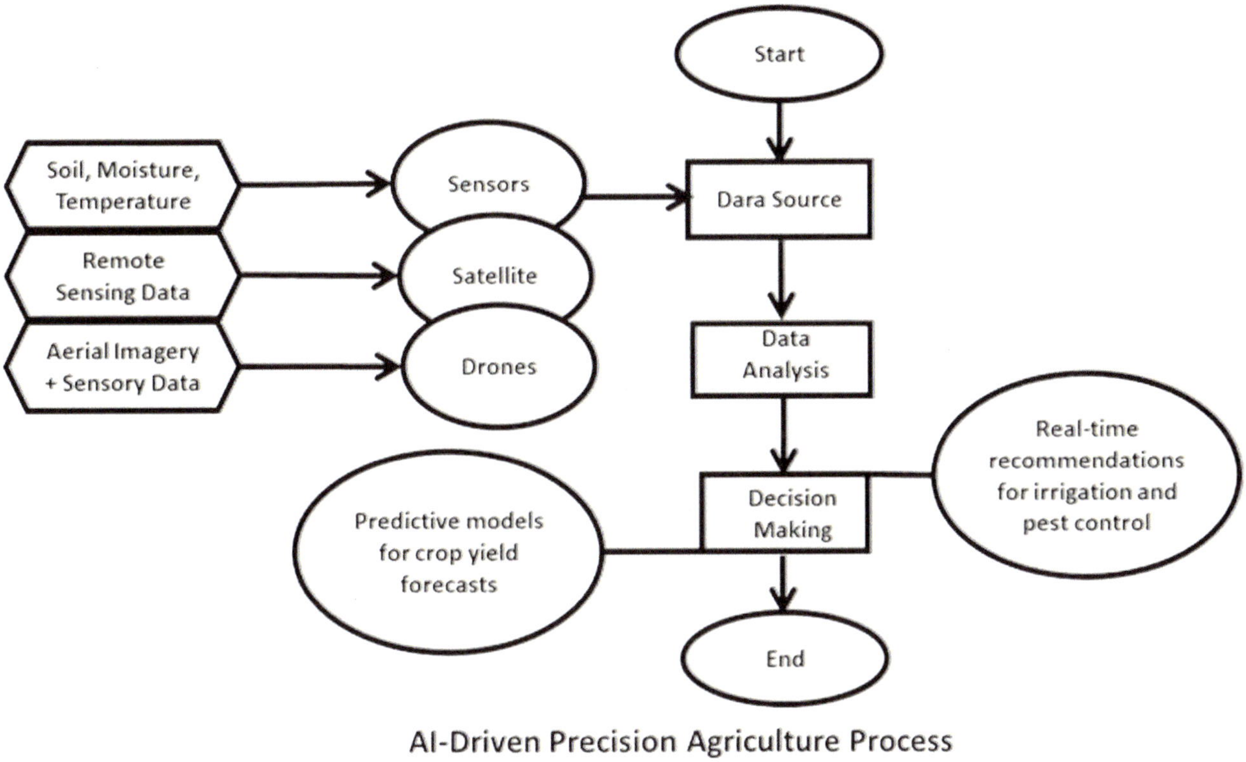

FIGURE 22.2 Applications of AI in agricultural improvements.

anticipating on-target and off-target effects with convolutional neural networks. It uses massive datasets to properly anticipate the efficiency and specificity of gRNAs, resulting in better CRISPR-based edits. The increasing precision and efficiency of CRISPR technology, assisted by DeepCRISPR, contribute greatly to food sustainability (Zhang et al., 2023). Furthermore, another tool, DANGER, has a critical impact on the CRISPR-based editing process as it uses advanced algorithms to design the gRNA that is highly specific to the target sequence. In addition, it also predicts the best-fit gRNA for the endonuclease using its deep learning as well as computational ability (Zhang et al., 2023). These tools not only increase the efficiency of CRISPR technology but also help the researchers to work in a more precise and user-friendly manner, as they rely on these tools before conducting the experiment, to check the feasibility of an optimal gRNA as well as overcoming the challenges that could result in experimental errors. Additionally, the researchers get help from these tools to develop more effective tools by analyzing their ability in trait modification and functional genomics in plants (Kianpour, 2023). In the study by Brazelton and colleagues, multiple CRISPR design tools were discussed and evaluated for their potential use in genome editing. Among them, the authors found the CRISPR genome analyst tool was the most effective due to its wide range of functions according to the need of researchers in selecting the target gene for CRISPR-based gene editing (Brazelton et al., 2015). Finally, these AI tools are most specific in the context of multiplexing, which use multiple gRNAs to produce multiple mutations in the genome, aiming for more accurate targeting in the gene (Minkenberg, Wheatley, and Yang, 2017). So, the CRISPR system in combination with AI will become more effective, precise, and less time-consuming in the context of genome editing as compared to manual use. In the following sections we will provide other examples of how together these two technologies are enhancing not only crop yield but also helping to manage diseases due to pest attacks, increase the nutritional contents, growth, and global access to food production (Minkenberg, Wheatley, and Yang, 2017).

22.5 DISEASE AND PEST MANAGEMENT

Management of pests and diseases is a crucial component in the field of agriculture as it is directly associated with yield as well as the security of crops. The two well-known programs, i.e. Integrated Disease Management (IDM) and Integrated Pest Control (IPC), are helping to reduce disease and control pest attack respectively. These programs are used to decrease the use of chemicals and emphasize ecological sustainability, using mechanical, genetically resistant cultivars, biocontrol, and a combination of cultural measures (Hartman et al., 2016).

Image processing and machine learning are two AI techniques that are revolutionizing the pest and disease identification processes and improving IDC and IDM programs significantly. These techniques help in the early detection of diseases as well as pathogen attacks which lead to on-time interventions for the early management of disease and pest attacks in plants (Susheel and Rajkumar, 2023). Additionally, these techniques have been shown to increase the functioning ability of Decision Support Systems (DSS) in managing disease and pests in agriculture. Support Vector Machine (SVM) and Artificial Neural Networks (ANN) are data mining methods that are successfully implicated in the detection of various kinds of diseases, soil parameters, and the growth of crops (Sabareeswaran and Robert, 2016; Reiger, 2022).

CRISPR-Cas technology has immense potential to control diseases and pests compared to traditional methods because it modifies the genome rather than applying chemicals that are toxic to plants as well as human health and the environment. This system has been successfully demonstrated to produce pest-resistant crops such as a late blight-resistant potato, which is resistant to the pathogen *Phytophthora infestans*, avoiding the application of chemical fungicides on potatoes (Bradshaw, 2021). Furthermore, CRISPR-Cas9 has also been used to decrease browning in potatoes, thus helping to improve the post-harvest trait of potatoes (González et al., 2020). In this way, it decreases the pressure on farmers in pest and disease management by producing pest- and disease-resistant varieties of crops that have a natural ability to fight diseases (Ahmad et al., 2020). CRISPR with AI in the management of disease and pests has game-changing potential which will increase the yield of crops by improving the development and management processes.

The combined application of AI and CRISPR techniques to protect crops from the devastating attack of pathogens is already delivering on its promise. AI enhances the performance of the CRISPR system by increasing its precision through a comprehensive literature review, on-target and off-target predictions, and helps researchers adopt smart research and technology to manage different diseases of crops (Afzal, Sirohi and Singh, 2020). Moreover, CRISPR, by producing genetically resistant varieties of crops, overcomes food security challenges and integration with AI has boosted these benefits tremendously by increasing the process of novel trait recognition, improving the genome editing, and overcoming regulatory oversights regarding transgene-free CRISPR crops (Ahmed et al., 2020; Rehman, 2022).

22.6 ENHANCING NUTRITIONAL CONTENT

Nutritional contents in food play a significant role in the growth and health of human beings. Enhancement of the nutritional content of food in underdeveloped areas is a global challenge that is associated with the health of the people and food security. Nutritional contents such as minerals and essential vitamins have been significantly improved by CRISPR-mediated genome editing. Golden rice is an example of a gene-edited crop using CRISPR/Cas system. The golden color of rice is due to the increased content of beta-carotene which is a precursor of Vitamin A. So gene-edited golden rice is a rich source of beta carotene that is being used to treat Vitamin A deficiency naturally (Datta, Majumder and Datta, 2021; Kumar, 2022).

AI, on the other hand, synergistically increases the efficiency of the CRISPR-Cas system in several ways by assessing and improving nutritional contents, revealing the role of nutrients in health, and decreasing the onset of diseases. Similarly, AI and CRISPR have tremendous roles in the future of human health and precision medicine. Recent research has shown AI to be a successful analyzer of nutritional content in fruits and vegetables and their impact on the health of humans (Salinari et al. 2023).

Convolutional Neural Networks (CNN) are providing vital information regarding nutritional deficiencies in various crops and classifying them based on nutritional contents. So, it is making it easy for researchers to select crop varieties that have nutritional deficiencies and edit them using the CRISPR-Cas system to enhance nutritional contents in a very precise manner (Sripada, 2023). Additionally, AI also helps to explain the role of nutrients with a thorough analysis of data regarding various nutrients and their role in health and disease conditions based on different databases. It provides information regarding the specific role of nutrients in the body, for example, the effect of vitamin A deficiency in human vision. AI provides information related to the formation of vitamin A, its precursors, its mechanism in vision, and its mode of production in fruits and vegetables. It not only provides metabolic information regarding the synthesis of nutrients but also classifies them according to their ability to produce these nutrients, making an easy way to target key components in the genome using CRISPR to improve nutritional contents (Côté and Lamarche, 2021).

Furthermore, together these two technologies are helping to improve the bio-fortification process by providing effective strategies and solutions to mitigate malnutrition due to deficiency of iron, zinc, and other essential nutrients in staple foods, addressing micro and macronutrients that affect the health of millions of people around the globe. This approach has an important role in decreasing the impact of various curable diseases associated with nutritional deficiencies (Sánchez, García-Bañuelos and Sida-Arreola, 2014; Kiran et al., 2022; Kumar et al., 2022). However, it is crucial to note that the use and advancement of these approaches must be linked to increase food security, sustainable resource management and achieve the UN's Sustainable Development Goals (Kumar et al., 2022).

22.7 SUSTAINABLE RESOURCE MANAGEMENT IN AGRICULTURE

Sustainable resource management in agriculture is the careful use of interventions or strategies to produce food that could be stored for a long time and made available for human consumption when needed. Climate change impacts crop growth and decreases productivity. To mitigate the harmful effects of climate change on crops, there is a need for sustainable resource management. It not only safeguards the environment but also increases the long-term productivity of crops. It also maintains a balance to meet the increasing demand for food around the world (Abbass et al., 2022). Careful use of AI and CRISPR is imperative for sustainable resource management in agriculture, due to their potential to analyze the vast amount of data and produce targeted genetic modifications to improve food crops, respectively. AI is helping in water and fertilizer resource management through the implementation of the Internet of Things (IoT), cloud computing, and machine learning. IoT uses various kinds of sensors that are embedded in the soil to monitor soil moisture among other parameters, such as the amount of nutrients in the soil. Still others detect and gather data about weather forecasts and warn farmers to maintain their resources according to the changing conditions of the environment. Cloud computing is specific in providing infrastructure for gathering and storing information that is detected and sent by the sensors of IoT. Machine learning algorithms process the stored data about soil moisture, the health of crops, and the nutrient contents and provide recommendations and strategies for efficient use of water and fertilizers (Virnodkar et al., 2020). Real-time data obtained from these sensors is being used to optimize the irrigation process, implementing timely decisions to reduce the excessive use of water through the efficient use of AI-based sensors (Cardoso et al., 2020; Phasinam et al., 2022).

Synergism between CRISPR and AI in resource management is becoming more evident due to these evolving technologies in agriculture. Using the information of AI in the field, one can make crops naturally resilient to changing environmental conditions by using CRISPR technology in a more accurate way, compared to traditional manual processes of data collection. There are several examples in the literature that justify this statement. For example, the study by Fang (2024) explored the leveraging effect of AI and CRISPR in agricultural resource management. In this study, researchers analyzed a wide range of data collected from fields using AI and selected genetic targets for salinity, drought, and heat. These insights helped to use CRIPSR for precise targeting of these genetic targets to produce crops that were tolerant of salt, drought, and heat. In this way, it is more convenient to produce genetically improved crops that can withstand harsh environmental conditions. Additionally, it also mitigates excessive use of chemical fertilizers to maintain nutritional contents and, as a result, helps to maintain sustainable resource management (Mohamed, 2023; Fang, 2024). Table 22.1 shows the United Nations Sustainable Development Goals (SDGs) that are directly relevant to food and agricultural production, as well as how CRISPR and artificial intelligence (AI) can be employed to accomplish them.

22.8 ETHICAL AND REGULATORY CONSIDERATIONS

CRISPR technology has revolutionized genome editing, but it also raises significant ethical concerns. CRISPR-Cas systems are incredibly powerful, but their use is subject to varying regulatory frameworks and guidelines. There is need for a clear and global regulatory framework for CRISPR-edited crops, given the technology's potential and the ethical and safety concerns it raises. This framework would emphasize the importance of resolving technical and regulatory uncertainties to fully benefit from CRISPR technologies in plant breeding. Different studies have discussed the differences in regulatory frameworks for CRISPR-edited crops, with some countries following process-based regulations and others using product-based regulations. These frameworks typically include risk assessment and toxicity studies to ensure the safety of CRISPR-edited crops (Gostimskaya, 2022).

The convergence of AI and CRISPR technology in agriculture also presents complex ethical and regulatory challenges that necessitate careful consideration. Both AI and CRISPR offer immense potential for advancing crop improvement, sustainability, and food security, but their use also raises ethical concerns related to data privacy, genetic modifications, editing outcomes, and equitable access to technology. In the context of CRISPR, ethical considerations revolve around the potential for unintended genetic consequences, such as off-target mutations, gene drives and their ecological impacts, as well as the ethical implications of gene-edited organisms. Meanwhile, AI algorithms used in conjunction with CRISPR for tasks like gene-editing design and off-target predictions raise issues of algorithmic bias, data transparency, and intellectual property rights. Regulatory frameworks must be developed to ensure that AI-driven CRISPR applications adhere to ethical guidelines, promote transparency, and protect against potential harms, while also fostering innovation and equitable access to agricultural technologies. By addressing these ethical and regulatory considerations comprehensively, stakeholders can harness the transformative potential of AI and CRISPR in agriculture while minimizing risks and maximizing societal benefits (Zafar et al., 2021; Srushti et al., 2022; Munawar et al., 2024).

22.9 CHALLENGES AND FUTURE DIRECTIONS

The integration of CRISPR and AI shows significant potential for the future, with significant advances in nucleic acid detection, genetic modification, and genome editing. The use of artificial intelligence in CRISPR technology can assist researchers to address issues such as off-target effects while also improving the system's precision and efficiency. The use of AI can also improve the efficiency as well as the accuracy of CRISPR-based diagnostics and medicines, resulting in more effective and tailored treatments. Furthermore, AI can help with the design and optimization of gRNA for CRISPR, potentially lowering costs, off-target effects and boosting the safety of the technology (Naeem et al., 2020; Ahmad et al., 2021; Liu et al., 2023). The integration of CRISPR with AI brings various challenges and constraints. These issues are amplified when AI is used, as the AI system must accurately forecast the effects of CRISPR-based editing, which is difficult due to the complexity of genetic connections. Furthermore, the use of AI in CRISPR design and analysis requires huge amounts of high-quality data, which are not always available (Mohamed, 2023; Dixit et al., 2024). Despite these obstacles, this combination has tremendous potential for treatment of plant diseases, improvements of traits, and increased yield of crops. However, additional research is required to solve the current limitations.

TABLE 22.1
UN Sustainable Development Goals and how CRISPR and AI can be employed to accomplish them

SDG number	SDG title	Description	Role of CRISPR	Role of AI	Reference
SDG1	No Poverty	Reduce poverty in all of its forms, particularly in rural areas by the development of agriculture	Promote high in yield, stress-resistant cultivars to boost farmer earnings	Improve the distribution of resources and farming practices to increase productivity and sustainability	(Ahmad et al., 2020); (Javaid et al., 2022); (Goralski et al., 2022)
SDG2	Zero Hunger	End hunger, increase nutritional and food security, and promote sustainable agriculture	Develop nutrient-dense, pest-resistant, and high-yielding crop types	Predict crop yields, optimize planting timelines, and coordinate supply chains to maintain food security	(Wani et al., 2022); (Javaid et al., 2022); (Husaini and Sohail, 2024)
SDG3	Good Health and Well-being	Promote healthy lives and foster well-being for people of all ages, including providing access to nutritional foods	Increase crop nutritional value (for example, through nutritional fortification includes vitamins as well as minerals)	Monitor the health effects of agricultural techniques and patterns of food consumption in order to enhance nutrition	(Mohamed et al., 2024); (Javaid et al., 2022); (Fernandez and Rosa Maria, 2020)
SDG6	Clean Water and Sanitation	Ensure universal access to and efficient management of water and hygiene, which is vital to agricultural output	Develop drought-resistant agricultural types to reduce water consumption	To conserve water resources, control irrigation systems and anticipate water requirements	(Pachiappan et al., 2024); (Joshi et al., 2020); (Dayıoğlu and Turker, 2021)
SDG12	Responsible Consumption and Production	Encourage equitable patterns of consumption and production, notably in the agriculture sector	Produce crops having longer shelf lives and less waste	Analyze patterns of consumption and manage food production to reduce waste	(Inoue, 2020); (Mondejar et al., 2021); (Ninama et al., 2024)
SDG13	Climate Action	Take strong steps to prevent climate change and its consequences, which are critical for food security and the sustainability of agriculture	Create crops that are resistant to climate change impacts such as extreme temperatures, drought, and floods	Analyze the effects of climate on agriculture and devise measures to prevent negative consequences	(Pachiappan et al., 2024); (Joshi et al., 2020); (Javaid et al., 2022); (Campbell et al., 2018)

22.10 CONCLUSION

Combining CRISPR and AI has enormous revolutionary potential for improving food security and sustainable growth. CRISPR's precise gene-editing capabilities enable crop modification for desired features such as higher yield, resistance to diseases, and nutritional value. When combined with AI, this technology improved the identification of target genes, forecast the results of genetic alterations, and optimized agricultural techniques for a variety of conditions.

CRISPR-AI solutions address food scarcity concerns by expediting the creation of resistant and high-yielding crop varieties, mitigating the effects of climate change on agriculture, and promoting sustainable farming methods. Furthermore, this synergy allows for tailored solutions to local needs, helping communities to achieve food sovereignty and lessen their dependency on outside sources. The combination of CRISPR and AI promises a potent toolkit for enhancing global food security and promoting sustainable development routes. Encouraging additional research, collaboration, and appropriate use of CRISPR and AI technologies is critical for realizing their full promise while minimizing hazards. Continued research for the safety, ethical implications, and long-term effects of gene editing and AI applications in agriculture is critical for decision making on adopting these technologies on a large scale. Collaboration among scientists, politicians, industry stakeholders, and civil society can promote knowledge sharing, multidisciplinary innovation, and the creation of strong regulatory frameworks. Increasing transparency, openness, inclusion, and public participation will pave the way so that advances in CRISPR and AI serve society fairly and are in accordance with ethical values. By developing a culture of responsible innovation and governance, we can traverse the complexity of these transformational technologies and lead them toward long-term food security and development results.

REFERENCES

Abbass, K., Qasim, M.Z., Song, H., Murshed, M., Mahmood, H. and Younis, I. (2022). A review of the global climate change impacts, adaptation, and sustainable mitigation measures. *Environmental Science and Pollution Research*, 29(1), 42539–42559. https://doi.org/10.1007/s11356-022-19718-6.

Afzal, S., Sirohi, P. and Singh, N.K. (2020). A review of CRISPR associated genome engineering: Application, advances and future prospects of genome targeting tool for crop improvement. *Biotechnology Letters*, 42(9), 1611–1632. https://doi.org/10.1007/s10529-020-02950-w.

Ahmad, A., Abd-Elsalam, K. A., Ali, M., Majeed, H. N., Arif, A., Munir, A., Khan, Z. and Munawar, N. (2024). Regulatory triggers of CRISPR-edited crops. In Elsevier EBooks (pp. 91–112). Elsevier. https://doi.org/10.1016/b978-0-443-18444-4.00016-8

Ahmad, A., Munawar, N., Khan, Z., Qusmani, A.T., Khan, S.H., Jamil, A., Ashraf, S., Ghouri, M.Z., Aslam, S., Mubarik, M.S., Munir, A., Sultan, Q., Abd-Elsalam, K.A. and Qari, S.H. (2021). An outlook on global regulatory landscape for genome-edited crops. *International Journal of Molecular Sciences,* 22(21), 11753. https://doi.org/10.3390/ijms222111753

Ahmad, S., Wei, X., Sheng, Z., Hu, P. and Tang, S. (2020). CRISPR/Cas9 for development of disease resistance in plants: Recent progress, limitations and future prospects. *Briefings in Functional Genomics*, 19(1), 26–39. https://doi.org/10.1093/bfgp/elz041.

Aslam, S., Munir, A., Aslam, H.M.U., Khan, S.H. and Ahmad, A. (2022). Genome editing advances in soybean improvement against biotic and abiotic stresses. In *Soybean Improvement: Physiological, Molecular and Genetic Perspectives* (pp. 241–274). Cham: Springer International Publishing. https://doi.org/10.1007/978-3-031-12232-3_13

Balogun, A.-L., Adebisi, N., Abubakar, I.R., Dano, U.L. and Tella, A. (2022). Digitalization for transformative urbanization, climate change adaptation, and sustainable farming in Africa: Trend, opportunities, and challenges. *Journal of Integrative Environmental Sciences*, 19(1), 1–21. https://doi.org/10.1080/1943815x.2022.2033791.

Bradshaw, J.L. (2021). Gene editing and genetic transformation of potatoes. In Potato breeding. Springer. https://doi.org/10.1007/978-3-030-64414-7_9.

Brazelton, V.A., Zarecor, S., Wright, D.A., Wang, Y., Liu, J., Chen, K., Yang, B. and Lawrence-Dill, C.J. (2015). A quick guide to CRISPR sgRNA design tools. *GM Crops & Food,* 6(4), 266–76. https://doi.org/10.1080/21645698.2015.1137690.

Campbell, B.M., Hansen, J., Rioux, J., Stirling, C.M., Twomlow, S. and Lini Wollenberg, E. (2018). Urgent action to combat climate change and its impacts (SDG 13): Transforming agriculture and food systems. *Current Opinion in Environmental Sustainability*, 34(1), 13–20. https://doi.org/10.1016/j.cosust.2018.06.005

Cardoso, J., Gloria, A. and Sebastiao, P. (2020). *Improve irrigation timing decision for agriculture using real time data and machine learning*. In 2020 International Conference on Data Analytics for Business and Industry: Way Towards a Sustainable Economy (ICDABI) (pp. 1–5). IEEE.

Chen, K., Wang, Y., Zhang, R., Zhang, H. and Gao, C. (2019). CRISPR/Cas genome editing and precision plant breeding in agriculture. *Annual Review of Plant Biology*, 70(1), 667–697. https://doi.org/10.1146/annurev-arplant-050718-100049.

Concordet, J.-P. and Haeussler, M. (2018). CRISPOR: Intuitive guide selection for CRISPR/Cas9 genome editing experiments and screens. *Nucleic Acids Research*, 46(W1), W242–W245. https://doi.org/10.1093/nar/gky354.

Côté, M. and Lamarche, B. (2021). Artificial intelligence in nutrition research: Perspectives on current and future applications. *Applied Physiology, Nutrition, and Metabolism,* 15, 1–8. https://doi.org/10.1139/apnm-2021-0448.

Datta, S.K., Majumder, S. and Datta, K. (2021). Molecular breeding for improved β-carotene synthesis in golden rice: Recent progress and future perspectives. In M.A. Hossain et al. (Eds.), Molecular breeding for rice abiotic stress tolerance and nutritional quality (pp. 287–303) Wiley. https://doi.org/10.1002/9781119633174.ch15.

Dayıoğlu, M.A. and Turker, U. (2021). Digital transformation for sustainable future – agriculture 4.0: A review. Tarım Bilimleri Dergisi, 27(4). https://doi.org/10.15832/ankutbd.986431

Dixit, S., Kumar, A., Srinivasan, K., Vincent, P.M. and Ramu Krishnan, N. (2024). Advancing genome editing with artificial intelligence: Opportunities, challenges, and future directions. *Frontiers in Bioengineering and Biotechnology,* 11, 1335901.

Elbasi, E., Mostafa, N., Al-Arnaout, Z., Zreikat, A. I., Cina, E., Varghese, G., ... & Zaki, C. (2023). Artificial intelligence technology in the agricultural sector: A systematic literature review. *IEEE Access,* 11, 171–202.

Fang, J. (2024). Breeding 5.0: AI-driven revolution in designed plant breeding. *Molecular Plant Breeding*, 15. Available at: https://genbreedpublisher.com/index.php/mpb/article/view/3993 [Accessed 19 Apr. 2024].

Fernandez, R.M. (2020). *SDG3 good health and well-being: Integration and connection with other SDGs*. United Nations.

Ghazali, M. H. M., Azmin, A. and Rahiman, W. (2022). Drone implementation in precision agriculture: A survey. *International Journal of Emerging Technology and Advanced Engineering*, 12(4), 67–77.

González, M.N., Massa, G.A., Andersson, M., Turesson, H., Olsson, N., Fält, A.S. and Feingold, S.E. (2020). Reduced enzymatic browning in potato tubers by specific editing of a polyphenol oxidase gene via ribonucleoprotein complexes delivery of the CRISPR/Cas9 system. *Frontiers in Plant Science,* 10, 497481.

Goralski, M.A. and Tan, T.K. (2022). Artificial intelligence and poverty alleviation: Emerging innovations and their implications for management education and sustainable development. *International Journal of Management Education*, 20(3), 100662.

Gostimskaya, I. (2022). CRISPR–Cas9: A history of its discovery and ethical considerations of its use in genome editing. *Biochemistry (Moscow)*, 87(8), 777–788. https://doi.org/10.1134/s0006297922080090.

Hartman, G.L., Pawlowski, M.L., Chang, H.-X. . and Hill, C.B. (2016). 3 – Successful technologies and approaches used to develop and manage resistance against crop diseases and pests. [online] ScienceDirect. Available at: www.sciencedirect.com/science/article/pii/B9781782423355000032.

Husaini, A.M. and Sohail, M. (2024). Agrochemical-free genetically modified and genome-edited crops – towards a "greener" green revolution and achieving the United Nations' Sustainable Development Goals. *Journal of Biotechnology*, 389, 68–77. https://doi.org/10.1016/j.jbiotec.2024.04.015

Inoue, Y. (2020). Satellite- and drone-based remote sensing of crops and soils for smart farming: A review. *Soil Science and Plant*

Nutrition, 66(6), 798–810. https://doi.org/10.1080/00380768.2020.1738899

Javaid, M., Haleem, A., Khan, I.H. and Suman, R. (2022). Understanding the potential applications of Artificial Intelligence in agriculture sector. *Advanced Agrochem*, 2(1), 15–30. https://doi.org/10.1016/j.aac.2022.10.001.

Joshi, R.K., Bharat, S.S. and Mishra, R. (2020). Engineering drought tolerance in plants through CRISPR/Cas genome editing. 3 Biotech, 10(9). https://doi.org/10.1007/s13205-020-02390-3

Khanzadi, M.N. and Khan, A.A. (2019). CRISPR/Cas9: Nature's gift to prokaryotes and an auspicious tool in genome editing. *Journal of Basic Microbiology*, 60(2), 91–102. https://doi.org/10.1002/jobm.201900420.

Kianpour, N. (2023). Identification of optimal sgRNA candidates for mutagenesis of the ALS1 gene in Ipomoea batatas using in-vitro cleavage assay and CRISPR/Cas9 technology. *[online]. Available at:* https://brage.inn.no/inn-xmlui/handle/11250/3109914 *(accessed* 30 March 2024).

Kiran, A., Wakeel, A., Mahmood, K., Mubaraka, R., Hafsa and Haefele, S.M. (2022). Biofortification of staple crops to alleviate human malnutrition: Contributions and potential in developing countries. *Agronomy*, 12(2), 452. https://doi.org/10.3390/agronomy12020452.

Kumar, D., Yadav, A., Ahmad, R., Dwivedi, U.N. and Yadav, K. (2022). CRISPR-based genome editing for nutrient enrichment in crops: A promising approach toward global food security. *Frontiers in Genetics,* 13, 932859.

Li, C., Chu, W., Gill, R.A., Sang, S., Shi, Y., Hu, X., Yang, Y., Zaman, Q.U. and Zhang, B. (2022). Computational tools and resources for CRISPR/Cas genome editing. Genomics, Proteomics & Bioinformatics. https://doi.org/10.1016/j.gpb.2022.02.006

Liu, F.X., Cui, J.Q., Wu, Z. and Yao, S. (2023). Recent progress in nucleic acid detection with CRISPR. *Lab on a Chip*, 23(6), 1467–1492. https://doi.org/10.1039/D2LC00928E

Maraveas, C. (2022). Incorporating artificial intelligence technology in smart greenhouses: Current state of the art. *Applied Sciences*, 13(1), 14. https://doi.org/10.3390/app13010014

Maximiano, M.R., Távora, F.T.P.K., Prado, G.S., Dias, S.C., Mehta, A. and Franco, O.L. (2021). CRISPR genome editing technology: A powerful tool applied to developing agribusiness. *Journal of Agricultural and Food Chemistry*, 69(23), 6379–6395. https://doi.org/10.1021/acs.jafc.1c01062

Minkenberg, B., Wheatley, M. and Yang, Y. (2017). CRISPR/Cas9-enabled multiplex genome editing and its application. [online]. ScienceDirect. Available at: www.sciencedirect.com/science/article/pii/S1877117317300698

Mohamed, H. I., Khan, A. and Basit, A. (2024). CRISPR-Cas9 system mediated genome editing technology: An ultimate tool to enhance abiotic stress in crop plants. *Journal of Soil Science and Plant Nutrition,* 24, 1799–1822. https://doi.org/10.1007/s42729-024-01778-x

Mohamed, M. (2023a). Agricultural sustainability in the age of deep learning: Current trends, challenges, and future trajectories. Sustainable Machine Intelligence Journal, 4. https://doi.org/10.61185/smij.2023.44102

Mondejar, M.E., Avtar, R., Diaz, H.L.B., Dubey, R.K., Esteban, J., Gómez-Morales, A., Hallam, B., Mbungu, N.T., Okolo, C.C., Prasad, K.A., She, Q. and Garcia-Segura, S. (2021). Digitalization to achieve sustainable development goals: Steps towards a smart green planet. *Science of the Total Environment,* 794, 148539. https://doi.org/10.1016/j.scitotenv.2021.148539

Mrabet, R. (2023). Sustainable agriculture for food and nutritional security. [online]. ScienceDirect. Available at: www.sciencedirect.com/science/article/pii/B9780323905008000130

Munawar, N., Faheem, M., Niamat, A., Munir, A., Habibullah Khan, S., Zahoor, M.K., Aslam, A. and Ahmad, A. (2024). Regulatory, ethical, social, and biosafety concerns in genome-edited horticultural crops. In K.A. Abd-Elsalam, A. Ahmad, and B. Zhang (Eds.), CRISPR horticultural crops. ScienceDirect. Available at: www.sciencedirect.com/science/article/pii/B9780443132292000260 (accessed 15 April 2024).

Munir, A., Amin, I., Zahoor, M.K., Majeed, H.N., Almoammar, H., Ghaffar, A. and Ahmad, A. (2024). Multiplex genome editing in plants through CRISPR-Cas. In Elsevier EBooks (pp. 127–142). Elsevier. https://doi.org/10.1016/b978-0-443-13229-2.00012-0

Naeem, M., Majeed, S., Hoque, M.Z. and Ahmad, I. (2020). Latest developed strategies to minimize the off-target effects in CRISPR-Cas-mediated genome editing. *Cells*, 9(7), 1608. https://doi.org/10.3390/cells9071608.

Ninama, N., Gangal, L., Khayum, A., Harshitha, S.B., Divya, C., Sowmya, H.M., Singh, A. and Anushi. (2024). Post-harvest biotechnology or genetic engineering solutions: Extending shelf life and reducing food waste. *Journal of Advances in Biology & Biotechnology*, 27(4), 1–26. https://doi.org/10.9734/jabb/2024/v27i47oo36

Pachiappan, K., Anitha, K., Pitchai, R., Sangeetha, S., Satyanarayana, T. V. V. and Boopathi, S. (2024). *Intelligent machines, IoT, and AI in revolutionizing agriculture for water processing.* Available at: www.igi-Global.com; IGI Global. www.igi-global.com/chapter/intelligent-machines-iot-and-ai-in-revolutionizing-agriculture-for-water-processing/334480

Patil, A. (2023). Use of Artificial Intelligence to hasten progress in plant genetics. [online]. Preprints.org. https://doi.org/10.20944/preprints202309.1067.v1.

Phasinam, K., Kassanuk, T., Shinde, P.P., Thakar, C.M., Sharma, D.K., Mohiddin, Md.K. and Rahmani, A.W. (2022). Application of IoT and cloud computing in automation of agriculture irrigation. *Journal of Food Quality,* 2022, 1–8. https://doi.org/10.1155/2022/8285969.

Rehman, F., Gong, H., Bao, Y., Zeng, S., Huang, H. and Wang, Y. (2022). CRISPR gene editing of major domestication traits accelerating breeding for Solanaceae crops improvement. *Plant Molecular Biology*, 108(3), 157–173. https://doi.org/10.1007/s11103-021-01229-6.

Reiger, B. (2022). Adoption of Artificial Intelligence-based technologies in Sub-saharan African agriculture. [online]. Available at: https://run.unl.pt/bitstream/10362/145589/1/TGI1447.pdf

Sabareeswaran, D. and Robert, A.E. (2016). A survey on data mining techniques in agriculture. Indian Journal of Innovations and Developments, pp. 1–6. Available at: https://ischolar.sscldl.in/index.php/ijid/article/view/137247 (accessed 11 April 2024).

Salinari, A., Machì, M., Armas Diaz, Y., Cianciosi, D., Qi, Z., Yang, B., Ferreiro Cotorruelo, M. S., Villar, S. G., Dzul Lopez, L. A., Battino, M., & Giampieri, F. (2023). The application of digital technologies and artificial intelligence in healthcare: An overview on nutrition assessment. *Diseases*, 11(3), 97. https://doi.org/10.3390/diseases11030097

Sampath, V., Rangarajan, N., C. H., S., Deori, M., Veeraragavan, M., Ghodake, B.D. and Kaushal, K. (2023). Advancing crop improvement through crispr technology in precision agriculture trends: A review. *International Journal of Environment and*

Climate Change, 13(11), 4683–4694. https://doi.org/10.9734/ijecc/2023/v13i113647.

Sánchez, E., García-Bañuelos, M.L. and Sida-Arreola, J.P. (2014). Biofortification – promising approach to increasing the content of iron and zinc in staple food crops. Journal of Elementology. https://doi.org/10.5601/jelem.2014.19.3.708.

Sarker, I.H. (2022). AI-based modeling: Techniques, applications and research issues towards automation, intelligent and smart systems. SN Computer Science, 3(2). https://doi.org/10.1007/s42979-022-01043-x

Sripada, N.K., Challa, S.C. and Kothakonda, S. (2023, *October). AI-Driven nutritional assessment improving diets with machine learning and Deep Learning for food image classification*. In 2023 International Conference on Self Sustainable Artificial Intelligence Systems (ICSSAS) (pp. 138–144). IEEE.

Srivastav, A.L., Dhyani, R., Ranjan, M., Madhav, S. and Sillanpää, M. (2021). Climate-resilient strategies for sustainable management of water resources and agriculture. *Environmental Science and Pollution Research*, 28(31), 41576–41595. https://doi.org/10.1007/s11356-021-14332-4.

Srushti, S.C., Biradar, S., Shreedevi, S.J., Naveen Kumar, M., Mondal, S. and Babu, R.L. (2022). Ethical aspects and public perception on plant genomics In R.L. Singh et al. (Eds.), Plant genomics for sustainable agriculture (pp. 371–399). Springer. https://doi.org/10.1007/978-981-16-6974-3_15

Sudduth, K. A., Woodward-Greene, M. J., Penning, B. W., Locke, M. A., Rivers, A. R., & Veum, K. S. (2020). AI down on the farm. *IT Professional*, 22(3), 22–26. https://doi.org/10.1109/mitp.2020.2986104

Sultan, Q., Ashraf, S., Munir, A., Khan, S. H., Munawar, N., Abd-Elsalam, K. A. and Ahmad, A. (2022). Beyond genome editing: CRISPR approaches. In A. Ahmad et al. (Eds.), The CRISPR/Cas tool kit for genome editing (pp.187–218). Springer. https://doi.org/10.1007/978-981-16-6305-5_6

Susheel, K. and Rajkumar, R. (2023). A review on cutting edge technologies in crop pests and diseases detection. *Journal of Data Acquisition and Processing*, 38(3), 640. https://doi.org/10.5281/zenodo.7922968.

Virnodkar, S.S., Pachghare, V.K., Patil, V.C. and Jha, S.K. (2020). Remote sensing and machine learning for crop water stress determination in various crops: A critical review. Precision Agriculture. https://doi.org/10.1007/s11119-020-09711-9

Wani, S.H., Choudhary, M., Barmukh, R., Bagaria, P.K., Samantara, K., Razzaq, A., Jaba, J., Ba, M.N. and Varshney, R.K. (2022). Molecular mechanisms, genetic mapping, and genome editing for insect pest resistance in field crops. Theoretical and Applied Genetics. https://doi.org/10.1007/s00122-022-04060-9

Zafar, I., Rafique, A., Fazal, J., Manzoor, M., Ain, Q. U. and Rayan, R. A. (2021). Genome and gene editing by artificial intelligence programs. In L. Gaur et al. (Eds.), *Advanced AI techniques and applications in bioinformatics* (pp. 165–188). CRC Press.

Zhang, G., Luo, Y., Dai, X. and Dai, Z. (2023). Benchmarking deep learning methods for predicting CRISPR/Cas9 sgRNA on- and off-target activities. Briefings in Bioinformatics, 24(6). https://doi.org/10.1093/bib/bbad333

Zhu, H., Li, C. and Gao, C. (2020). Applications of CRISPR–Cas in agriculture and plant biotechnology. *Nature Reviews Molecular Cell Biology*, 21(11), 661–677. https://doi.org/10.1038/s41580-020-00288-9

23 The Genome-Editing Toolkit for Food Security

Challenges and Opportunities

Waqar Majeed, Noreen Fatima, Aqsa Riaz, Uzma Ramzan, Hammad Ahmad Khan, Muhammad Sarfraz Ahmed, Asma Ayub, and Naureen Rana

23.1 INTRODUCTION

In contemporary agricultural practices, genome engineering is swiftly evolving as a groundbreaking technology with the potential to revolutionize our methodologies in crop production and sustainability. This scientific progression, marked by the meticulous and targeted modification of an organism's genetic material, transcends the mere augmentation of crop traits. It embodies a pivotal solution to paramount global challenges, encompassing food security, environmental sustainability, and the imperative need for adaptation to the swiftly evolving climate change.

The importance of genome engineering in agriculture is emphasized by its capacity to tackle the pressing concern of food security, a matter of considerable significance as the global population is anticipated to approach nearly 10 billion by 2050. Conventional breeding techniques and agricultural approaches may prove insufficient in meeting the escalating demand for food. Genome engineering presents a propitious substitute, facilitating the cultivation of high-yield, nutrient-enriched crops adaptable to diverse climates and soil conditions (Qaim, 2010).

Environmental sustainability stands out as another pivotal domain wherein genome engineering has potential for substantial influence. Traditional agricultural methodologies have frequently resulted in soil degradation, water scarcity, and the depletion of biodiversity. Through genome engineering, crops can be meticulously crafted to necessitate diminished quantities of water, fertilizer, and pesticides, thereby mitigating the ecological impact associated with agricultural practices (Chen et al., 2019). For instance, developing nitrogen-efficient crops could significantly minimize fertilizer use, thus mitigating soil and water pollution (Kamburova et al., 2017). Furthermore, the menace of climate change poses a substantial challenge to agricultural productivity, characterized by shifting weather patterns and the occurrence of extreme conditions such as droughts and floods. Genome engineering emerges as a crucial tool in the creation of crops that exhibit enhanced resilience to these environmental stressors. An illustrative instance involves the incorporation of drought-tolerant traits into crops such as maize and rice, thereby securing stability in food production amidst climate variability (Henry et al., 2016).

In summary, genome engineering within the realm of agriculture signifies an innovative scientific endeavor and an indispensable response to worldwide challenges. Its capacity to elevate crop productivity, ensure environmental sustainability, and address climate change positions it as a foundational technology for the future of agriculture. Nevertheless, it is imperative to navigate this domain with careful consideration, taking into account ethical, regulatory, and societal implications, to optimize its benefits for humanity.

23.2 THE IMPORTANCE OF GENETIC MODIFICATION IN AGRICULTURAL PRACTICES

Genome engineering in agriculture goes beyond the limits of conventional breeding methods and provides a more accurate and successful approach to improving crop attributes. This technology is essential to enhancing agricultural output, resistance to pests and diseases, tolerance of stress, and nutritional value. The future setting of a rapidly expanding global population and mounting environmental issues gives rise to an intensification of the relevance of this issue. Recent studies have underlined that the goal of genome engineering goes beyond just increasing agricultural production. Its purpose is to provide plant life with the capacity to withstand and adapt to a diverse range of environmental circumstances that are constantly changing. This resilience is essential for guaranteeing food security in the face of difficulties such as climate change, the deterioration of soil, and a lack of water availability. Genome engineering, which makes it possible to create targeted genetic modifications, may be used to develop crop varieties that can better endure these problems and thus contribute to sustainable agricultural practices (Gao, 2018; Jansing et al., 2019).

23.3 GENOME EDITING IN PLANTS: POTENTIAL DNA MODIFICATIONS

The process of editing plant genomes covers a wide range of DNA alterations, each of which has unique uses and the ability to influence the crop's characteristics.

DOI: 10.1201/9781003500933-26

23.3.1 Single-Nucleotide Polymorphisms

In producing genetic variants that might dramatically improve agricultural attributes, single-nucleotide polymorphisms, often SNPs, play an essential part. These polymorphisms entail changes made to a single nucleotide. According to a study by Kartseva et al. (2023), targeted SNPs in wheat have been connected to higher grain protein content. This suggests that there may be possible remedies to the problems caused by nutritional deficiencies. This sort of precise genetic alteration may lead to the production of crops with enhanced nutritional profiles, which is one way to address concerns relating to malnutrition and food quality.

23.3.2 Chromosomal Restricting

Significant advances have been made in the field of plant gene editing in recent years. By the production of single site-specific double-strand breaks (DSBs), the knockout of genes by non-homologous end joining has become routine in many plant species. Recently, there has also been a significant improvement in the effectiveness of homologous recombination in causing pre-planned alterations. To date, however, relatively little work has been done to induce several DSBs simultaneously to produce more complicated modifications in plant genomes. Several studies on the effective induction of deletions have been released. Nevertheless, reports of CRISPR/Cas-induced intrachromosomal inversions and inter-chromosomal recombination are quite recent (Schmidt et al., 2020).

CRs in plants are associated with evolutionary processes such as adaptation and species formation in a much broader way compared to mammals. An obvious application for the targeted modification of CRs would be the targeted breaking or stabilization of genetic linkages between quantitative trait loci (QTLs), resistance markers or other attractive traits. First attempts to induce CR events in plants artificially were made with the use of TEs, site-specific recombinases and mega nucleases (Wellenreuther and Bernatchez, 2018; Lee et al., 2017). A rather obvious CR desirable for plant breeding is the induction of crossovers between homologous chromosomes. This kind of homologous recombination reaction occurs in a programmed way during meiosis (Wang and Copenhaver, 2018). Meiotic HR allows for the creation of new allelic combinations and is the source of genetic diversity. The random nature of meiotic recombination decreases the possibility of favorable results and the presence of hot and cold spots further restricts viable outcomes, due to the inaccessibility of genomic regions (Choi, 2017).

23.3.3 Gene Regulation

There are many ways in which plants and animals are different, but one that matters is that plants are more susceptible to environmental change than animals. To adapt to changes in their surroundings, plants have developed several sophisticated processes during their long evolution and artificial domestication. The most crucial foundation is the molecular biology-related apparatus. The clarification of it will greatly and deliberately encourage the sustainable use of plant resources and maximize its potential at various scales. At the very least, this molecular mechanism consists of the following: the recognition of drought signals (input), their transduction (which involves numerous cascade biochemical reactions), their output, their responses, and the realization of their phenotypes, which is a multi-dimensional network system with numerous levels of gene expression and regulation. We conclude that biological measures are the fundamental answer to a variety of problems pertaining to sustainable development, and that plant measures are the best course of action (Ni et al., 2009).

According to Robertson et al. (2009), plants must coordinate a variety of systems to adapt to the constantly changing soil, plant, and atmosphere (SPAC) environment in which they reside. The plant gene regulatory network system, which is primarily made up of inducible genes (environmental factors and developmental cues), their expression programming and regulatory elements (cis-element and trans-element), corresponding biochemical pathways, and diverse signal factors, has been better understood thanks to recent advances in molecular biology and bioinformatics (especially, DNA microarray technology), genomics, proteomics, metabolomics, and transcriptomics (Sakuma et al., 2006).

23.3.4 Indels in DNA

Indels, which are short DNA sequences that may be inserted into or deleted from one another, can change or impair the activity of genes. According to Lawrenson et al. (2015), this may result in favorable features like increased drought resistance or improved metabolic pathways, which contribute to the production of crops that are better adapted for various environmental situations. These alterations are especially crucial for adapting crops to the pressures associated with climate change, including fluctuating rainfall patterns and severe weather events.

23.3.5 Large Deletions in DNA

Large deletions require the removal of substantial portions of DNA. They may play an essential role in the process of eradicating unwanted characteristics as well as in the study of functional genomics (Zhou et al., 2015). The purpose of these investigations is to understand the functions that genes play in plants' growth, development, and stress response. Scientists can design crops more resistant to abiotic stressors such as salt, temperature extremes, and water shortages if they first understand and then manipulate the genetic components contributing to this resistance.

23.3.6 Insertions in DNA

Introducing DNA sequences into the plant genome is known as insertion. According to Baltes et al. (2015), this strategy has the potential to introduce new features, such as better photosynthetic efficiency or resistance to viral infections. This, in turn, may enhance crop resilience and production. These developments are essential for expanding agricultural output in an environmentally friendly way, which is necessary to guarantee that food production can keep up with the rising needs of the world's population.

In conclusion, the possible DNA alterations achieved in plants via genome editing constitute a big step forward in agricultural biotechnology. These alterations provide the option of generating crops that are both high-yielding and nutrient-dense and robust to the difficulties of a changing environment. This will play a significant part in the future of sustainable agriculture since it is one of the most critical aspects of industry.

23.4 THE TOOLS

CRISPR/Cas9 and TALENs are the principal tools used for editing genomes, and because of their pinpoint accuracy and high-efficiency level, they have completely transformed the field. CRISPR/Cas9 has been a game-changer since it makes it possible to modify plant genomes in a targeted manner with an accuracy that has never been seen before (Doudna and Charpentier, 2014). These technologies have made the generation of crops with better characteristics and much more straightforward, leading to new doors in agricultural biotechnology.

Using genome engineering in agricultural production represents a huge step forward regarding our potential to improve crop performance and sustainability. The uses of this technology in the production of genetically modified crops are essential in tackling the difficulties of ensuring food security and maintaining environmental sustainability. It is vital to approach this technology with a balanced viewpoint, considering the ethical, ecological, and social ramifications.

TABLE 23.1
Tools (webservers) helpful for the guide RNA design, off-target and gRNA prediction, respectively

Guide RNA design	
1. ATUM	2. Benchling
3. Breaking-Cas	4. Cas-Designer
5. Cas13design	6. CGAT
7. CHOPCHOPv3	8. CLD
9. Crisflash	10. CRISpick
11. CRISPETa	12. CRISPOR
Off target prediction	
13. Cas-OFFinder	14. CCtop
15. CHOPCHOP	16. CRISPR RGEN Tools
17. CRISPRTarget	18. flyCRISPR
19. Geneious	20. GT-Scan
21. GT-Scan2	22. sgRNAcas9
gRNA prediction	
23. AmpliconDIVider	24. BATCH-GE
25. Cas-analyze	26. CRIS.py
27. crispr-DART	28. CRISPR-DAV
29. CRISPR-GA	30. CRISPResso2
31. CRISPRmatch	32. CrispRVariants

23.4.1 Guide RNA Design

gRNA (guide RNA) design is a critical step in CRISPR-based genome editing, and several computational tools are available for gRNA design, each with its unique features. When designing gRNAs, several criteria should be taken into consideration, including GC content (guanine-cytosine content), length and potential off-target sites of the designed gRNA (Bergenholm et al., 2021). Additionally, integrating transcription factor binding data has been found to be useful for designing guide RNAs. These tools play a crucial role in enabling efficient and precise genome editing by facilitating the design of gRNAs tailored to specific editing needs. Some of these tools (webservers) are given in Table 23.1.

23.4.2 Off-Target Prediction

The importance of avoiding off-target effects in gRNA design cannot be overstated. Off-target activity of gRNAs can lead to unintended genetic modifications, potentially causing undesirable effects. Therefore, careful consideration of various design criteria is essential to minimize off-target effects. Factors such as GC content, gRNA length, and the presence of potential off-target sites should be considered during the design process. For instance, the GC content of the gRNA sequence is crucial, as a higher GC content can stabilize the RNA:DNA duplex while destabilizing off-target hybridization. Additionally, the length of the gRNA sequence is adjustable and can range from 17–24 base pairs, with shorter sequences leading to minimized off-target effects. Furthermore, advanced algorithmically designed scoring-based models can aid in picking the most appropriate gRNA for an experiment based on identified off-target sites. Careful design, engineering, and modification of gRNAs are necessary to circumvent off-target mutations and ensure the specificity and efficacy of CRISPR-based genome editing (Bergenholm et al., 2021; Naeem et al., 2020). These considerations are vital for maximizing on-target activity while minimizing off-target activity, ultimately contributing to the success and accuracy of genome-editing experiments. Some selected tools and webservers that are important for the prediction of off target are given in Table 23.1.

23.4.3 Genome Editing of Outcomes and Predictions of GRNA

The outcome of genome-editing experiments can be influenced by several factors, including the efficiency of the gRNA used. Therefore, predicting the efficiency of gRNAs is crucial when designing the CRISPR-Cas9 system. Several computational tools have been developed to predict gRNA efficiency, including CRISPRedict, which is an interpretable gRNA efficiency prediction model for CRISPR-Cas9 gene editing (Bergenholm et al., 2021; Konstantakos et al., 2022).

These tools use various features, such as GC content, gRNA length, and potential off-target sites, to predict the efficiency of gRNAs (Zhong et al., 2023). Accurate prediction of gRNA efficiency is essential for maximizing on-target activity while minimizing off-target activity, ultimately contributing to the success and accuracy of genome-editing experiments (Xiang et al., 2021). Some of the noteworthy tools (webservers) for this purpose are given in Table 23.1.

23.5 AGRICULTURAL DEMANDS AND GENOME-EDITING SUCCESSES

23.5.1 To Increase Food Production

When it comes to employing sustainable practices to deliver food and ensure proper nutrition, agriculture faces enormous challenges. Regarding sustainability, sustainable agriculture is defined as a method for growing crops in both the short and long term without endangering the ecosystem (Tripathi et al., 2022). The main objectives of sustainable agriculture are to increase the amount of nutritious crop products produced, to use ecological resources efficiently and with minimal harm, to raise the standard of living through the fair distribution of food, and to benefit farmers economically (Tseng et al., 2020). Since it is difficult to produce large amounts of food with minimal environmental damage, these goals have acquired popularity in the agricultural sector and have been widely acknowledged in scientific publications (Abdul Aziz et al., 2022).

Genetic modification of plants has produced significant advances in the field of agriculture. Plant biotechnology has facilitated the development of cultivars that enable the agricultural industry to enhance crop productivity in an ecologically sustainable manner. Production has experienced a substantial increase, reaching the same levels as the Green Revolution in the early 1970s (Raman, 2017). A genetically modified (GM) crop refers to a plant that has undergone deliberate genetic manipulation using advanced techniques, resulting in genetic material alterations that do not occur naturally (Sendhil et al., 2022). The agricultural industry has been at the forefront of substantial investments in the use of genetic modifications (Raman, 2017). Extensive research into agricultural biotechnology has made it easier to cultivate plants with the desired traits in order to provide food. The use of such techniques to incorporate exogenous genes, as well as to suppress and activate certain genes in plants, has led to a significant proliferation of GM crops (Kumar ct al., 2020). The advance has resulted in the proliferation of crops that possess resistance to diseases, tolerance to environmental stress, and an enhanced nutritious content for consumers (Batista et al., 2017).

23.5.2 To Increase Crop Productivity

According to the Food and Agriculture Organization (FAO) in 2017, it is anticipated that the global population would reach 10 billion by the year 2050. One of the significant contemporary challenges is to effectively nourish the growing population. Because of the considerable influence of advanced plant breeding techniques and the Green Revolution, the majority of the global population can receive an adequate supply of food from current crop yields. Nevertheless, agricultural production is showing signs of reaching a plateau and even decreasing because of lack of arable land and fluctuations in climate globally. To sustain a worldwide population of 10 billion individuals, a 60% rise in production yields would be necessary (Springmann et al., 2018). Hence, enhancing agricultural productivity and sustainability is imperative on a global scale. To ensure future global food security, scientific breakthroughs and technical advances in crop production are desperately needed (Gao, 2021).

Genome-editing techniques have been developed to modify the genetic composition of plants precisely and reliably to attain the desired traits. These strategies are facilitating the advance of accurate breeding methodologies that are reshaping the prospects of plant breeding. The CRISPR-Cas, which stands for clustered regularly interspaced short palindromic repeats-CRISPR associated, has become an innovative technique for modifying genetic composition of crops (Shan et al., 2013). Important cereal crops like wheat, maize, and rice, as well as other vital crops for food security like potatoes (*Solanum tuberosum*) and cassava (*Manihot esculenta*) have experienced a significant increase through the use of this technique (Zhu et al., 2020). Furthermore, the development of CRISPR-related technologies, such as base and prime editors, has significantly broadened the range of possible genome-editing applications. These methods make it easier to add and remove specific DNA segments as well as create accurate nucleotide substitutions. Efforts to improve crops will be significantly impacted by the integration of CRISPR-Cas technology into contemporary breeding processes (Anzalone et al., 2020).

Sequence-specific nucleases (SSNs) that are programmed are used to modify the plant genome. Zinc finger nucleases (ZFNs), transcription activator-like effector nucleases (TALENs), zinc-finger nucleases (ZFNs), and the CRISPR-Cas system are the components of SSNs. Enzymatic proteins called nucleases cause double-stranded DNA to cleave at certain locations, or "target sites." DNA repair processes are utilized to achieve genomic changes with precision (Voytas and Gao, 2014). While ZFNs, TALENs, and mega nucleases identify target sequences by interactions between proteins and DNA, the CRISPR-Cas system uses the complementary base pairing between the target DNA and a programmable "guide" RNA to target specific DNA sequences. Because it is inexpensive, easy to use, and highly effective, CRISPR-Cas technology has gained a lot of traction in the field of plant genome editing (Yin et al., 2017) (Figure 23.1).

DuPont Pioneer achieved effective development of a highly productive economic maize variety by employing the CRISPR/Cas9 knockdown technology on the waxy maize line (Waltz, 2016). Genome editing techniques are used to develop

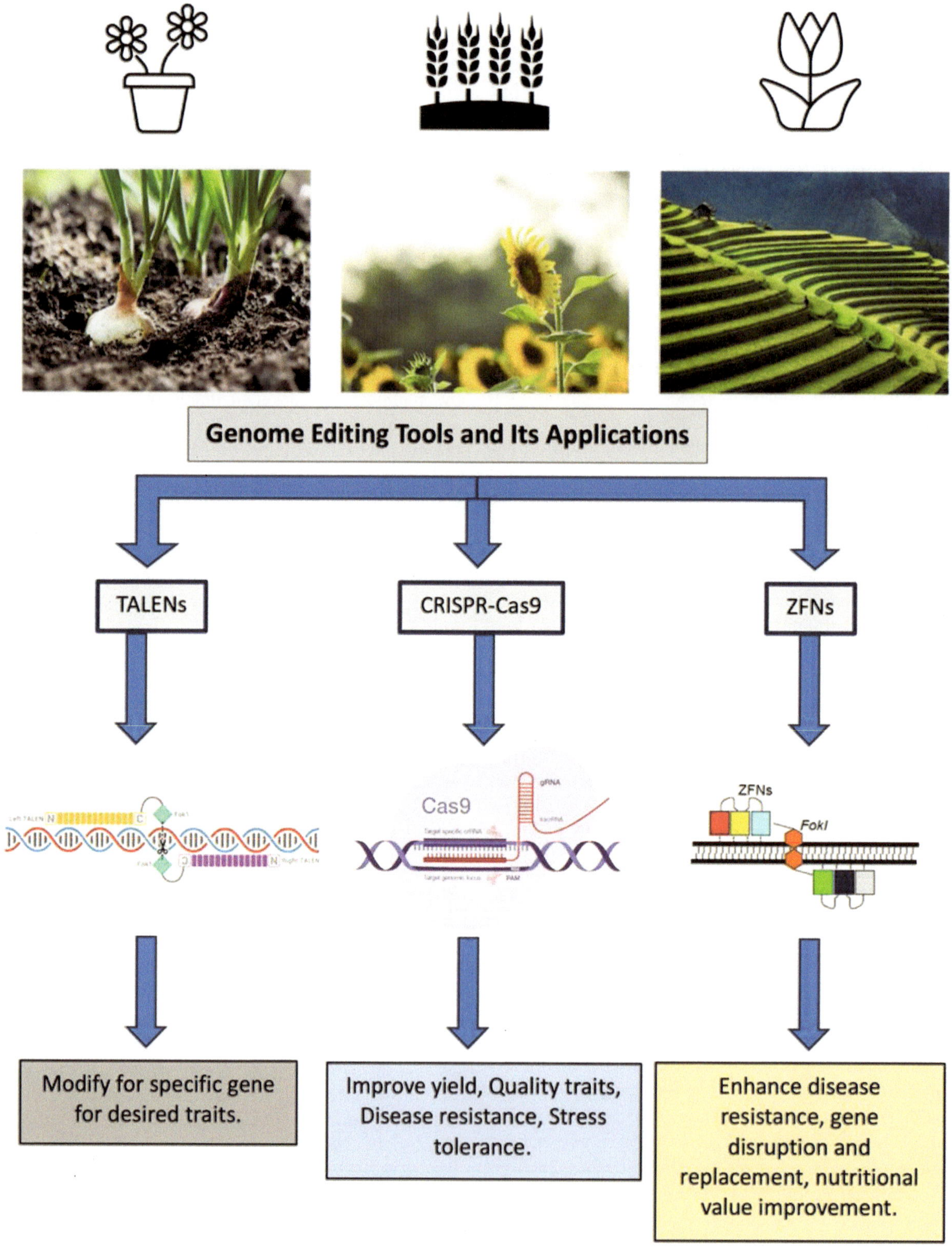

FIGURE 23.1 Genome editing success in agriculture. (Modified from the source Daniel et al., 2023.)

semi-dwarf corn varieties that exhibit improved yields and decreased height. This, in turn, leads to a decrease in the nutrient and moisture requirements of maize (Bage et al., 2020). Additionally, by generating sensitive promoter alleles of CLE genes, the use of CRISPR/Cas9 technology in maize has improved a number of grain yield features. A null allele of a recently discovered partially duplicated compensatory CLE gene was also generated. The current gene-editing research is primarily directed to augmenting the efficiency, size, and mass of wheat seeds (Li et al., 2021).

23.5.3 To Increase Resistance to Plant Pathogens

The advance of genome-editing technologies has been rapid, making them a crucial genetic tool for introducing disease resistance in plants. The use of certain modification techniques, such as zinc finger nucleases (ZFNs), transcription activator-like effector nucleases (TALENs), mega nucleases, and clustered regularly interspaced short palindrome repeats (CRISPR)/CRISPR-associated protein 9 (Cas9), has increased in recent years. CRISPR/Cas9 has become dominant in genome-editing technology because it offers easier design and implementation, higher success rates, greater versatility and lower costs compared to other methods (Figure 23.1). Scientists have achieved viral disease resistance by employing specific strategies in model species like *Arabidopsis* and *Nicotiana benthamiana*. Combining CRISPR-encoding sequences that target and disrupt the viral genome with CRISPR-mediated targeted mutation induction in the host plant genome. Furthermore, a great deal of research is being done on the application of CRISPR/Cas9 targeted modification of susceptibility genes in crop species like citrus, rice, tomato, and wheat to improve resistance to bacterial and fungal diseases.

Many commercially significant staple crops are at serious risk from plant viruses. Based on their genetic composition, they are divided into different groups: reverse-transcribing viruses, double-stranded DNA (dsDNA) viruses, negative sense single-stranded RNA (ssRNA-) viruses, and positive sense single-stranded RNA (ssRNA+) viruses (Roossinck et al., 2015). Double-stranded DNA (dsDNA) viruses are distinct from plant viruses. The majority of studies on ssDNA geminivirus genomes for virus resistance in CRISPR-edited plants have been conducted (Ali et al., 2016; Ji et al., 2015).

Furthermore, the MLO loci of three different plant species (tomato, grapevine, and bread wheat) have been targeted by RNA-guided Cas9 endonuclease (Nekrasov et al., 2017). The protein that the MLO gene encodes has seven transmembrane domains. According to Acevedo-Garcia et al. (2014), this protein can be found in the plasma membrane of both monocots and dicots. Prior research revealed that MLO genes were susceptible (S) genes and that in barley, *Arabidopsis*, and tomato, homozygous loss-of-function mutants showed a significantly higher resistance to powdery mildew (Bai et al., 2008).

23.5.4 Management of Weeds

Since cultivated plants were first domesticated, weeds have continuously impeded them. This has led to a greater decrease in crop yield than diseases or pests, necessitating the use of weed control techniques. To ensure there is a sufficient supply of food to fulfill the demands of a rapidly expanding human population, effective weed management is crucial. The most effective and reliable method for managing weeds is to combine the application of herbicides with integrated weed management (IWM) techniques. Herbicide-resistant (HR) crops must be developed in order to combat weeds using herbicides. Recent advances in genome-editing technology, mainly with respect to CRISPR-Cas9, have opened up new avenues for the agriculture industry to pursue sustainable farming practices. Currently, numerous herbicide-resistant (HR) crops that have not been genetically modified (GM) have been created using genome-editing techniques. These crops potentially will play a significant role in addressing weed issues and simultaneously enhancing crop yield to satisfy the growing global food demand (Hussain et al., 2021).

Genome-editing techniques have successfully been used in several crop species to specifically target genes with the aim of enhancing the average crop production, to address the growing needs of the current global food scarcity. Shah et al. (2018) proposes an environmentally sustainable and practical agricultural plan to enhance crop varieties for improved quality, increased production, and resistance to diseases. In recent years, this technique has significantly transformed the field of crop breeding due to its remarkable effectiveness, adaptability, ease of use, and reliability (Mao et al., 2019). For the advance of homologous recombination in plants, a variety of modern genome-editing techniques have been used, such as transcription activator-like effector nucleases (TALENS), zinc finger nucleases (ZFNS), clustered regulatory interspaced short palindromic repeats (CRISPR), and CRISPR-associated (Cas) methods (Li et al., 2016a) (Figure 23.1).

23.5.5 To Increase Resistance to Abiotic Stress

Genome editing with CRISPR/Cas9 is a promising method for creating cultivars that are tolerant to abiotic stresses, hence supporting sustainable agricultural output (Figure 23.1). The CRISPR/Cas9 method has been widely used to study the effects of temperature, salt, and drought as abiotic stresses (Wang et al., 2020). Using the CRISPR/Cas9 system, two tomato drought stress response genes were removed: mitogen-activated protein kinase 3 (SlMAPK3) and non-expressor of pathogenesis-related gene 1 (SlNPR1). However, no enhancement in drought tolerance was detected because of this genetic modification (Wang et al., 2017). Functional analyses of genes linked to the receipt of salt stress revealed salt-tolerant alleles. Tomatoes' tolerance for high salt levels was reduced when the Salt Overly Sensitive 1 (SOS1) gene, which functions as a transporter to control sodium levels in the outer layer of the roots, was blocked (Wang et al., 2020). Furthermore, the ability of modified tomato plants to withstand high salinity levels was enhanced by the targeted mutation of protein domains in tomato hybrid proline-rich protein1 (HyPRP1), a protein that reduces salt stress responses (Tran et al., 2021).

Fluctuations in climate, coupled with significant temperature differences, have an impact on agricultural production. In order to design and breed crops that can withstand temperature stress, understanding the role that genes play in these responses is essential. Two particular genes, tomato C-repeat binding factor 1 (SlCBF1) and Brassinazole Resistant 1 (SlBZR1), which are linked to cold tolerance and heat sensitivity, respectively, were disrupted using the

CRISPR/Cas9 technique. The study demonstrated a strong correlation between these genes and temperature tolerance. Specifically, the modified versions of SlCBF1 and SlBZR1 genes exhibited reduced ability to withstand cold and heat stress (Yin et al., 2018).

The use of CRISPR/Cas9 technology to simultaneously deactivate the pyl1, pyl4, and pyl6 genes in the OsPYL abscisic acid receptor gene family resulted in enhanced resistance to high temperatures, improved grain production and reduced preharvest sprouting when compared to the wild type (Miao et al., 2018). Field trial and experimental results provide evidence that genome-edited crops enhance water and nitrogen use efficiency (Wen et al., 2021). Peng et al. (2021) showed that using genome-editing technologies to promote drought tolerance in cotton resulted in a reduced environmental impact of cotton production. Arable areas are significantly polluted with metal toxicity. Moreover, CRISPR/Cas9 was successfully used to eliminate the genes OsHAK1, OsARM1, and OsNramp5 in rice, resulting in variants that had lower levels of arsenic, radioactive cesium, and cadmium (Tang et al., 2017).

23.6 PRODUCTION OF HEALTHIER AND NUTRITIOUS FOOD

23.6.1 Oil Composition Improvements

The main component of vegetable oils is triacylglycerol (TAG), which is also the main component used for storage in the seeds or fruits of the majority of plants that produce oil. Plant-derived oils are widely used in many different industries, including food, feed, medicines, lubricants, emulsifiers, detergents, adhesives, cosmetics, and oleochemicals. They also have significant economic value (Rahman and de Jiménez, 2016). With an almost 40% share, oil palm (*Elaeis guineensis*) is a significant contributor to the world's vegetable oil supply. Together, rapeseed/canola (*Brassica napus*) and soybeans (*Glycine max*) produce around 40% of the vegetable oil produced worldwide. The remaining vegetable oil sources are primarily composed of sunflower (*Helianthus annuus*), peanut (*Arachis hypogaea*), cotton (*Gossypium hirsutum*), coconut (*Coco nucifera*), and olive (*Olea europaea*).

An increasing amount of research is being done to improve the fatty acid composition and seed oil output of oilseed crops due to the expanded demand for vegetable oils, which are mainly composed of the storage lipid triacylglycerol. Thus far, progress in this area has been achieved by the application of both conventional breeding practices and transgenic approaches. However, the polyploid nature of many oilseed crops combined with the short time frame present obstacles to further advances using traditional breeding methods. Additionally, the commercialization of genetically modified oilseed crops is hampered by public opinion and the high costs associated with regulatory processes. CRISPR/Cas genome editing is emerging as a cutting-edge breeding method that can satisfy growing demand while alleviating regulatory restrictions (Subedi et al., 2020).

23.6.2 Reduction of Acrylamide in Potatoes

After rice and wheat, the cultivated potato, formally known as *Solanum tuberosum* L., is the third most important staple food crop in the world and is widely grown (FAO, 2024). The main reason potatoes are grown is for human consumption; in the USA and Europe, 80–85% of the harvest is used to produce frozen foods (Devaux et al., 2021). Over 70 countries produced 8.9 million tons of processed potato products in 2019, up from 7.4 million tonnes in 2015 (FAO, 2024). The items that are listed include crisps, chips, and French fries. These products are frequently made from potato tubers after they have been refrigerated for a while.

Cold-induced sweetening (CIS), which occurs when potato tubers are kept in cold storage, is caused by plant invertases breaking down stored sucrose into glucose and fructose (Sowokinos, 2001). The hexose sugars in cold-stored tubers react chemically with free amino acids, especially asparagine, when they are heated to high temperatures, as in frying. This process is known as the Maillard reaction. The potato product browns because of this process. Usually, the reaction yields organic compounds that improve processed foods' flavor and scent (Mottram et al., 2002). Nevertheless, the Maillard reaction has the capability to produce acrylamide, a substance that has the potential to cause cancer and damage the nervous system, as well as several simple carbon compounds that are linked to the bitter flavor and darker appearance observed in potato goods when subjected to high-temperature processing (Maan et al., 2022). Low amounts of hexose sugars and non-essential amino acids, which accumulate in cold-stored tubers, can reduce the amount of browning and acrylamide in processed potato products (Shepherd et al., 2010). Potato processors look for cultivars with a lesser propensity to collect reducing sugars because they must store potato tubers at low temperatures in order to guarantee a steady supply and the production of processed potato products (Powers et al., 2021). It has been shown that employing management strategies to reduce acrylamide formation and discoloration during high-temperature processing is impractical (Palermo et al., 2016). More than 4780 potato accessions are kept in storage by the International Potato Centre and other germplasm collections, providing a wealth of resources for the development of cultivars with enhanced qualities for high-temperature processing. However, conventional breeding techniques have not yet yielded commercially viable cultivars with all the necessary traits for high-temperature processing. CV Atlantic is a popular potato cultivar in North America that accumulates reducing sugars even after being stored for long periods of time (Richael, 2021). Therefore, it is essential to investigate new methods for growing superior potato varieties that are appropriate for processing at high temperatures.

Efforts have been made to use molecular tools to introduce characteristics into potatoes that result in heat-processed products with a low potential for generating acrylamide and reduced browning. These methods have been focused on blocking the function of the vital enzyme vacuolar invertase (VINV), which converts sucrose into glucose and fructose in tubers kept at low temperatures. Furthermore, also targeted have been the two isoforms of asparagine synthetase (AS1 and AS2), which catalyze the conversion of aspartate and glutamine into asparagine (Bhaskar et al., 2010). For example, it has been discovered that RNA interference (RNAi) against the AS1 gene reduces asparagine production. This has resulted in certain potato varieties exhibiting up to a 70% reduction in acrylamide content compared to the natural type. Conversely, RNAi targeting the AS2 gene does not seem to affect the levels of asparagine synthetase and may have adverse effects on plant growth (Wu et al., 2011).

23.6.3 Increased Fruits' Antioxidants

Horticultural crops, such as vegetables, fruits, and ornamental plants are essential for sustaining human existence and are crucial for agricultural productivity. Due to a significant rise in the global human population and the resulting demand for additional food, it has become imperative to enhance agricultural output to ensure food security. Traditional breeding methods have supported the creation of better varieties but, to further boost agricultural production, it is necessary to acquire new breeding procedures. The CRISPR-Cas9 system is a remarkable and powerful genome-editing tool that makes precise modifications to DNA possible. Inspired by the bacterial adaptive immune system, this method uses an endonuclease driven by a single guide RNA to produce double-stranded breaks (DSBs) at specified target loci. A cellular healing mechanism that introduces small insertions and deletions (indels) at the sites of the wounds can mend double-strand breaks (DSBs). Due to its ease of usage, simplicity, and low off-target effects, CRISPR, a cas-based editing tool, has become increasingly popular in comparison to other editing techniques like ZFN, TALENs, and mega nucleases. The CRISPR technology has been effective in improving stress tolerance, shelf life, nutritional content, flavor, and metabolites in several horticulture and industrial crops. The CRISPR-based technique is the most suitable option for achieving the goal of producing non-transgenic crops and eliminating regulatory challenges associated with introducing genetically modified crops to the market. Despite the existing hurdles in editing horticultural, industrial, and ornamental crops, the introduction of this innovative nuclease, which can be specifically applied to crops, renders it a powerful instrument for enhancing crop quality (Sharma et al., 2023).

23.7 CHALLENGES AND FUTURE OUTLOOK

23.7.1 Challenges in Regulating Gene-Edited Crops

- *Ethical concern*: Gene editing prompts ethical inquiries regarding the manipulation of life's core elements. Questions arise about unintended repercussions, potential misuse, and the ethical ramifications of assuming the role of "nature's architect," demanding meticulous examination.
- *Environmental impact*: Modifying the genetic composition of crops can have unforeseen impacts on ecosystems. Introducing gene-edited organisms into the environment might result in ecological disturbances, influencing biodiversity and possibly engendering invasive species.
- *Safety for human consumption*: Stringent safety evaluations are indispensable to guarantee the harmlessness of gene-edited crops intended for human consumption. Devising standardized protocols for testing and supervising these crops involves a complex undertaking.
- *Regulatory structures*: Current regulations might not adequately address the distinctive facets of gene editing. Formulating new, adaptable frameworks that strike a balance between fostering innovation and ensuring safety is an arduous yet imperative task.
- *International consistency*: Securing worldwide consensus on regulating gene-edited crops is formidable. Differing regulatory approaches across nations could erect trade barriers and complicate the global movement of agricultural goods.
- *Public perception and acceptance*: Cultivating public trust is paramount. Doubt and apprehensions about the safety of gene-edited crops can influence their reception. Effective communication and transparent regulatory processes are vital to allay public concerns.
- *Long-term ramifications*: Evaluating the enduring effects of gene-edited crops on soil health, crop sustainability, and other agricultural factors demands continual research. Crafting methodologies to monitor and manage these effects poses a challenge for regulators.

Addressing these obstacles necessitates cooperation among scientists, policymakers, and the public to establish resilient regulatory frameworks that harmonize innovation with safety and ethical considerations.

23.7.2 Genome Editing: Addressing the Food System Challenges

Genome editing plays a pivotal role in enhancing crop genetics through two distinct avenues. First, tools like CRISPR-Cas facilitate research into intricate gene functionalities within complex biological processes. Second, these tools enable the precise introduction of desired genetic modifications into plants.

Researchers leverage genome editing to optimize the utilization of genetic diversity, probe gene functionalities, and uncover key genes and genetic variants governing desirable traits. However, the knowledge derived from genome-editing research need not exclusively culminate in the creation of genome-edited crops; it can also inform conventional breeding methodologies. Plant breeders choose breeding tools based on their suitability for the specific issue at hand. Each tool possesses unique advantages in terms of efficacy, swiftness, and precision. While some breeding objectives may be achievable through older tools, they often sacrifice speed and precision. Conventional breeding typically spans nine to eleven years for a new variety to reach the market, with fruit tree breeding requiring an even longer duration. Genome editing significantly truncates this timeline, expediting the release of new varieties (Kaiser et al., 2020; Watson et al., 2017; Wolter et al., 2019).

Moreover, genome editing allows for the targeted introduction of specific genetic alterations into elite plant varieties, excluding the simultaneous transfer of undesired genetic changes (Wolter et al., 2019).

23.7.3 Genetically Modified Organisms (GMOs) for the Climate Change Challenge

The integration of genetically modified organisms (GMOs) presents a potential avenue for addressing significant challenges in various human activities. However, their development and implementation are markedly influenced by regulatory frameworks and public perceptions. This is particularly evident in the application of GMOs to tackle climate change, where biotech crops are proposed as solutions to mitigate greenhouse gas (GHG) emissions. Biotech crops, throuh use of land optimization, increasing crop yields, and reducing energy, chemical and transportation resources in agricultural production (Kovak et al., 2022), have shown promise in reducing global pesticide usage by an estimated 8.3% compared to conventional crops (Brookes and Barfoot, 2020). Notably, insect- and herbicide-resistant characters have facilitated diminished tillage between agricultural cycles, thereby promoting carbon sequestration by enhancing atmospheric CO_2 fluxes toward the soil (Sutherland et al., 2021). Additionally, insect-resistant traits have led to a reduction in insecticide spraying, subsequently decreasing fuel consumption globally. Furthermore, the cultivation of genetically modified maize, soybean, or their rotation, has shown carbon sequestration potential, 102–250 kg of carbon hectare^{-1} year^{-1} in different regions (Brookes and Barfoot, 2020).

Apart from conventional breeding, genetic modification has emerged as a viable approach to increase the plant resistance to drought and heat stress. Drought tolerance has emerged as a pivotal trait in crop development, with genetic modifications aiming to express the sunflower Hahb-4 TF (transcription factor) associated with drought stress responses in crops like soybean and wheat (Dezar et al., 2005). These transgenic crops, known as HB4 crops, now available commercially, exhibit no differences in nutritional content compared to their non-GMO counterparts (Ayala et al., 2019; Chiozza et al., 2020; Miranda et al., 2022). Under field conditions, HB4 soybean crop has improved the water-use efficiency and seed yield in dry climatic conditions as compared to non-GMO crops (Ribichich et al., 2020). Experience in the USA with drought-tolerant maize has showcased increased grain yield in water-stress environments compared to conventional hybrids, with yield variations ranging from 1% to 9.7% (Cooper et al., 2014; Gaffney et al., 2015). The understanding of water management and root systems in plant biology stands as a crucial part for developing this trait (Bailey-Serres et al., 2019). For instance, modification of the rice root architecture related locus Dro1 has shown to increase root depth and enhance grain yield under water stress environments (Arai-Sanoh et al., 2014; Uga et al., 2013). Another avenue to stress resistance and drought tolerance revolves around the abscisic acid (ABA) hormone signaling pathway, representing a potential target for plant genetic modification to produce new crop varieties (Bailey-Serres et al., 2019). GMO canola incorporating an antisense construct against farnesyltransferase (ERA1), a down-regulator of ABA signaling, has demonstrated increased seed yield under drought condition during flowering (Wang et al., 2005). Indeed, complementary approaches focusing on modifying root systems, ABA signaling, and early-flowering strategies could prove instrumental in coping with warmer seasons, thereby averting exposure to heat stress and reduced water levels in drought-prone regions (Bailey-Serres et al., 2019; Ortega et al., 2019; Shavrukov et al., 2017; Wang et al., 2005).

23.7.4 Challenges in Delivering Genome-Editing Reagents

A critical aspect in the application of genome editing in crops, albeit a challenging one, revolves around delivering genome-editing agents and subsequently generating modified plants. Regrettably, the existing methods for transformation and regeneration in most crops exhibit low efficiency, demand considerable time and labor, or are altogether absent. These deficiencies have posed significant hurdles in the widespread adoption of genome editing in crop cultivation. Moreover, each transformation method comes with inherent limitations and unfavorable attributes that impede the attainment of appropriately modified plants. The prevalent delivery techniques—Agrobacterium, biolistics, and protoplast transformation—have distinct strengths and drawbacks, necessitating a comprehensive understanding for the effective implementation of genome editing in crops.

23.7.5 Agrobacterium Transformation

The use of Agrobacterium-mediated transformation stands as a prominent approach for transporting genome-editing reagents into crops. This method initiates when plant cells come into contact with Agrobacterium, prompting the injection of T-DNA/protein complexes into the plant cell's cytoplasm. Subsequently, these complexes traverse into the cell nucleus. Within the cell nucleus, the fate of the T-DNA/protein complex varies; it might either randomly integrate into the host genome or persist extra chromosomally. The latter scenario facilitates transient gene expression. However, for most genome engineering objectives, stable integration of T-DNA into the plant genome is sought. Instances of successful genome editing through stable T-DNA integration are evident in various crops such as potato, soybean, barley, *Brassica oleracea*, corn, and rice (Char et al., 2015; Gurushidze et al., 2014).

Agrobacterium's application in delivering genome-editing agents presents several advantages, notably a heightened likelihood of achieving single-copy insertions, coupled with established and refined protocols adaptable for transforming diverse crops. Yet, challenges persist, including prolonged tissue culture requirements for plant recovery, a low frequency of stable transformation, limitations concerning the range of genotypes transformable within a crop species, and constraints related to specific Agrobacterium species' host-range (Nam et al., 1997). Moreover, there exists an inverse relationship between the efficiency of transformation and the capacity of target cells to regenerate (Altpeter et al., 2016). Exploring methods to broaden the scope of transformable genotypes and reducing reliance on tissue culture could enhance Agrobacterium's utility in transporting genome-editing reagents to crops.

23.7.6 A Biolistics Approach

The biolistic approach is commonly known as particle bombardment, which serves as a prevalent technique for effecting nuclear plant transformation, particularly when DNA delivery to chloroplasts and mitochondria is necessary. This method revolves around the introduction of DNA into plant cells directly using gold or tungsten particles (Sanford, 1988). These particles, coated with DNA, are propelled at high speed toward plant tissue, effectively lodging themselves inside the plant cells. Once within the cell, the DNA separates from particles and can exhibit transient expression or integrate durably into the target genome.

The physical transfer of DNA via biolistics bypasses the limitations associated with host-range, a challenge occasionally encountered in Agrobacterium-based methods. Biolistics has proven successful in delivering genome-editing elements in various crop plants, for example, in cotton, corn, wheat, and soybean (Ainley et al., 2013; D'Halluin et al., 2013; Jacobs et al., 2015; Li et al., 2015; Wang et al., 2014).

Biolistics offers several advantages: it enables transformation of diverse tissues and cell types, eliminates the necessity for a binary vector, facilitates high-frequency co-transformation of multiple plasmids, boasts a straightforward transformation protocol, allows delivery of large DNA fragments, and is adaptable to convey mRNA or protein (Svitashev et al., 2016; Zhang et al., 2016). Nonetheless, drawbacks include erratic patterns of integration, relatively limited material, high costs, and the inability to precisely control the cellular target.

23.7.7 Protoplast Transformation

This shows the direct transfer of DNA into the plant cells using polyethylene glycol (PEG) or electroporation. Once inside the cell nucleus, the DNA shows the potential for either transient expression or stable integration into the target genome. Among the three procedures delineated, protoplast transformation possibly boasts the highest efficacy, capable of achieving transformation frequencies surpassing 70%. This method has successfully delivered genome-editing reagents in many crop plants, including flax, rice, wheat, and potato (Clasen et al., 2016; Liang et al., 2014; Sauer et al., 2016).

Advantages inherent in protoplast transformation encompass the capacity to deliver multiple plasmids with substantial co-transformation rates, absence of the necessity for a binary vector, high-frequency transformation, and the general amenability of many plant species to protoplast separation and transformation. Nevertheless, limitations involve the restricted number of plant species conducive to regeneration from protoplasts. Additionally, this process tends to be laborious.

23.8 OVERCOMING DELIVERY LIMITATIONS

Most of the transformation approaches encounter limitations, typically restricted to only a handful of genotypes per crop due to the limited amenability of certain genotypes to regeneration. Unfortunately, this compatibility with transformation and regeneration does not consistently align with desirable agronomic performance. To overcome these constraints, Scientists proposed a novel approach employing the *Baby boom (Bbm)* and *Wuschel2 (Wus2)* transcription factors (Lowe et al., 2016). This method, using Agrobacterium and biolistic-mediated transformation, targeted tissues from sorghum, rice, sugarcane, and maize varieties. Transformation frequencies notably increased in rice, sugarcane, and sorghum, elevating from 0 to 3% to 15 to 885%. While this strategy hasn't achieved a worldwide genotype-independent transformation system for all crops, it represents a significant advance to this goal.

An alternate strategy to modify recalcitrant crops involves employing virus-based *in planta* delivery systems. In this method, the plant is infested with a genetically engineered virus carrying genome-editing components. Within the plant as the virus spreads, these reagents are expressed, inducing genetic modifications. Ideally, the virus reaches the meristematic tissue, facilitating the introduction of heritable genetic alterations. For instance, the geminivirus vector, cabbage leaf curl virus, were used to transfer gRNAs for Cas9 in plants (Yin et al., 2015), resulting in high mutation rate in somatic tissues but lacking heritable mutations due to exclusion from meristematic cells. In contrast, the *Tobacco rattle virus* (TRV), altered to deliver genome-editing components to *N. benthamiana* and in petunia (Ali et al., 2015; Marton et al., 2010), exhibited the ability to enter meristematic cells (Martín-Hernández and Baulcombe, 2008). Consequently, mutations' lower frequencies were detected in seeds from plants infected with engineered TRV, evidencing its capability to access meristem cells and introduce heritable DNA alterations. Further refinements in virus-based delivery systems hold promise to expand the range of editable crops and enhance the frequency of heritable mutations.

23.9 NON-GMO GENOME ENGINEERING

When selecting a transfer method, it is crucial to consider governmental principles that might govern the deployment of genome-edited plants in field settings. Many jurisdictions have established regulations triggered by the transfer of foreign DNA into a cell (Voytas and Gao, 2014). Moreover, the delivery process itself may inadvertently lead to unintended off-target alterations owing to the random addition of nuclease-encoding DNA. Editing plant genomes through transient delivery of DNA into cells, or by conveying mRNA/protein, presents a potential strategy to alleviate concerns related to regulatory compliance and off-target effects associated with genome-edited plants.

Numerous strategies have emerged to tackle these concerns, including the delivery of sequence-specific nucleases in the form of mRNA (Stoddard et al., 2016), protein (Luo et al., 2015), mRNA or protein Cas9-gRNA complexes (Svitashev et al., 2015; Woo et al., 2015), or transient transfer of plasmid DNA or RNA (Clasen et al., 2016; Zhang et al., 2016). Results from these studies offer innovative pathways to diminish the possibility of undesired random DNA incorporation, thus addressing regulatory and off-target concerns associated with genome editing in plants.

23.9.1 Identification of Trait Gene/Locus

A pivotal challenge in agricultural genome editing lies in determining the specific genetic modifications essential to manifest a desired trait. This process entails pinpointing target genes or DNA regions, demanding comprehensive comprehension of both crop genomes and gene functionalities. But numerous characters and target identifications derive from gene-function analyses conducted primarily in model plants. To enrich our understanding of crop genetic composition, extensive genome sequencing efforts have been undertaken or are ongoing for various crops such as bread wheat, rice, soybean, corn, and potato (Mayer et al., 2014; Schmutz et al., 2010). Nevertheless, these genome sequences predominantly represent a singular variety within each crop.

The existence of genomic disparities among varieties necessitates resequencing target genes or DNA regions, as they may exhibit variations such as indels, SNPs, or significant chromosomal abnormalities. This resequencing imperative is fundamental for precise genome editing, as the accuracy of genome-editing tools depends on possessing the exact sequence of the target region within the modified plant. Additionally, the profound challenges associated with transformation and regeneration in many crops have limited the molecular characterization of gene functionalities. Addressing these challenges warrants a notable advance in our comprehension of gene functions and the broader systems biology governing crop plants. This advance is crucial to enhance our ability to decipher gene function accurately and comprehensively in diverse crop species.

23.9.2 Frequency of Introducing a Candidate Gene Edit

Another significant challenge in applying genome editing lies in the effectiveness of introducing specific DNA edits into genetic material. Among the edits facilitated by nuclease technology, the simplest alteration involves indels (insertions or deletions). These indels are a result of nucleases creating targeted DNA double-strand breaks (DSBs) in plant cells, followed by imprecise repairs via the Non-Homologous End Joining (NHEJ) pathway. The frequency of inducing indels can be notably high, exceeding 70% (Li et al., 2016b). However, indels represent just one type of potential genetic alteration among many.

Other types of edits contain large DNA deletions, single-nucleotide polymorphisms (SNPs), gene additions, replacements, and inversions. Yet, the occurrences of generating these edits are usually minor compared to the high frequencies observed with indels. Among the common edits besides indels, large deletions stand out, where two DSBs are induced within the same chromosome, resulting in the removal of intervening sequences. In a proof-of-concept study in rice, scientists successfully induced two DSBs by using CRISPR/Cas to delete sizable DNA segments, about 245 kb, at a frequency of around 4 out of 24 transgenic calli (Zhou et al., 2014). However, the efficiency of deletion was shown to inversely correlate with its size, with smaller deletions (<100 bp) occurring at relatively higher frequencies (~10%) compared to larger deletions (5–120 kb) which dropped below 1% (Ordon et al., 2017).

The challenge of low editing frequencies extends to introducing SNPs, gene substitutions, and additions, as these edits rely on Homology-Directed Repair (HR). HR involves providing both nucleases and the donor fragments to plant cells; the resultant DSBs are then fixed using the donor molecule as a template. Nevertheless, the efficiency of HR is significantly lower than that of NHEJ, (Steinert et al., 2016) with reported occurrences ranging from ~10.00% to 0.08% of transformed cells (Wright et al., 2005). This wide range of gene targeting occurrences stems from multiple contributing factors including nuclease efficiency, plant species, cell type, donor molecule structure, and various other molecular and cellular factors.

To address the challenge of low gene targeting frequencies, alternative approaches have been explored. For instance, scientists confirmed in Arabidopsis that knockout of genes play role in classical NHEJ, *ku70, or lig4*, resulted in three-fold to 16-fold improvement in HR-based gene targeting frequencies(Qi et al., 2013). Similarly, in rice, simultaneous disruption of DNA ligase-4 together with Cas9 and donor molecule delivery led to increased HR frequencies (Endo et al., 2016).

23.9.3 Genome-Edited Crops with Breeding Challenges

Due to the partial amenability of only a few genotypes within a given species to transformation, most gene edits are transferred into varieties that often possess suboptimal agronomic traits. This situation necessitates the integration of genome-edited crops into breeding agendas or the allocation of the mutations into elite varieties. However, employing breeding methods for this purpose comes with several inherent challenges. These include significant time investments, spanning multiple years for backcrossing, especially in cases involving numerous unlinked gene edits. Furthermore, certain crops require labor-intensive crossing protocols, continuous genotyping incurs costs, and relocating plants to varied environments (e.g., winter and summer nurseries) adds potential expenses and labor.

While breeding remains a workable option for transferring a limited number of mutations, its feasibility diminishes as the number of unlinked gene edits increases. For instance, when selfing a plant heterozygous for a single gene edit, the occurrence of obtaining offspring with homozygous edits is relatively high (1 out of every 4 plants). However, if the plant carries four gene edits in a heterozygous state, the frequency declines substantially to 1 out of every 256 plants. Managing such low frequencies becomes impractical for many breeding platforms. Moreover, increase in the number of the gene edits concurrently heightens the likelihood of an edit occurring near a locus that adversely affects the complete performance of the plant. This scenario results in linkage drag when crossed with elite varieties, potentially offsetting their favorable traits (Peleman and van der Voort, 2003).

Lastly, when evaluating traits like yield, careful consideration is necessary to compare the value of a product produced through backcrossing with one that might be developed via traditional breeding and the selection programs. Assessing the annual yield increases related with such programs becomes critical in determining the optimal approach for trait enhancement.

23.10 CONCLUSION

Genome editing holds the promise of revolutionizing agriculture in the twenty-first century and beyond. The initial strides have already resulted in the emergence of genome-edited foodstuffs, spanning from the soybean exhibiting better-quality oil characteristics to water stress tolerant maize. However, as genome editing continues to be applied in crop cultivation, it becomes imperative to comprehensively grasp the constraints and limitations inherent in this technology. Addressing these limitations is crucial to unlock the complete potential of genome editing in the agriculture. Notably, our current limitations lie in our understanding of the plant systems biology and our capacity to effectively alter and redevelop elite crop varieties. Overcoming or mitigating these restrictions stands as a pivotal step toward achieving the goal of genome editing in agriculture: the ability to intricately engineer crop plants from the foundational level upward.

REFERENCES

Abdul, M., Brini, F., Rouached, H., et al. 2022. Genetically engineered crops for sustainably enhanced food production systems. *Frontiers in Plant Science,* 13, 1027828.

Acevedo-Garcia, J., Kusch, S., Panstruga, R. 2014. Magical mystery tour: MLO proteins in plant immunity and beyond. *New Phytologist,* 204: 273–281.

Ainley, W.M., Sastry-Dent, L., Welter, M.E., et al. 2013. Trait stacking via targeted genome editing. *Plant Biotechnology Journal,* 11: 1126–1134.

Ali, Z,. Abul-Faraj, A., Piatek, M., et al. 2015. Activity and specificity of TRV-mediated gene editing in plants. *Plant Signaling & Behavior,* 10: 1–4.

Ali, Z., Ali, S., Tashkandi, M., et al. 2016. CRISPR/Cas9-mediated immunity to geminiviruses: Differential interference and evasion. *Science Reports*, 2016; 6: 26912.

Altpeter, F., Springer. N.M, Bartley, L.E., et al. 2016. Advancing crop transformation in the era of genome editing. *Plant Cell,* 28: 1510–1520.

Anzalone, A.V., Koblan, L.W., Liu, D.R. 2020. Genome editing with CRISPR–Cas nucleases, base editors, transposases and prime editors. *Nature Biotechnology,* 38: 824–844.

Arai-Sanoh, Y., Takai, T., Yoshinaga, S., et al. 2014. Deep rooting conferred by DEEPER ROOTING 1 enhances rice yield in paddy fields. *Science Reports*, 4: 1–6.

Ayala, F., Fedrigo, G. V., Burachik, M., et al. 2019. Compositional equivalence of event IND-ØØ412-7 to non-transgenic wheat. *Transgenic Research,* 28: 165–176.

Bage, S.A., Barten, T.J., Brown, A.N., et al. 2020. Genetic characterization of novel and CRISPR-Cas9 gene edited maize brachytic 2 alleles. *Plant Gene*, 21: 100198.

Bai, Y., Pavan, S., Zheng, Z., et al. 2008. Naturally occurring broad-spectrum powdery mildew resistance in a Central American tomato accession is caused by loss of Mlo function. *Molecular Plant-Microbe Interactions,* 21: 30–39.

Bailey-Serres, J., Parker, J.E., Ainsworth, E.A., et al. 2019. Genetic strategies for improving crop yields. *Nature,* 575: 109–118.

Baltes, N.J., Hummel, A.W., Konecna, E., et al. 2015. Conferring resistance to geminiviruses with the CRISPR–Cas prokaryotic immune system. *Nature Plants,* 1: 1–4.

Batista, R., Fonseca, C., Planchon, S., et al. 2017. Environmental stress is the major cause of transcriptomic and proteomic changes in GM and non-GM plants. *Science Reports*, 7: 10624.

Bergenholm, D., Dabirian, Y., Ferreira, R., et al. 2021. Rational gRNA design based on transcription factor binding data. *Synthetic Biology*, 6: ysab014.

Bhaskar, P.B., Wu, L., Busse, J.S., et al. 2010. Suppression of the vacuolar invertase gene prevents cold-induced sweetening in potato. *Plant Physiology,* 154: 939–948.

Brookes, G,. Barfoot, P.. 2020. Environmental impacts of genetically modified (GM) crop use 1996–2018: Impacts on pesticide use and carbon emissions. *GM Crops & Food*, 11: 215–241.

Char, S.N., Unger-Wallace, E., Frame, B., et al. 2015. Heritable site-specific mutagenesis using TALENs in maize. *Plant Biotechnology Journal,* 13: 1002–1010.

Chen, K., Wang, Y., Zhang, R., et al. 2019. CRISPR/Cas genome editing and precision plant breeding in agriculture. *Annual Review of Plant Biology,* 70: 667–697.

Chiozza, M.V., Burachik, M., Miranda, P.V. 2020. Compositional analysis of soybean event IND-ØØ41Ø-5. *GM Crops & Food,* 11: 154–163.

Choi, K. 2017. Advances towards controlling meiotic recombination for plant breeding. *Molecular Cells*, 40: 814.

Clasen, B.M., Stoddard, T.J., Luo, S., et al. 2016. Improving cold storage and processing traits in potato through targeted gene knockout. *Plant Biotechnology Journal,* 14: 169–176.

Cooper, M., Gho, C., Leafgren, R., et al. 2014. Breeding drought-tolerant maize hybrids for the US corn-belt: Discovery to product. *Journal of Experimental Botany,* 65: 6191–6194.

Daniel, M.A., Sebastin, R., Yu, J-K., et al. 2023. Enhancing horticultural crops through genome editing: Applications, benefits, and considerations. *Horticulturae*, 9: 884.

Devaux, A., Goffart, J-P., Kromann, P., et al. 2021. The potato of the future: Opportunities and challenges in sustainable agri-food systems. *Potato Research,* 64: 681–720.

Dezar, C.A., Gago, G.M., González, D.H., et al. 2005. Hahb-4, a sunflower homeobox-leucine zipper gene, is a developmental regulator and confers drought tolerance to Arabidopsis thaliana plants. *Transgenic Research,* 14: 429–440.

D'Halluin, K., Vanderstraeten, C., Van Hulle, J., et al. 2013. Targeted molecular trait stacking in cotton through targeted double-strand break induction. *Plant Biotechnology Journal,* 11: 933–941.

Doudna, J.A., Charpentier, E. 2014. *The new frontier of genome engineering with CRISPR-Cas9.* Science (1979), 346: 1258096.

Endo, M., Mikami, M., Toki, S. 2016. Biallelic gene targeting in rice. *Plant Physiology,* 170: 667–677.

FAO. 2024. FAOSTAT Cropland nutrient budget. www.fao.org/faostat *(accessed* 12 May 2024).

Gaffney, J., Schussler, J., Löffler, C., et al. 2015. Industry-scale evaluation of maize hybrids selected for increased yield in drought-stress conditions of the US corn belt. *Crop Science,* 55: 1608–1618.

Gao, C. 2018. The future of CRISPR technologies in agriculture. *Nature Reviews Molecular Cell Biology,* 19: 275–276.

Gao, C. 2021. Genome engineering for crop improvement and future agriculture. *Cell,* 184: 1621–1635.

Gurushidze, M., Hensel, G., Hiekel, S., et al. 2014. True-breeding targeted gene knock-out in barley using designer TALE-nuclease in haploid cells. *PLoS One,* 9: 1–9.

Henry, R.J., Rangan, P., Furtado, A. 2016. Functional cereals for production in new and variable climates. *Current Opinion in Plant Biology,* 30: 11–18.

Hussain, A., Ding, X., Alariqi, M., et al. 2021. Herbicide resistance: Another hot agronomic trait for plant genome editing. *Plants,* 10: 621.

Jacobs, T.B., LaFayette, P.R., Schmitz, R.J., et al. 2015. Targeted genome modifications in soybean with CRISPR/Cas9. *BMC Biotechnology,* 15: 1–10.

Jansing, J., Schiermeyer, A., Schillberg, S., et al. 2019. Genome editing in agriculture: Technical and practical considerations. *International Journal of Molecular Sciences,* 20: 2888.

Ji, X., Zhang, H., Zhang, Y., et al. 2015. Establishing a CRISPR–Cas-like immune system conferring DNA virus resistance in plants. *Nature Plants,* 1: 1–4.

Kaiser, N., Douches, D., Dhingra, A., et al. 2020. The role of conventional plant breeding in ensuring safe levels of naturally occurring toxins in food crops. *Trends in Food Science & Technology,* 100: 51–66.

Kamburova, V.S., Nikitina, E.V., Shermatov, S.E., et al. 2017. Genome editing in plants: An overview of tools and applications. *International Journal of Agronomy,* 78,: 7315351.

Kartseva, T., Alqudah, A.M,. Aleksandrov, V., et al. 2023. Nutritional genomic approach for improving grain protein content in wheat. *Foods*, 12: 1399.

Konstantakos, V., Nentidis, A., Krithara, A., et al. 2022. CRISPRedict: A CRISPR-Cas9 web tool for interpretable efficiency predictions. *Nucleic Acids Research*, 50: W191–W198.

Kovak, E., Blaustein-Rejto, D., Qaim, M. 2022. Genetically modified crops support climate change mitigation. *Trends in Plant Science,* 27: 627–629.

Kumar, K., Raina, S.K., Sultan, S.M. 2020. Arabidopsis MAPK signaling pathways and their cross talks in abiotic stress response. *Journal of Plant Biochemistry and Biotechnology,* 29: 700–714.

Lawrenson, T., Shorinola, O., Stacey, N., et al. 2015. Induction of targeted, heritable mutations in barley and Brassica oleracea using RNA-guided Cas9 nuclease. *Genome Biology,* 16: 1–13.

Lee, C-R., Wang, B., Mojica, J.P., et al. 2017.Young inversion with multiple linked QTLs under selection in a hybrid zone. *Nature Ecology & Evolution*, 1: 0119.

Li, J., Li, Y., Ma, L. 2021. Recent advances in CRISPR/Cas9 and applications for wheat functional genomics and breeding. *aBIOTECH,* 1–11.

Li, J., Stoddard, T.J., Demorest, Z.L., et al. 2016a. Multiplexed, targeted gene editing in Nicotiana benthamiana for glyco-engineering and monoclonal antibody production. *Plant Biotechnology Journal,* 14: 533–542.

Li, M., Li, X., Zhou, Z., et al. 2016b. Reassessment of the four yield-related genes Gn1a, DEP1, GS3, and IPA1 in rice using a CRISPR/Cas9 system. *Frontiers in Plant Science,* 7: 377.

Li, Z., Liu, Z.B., Xing, A., et al. 2015. Cas9-guide RNA directed genome editing in soybean. *Plant Physiology,* 169: 960–970.

Liang, Z., Zhang, K., Chen, K., et al. 2014. Targeted mutagenesis in Zea mays using TALENs and the CRISPR/Cas system. *Journal of Genetics and Genomics,* 41: 63–68.

Lowe, K., Wu, E., Wang, N., et al. 2016. Morphogenic regulators Baby boom and Wuschel improve monocot transformation. *Plant Cell,* 28: 1998–2015.

Luo, S., Li, J., Stoddard, T.J., et al. 2015. Non-transgenic plant genome editing using purified sequence-specific nucleases. *Molecular Plant,* 8: 1425–1427.

Maan, A.A., Anjum, M.A., Khan, M.K.I., et al. 2022. Acrylamide formation and different mitigation strategies during food processing: A review. *Food Reviews International,* 38: 70–87.

Mao, Y., Botella, J.R., Liu, Y., et al. 2019. Gene editing in plants: Progress and challenges. *National Science Review*, 6: 421–437.

Martín-Hernández, A.M., Baulcombe, D.C. 2008. Tobacco rattle virus 16-Kilodalton protein encodes a suppressor of RNA silencing that allows transient viral entry in meristems. *Journl of Virology,* 82: 4064–4071.

Marton, I., Zuker, A., Shklarman, E. et al. 2010. Nontransgenic genome modification in plant cells. *Plant Physiology,* 154: 1079–1087.

Mayer, K.F.X., Rogers, J., Doleel, J.P.C., et al. 2014. *A chromosome-based draft sequence of the hexaploid bread wheat (Triticum aestivum) genome.* Science (1979), 345: 1251788.

Miao, C., Xiao, L., Hua, K., et al. 2018. Mutations in a subfamily of abscisic acid receptor genes promote rice growth and productivity. *Proceedings of the National Academy of Sciences,* 115: 6058–6063.

Miranda, P.V., Iglesias, B.F., Charriere, M.V., et al. 2022. Drought tolerant wheat IND-ØØ412-7 is nutritionally equivalent to its Non-Transgenic Comparator. *GM Crops & Food,* 13: 119–125.

Mottram, D.S., Wedzicha, B.L., Dodson, A.T. 2002. Acrylamide is formed in the Maillard reaction. *Nature,* 419: 448–449.

Naeem, M., Majeed, S., Hoque, M.Z., et al. 2020. Latest developed strategies to minimize the off-target effects in CRISPR-Cas-mediated genome editing. *Cells*, 9: 1608.

Nam, J., Matthysse, A.G., Gelvin, S.B. 1997. Differences in susceptibility of arabidopsis ecotypes to crown gall disease may result from a deficiency in T-DNA integration. *Plant Cell,* 9: 317–333.

Nekrasov, V., Wang, C., Win, J., et al. 2017. Rapid generation of a transgene-free powdery mildew resistant tomato by genome deletion. *Science Reports*, 7: 482.

Ni, F-T., Chu, L-Y., Shao, H-B., et al. 2009. Gene expression and regulation of higher plants under soil water stress. *Current Genomics,* 10: 269–280.

Ordon, J., Gantner, J., Kemna, J., et al. 2017. Generation of chromosomal deletions in dicotyledonous plants employing a user-friendly genome editing toolkit. *Plant Journal,* 89: 155–168.

Ortega, A., de Marcos, A., Illescas-Miranda, J., et al. 2019. The tomato genome encodes SPCH, MUTE, and FAMA candidates that can replace the endogenous functions of their Arabidopsis orthologs. *Frontiers in Plant Science,* 10: 1–18.

Palermo, M., Gökmen, V., De Meulenaer, B., et al. 2016. Acrylamide mitigation strategies: Critical appraisal of the FoodDrinkEurope toolbox. *Food & Function,* 7: 2516–2525.

Peleman, J.D., van der Voort, J.R., et al. 2003. The challenges in Marker Assisted Breeding. *CGN,* 125: 125–130.

Peng, R., Jones, D.C., Liu, F., et al. 2021. From sequencing to genome editing for cotton improvement. *Trends in Biotechnology,* 39: 221–224.

Powers, S.J., Mottram, D.S., Curtis, A., et al. 2021. Progress on reducing acrylamide levels in potato crisps in Europe, 2002 to 2019. Food Additives and Contaminants, *Part A,* 38: 782–806.

Qaim, M. 2010. Benefits of genetically modified crops for the poor: Household income, nutrition, and health. *Nature Biotechnology,* 27: 552–557.

Qi ,Y., Zhang, Y., Zhang, F., et al. 2013. Increasing frequencies of site-specific mutagenesis and gene targeting in Arabidopsis by manipulating DNA repair pathways. *Genome Research,* 23: 547–554.

Rahman, M., de Jiménez, M.M. 2016. Designer oil crops. In S.K. Gupta (Ed.), *Breeding oilseed crops for sustainable production* (pp. 361–376). Elsevier.

Raman, R. 2017. The impact of Genetically Modified (GM) crops in modern agriculture: A review. *GM Crops & Food,* 8: 195–208.

Ribichich, K.F., Chiozza, M., Ávalos-Britez, S., et al. 2020. Successful field performance in warm and dry environments of soybean expressing the sunflower transcription factor HB4. *Journal of Experimental Botany,* 71: 3142–3156.

Richael, C.M. 2021. Development of the genetically modified Innate® potato. *Plant Breedng Reviews,* 44: 57–78.

Robertson, F.C., Skeffington A.W., Gardner M.J., et al. 2009. Interactions between circadian and hormonal signalling in plants. *Plant Molecular Biology,* 69: 419–427.

Roossinck, M.J., Martin D.P., Roumagnac P. 2015. Plant virus metagenomics: advances in virus discovery. *Phytopathology,* 105: 716–727.

Sakuma, Y., Maruyama, K., Osakabe, Y., et al. 2006. Functional analysis of an Arabidopsis transcription factor, DREB2A, involved in drought-responsive gene expression. *Plant Cell,* 18: 1292–1309.

Sanford, J.C. 1988. The biolistic process. *Trends in Biotechnology,* 6, 299–302.

Sauer, N.J., Narváez-Vásquez, J., Mozoruk, J., et al. 2016. Oligonucleotide-mediated genome editing provides precision and function to engineered nucleases and antibiotics in plants. *Plant Physiology,* 170: 1917–1928.

Schmidt, C., Schindele, P., Puchta, H. 2020. From gene editing to genome engineering: Restructuring plant chromosomes via CRISPR/Cas. *aBIOTECH,* 1: 21–31.

Schmutz, J., Cannon, S.B., Schlueter, J., et al. 2010. Genome sequence of the palaeopolyploid soybean. *Nature,* 463: 178–183.

Sendhil, R., Nyika, J., Yadav, S., et al. 2022. Genetically modified foods: Bibliometric analysis on consumer perception and preference. *GM Crops & Food,* 13: 65.

Shah, T., Andleeb, T., Lateef, S., et al. 2018. Genome editing in plants: Advancing crop transformation and overview of tools. *Plant Physiology and Biochemistry,* 131: 12–21.

Shan, Q., Wang, Y., Li, J., et al. 2013. Targeted genome modification of crop plants using a CRISPR-Cas system. *Nature Biotechnology,* 31: 686–688.

Sharma, P., Pandey, A., Malviya, R., et al. 2023. Genome editing for improving nutritional quality, post-harvest shelf life and stress tolerance of fruits, vegetables, and ornamentals. *Frontiers in Genome Editing*, 5: 1094965.

Shavrukov, Y., Kurishbayev, A., Jatayev, S., et al. 2017. Early flowering as a drought escape mechanism in plants: How can it aid wheat production? *Frontiers in Plant Science,* 8: 1–8.

Shepherd, L.V.T., Bradshaw, J.E., Dale, M.F.B., et al. 2010. Variation in acrylamide producing potential in potato: segregation of the trait in a breeding population. *Food Chemistry,* 123: 568–573.

Sowokinos, J.R. 2001. Biochemical and molecular control of cold-induced sweetening in potatoes. *American Journal of Potato Research,* 78: 221–236.

Springmann, M., Clark, M., Mason-D'Croz, D., et al. 2018. Options for keeping the food system within environmental limits. *Nature,* 562: 519–525.

Steinert, J., Schiml, S., Puchta, H. 2016. Homology-based double-strand break-induced genome engineering in plants. *Plant Cell Reports,* 35: 1429–1438.

Stoddard, T.J., Clasen, B.M., Baltes, N.J., et al. 2016. Targeted mutagenesis in plant cells through transformation of sequence-specific nuclease mRNA. *PLoS One,* 11: 1–10.

Subedi, U., Jayawardhane, K.N., Pan, X., et al. 2020. The potential of genome editing for improving seed oil content and fatty acid composition in oilseed crops. *Lipids,* 55: 495–512.

Sutherland, C., Gleim, S., Smyth S.J. 2021; Correlating genetically modified crops, glyphosate use, and increased carbon sequestration. *Sustainability (Switzerland)*, 13. DOI: 10.3390/su132111679.

Svitashev, S., Schwartz, C., Lenderts, B., et al. 2016. Genome editing in maize directed by CRISPR-Cas9 ribonucleoprotein complexes. *Nature Communications,* 7: 1–7.

Svitashev, S., Young, J.K., Schwartz, C., et al. 2015. Targeted mutagenesis, precise gene editing, and site-specific gene insertion in maize using Cas9 and guide RNA. *Plant Physiology,* 169: 931–945.

Tang, L., Mao, B., Li, Y., et al. 2017. Knockout of OsNramp5 using the CRISPR/Cas9 system produces low Cd-accumulating indica rice without compromising yield. *Science Reports*, 7: 14438.

Tran, M.T., Doan, D.T.H., Kim, J., et al. 2021. CRISPR/Cas9-based precise excision of SlHyPRP1 domain (s) to obtain salt stress-tolerant tomato. *Plant Cell Reports,* 40: 999–1011.

Tripathi, R., Tewari, R., Singh, K.P., et al. 2022. Plant mineral nutrition and disease resistance: A significant linkage for sustainable crop protection. *Frontiers in Plant Science,* 13, 3116.

Tseng, M-C., Roel, A., Deambrosi, E., et al. 2020. Towards actionable research frameworks for sustainable intensification in high-yielding rice systems. *Science Reports*, 10: 9975.

Uga, Y., Sugimoto, K., Ogawa, S., et al. 2013. Control of root system architecture by DEEPER ROOTING 1 increases rice yield under drought conditions. *Nature Genetics,* 45: 1097–1102.

Voytas, D.F., Gao, C. 2014. Precision genome engineering and agriculture: opportunities and regulatory challenges. *PLoS Biology,* 12, e1001877.

Waltz, E. 2016. CRISPR-edited crops free to enter market, skip regulation. *Nature Biotechnology,* 34: 582–583.

Wang, L., Chen, L., Li, R., et al. 2017. Reduced drought tolerance by CRISPR/Cas9-mediated SlMAPK3 mutagenesis in tomato plants. *Journal of Agriculture and Food Chemistry,* 65: 8674–8682.

Wang, Y., Cheng, X., Shan, Q., et al. 2014. Simultaneous editing of three homoeoalleles in hexaploid bread wheat confers heritable resistance to powdery mildew. *Nature Biotechnology,* 32: 947–951.

Wang, Y., Copenhaver, G.P. 2018. Meiotic recombination: mixing it up in plants. *Annual Review of Plant Biology,* 69: 577–609.

Wang, Y., Ying, J., Kuzma M., et al. 2005. Molecular tailoring of farnesylation for plant drought tolerance and yield protection. *Plant Journal,* 43: 413–424.

Wang, Z., Hong, Y., Zhu, G., et al. 2020. Loss of salt tolerance during tomato domestication conferred by variation in a Na+/K+ transporter. *EMBO Journal,* 39: e103256.

Watson Amy, S.G., Williams, M.J., Cuddy, W.S., et al. 2017. Speed breeding: A powerful tool to accelerate crop research and breeding. *Nature Plants,* 4: 23–29.

Wellenreuther, M., Bernatchez, L. 2018. Eco-evolutionary genomics of chromosomal inversions. *Trends in Ecology & Evolution,* 33: 427–440.

Wen, A., Havens, K.L., Bloch, S.E., et al. 2021. Enabling biological nitrogen fixation for cereal crops in fertilized fields. *ACS Synthetic Biology,* 10: 3264–3277.

Wolter, F., Schindele, P., Puchta, H. 2019. Plant breeding at the speed of light: The power of CRISPR/Cas to generate directed genetic diversity at multiple sites. *BMC Plant Biology,* 19: 1–8.

Woo, J.W., Kim, J., Kwon, S. II, et al. 2015. DNA-free genome editing in plants with preassembled CRISPR-Cas9 ribonucleoproteins. *Nature Biotechnology,* 33: 1162–1164.

Wright, D.A., Townsend, J.A., Winfrey, R.J., et al. 2005. High-frequency homologous recombination in plants mediated by zinc-finger nucleases. *Plant Journal,* 44: 693–705.

Wu, L., Bhaskar, P.B., Busse, J.S., et al. 2011. Developing cold-chipping potato varieties by silencing the vacuolar invertase gene. *Crop Science,* 51: 981–990.

Xiang, X., Corsi, G.I., Anthon, C., et al. 2021. Enhancing CRISPR-Cas9 gRNA efficiency prediction by data integration and deep learning. *Nature Communications,* 12: 3238.

Yin, K., Gao, C., Qiu, J-L. 2017. Progress and prospects in plant genome editing. *Nature Plants,* 3: 1–6.

Yin, K., Han, T., Liu, G., et al. 2015. A geminivirus-based guide RNA delivery system for CRISPR/Cas9 mediated plant genome editing. *Science Reports,* 5: 1–10.

Yin, Y., Qin, K., Song, X., et al. 2018. BZR1 transcription factor regulates heat stress tolerance through FERONIA receptor-like kinase-mediated reactive oxygen species signaling in tomato. *Plant Cell Physiology,* 59: 2239–2254.

Zhang, Y., Liang, Z., Zong, Y., et al. 2016. Efficient and transgene-free genome editing in wheat through transient expression of CRISPR/Cas9 DNA or RNA. *Nature Communications,* 7: 1–8.

Zhong, Z., Li, Z., Yang, J., et al. 2023. *Unified model to predict gRNA efficiency across diverse cell lines and CRISPR-Cas9 systems. Journal of Chemical Information and Modeling.* DOI: 10.1021/acs.jcim.3c01339.

Zhou, H., Liu, B., Weeks, D.P., et al. 2014. Large chromosomal deletions and heritable small genetic changes induced by CRISPR/Cas9 in rice. *Nucleic Acids Research,* 42: 10903–10914.

Zhou, J., Peng, Z., Long, J., et al. 2015. Gene targeting by the TAL effector PthXo2 reveals cryptic resistance gene for bacterial blight of rice. *Plant Journal,* 82: 632–643.

Zhu, H., Li, C., Gao, C. 2020. Applications of CRISPR–Cas in agriculture and plant biotechnology. *Nature Reviews Molecular Cell Biology,* 21: 661–677.

Index

Note: Figures are in *italics* and tables in **bold** type. The acronyms AI, CRISPR, DNA, GM, GMO, gRNA, and RNA are used for artificial intelligence, Clustered Regularly Interspaced Short Palindromic Repeats, deoxyribonucleic acid, genetically modified, genetically modified organisms, guide ribonucleic acid, and ribonucleic acid respectively.